W9-CBA-991

INTRODUCTION TO TRIBOLOGY

Tribology Series

Bhushan	Introduction to Tribology, 2nd Edition	March 2013
Bhushan	Principles and Applications to Tribology, 2nd Edition	March 2013
Lugt	Grease Lubrication in Rolling Bearings	January 2013
Honary and Richter	Biobased Lubricants and Greases: Technology and Products	April 2011
Martin and Ohmae	Nanolubricants	April 2008
Khonsari and Booser	Applied Tribology: Bearing Design and Lubrication, 2nd Edition	April 2008
Stachowiak (ed)	Wear: Materials, Mechanisms and Practice	November 2005
Lansdown	Lubrication and Lubricant Selection: A Practical Guide, 3rd Edition	November 2003
Cartier	Handbook of Surface Treatment and Coatings	May 2003
Sherrington, Rowe and Wood (eds)	Total Tribology: Towards an Integrated Approach	December 2002
Kragelsky and	Tribology: Lubrication, Friction and Wear	April 2001
Stolarski and Tobe	Rolling Contacts	December 2000
Neale and Gee	Guide to Wear Problems and Testing for Industry	October 2000

INTRODUCTION TO TRIBOLOGY

SECOND EDITION

Bharat Bhushan

Ohio Eminent Scholar and the Howard D. Winbigler Professor
Director, Nanoprobe Laboratory for Bio- & Nanotechnology and Biomimetics
The Ohio State University
Columbus, Ohio
USA

A John Wiley & Sons, Ltd., Publication

This edition first published 2013
© 2013 John Wiley & Sons, Ltd

First Edition published in 2002
©2002, John Wiley & Sons, Inc., New York. All rights reserved.

Registered office
John Wiley & Sons Ltd, The Atrium, Southern Gate, Chichester, West Sussex, PO19 8SQ, United Kingdom

For details of our global editorial offices, for customer services and for information about how to apply for permission to reuse the copyright material in this book please see our website at www.wiley.com.

The right of the author to be identified as the author of this work has been asserted in accordance with the Copyright, Designs and Patents Act 1988.

All rights reserved. No part of this publication may be reproduced, stored in a retrieval system, or transmitted, in any form or by any means, electronic, mechanical, photocopying, recording or otherwise, except as permitted by the UK Copyright, Designs and Patents Act 1988, without the prior permission of the publisher.

Wiley also publishes its books in a variety of electronic formats. Some content that appears in print may not be available in electronic books.

Designations used by companies to distinguish their products are often claimed as trademarks. All brand names and product names used in this book are trade names, service marks, trademarks or registered trademarks of their respective owners. The publisher is not associated with any product or vendor mentioned in this book. This publication is designed to provide accurate and authoritative information in regard to the subject matter covered. It is sold on the understanding that the publisher is not engaged in rendering professional services. If professional advice or other expert assistance is required, the services of a competent professional should be sought.

Library of Congress Cataloging-in-Publication Data

Bhushan, Bharat, 1949–
Introduction to tribology / Bharat Bhushan. – Second edition.
pages cm
Includes bibliographical references and index.
ISBN 978-1-119-94453-9 (cloth)
1. Tribology. I. Title.
TJ1075.B472 2013
621.8′9–dc23

2012031551

A catalogue record for this book is available from the British Library.

ISBN: 978-1-119-94453-9

Typeset in 10/12pt Times by Aptara Inc., New Delhi, India

To my wife Sudha, my son Ankur and my daughter Noopur

Contents

About the Author

Dr Bharat Bhushan received an MS in mechanical engineering from the Massachusetts Institute of Technology in 1971, an MS in mechanics and a PhD in mechanical engineering from the University of Colorado at Boulder in 1973 and 1976, respectively, an MBA from Rensselaer Polytechnic Institute at Troy, NY in 1980, Doctor Technicae from the University of Trondheim at Trondheim, Norway in 1990, a Doctor of Technical Sciences from the Warsaw University of Technology at Warsaw, Poland in 1996, and Doctor Honouris Causa from the National Academy of Sciences at Gomel, Belarus in 2000 and University of Kragujevac, Serbia in 2011. He is a registered professional engineer. He is presently an Ohio Eminent Scholar and The Howard D. Winbigler Professor in the College of Engineering, and the Director of the Nanoprobe Laboratory for Bio- & Nanotechnology and Biomimetics (NLB^2) at the Ohio State University, Columbus, Ohio. His research interests include fundamental studies with a focus on scanning probe techniques in the interdisciplinary areas of bio/nanotribology, bio/nanomechanics and bio/nanomaterials characterization and applications to bio/nanotechnology, and biomimetics. He is an internationally recognized expert of bio/nanotribology and bio/nanomechanics using scanning probe microscopy, and is one of the most prolific authors. He is considered by some a pioneer of the tribology and mechanics of magnetic storage devices. He has authored 8 scientific books, 90+ handbook chapters, 700+ scientific papers (h-index – 57+; ISI Highly Cited in Materials Science, since 2007; ISI Top 5% Cited Authors for Journals in Chemistry since 2011), and 60+ technical reports. He has also edited 50+ books and holds 17 US and foreign patents. He is co-editor of the Springer NanoScience and Technology Series and co-editor of *Microsystem Technologies*. He has given more than 400 invited presentations on 6 continents and more than 200 keynote/plenary addresses at major international conferences.

Dr Bhushan is an accomplished organizer. He organized the 1st Symposium on Tribology and Mechanics of Magnetic Storage Systems in 1984 and the 1st International Symposium on Advances in Information Storage Systems in 1990, both of which are now held annually. He is the founder of an ASME Information Storage and Processing Systems Division founded in 1993 and served as the founding chair during 1993–1998. His biography has been listed in over two dozen Who's Who books including *Who's Who in the World* and has received more than two dozen awards for his contributions to science and technology from professional societies, industry, and US government agencies. He is also the recipient of various international fellowships including the Alexander von Humboldt Research Prize for Senior Scientists, Max Planck Foundation Research Award for Outstanding Foreign Scientists, and the Fulbright Senior Scholar Award. He is a foreign member of the International Academy of Engineering (Russia), Byelorussian Academy of Engineering and Technology and the Academy of Triboengineering of Ukraine, an honorary member of the Society of Tribologists of Belarus, a fellow of ASME, IEEE, STLE, and the New York Academy of Sciences, and a member of ASEE, Sigma Xi and Tau Beta Pi.

Dr Bhushan has previously worked for Mechanical Technology Inc., Latham, NY; SKF Industries Inc., King of Prussia, PA; IBM, Tucson, AZ; and IBM Almaden Research Center, San Jose, CA. He has held visiting professorship at the University of California at Berkeley, the University of Cambridge, UK, the Technical University Vienna, Austria, the University of Paris, Orsay, ETH Zurich, and EPFL Lausanne. He is currently a visiting professor at KFUPM, Saudi Arabia, the Harbin Institute, China, the University of Kragujevac, Serbia and the University of Southampton, UK.

Foreword

The concept of Tribology was enunciated in 1966 in a report of the UK Department of Education and Science. It encompasses the interdisciplinary science and technology of interacting surfaces in relative motion and associated subjects and practices. It includes parts of physics, chemistry, solid mechanics, fluid mechanics, heat transfer, materials science, lubricant rheology, reliability, and performance.

Although the name tribology is new, the constituent parts of tribology – encompassing friction and wear – are as old as history. The economic aspects of tribology are significant. Investigations by a number of countries arrived at figures of savings of 1.0% to 1.4% of the GNPs, obtainable by the application of tribological principles, often for proportionally minimal expenditure in Research and Development.

Being an interdisciplinary area, the important aspects of tribology have been difficult to cover in a single book of interest to readers ranging from students to active researchers in academia and industry.

To prepare such a wide-ranging book on tribology, Professor Bhushan has harnessed the knowledge and experience gained by him in several industries and universities. He has set out to cover not only the fundamentals of friction, wear, and lubrication, friction and wear test methods and industrial applications, but also includes a chapter on the field of micro/nanotribology, which may be of special interest in the light of the emergence of proximal probes and computational techniques for simulating tip surface interactions and interface properties.

Professor Bharat Bhushan's comprehensive book is intended to serve both as a textbook for university courses as well as a reference for researchers. It is a timely addition to the literature on tribology and I hope that it will stimulate and further the interest of tribology and be found useful by the international scientific and industrial community.

Professor H. Peter Jost
President, International Tribology Council
Angel Lodge Laboratories & Works
London, UK
July, 1998

Series Preface

This Second Edition of the successful *Introduction to Tribology* published in 1999 promises to deliver much more than its earlier version. Over the last few decades, since the concept of 'tribology' was introduced by Peter Jost in 1966, the industry has gone through dramatic changes. These changes were dictated by demands for new, more reliable products and for improving the quality of life. To fulfill these demands, new technologies have emerged. Much has changed in many areas of science over the last decade and the tribology is not an exception. Improved materials and surface treatments were developed, novel lubricants were introduced and new insights into the mechanisms of contacting surfaces were gained. Nowadays, humanity is facing new challenges such as sustainability, climate change, and gradual degradation of the environment. There are also concerns about providing enough food and clean water to the human population and issues associated with supplying enough energy to allow people to pursue a civilized life. Tribology makes vital contribution to the resolution of these problems. As is any other field of science, tribology is continuously evolving to stay at the forefront of the emerging technologies.

As tribology is an interdisciplinary area of science, knowledge from chemistry, physics, material science, engineering, computational science, and many others is required to allow for the understanding of the tribological phenomena. This book provides a comprehensive account of the field of tribology and this edition includes the latest developments in the understanding and interpretation of friction, wear, and lubrication. It introduces tribology at the nano- and micro-level, i.e. nanotribology, tribology in MEMS and magnetic surface storage devices. This approach demonstrates to the reader that tribology continuously evolves and adapts and remains relevant to the modern industry. This is a much-welcomed edition to the tribology book series as tribology provides badly needed answers to many problems. The book is recommended for both under- and postgraduate students and engineers.

Gwidon Stachowiak
University of Western Australia

Preface to the Second Edition

Tribology is an important interdisciplinary field. It involves the design of components with static and dynamic contacts for a required performance and reliability. The second edition of the book is thoroughly updated. Notable additions include an updated chapter on nanotribology, introduction to nanotechnology (MEMS/NEMS), and a new chapter on green tribology and biomimetics.

Modern tools and techniques as well as computational modeling have allowed systematic investigations of interfacial phenomena down to atomic scales. These developments have led to the development of the field of nanotribology and nanomechanics. These studies are needed to develop a fundamental understanding of the interface of science and technology.

The advances in micro/nanofabrication processes have led to the development of micro/nanoelectromechanical systems (MEMS/NEMS) used in various electro/mechanical, chemical, optical, and biological applications. These devices are expected to have a major impact on our lives, comparable to that of semiconductor technology, information technology, or cellular or molecular biology.

Ecological, or green, tribology is a relatively new field. It is defined as the science and technology of the tribological aspects of ecological balance and of environmental and biological impacts. This includes tribological components and materials and surfaces that mimic nature (biomimetic surfaces) and the control of friction and wear that is important for alternative energy production.

The author hopes that the second edition will be a useful addition to the interface between science and technology. Thanks are due to Megan BeVier for typing the manuscript.

A Power Point presentation of the entire book for a semester course is available from the author. A solution manual is also available from the author. Both Power Point presentation and the solution manual will be shipped to those who are using the book as a textbook for a class of a minimum of six students.

Professor Bharat Bhushan
Powell, Ohio
May, 2012

Preface to the First Edition

Tribology is the science and technology of interacting surfaces in relative motion and of related subjects and practices. Its popular English language equivalent is friction, wear, and lubrication, or lubrication science. The nature and consequence of the interactions that take place at the interface control its friction, wear, and lubrication behavior. During these interactions, forces are transmitted, mechanical energy is converted, the physical and the chemical nature, including surface topography, of the interacting materials are altered. Understanding the nature of these interactions and solving the technological problems associated with the interfacial phenomena constitute the essence of tribology.

Sliding and rolling surfaces represent the key to much of our technological society. An understanding of tribological principles is essential for the successful design of machine elements. When two nominally flat surfaces are placed in contact, surface roughness causes contact to occur at discrete contact spots and interfacial adhesion occurs. Friction is the resistance to motion that is experienced whenever one solid body moves over another. Wear is the surface damage or removal of material from one or both of two solid surfaces in a moving contact. Materials, coatings, and surface treatments are used to control friction and wear. One of the most effective means of controlling friction and wear is by proper lubrication, which provides smooth running and satisfactory life for machine elements. Lubricants can be liquid, solid, or gas. The role of surface roughness, mechanisms of adhesion, friction, and wear, and physical and chemical interactions between the lubricant and the interacting surfaces must be understood for optimum performance and reliability. The importance of friction and wear control cannot be overemphasized for economic reasons and long-term reliability. The savings can be substantial, and these savings can be obtained without the deployment of investment.

The recent emergence and proliferation of proximal probes, in particular tip-based microscopies (the scanning tunneling microscope and the atomic force microscope) and the surface force apparatus, and of computational techniques for simulating tip–surface interactions and interfacial properties, have allowed systematic investigations of interfacial problems with high resolution as well as ways and means for modifying and manipulating nanoscale structures. These advances provide the impetus for research aimed at developing a fundamental understanding of the nature and consequences of the interactions between materials on the atomic scale, and they guide the rational design of material for technological applications. In short, they have led to the appearance of the new field of micro/nanotribology, which pertains to experimental and theoretical investigations of interfacial processes on scales ranging from the atomic and molecular to the microscale. Micro/nanotribological studies are valuable in gaining a fundamental understanding of interfacial phenomena to provide a bridge between science and engineering.

There is a concern that some of today's engineering and applied science students may not be learning enough about the fundamentals of tribology. No single, widely accepted textbook exists for a comprehensive course on tribology. Books to date are generally based on their authors' own expertise in narrow aspects of tribology. A broad-based textbook is needed. This book is a condensed version of the comprehensive book titled *Principles and Applications of Tribology* published by Wiley first in 1999. The purpose of this book is to present the principles of tribology and the tribological understanding of the most common industrial applications. The book is based on the author's broad experience in research and teaching in the area of tribology, mechanics, and materials science for more than 30 years. The emphasis is on contemporary knowledge of tribology, and includes the emerging field of micro/nanotribology. The book integrates the knowledge of tribology from mechanical engineering, mechanics, and materials science points of view. The organization of the book is straightforward. The first part of the book starts with the principles of tribology and prepares students to understand the tribology of industrial applications. The principles of tribology follow with the emerging field of micro/nanotribology. The last chapter describes the tribological components and applications.

The book should serve as an excellent text for a one semester graduate course in tribology as well as for a senior level undergraduate course of mechanical engineering, materials science, or applied physics. The book is also intended for use by research workers who are active or intend to become active in this field, and practicing engineers who have encountered a tribology problem and hope to solve it as expeditiously as possible.

A Power Point presentation of the entire book for a semester course is available from the author. A solution manual is also available from the author. Both Power Point presentation and the solution manual will be shipped to those who are using the book as textbook for a class of a minimum of six students.

I wish to thank all of my former and present colleagues and students who have contributed to my learning of tribology. I was introduced to the field of tribology via a graduate course in Tribology in Fall 1970 from Profs. Brandon G. Rightmyer and Ernest Rabinowicz at Massachusetts Institute of Technology. I learnt a great deal from Prof. Nathan H. Cook, my MS thesis supervisor. My real learning started at the R& D Division of Mechanical Technology Inc., Latham, New York with the guidance from Dr Donald F. Wilcock, Dr Jed A. Walowit and Mr Stanley Gray, and at Technology Services Division of SKF Industries Inc., King of Prussia, Pennsylvania with the guidance from Dr Tibor Tallian. I immensely benefited from many colleagues at General Products Division of IBM Corporation, Tucson, Arizona and at Almaden Research Center of IBM Corporate Research Division, San Jose, California. Dr Kailash C. Joshi helped me in establishing at IBM Tucson and Dr Barry H. Schechtman mentored me at IBM Almaden, San Jose and helped me immensely. Prof. Bernard H. Hamrock at The Ohio State University has provided a nice companionship. Since 1991, I have offered many graduate and undergraduate tribology courses at The Ohio State University as well as many on-site short tribology courses in the United States and overseas. The book is based on the class notes used for various courses taught by me.

My special thanks go to my wife Sudha, my son Ankur and my daughter Noopur, who have been forebearing during the years when I spent long days and nights in conducting the research and keeping up with the literature and preparation of this book. They provided the lubrication necessary to minimize friction and wear at home.

Professor Bharat Bhushan
Powell, Ohio
August, 2001

1

Introduction

In this introductory chapter, the definition and history of tribology and their industrial significance are described, followed by the origins and significance of an emerging field of micro/nanotribology. In the last section the organization of the book is presented.

1.1 Definition and History of Tribology

The word tribology was first reported in a landmark report by Jost (1966). The word is derived from the Greek word *tribos* meaning rubbing, so the literal translation would be "the science of rubbing." Its popular English language equivalent is friction and wear or lubrication science, alternatively used. The latter term is hardly all-inclusive. Dictionaries define tribology as the science and technology of interacting surfaces in relative motion and of related subjects and practices. Tribology is the art of applying operational analysis to problems of great economic significance, namely, reliability, maintenance, and wear of technical equipment, ranging from spacecraft to household appliances. Surface interactions in a tribological interface are highly complex, and their understanding requires knowledge of various disciplines, including physics, chemistry, applied mathematics, solid mechanics, fluid mechanics, thermodynamics, heat transfer, materials science, rheology, lubrication, machine design, performance, and reliability.

It is only the name tribology that is relatively new, because interest in the constituent parts of tribology is older than recorded history (Dowson, 1998). It is known that drills made during the Paleolithic period for drilling holes or producing fire were fitted with bearings made from antlers or bones, and potters' wheels or stones for grinding cereals, etc., clearly had a requirement for some form of bearings (Davidson, 1957). A ball thrust bearing dated about AD 40 was found in Lake Nimi near Rome.

Records show the use of wheels from 3500 BC, which illustrates our ancestors' concern with reducing friction in translationary motion. Figure 1.1.1 shows a two wheeled harvest cart with studded wheels, circa 1338 AD. The transportation of large stone building blocks and monuments required the know-how of frictional devices and lubricants, such as water-lubricated sleds. Figure 1.1.2 illustrates the use of a sledge to transport a heavy statue

Introduction to Tribology, Second Edition. Bharat Bhushan.
© 2013 John Wiley & Sons, Ltd. Published 2013 by John Wiley & Sons, Ltd.

Figure 1.1.1 Drawing of two-wheeled harvest cart with studded wheels. Luttrell Psalter (folio 173v), circa 1338 AD.

by the Egyptians, circa 1880 BC (Layard, 1853). In this transportation, 172 slaves are being used to drag a large statue weighing about 600 kN along a wooden track. One man, standing on the sledge supporting the statue, is seen pouring a liquid (most likely water) into the path of motion; perhaps he was one of the earliest lubrication engineers. Dowson (1998) has estimated that each man exerted a pull of about 800 N. On this basis, the total effort, which must at least equal the friction force, becomes 172 × 800 N. Thus, the coefficient of friction is about 0.23. A tomb in Egypt that was dated several thousand years BC provides the evidence of use of lubricants. A chariot in this tomb still contained some of the original animal-fat lubricant in its wheel bearings.

During and after the Roman Empire, military engineers rose to prominence by devising both war machinery and methods of fortification, using tribological principles. It was the Renaissance engineer-artist Leonardo da Vinci (1452–1519), celebrated in his day for his genius in military construction as well as for his painting and sculpture, who first postulated a scientific approach to friction. Da Vinci deduced the rules governing the motion of a rectangular

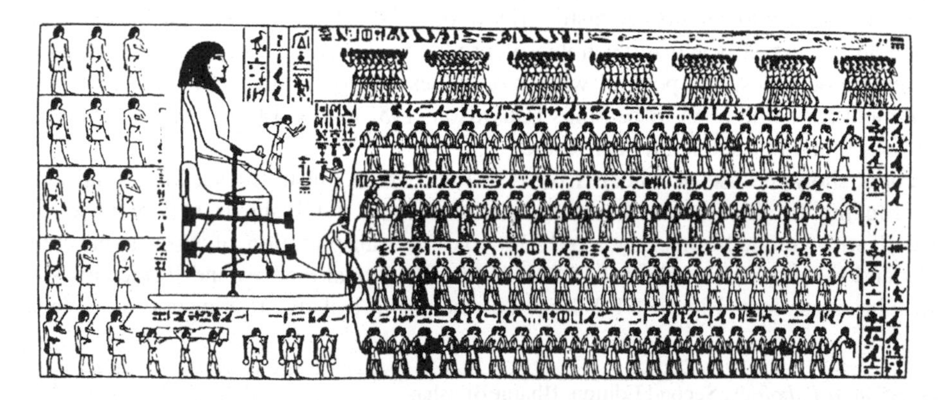

Figure 1.1.2 Egyptians using lubricant to aid movement of colossus, El-Bersheh, circa 1880 BC.

block sliding over a flat surface. He introduced the concept of the coefficient of friction as the ratio of the friction force to normal load. His work had no historical influence, however, because his notebooks remained unpublished for hundreds of years. In 1699, the French physicist Guillaume Amontons rediscovered the rules of friction after he studied dry sliding between two flat surfaces (Amontons, 1699). First the friction force that resists sliding at an interface is directly proportional to the normal load. Second the amount of friction force does not depend on the apparent area of contact. These observations were verified by the French physicist Charles-Augustin Coulomb (better known for his work on electrostatics [Coulomb, 1785]). He added a third law that the friction force is independent of velocity once motion starts. He also made a clear distinction between static friction and kinetic friction.

Many other developments occurred during the 1500s, particularly in the use of improved bearing materials. In 1684, Robert Hooke suggested the combination of steel shafts and bell-metal bushes would be preferable to wood shod with iron for wheel bearings. Further developments were associated with the growth of industrialization in the latter part of the eighteenth century. Early developments in the petroleum industry started in Scotland, Canada, and the United States in the 1850s (Parish, 1935; Dowson, 1998).

Though essential laws of viscous flow were postulated by Sir Isaac Newton in 1668, scientific understanding of lubricated bearing operations did not occur until the end of the nineteenth century. Indeed, the beginning of our understanding of the principle of hydrodynamic lubrication was made possible by the experimental studies of Beauchamp Tower (1884) and the theoretical interpretations of Osborne Reynolds (1886) and related work by N.P. Petroff (1883). Since then, developments in hydrodynamic bearing theory and practice have been extremely rapid in meeting the demand for reliable bearings in new machinery.

Wear is a much younger subject than friction and bearing development, and it was initiated on a largely empirical basis. Scientific studies of wear scarcely developed until the mid-twentieth century. Ragnar Holm made one of the earliest substantial contributions to the study of wear (Holm, 1946).

In the West, the Industrial Revolution (AD 1750–1850) is recognized as the period of rapid and impressive development of the machinery of production. The use of steam power and the subsequent development of the railways in the 1830s, automobiles in the early 1900s and aircraft in the 1940s led to the need for reliable machine components. Since the beginning of the twentieth century, from enormous industrial growth leading to demand for better tribology, knowledge in all areas of tribology has expanded tremendously (Holm, 1946; Bowden and Tabor, 1950, 1964; Bhushan, 1996, 2001a; Bhushan and Gupta, 1997; Nosonovsky and Bhushan, 2012).

1.2 Industrial Significance of Tribology

Tribology is crucial to modern machinery which uses sliding and rolling surfaces. Examples of productive friction are brakes, clutches, driving wheels on trains and automobiles, bolts, and nuts. Examples of productive wear are writing with a pencil, machining, polishing, and shaving. Examples of unproductive friction and wear are internal combustion and aircraft engines, gears, cams, bearings, and seals.

According to some estimates, losses resulting from ignorance of tribology amount in the United States to about 4% of its gross national product (or about $200 billion dollars per year in 1966), and approximately one-third of the world's energy resources in present use appear as friction in one form or another. Thus, the importance of friction reduction and wear control cannot be overemphasized for economic reasons and long-term reliability. According to Jost (1966, 1976), savings of about 1% of gross national product of an industrial nation can be realized by better tribological practices. According to recent studies, expected savings are expected to be of the order of 50 times the research costs. The savings are both substantial and significant, and these savings can be obtained without the deployment of large capital investment.

The purpose of research in tribology is understandably the minimization and elimination of losses resulting from friction and wear at all levels of technology where the rubbing of surfaces is involved. Research in tribology leads to greater plant efficiency, better performance, fewer breakdowns, and significant savings.

Since the 1800s, tribology has been important in numerous industrial applications requiring relative motion, for example, railroads, automobiles, aircraft, and the manufacturing process of machine components. Some of the tribological machine components used in these applications include bearings, seals, gears, and metal cutting (Bhushan, 2001a). Since the 1980s, other applications have included magnetic storage devices, and micro/nanoelectromechanical systems (MEMS/NEMS) as well as biomedical and beauty care products (Bhushan, 1996, 1998, 1999, 2000, 2001a, 2001b, 2010a, 2010b, 2011, 2012b). Since the 2000s, bioinspired structures and materials, some of which are eco-friendly, have been developed and exploited for various applications (Nosonovsky and Bhushan, 2008, 2012; Bhushan, 2012a).

Tribology is not only important to heavy industry, it also affects our day-to-day life. For example, writing is a tribological process. Writing is accomplished by the controlled transfer of lead (pencil) or ink (pen) to the paper. During writing with a pencil there should be good adhesion between the lead and the paper so that a small quantity of lead transfers to the paper and the lead should have adequate toughness/hardness so that it does not fracture/break. The objective when shaving is to remove hair from the body as efficiently as possible with minimum discomfort to the skin. Shaving cream is used as a lubricant to minimize friction between the razor and the skin. Friction is helpful during walking and driving. Without adequate friction, we would slip and a car would skid! Tribology is also important in sports. For example, a low friction between the skis and the ice is desirable during skiing. Fabric fibers should have low friction when touching human skin.

Body joints need to be lubricated for low friction and low wear to avoid osteoarthritis and joint replacement. The surface layer of cartilage present in the joint provides the bearing surface and is lubricated with a joint fluid consisting of lubricin, hyaluronic acid (HA) and lipid. Hair conditioner coats hair in order to repair hair damage and lubricate it. It contains silicone and fatty alcohols. Low friction and adhesion provide a smooth feel in wet and dry environments, reduce friction between hair fibers during shaking and bouncing, and provide easy combing and styling. Skin creams and lotions are used to reduce friction between the fingers and body skin. Saliva and other mucous biofluids lubricate and facilitate the transport of food and soft liquids through the body. The saliva in the mouth interacts with food and influences the taste–mouth feel.

1.3 Origins and Significance of Micro/Nanotribology

At most interfaces of technological relevance, contact occurs at numerous levels of asperity. Consequently, the importance of investigating a single asperity contact in studies of the

Macrotribology	Micro/nanotribology
Large Mass Heavy Load	Small Mass (μg) Light Load (μg to mg)
↓	↓
Wear (Inevitable)	No Wear (Few atomic layers)
Bulk Material	Surface (few atomic layers)

Figure 1.3.1 Comparisons between macrotribology and micro/nanotribology.

fundamental tribological and mechanical properties of surfaces has long been recognized. The recent emergence and proliferation of proximal probes, in particular tip-based microscopies (the scanning tunneling microscope and the atomic force microscope) and of computational techniques for simulating tip-surface interactions and interfacial properties, have allowed systematic investigations of interfacial problems with high resolution as well as ways and means of modifying and manipulating nanoscale structures. These advances have led to the development of the new field of microtribology, nanotribology, molecular tribology, or atomic-scale tribology (Bhushan *et al.*, 1995; Bhushan, 1997, 1998, 2001b, 2010a, 2011). This field is concerned with experimental and theoretical investigations of processes ranging from atomic and molecular scales to microscales, occurring during adhesion, friction, wear, and thin-film lubrication at sliding surfaces.

The differences between the conventional or macrotribology and micro/nanotribology are contrasted in Figure 1.3.1. In macrotribology, tests are conducted on components with relatively large mass under heavily loaded conditions. In these tests, wear is inevitable and the bulk properties of mating components dominate the tribological performance. In micro/nanotribology, measurements are made on components, at least one of the mating components, with relatively small mass under lightly loaded conditions. In this situation, negligible wear occurs and the surface properties dominate the tribological performance.

The micro/nanotribological studies are needed to develop a fundamental understanding of interfacial phenomena on a small scale and to study interfacial phenomena in micro- and nano structures used in magnetic storage systems, micro/nanoelectromechanical systems (MEMS/NEMS), and other industrial applications. The components used in micro- and nano structures are very light (of the order of few micrograms) and operate under very light loads (of the order of a few micrograms to a few milligrams). As a result, friction and wear (on a nanoscale) of lightly-loaded micro/nano components are highly dependent on the surface interactions (few atomic layers). These structures are generally lubricated with molecularly-thin films. Micro- and nanotribological techniques are ideal ways to study the friction and wear processes of micro- and nanostructures. Although micro/nanotribological studies are critical to study micro- and nanostructures, these studies are also valuable in the fundamental understanding of interfacial phenomena in macrostructures to provide a bridge between science and engineering.

The scanning tunneling microscope, the atomic force and friction force microscopes, and the surface force apparatus are widely used for micro/nanotribological studies (Bhushan *et al.*, 1995; Bhushan, 1997, 1999). To give a historical perspective of the field, the scanning tunneling microscope (STM) developed by Doctors Gerd Binnig and Heinrich Rohrer and their colleagues in 1981 at the IBM Zurich Research Laboratory, the Forschungslabor, is the first instrument capable of directly obtaining three-dimensional (3D) images of solid surfaces with atomic resolution (Binnig *et al.*, 1982). STMs can only be used to study surfaces which are electrically conductive to some degree. Based on their design of the STM, in 1985, Binnig *et al.* (1986, 1987) developed an atomic force microscope (AFM) to measure ultrasmall forces (less than 1 μN) present between the AFM tip surface and the sample surface. AFMs can be used in the measurement of all engineering surfaces which may be either electrically conducting or insulating. AFM has become a popular surface profiler for topographic measurements on the micro- to nanoscale. AFMs modified to measure both normal and friction forces, generally called friction force microscopes (FFMs) or lateral force microscopes (LFMs), are used to measure friction on the micro- and nanoscales. AFMs are also used for studies of adhesion, scratching, wear, lubrication, surface temperatures, and for the measurement of elastic/plastic mechanical properties (such as indentation hardness and modulus of elasticity). Surface force apparatuses (SFAs), first developed in 1969, are used to study both static and dynamic properties of the molecularly thin liquid films sandwiched between two molecularly-smooth surfaces (Tabor and Winterton, 1969; Bhushan, 1999).

Meanwhile, significant progress in understanding the fundamental nature of bonding and interactions in materials, combined with advances in computer-based modeling and simulation methods, have allowed theoretical studies of complex interfacial phenomena with high resolution in space and time (Bhushan, 1999, 2001b, 2011). Such simulations provide insights into the atomic-scale energetics, structure, dynamics, thermodynamics, transport and rheological aspects of tribological processes. Furthermore, these theoretical approaches guide the interpretation of experimental data and the design of new experiments, and enable the prediction of new phenomena based on atomistic principles.

1.4 Organization of the Book

The friction, wear, and the lubrication behavior of interfaces is very dependent upon the surface material, the shape of mating surfaces and the operating environment. A surface film may change the physical and chemical properties of the first few atomic layers of material through interaction with the environment. Following this introductory, Chapter 2 includes a discussion on solid surface characterization. Chapter 2 includes a discussion on the nature of surfaces, the physico-chemical characteristics of solid surfaces, the statistical analysis of surface roughness, and the methods of characterization of solid surfaces. Chapter 3 is devoted to the elastic and plastic real area of contacts that occur when two solid surfaces are placed in contact. Statistical and numerical analyses and measurement techniques are presented. Chapter 4 covers various adhesion mechanisms in dry and wet conditions. Various analytical and numerical models to predict liquid-mediated adhesion are described. When the two surfaces in contact slide or roll against each other friction is encountered, thus, various friction mechanisms, the physical and chemical properties that control friction, and the typical friction data of materials are discussed in Chapter 5. Chapter 6 is devoted to the interface temperatures generated from the dissipation of the frictional energy input. Analysis and measurement techniques for interface temperatures and the impact of a temperature rise on an interface performance are discussed.

Repeated sliding or rolling results in wear. In Chapter 7, various wear mechanisms, types of particles present in wear debris, and representative data for various materials of engineering interest are presented. Chapter 8 reviews various the various regimes of lubrication, the theories of hydrostatic, hydrodynamic and elastohydrodynamic lubrication and various designs of bearings. In Chapter 9, mechanisms of boundary lubrication and the description of various liquid lubricants and additives and greases are presented. In Chapter 10, various experimental techniques and molecular dynamics computer simulation techniques used for micro/nanotribological studies and state-of-the art techniques and their applications are described and relevant data are presented. In Chapter 11, the design methodology and typical test geometries for friction and wear test methods are described.

In Chapter 12, descriptions, relevant wear mechanisms and commonly used materials for standard tribological components, microcomponents, material processing and industrial applications are presented. In Chapter 13, the fields of green tribology and biomimetics are introduced and various examples in each field are presented.

References

Amontons, G. (1699), "De la resistance causée dans les Machines," *Mémoires de l'Academic Royale, A*, 257–282.

Bhushan, B. (1996), *Tribology and Mechanics of Magnetic Storage Devices*, Second edition, Springer-Verlag, New York.

Bhushan, B. (1997), *Micro/Nanotribology and its Applications*, NATO ASI Series E: Applied Sciences, Vol. 330, Kluwer Academic, Dordrecht, The Netherlands.

Bhushan, B. (1998), *Tribology Issues and Opportunities in MEMS*, Kluwer Academic Publishers, Dordrecht, Netherlands.

Bhushan, B. (1999), *Handbook of Micro/Nanotribology*, Second edition, CRC Press, Boca Raton, Florida.

Bhushan, B. (2000), *Mechanics and Reliability of Flexible Magnetic Media*, Second edition, Springer-Verlag, New York.

Bhushan, B. (2001a), *Modern Tribology Handbook, Vol. 1 – Principles of Tribology; Vol. 2 – Materials, Coatings, and Industrial Applications*, CRC Press, Boca Raton, Florida.

Bhushan, B. (2001b), *Fundamentals of Tribology and Bridging the Gap Between the Macro- and Micro/Nanoscales*, NATO Science Series II: Mathematics, Physics and Chemistry – Vol. 10, Kluwer Academic Publishers, Dordrecht, Netherlands.

Bhushan, B. (2010a), *Springer Handbook of Nanotechnology*, Third edition, Springer-Verlag, Heidelberg, Germany.

Bhushan, B. (2010b), *Biophysics of Human Hair: Structural, Nanomechanical and Nanotribological Studies*, Springer-Verlag, Heidelberg, Germany.

Bhushan, B. (2011), *Nanotribology and Nanomechanics I – Measurement Techniques and Nanomechanics, II – Nanotribology, Biomimetics, and Industrial Applications*, Third edition, Springer-Verlag, Heidelberg, Germany.

Bhushan, B. (2012a), *Biomimetics: Bioinspired Hierarchical-Structured Surfaces for Green Science and Technology*, Springer-Verlag, Heidelberg, Germany.

Bhushan, B. (2012b), "Nanotribological and Nanomechanical Properties of Skin with and without Cream Treatment Using Atomic Force Microscopy and Nanoindentation (Invited Feature Article)," *Journal of Colloid and Interface Science* **367**, 1–33.

Bhushan, B. and Gupta, B.K. (1997), *Handbook of Tribology: Materials, Coatings and Surface Treatments*, McGraw-Hill, New York (1991); Reprinted with corrections, Krieger Publishing Co., Malabar, Florida.

Bhushan, B., Israelachvili, J.N., and Landman, U. (1995), "Nanotribology: Friction, Wear and Lubrication at the Atomic Scale," *Nature* **374**, 607–616.

Binnig, G., Rohrer, H., Gerber, Ch., and Weibel, E. (1982), "Surface Studies by Scanning Tunneling Microscopy," *Phys. Rev. Lett.* **49**, 57–61.

Binnig, G., Quate, C.F., and Gerber, Ch. (1986), "Atomic Force Microscope," *Phys. Rev. Lett.* **56**, 930–933.

Binnig, G., Gerber, Ch., Stoll, E., Albrecht, T.R., and Quate, C.F. (1987), "Atomic Resolution with Atomic Force Microscope," *Europhys. Lett.* **3**, 1281–1286.

Bowden, F.P. and Tabor, D. (1950), *The Friction and Lubrication of Solids*, Part I, Clarendon Press, Oxford, UK; Revised edition (1954); Paperback edition (1986).

Bowden, F.P. and Tabor, D. (1964), *The Friction and Lubrication of Solids*, Part II, Clarendon, Press, Oxford, UK.
Coulomb, C.A. (1785), "Théorie des Machines Simples, en ayant regard au Frottement de leurs Parties, et à la Roideur des Cordages," *Mem. Math. Phys.*, X, Paris, 161–342.
Davidson, C.S.C. (1957), "Bearings Since the Stone Age," *Engineering* **183**, 2–5.
Dowson, D. (1998), *History of Tribology*, Second edition, Instn Mech. Engrs, London, UK.
Holm, R. (1946), *Electrical Contacts*, Springer-Verlag, New York.
Jost, P. (1966), *Lubrication (Tribology) – A Report on the Present Position and Industry's Needs*, Dept. of Education and Science, H.M. Stationary Office, London.
Jost, P. (1976), "Economic Impact of Tribology," *Proc. Mechanical Failures Prevention Group*, NBS Spec. Pub. 423, Gaithersburg, Maryland.
Layard, A.G. (1853), *Discoveries in the Ruins of Nineveh and Babylon*, I and II, John Murray, Albemarle Street, London, UK.
Nosonovsky, M. and Bhushan, B. (2008), *Multiscale Dissipative Mechanisms and Hierarchical Surfaces: Friction, Superhydrophobicity, and Biomimetics*, Springer-Verlag, Heidelberg, Germany.
Nosonovsky, M. and Bhushan, B. (2012), *Green Tribology: Biomimetics, Energy Conservation and Sustainability*, Springer-Verlag, Heidelberg, Germany.
Parish, W.F. (1935), "Three Thousand Years of Progress in the Development of Machinery and Lubricants for the Hand Crafts," *Mill and Factory* **16** and **17**.
Petroff, N.P. (1883), "Friction in Machines and the Effects of the Lubricant," *Engng. J.* (in Russian), St Petersburg, 71–140, 228–279, 377–436, 535–564.
Reynolds, O.O. (1886), "On the Theory of Lubrication and its Application to Mr. Beauchamp Tower's Experiments," *Phil. Trans. R. Soc. Lond.* **177**, 157–234.
Tabor, D. and Winterton, R.H.S. (1969), "The Direct Measurement of Normal and Retarded van der Waals Forces," *Proc. R. Soc. Lond.* A **312**, 435–450.
Tower, B. (1884), "Report on Friction Experiments," *Proc. Inst. Mech. Engrs* **632**, 29–35.

2

Solid Surface Characterization

2.1 The Nature of Surfaces

A solid surface, or more exactly a solid–gas or solid–liquid interface, has a complex structure and complex properties dependent upon the nature of solids, the method of surface preparation, and the interaction between the surface and the environment. Properties of solid surfaces are crucial to surface interaction because surface properties affect real area of contact, friction, wear, and lubrication. In addition to tribological functions, surface properties are important in other applications, such as optical, electrical and thermal performance, painting, and appearance.

Solid surfaces, irrespective of the method of formation, contain irregularities or deviations from the prescribed geometrical form (Whitehouse, 1994; Bhushan, 1996; Thomas, 1999). The surfaces contain irregularities of various orders ranging from shape deviations to irregularities of the order of interatomic distances. No machining method, however precise, can produce a molecularly flat surface on conventional materials. Even the smoothest surfaces, such as those obtained by cleavage of some crystals, contain irregularities the heights of which exceed the interatomic distances. For technological applications, both macro- and micro/nanotopography of the surfaces (surface texture) are important.

In addition to surface deviations, the solid surface itself consists of several zones having physico-chemical properties peculiar to the bulk material itself (Figure 2.1.1) (Gatos, 1968; Haltner, 1969; Buckley, 1981). As a result of the forming process in metals and alloys, there is a zone of work-hardened or deformed material. Deformed layers would also be present in ceramics and polymers. These layers are extremely important because their properties, from a surface chemistry point of view, can be entirely different from the annealed bulk material. Likewise, their mechanical behavior is also influenced by the amount and depth of deformation of the surface layers.

Many of the surfaces are chemically reactive. With the exception of noble metals, all metals and alloys and many nonmetals form surface oxide layers in air, and in other environments they are likely to form other layers (for example, nitrides, sulfides, and chlorides). Besides the chemical corrosion film, there are also adsorbed films that are produced either by physisorption or chemisorption of oxygen, water vapor, and hydrocarbons, from the environment.

Introduction to Tribology, Second Edition. Bharat Bhushan.
© 2013 John Wiley & Sons, Ltd. Published 2013 by John Wiley & Sons, Ltd.

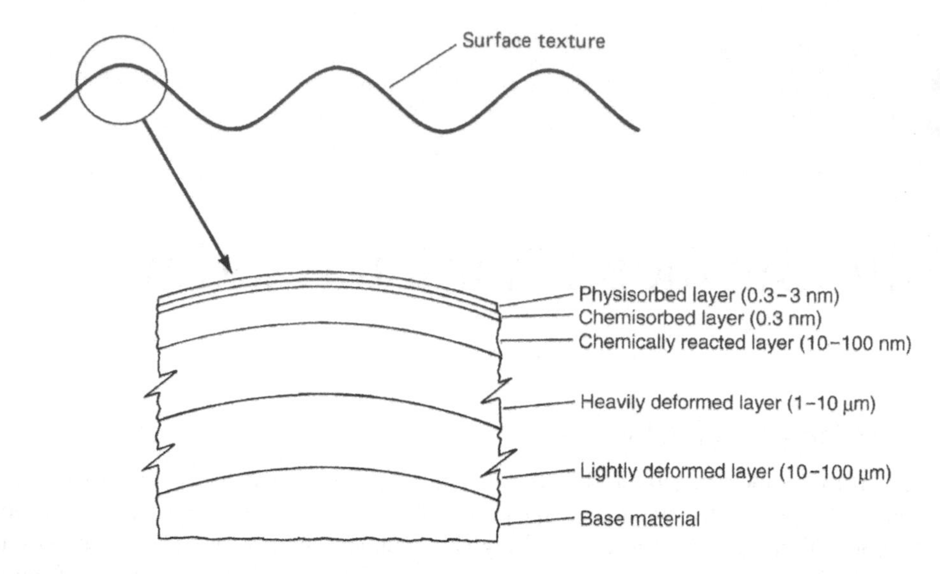

Figure 2.1.1 Solid surface details: surface texture (vertical axis magnified) and typical surface layers.

Occasionally, there will be a greasy or oily film derived from the environment. These films are found both on the metallic and nonmetallic surfaces.

The presence of surface films affects friction and wear. The effect of adsorbed films, even a fraction of a monolayer, is significant on the surface interaction. Sometimes, the films are worn out in the initial period of running and subsequently have no effect. The effect of greasy or soapy film, if present, is more marked; it reduces the severity of surface interaction often by one or more orders of magnitude.

Besides the chemical reactivity of the surfaces and the tendency of molecules to adsorb on it, which are regarded as extrinsic properties of the surface, an important property that must be considered is surface tension or surface free energy. This affects the adsorption behavior of the surfaces. Details on different surface layers will be presented next followed by the analysis of surface roughness and measurement of surface roughness.

2.2 Physico-Chemical Characteristics of Surface Layers

2.2.1 Deformed Layer

The metallurgical properties of the surface layer of a metal, alloy or a ceramic can vary markedly from the bulk of the material as a result of the forming process with which the material surface was prepared. For example, in grinding, lapping, machining, or polishing, the surface layers are plastically deformed with or without a temperature gradient and become highly strained. Residual stresses may be released of sufficient magnitude to affect dimensional stability. The strained layer is called the deformed (or work hardened) layer and is an integral part of the material itself in the surface region (Samuels, 1960; Bhushan, 1996; Shaw, 1997). The deformed layer can also be produced during the friction process (Cook and Bhushan, 1973).

The amount of the deformed material present and the degree of deformation that occurs are functions of two factors: (1) the amount of work or energy that was put into the deformation process; and (2) the nature of the material. Some materials are much more prone to deformation and work hardening than are others. The deformed layer would be more severely strained near the surface. The thickness of the lightly and heavily deformed layers typically ranges from 1 to 10 and 10 to 100 μm, respectively.

We generally find smaller grains in the deformed zone from recrystallization of the grains. In addition, the individual crystallite or grains with interface rubbing can orient themselves at the surface. The properties of the deformed layers can be entirely different from the annealed bulk material. Likewise, their mechanical behavior is also influenced by the amount and the depth of deformation of the surface layers.

2.2.2 *Chemically Reacted Layer*

With the exception of some noble metals (such as gold and platinum), all metals and alloys react with oxygen and form oxide layers in air; however, in other environments, they are quite likely to form other layers (for example, nitrides, sulfides, and chlorides) (Kubaschewski and Hopkins, 1953), Figure 2.2.1. With many non-oxide nonmetals, the oxide and other chemically reacted layers may also be present. For example, silicon exposed to air readily forms a silicon dioxide layer. In the case of oxides, for example, aluminum oxide, oxygen is an integral part of the structure, so an oxide layer is not expected. Polymers generally do not form an oxide layer. Interaction of surfaces with gases does not necessarily cease with the formation

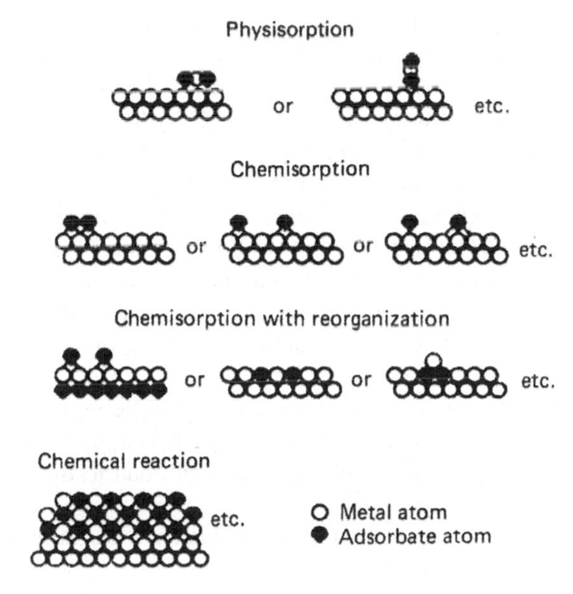

Figure 2.2.1 Schematic diagrams of physisorption, chemisorption, and a chemical reaction. Reproduced with permission from Buckley, D.H. (1981), *Surface Effects in Adhesion, Friction, Wear and Lubrication*, Elsevier, Amsterdam. Copyright 1981. Elsevier.

of an adsorbed monolayer. If a mechanism is available for the continuous exposure of new surface, the interaction with the ambient proceeds, leading to the formation of a thick film. The thickness of the oxide and other chemically reacted layers depends on the reactivity of the materials to the environment, reaction temperature, and reaction time. Typical thicknesses of these layers range from 10 to 100 nm, although much thicker layers can be formed.

Oxide layers can also be produced during the machining or the friction process. The heat released by almost all processing methods increases the rate of oxidation and leads to several types of oxides. During the friction process, because of a rise in temperature, the chemical reaction with the environment is accelerated. When a metal friction pair operates in air, the reaction may take place between the oxide layers of the two surfaces. The presence of lubricant and additives causes the formation of solid reaction layers that are important in surface protection.

Oxide layers may be of one or more elemental oxides. For example, on iron it may be iron oxide, or the film may contain a mixture of oxides such as Fe_2O_3, Fe_2O_4, and an innermost layer of FeO. With alloys, the surface oxides may consist of a mixture of oxides. For example, on stainless steels, the oxides may be a mixture of iron oxide and chromium oxide (Cr_2O_3).

With some materials, the oxides that are formed are very tenacious, very thin films form on the materials, and the surface becomes passivated with no further oxidation taking place: for example, aluminum and titanium surfaces. With some metals, however, the oxide can continue to grow; for example, Fe_2O_3 continues to grow in a humid air environment.

2.2.3 *Physisorbed Layer*

Besides the chemically reacted layer that forms on metals in reactive environments, adsorbed layers may be formed from the environment both on metallic or nonmetallic surfaces. For example, the admission of an inert gas, such as argon or krypton, to the surface can produce the physical adsorption of the argon to the clean surface. The most common constituents of adsorbate layers are molecules of water vapor, oxygen, or hydrocarbons from the environment that may be condensed and become physically adsorbed to the solid surface (Haltner, 1969). This layer can be either monomolecular (about 0.3 nm thick) or polymolecular.

With physisorption, no exchange of electrons takes place between the molecules of the adsorbate and those of the adsorbent. The physisorption process typically involves van der Waals forces, which are relatively weak compared to the forces acting in the liquefication of inert gases. It takes very little energy (1 to 2 kcal/mol) to remove physisorbed species from a solid surface, and all surfaces in high vacuum ($\sim 10^{-8}$ Pa *or* $\sim 10^{-10}$ Torr) are free of physisorbed species.

An example of physisorption is shown in Figure 2.2.1. The molecule depicted, bonding itself to the surface, is shown as a diatomic molecule, such as might occur in oxygen. In such a case, both oxygen atoms of the diatomic molecule can bond to the already contaminated surface.

Occasionally, there will also be greasy or oily film, which may partially displace the adsorbed layer derived from the environment. This greasy film may be derived from a variety of sources, such as the oil drops found in most industrial environments, the lubricants that were applied while the surface was being prepared, or natural greases from the fingers of people who handled the solid. The thickness of greasy films could be as small as 3 nm.

2.2.4 Chemisorbed Layer

In chemisorption, in contrast to physisorption, there is an actual sharing of electrons or electron interchange between the chemisorbed species and the solid surface. In chemisorption, the solid surface very strongly bonds to the adsorption species through covalent bonds; it therefore requires a great deal of energy comparable to those associated with chemical bond formation (10–100 kcal/mol) to remove the adsorbed species, the energy being a function of the solid surface to which the adsorbing species attaches itself and the character of the adsorbing species as well (Trapnell, 1955).

In chemisorption, the chemisorbing species, while chemically bonding to the surface, retain their own individual identity so that we can, by proper treatment of the surfaces, recover the initial adsorbing species. The chemisorbed layer is limited to a monolayer. This is a distinction between chemisorption and chemical reaction. Once the surface is covered with a layer, chemisorption ceases; any subsequent layer formation is either by physisorption or chemical reaction.

A series of qualitative criteria are available for establishing the difference between the two types of adsorption. A first criterion is the value of heat of adsorption. As chemical bonds are stronger than physical bonds, the heat of chemisorption will be greater than the heat of adsorption. Typical physisorption values range from 1 to 2 kcal/mol but typical chemisorption values range from 10 to 100 kcal/mol (1 kcal/mol = 4.187 kJ/mol = 0.1114 eV/atom).

Another criterion for differentiating between the two types of adsorption is the temperature range in which the process may take place. As distinguished from physisorption, chemisorption can also take place at temperatures much higher than the boiling point of the adsorbate. If adsorption takes place at a certain temperature and pressure (p) at which the pressure of the saturated vapors is p_0, then physisorption generally does not take place until the ratio p/p_0 reaches the value 0.01. This criterion cannot be considered absolute as for some active adsorbents, particularly those with a fine porous structure; gases and vapors can be adsorbed even at values of $p/p_0 = 10^{-8}$.

Another criterion used for distinguishing chemisorption from physisorption is the activation energy. For a high rate of chemisorption, a certain activation energy is necessary. This may be due to the existence of a temperature threshold below which chemisorption does not take place. As physical adsorption needs no activation energy, it will take place at a certain rate at any temperature, namely, at the rate at which the adsorbate reaches the solid surface. Likewise, chemisorption, as distinguished from physisorption, depends on the purity of the adsorbent surface. On the contrary, physisorption takes place on all surfaces.

Another difference between the two types of adsorption is the thickness of the adsorbed layer. While the chemisorption layer is always monomolecular, physisorbed layers may be either monomolecular or polymolecular.

A schematic diagram comparing physisorption, chemisorption, and a chemical reaction is shown in Figure 2.2.1.

2.2.5 Methods of Characterization of Surface Layers

Numerous surface analytical techniques that can be used for the characterization of surface layers are commercially available (Buckley, 1981; Bhushan, 1996). The metallurgical properties (grain structure) of the deformed layer can be determined by sectioning the surface and examining the cross section by a high-magnification optical microscope or a scanning electron microscope (SEM). Microcrystalline structure and dislocation density can be studied by preparing thin samples (a few hundred nm thick) of the cross section and examining them with a transmission electron microscope (TEM). The crystalline structure of a surface layer can also be studied by X-ray, high-energy or low-energy electron diffraction techniques. An elemental analysis of a surface layer can be performed by an X-ray energy dispersive analyzer (X-REDA) available with most SEMs, an Auger electron spectroscope (AES), an electron probe microanalyzer (EPMA), an ion scattering spectrometer (ISS), a Rutherford backscattering spectrometer (RBS), or by X-ray fluorescence (XRF). The chemical analysis can be performed using X-ray photoelectron spectroscopy (XPS) and secondary ion mass spectrometry (SIMS). The thickness of the layers can be measured by depth-profiling a surface, while simultaneously conducting surface analysis. The thickness and severity of deformed layer can be measured by measuring residual stresses in the surface.

The chemical analysis of adsorbed organic layers can be conducted by using surface analytical tools, such as mass spectrometry, Fourier transform infrared spectroscopy (FTIR), Raman scattering, nuclear magnetic resonance (NMR) and XPS. The most commonly used techniques for the measurement of organic layer (including lubricant) thickness are depth profiling using XPS and ellipsometry.

2.3 Analysis of Surface Roughness

Surface texture is the repetitive or random deviation from the nominal surface that forms the three-dimensional topography of the surface. Surface texture includes: (1) roughness (nano- and microroughness); (2) waviness (macroroughness); (3) lay; and (4) flaws. Figure 2.3.1 is a pictorial display of surface texture with unidirectional lay.

Nano- and microroughness are formed by fluctuations in the surface of short wavelengths, characterized by hills (asperities) (local maxima) and valleys (local minima) of varying amplitudes and spacings, and these are large compared to molecular dimensions. Asperities are referred to as peaks in a profile (two dimensions) and summits in a surface map (three dimensions). Nano- and microroughness include those features intrinsic to the production process. These are considered to include traverse feed marks and other irregularities within the limits of the roughness sampling length. Waviness is the surface irregularity of longer wavelengths and is referred to as macroroughness. Waviness may result from such factors as machine or workpiece deflections, vibration, chatter, heat treatment, or warping strains. Waviness includes all irregularities whose spacing is greater than the roughness sampling length and less than the waviness sampling length. Lay is the principal direction of the predominant surface pattern, ordinarily determined by the production method. Flaws are unintentional, unexpected, and unwanted interruptions in the texture. In addition, the surface may contain gross deviations from nominal shape of very long wavelength, which is known as error of form. They are not normally considered part of the surface texture. A question often asked is whether various geometrical features should be assessed together or separately. What features are included together depends on the applications. It is generally not possible to measure all the features at the same time.

A very general typology of a solid surface is seen in Figure 2.3.2. Surface textures that are deterministic may be studied by relatively simple analytical and empirical methods; their

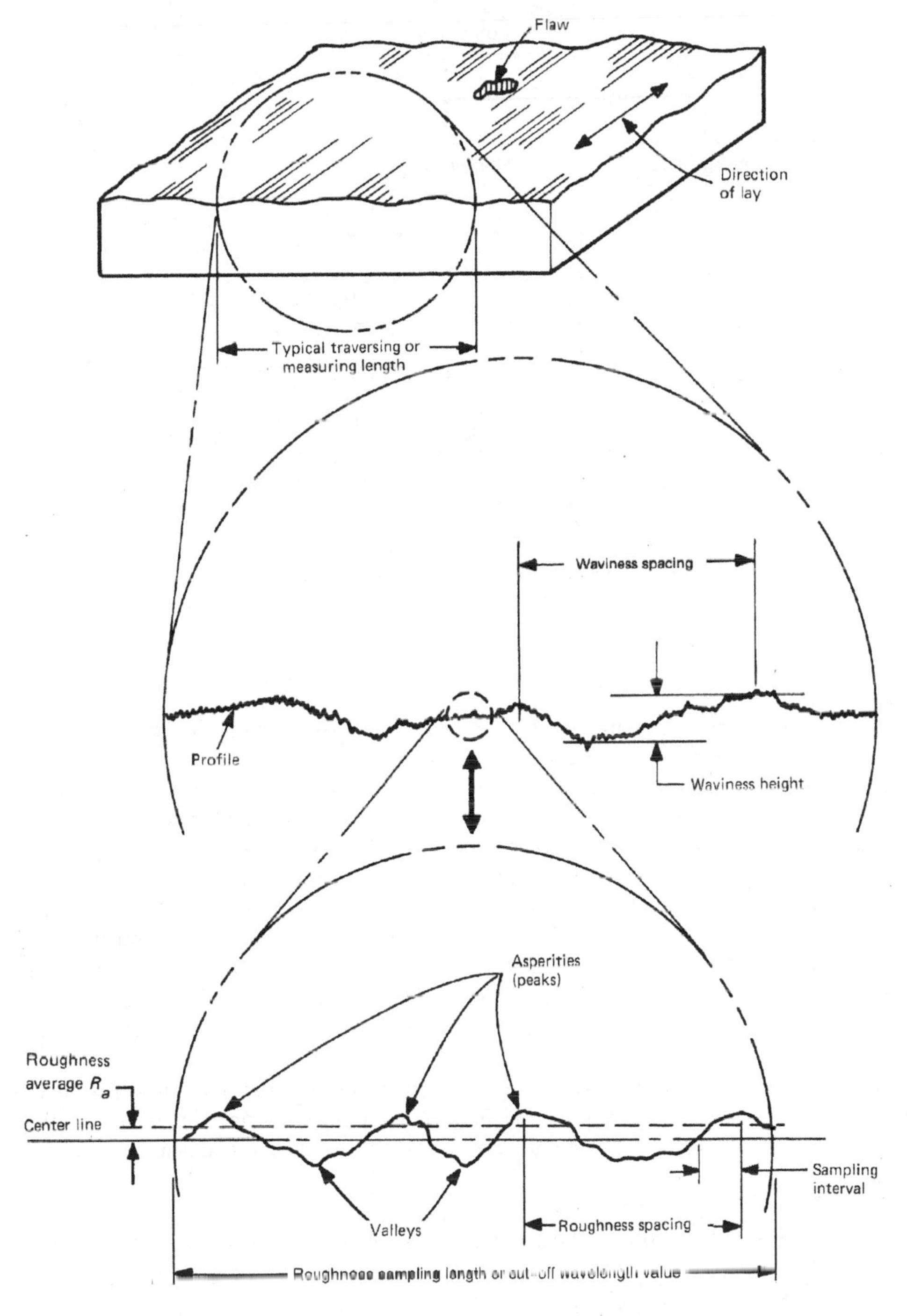

Figure 2.3.1 Pictorial display of surface texture. (*Source*: Anonymous, 1985). Reproduced from ASME B46.1-1985, by permission of The American Society of Mechanical Engineers. All rights reserved. No further copies can be made without written permission.

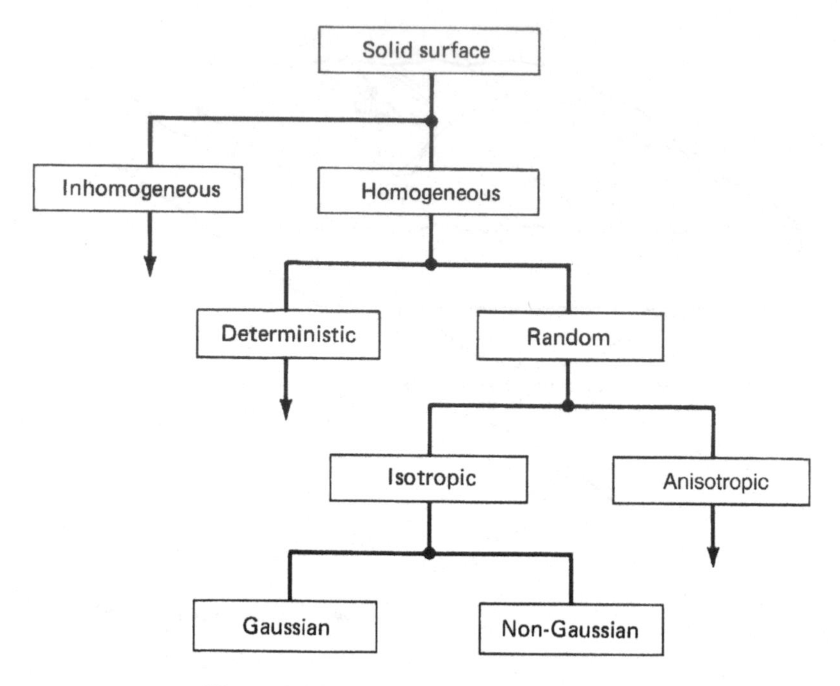

Figure 2.3.2 General typology of surfaces.

detailed characterization is straightforward. However, the textures of most engineering surfaces are random, either isotropic or anisotropic, and either Gaussian or non-Gaussian. Whether the surface height distribution is isotropic or anisotropic and Gaussian or non-Gaussian depends upon the nature of the processing method. Surfaces that are formed by so called cumulative processes (such as peening, electropolishing and lapping) in which the final shape of each region is the cumulative result of a large number of random discrete local events and irrespective of the distribution governing each individual event, will produce a cumulative effect that is governed by the Gaussian form; it is a direct consequence of the central limit theorem of statistical theory. Single-point processes (such as turning and shaping) and extreme-value processes (such as grinding and milling) generally lead to anisotropic and non-Gaussian surfaces. The Gaussian (normal) distribution has become one of the mainstays of surface classification.

In this section, we first define average roughness parameters followed by statistical analyses and fractal characterization of surface roughness that are of importance in contact problems. Emphasis is placed on random, isotropic surfaces that follow a Gaussian distribution.

2.3.1 Average Roughness Parameters

2.3.1.1 Amplitude Parameters

Surface roughness most commonly refers to the variations in the height of the surface relative to a reference plane. It is measured either along a single line profile or along a set of parallel

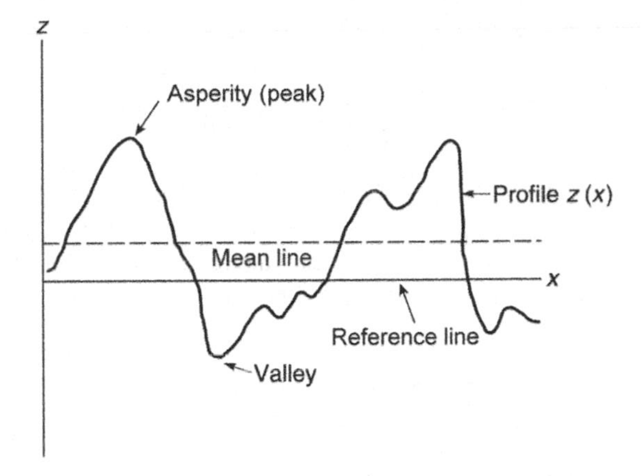

Figure 2.3.3 Schematic of a surface profile z(x).

line profiles (surface maps). It is usually characterized by one of the two statistical height descriptors advocated by the American National Standards Institute (ANSI) and the International Standardization Organization (ISO) (Anonymous, 1975, 1985). These are (1) R_a, CLA (center-line average), or AA (arithmetic average) and (2) the standard deviation or variance (σ), R_q or root mean square (*RMS*). Two other statistical height descriptors are skewness (*Sk*) and kurtosis (*K*); these are rarely used. Another measure of surface roughness is an extreme-value height descriptor (Anonymous, 1975, 1985) R_t(or R_y, $R_{\max}$, or maximum peak-to-valley height or simply *P-V* distance). Four other extreme-value height descriptors in limited use, are: R_p (maximum peak height, maximum peak-to-mean height or simply *P-M* distance), R_v (maximum valley depth or mean-to-lowest valley height), R_z (average peak-to-valley height) and R_{pm} (average peak-to-mean height).

We consider a profile, $z(x)$ in which profile heights are measured from a reference line, Figure 2.3.3. We define a center line or mean line as the line such that the area between the profile and the mean line above the line is equal to that below the mean line. R_a, *CLA* or *AA* is the arithmetic mean of the absolute values of vertical deviation from the mean line through the profile. The standard deviation σ is the square root of the arithmetic mean of the square of the vertical deviation from the mean line.

In mathematical form, we write

$$R_a = CLA = AA = \frac{1}{L}\int_0^L |z - m|dx \tag{2.3.1a}$$

and

$$m = \frac{1}{L}\int_0^L z dx \tag{2.3.1b}$$

where L is the sampling length of the profile (profile length).

The variance is given as

$$\sigma^2 = \frac{1}{L}\int_0^L (z-m)^2 dx \tag{2.3.2a}$$

$$= R_q^2 - m^2 \tag{2.3.2b}$$

where σ is the standard deviation and R_q is the square root of the arithmetic mean of the square of the vertical deviation from a reference line, or

$$R_q^2 = RMS^2 = \frac{1}{L}\int_0^L (z^2)\, dx \tag{2.3.3a}$$

For the special case where m is equal to zero,

$$R_q = \sigma \tag{2.3.3b}$$

In many cases, R_a and σ are interchangeable, and for Gaussian surfaces,

$$\sigma \sim \sqrt{\frac{\pi}{2}} R_a \sim 1.25\, R_a \tag{2.3.4}$$

The value of R_a is an official standard in most industrialized countries. Table 2.3.1 gives internationally adopted R_a values together with the alternative roughness grade number. The standard deviation σ is most commonly used in statistical analyses.

The skewness and kurtosis in the normalized form are given as

$$Sk = \frac{1}{\sigma^3 L}\int_0^L (z-m)^3 dx \tag{2.3.5}$$

Table 2.3.1 Center-line average and roughness grades.

R_a values up to μm	Roughness grade number
0.025	N1
0.05	N2
0.1	N3
0.2	N4
0.4	N5
0.8	N6
1.6	N7
3.2	N8
6.3	N9
12.5	N10
25.0	N11

and

$$K = \frac{1}{\sigma^4 L} \int_0^L (z - m)^4 dx \tag{2.3.6}$$

More discussion of these two descriptors will be presented later.

Five extreme-value height descriptors are defined as follows: R_t is the distance between the highest asperity (peak or summit) and the lowest valley; R_p is defined as the distance between the highest asperity and the mean line; R_v is defined as the distance between the mean line and the lowest valley; R_z is defined as the distance between the averages of five highest asperities and the five lowest valleys; and R_{pm} is defined as the distance between the averages of five highest asperities and the mean line. The reason for taking an average value of asperities and valleys is to minimize the effect of unrepresentative asperities or valleys which occasionally occur and can give an erroneous value if taken singly. R_z and R_{pm} are more reproducible and are advocated by ISO. In many tribological applications, height of the highest asperities above the mean line is an important parameter because damage of the interface may be done by the few high asperities present on one of the two surfaces; on the other hand, valleys may affect lubrication retention and flow.

The height parameters R_a (or σ in some cases) and R_t (or R_p in some cases) are most commonly specified for machine components. For the complete characterization of a profile or a surface, any of the parameters discussed earlier are not sufficient. These parameters are seen to be primarily concerned with the relative departure of the profile in the vertical direction only; they do not provide any information about the slopes, shapes, and sizes of the asperities or about the frequency and regularity of their occurrence. It is possible, for surfaces of widely differing profiles with different frequencies and different shapes to give the same R_a or σ (R_q) values (Figure 2.3.4). These single numerical parameters are mainly useful for classifying surfaces of the same type that are produced by the same method.

Average roughness parameters for surface maps are calculated using the same mathematical approach as that for a profile presented here.

Example Problem 2.3.1

Consider two sinusoidal profiles with wavelengths λ *and* 2λ and a maximum amplitude A_0. Show that (a) R_a and (b) σ for the two profiles are the same.

Solution

The expression for a sinusoidal profile of wavelength λ is

$$z(x) = A_0 \sin\left(\frac{2\pi}{\lambda}x\right)$$

and

$$\mathrm{m} = 0 \tag{2.3.7}$$

One can select any profile length with multiples of the length of the repeated wave structure in terms of height (quarter of the wavelength for a sine or a cosine wave). Here, we select two

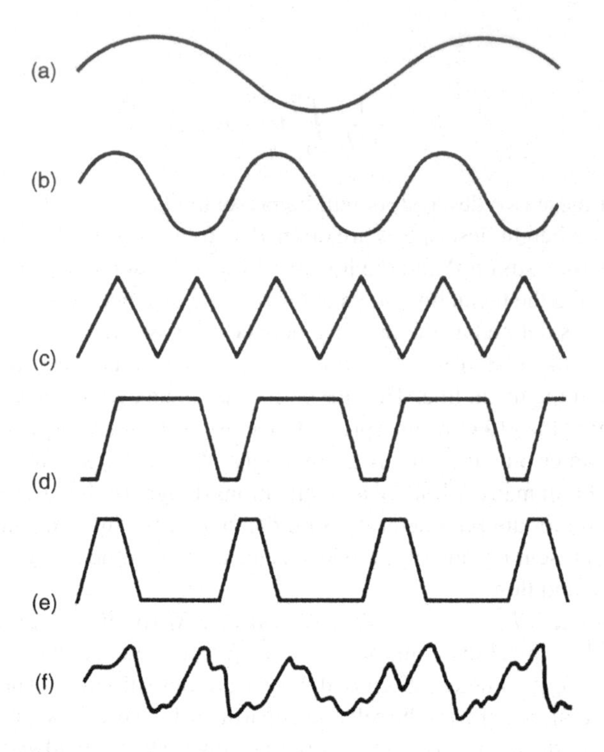

Figure 2.3.4 Various surface profiles having the same R_a value.

profile lengths of quarter and one wavelength for demonstration that one gets the same results irrespective of the differences in the profile length.

(a) If the profile length is $\lambda/4$,

$$\begin{aligned} R_a &= \frac{1}{L}\int_0^L |z-m|\,dx = \frac{4}{\lambda}\int_0^{\lambda/4} A_0 \sin\left(\frac{2\pi}{\lambda}x\right)dx \\ &= -\left(\frac{2A_0}{\pi}\right)\cos\left(\frac{2\pi}{\lambda}x\right)\Big|_0^{\lambda/4} \\ &= \frac{2A_0}{\pi} \end{aligned} \tag{2.3.8a}$$

If the profile length is λ,

$$\begin{aligned} R_a &= \frac{1}{\lambda}\left[\int_0^{\lambda/2} A_0 \sin\left(\frac{2\pi}{\lambda}x\right)dx - \int_{\lambda/2}^{\lambda} A_0 \sin\left(\frac{2\pi}{\lambda}x\right)dx\right] \\ &= \frac{2A_0}{\pi} \end{aligned} \tag{2.3.8b}$$

As expected, the value of R_a is independent of the profile length. Furthermore, R_a is independent of the wavelength.

(b) For a profile length of quarter wavelength,

$$\begin{aligned}\sigma^2 &= \frac{1}{L}\int_0^L (z-m)^2\,dx = \frac{4}{\lambda}\int_0^{\lambda/4} A_0^2 \sin^2\left(\frac{2\pi}{\lambda}x\right)dx \\ &= \frac{4}{\lambda}\int_0^{\lambda/4} A_0^2\left[\frac{1}{2}-\frac{1}{2}\cos\left(\frac{4\pi}{\lambda}x\right)\right]dx \\ &= \frac{2A_0^2}{\lambda}\left[x-\frac{\lambda}{4\pi}\sin\left(\frac{4\pi}{\lambda}x\right)\right]_0^{\lambda/4} \\ &= \frac{A_0^2}{2}\end{aligned}$$

Therefore,

$$\sigma = \frac{A_0}{\sqrt{2}} \tag{2.3.9}$$

The preceding expression for σ^2 can be used for a profile length that is a multiple of $\lambda/4$. Again σ is independent of the wavelength.

Example Problem 2.3.2

Consider a sinusoidal and two triangular profiles with wavelength λ as shown in Figure 2.3.5. Calculate the relationships between the maximum amplitudes of the two profiles which give the same values of R_a and σ.

Solution

Expressions of R_a and σ for a sinusoidal profile have been obtained in the Example Problem 2.3.1. We calculate expressions for two triangular profiles of maximum amplitude A_1. Expression for the triangular profile shown in Figure 2.3.5b is given as

$$\begin{aligned}z &= \frac{4A_1}{\lambda}x, \quad x \le \lambda/4 \\ &= 2A_1\left[1-\frac{2}{\lambda}x\right], \quad \frac{\lambda}{4} \le x \le \frac{3\lambda}{4} \\ &= 4A_1\left[-1+\frac{1}{\lambda}x\right], \quad \frac{3\lambda}{4} \le x \le \lambda\end{aligned}$$

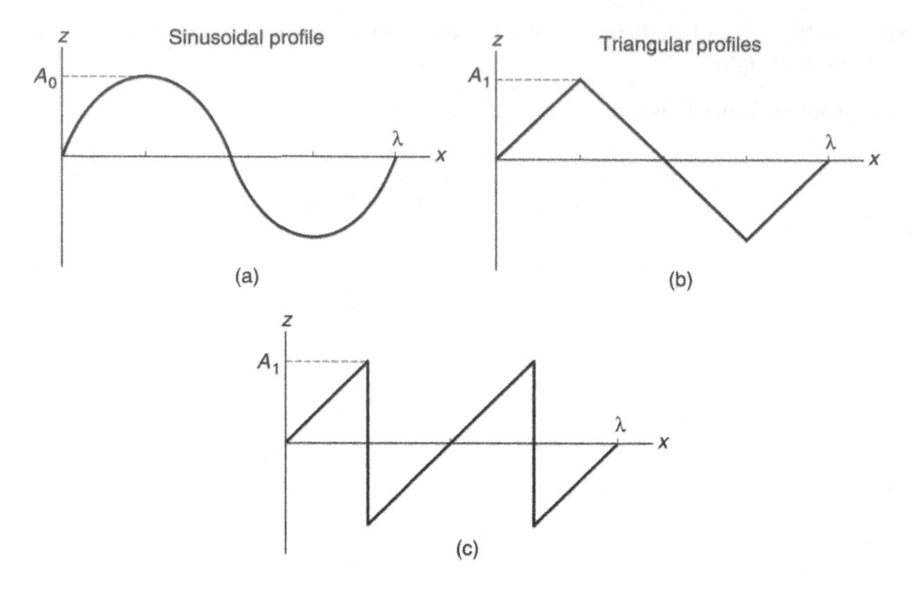

Figure 2.3.5 Schematics of (a) a sinusoidal and (b, c) two triangular profiles.

We only need to consider a profile length of $\lambda/4$. For this profile,

$$\begin{aligned} R_a &= \frac{4}{\lambda}\int_0^{\lambda/4} \frac{4A_1}{\lambda} x\, dx \\ &= \frac{A_1}{2} \\ \sigma^2 &= \frac{4}{\lambda}\int_0^{\lambda/4} \frac{16A_1^2}{\lambda^2} x^2 dx \\ &= \frac{A_1^2}{3} \end{aligned} \tag{2.3.10.a}$$

Therefore,

$$\sigma = \frac{A_1}{\sqrt{3}} \tag{2.3.10.b}$$

Next, we calculate the relationships between the maximum amplitudes of the sinusoidal profile and the triangular profile (b), using Equations (2.3.8) to (2.3.10).

$$\text{For the same } R_a, \quad A_0 = \frac{\pi}{4} A_1 \tag{2.3.11a}$$

$$\text{For the same } \sigma, \quad A_0 = \sqrt{\frac{2}{3}} A_1 \tag{2.3.11b}$$

Finally we consider the second triangular profile (c). Expressions for R_a *and* σ are the same as that for the triangular profile (b).

2.3.1.2 Spacing (or Spatial) Parameters

One way to supplement the amplitude (height) information is to provide some index of crest spacing or wavelength (which corresponds to lateral or spatial distribution) on the surface. Two parameters occasionally used are the peak (or summit) density, N_p (η), and zero crossings density, N_0. N_p is the density of peaks (local maxima) of the profile in number per unit length and η is the density of summits of the surface in number per unit area. N_p and η are just a measure of maxima irrespective of height. This parameter is in some use. N_0 is the zero crossings density defined as the number of times the profile crosses the mean line per unit length. From Longuet-Higgins (1957a), the number of surface zero crossings per unit length is given by the total length of the contour where the autocorrelation function (to be described later) is zero (or 0.1) divided by the area enclosed by the contour. This count N_0 is rarely used.

A third parameter – mean peak spacing (A_R) is the average distance between measured peaks. This parameter is merely equal to $\left(1/N_p\right)$. Other spacial parameters rarely used are the mean slope and mean curvature which are the first and second derivative of the profile/surface, respectively.

2.3.2 Statistical Analyses

2.3.2.1 Amplitude Probability Distribution and Density Functions

The cumulative probability distribution function or simply cumulative distribution function (CDF), $P(h)$ associated with the random variable $z(x)$, which can take any value between $-\infty$ *and* ∞ *or* $z_{\min}$ *and* $z_{\max}$, is defined as the probability of the event $z(x) \leq h$ and is written as (McGillem and Cooper, 1984; Bendat and Piersol, 1986)

$$P(h) = Prob(z \leq h) \tag{2.3.12}$$

with $P(-\infty) = 0$ *and* $P(\infty) = 1$.

It is common to describe the probability structure of random data in terms of the slope of the distribution function given by the derivative

$$p(z) = \frac{dP(z)}{dz} \tag{2.3.13a}$$

where the resulting function $p(z)$ is called the probability density function (PDF). Obviously, the cumulative distribution function is the integral of the probability density function $p(z)$, that is,

$$P(z \leq h) = \int_{-\infty}^{h} p(z)\,dz = P(h) \tag{2.3.13b}$$

and

$$P(h_1 \leq z \leq h_2) = \int_{h_1}^{h_2} p(z)\,dz = P(h_2) - P(h_1) \tag{2.3.13c}$$

Furthermore, the total area under the probability density function must be unity; that is, it is certain that the value of z at any x must fall somewhere between plus and minus infinity or $z_{\max}$ *and* $z_{\min}$.

The data representing a wide collection of random physical phenomenon in practice tend to have a Gaussian or normal probability density function,

$$p(z) = \frac{1}{\sigma(2\pi)^{1/2}} \exp\left[-\frac{(z-m)^2}{2\sigma^2}\right] \tag{2.3.14a}$$

where σ is the standard deviation and m is the mean.

For convenience, the Gaussian function is plotted in terms of a normalized variable,

$$z^* = (z - m)/\sigma \tag{2.3.14b}$$

which has zero mean and unity standard deviation. With this transformation of variables, Equation (2.3.14a) becomes

$$p(z^*) = \frac{1}{(2\pi)^{1/2}} \exp\left[\frac{-(z^*)^2}{2}\right] \tag{2.3.14c}$$

which is called the standardized Gaussian or normal probability density function. To obtain $P(h)$ from $p(z^*)$ of Equation (2.3.14c), the integral cannot be performed in terms of the common functions, and the integral is often listed in terms of the "error function" and its values are listed in most statistical text books. The error function is defined as

$$erf(h) = \frac{1}{(2\pi)^{1/2}} \int_0^h \exp\left[\frac{-(z^*)^2}{2}\right] dz^* \tag{2.3.15}$$

An example of a random variable $z^*(x)$ with its Gaussian probability density and corresponding cumulative distribution functions are shown in Figure 2.3.6. Examples of $P(h)$ and $P(z^* = h)$ are also shown. The probability density function is a bell-shaped and the cumulative distribution function is an S-shaped appearance.

We further note that for a Gaussian function

$$P(-1 \le z^* \le 1) = 0.682$$
$$P(-2 \le z^* \le 2) = 0.954$$
$$P(-3 \le z^* \le 3) = 0.999$$

and

$$P(-\infty \le z^* \le \infty) = 1$$

which implies that the probabilities of some number that follows a Gaussian distribution is within the limits of $\pm 1\sigma$, $\pm 2\sigma$, and $\pm 3\sigma$ are 68.2, 95.4, and 99.9%, respectively.

A convenient method for testing for Gaussian distribution is to plot the cumulative distribution function on a probability graph paper to show the percentage of the numbers below a given number; this is scaled such that a straight line is produced when the distribution is Gaussian (typical data to be presented later). To test for Gaussian distribution, a straight line corresponding to a Gaussian distribution is drawn on the plot. The slope of the straight line

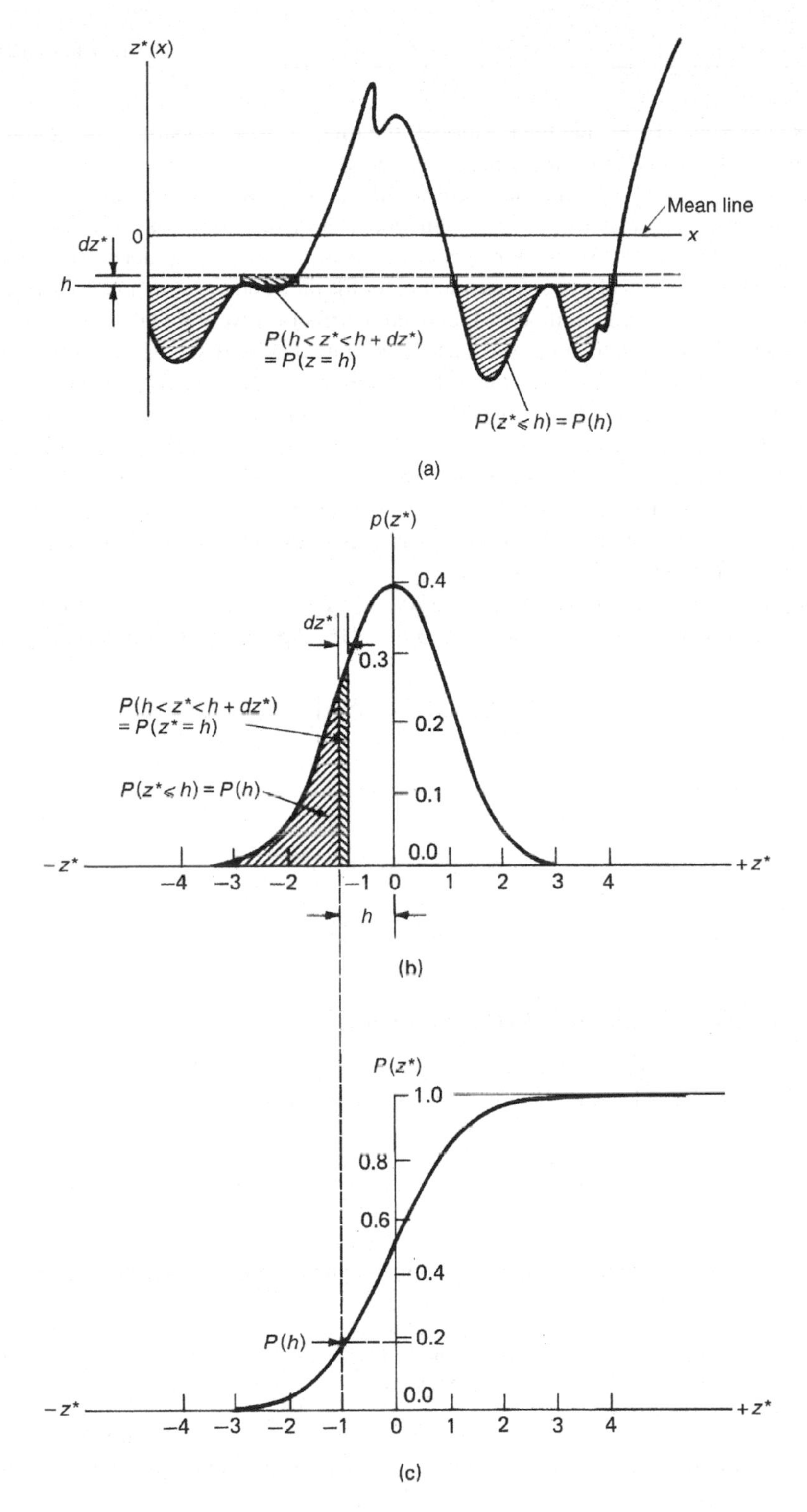

Figure 2.3.6 (a) Random function $z^*(x)$, which follows Gaussian probability functions, (b) Gaussian probability density function $p(z^*)$, and (c) Gaussian probability distribution function $P(z^*)$.

portion is determined by σ, and the position of the line for 50% probability is set at the mean value (which is typically zero for surface height data).

The most practical method for the goodness of the fit between the given distribution and the Gaussian distribution is to use the Kolmogorov–Smirnov test (Smirnov, 1948; Massey, 1951; Siegel, 1956). In the Kolmogorov–Smirnov test, the maximum departure between the percentage of the numbers above a given number for the data and the percentage of the numbers that would be above a given number if the given distribution were a Gaussian distribution is first calculated. Then, a calculation is made to determine if indeed the distribution is Gaussian. The level of significance, P, is calculated; this gives the probability of mistakenly or falsely rejecting the hypothesis that the distribution is a Gaussian distribution. Common minimum values for P for accepting the hypothesis are 0.01–0.05 (Siegel, 1956). The chi-square test (Siegel, 1956) can also be used to determine how well the given distribution matches a Gaussian distribution. However, the chi-square test is not very useful because the goodness of fit calculated depends too much upon how many bins or discrete cells the surface height data are divided into (Wyant *et al.*, 1986).

For the sake of mathematical simplicity in some analyses, sometimes an exponential distribution is used instead of the Gaussian distribution. The exponential distribution is given as

$$p(z) = \frac{1}{\sigma} \exp\left[-\frac{(z-m)}{\sigma}\right], \quad z \geq m \tag{2.3.16a}$$

or

$$p(z^*) = \exp(-z^*) \tag{2.3.16b}$$

In this function, m is the minimal value of the variable.

2.3.2.2 Moments of Amplitude Probability Functions

The shape of the probability density function offers useful information on the behavior of the process. This shape can be expressed in terms of moments of the function,

$$m_n = \int_{-\infty}^{\infty} z^n p(z)\, dz \tag{2.3.17}$$

m_n is called the nth moment. Moments about the mean are referred to as central moments,

$$m_n^c = \int_{-\infty}^{\infty} (z-m)^n p(z)\, dz \tag{2.3.18}$$

The zeroth moment ($n = 0$) is equal to 1. The first moment is equal to m, mean value of the function $z(x)$, whereas the first central moment is equal to zero. For completeness we note that,

$$R_a = \int_{-\infty}^{\infty} |z-m|\, p(z)\, dz \tag{2.3.19}$$

The second moments are

$$m_2 = \int_{-\infty}^{\infty} z^2 p(z)\, dz = R_q^2 \tag{2.3.20}$$

and

$$m_2^c = \int_{-\infty}^{\infty} (z - m)^2 p(z)\, dz = \sigma^2 \tag{2.3.21a}$$

$$= R_q^2 - m^2 \tag{2.3.21b}$$

The third moment m_3^c is the skewness (Sk), a useful parameter in defining variables with an asymmetric spread and represents the degree of symmetry of the density function, Figure 2.3.7. It is usual to normalize the third central moment as

$$Sk = \frac{1}{\sigma^3} \int_{-\infty}^{\infty} (z - m)^3 p(z)\, dz \tag{2.3.22}$$

Symmetrical distribution functions, including Gaussian, have zero skewness.

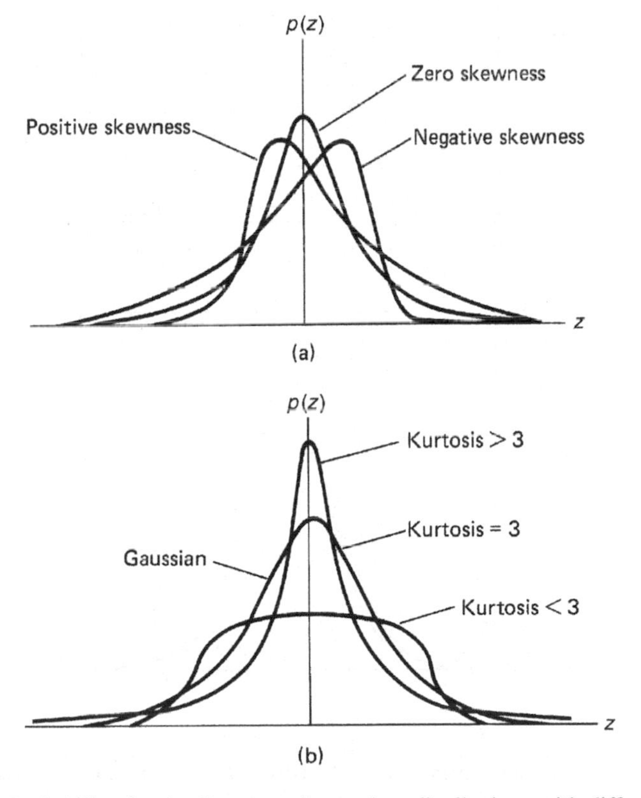

Figure 2.3.7 (a) Probability density functions for random distributions with different skewness, and for (b) symmetrical distributions (zero skewness) with different kurtosis.

The fourth moment m_4^c is the kurtosis (K) and represents the peakedness of the density and is a measure of the degree of pointedness or bluntness of a density function, Figure 2.3.7. Again, it is usual to normalize the fourth central moment as

$$K = \frac{1}{\sigma^4} \int_{-\infty}^{\infty} (z - m)^4 p(z) \, dz \tag{2.3.23}$$

Note that the symmetric Gaussian distribution has a kurtosis of 3. Distributions with $K > 3$ are called leptokurtic and those with $K < 3$ are called platykurtic.

Kotwal and Bhushan (1996) developed an analytical method to generate probability density functions for non-Gaussian distributions using the so-called Pearson system of frequency curves based on the methods of moments. Chilamakuri and Bhushan (1998) generated non-Gaussian distributions on the computer. The probability density functions are plotted in Figure 2.3.8. From this figure, it can be seen that a Gaussian distribution with zero skewness

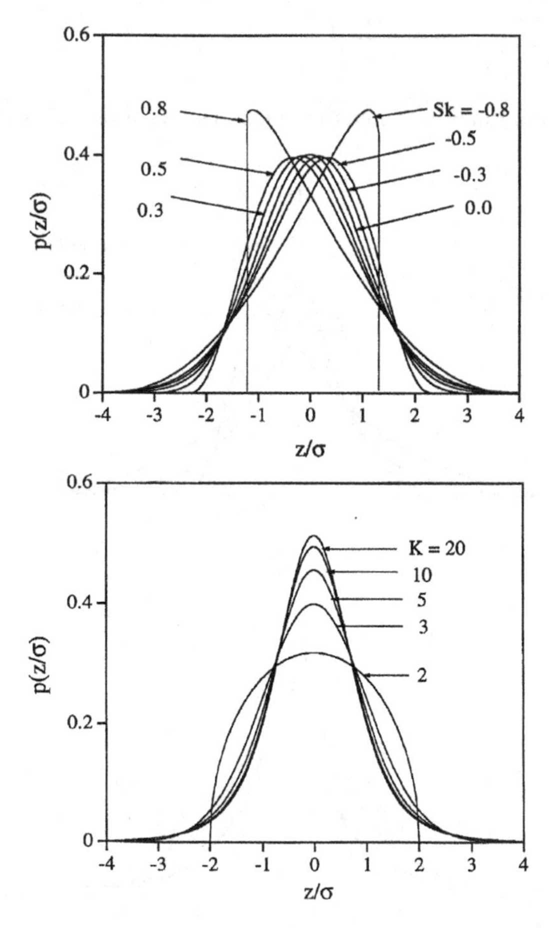

Figure 2.3.8 Probability density function for random distributions with selected skewness and kurtosis values.

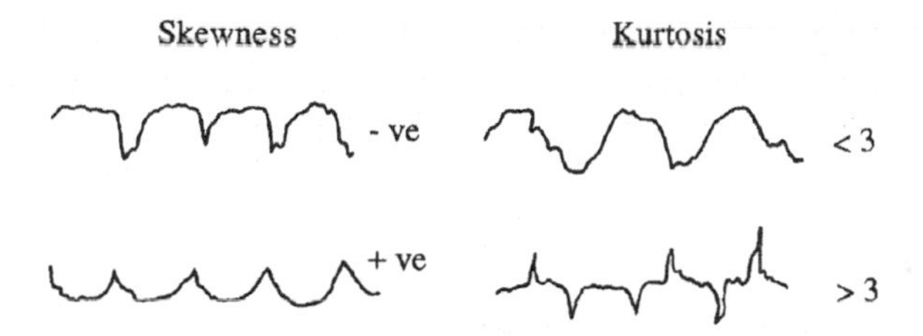

Figure 2.3.9 Schematic illustration for random functions with various skewness and kurtosis values.

and a kurtosis of three has an equal number of local maxima and minima at a certain height above and below the mean line. A surface with a high negative skewness has a larger number of local maxima above the mean as compared to a Gaussian distribution; for a positive skewness, the converse is true, Figure 2.3.9. Similarly, a surface with a low kurtosis has a larger number of local maxima above the mean as compared to that of a Gaussian distribution; again, for a high kurtosis, the converse is true, Figure 2.3.9.

In practice, many engineering surfaces have symmetrical Gaussian height distribution. Experience with most engineering surfaces shows that the height distribution is Gaussian at the high end, but at the lower end, the bottom 1–5% of the distribution is generally found to be non-Gaussian (Williamson, 1968). Many of the common machining processes produce surfaces with non-Gaussian distribution, Figure 2.3.10. Turning, shaping and electrodischarge

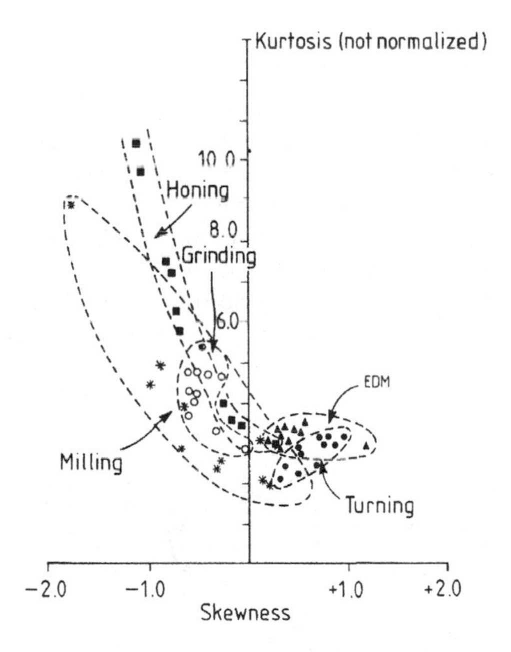

Figure 2.3.10 Typical skewness and kurtosis envelopes for various manufacturing processes. Reproduced with permission from Whitehouse, D.J. (1994), *Handbook of Surface Metrology*, Institute of Physics Publishing, Bristol. Copyright 1994. Taylor and Francis.

machining (EDM) processes produce surfaces with positive skewness. Grinding, honing, milling and abrasion processes produce grooved surfaces with negative skewness but high kurtosis values. Laser polishing produces surfaces with high kurtosis.

Example Problem 2.3.3

Write an expression for *Sk* in terms of moments.

Solution

$$\begin{aligned} Sk &= \frac{1}{\sigma^3}\int_{-\infty}^{\infty}(z-m)^3\,p(z)\,dz \\ &= \frac{1}{\sigma^3}\int_{-\infty}^{\infty}\left(z^3-m^3-3mz^2+3m^2z\right)p(z)\,dz \\ &= \frac{1}{\sigma^3}\left[m_3-3m\,m_2+2m^3\right] \\ &= \frac{1}{\sigma^3}\left[m_3-3m\,\sigma^2-m^3\right] \end{aligned}$$

where m is the first moment equal to the mean value of the function z.

2.3.2.3 Surface Height Distribution Functions

If the surface or profile heights are considered as random variables, then their statistical representation in terms of the probability density function $p(z)$ is known as the height distribution or a histogram. The height distribution can also be represented as cumulative distribution function $P(z)$. For a digitized profile, the histogram is constructed by plotting the number or fraction of surface heights lying between two specific heights as a function of height, Figure 2.3.11. The interval between two such heights is termed the class interval and is shown as dz in Figure 2.3.11. It is generally recommended to use 15–50 class intervals for general random data, but choice is usually a trade-off between accuracy and resolution. Similarly,

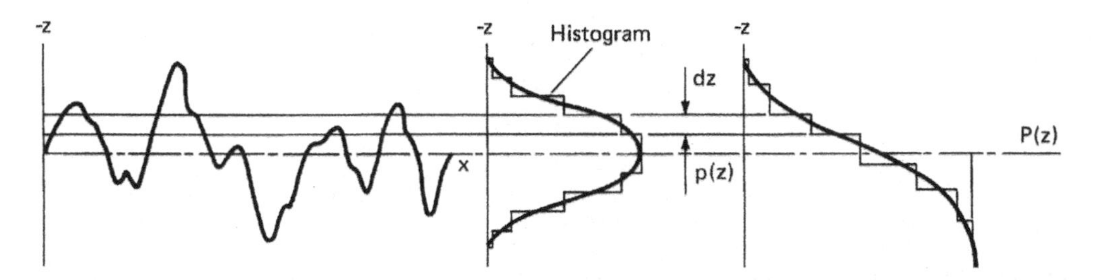

Figure 2.3.11 Method of deriving the histogram and cumulative distribution function from a surface height distribution.

from the surface or profile height distribution, the cumulative distribution function is derived. It is constructed by plotting the cumulative number or proportion of the surface height lying at or below a specific height as a function of that height, Figure 2.3.11. An example of a profile and corresponding histogram and cumulative height distribution on a probability paper for a lapped nickel-zinc ferrite is given in Figure 2.3.12.

Probability density and distribution curves can also be obtained for the slope and curvature of the surface or the profile. If the surface, or profile height, follows a Gaussian distribution, then its slope and curvature distribution also follow a Gaussian distribution. Because it is known that if two functions follow a Gaussian distribution, their sum and difference also follows a Gaussian distribution. Slope and curvatures are derived by taking the difference in a height distribution, and therefore slope and curvatures of a Gaussian height distribution would be Gaussian.

For a digitized profile of length L with heights z_i, $i = 1$ to N, at a sampling interval $\Delta x = L/(N-1)$, where N represents the number of measurements, average height parameters are given as

$$R_a = \frac{1}{N}\sum_{i=1}^{N}|z_i - m| \tag{2.3.24a}$$

$$\sigma^2 = \frac{1}{N}\sum_{i=1}^{N}(z_i - m)^2 \tag{2.3.24b}$$

$$Sk = \frac{1}{\sigma^3 N}\sum_{i=1}^{N}(z_i - m)^3 \tag{2.3.24c}$$

$$K = \frac{1}{\sigma^4 N}\sum_{i=1}^{N}(z_i - m)^4 \tag{2.3.24d}$$

and

$$m = \frac{1}{N}\sum_{i=1}^{N} z_i \tag{2.3.24e}$$

Two average spacing parameters, mean of profile slope $(\partial z/\partial x)$ and profile curvature $(-\partial^2 z/\partial x^2)$ of a digitized profile are given as

$$\text{mean slope} = \frac{1}{N-1}\sum_{i=1}^{N-1}\left(\frac{z_{i+1} - z_i}{\Delta x}\right) \tag{2.3.25a}$$

and

$$\text{mean curvature} = \frac{1}{N-2}\sum_{i=2}^{N-1}\left(\frac{2z_i - z_{i-1} - z_{i+1}}{\Delta x^2}\right) \tag{2.3.25b}$$

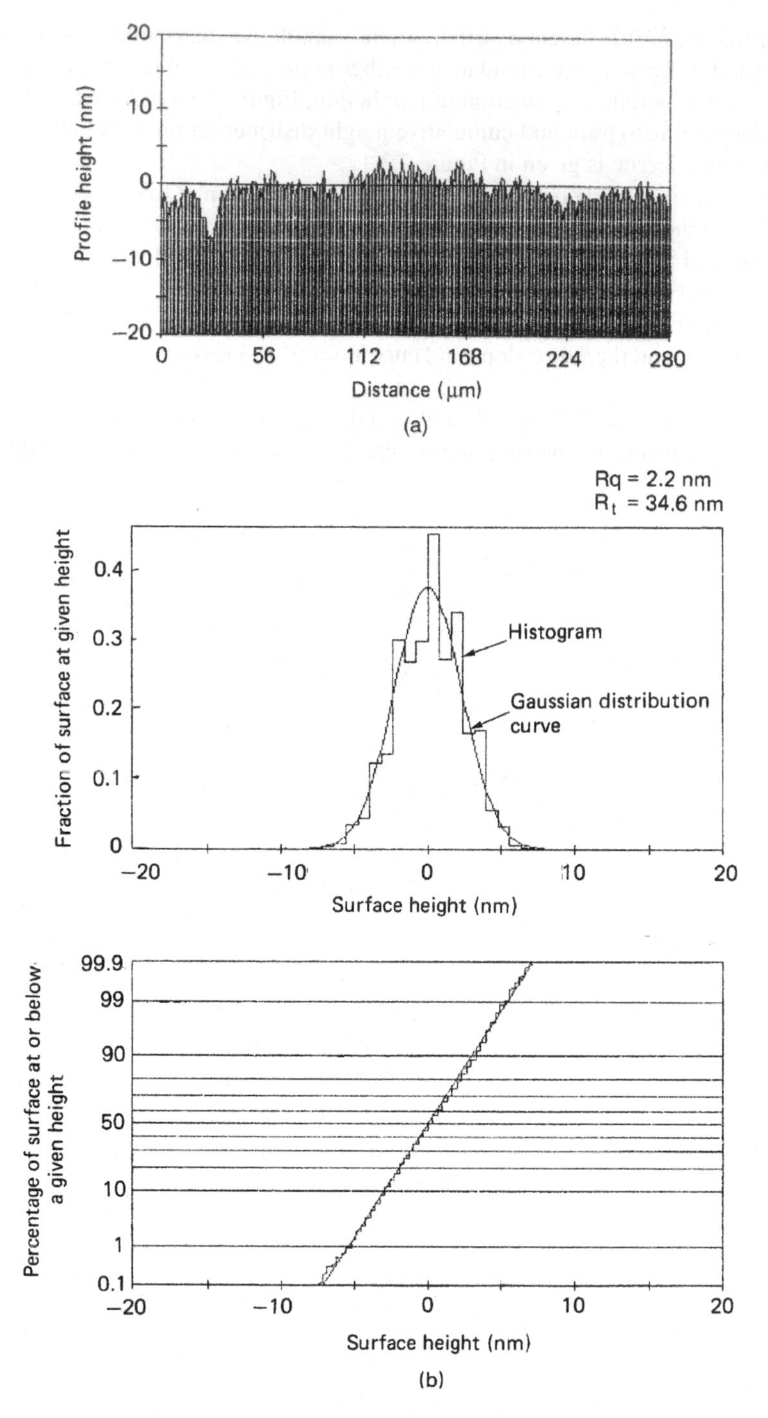

Figure 2.3.12 (a) Profile and (b) corresponding histogram and distribution of profile heights of lapped nickel-zinc ferrite.

The surface slope at any point on a surface is obtained by finding the square roots of the sum of the squares of the slopes in two orthogonal (x and y) axes. The curvature at any point on the surface is obtained by finding the average of the curvatures in two orthogonal (x and y) axes (Nayak, 1971).

Before calculation of roughness parameters, the height data are fitted in a least-square sense to determine the mean height, tilt, and curvature. The mean height is always subtracted and usually the tilt is also subtracted. In some cases, curvature needs to be removed as well. Spherical and cylindrical radii of curvature are removed for spherical and cylindrical surfaces, respectively (e.g., balls and cylinders) before roughness parameters are calculated.

Example Problem 2.3.4

A surface profile is sinusoidal, with an amplitude A_0 and wavelength λ. The profile is sampled at equal intervals, with the origin on the center line at a position of maximum amplitude. Calculate the σ and P-V values for this profile for sampling intervals of $\lambda/2$, $\lambda/4$, $\lambda/8$ and $\lambda/16$. Also calculate the σ and P-V distance values derived from the analog signal for the same profile.

Solution

For the sinusoidal wave

$$z(x) = A_0 \cos\left(\frac{2\pi}{\lambda}x\right)$$

For a sampling interval $\frac{\lambda}{2}$,

$$\sigma^2 = \frac{1+1}{2}A_0^2, \quad \sigma = A_0$$
$$P - V = 2A_0$$

For a sampling interval $\frac{\lambda}{4}$,

$$\sigma^2 = A_0^2\frac{1+0+1+0}{4}, \quad \sigma = \frac{A_0}{\sqrt{2}}$$
$$P - V = 2A_0$$

For $\frac{\lambda}{8}$,

$$\sigma^2 = A_0^2\frac{1+\frac{1}{2}+0+\frac{1}{2}}{4}, \quad \sigma = \frac{A_0}{\sqrt{2}}$$
$$P - V = 2A_0$$

For $\frac{\lambda}{16}$,

$$\sigma^2 = A_0^2 \frac{1 + 0.9238^2 + \frac{1}{2} + 0.3826^2 + 0 + 0.3826^2 + \frac{1}{2} + 0.9238^2 + 1}{8}, \quad \sigma = \frac{A_0}{\sqrt{2}}$$

$$P - V = 2A_0$$

For analog signal,

$$\sigma^2 = \frac{2A_0^2}{\pi} \int_0^{\pi/2} \cos^2\left(\frac{2\pi}{\lambda}x\right) dx$$

Note:

$$\cos^2\theta = \frac{1 + \cos 2\theta}{2}$$

Therefore,

$$\begin{aligned}\sigma^2 &= \frac{A_0^2}{\pi} \int_0^{\pi/2} \left[1 + \cos\left(\frac{4\pi}{\lambda}x\right)\right] dx \\ &= \frac{A_0^2}{\pi} \left[x + \frac{\lambda}{4\pi} \sin\left(\frac{4\pi}{\lambda}x\right)\right]_0^{\pi/2} \\ &= \frac{A_0^2}{\pi} \left[\frac{\pi}{2} + 0\right] \\ &= \frac{A_0^2}{2}\end{aligned}$$

or

$$\sigma = \frac{A_0}{\sqrt{2}}$$

and

$$P - V = 2A_0$$

A sampling interval of $\lambda/2$ gives erroneous values whereas sampling intervals ranging from $\lambda/4$ to $\lambda/16$ give exact values. Thus, the results show the importance of selection of a suitable sampling interval.

2.3.2.4 Bearing Area Curves

The real area of contact (to be discussed in the next chapter) is known as the bearing area and may be approximately obtained from a surface profile or a surface map. The bearing area curve (BAC) first proposed by Abbott and Firestone (1933) is also called the Abbott–Firestone curve or simply Abbott curve. It gives the ratio of material total length at any level, starting at the highest peak, called the bearing ratio or material ratio, as a function of level.

To produce a BAC from a surface profile, some distance from a reference (or mean) line a parallel line (bearing line) is drawn. The length of each material intercept (land) along the line is measured and these lengths are summed together. The proportion of this sum to the total length, the bearing length ratio (t_p), is calculated. This procedure is repeated along a number of bearing lines starting at the highest peak to the lowest valley and the fractional land length (bearing length ratio) as a function of the height of each slice from the highest peak (cutting depth) is plotted, Figure 2.3.13. For a Gaussian surface, the BAC has an S-shaped appearance. In the case of a surface map, bearing planes are drawn and the area of each material intercept is measured. For a random surface, the bearing length and bearing area fractions are numerically identical.

The BAC is related to the CDF. The fraction of heights lying above a given height z (i.e. the bearing ratio at height h) is given by

$$\text{Prob}\,(z \geq h) = \int_h^\infty p(z)\;dz \tag{2.3.26a}$$

which is $1 - P(h)$, where $P(h)$ is the cumulative distribution function at $z \leq h$, Figure 2.3.6. Therefore, the BAC can be obtained from the height distribution histogram. The bearing ratio histograph at height h is simply the progressive addition of all the values of $p(z)$ starting at the

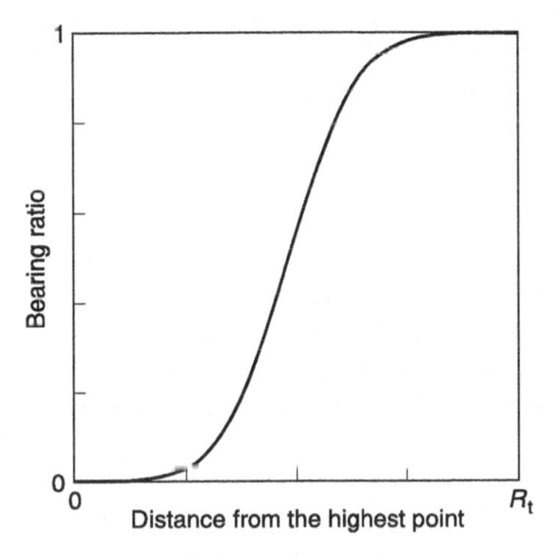

Figure 2.3.13 Schematic of bearing area curve.

highest point and working down to the height $z = h$, and this cumulative sum multiplied by the class interval Δz is

$$P(z \geq h) = \Delta z \sum_{z=h}^{\infty} p(z) \tag{2.3.26b}$$

The relationship of bearing ratio to the fractional real area of contact is highly approximate as material is sliced off in the construction of BAC and the material deformation is not taken into account.

2.3.2.5 Spatial Functions

Consider two surfaces with sine wave distributions with the same amplitude but different frequencies. We have shown that these will have the same R_a and σ, but with different spatial arrangements of surface heights. Slope and curvature distributions are not, in general, sufficient to represent the surface, as they refer only to one particular spatial size of features. The spatial functions (McGillem and Cooper, 1984; Bendat and Piersol, 1986), namely the autocovariance (or autocorrelation) function (ACVF), structure function (SF), or power spectral (or autospectral) density function (PSDF), offer a means of representing the properties of all wavelengths, or spatial sizes of the feature; these are also known as surface texture descriptors.

ACVF has been the most popular way of representing spatial variation. The ACVF of a random function is most directly interpreted as a measure of how well future values of the function can be predicted based on past observations. SF contains no more information than the ACVF. The PSDF is interpreted as a measure of frequency distribution of the mean square value of the function, that is the rate of change of the mean square value with frequency. In this section, we will present the definitions for an isotropic and random profile z(x). The definitions of an isotropic surface z(x,y) can be found in a paper by Nayak (1971). Analysis of an anisotropic surface is considerably complicated by the number of parameters required to describe the surface. For example, profile measurements along three different directions are needed for complete surface characterization of selected anisotropic surfaces. For further details on anisotropic surfaces, see Longuet-Higgins (1957a), Nayak (1973), Bush *et al.* (1979), and Thomas (1999).

Autocovariance and Autocorrelation Functions

For a function $z(x)$, the ACVF for a spatial separation of τ is an average value of the product of two measurements taken on the profile a distance τ apart, $z(x)$ and $z(x + \tau)$. It is obtained by comparing the function $z(x)$ with a replica of itself where the replica is shifted an amount τ (see Figure 2.3.14),

$$R(\tau) = \lim_{L \to \infty} \frac{1}{L} \int_0^L z(x)\, z(x + \tau)\, dx \tag{2.3.27a}$$

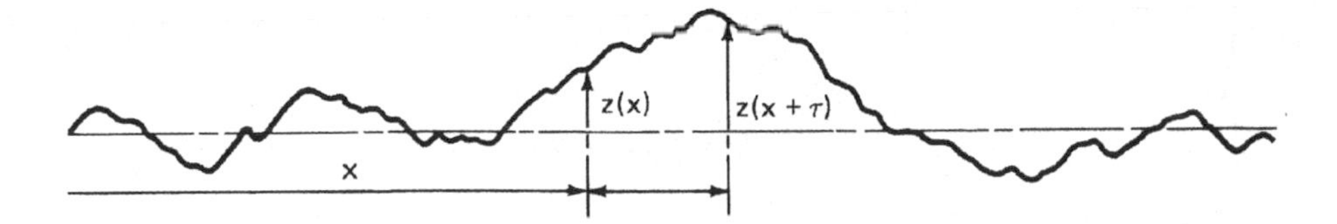

Figure 2.3.14 Construction of the autocovariance function.

where L is the sampling length of the profile. From its definition, ACVF is always an even function of τ, that is,

$$R(\tau) = R(-\tau) \tag{2.3.27b}$$

The values of ACVF at $\tau = 0$ and ∞ are,

$$R(0) = R_q^2 = \sigma^2 + m^2 \tag{2.3.27c}$$

and

$$R(\infty) = m^2 \tag{2.3.27d}$$

The normalized form of the ACVF is called the autocorrelation function (ACF) and is given as

$$C(\tau) = \lim_{L\to\infty} \frac{1}{L\sigma^2} \int_0^L [z(x) - m][z(x+\tau) - m]\, dx - \left[R(\tau) - m^2\right]/\sigma^2 \tag{2.3.28}$$

For a random function, $C(\tau)$ would be maximum (= 1) at $\tau = 0$. If the signal is periodic, $C(\tau)$ peaks whenever τ is a multiple of wavelength. Many engineering surfaces are found to have an exponential ACF,

$$C(\tau) = \exp(-\tau/\beta) \tag{2.3.29}$$

The measure of how quickly the random event decays is called the correlation length. The correlation length is the length over which the autocorrelation function drops to a small fraction of its value at the origin, typically 10% of its original value. The exponential form has a correlation length of $\beta^*[C(\tau) = 0.1]$ equal to 2.3 β, Figure 2.3.15. Sometimes, correlation length is defined as the distance at which value of the autocorrelation function is $1/e$, that is 37%, which is equal to β for exponential ACF. The correlation length can be taken as that at which two points on a function have just reached the condition where they can be regarded as being independent. This follows from the fact that when $C(\tau)$ is close to unity, two points on the function at a distance τ are strongly interdependent. However, when $C(\tau)$ attains values close to zero, two points on the function at a distance τ are weakly correlated. The correlation length, β^*, can be viewed as a measure of randomness. The degree of randomness of a surface decreases with an increase in the magnitude of β^*.

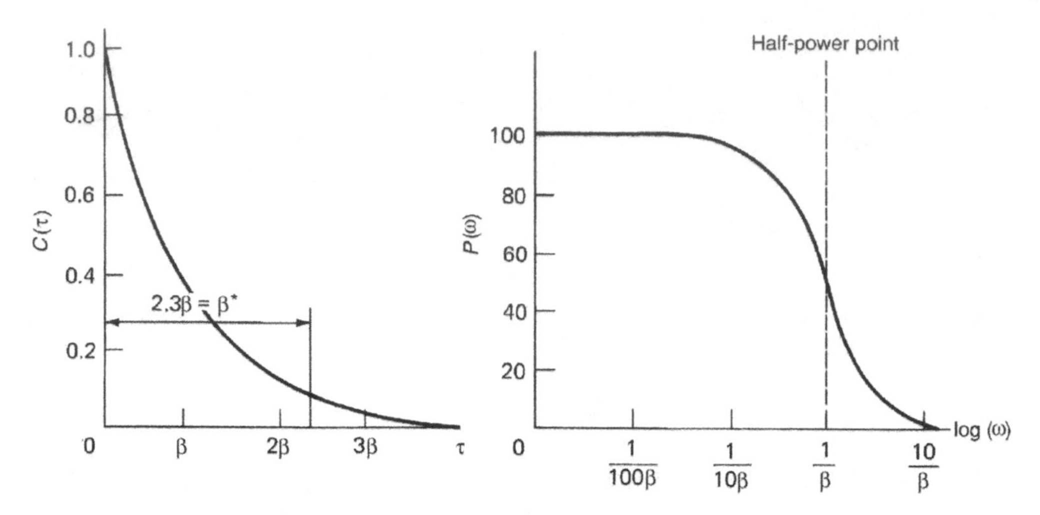

Figure 2.3.15 An exponential autocorrelation function and corresponding power spectral density function.

The directionality of a surface can be found from its autocorrelation function. By plotting the contours of equal autocorrelation values, one can obtain contours to reveal surface structure. The anisotropy of the surface structure is given as the ratio between the longer and shorter axes of the contour (Wyant *et al.*, 1986; Bhushan, 1996). For a theoretically isotropic surface structure, the contour would have a constant radius, that is, it would be a circle.

The autocorrelation function can be calculated either by using the height distribution of the digitized profile or the fast Fourier transform (FFT) technique. In the FFT technique, the first PSDF (described later) is obtained by taking an FFT of the surface height and squaring the results, then an inverse FFT of the PSDF is taken to get ACVF.

Structure Function (SF)

The structure function (SF) or variance function (VF) in an integral form for a profile $z(x)$ is

$$S(\tau) = \lim_{L\to\infty} \frac{1}{L} \int_0^L [z(x) - z(x+\tau)]^2 dx \tag{2.3.30}$$

The function represents the mean square of the difference in height expected over any spatial distance τ. For stationary structures, it contains the same information as the ACVF. The two principal advantages of SF are that its construction is not limited to the stationary case, and it is independent of the mean plane.

Structure function is related to ACVF and ACF as

$$S(\tau) = 2\left[\sigma^2 + m^2 - R(\tau)\right] \tag{2.3.31a}$$

$$= 2\sigma^2 [1 - C(\tau)] \tag{2.3.31b}$$

Power Spectral Density Function (PSDF)

The PSDF is another form of spatial representation and provides the same information as the ACVF or SF, but in a different form. The PSDF is the Fourier transform of the ACVF,

$$\begin{aligned} P(\omega) = P(-\omega) &= \int_{-\infty}^{\infty} R(\tau)\exp(-i\omega\tau)\ d\tau \\ &= \int_{-\infty}^{\infty} \sigma^2 C(\tau)\exp(-i\omega\tau)\ d\tau + m^2\delta(\omega) \end{aligned} \tag{2.3.32}$$

where ω is the angular frequency in length^{-1} (=2π f or $2\pi/\lambda$, f is frequency in cycles/length and λ is wavelength in length per cycle) and $\delta(\omega)$ is the delta function. $P(\omega)$ is defined over all frequencies, both positive and negative, and is referred to as a two sided spectrum. $G(\omega)$ is a spectrum defined over nonnegative frequencies only and is related to $P(\omega)$ for a random surface by

$$\begin{aligned} G(\omega) &= 2P(\omega), \quad && \omega \geq 0 \\ &= 0, && \omega < 0 \end{aligned} \tag{2.3.33a}$$

Since the ACVF is an even function of τ, it follows that the PSDF is given by the real part of the Fourier transform in Equation (2.3.32). Therefore,

$$P(\omega) = \int_{-\infty}^{\infty} R(\tau)\ \cos(\omega\tau)\ d\tau = 2\int_{0}^{\infty} R(\tau)\ \cos(\omega\tau)\ d\tau \tag{2.3.33b}$$

Conversely, the ACVF is given by the inverse Fourier transform of the PSDF,

$$R(\tau) = \frac{1}{2\pi}\int_{-\infty}^{\infty} P(\omega)\ \exp(i\omega\tau)\ d\omega = \frac{1}{2\pi}\int_{-\infty}^{\infty} P(\omega)\ \cos(\omega\tau)\ d\omega \tag{2.3.34}$$

$$\text{For } \tau = 0,\quad R(0) = R_q^2 = \frac{1}{2\pi}\int_{-\infty}^{\infty} P(\omega)\ d\omega \tag{2.3.35}$$

The equation shows that the total area under the PSDF curve (when frequency is in cycles/length) is equal to R_q^2. The area under the curve between any frequency limits gives the mean square value of the data within that frequency range.

The PSDF can also be obtained directly in terms of the Fourier transform of the profile data z(x) by taking an FFT of the profile data and squaring the results, as follows:

$$P(\omega) = \lim_{L\to\infty} \frac{1}{L}\left[\int_0^L z(x)\ \exp(-i\omega x)\ dx\right]^2 \tag{2.3.36}$$

The PSDF can be evaluated from the data either via the ACVF using Equation (2.3.33) or the Fourier transform of the data, Equation (2.3.36). Note that the units of the one-dimensional PSDF are in terms of length to the third power and for the two-dimensional case, it is the length to the fourth power.

Figure 2.3.15 shows the PSDF for an exponential ACF previously presented in Equation (2.3.29). The magnitude of the $P(\omega)$ at $\omega = 1/\beta$ is known as the half-power point. For an exponential ACF, the PSDF is represented by white noise in the upper frequencies. The physical meaning of the model is that the main components of the function consist of a band covering the lower frequencies (longer wavelengths). Shorter wavelength components exist but their magnitude declines with increasing frequency so that, in this range, the amplitude is proportional to wavelength. To cover a large spatial range, it is often more convenient with surface data to represent ACF, SF and PSDF on a log-log scale.

Figure 2.3.16a shows examples of selected profiles. Figures 2.3.16b and 2.3.16c show the corresponding ACVF and PSDF (Bendat and Piersol, 1986). (For calculation of ACVF and PSDF, profile length of multiple of wavelengths (a minimum of one wavelength) needs to be used.) The ACVF of a sine wave is a cosine wave. The envelope of the sine wave covariance function remains constant over all time delays, suggesting that one can predict future values of the data precisely based on past observations. Looking at the PSDF of the sine wave, we note that the total mean square value of the sine wave is concentrated at the single frequency, ω_0. In all other cases, because of the erratic character of $z(x)$ in Figure 2.3.16a, the past record does not significantly help one predict future values of the data beyond the very near future. To calculate the autocovariance function for (iii) to (iv) profiles, the power spectrum of the data is considered uniform over a wide bandwidth B. ACVF and PSDF of a sine wave plus wide-band random noise is simply the sum of the functions of the sine wave and wide-band random noise.

The moments of the PSDF are defined as

$$M_n = \frac{1}{2\pi} \int_{-\infty}^{\infty} \left[P(\omega) - m^2 \delta(\omega)\right] \omega^n \, d\omega \tag{2.3.37}$$

where M_n are known as the spectral moments of the nth order. We note for a Gaussian function (Nayak, 1971),

$$M_0 = \sigma^2 = \frac{1}{L} \int_0^L (z - m)^2 \, dx \tag{2.3.38a}$$

$$M_2 = \left(\sigma'\right)^2 = \frac{1}{L} \int_0^L \left(dz/dx\right)^2 dx \tag{2.3.38b}$$

and

$$M_4 = \left(\sigma''\right)^2 = \frac{1}{L} \int_0^L \left(d^2z/dx^2\right)^2 dx \tag{2.3.38c}$$

where σ' and σ'' are the standard deviations of the first and second derivatives of the functions. For a surface/profile height, these are the surface/profile slope and curvature, respectively.

According to Nayak (1971), a random and isotropic surface with a Gaussian height distribution can be adequately characterized by the three-zeroth (M_0), second (M_2) and fourth moments (M_4) of the power spectral density function. Based on the theory of random processes, a random and isotropic surface can be completely characterized in a statistical sense (rather than a deterministic sense) by two functions: the height distribution and the autocorrelation

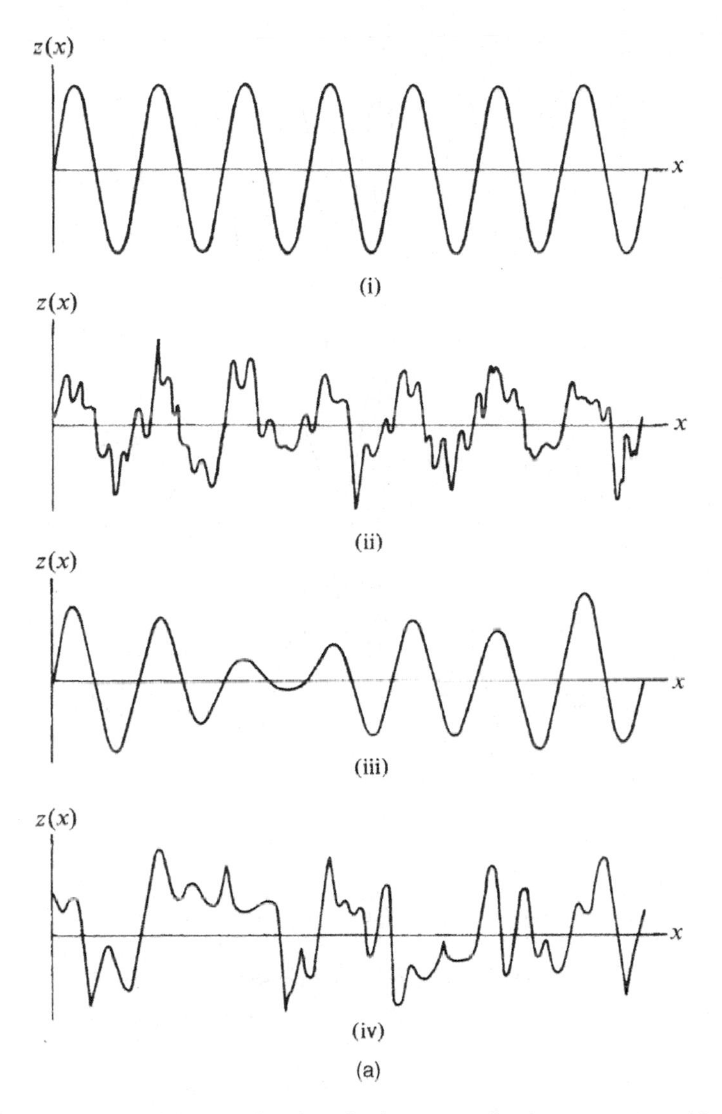

Figure 2.3.16 (a) Four special time histories: (i) sine wave, (ii) sine wave plus wide-band random noise, (iii) narrow-band random noise, and (iv) wide-band random noise. (b) corresponding idealized autocovariance functions, and (c) corresponding power spectral density functions. Reproduced with permission from Bendat, J.S. and Piersol, A.G. (1986), *Engineering Applications of Correlation and Spectral Analysis*, Second edition, Wiley, New York. Copyright 1986. Wiley. (*Continued*)

function. A random surface with Gaussian height distribution and exponential autocorrelation function can then simply be characterized by two parameters, two lengths: standard deviation of surface heights (σ) and the correlation distance (β^*) (Whitehouse and Archard, 1970). For characterization of a surface with a discrete, arbitrary autocorrelation function, three points $C(0)$, $C(h)$ and $C(2h)$ for a profile, where h is an arbitrary distance and four or more points are

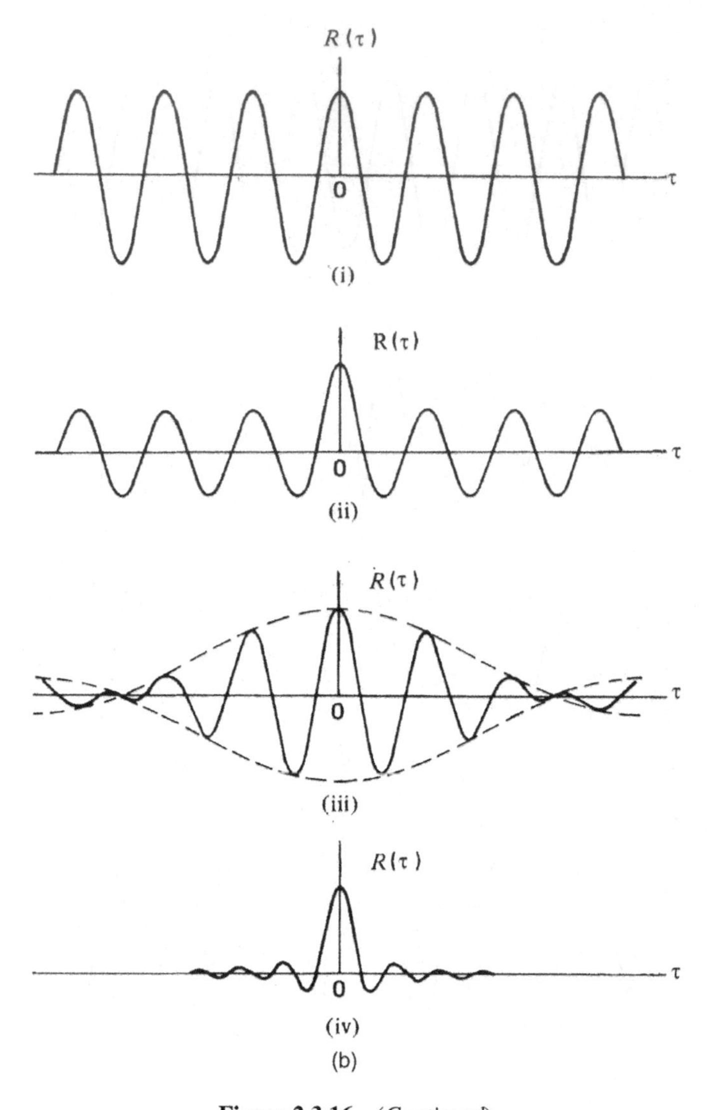

Figure 2.3.16 *(Continued)*

needed on the $C(\tau)$, depending upon the type of the surface (Whitehouse and Phillips, 1978, 1982).

2.3.2.6 Probability Distribution of the Asperities and Valleys

Surfaces consist of hills (asperities) of varying heights and spacing and valleys of varying depths and spacing. For a two-dimensional profile, the peak is defined as a point higher than its two adjacent points greater than a threshold value. For a three-dimensional surface map, the summit is defined as a point higher than its four adjacent points greater than a threshold value.

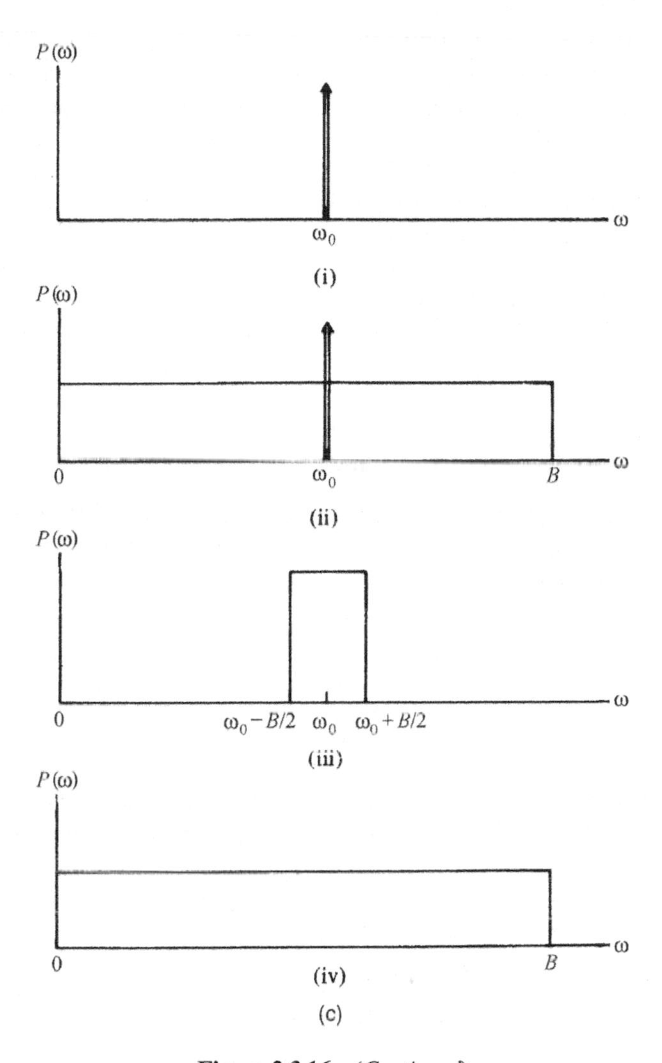

Figure 2.3.16 *(Continued)*

A valley is defined in the same way as a peak/summit but in a reversed order. A threshold value is introduced to reduce the effect of noise in the measured data and ensure that every peak/summit identified is truly substantial. Based on analysis of roughness data of variety of smooth samples, Poon and Bhushan (1995a) recommend a threshold value as one-tenth of the σ roughness of smooth surfaces (with σ less than about 50 nm); it should be lower than 10% of the σ value for rougher surfaces.

Gaussian surfaces might be considered as comprising a certain number of hills (asperities) and an equal number of valleys. These features may be assessed and represented by their appropriate distribution curves, which can be described by the same sort of characteristics as were used previously for the surface height distributions. Similar to surface height distributions, the height distributions of peaks (or summits) and valleys often follow the Gaussian curve

(Greenwood, 1984; Wyant *et al.*, 1986; Bhushan, 1996). Distribution curves can also be obtained for the absolute values of slope and for the curvature of the peaks (or summits) and valleys. Distributions of peak (or summit) curvature follow a log normal distribution (Gupta and Cook 1972; Wyant *et al.*, 1986; Bhushan, 1996). The mean of the peak curvature increases with the peak height for a given surface (Nayak, 1971).

The parameters of interest in some analytical contact models of two random rough surfaces to be discussed in the next chapter are the density of summits (η), the standard deviation of summit heights (σ_p), and the mean radius (R_p) (or curvature, κ_p) of the summit caps or $\eta, \sigma,$ *and* β^*. The former three roughness parameters (η, σ_p, R_p) can be related to other easily measurable roughness parameters using the theories of Longuet-Higgins (1957a, 1957b), Nayak (1971) and Whitehouse and Phillips (1978, 1982).

2.3.2.7 Composite Roughness of Two Random Rough Surfaces

For two random rough surfaces in contact, the composite roughness of interest is defined as the sum of two roughness processes obtained by adding together the local heights (z), the local slope (θ) and local curvature (κ)

$$z = z_1 + z_2$$
$$\theta = \theta_1 + \theta_2 \tag{2.3.39}$$
$$\kappa = \kappa_1 + \kappa_2$$

For two random rough surfaces in contact, an equivalent rough surface can be described of which the values of $\sigma, \sigma', \sigma'', R(\tau), P(\omega)$ *and* $M_0, M_2,$ *and* M_4 are summed for the two rough surfaces, that is,

$$\sigma^2 = \sigma_1^2 + \sigma_2^2$$
$$\sigma'^2 = \sigma_1'^2 + \sigma_2'^2$$
$$\sigma''^2 = \sigma_1''^2 + \sigma_2''^2$$
$$R(\tau) = R_1(\tau) + R_2(\tau)$$
$$P(\omega) = P_1(\omega) + P_2(\omega)$$

and

$$M_i = (M_i)_1 + (M_i)_2 \tag{2.3.40a}$$

where $i = 0, 2, 4$. These equations state that variances, autocovariance function and power spectra are simply additive. Since autocovariance functions of two functions are additive, simple geometry shows that correlation lengths of two exponential ACVFs are related as

$$\frac{1}{\beta^*} = \frac{1}{\beta_1^*} + \frac{1}{\beta_2^*} \tag{2.3.40b}$$

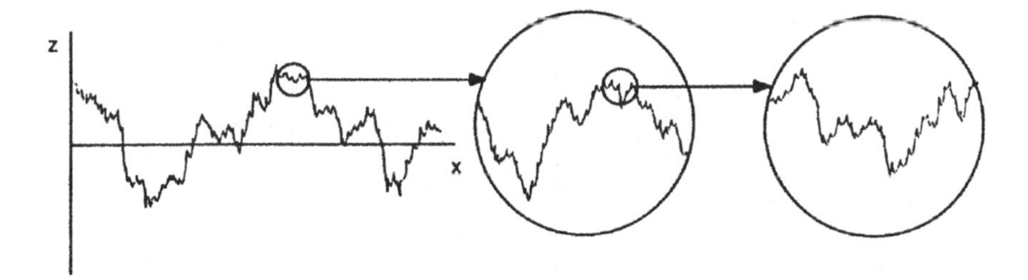

Figure 2.3.17 Qualitative description of statistical self-affinity for a surface profile.

2.3.3 Fractal Characterization

A surface is composed of a large number of length scales of roughness that are superimposed on each other. As stated earlier, surface roughness is generally characterized by the standard deviation of surface heights. However, due to the multiscale nature of the surface, it is known that the variances of surface height and its derivatives and other roughness parameters depend strongly on the resolution of the roughness measuring instrument or any other form of filter, hence they are not unique for a surface (Ganti and Bhushan, 1995; Poon and Bhushan, 1995a). Therefore, rough surfaces should be characterized in a way such that the structural information of roughness at all scales is retained. It is necessary to quantify the multiscale nature of surface roughness.

A unique property of rough surfaces is that if a surface is repeatedly magnified, increasing details of roughness are observed right down to nanoscale. In addition, the roughness at all magnifications appear quite similar in structure as qualitatively shown in Figure 2.3.17. The statistical self-affinity is due to similarity in appearance of a profile under different magnifications. Such a behavior can be characterized by fractal geometry (Majumdar and Bhushan, 1990; Ganti and Bhushan, 1995; Bhushan, 1999). The fractal approach has the ability to characterize surface roughness by scale-independent parameters and provides information of the roughness structure at all length scales that exhibit the fractal behavior. Surface characteristics can be predicted at all length scales within the fractal regime by making measurements at one scan length.

The structure function and power spectrum of a self-affine fractal surface follow a power law and can be written as (Ganti and Bhushan model)

$$S(\tau) = C\eta^{(2D-3)}\tau^{(4-2D)} \tag{2.3.41}$$

$$P(\omega) = \frac{c_1\eta^{(2D-3)}}{\omega^{(5-2D)}} \tag{2.3.42a}$$

and

$$c_1 = \frac{\Gamma\,(5-2D)\,\sin\,[\pi\,(2-D)]}{2\pi}C \tag{2.3.42b}$$

The fractal analysis allows the characterization of surface roughness by two parameters D and C which are instrument-independent and unique for each surface. The parameter D (ranging

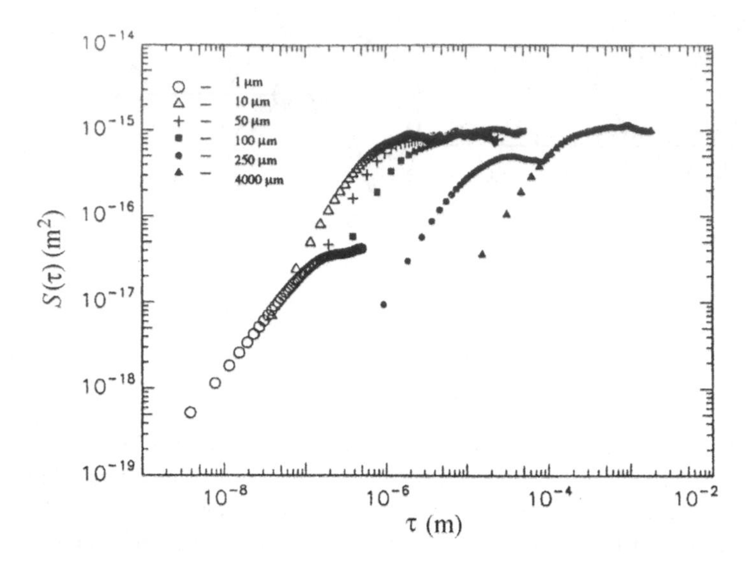

Figure 2.3.18 Structure functions for the roughness data measured using AFM and NOP, for a thin-film magnetic rigid disk. Reproduced with permission from Ganti, S. and Bhushan, B. (1996), "Generalized Fractal Analysis and its Applications to Engineering Surfaces," *Wear* **180**, 17–34. Copyright 1995. Elsevier.

from 1 to 2 for a surface profile) primarily relates to the relative power of the frequency contents, and C to the amplitude of all frequencies. η is the lateral resolution of the measuring instrument, τ is the size of the increment (distance), and ω is the frequency of the roughness. Note that if $S(\tau)$ or $P(\omega)$ are plotted as a function of ω or τ, respectively, on a log-log plot, then the power-law behavior results in a straight line. The slope of the line is related to D and the location of the spectrum along the power axis is related to C.

Figure 2.3.18 presents the structure functions of a thin-film magnetic rigid disk measured using an atomic force microscope (AFM) and noncontact optical profiler (NOP). A horizontal shift in the structure functions from one scan to another arises from the change in the lateral resolution. The D and C values for various scan lengths are listed in Table 2.3.2. Note

Table 2.3.2 Surface roughness parameters for a polished thin-film rigid disk.

Scan size (µm × µm)	σ (nm)	D	C (nm)
1(AFM)	0.7	1.33	9.8×10^{-4}
10(AFM)	2.1	1.31	7.6×10^{-3}
50(AFM)	4.8	1.26	1.7×10^{-2}
100(AFM)	5.6	1.30	1.4×10^{-2}
250(NOP)	2.4	1.32	2.7×10^{-4}
4000(NOP)	3.7	1.29	7.9×10^{-5}

AFM - Atomic force microscope.
NOP - Noncontact optical profiler.

that fractal dimension of the various scans is fairly constant (1.26 to 1.33); however, C increases/decreases monotonically with σ for the AFM data. The error in estimation of η is believed to be responsible for the variation in C. These data show that the disk surface follows a fractal structure for three decades of length scales.

2.3.4 Practical Considerations in Measurement of Roughness Parameters

2.3.4.1 Short- and Long-Wavelength Filtering

Engineering surfaces cover a broad bandwidth of wavelengths, and samples, however large, often exhibit nonstationary properties (in which the roughness is dependent upon the sample size). Surface roughness is intrinsic, however, measured roughness is a function of the bandwidth of the measurement and thus is not an intrinsic property. Instruments using different sampling intervals measure features with different length scales. Roughness is found at scales ranging from millimeter to nanometer (atomic) scales. A surface is composed of a large number of length scales of roughness that are superimposed on each other. Therefore, on a surface, it is not that different asperities come in different sizes but that one asperity comes in different sizes. Distribution of size and shape of asperities is dependent on the short-wave length limit or the sampling interval of the measuring instrument. When the sampling interval at which the surface is sampled is reduced, the number of asperities detected and their curvature appear to rise without limit down to atomic scales. This means that asperity is not a "definite object." Attempts are made to identify a correct sampling interval which yields the relevant number of asperities for a particular application. An asperity relevant for contact mechanics is defined as that which makes a contact in a particular application (contacting asperity) and carries some load.

The short-wavelength limit or the sampling interval affects asperity statistics. The choice of short-wavelength limit depends on the answer to the following question: what is the smallest wavelength that will affect the interaction? It is now known that it is the asperities on a nanoscale which first come into contact and plastically deform instantly, and subsequently the load is supported by the deformation of larger-scale asperities (Bhushan and Blackman, 1991; Poon and Bhushan, 1996). Since plastic deformation in most applications is undesirable, asperities on a nanoscale need to be detected. (See Chapter 3 for more discussion.) Therefore, the short-wavelength limit should be as small as possible.

The effect of the short-wavelength limit on a roughness profile can be illustrated by a sinusoidal profile represented by different numbers of sampling points per wavelength as shown in Figure 2.3.19. The waveform of the sinusoidal profile is distorted when the number of sampling points decreases. The profile parameters do not change significantly with sampling points equal to 6 or greater per wavelength. Therefore, the minimum number of sampling points required to represent a wavelength structure may be set to 6, i.e., the optimum sampling interval is $\lambda/6$, where λ is the wavelength of the sinusoidal profile. By analogy, the suitable sampling interval should be related to the main wavelength structure of a random profile which is represented by β^*. However, β^* is a function of the bandwidth of the measurement and thus is not an intrinsic property. It is reasonable to select a sampling interval a fraction of β^* measured at the long-wavelength limit, say 0.25 β^* to 0.5 β^* (Poon and Bhushan, 1995a).

Figure 2.3.20 demonstrates how the long wavelength limit, also called the cutoff wavelength or sampling length (size), can affect the measured roughness parameters (Anonymous, 1985).

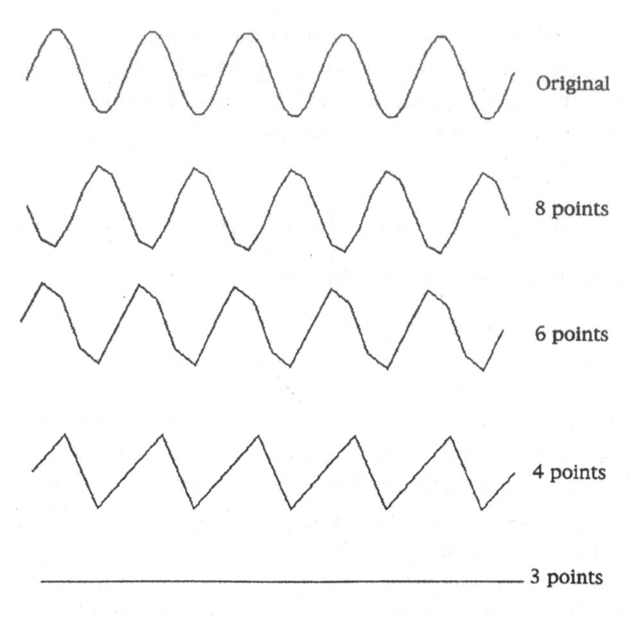

Figure 2.3.19 Sinusoidal profiles with different number of sampling points per wavelength.

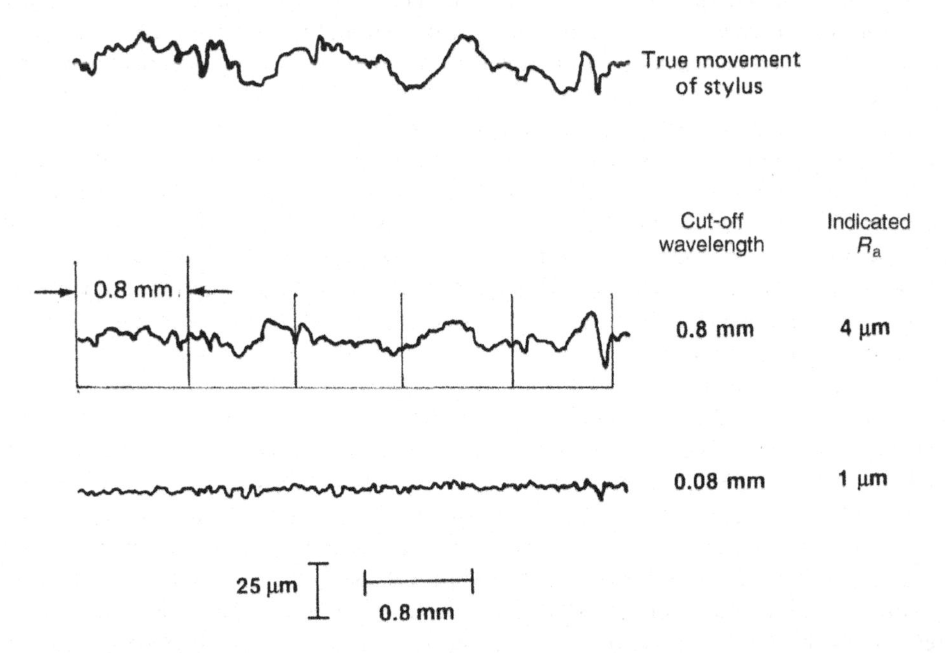

Figure 2.3.20 The effect of the cutoff wavelength is to remove all components of the total profile that have wavelengths greater than cutoff value.

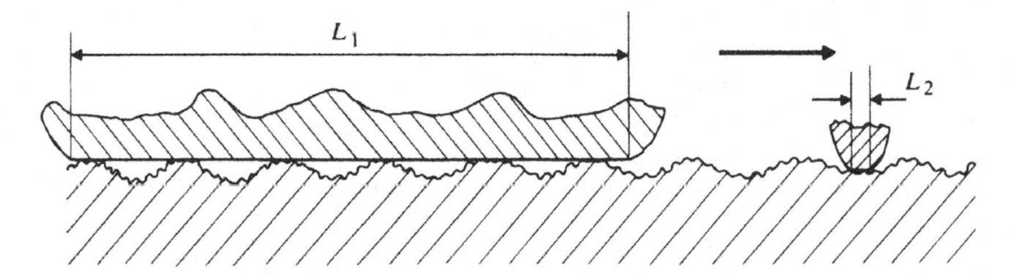

Figure 2.3.21 Contact size of two moving components of different lengths L_1 and L_2 on the same rough surface.

The top profile represents the actual movement of the stylus on a surface. The lower ones show the same profile using cutoff wavelength values of 0.8 and 0.08 mm. A small cutoff value would isolate the waviness while a large cutoff value would include the waviness. Thomas (1999) has shown that the standard deviation of surface roughness, σ, will increase with an increase in the cutoff wavelength or sampling length L, as given by the following relation,

$$\sigma \propto L^{1/2} \tag{2.3.43}$$

Ganti and Bhushan (1995) and Poon and Bhushan (1995a) have reported that σ and other roughness parameters initially increase with L and then reach a constant value because engineering surfaces seem to have a long wavelength limit. Thus, before the surface roughness can be effectively quantified, an application must be defined. Having a knowledge of the application enables a measurement to be planned and in particular for it to be decided to what bandwidth of surface features the information collected should refer. Features that appear as roughness in one application of a surface may well constitute waviness in another.

The long-wavelength limit (which is the same as scan size in many instruments) in contact problems is set by the dimensions of the nominal contact area (Figure 2.3.21). This is simply to say that a wavelength much longer than the nominal contact area will not affect what goes on inside it. In addition, the long-wavelength limit of the surface roughness in the nominal contact area, if it exists, should be obtained. The long-wavelength limit can be chosen to be twice the nominal contact size or the long-wavelength limit of the roughness structure in the nominal contact size, if it exists, whichever is smaller.

To provide a basis of instrumentation for roughness measurement, a series of cutoff wavelength values has been standardized in a British standard, BS1134-1972, an ANSI/ASME B46.1-1985, and an ISO Recommendation, R468. The international standard cutoff values are 0.08, 0.25, and 0.8 mm. The preferred value of 0.8 mm is assumed unless a different value is specified. Note that waviness measurements are made without long-wavelength filtering.

Long- and short-wavelength filtering in measuring instruments are most commonly accomplished by digital filtering. For example, in a fast Fourier transform (FFT) technique, the FFT of the raw data is taken, the appropriate frequency contents are removed and the inverse FFT is taken to obtain the filtered data. However, this technique is slow and one method commercially used is the Finite Impulse Response (FIR) technique. The FIR technique filters by convoluting the trace data with the impulse response of the filter specification. The impulse response is obtained by taking the inverse FFT of the filter specification.

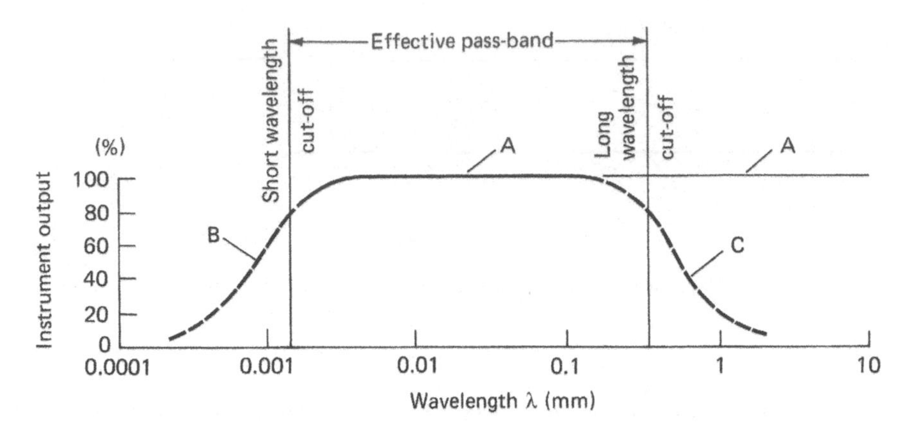

Figure 2.3.22 Transmission characteristics of a profiler with low bandpass and high bandpass filters.

Anonymous (1985) also describes the electronic filtering method for short- and long-wavelength filtering, which is accomplished by passing the alternating voltage representing the profile through an electrical wave filter, such as the standard RC filter. The electronic filtering is generally used to filter out short-wavelength electronic noise (low band pass filtering).

Mechanical short-wavelength filtering also results from the design and construction of a measuring instrument. For example in the stylus instrument or the atomic force microscope, the stylus removes certain short wavelengths on the order of the stylus tip radius, which is referred to as lateral resolution of the instrument. The stylus is not able to enter the grooves. As the spacing between grooves increases, the stylus displacement will rise, but once it has become sufficient for the stylus to reach to the bottom, there will be a full indication. In a digital optical profiler, lateral resolution is controlled by the physical size of the charge-coupled device (CCD) image sensors at the microscope objective magnifications. A short-wavelength limit, if selected, should be at least twice the lateral resolution of the instrument.

For the instrument in which a short-wavelength filter is introduced, the output will tend to fall off above a certain frequency, that is below a certain wavelength, for example, as shown by the dotted curve B in Figure 2.3.22, even though the stylus continues to rise and fall over the irregularities. Dotted curve C in Figure 2.3.22 also shows the fall-off of instrument output at longer wavelength. Only within the range of wavelengths for which the curve is substantially level will the indication be a measure solely of the amplitude and be independent of wavelength curve A in Figure 2.3.22.

2.3.4.2 Scan Size

After the short-wavelength and long-wavelength limits are selected, the roughness measurement must be made on a length large enough to provide a statistically significant value for the chosen locality. The total length involved is called the measuring length, evaluation length, traversing length or scan length. In some cases, a length of several individual scan lengths (say five) is chosen (Whitehouse, 1994). In most measurements, scan length is the same as the long-wavelength limit. For two-dimensional measurement, a certain area is measured rather than a length.

Wyant *et al.* (1984) and Bhushan *et al.* (1985) have suggested that in measurement of a random surface, a scan length equal to or greater than 200 β^* should be used.

2.4 Measurement of Surface Roughness

A distinction is made between methods of evaluating the nanoscale to atomic scale and microscale features of surface roughness. Physicists and physical chemists require fine-scale details of surfaces and often details of molecular roughness. These details are usually provided using methods such as low-energy electron diffraction, molecular-beam methods, field-emission and field-ion microscopy, scanning tunneling microscopy, and atomic force microscopy. On the other hand, for most engineering and manufacturing surfaces, microscopic methods suffice, and they are generally mechanical or optical methods. Some of these methods can also be used to measure geometrical parameters of surfaces (Bhushan, 1996, 1999, 2011, 2013).

Various instruments are available for the roughness measurement. The measurement technique can be divided into two broad categories: (a) a contact type in which during measurement a component of the measurement instrument actually contacts the surface to be measured; and (2) a noncontact type. A contact-type instrument may damage surfaces, when used with sharp stylus tip, particularly soft surfaces, Figure 2.4.1. For these measurements, the normal loads have to be low enough such that the contact stresses do not exceed the hardness of the surface to be measured.

The first practical stylus instrument was developed by Abbott and Firestone (1933). In 1939, Rank Taylor Hobson in Leicester, England introduced the first commercial instrument called Talysurf. Today, contact-type stylus instruments using electronic amplification are the most popular. The stylus technique, recommended by the ISO, is generally used for reference purposes. In 1983, a noncontact optical profiler based on the principle of two-beam optical interferometry was developed and is now widely used in the electronics and optical industries to measure smooth surfaces (Wyant *et al.*, 1984). In 1985, an atomic force microscope was developed which is basically a nano-profiler operating at ultra-low loads (Binnig *et al.*, 1986). It can be used to measure surface roughness with lateral resolution ranging from microscopic to atomic scales. This instrument is commonly used in research to measure roughness with extremely high lateral resolution, particularly nanoscale roughness.

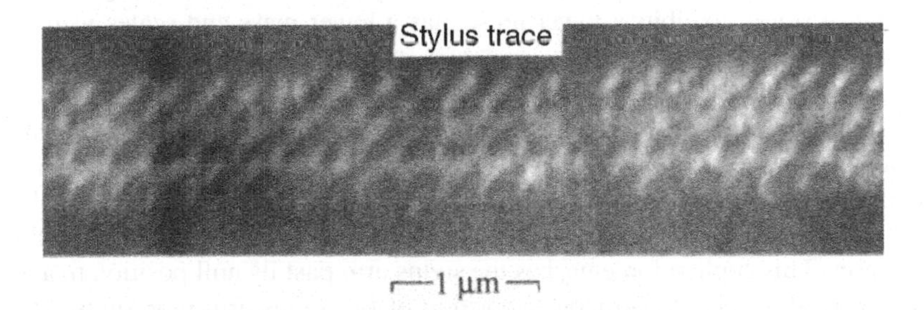

Figure 2.4.1 SEM micrograph of a trace made by a stylus instrument showing surface damage of electroless Ni-P coating (stylus material, diamond; stylus radius = 0.1 μm; and stylus load = 10 μN or 1 mg). Reproduced with permission from Poon, C.Y. and Bhushan, B. (1995a), "Comparison of Surface Roughness Measurements by Stylus Profiler, AFM and Non-Contact Optical Profiler," *Wear* **190**, 76–88. Copyright 1995. Elsevier.

There are a number of other techniques that have been either demonstrated in the laboratory and never commercially used or used in specialized applications. We will divide the different techniques into six categories based on the physical principle involved: mechanical stylus method, optical methods, scanning probe microscopy (SPM) methods, fluid methods, electrical method, and electron microscopy methods. Descriptions of these methods are presented, and the detailed descriptions of only three stylus, optical (based on optical interferometry) and AFM techniques, are provided. We will conclude this section by comparing various measurement methods.

2.4.1 Mechanical Stylus Method

This method uses an instrument that amplifies and records the vertical motions of a stylus displaced at a constant speed by the surface to be measured. Commercial profilers include: Rank Taylor Hobson (UK) Talysurf profilers, KLA-Tencor Corporation Alpha-Step and P-series profilers, Bruker Instruments Dektak profilers, and Kosaka Laboratory, Tokyo (Japan) profilers. The stylus is mechanically coupled mostly to a linear variable differential transformer (LVDT), to an optical or a capacitance sensor. The stylus arm is loaded against the sample and either the stylus is scanned across the stationary sample surface using a traverse unit at a constant speed or the sample is transported across an optical flat reference. As the stylus or sample moves, the stylus rides over the sample surface detecting surface deviations by the transducer. It produces an analog signal corresponding to the vertical stylus movement. This signal is then amplified, conditioned and digitized (Bhushan, 1996; Thomas, 1999).

In a profiler, as is shown in Figure 2.4.2a, the instrument consists of stylus measurement head with a stylus tip and a scan mechanism. The measurement head houses a stylus arm with a stylus, sensor assembly, and the loading system. The stylus arm is coupled to the core of an LVDT to monitor vertical motions. The core of a force solenoid is coupled to the stylus arm and its coil is energized to load the stylus tip against the sample. A proximity probe (photo optical sensor) is used to provide a soft limit to the vertical location of the stylus with respect to the sample. The sample is scanned under the stylus at a constant speed. In high precision ultralow load profilers, shown in Figures 2.4.2b and 2.4.2c, the vertical motion is sensed using a capacitance sensor and a precision stage transports the sample during measurements. The capacitance sensor exhibits a lower noise, has a lower mass and scales well to smaller dimensions as compared to LVDTs.

In order to track the stylus across the surface, force is applied to the stylus. The ability to accurately apply and control this force is critical to the profiler performance. The measurement head uses a wire coil to set a programmable stylus load as low as 0.05 mg. Attached above the stylus flexure pivot is an arm with a magnet mounted to the end. The magnet is held in close proximity to the wire coil, and the coil, when energized, produces a magnetic field that moves the magnet arm. This applied force pushes the stylus arm past its null position to a calibrated force displacement, where the horizontal position of the stylus arm represents zero applied force to the stylus. The force coil mechanism and a sophisticated digital signal processor are used to maintain a constant applied force to the stylus.

The scan mechanism shown in Figure 2.4.2c, holds the sensor assembly (Figure 2.4.2.b) stationary while the sample stage is moved with a precision lead screw drive mechanism. This drive mechanism, called the X drive, uses a motor to drive the lead screw, which then moves the

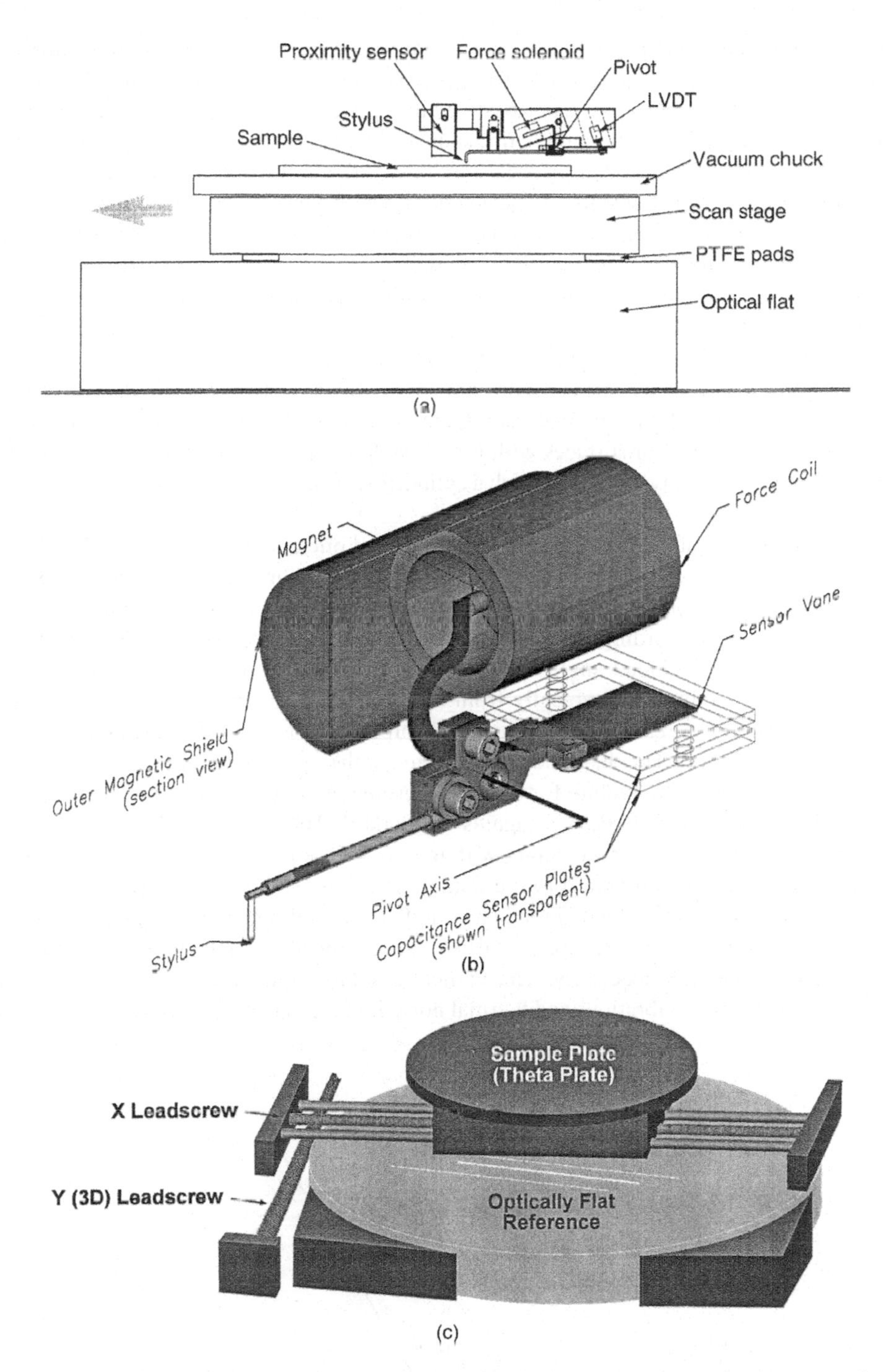

Figure 2.4.2 Schematics of (a) stylus measurement head with loading system and scan mechanism used in Veeco/Sloan Dektak profilers (Courtesy of Bruker Instruments, Santa Barbara, CA), (b) stylus measurement head with loading system and (c) scan mechanism used in Tencor P-series profilers (Courtesy of KLA-Tencor Corporation, Milpitas, CA).

sample stage with guide wires along an optical flat via PTFE skids. The motion is monitored by an optical encoder and is accurate to 1–2 μm. The optical flat ensures a smooth and stable movement of the stage across the scan length, while a guide bar provides a straight, directional movement. This scanning of the sample limits the measurement noise from the instrument, by decoupling the stage motion from vertical motions of the stylus measured using the sensor. Surface topography measurements can be acquired with high sensitivity over a 205 mm scan. Three-dimensional images can be obtained by acquiring two-dimensional scans in the X direction while stepping in the Y direction by 5 μm with the Y lead screw used for precise sample positioning. When building a surface map by parallel traversing, it is essential to maintain a common origin for each profile. This can be achieved by a flattening procedure in which the mean of each profile is calculated separately and these results are spliced together to produce an accurate surface map. The surface maps are generally presented such that the vertical axis is magnified by three to four orders of magnitude as compared to the horizontal scan axis.

Measurements on circular surfaces with long scan lengths can be performed by a modified stylus profiler (such as Talyround) in which a cylindrical surface is rotated about an axis during measurement.

Styli are made of diamond. The shapes can vary from one manufacturer to another. Chisel-point styli with tips (e.g., 0.25 μm × 2.5 μm) may be used for detection of bumps or other special applications. Conical tips are almost exclusively used for microroughness measurements, Figure 2.4.3. According to the international standard (ISO 3274-1975), a stylus is a cone of a 60° to 90° included angle and a (spherical) tip radius of curvature of 2, 5, or 10 μm. The radius of styli ranges typically from 0.1-0.2 μm to 25 μm with the included angle ranging from 60° to 80°. The stylus is a diamond chip tip that is braised to a stainless steel rod mounted to a stylus arm. The diamond chip is cleaved, then ground and polished to a specific dimension. The radius of curvature for the sub-micrometer stylus tip, which is assumed to be spherical, is measured with a SEM, or against a standard. The portion of the stylus tip that is in contact with the sample surface, along with the known radius of curvature, determines the actual radius of the tip with regard to the feature size. The stylus cone angle is determined from the cleave and grind of the diamond chip, and is checked optically or against a standard.

Maximum vertical and spatial (horizontal) magnifications that can be used are on the order of x100,000 and x100, respectively. The vertical resolution is limited by sensor response, background mechanical vibrations and thermal noise in the electronics. Resolution for smooth

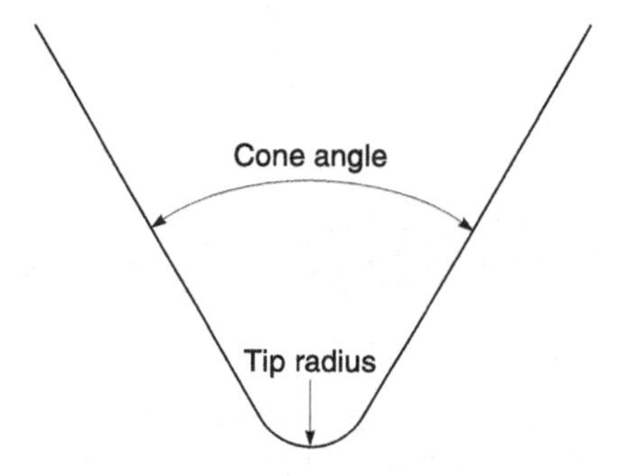

Figure 2.4.3 Schematic of a diamond conical stylus showing its cone angle and tip radius.

surfaces is as low as 0.1 nm and 1 nm for rough surfaces for large steps. Lateral resolution is on the order of the square root of the stylus radius. The step height repeatability is about 0.8 nm for a step height of 1 μm. The stylus load ranges typically from 0.05 to 100 mg. Long-wave cutoff wavelengths range typically from 4.5 μm to 25 mm. Short-wave cutoff wavelengths range typically from 0.25 μm to several mm. The scan lengths can be typically as high as 200 mm and for three-dimensional imaging, the scan areas can be as large as 5 mm $\times$ 5 mm. The vertical range ranges typically from 2 to 250 μm. The scan speed ranges typically from 1 μm/s to 25 mm/s. The sampling rate ranges typically from 50 Hz to 1 kHz.

2.4.1.1 Relocation

There are many situations where it would be very useful to look at a particular section of a surface before and after some experiment, such as grinding or run-in, to see what changes in the surface roughness have occurred. This can be accomplished by the use of a relocation table (Thomas, 1999). The table is bolted to the bed of the stylus instrument, and the specimen stage is kinematically located against it at three points and held in position pneumatically. The stage can be lowered and removed, an experiment of some kind performed on the specimen, and the stage replaced on the table. Relocation of the stylus then occurs to within the width of the original profile.

2.4.1.2 Replication

Replication is used to obtain measurements on parts that are not easily accessible, such as internal surfaces or underwater surfaces. It is used in compliant surfaces because direct measurement would damage or misrepresent the surface (Thomas, 1999). The principle is simply to place the surface to be measured in contact with a liquid that will subsequently set to a solid, hopefully faithfully reproducing the detail of the original as a mirror image or a negative. Materials such as plaster of paris, dental cement, or polymerizing liquids are used. The vital question is how closely the replica reproduces the features of the original. Lack of fidelity may arise from various causes.

2.4.1.3 Sources of Errors

A finite size of stylus tip distorts, a surface profile to some degree (Radhakrishnan, 1970; McCool, 1984). Figure 2.4.4 illustrates how the finite size of the stylus distorts the surface profile. The radius of curvature of a peak may be exaggerated and the valley may be represented as a cusp. A profile containing many peaks and valleys of radius of curvature of about 1 μm or less or many slopes steeper than 45° would probably be more or less badly misrepresented by a stylus instrument.

Another error source is due to stylus kinematics (McCool, 1984). A stylus of finite mass held in contact with a surface by a preloaded spring may, if traversing the surface at a high enough velocity, fail to maintain contact with the surface being traced. Where and whether this occurs depends on the local surface geometry, the spring constant to the mass ratio, and the tracing speed. It is clear that a trace for which stylus contact has not been maintained presents inaccurate information about the surface microroughness.

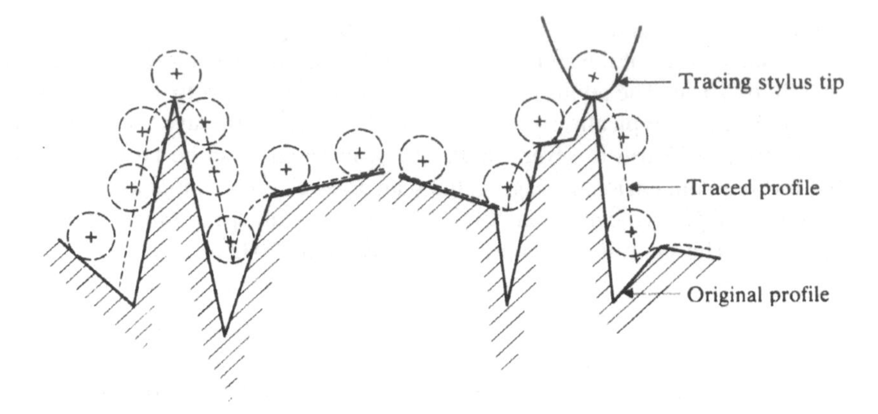

Figure 2.4.4 Distortion of profile due to finite dimensions of stylus tip (exaggerated).

Stylus load also introduces error. A sharp stylus even under low loads results in the area of contact so small that the local pressure may be sufficiently high to cause significant local elastic deformation of the surface being measured. In some cases, the local pressure may exceed the hardness of the material and plastic deformation of the surface may result. Styli generally make a visible scratch on softer surfaces, for example, some steels, silver, gold, lead and elastomers (Poon and Bhushan, 1995; Bhushan, 1996). The existence of scratches results in measurement errors and unacceptable damage. As shown in Figure 2.4.1 presented earlier, the stylus digs into the surface and the results do not truly represent the microroughness. It is important to select stylus loads low enough to minimize plastic deformation.

2.4.2 Optical Methods

When electromagnetic radiation (light wave) is incident on an engineering surface, it is reflected either specularly or diffusely or both, Figure 2.4.5. Reflection is totally specular when the angle of reflection is equal to the angle of incidence (Snell's law); it is true for perfectly smooth surfaces. Reflection is totally diffused or scattered when the energy in the incident beam is distributed as the cosine of the angle of reflection (Lambert's law). As roughness increases, the intensity of the specular beam decreases while the diffracted radiation increases in intensity and becomes more diffuse. In most real surfaces, reflections are neither completely specular nor completely diffuse. Clearly, the relationships between the wavelength of radiation and the surface roughness will affect the physics of reflection; thus, a surface that is smooth to radiation of one wavelength may behave as if it were rough to radiation of a different wavelength.

The reflected beams from two parallel plates placed normal to the incident beam interfere and result in the formation of the fringes (Figure 2.4.6). The fringe spacing is a function of the spacing of the two plates. If one of the plates is a reference plate and another is the engineering surface whose roughness is to be measured, fringe spacing can be related to the surface roughness. We have just described so-called two-beam optical interference. A number of other interference techniques are used for roughness measurement.

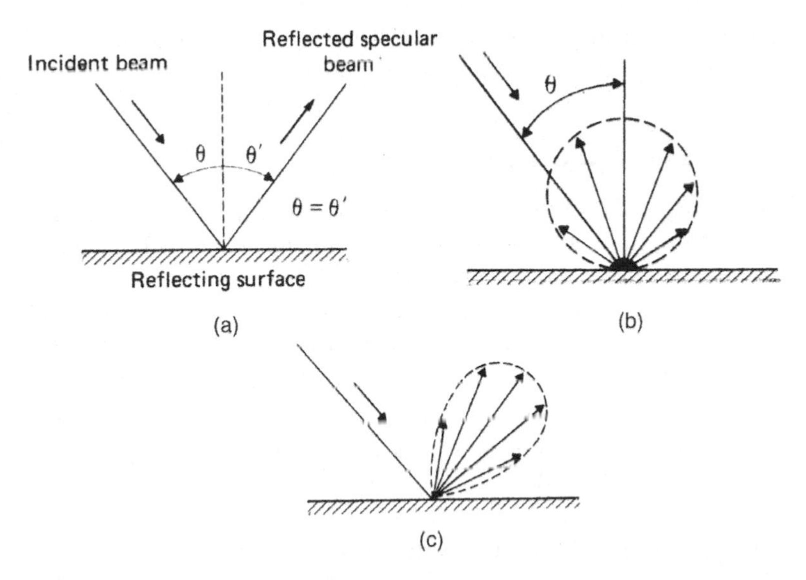

Figure 2.4.5 Modes of reflection of electromagnetic radiation from a solid surface, (a) specular only, (b) diffuse only, and (c) combined specular and diffuse. Reproduced with permission from Thomas, T.R. (1999), *Rough Surfaces*, Second edition, Imperial College Press, London, UK.

Numerous optical methods have been reported in the literature for measurement of surface roughness (Bhushan, 2013). Optical microscopy has been used for overall surveying, which only provides qualitative information. Optical methods may be divided into geometrical and physical methods (Thomas, 1999). Geometrical methods include taper-sectioning and light-sectioning methods. Physical methods include specular and diffuse reflections, speckle pattern,

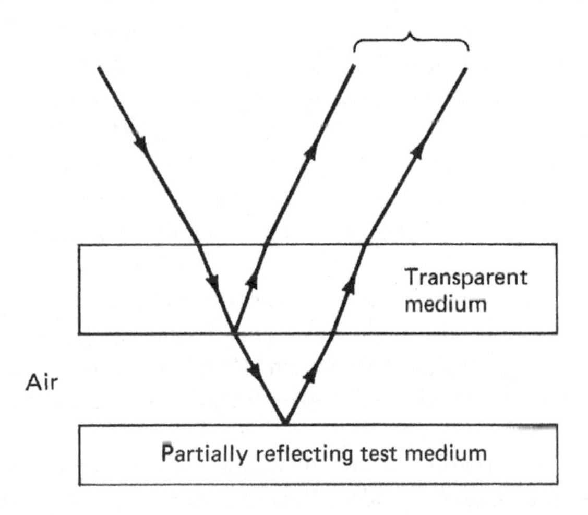

Figure 2.4.6 Schematic of two-beam interference.

and optical interference. In this section, we will describe only commonly used methods based on specular and diffuse reflections and optical interference.

2.4.2.1 Specular Reflection Methods

Gloss or specular reflectance (sometimes referred to as sheen or luster) is a surface property of the material, namely, the refractive index and surface roughness. Fresnel's equations provide a relationship between refractive index and reflectance. Surface roughness scatters the reflected light, thus affecting the specular reflectance. If the surface roughness σ is much smaller than the wavelength of the light (λ) and the surface has a Gaussian height distribution, the correlation between specular reflectance (R) and σ is described by (Beckmann and Spizzichino, 1963)

$$\frac{R}{R_0} = \exp\left[-\left(\frac{4\pi\sigma\cos\theta_i}{\lambda}\right)^2\right] \sim 1 - \left(\frac{4\pi\sigma\cos\theta_i}{\lambda}\right)^2 \tag{2.4.1}$$

where θ_i is the angle of incidence measured with respect to the sample normal and R_0 is the total reflectance of the rough surface and is found by measuring the total light intensity scattered in all directions including the specular direction. If roughness-induced, light-absorption processes are negligible, R_0 is equal to the specular reflectance of a perfectly smooth surface of the same material. For rougher surfaces ($\sigma \geq \lambda/10$), the true specular beam effectively disappears, so R is no longer measurable. Commercial instruments following the general approach are sometimes called specular glossmeters, Figure 2.4.7. The first glossmeter was used in the 1920s. A glossmeter detects the specular reflectance (or gloss) of the test surface (of typical size 50 mm $\times$ 50 mm), which is simply the fraction of the incident light reflected from a surface (Gardner and Sward, 1972). Measured specular reflectance is assigned a gloss number. The gloss number is defined as the degree to which the finish of the surface approaches that of the theoretical gloss standard, which is the perfect mirror, assigned a value of 1000.

The practical, primary standard is based on the black gloss (refractive index, $n = 1.567$) under angles of incidence of 20°, 60°, or 85°, according to ISO 2813 or American Society for Testing and Materials (ASTM) D523° standards. The specular reflectance of the black gloss at 60° for unpolarized radiation is 0.100 (Fresnel's equation, to be discussed later). By definition, the 60° gloss value of this standard is $1000 \times 0.10 = 100$. For 20° and 85°, Fresnel reflectances are 0.049 and 0.619, respectively, which are again by definition set to give a gloss value of 100. The glossmeter described by Budde (1980) operates over the wavelength range from 380 to 760 nm with a peak at 555 nm. There are five different angles of incidence that are commonly used – 20°, 45°, 60°, 75°, and 85°. Higher angles of incidence are used for rougher surfaces and vice versa.

Glossmeters are commonly used in the paint, varnish, and paper coating industries (Gardner and Sward, 1972). These are also used in magnetic tapes at 45° or 60° incident angles, depending upon the level of roughness (Bhushan, 1996). It is very convenient to measure the roughness of magnetic tape coatings during manufacturing by a glossmeter. The advantage of a glossmeter is its intrinsic simplicity, ease, and speed of analysis.

Other than accuracy and reproducibility, the major shortcoming of the gloss measurement is its dependence on the refractive index. Specular reflectance of a dielectric surface for unpolarized incident radiation increases with an increase in the refractive index. Use of a

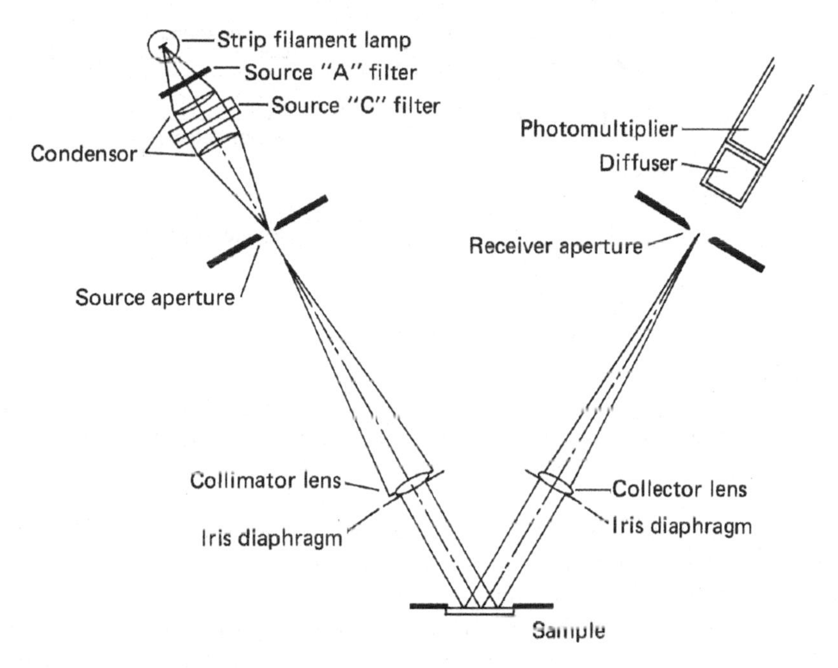

Figure 2.4.7 Schematic of a glossmeter. Reproduced with permission from Budde, W. (1980), "A Reference Instrument for 20°, 40°, and 85° Gloss Measurements," *Metrologia* **16**, 1–5. Copyright 1980. IOP Science.

glossmeter for roughness measurement is not appropriate; however, for luster or general appearance it may be acceptable.

2.4.2.2 Diffuse Reflection (Scattering) Methods

Vision depends on diffuse reflection or scattering. Texture, defects, and contamination causes scattering (Bennett and Mattson, 1989; Stover, 1995). It is difficult to obtain detailed roughness distribution data from the scattering measurements. Its spatial resolution is based on optical beam size typically 0.1 to 1 mm in diameter. Because scatterometers measure light reflectance rather than the actual physical distance between the surface and the sensor, they are relatively insensitive to changes in temperature and mechanical or acoustical vibrations, making them extremely stable and robust. To measure large surface areas, traditional methods scan the roughness of several, relatively small areas (or sometimes just a single scan line) at a variety of locations on the surface. On the other hand, with scatterometers, the inspection spot is quickly and automatically rastered over a large surface. The scattering is sometimes employed to measure surface texture. This technique is particularly suitable for on-line, roughness measurement during manufacture because it is continuous, fast, noncontacting, nondestructive, and relatively insensitive to the environment.

Three approaches used to measure defects and roughness by light scattering, include total integrated scatter, diffuseness of scattered light, and angular distribution (Bhushan, 2013). The

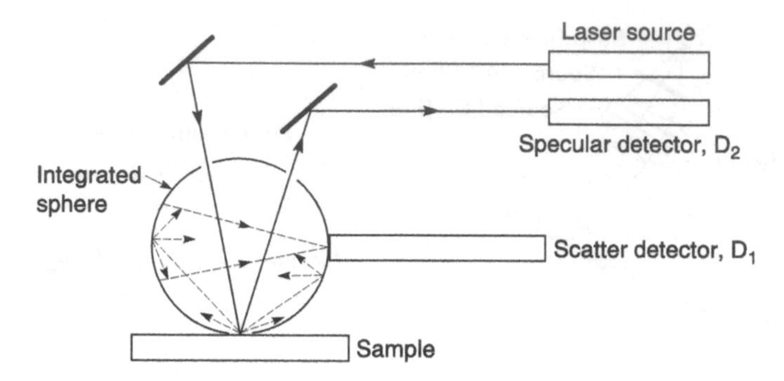

Figure 2.4.8 Schematic of the total integrated scatter apparatus with a diffuse integrated sphere. Reproduced with permission from Stover, J.C., Bernt, M., and Schiff, T. (1996), "TIS Uniformity Maps of Wafers, Disks and Other Samples," *Proc. Soc. Photo-Opt. Instrum. Eng.* **2541**, 21–25.

first of these three techniques is commercially used and is described here. The total integrated scatter (TIS) method is complementary to specular reflectance. Instead of measuring the intensity of the specularly reflected light, one measures the total intensity of the diffusely scattered light (Bennett, 1978; Stover, 1995). In the first TIS instrument, an aluminized, specular Coblentz sphere (90° integrating sphere) was used (Bennett and Porteus, 1961). Another method, shown in Figure 2.4.8 uses a high-reflectance diffuse integrated sphere. The incident laser beam travels through the integrated sphere, and strikes the sample port at a few degrees off-normal. The specular reflection traverses the sphere again and leaves through the exit port where it is measured by the specular detector, D_2. The inside of the sphere is covered with a diffuse white coating that rescatters the gathered sample scatter throughout the interior of the sphere. The sphere takes on a uniform glow regardless of the orientation of the scatter pattern. The scatter signal is measured by sampling this uniform glow with a scatter detector, D_1, located on the right side of the sphere. The TIS is then the ratio of the total light scattered by the sample to the total intensity of scattered radiation (both specular and diffuse). If the surface has a Gaussian height distribution and its standard deviation σ is much smaller than the wavelength of light (λ), the TIS can be related to σ as given by Equation (2.4.1) (Bennett, 1978):

$$TIS = \frac{R_0 - R}{R_0} = 1 - \exp\left[-\left(\frac{4\pi\sigma\,\cos\,\theta_i}{\lambda}\right)^2\right] \sim \left(\frac{4\pi\sigma\,\cos\,\theta_i}{\lambda}\right)^2 \quad (2.4.2a)$$

$$= \left(\frac{4\pi\sigma}{\lambda}\right)^2, \quad if\theta_i = 0 \quad (2.4.2b)$$

Samples of known specular reflectance are used to calibrate the reflected power (R_0) signals. The same samples, used to reflect the beam onto the sphere interior, can be used to calibrate the scattered power ($R_0 - R$) measurement signals (Stover *et al.*, 1996).

The scattering method is generally limited by available theories to studies of surface whose σ is much less than λ. With a He-Ne laser as the light source, the preceding constraint means that these techniques have been used mainly on optical quality surfaces where $\sigma < 0.1$ μm. Within that limited regime, they can provide high-speed, quantitative measurements of the roughness of both isotropic surfaces and those with a pronounced lay. The ultimate vertical resolution is 1 nm or better but the horizontal range is limited to fairly short surface wavelengths. Both the vertical and horizontal ranges can be increased by using long wavelength (infrared) radiation, but there is an accompanying loss of vertical and horizontal resolution.

Several commercial instruments, such as a Surfscan (KLA-Tencor Corporation, Mountain View, CA), Diskan (GCA Corp., Bedford, MA), and Dektak TMS-2000 (Bruker Instruments, Santa Barbara, CA) are built on this principle. In these instruments, to map a surface, either the sample moves or the light beam raster scans the sample. These instruments are generally used to generate maps of asperities, defects or particles rather than microroughness distribution.

2.4.2.3 Optical Interference Methods

Optical interferometry is a valuable technique for measuring surface shape, on both a macroscopic and microscopic scale (Tolansky, 1973). The traditional technique involves looking at the interference fringes and determining how much they depart from going straight and equally spaced. With suitable computer analysis, these can be used to completely characterize a surface. Both the differential interference contrast (DIC) and the Nomarski polarization interferometer techniques (Francon, 1966; Francon and Mallick, 1971) are commonly used for qualitative assessment of surface roughness. While those interferometers are very easy to operate, and they are essentially insensitive to vibration, they have the disadvantage that they measure what is essentially the slope of the surface errors, rather than the surface errors themselves. A commercial Nomarski type profiler based on the linearly polarized laser beam is made by Chapman Instruments, Rochester, New York.

The Tolansky or multiple-beam interferometer is another common interferometer used with a microscope. The surface being examined must have a high reflectivity and must be in near contact with the interferometer reference surface, which can scratch the surface under test.

One of the most common optical methods for the quantitative measurement of surface roughness is to use a two-beam interferometer. The actual sample can be measured directly without applying a high-reflectivity coating. The surface-height profile itself is measured. The option of changing the magnification can be used to obtain different values of lateral resolution and different fields of view. Short-wavelength visible-light interferometry and computerized phase-shifting techniques can measure surface-height variations with resolutions better than 1/100 of a wavelength of light. The short wavelength of visible light is a disadvantage, however, when measuring large surface-height variations and slopes. If a single wavelength is used to make a measurement and the surface-height difference between adjacent measurement points is greater than one-quarter wavelength, height errors of multiple half-wavelengths, may be introduced. The use of white light, or at least a few different wavelengths for the light source can solve this height ambiguity problem. Two techniques can extend the range of measurement of surface microstructure where the surface slopes are large. One technique, measuring surface heights at two or more visible wavelengths, creates a much longer nonvisible synthetic wavelength, which increases the dynamic range of the measurement by the ratio of the synthetic wavelength to the visible wavelength. Increases in the dynamic range by factors of 50 to 100 are possible. Another more powerful method uses a white-light scanning interferometer which involves measuring the degree of fringe modulation or coherence, instead of the phase of the interference fringes. Surface heights are measured by changing the path length of the sample arm of the interferometer to determine the location of the sample for which the white-light fringe with the best contrast is obtained. Vertical position at each location gives the surface height map. Various commercial instruments based on optical phase-shifting and vertical scanning interferometry are available (Bruker AXS (Wyko), Tucson, AZ; Zygo Corp., Middlefield, CT; and Phase Shift Technology (subsidiary of KLA-Tencor), Tucson, AZ).

Next, we first describe the principles of operation followed by a description of a typical commercial optical profiler.

Phase Shifting Interferometry

Several phase-measurement techniques (Wyant, 1975; Bruning, 1978; Wyant and Koliopoulos, 1981; Creath, 1988) can be used in an optical profiler to give more accurate height measurements than is possible by simply using the traditional technique of looking at the interference fringes and determining how much they depart from going straight and equally spaced. One mode of operation used in commercial profilers is the so-called integrated-bucket phase-shifting technique (Wyant *et al.*, 1984, 1986; Bhushan *et al.*, 1985).

For this technique, the phase difference between the two interfering beams is changed at a constant rate as the detector is read out. Each time the detector array is read out, the time variable phase, $\alpha(t)$, has changed by 90° for each pixel. The basic equation for the irradiance of a two-beam interference pattern is given by:

$$I = I_1 + I_2 \cos[\phi(x, y) + \alpha(t)] \tag{2.4.3}$$

where the first term is the average irradiance, the second term is the interference term, and $\phi(x, y)$ is the phase distribution being measured. If the irradiance is integrated while $\alpha(t)$ varies from 0 to $\pi/2, \pi/2$ *to* π, *and* π *to* $3\pi/2$, the resulting signals at each detected point are given by

$$\begin{aligned} A(x, y) &= I_1' + I_2'[\cos\phi(x\ y) - \sin\phi(x, y)] \\ B(x, y) &= I_1' + I_2'[-\cos\phi(x\ y) - \sin\phi(x, y)] \\ C(x, y) &= I_1' + I_2'[-\cos\phi(x\ y) + \sin\phi(x, y)] \end{aligned} \tag{2.4.4}$$

From the values of *A*, *B*, and *C*, the phase can be calculated as

$$\phi(x, y) = \tan^{-1}[(C(x, y) - B(x, y))/(A(x, y) - B(x, y))] \tag{2.4.5}$$

The subtraction and division cancel out the effects of fixed-pattern noise and gain variations across the detector, as long as the effects are not as large as to make the dynamic range of the detector too small to be of use.

Four frames of intensity data are measured. The phase $\phi(x, y)$ is first calculated, by means of Equation (2.4.5), using the first three of the four frames. It is then similarly calculated using the last three of the four frames. These two calculated phase values are then averaged to increase the accuracy of the measurement.

Because Equation (2.4.5) gives the phase modulo 2π, there may be discontinuities of 2π present in the calculated phase. These discontinuities can be removed as long as the slopes on the sample being measured are limited so that the actual phase difference between adjacent pixels is less than π. This is done by adding or subtracting a multiple of 2π to a pixel until the difference between it and its adjacent pixel is less than π.

Once the phase $\phi(x, y)$ is determined across the interference field, the corresponding height distribution $h(x, y)$ is determined by the equation

$$h(x, y) = \left(\frac{\lambda}{4\pi}\right)\phi(x, y) \tag{2.4.6}$$

Phase-shifting interferometry using a single wavelength has limited dynamic range. The height difference between two consecutive data points must be less than $\lambda/4$, where λ is the wavelength of the light used. If the slope is greater than $\lambda/4$ per detector pixel then height ambiguities of multiples of half-wavelengths exist. One technique that has been very successful in overcoming these slope limitations is to perform the measurement using two or more wavelengths λ_1 *and* λ_2, and then to subtract the two measurements. This results in the limitation in height difference between two adjacent detector points of one quarter of a synthesized equivalent wavelength λ_{eq}:

$$\lambda_{eq} = \frac{\lambda_1\lambda_2}{|\lambda_1 - \lambda_2|} \tag{2.4.7}$$

Thus, by carefully selecting the two wavelengths it is possible to greatly increase the dynamic range of the measurement over what can be obtained using a single wavelength (Cheng and Wyant, 1985).

While using two wavelength phase-shifting interferometry works very well with step heights, it does not work especially well with rough surfaces. A much better approach is to use a broad range of wavelengths and the fringe modulation or coherence peak sensing approach whose description follows.

Vertical Scanning Coherence Peak Sensing

In the vertical scanning coherence peak sensing mode of operation, a broad spectral white light source is used. Due to the large spectral bandwidth of the source, the coherence length of the source is short, and good contrast fringes will be obtained only when the two paths of the interferometer are closely matched in length. Thus, if in the interference microscope the path length of the sample arm of the interferometer is varied, the height variations across the sample can be determined by looking at the sample position for which the fringe contrast is a maximum. In this measurement there are no height ambiguities and since in a properly adjusted interferometer the sample is in focus when the maximum fringe contrast is obtained, there are no focus errors in the measurement of surface texture (Davidson *et al.*, 1987). Figure 2.4.9 shows the irradiance at a single sample point as the sample is translated through focus. It should be noted that this signal looks a lot like an amplitude modulated (AM) communication signal.

The major drawback of this type of scanning interferometer measurement is that only a single surface height is being measured at a time and a large number of measurements and calculations are required to determine a large range of surface height values. One method for processing the data that gives both fast and accurate measurement results is to use conventional

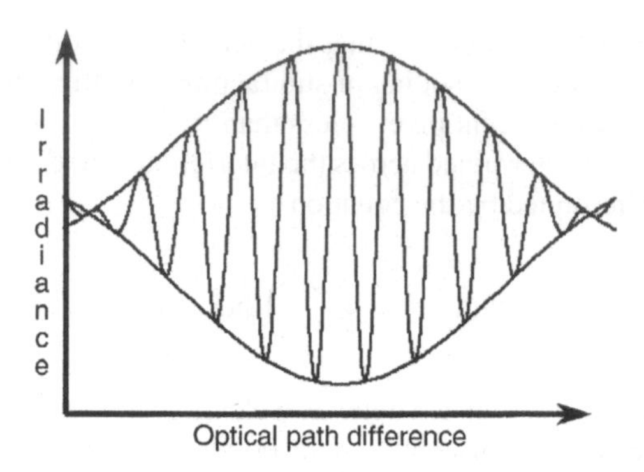

Figure 2.4.9 Irradiance at a single sample point as the sample is translated through focus. Reproduced with permission from Caber, P. (1993), "An Interferometric Profiler for Rough Surfaces," *Appl. Opt.* **32**, 3438–3441. Copyright 1993. Optical Society.

communication theory and digital signal processing (DSP) hardware to demodulate the envelope of the fringe signal to determine the peak of the fringe contrast (Caber, 1993). This type of measurement system produces fast, noncontact, true three-dimensional area measurements for both large steps and rough surfaces to nanometer precision.

A Commercial Digital Optical Profiler

Figure 2.4.10 shows a schematic of a commercial phase-shifting/vertical sensing interference microscope (Wyant, 1995). For smooth surfaces, the phase-shifting mode is used since it gives subnanometer height resolution capability. For rough surfaces and large steps, up to 500 μm surface height variations, the vertical scanning coherence sensing technique is used which gives an approximately 3 nm height resolution. The instrument operates with one of several interchangeable magnification objectives. Each objective contains an interferometer, consisting

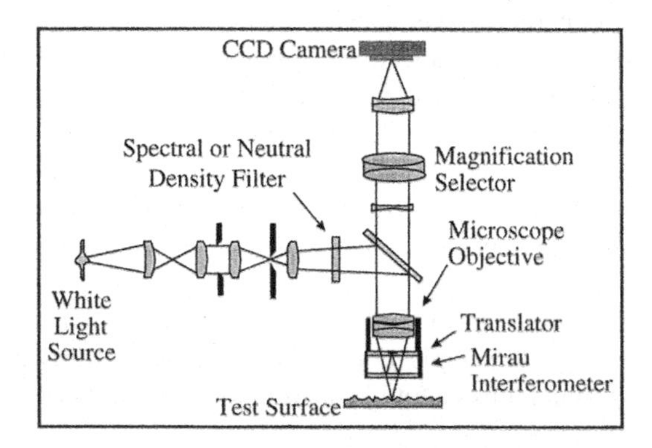

Figure 2.4.10 Optical schematic of the three-dimensional digital optical profiler based on phase-shifting/vertical sensing interferometer, Wyko HD-2000. Reproduced with permission from Wyant, J.C. (1995), "Computerized Interferometric Measurement of Surface Microstructure," *Proc. Soc. Photo-Opt. Instrum. Eng.* **2576**, 122–130. Copyright 1995. SPIE.

of a reference mirror and beams splitter, which produces interference fringes when light reflected off the reference mirror recombines with light reflected off the sample. Determination of surface height using phase-shifting interferometry typically involves the sequential shifting of the phase of one beam of the interferometer relative to another beam by known amounts, and measuring the resulting interference pattern irradiance. Using a minimum of three frames of intensity data, the phase is calculated which is then used to calculate the surface height variations over a surface. In vertical scanning interferometry when short coherence white light is used, these interference fringes are present only over a very shallow depth on the surface. The surface is profiled vertically so that each point on the surface produces an interference signal and then locating the exact vertical position where each signal reaches its maximum amplitude. To obtain the location of the peak, and hence the surface height information, this irradiance signal is detected using a CCD array. The instrument starts the measurement sequence by focusing above the top of the surface being profiled and quickly scanning downward. The signal is sampled at fixed intervals, such as every 50 to 100 nm, as the sample path is varied. The motion can be accomplished using a piezoelectric transducer. Low-frequency and DC signal components are removed from the signal by digital high-bandpass filtering. The signal is next rectified by square-law detection and digitally lowpass filtered. The peak of the lowpass filter output is located and the vertical position corresponding to the peak is noted. Frames of interference data imaged by a video camera are captured and processed by high-speed digital signal-processing hardware. As the system scans downward, an interference signal for each point on the surface is formed. A series of advanced algorithms are used to precisely locate the peak of the interference signal for each point on the surface. Each point is processed in parallel and a three-dimensional map is obtained.

The configuration shown in Figure 2.4.10 utilizes a two-beam Mirau interferometer at the microscope objective. Typically the Mirau interferometer is used for magnifications between 10 and 50x, a Michelson interferometer is used for low magnifications (between 1.5 and 5x) and the Linnik interferometer is used for high magnifications (between 100 and 200x), Figure 2.4.11. A separate magnification selector is placed between the microscope objective and the CCD camera to provide additional image magnifications. High magnifications are used for roughness measurement (typically 40x) and low magnifications (typically 1.5x) are used for geometrical parameters. A tungsten halogen lamp is used as the light source. In the phase shifting mode of operation a spectral filter of 40 nm bandwidth centered at 650 nm is used to increase the coherence length. For the vertical scanning mode of operation the spectral filter is not used. Light reflected from the test surface interferes with light reflected from the reference. The resulting interference pattern is imaged onto the CCD array, with a size of about 736 × 480 and pixel spacing of about 8 μm. The output of the CCD array can be viewed on the TV monitor as well as is digitized and read by the computer. The Mirau interferometer is mounted on either a piezoelectric transducer (PZT) or a motorized stage so that it can be moved at constant velocity. During this movement, the distance from the lens to the reference surface remains fixed. Thus, a phase shift is introduced into one arm of the interferometer. By introducing a phase shift into only one arm while recording the interference pattern that is produced, it is possible to perform either phase-shifting interferometry or vertical scanning coherence peak sensing interferometry.

Major advantages of this technique are that it is noncontact and three-dimensional measurements can be made rapidly without moving the sample or the measurement tool. One of the limitations of these instruments is that they can be used for surfaces with similar optical

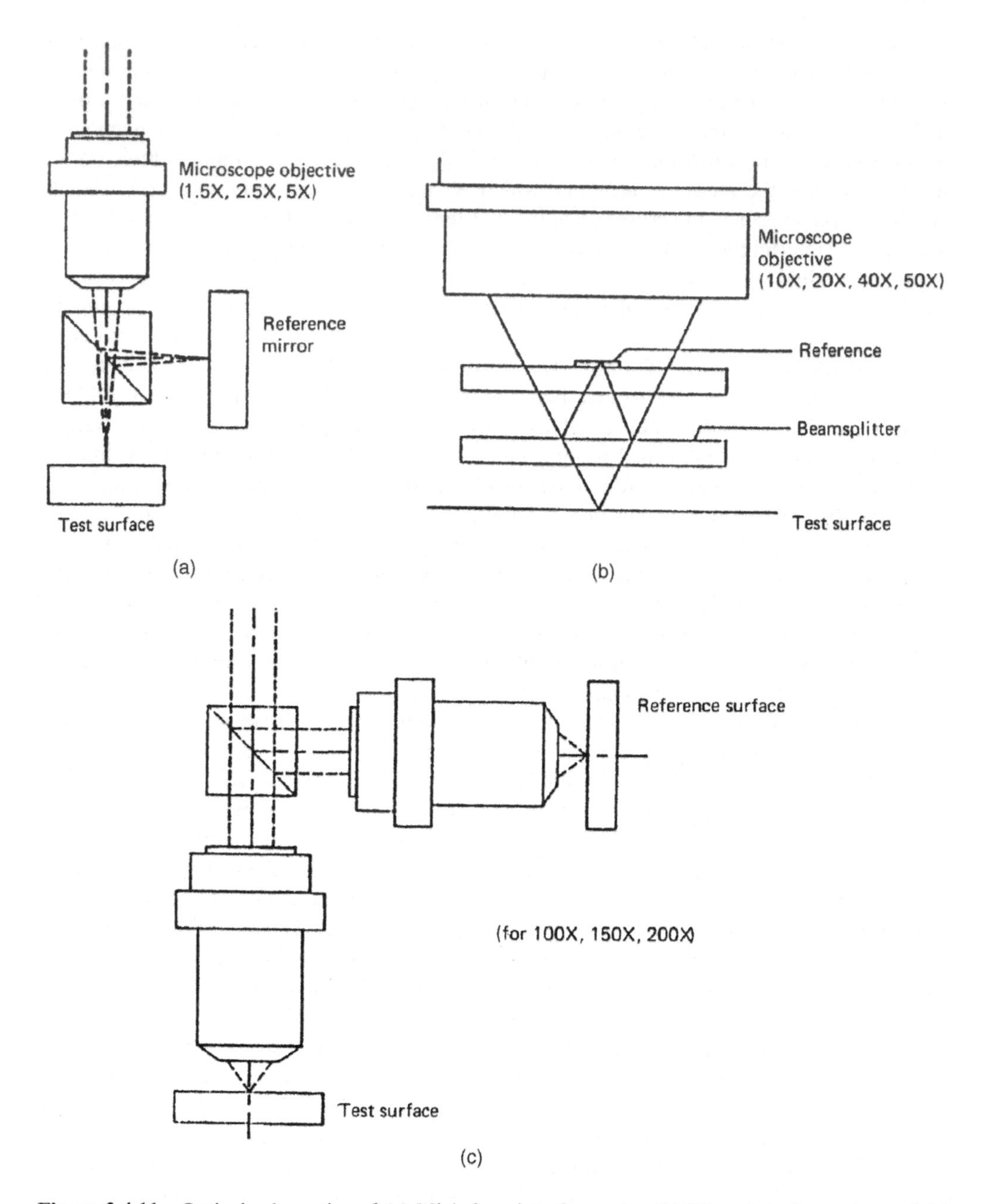

Figure 2.4.11 Optical schematics of (a) Michelson interferometer, (b) Mirau interferometer, and (c) Linnik interferometer.

properties. When dealing with thin films, incident light may penetrate the film and can be reflected from the film-substrate interface. This reflected light wave would have a different phase from that reflected from the film surface.

The smooth surfaces using the phase measuring mode can be measured with a vertical resolution as low as 0.1 nm. The vertical scanning mode provides a measurement range to about 500 μm. The field of view depends on the magnification, up to 10 mm × 10 mm. The lateral sampling interval is given by the detector spacing divided by the magnification; it is about 0.15 μm at 50x magnification. The optical resolution, which can be thought of as the closest distance between two features on the surface such that they remain distinguishable, is given by 0.61 $\lambda/(NA)$, where λ is the wavelength of the light source and NA is the numerical aperture of the objective (typically ranging from 0.036 for 1.5x to 0.5 for 40x). In practice, because of aberrations in the optical system, the actual resolution is slightly worse than the optical resolution. The best optical resolution for a lens is on the order of 0.5 μm. The scan speed is typically up to about 7 μm/s. The working distance, which is the distance between the last element in the objective and the sample, is simply a characteristic of the particular objective used.

Church *et al.* (1985) measured a set of precision-machined smooth optical surfaces by a mechanical-stylus profiler and an optical profiler in phase-measuring mode. They reported an excellent quantitative agreement between the two profilers. Boudreau *et al.* (1995) measured a set of machined (ground, milled, and turned) steel surfaces by a mechanical-stylus profiler and an optical profiler in the vertical scanning mode. Again, they reported an excellent quantitative agreement between the two profilers.

Typical roughness data using a digital optical profiler can be found in Wyant *et al.* (1984, 1986); Bhushan *et al.* (1985, 1988); Lange and Bhushan (1988), Caber (1993), and Wyant (1995).

2.4.3 Scanning Probe Microscopy (SPM) Methods

The family of instruments based on scanning tunneling microscopy (STM) and atomic force microscopy (AFM) are called scanning probe microscopies (SPM) (Bhushan, 2011).

2.4.3.1 Scanning Tunneling Microscopy (STM)

The principle of electron tunneling was proposed by Giaever (1960). He envisioned that if a potential difference is applied to two metals separated by a thin insulating film, a current will flow because of the ability of electrons to penetrate a potential barrier. To be able to measure a tunneling current, the two metals must be spaced no more than 10 nm apart. In 1981, Gerd Binnig, Heinrich Rohrer and their colleagues introduced vacuum tunneling combined with lateral scanning (Binnig *et al.*, 1982; Binnig and Rohrer, 1983). Their instrument is called the scanning tunneling microscope (STM). The vacuum provides the ideal barrier for tunneling. The lateral scanning allows one to image surfaces with exquisite resolution, laterally less than 1 nm and vertically less than 0.1 nm, sufficient to define the position of single atoms. The very high vertical resolution of the STM is obtained because the tunnel current varies exponentially with the distance between the two electrodes, that is, the metal tip and the scanned surface. Very high lateral resolution depends upon the sharp tips. Commercial STMs have been developed for operation in ambient air as well. An excellent review on this subject is presented by Bhushan (1999, 2011).

The principle of STM is straightforward. A sharp metal tip (one electrode of the tunnel junction) is brought close enough (0.3–1 nm) to the surface to be investigated (second electrode) so that, at a convenient operating voltage (10 mV–2 V), the tunneling current varies from 0.2 to 10 nA, which is measurable. The tip is scanned over a surface at a distance of 0.3 to 1 nm, while the tunnel current between it and the surface is sensed. The tunnel current J_T is a sensitive function of the gap width d, that is, $J_T \propto V_T \exp(-A\phi^{1/2}d)$, where V_T is the bias voltage, ϕ is the average barrier height (work function) and $A \sim 1$ if ϕ is measured in eV and d in Å. With a work function of a few eV, J_T changes by an order of magnitude for every angstrom change of h. If the current is kept constant to within, for example, 2%, then the gap h remains constant to within 1 pm. For operation in the constant current mode, the control unit (CU) applies a voltage V_z to the piezo P_z such that J_T remains constant when scanning the tip with $P_y$and P_x over the surface. At the constant work function ϕ, $V_z(V_x, V_y)$ yields the roughness of the surface $z(x, y)$ directly, as illustrated at a surface step at A. Smearing of the step, δ (lateral resolution) is on the order of $(R)^{1/2}$, where R is the radius of the curvature of the tip. Thus, a lateral resolution of about 2 nm requires tip radii on the order of 10 nm. A 1-mm-diameter solid rod ground at one end at roughly 90° yields overall tip radii of only a few hundred nanometers, but with closest protrusion of rather sharp microtips on the relatively dull end yields a lateral resolution of about 2 nm. In-situ sharpening of the tips by gently touching the surface brings the resolution down to the 1-nm range; by applying high fields (on the order of 10^8 V/cm) during, for example, half an hour, resolutions considerably below 1 nm could be reached.

There are a number of commercial STMs available on the market. Digital Instruments (now Bruker Instruments) introduced the first commercial STM, the Nanoscope I, in 1987. In the Nanoscope IV STM for operation in ambient air, the sample is held in position while a piezoelectric crystal in the form of a cylindrical tube scans the sharp metallic probe over the surface in a raster pattern while sensing and outputting the tunneling current to the control station, Figure 2.4.12. The digital signal processor (DSP) calculates the desired separation of the tip from the sample by sensing the tunneling current flowing between the sample and the tip. The bias voltage applied between the sample and the tip encourages the tunneling current

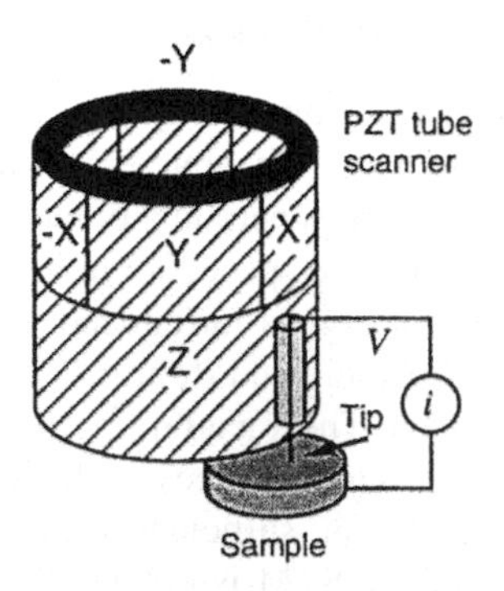

Figure 2.4.12 Principle of operation of a commercial STM; a sharp tip attached to a piezoelectric tube scanner is scanned on a sample.

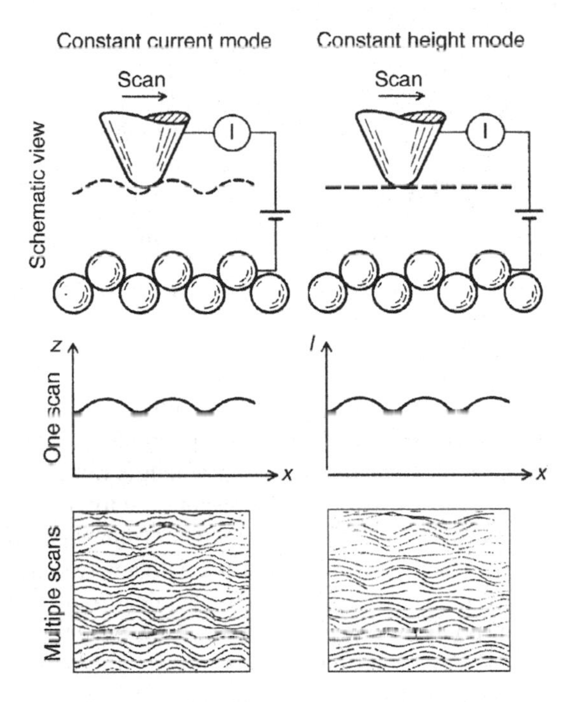

Figure 2.4.13 Scanning tunneling microscope can be operated in either the constant current or the constant height mode. The images are of graphite in air.

to flow. The DSP completes the digital feedback loop by outputting the desired voltage to the piezoelectric tube. The STM operates in both the "constant height" and "constant current" modes depending on a parameter selection in the control panel. In the constant current mode, the feedback gains are set high, the tunneling tip closely tracks the sample surface, and the variation in the tip height required to maintain constant tunneling current is measured by the change in the voltage applied to the piezo tube, Figure 2.4.13. In the constant height mode, the feedback gains are set low, the tip remains at a nearly constant height as it sweeps over the sample surface, and the tunneling current is imaged, Figure 2.4.13. A current mode is generally used for atomic-scale images. This mode is not practical for rough surfaces. A three-dimensional picture [$z(x, y)$] of a surface consists of multiple scans [$z(x)$] displayed laterally from each other in the y direction. Note that if atomic species are present in a sample, the different atomic species within a sample may produce different tunneling currents for a given bias voltage. Thus the height data may not be a direct representation of the texture of the surface of the sample.

Samples to be imaged with STM must be conductive enough to allow a few nanoAmperes of current to flow from the bias voltage source to the area to be scanned. In many cases, nonconductive samples can be coated with a thin layer of a conductive material to facilitate imaging. The bias voltage and the tunneling current depend on the sample. The scan size ranges from a fraction of a nm x fraction of a nm to about 125 μm × 125 μm. A maximum scan rate of 122 Hz can be used. Typically, 256 × 256 data formats are used. The lateral resolution at larger scans is approximately equal to scan length divided by 256.

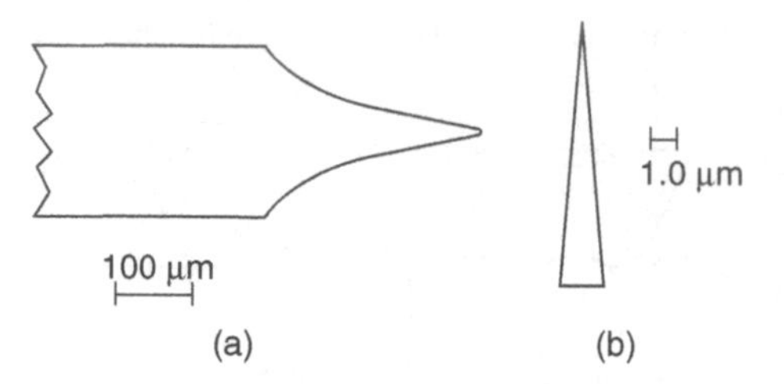

Figure 2.4.14 Schematics of (a) a typical tungsten cantilever with a sharp tip produced by electrochemical etching, and (b) CG Pt/Ir.

The STM cantilever should have a sharp metal tip with a low aspect ratio (tip length/tip shank) to minimize flexural vibrations. Ideally, the tip should be atomically sharp, but in practice, most tip preparation methods produce a tip which is rather ragged and consists of several asperities with the one closest to the surface responsible for tunneling. STM cantilevers with sharp tips are typically fabricated from metal wires of tungsten (W), platinum-iridium (Pt-Ir), or gold (Au) and sharpened by grinding, cutting with a wire cutter or razor blade, field emission/evaporator, ion milling, fracture, or electrochemical polishing/etching (Ibe *et al.*, 1990). The two most commonly used tips are made from either a Pt-Ir (80/20) alloy or tungsten wire. Iridium is used to provide stiffness. The Pt-Ir tips are generally mechanically formed and are readily available. The tungsten tips are etched from tungsten wire with an electrochemical process. The wire diameter used for the cantilever is typically 250 μm with the radius of curvature ranging from 20 to 100 nm and a cone angle ranging from 10° to 60°, Figure 2.4.14a. For calculations of normal spring constant and natural frequency of round cantilevers, see Sarid and Elings (1991).

Controlled Geometry (CG) Pt-Ir probes are commercially available, Figure 2.4.14b. These probes are electrochemically etched from Pt-Ir (80/20) wire and polished to a specific shape which is consistent from tip to tip. Probes have a full cone angle of approximately 15° and a tip radius of less than 50 nm. For imaging of deep trenches (> 0.25 μm) and nanofeatures, focused ion beam (FIB) milled CG milled probes with an extremely sharp tip radius (< 5 nm) are used. For electrochemistry, Pt-Ir probes are coated with a nonconducting film (not shown in the figure).

2.4.3.2 Atomic Force Microscopy (AFM)

STM requires that the surface to be measured is electrically conductive. In 1985, Gerd Binnig and his colleagues developed an instrument called the atomic force microscope, capable of investigating surfaces of both conductors and insulators on an atomic scale (Binnig *et al.*, 1986). Like the STM, the AFM relies on a scanning technique to produce very high resolution, three-dimensional images of sample surfaces. AFM measures ultrasmall forces (less than 1 nN) present between the AFM tip surface and a sample surface. These small forces are measured by measuring the motion of a very flexible cantilever beam having an ultrasmall mass. In the operation of high-resolution AFM, the sample is generally scanned instead of the tip as an STM, because AFM measures the relative displacement between the cantilever surface and reference surface, and any cantilever movement would add vibrations. However, AFMs are

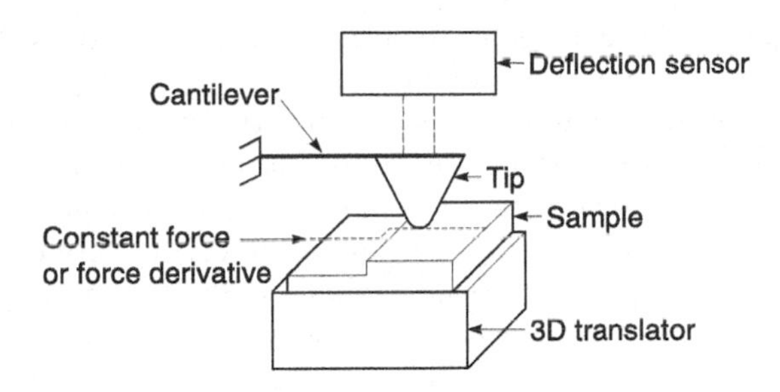

Figure 2.4.15 Principle of operation of the atomic force microscope.

now available where the tip is scanned and the sample is stationary. As long as the AFM is operated in the so-called contact mode, little if any vibration is introduced.

The AFM combines the principles of the STM and the stylus profiler, Figure 2.4.15. In the AFM, the force between the sample and tip is detected rather than the tunneling current to sense the proximity of the tip to the sample. A sharp tip at the end of a cantilever is brought with contact with a sample surface by moving the sample with piezoelectric scanners. During initial contact, the atoms at the end of the tip experience a very weak repulsive force due to electronic orbital overlap with the atoms in the sample surface. The force acting on the tip causes a lever deflection which is measured by tunneling, capacitive, or optical detectors such as laser interferometry. The deflection can be measured to within $\pm$0.02 nm, so for a typical lever force constant at 10 N/m a force as low as 0.2 nN (corresponding normal pressure ~200 MPa for an Si3N4 tip with a radius of about 50 nm against single-crystal silicon) could be detected. This operational mode is referred to as "repulsive mode" or "contact mode" (Binnig *et al.*, 1986). In the dynamic mode of operation, also referred to as "attractive force imaging" or "noncontact imaging" mode, the tip is brought into close proximity (within a few nanometers) to, and not in contact with, the sample (Martin *et al.*, 1987). The cantilever is vibrated in either amplitude modulation (AM) or frequency modulation (FM) mode. Very weak van der Waals attractive forces are present at the tip–sample interface. Although in this technique the normal pressure exerted at the interface is zero (desirable to avoid any surface deformation), it is slow and difficult to use and is rarely used outside research environments. In either mode, surface topography is measured by laterally scanning the sample under the tip while simultaneously measuring the separation-dependent force or force gradient (derivative) between the tip and the surface. The force gradient is obtained by vibrating the cantilever (Martin *et al.*, 1987; Sarid and Elings, 1991) and measuring the shift of resonance frequency of the cantilever. To obtain topographic information, the interaction force is either recorded directly or used as a control parameter for a feedback circuit that maintains the force or force derivative at a constant value. Force derivative is normally tracked in noncontact imaging.

With AFM operated in the contact mode, topographic images with a vertical resolution of less than 0.1 nm (as low as 0.01 nm) and a lateral resolution of about 0.2 nm have been obtained. With a 0.01 nm displacement sensitivity, 10 nN to 1 pN forces are measurable. These forces are comparable to the forces associated with chemical bonding, e.g., 0.1 μN for an ionic bond and 10 pN for a hydrogen bond (Binnig *et al.*, 1986). For further reading, see Bhushan (1999, 2011).

STM is ideal for atomic-scale imaging. To obtain atomic resolution with AFM, the spring constant of the cantilever should be weaker than the equivalent spring between atoms on the order of 10 N/m. Tips have to be as sharp as possible. Tips with a radius ranging from 5 to 50 nm are commonly available. "Atomic resolution" cannot be achieved with these tips at the normal force in the nanoNewton range. Atomic structures obtained at these loads have been obtained from lattic imaging or by imaging of the crystal periodicity. Reported data show either perfectly ordered periodic atomic structures or defects on a large lateral scale, but no well-defined, laterally resolved atomic-scale defects like those seen in images routinely obtained with STM. Interatomic forces with one or several atoms in contact are 20–40 or 50–100 pN, respectively. Thus, atomic resolution with AFM is only possible with a sharp tip on a flexible cantilever at a net repulsive force of 100 pN or lower.

The first commercial AFM was introduced in 1989 by Digital Instruments (now Bruker Instruments). Now there are a number of commercial AFMs available on the market. Major manufacturers of AFMs for use in an ambient environment are as follows: Bruker Instruments (formally Digital Instruments and Veeco Metrology), Santa Barbara, CA; Agilent Technologies, Chandler, Arizona; ND-MDT, Russia; JPK Instruments, Berlin, Germany; Park Systems, SuWon, Korea; Asylum Research, Santa Barbara, CA; and KLA-Tencor and Seiko Instruments. Ultra-high vacuum (UHV) AFM/STMs are manufactured by Omicron Vakuumphysik GmbH, Germany. Low temperature AFMs are manufactured by Nanonics Imaging, Jerusalem, Israel. Personal STMs and AFMs for ambient environment and UHV/STMs are manufactured by various manufactures including Nanosurf AG, Liestal, Switzerland and Nanonics Imaging, Jerusalem, Israel.

We describe here a commercial AFM called Nanoscope IV from Bruker Instruments for operation in ambient air, with scanning lengths ranging from about 0.7 μm (for atomic resolution) to about 125 μm, Figure 2.4.16a. This is the most commonly used design and the multimode AFM comes with many capabilities. In this AFM, the sample is mounted on a PZT tube scanner which consists of separate electrodes to scan precisely the sample in the *X-Y* plane in a raster pattern as shown in Figure 2.4.16b and to move the sample in the vertical (*Z*) direction. A sharp tip at the end of a flexible cantilever is brought into contact with the sample. Normal and frictional forces (to be discussed in Chapter 10) being applied at the tip–sample interface are measured using a laser beam deflection technique. A laser beam from a diode laser is directed by a prism onto the back of a cantilever near its free end, tilted downward at about 10° with respect to a horizontal plane. The reflected beam from the vertex of the cantilever is directed through a mirror onto a quad photodetector (split photodetector with four quadrants). The differential signal from the top and bottom photodiodes provides the AFM signal, which is a sensitive measure of the cantilever vertical deflection. Topographic features of the sample cause the tip to deflect in the vertical direction as the sample is scanned under the tip. This tip deflection will change the direction of the reflected laser beam, changing the intensity difference between the top and bottom photodetector (AFM signal). In the AFM operating mode of the "height mode," for topographic imaging, or for any other operation in which the applied normal force is to be kept a constant, a feedback circuit is used to modulate the voltage applied to the PZT scanner to adjust the height of the PZT, so that the cantilever vertical deflection (given by the intensity difference between the top and bottom detector) will remain almost constant during scanning. The PZT height variation is thus a direct measure of surface roughness of the sample.

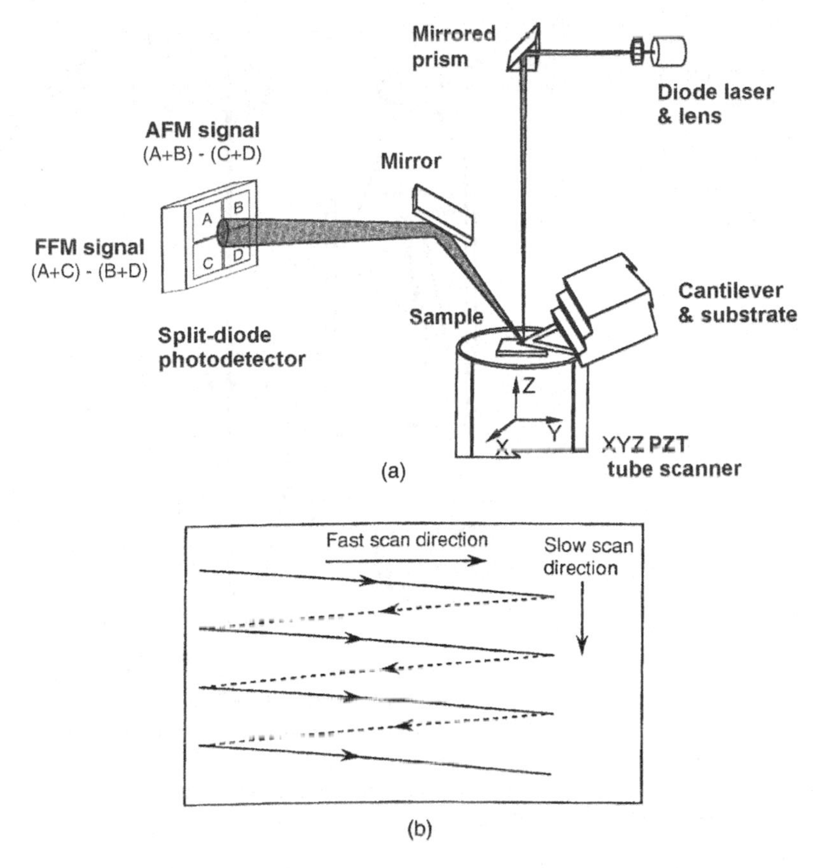

Figure 2.4.16 (a) Principle of operation of a commercial atomic force/friction force microscope, sample mounted on a piezoelectric tube scanner is scanned against a sharp tip and the cantilever deflection is measured using a laser beam deflection technique and (b) schematic of triangular pattern trajectory of the AFM tip as the sample is scanned in two dimensions. During imaging, data are recorded only during scans along the solid scan lines.

This AFM can be used for roughness measurements in the "tapping mode," also referred to as dynamic force microscopy. In the tapping mode, during scanning over the surface, the cantilever is vibrated by a piezo mounted above it, and the oscillating tip slightly taps the surface at the resonant frequency of the cantilever (70-400 kHz) with a 20-100 nm oscillating amplitude introduced in the vertical direction with a feedback loop keeping the average normal force constant. The oscillating amplitude is kept large enough so that the tip does not get stuck to the sample because of adhesive attraction. The tapping mode is used in roughness measurements to minimize the effects of friction and other lateral forces and to measure the roughness of soft surfaces.

There are several AFM designs in which both force sensors using optical beam deflection method and scanning unit are mounted on the microscope head; then these AFMs can be used

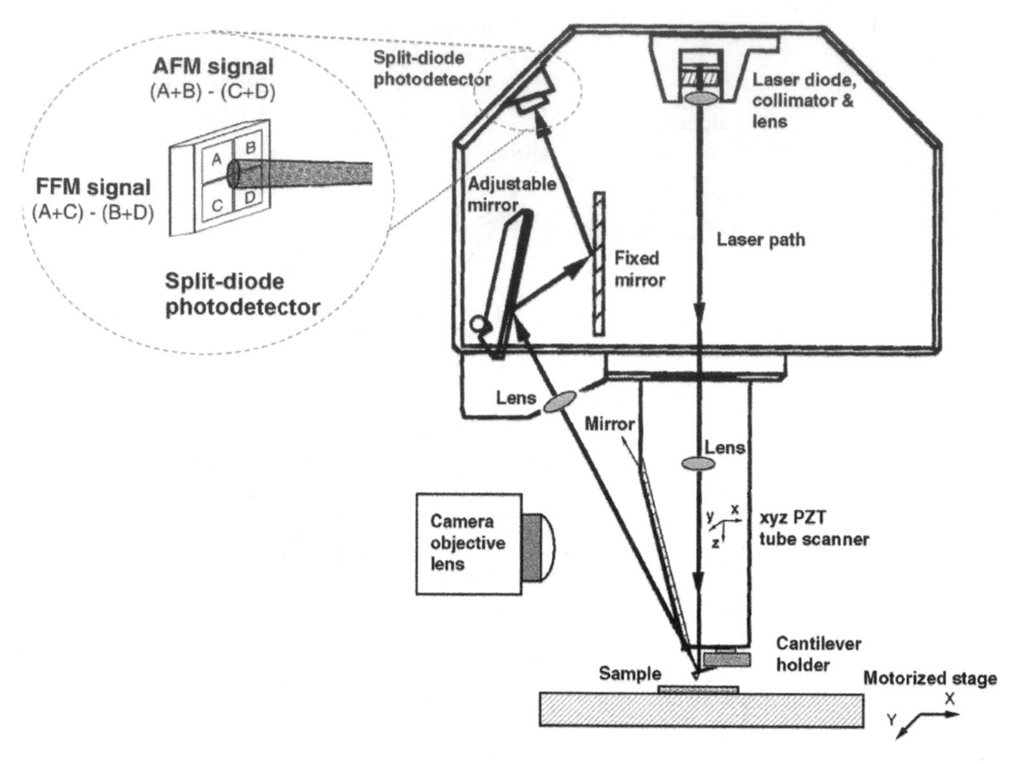

Figure 2.4.17 Principle of operation of a commercial atomic force/friction force microscope, the head scans as well as generates the cantilever deflection.

to image large samples. Schematic of one such design called Dimension 3000 from Bruker Instruments is shown in Figure 2.4.17. The head scans as well as generates the cantilever deflection. The beam emitted by the laser diode reflects off the cantilever and is detected by a quad photodetector.

Roughness measurements are typically made using a sharp tip on a cantilever beam at a normal load on the order of 10 nN. The tip is scanned in such a way that its trajectory on the sample forms a triangular pattern. Scanning speeds in the fast and slow scan directions depend on the scan area and scan frequency. The scan sizes available for this instrument range from 0.7 μm × 0.7 μm to 125 μm × 125 μm. A maximum scan rate of 122 Hz can typically be used. Higher scan rates are used for small scan length. 256 × 256 data points are taken for each image. For example, scan rates in the fast and slow scan directions for an area of 10 μm × 10 μm scanned at 0.5 Hz are 10 μm/s and 20 nm/s, respectively. The lateral resolution at larger scans is approximately equal to scan length divided by 256. At a first instance, scanning angle may not appear to be an important parameter for roughness measurements. However, the friction force between the tip and the sample will affect the roughness measurements in a parallel scan (scanning along the long axis of the cantilever). Therefore, a perpendicular scan may be more desirable. Generally, one picks a scanning angle which gives the same roughness data in both directions; this angle may be slightly different than that for the perpendicular scan.

The most commonly used cantilevers for roughness measurements in contact AFM mode, are microfabricated plasma enhanced chemical vapor deposition (PECVD) silicon nitride triangular beams with integrated square pyramidal tips with a radius on the order of 30–50 nm. Four cantilevers with different sizes and spring stiffnesses (ranging from 0.06 to 0.6 N/m) on each cantilever substrate made of boron silicate glass are shown in Figure 2.4.18a. Etched single-crystal n-type silicon rectangular cantilevers with square pyramidal tips with a radius of about 10 nm are used for contact and tapping modes, Figure 2.4.18b. The cantilevers used for contact mode are stiff. For imaging within trenches by AFM, high-aspect ratio tips (HART) are used. An example of a probe is shown in Figure 2.4.18c. The probe is approximately 1 μm long and 0.1 μm in diameter. It tapers to an extremely sharp point with the radius better than few nanometers. Carbon nanotube tips with small diameters and high aspect ratios are also used for high-resolution imaging of surfaces and of deep trenches.

For scratching, wear and indentation studies, single-crystal natural diamond tips ground to the shape of a three-sided pyramid with an apex angle of either 60° or 80° and a point sharpened to a radius of 50-100 nm are commonly used.

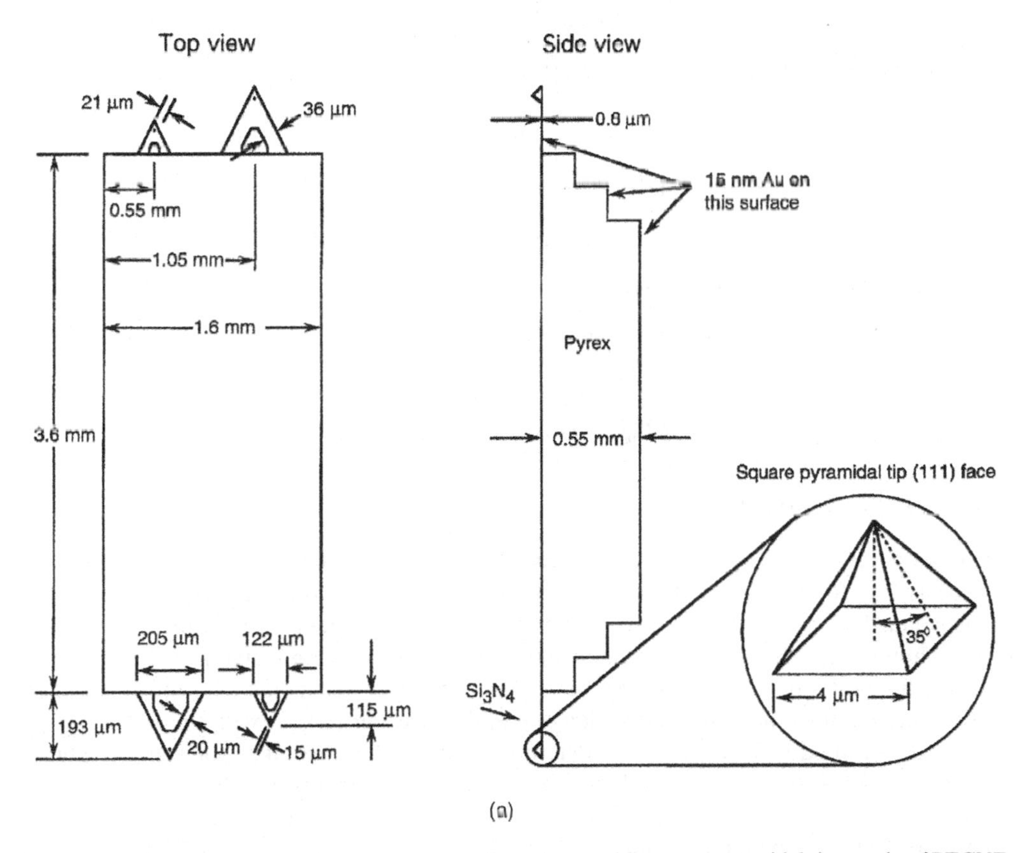

Figure 2.4.18 Schematics of (a) triangular cantilever beam with square pyramidal tips made of PECVD Si_3N_4, (b) rectangular cantilever beams with square pyramidal tips made of single-crystal silicon, and (c) high-aspect ratio Si_3N_4 probe. (*Continued*)

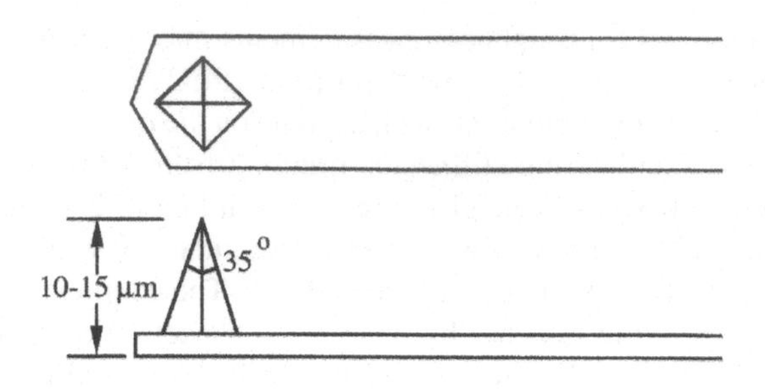

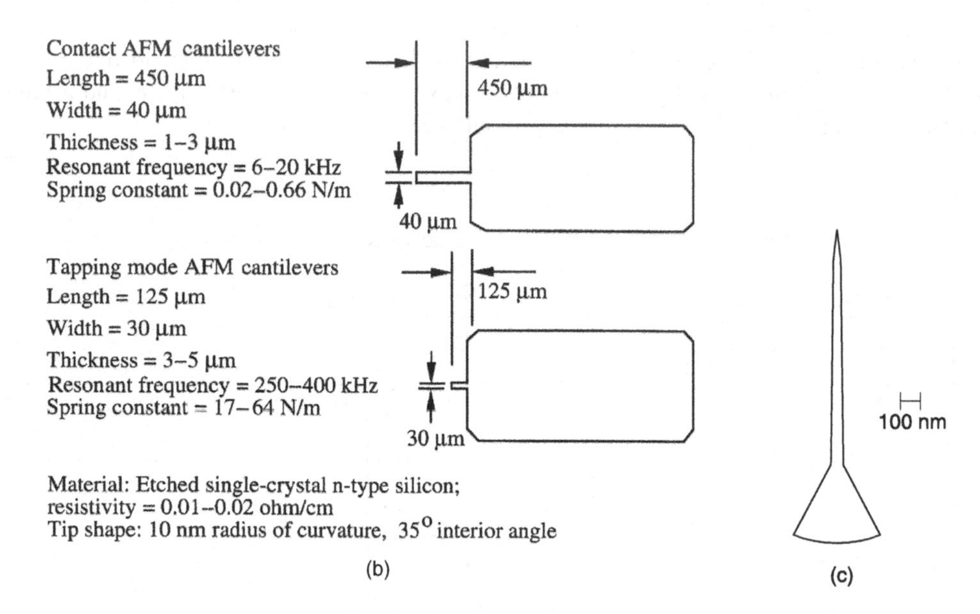

Figure 2.4.18 *(Continued)*

2.4.4 Fluid Methods

Such techniques are mainly used for continuous inspection (quality control) procedures in service as they function without contact with the surface and are very fast. These provide numerical data that can only be correlated empirically to the roughness. The two most commonly used techniques are the hydraulic method and the pneumatic gaging method.

In the hydraulic method, sometimes called the outflow meter method, an open-bottomed vessel with a compliant annulus at its lower end is placed in contact with the surface to be measured and filled with water to a predetermined level. The time taken for a given volume of water to escape through the gap between the compliant annulus and the rough surface is measured (Thomas, 1999). A simple relationship exists between the standard deviation of asperity heights, σ_p and the flow time t,

$$\sigma_p = at^n \tag{2.4.8}$$

where a and n are constants determined by the characteristics of the method employed. This method was initially developed to measure road surfaces but can be used for any large roughness pattern.

The pneumatic gaging method is used for finer scale roughness, such as machined metal surfaces. An outflow meter is used with air rather than water as the working medium and surface roughness is measured by means of pneumatic resistance between the compliant annulus and the surface. For a constant rate of air flow, the pressure drop is determined by the overall surface roughness (Thomas, 1999).

2.4.5 Electrical Method

An electrical method used is the capacitance method based on the parallel capacitor principle. The capacitance between two conducting elements is directly proportional to their area and the dielectric constant of the medium between them and inversely proportional to their separation. If a rough surface is regarded as the sum of a number of small elemental areas at different heights, it is fairly easy to work out the effective capacitance between it and a smooth surface disk for various deterministic models. The capacitance between a smooth disk surface and the surface to be measured is a function of the surface roughness. A commercial instrument is available based on this principle (Brecker *et al.*, 1977). The capacitance method is also used for the continuous inspection procedures (quality control).

2.4.6 Electron Microscopy Methods

Electron microscopy, both reflection and replica, can reveal both macroscopic and microscopic surface features (Halliday, 1955). But they have two major limitations: first, it is difficult to derive quantitative data; and second, because of their inherent limited field of view, they show only few asperities, whereas in fact the salient point about surface contact is that it involves whole populations of contacting asperities. Sato and O-Hori (1982) have shown that the profile of a surface can be obtained by processing backscattered electron signals (BES) using a computer connected to a scanning electron microscope (SEM). A backscattered electron image is produced by a BES, which is proportional to the surface inclination along the electron beam scanning.

The use of SEM requires placing specimens in a vacuum. In addition, for insulating specimens, a conductive coating (e.g., gold or carbon) is required.

The application of stereomicroscopy to obtain surface roughness information is based on the principle of stereo effects (Bhushan, 1999, 2013). The stereo effects can be obtained by preparing two images of the same surface with slightly different angular views (typically less than 10°). The result is a parallax shift between two corresponding image points of the same feature relative to some reference point, due to a difference in the elevation between the feature and the reference point (Boyde, 1970). By measuring the parallax shift, one can extract the height information from these stereo-pair images. Since an SEM is typically used to obtain the pair of stereo images, the lateral resolution is limited by the electron beam size, which is typically 5 nm. Vertical resolution is a function of lateral parallax resolution and the angle θ.

2.4.7 Analysis of Measured Height Distribution

The measured height distribution across the sample can be analyzed to determine surface roughness statistics and geometrical parameters of a surface. The following surface roughness statistics can be obtained from the height distribution data: surface height distributions; surface slope and curvature distributions in x, y, and radial directions; heights, absolute slopes, and curvatures of all summits and the upper 25% summits; summit density and the upper 25% summit density; number of zero crossings per unit length in x, y, and two dimensions; and a three-dimensional plot of the autocovariance function with a contour of the autocovariance function at 0 and 0.1 (Wyant *et al.*, 1986; Bhushan, 1996). The following geometrical parameters of a surface can be measured, for example, the radii of spherical curvature and cylindrical curvature by fitting spherical and cylindrical surfaces, respectively.

2.4.8 Comparison of Measurement Methods

Comparison of the various methods of roughness measurement may be made based on a number of grounds, such as ease of use, whether quantitative information can be obtained, whether three-dimensional data of topography can be obtained, lateral and vertical resolutions, cost, and on-line measurement capability. Table 2.4.1 summarizes the comparison of the relevant information.

The final selection of the measurement method depends very much on the application that the user has in mind. For in-process inspection procedures, measurement methods employing specular reflection, diffuse reflection, or speckle pattern are used. For continuous inspection (quality control) procedures requiring limited information, either fluid or electrical methods can be used. For procedures requiring detailed roughness data, either the stylus profiler, digital optical profiler or atomic force microscope is used. For a soft or superfinished surface, the digital optical profiler or AFM is preferred.

Roughness plots of a disk measured using an atomic force microscope or AFM (spatial resolution $\sim$ 15 nm), noncontact optical profiler or NOP (spatial resolution $\sim$1 μm) and a stylus profiler or SP (spatial resolution $\sim$0.2 μm), are shown in Figure 2.4.19. The figure shows that roughness is found at scales ranging from millimeter to nanometer scales. The measured roughness profile is dependent on the spatial and normal resolutions of the measuring instrument. Instruments with different lateral resolutions measure features with different length scales. It can be concluded that a surface is composed of a large number of length scales of roughness that are superimposed on each other. Figure 2.4.20 shows the comparison of AFM, SP, and NOP profiles extracted from the measurements with about the same profile lengths and sampling intervals. The roughness measurements are affected by the spatial (lateral) resolution of the measuring instrument. It refers to the stylus size of AFM and stylus profiler and the pixel size used in NOP for roughness measurement. For AFM and stylus profiler instruments, the ability of the stylus to reproduce the original surface features depends on the stylus size. The smaller the stylus size, the closer it will follow the original profile. The stylus tip radius of AFM is smaller than SP and therefore the AFM measurement is expected to be more accurate. A profile measured by AFM is used to assess the effect of the stylus size on the accuracy of roughness measurements (Poon and Bhushan, 1995a, 1995b). Figure 2.4.21 shows the loci of different stylus radii on an AFM profile. By increasing the stylus size, the original profile is distorted resulting in the underestimation of σ and the overestimation of β^*. σ drops from

Table 2.4.1 Comparison of roughness measurement methods.

Method	Quantitative information	Three-dimensional data	Resolution (nm)		On-line measurement capability	Limitations
			Spatial	Vertical		
Stylus instrument	Yes	Yes	15–100	0.1–1	No	Contact type can damage the sample, slow measurement speed in 3D mapping
Optical methods						
Specular reflection	No	No	10^5–10^6	0.1–1	Yes	Semiquantitative
Diffuse reflection (scattering)	Limited	Yes	10^5–10^6	0.1–1	Yes	Smooth surfaces (<100 nm)
Optical interference	Yes	Yes	500–1000	0.1–1	No	
Scanning tunneling microscopy	Yes	Yes	0.2	0.02	No	Requires a conducting surface; scans small areas
Atomic force microscopy	Yes	Yes	0.2–1	0.02	No	Scans small areas
Fluid/electrical	No	No			Yes	Semiquantiative
Electron microscopy	Yes	Yes	5	50	No	Expensive instrumentation, tedious, limited data, requires a conducting surface, scans small areas

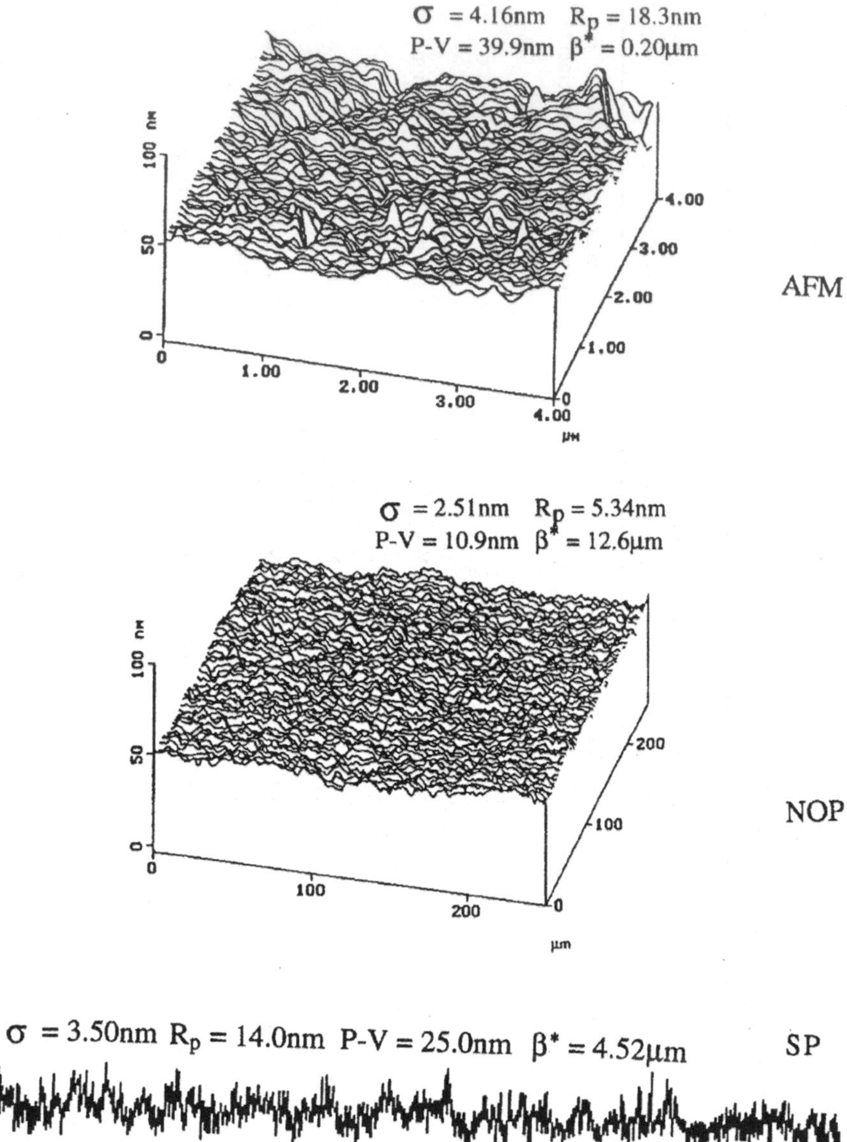

Figure 2.4.19 Surface roughness plots of a glass-ceramic disk measured using an atomic force microscope (spatial resolution ~15 nm), noncontact optical profiler (spatial resolution ~1 μm), and stylus profiler (tip radius ~0.2 μm).

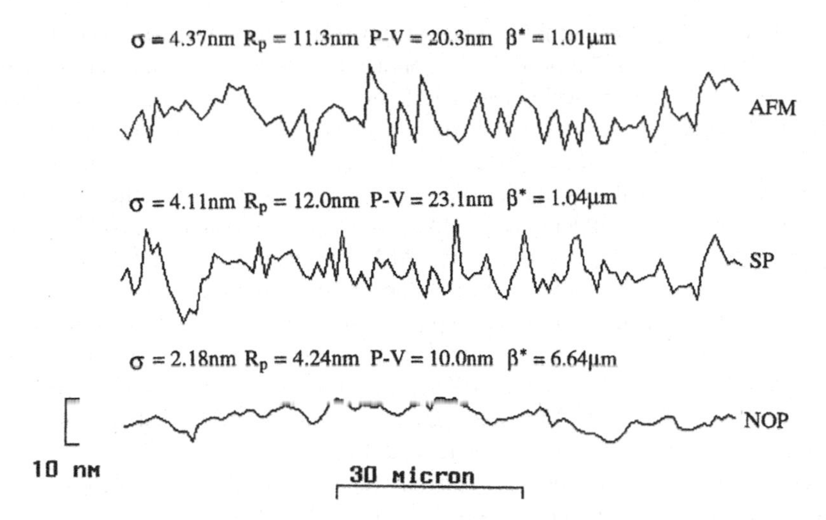

Figure 2.4.20 Comparison of surface plots of a glass-ceramic disk measured using AFM (~0.16 μm), SP (~0.2 μm) and NOP (~1 μm) drawn on a same scale.

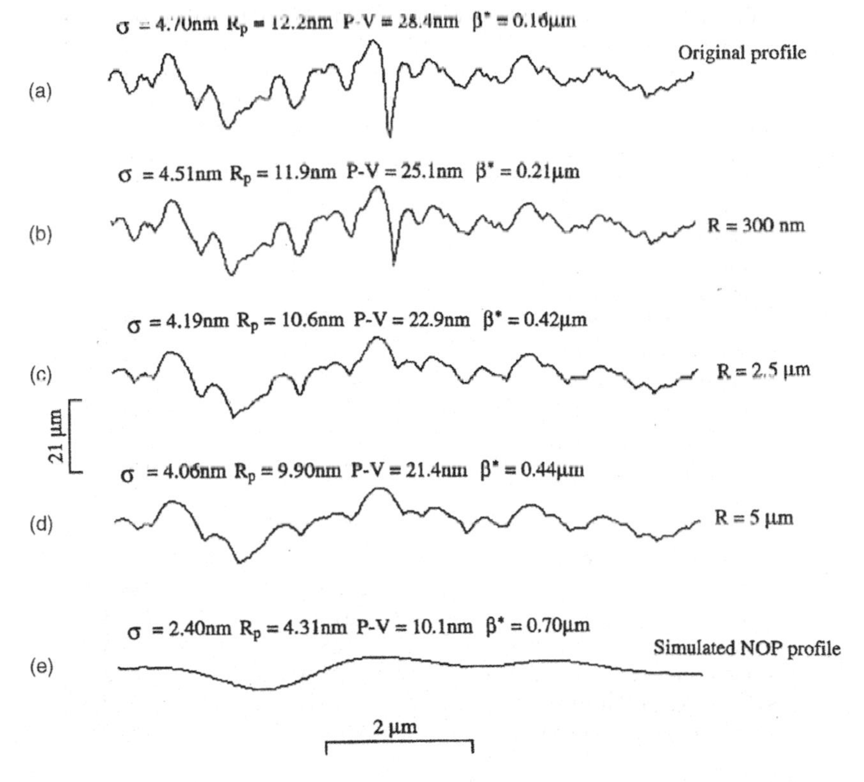

Figure 2.4.21 Simulated profiles of different stylus sizes sliding on the original AFM profile and the simulated NOP profile. Reproduced with permission from Poon, C.Y. and Bhushan, B. (1995a), "Comparison of Surface Roughness Measurements by Stylus Profiler, AFM and Non-Contact Optical Profiler," *Wear* **190**, 76–88. Copyright 1995. Elsevier.

4.70 nm to 4.06 nm by 14% and β^* increases from 0.16 μm to 0.44 μm by 175% when the stylus tip radius increases to 5 μm. NOP is an optical technique to measure surface roughness using the optical interference technique. The light intensity of the fringes is related to the surface height. In the optical system, the fringe pattern is discretized into pixels. Within one pixel or one sampling interval, the light intensity represents the averaged value of surface heights within the pixel. Effectively, the optical probe acts as an optical filter to remove high-frequency details using a cutoff length equal to the sampling interval. In the NOP measurement, the sampling interval is 1 μm. Therefore, the AFM profile in Figure 2.4.21a can be used to simulate the profile given by NOP by splitting the profile into number of cutoff lengths equal to 1 μm. The mean of each cutoff length represents the surface height measured by NOP. A cubic spline curve is obtained to go through the mean points and shown in Figure 2.4.21e. σ for the simulated NOP profile is about 50% underestimated and β^* is 45% overestimated as compared with the AFM profile. Various roughness parameters of the disk measured using the AFM with two scan sizes are presented in Table 2.4.2.

As stated earlier, surface roughness is generally characterized by σ, sometimes along with other parameters. From the profiles in Figs. 2.4.20 and 2.4.21, vertical roughness parameters

Table 2.4.2 Various roughness parameters of a glass-ceramic disk measured using AFM at two scan sizes.

	Scan size (μm^2)	
Roughness parameters	8 × 8	32 × 32
σ, surface height (nm)	5.13	5.42
Skewness	−0.24	0.24
Kurtosis	6.01	4.1
σ, profile slope x (mrad)	53.5	22
σ, profile slope y (mrad)	67.7	25.2
σ, surface slope (mrad)	86.3	33.5
σ, profile curvature x (mm^{-1})	1635	235.5
σ, profile curvature y (mm^{-1})	3022	291.2
σ, surface curvature (mm^{-1})	1950	228.3
Summit height (nm)		
Mean	2.81	4.26
σ	5.56	5.08
Summit curvature (mm^{-1})		
Mean	3550	384
σ	1514	225.5
Summit-valley distance (nm)	45.9	48.5
Summit-mean distance (nm)	22.9	24.2
Summit density (μm^{-2})	15.6	2.97
Profile zero crossing x (mm^{-1})	2794	1279
Profile zero crossing y (mm^{-1})	4157	1572
Mean correlation length (μm)	0.32	0.67

x and y are along radial and tangential directions, respectively; summit threshold is taken as 0.5 nm.

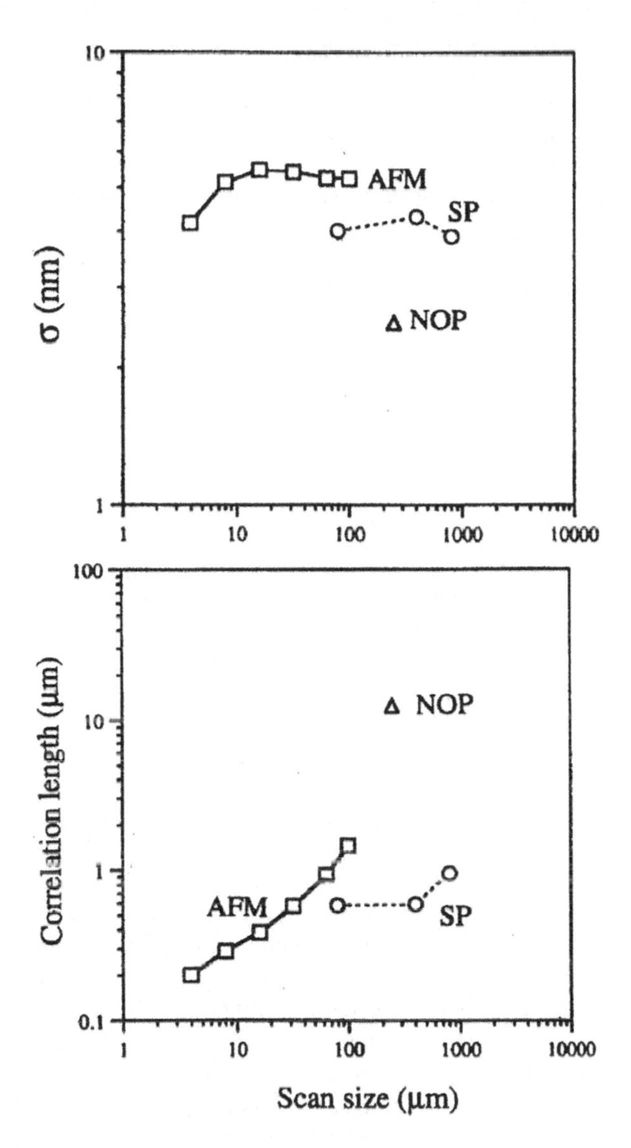

Figure 2.4.22 Variation of σ and β^* with scan size for a glass-ceramic disk measured using AFM (scan length/256 data points), NOP (~1 μm) and SP (~0.2 μm).

σ, R_p *and* $P - V$ are seen to increase with the measuring instruments in the following order NOP <SP <AFM. On the other hand, the spatial parameter β^* is seen to increase in the reverse order, i.e., AFM <SP <NOP. σ and β^* as a function of scan size for three instruments shown in Figure 2.4.22 show a similar trend and are related to different instrument spatial resolutions. We also note that the σ initially increases with the scan size and then approaches a constant value, whereas β^* increases monotonically with the scan size. The result of σ as a function of scan size suggests that the disk has a long-wavelength limit. It is expected that β^*,

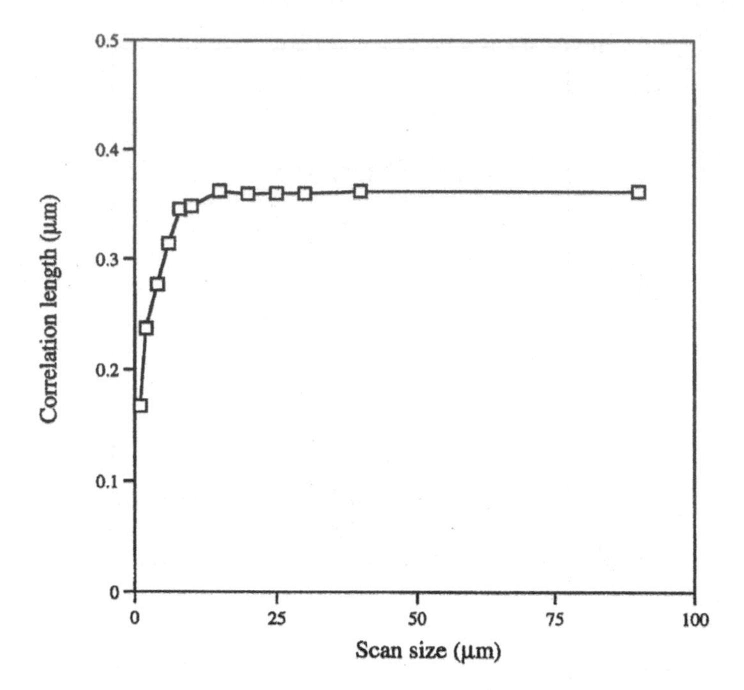

Figure 2.4.23 Variation of correlation length with scan size with a constant sampling interval (40 nm) for a glass-ceramic disk measured using AFM.

which is a measure of wavelength structure, should also approach a constant value. In contrast, β^* generally increases with the scan size. As the sampling interval increases with increasing scan size, high-frequency details of the original profile gradually disappear resulting in high β^*. σ is a vertical parameter not sensitive to sampling interval but generally it increases with scan size. β^* is a spatial parameter affected by both sampling interval and scan length. If the sampling interval can be kept the same for all scan sizes, β^* will be expected to approach a constant value, Figure 2.4.23 (Poon and Bhushan, 1995a).

The question often asked is what instrument should one use for roughness measurement? For a given instrument, what scan size and sampling interval should one use? Deformation of asperities is dependent on the roughness, mechanical properties, and loading. It will be shown in the next chapter, nanoasperities deform by plastic deformation which is undesirable (Bhushan and Blackman, 1991; Poon and Bhushan, 1996). Therefore, an instrument that can measure high-frequency data, such as in AFM, should be used, particularly in low-load conditions. As stated earlier, a sampling interval equal to 0.25 and 0.50 times the correlation length at the selected scan size should be selected. A scan size equal to or greater than the value at which σ approaches a constant value, or twice the nominal contact size of the physical problem, whichever is smaller, should be used.

2.5 Closure

Solid surfaces, irrespective of the method of formation contain deviations from the prescribed geometrical form, ranging from macro- to nanoscales. In addition to surface deviations, the solid surface consists of several zones having physico-chemical properties specular to the bulk material itself.

Surface texture, repetitive deviation from the nominal surface, includes roughness (nano- and microroughness, waviness or macroroughness and lay). Surface roughness is most commonly characterized with two average amplitude parameters: R_a or R_q (σ) and R_t (maximum peak-to-valley height). However, the amplitude parameters alone are not sufficient for complete characterization of a surface and spatial parameters are required as well. A random and isotropic surface can be completely characterized by two functions – the height distribution and autocorrelation functions. A random surface with Gaussian height distribution and exponential autocorrelation function can be completely characterized by two parameters σ and β^*; these parameters can be used to predict other roughness parameters.

A surface is composed of a large number of length scales of roughness superimposed on each other. Hence, commonly measured roughness parameters depend strongly on the resolution of the measuring instrument and are not unique for a surface. The multi-scale nature of rough surfaces can be characterized using a fractal analysis for fractal surfaces.

Various measurement techniques are used for off-line and on-line measurements of surface roughness. Optical techniques, such as specular reflection and scattering, are commonly used for on-line semiquantitative measurements. Commonly used techniques for off-line measurements are either contact profilers – stylus profilers and atomic force microscopes – or noncontact profilers – optical profilers based on two-beam interference. Contact - stylus based – profilers are the oldest form of measuring instruments and are most commonly used across the industry. However, the stylus tip can scratch the delicate surface during the course of the measurement. They also suffer from slow measurement speed, where three-dimensional mapping of the surfaces is required. Optical profilers are noncontact and can produce three-dimensional profiles rapidly and without any lateral motion between the optical head and the sample. Optical profilers can be used for surfaces with homogeneous optical properties, otherwise they need to be coated with a 10–20 nm thick reflective coating (e.g., gold) before measurement. Lateral resolutions of profilers with sharp tips are superior to optical profilers. Nanoscale roughness with atomic-scale resolutions can be measured using atomic force microscopes which are used at ultralow loads. However, these are more complex to use.

Three-dimensional roughness height data can be processed to calculate a variety of amplitude and spatial functions and parameters. Without the use of long-wavelength filtering, waviness data can be obtained and analyzed.

Problems

2.1 Consider a sinusoidal profile with wavelengths λ and a maximum amplitude A_0. Calculate (a) $R(\tau)$ and (b) $P(\omega)$.

2.2 A surface profile is sinusoidal with wavelength λ and maximum amplitude of unity. The profile is sampled at equal intervals, with the origin on the center line at a position of zero amplitude. (a) Calculate the R_a, σ and P-V distance values for this profile for sampling intervals of λ, $\lambda/2$, $\lambda/4$, $\lambda/8$, and $\lambda/16$. (b) Calculate the R_a and σ values derived from the analog signal for the same profile.

2.3 A surface profile consists of 10 triangular asperities having a constant flank angle θ and peak-to-valley heights of 1, 2, 3, 4, 5, 6, 7, 8, 9, 10, see Figure P2.1. If all of the valleys

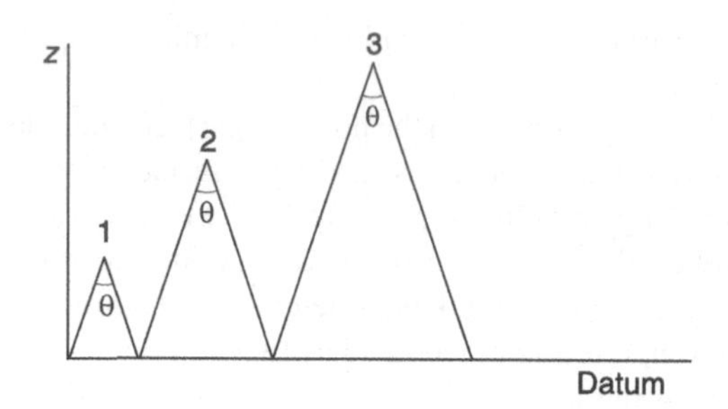

Figure P2.1

are at an arbitrary datum, calculate (a) the position of the mean line relative to the datum and (b) the R_a value of the profile.

2.4 For a sinusoidal profile of wavelength $2l$ and amplitude A, calculate (a) RMS, (b) CLA, (c) peak-to-valley distance, and (d) asperity tip radius.

References

Abbott, E.J. and Firestone, F.A. (1933) "Specifying Surface Quality," *Mech. Eng.* **55**, 569–572.

Anonymous (1975), "Instruments for the Measurement of Surface Roughness by Profile Methods," IS03274, International Standardization Organization.

Anonymous (1985), "Surface Texture (Surface Roughness, Waviness, and Lay)," ANSI/ASME B46.1, ASME, New York.

Beckmann, P. and Spizzichino, A. (1963), *The Scattering of Electromagnetic Waves from Rough Surfaces*, Chapter 5, Pergamon, New York.

Bendat, J.S. and Piersol, A.G. (1986), *Engineering Applications of Correlation and Spectral Analysis*, Second edition, Wiley, New York.

Bennett, H.E. (1978), "Scattering Characteristics of Optical Materials," *Opt. Eng.* **17**, 480–488.

Bennett, H.E. and Porteus, J.O. (1961), "Relation Between Surface Roughness and Specular Reflectance at Normal Incidence," *J. Opt. Soc. Amer.* **51**, 123–129.

Bennett, J.M. and Mattson, L. (1989), *Introduction to Surface Roughness and Scattering*, Opt. Soc. of Am., Washington, D.C.

Bhushan, B. (1996), *Tribology and Mechanics of Magnetic Storage Devices*, Second edition, Springer, New York.

Bhushan, B. (1999), *Handbook of Micro/Nanotribology*, Second edition, CRC, Boca Raton, Florida.

Bhushan, B. (2011), *Nanotribology and Nanomechanics I & II*, Third edition, Springer-Verlag, Heidelberg, Germany.

Bhushan, B. (2013), *Principles and Applications of Tribology*, Second edition, Wiley, New York.

Bhushan, B. and Blackman, G.S. (1991), "Atomic Force Microscopy of Magnetic Rigid Disks and Sliders and its Applications to Tribology," *ASME J. Trib.* **113**, 452–457.

Bhushan, B., Wyant, J.C., and Koliopoulos, C.L. (1985), "Measurement of Surface Topography of Magnetic Tapes by Mirau Interferometry," *Appl. Opt.* **24**, 1489–1497.

Bhushan, B., Wyant, J.C., and Meiling, J. (1988), "A New Three-Dimensional Digital Optical Profiler," *Wear* **122**, 301–312.

Binnig, G. and Rohrer, H. (1983), "Scanning Tunneling Microscopy," *Surface Sci.* **126**, 236–244.

Binnig, G., Rohrer, H., Gerber, Ch., and Weibel, E. (1982), "Surface Studies by Scanning Tunneling Microscopy," *Phys. Rev. Lett.* **49**, 57–61.

Binnig, G., Quate, C.F., and Gerber, Ch. (1986), "Atomic Force Microscope," *Phys. Rev. Lett.* **56**, 930–933.

Boudreau, B.D., Raja, J., Sannareddy, H., and Caber, P.J. (1995), "A Comparative Study of Surface Texture Measurement Using White Light Scanning Interferrometry and Contact Stylus Techniques," *Proc. Amer. Soc. Prec. Eng.* **12**, 120–123.

Boyde, A. (1970), "Practical Problems and Methods in the Three-Dimensional Analysis of Scanning Electron Microscope Images," *Scanning Electron Microscopy*, Proc. of the Third Annual SEM Symposium, pp. 105–112, IITRI, Chicago.

Brecker, J.N., Fromson, R.E., and Shum, L.Y. (1977), "A Capacitance Based Surface Texture Measuring System," *Annals CIRP* **25**, 375–377.

Bruning, J.H. (1978), "Fringe Scanning Interferometers," in *Optical Shop Testing* (D. Malacara, ed.), pp. 409–437, Wiley, New York.

Buckley, D.H. (1981), *Surface Effects in Adhesion, Friction, Wear and Lubrication*, Elsevier, Amsterdam.

Budde, W. (1980), "A Reference Instrument for 20°, 40°, and 85° Gloss Measurements," *Metrologia* **16**, 1–5.

Bush, A.W., Gibson, R.D., and Keogh, G.P. (1979), "Strongly Anisotropic Rough Surfaces," *ASME J. Trib.* **101**, 15–20.

Caber, P. (1993), "An Interferometric Profiler for Rough Surfaces," *Appl. Opt.* **32**, 3438–3441.

Cheng, Y.Y. and Wyant, J.C. (1985), "Multiple-Wavelength Phase-Shifting Interferometry," *Appl. Opt.* **24**, 804–807.

Chilamakuri, S. and Bhushan, B. (1998), "Contact Analysis of Non-Gaussian Random Surfaces," *Proc. Instn. Mech. Engrs., Part J: J. Eng. Trib.* **212**, 19–32.

Church, E.L., Vorburger, T.V., and Wyant, J.C. (1985), "Direct Comparison of Mechanical and Optical Measurements of the Finish of Precision-Machined Optical Surfaces," *Opt. Eng.* **24**, 388–395.

Cook, N.H. and Bhushan, B. (1973), "Sliding Surface Interface Temperatures," *ASME J. Lub. Tech.* **95**, 59–64.

Creath, K. (1988), "Phase-Shifting Interferometry Techniques," in *Progress in Optics*, **26** (E. Wolf, ed.), pp. 357–373, Elsevier, New York.

Davidson, M., Kaufman, K., Mazor, I., and Cohen, F. (1987), "An Application of Interference Microscopy to Integrated Circuit Inspection and Metrology," *Proc. Soc. Photo Opt. Instrum. Eng.* **775**, 233–247.

Francon, F. (1966), *Optical Interferometry*, Academic Press, San Diego, California.

Francon, F. and Mallick, S. (1971), *Polarization Interferometers*, Wiley (Interscience), New York.

Ganti, S. and Bhushan, B. (1995), "Generalized Fractal Analysis and its Applications to Engineering Surfaces," *Wear* **180**, 17–34.

Gardner, H.A. and Sward, G.G. (1972), *Paint Testing Manual, Physical and Chemical Examination: Paints, Varnishes, Lacquers and Colors*, 13th ed. ASTM Special Pub. 500, Philadelphia, Pennsylvania.

Gatos, H.C. (1968), "Structure of Surfaces and Their Interactions," in *Interdisciplinary Approach to Friction and Wear* (P.M. Ku, ed.), SP-181, pp. 7–84, NASA, Washington, DC.

Giaever, I. (1960), "Energy Gap in Superconductors Measured by Electron Tunnelling," *Phys. Rev. Lett.* **5**, 147–148.

Greenwood, J.A. (1984), "A Unified Theory of Surface Roughness," *Proc. Roy. Soc. Lond. A* **393**, 133–157.

Gupta, P.K. and Cook, N.H. (1972), "Statistical Analysis of Mechanical Interaction of Rough Surfaces," *ASME J. Lub. Tech.* **94**, 19–26.

Halliday, J.S. (1955), "Surface Examination by Reflection Electron Microscopy," *Proc. Instn Mech. Engrs* **109**, 777–781.

Haltner, A.J. (1969), "The Physics and Chemistry of Surfaces: Surface Energy, Wetting and Adsorption," in *Boundary Lubrication* (F.F. Linget al., eds), pp. 39–60, *ASME*, New York.

Ibe, J.P., Bey, P.P., Brandon, S.L., Brizzolara, R.A., Burnham, N.A., DiLella, D.P., Lee, K.P., Marrian, C.R.K., and Colton, R.J. (1990), "On the Electrochemical Etching of Tips for Scanning Tunneling Microscopy," *J. Vac. Sci. Technol.* A **8**, 3570–3575.

Kotwal, C.A. and Bhushan, B. (1996), "Contact Analysis of Non-Gaussian Surfaces for Minimum Static and Kinetic Friction and Wear," *Tribol. Trans.* **39**, 890–898.

Kubaschewski, O. and Hopkins (1953), *Oxidation of Metals and Alloys*, Butterworths, London, UK.

Lange, S.R. and Bhushan, B. (1988), "Use of Two- and Three-Dimensional, Noncontact Surface Profiler for Tribology Applications," *Surface Topography* **1**, 277–290.

Longuet-Higgins, M.S. (1957a), "The Statistical Analysis of a Random, Moving Surface," *Phil. Trans. R. Soc. Lond. A* **249**, 321–387.

Longuet-Higgins, M.S. (1957b), "Statistical Properties of an Isotropic Random Surface," *Phil. Trans. R. Soc. Lond. A* **250**, 157–174.

McCool, J.I. (1984), "Assessing the Effect of Stylus Tip Radius and Flight on Surface Topography Measurements," *ASME J. Trib.* **106**, 202–210.

McGillem, C.D. and Cooper, G.R. (1984), *Continuous and Discrete Signal and System Analysis*, Holt, Rinhart & Winston, New York.

Majumdar, A. and Bhushan, B. (1990), "Role of Fractal Geometry in Roughness Characterization and Contact Mechanics of Surfaces," *ASME J. Trib.* **112**, 205–216.

Martin, Y., Williams, C.C., and Wickramasinghe, H.K. (1987), "Atomic Force Microscope-Force Mapping Profiling on a sub 100-A Scale," *J. Appl. Phys.* **61**, 4723–4729.

Massey, F.J. (1951), "The Kolmogorov-Smirnov Test for Goodness of Fit," *J. Amer. Statist. Assoc.* **46**, 68–79.

Nayak, P.R. (1971), "Random Process Model of Rough Surfaces," *ASME J. Lub. Tech.* **93**, 398–407.

Nayak, P.R. (1973), "Some Aspects of Surface Roughness Measurement," *Wear* **26**, 165–174.

Poon, C.Y. and Bhushan, B. (1995a), "Comparison of Surface Roughness Measurements by Stylus Profiler, AFM and Non-Contact Optical Profiler," *Wear* **190**, 76–88.

Poon, C.Y. and Bhushan, B. (1995b), "Surface Roughness Analysis of Glass-Ceramic Substrates and Finished Magnetic Disks, and Ni-P Coated Al-Mg and Glass Substrates," *Wear* **190**, 89–109.

Poon, C.Y. and Bhushan, B. (1996), "Nano-Asperity Contact Analysis and Surface Optimization for Magnetic Head Slider/Disk Contact," *Wear* **202**, 83–98.

Radhakrishnan, V. (1970), "Effects of Stylus Radius on the Roughness Values Measured with Tracing Stylus Instruments," *Wear* **16**, 325–335.

Samuels, L.E. (1960), "Damaged Surface Layers: Metals," in *The Surface Chemistry of Metals and Semiconductors* (H.C. Gatos, ed.), pp. 82–103, Wiley, New York.

Sarid, D. and Elings, V. (1991), "Review of Scanning Force Microscopy," *J. Vac. Sci. Technol. B* **9**, 431–437.

Sato, H. and O-Hori, M. (1982), "Surface Roughness Measurement by Scanning Electron Microscope," *Annals CIRP* **31**, 457–462.

Shaw, M.C. (1997), *Metal Cutting Principles*, Second edition, Clarendon Press, Oxford, UK.

Siegel, S. (1956), *Nonparametric Statistics for the Behavioral Sciences*, McGraw-Hill, New York.

Smirnov, N. (1948), "Table for Estimating the Goodness of Fit of Empirical Distributions," *Annals of Mathematical Statistics* **19**, 279–281.

Stover, J.C. (1995), *Optical Scattering: Measurement and Analysis*, Second edition, SPIE Optical Engineering Press, Bellingham, Washington.

Stover, J.C., Bernt, M., and Schiff, T. (1996), "TIS Uniformity Maps of Wafers, Disks and Other Samples," *Proc. Soc. Photo-Opt. Instrum. Eng.* **2541**, 21–25.

Thomas, T.R. (1999), *Rough Surfaces*, Second edition, Imperial College Press, London, UK.

Tolansky, S. (1973), *Introduction to Interferometers*, Wiley, New York.

Trapnell, B.M.W. (1955), *Chemisorption*, Butterworths, London, UK.

Whitehouse, D.J. (1994), *Handbook of Surface Metrology*, Institute of Physics Publishing, Bristol, UK.

Whitehouse, D.J. and Archard, J.F. (1970), "The Properties of Random Surfaces of Significance in Their Contact," *Proc. Roy. Soc. Lond. A* **316**, 97–121.

Whitehouse, D.J. and Phillips, M.J. (1978), "Discrete Properties of Random Surfaces," *Phil. Trans. R. Soc. Lond. A* **290**, 267–298.

Whitehouse, D.J. and Phillips, M.J. (1982), "Two-Dimensional Discrete Properties of Random Surfaces," *Phil. Trans. R. Soc. Lond. A* **305**, 441–468.

Williamson, J.B.P. (1968), "Topography of Solid Surfaces," in *Interdisciplinary Approach to Friction and Wear* (P.M. Ku, ed.), SP-181 pp. 85–142, NASA Special Publication, NASA, Washington, D.C.

Wyant, J.C. (1975), "Use of an AC Heterodyne Lateral Shear Interferometer with Real Time Wavefront Corrections Systems," *Appl. Opt.* **14**, 2622–2626.

Wyant, J.C. (1995), "Computerized Interferometric Measurement of Surface Microstructure," *Proc. Soc. Photo-Opt. Instrum. Eng.* **2576**, 122–130.

Wyant, J.C. and Koliopoulos, C.L., (1981), "Phase Measurement System for Adaptive Optics," *Agard Conference Proceedings*, No. **300**, 48.1–48.12.

Wyant, J.C., Koliopoulos, C.L., Bhushan, B., and George, O.E. (1984), "An Optical Profilometer for Surface Characterization of Magnetic Media," *ASLE Trans.* **27**, 101–113.

Wyant, J.C., Koliopoulos, C.L., Bhushan, B., and Basila, D. (1986), "Development of a Three-Dimensional Noncontact Digital Optical Profiler," *ASME J. Trib.* **108**, 1–8.

Further Reading

Anonymous (1975), "Instruments for the Measurement of Surface Roughness by Profile Methods," IS03274, International Standardization Organization.

Anonymous (1979), *Wear* **57**.
Anonymous (1982), *Wear* **83**.
Anonymous (1985), "Surface Texture (Surface Roughness, Waviness, and Lay)," ANSI/ASME B46.1, ASME, New York.
Bhushan, B. (1996), *Tribology and Mechanics of Magnetic Storage Devices*, Second edition, Springer, New York.
Bhushan, B. (2011), *Nanotribology and Nanomechanics I & II*, Third edition, Springer-Verlag, Heidelberg, Germany.
Bhushan, B. (2013), *Principles and Applications of Tribology*, Second edition, Wiley, New York.
Buckley, D.H. (1981), *Surface Effects in Adhesion, Friction, Wear and Lubrication*, Elsevier, Amsterdam.
Gatos, H.C. (1968), "Structure of Surfaces and Their Interactions," in *Interdisciplinary Approach to Friction and Wear* (P.M. Ku, ed.), SP-181, pp. 7–84, NASA, Washington, D.C.
Thomas, T.R. (1999), *Rough Surfaces*, Second edition, Imperial College Press, London, UK.
Whitehouse, D.J. (1994), *Handbook of Surface Metrology*, Institute of Physics Publishing, Bristol, UK.

3

Contact Between Solid Surfaces

3.1 Introduction

When two nominally flat surfaces are placed in contact, surface roughness causes contact to occur at discrete contact spots (junctions), Figure 3.1.1. The sum of the areas of all the contact spots constitutes the real (true) area of contact or simply contact area, and for most materials with applied load, this will be only a small fraction of the apparent (nominal) area of contact (that which would occur if the surfaces were perfectly smooth). The real area of contact is a function of the surface texture, material properties and interfacial loading conditions. The proximity of the asperities results in adhesive contacts caused by interatomic interactions. When two surfaces move relative to each other, the friction force is contributed by adhesion of these asperities and other sources of surface interactions. Repeated surface interactions and surface and subsurface stresses, developed at the interface, result in formation of wear particles and eventual failure. A smaller real area of contact results in a lower degree of interaction, leading generally to lower wear. The problem of relating friction and wear to the surface texture and material properties generally involves the determination of the real area of contact. Therefore, understanding of friction and wear requires understanding of the mechanics of contact of solid bodies.

During the contact of two surfaces, contact will initially occur at only a few points to support the normal load (force). As the normal load is increased, the surfaces move closer together, a larger number of higher asperities on the two surfaces come into contact, and existing contacts grow to support the increasing load. Deformation occurs in the region of the contact spots, establishing stresses that oppose the applied load. The mode of surface deformation may be elastic, plastic, viscoelastic or viscoplastic, and depends on nominal normal and shear stresses (load/apparent contact area), surface roughness, and material properties. The local stresses at the contact spots are much higher than the nominal stresses. Although nominal stresses may be in the elastic range, the local stresses may exceed the elastic limit (yield strength) and the contact will yield plastically. In most contact situations, some asperities are deformed elastically, while others are deformed plastically; the load induces a generally elastic deformation of the solid bodies but at the tips of the asperities, where the actual contact occurs, local plastic deformation may take place.

Introduction to Tribology, Second Edition. Bharat Bhushan.
© 2013 John Wiley & Sons, Ltd. Published 2013 by John Wiley & Sons, Ltd.

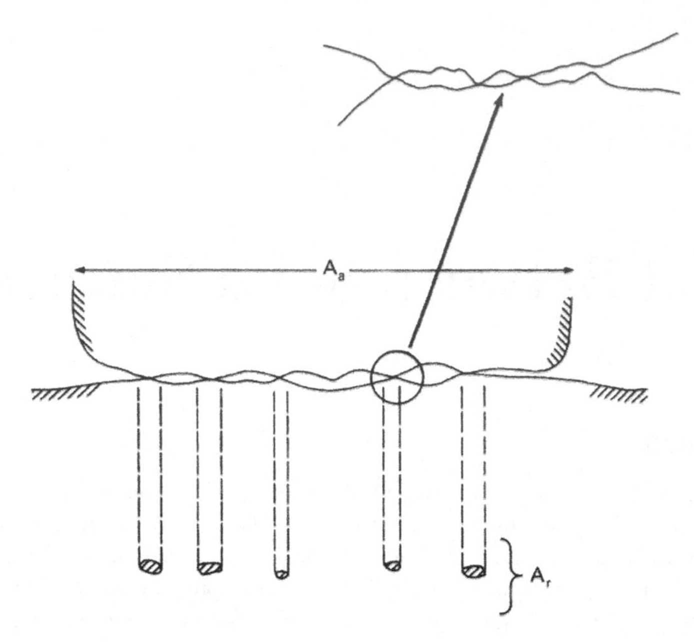

Figure 3.1.1 Schematic representation of an interface, showing the apparent and real areas of contact. Typical size of an asperity contact is from submicron to a few microns. Inset shows the details of a contact on a submicron scale.

In the contact of two rough surfaces, a large number of asperities of different shapes and sizes are pressed against each other (Bhushan, 1996a, 1998). Tips of surface asperities on solid bodies are sometimes considered spherically shaped so that the contact of two macroscopically flat bodies can be reduced to the study of an array of spherical contacts deforming at their tips. We shall consider the simpler idealized case of a single asperity loaded on homogeneous and layered elastic and elastic-plastic solids with and without sliding. Then we will consider the analysis of multiple asperity contacts. Next, methods of measuring the real area of contact in the static conditions are described and typical data are presented.

3.2 Analysis of the Contacts

3.2.1 Single Asperity Contact of Homogeneous and Frictionless Solids

A single asperity contact reduces to a problem of deformation of two curved bodies in contact. For the analysis of a single asperity contact, it is convenient to model an asperity as a small spherically shaped protuberance. For surfaces with anisotropic roughness distribution, asperities may be modeled with curved bodies of specific geometries.

3.2.1.1 Elastic Contact

The first analysis of the deformation and pressure at the contact of two elastic solids with geometries defined by quadratic surfaces is due to Hertz (1882) and such contacts are referred

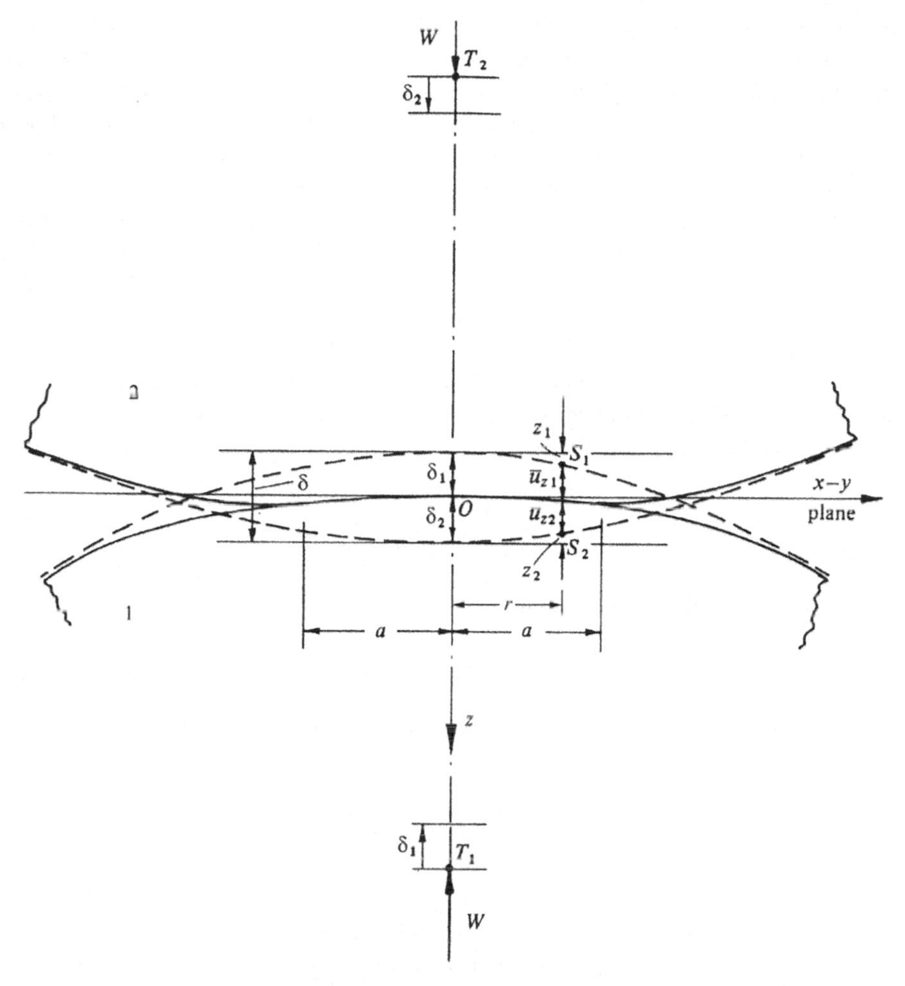

Figure 3.2.1 Schematic of two frictionless solids of general shape (but chosen convex for convenience) in static contact.

to as Hertzian contact. His analysis is based on the following assumptions: (1) the surfaces are continuous, smooth and nonconforming; (2) the strains are small; (3) each solid can be considered as an elastic half-space in the proximity of the contact region; and (4) the surfaces are frictionless. Two solids of general shape (but chosen convex for convenience) loaded together are shown in cross section after deformation in Figure 3.2.1. The x-y plane is the contact plane. The point of first contact is taken as the origin of a Cartesian coordinate system in which the x-y plane is the common tangent plane to the two surfaces and the z axis lies along the common normal directed positively into the lower solid. The separation between the two surfaces at radius r before loading is $z_1 + z_2$. During the compression by a normal force W, distant points in the two bodies T_1 and T_2 move towards O, parallel to the z axis, by vertical displacements δ_1 and δ_2, respectively. If the solids did not deform their profiles would overlap as shown by the dotted lines in Figure 3.2.1. The elastic deformation results in displacement of the surface outside the footprint such that the contact size (2a) is less than the overlap length resulting from intersection of the dotted lines. Due to the contact pressure the surface of each body is displaced parallel to Oz by an amount $\bar{u}_{z1}$ and $\bar{u}_{z2}$ (measured positive into each body), relative to the distant points T_1 and T_2, and after displacement points S_1 and S_2 become coincident. The total displacement $\delta = \delta_1 + \delta_2$ is called total interference or normal approach which is defined as the distance by which points on the two bodies remote from the deformation zone move together on application of a normal load; it arises from the flattening and displacement of the surface within the deformation zone. If the two bodies are solids of revolution, then from polar symmetry, the contact area will be circular, centered at O.

We now consider the problem of elastic deformation of two spheres of radii R_1 and R_2 in solid contact with an applied normal load W. The contact area is circular, having a radius a and the contact pressure is elliptical with $p(r)$ at a radius r in the contact zone. From Hertz analysis, we have the contact radius

$$a = \frac{\pi p_0 R}{2E^*} = \left(\frac{3WR}{4E^*}\right)^{1/3} \tag{3.2.1a}$$

The area of contact for the elastic case is

$$A_{re} = \pi a^2 = \pi R\delta \tag{3.2.1b}$$

The displacements within the contact case can be expressed as

$$\overline{u}_{z1} + \overline{u}_{z2} = \delta - z_1 - z_2 = \delta - \frac{r^2}{2R} \tag{3.2.2a}$$

and

$$\delta = \frac{a^2}{R} = \left(\frac{\pi p_0}{2E^*}\right)^2 R = \left(\frac{9W^2}{16RE^{*^2}}\right)^{1/3} \tag{3.2.2b}$$

The pressure distribution is elliptical with the maximum pressure at the contact center,

$$p = p_0\left\{1 - (r/a)^2\right\}^{1/2} \tag{3.2.3a}$$

with the maximum contact pressure p_0 being 3/2 times the mean contact pressure, p_m, given as,

$$p_0 = \frac{3}{2}p_m = \frac{3W}{2\pi a^2} = \left(\frac{6WE^{*^2}}{\pi^3 R^2}\right)^{1/3} \tag{3.2.3b}$$

where the composite or effective modulus

$$\frac{1}{E^*} = \frac{1 - \nu_1^2}{E_1} + \frac{1 - \nu_2^2}{E_2} \tag{3.2.4}$$

and the composite or effective curvature,

$$\frac{1}{R} = \frac{1}{R_1} + \frac{1}{R_2} \tag{3.2.5}$$

The parameters E and ν are Young's modulus of elasticity and the Poisson's ratio, respectively; subscripts 1 and 2 refer to the two bodies. Note that the real area of contact in Equation 3.2.1b is exactly half the area covered by intersection of dotted lines ($=2\pi\ R\delta$). From Equation 3.2.1b also note that the area of contact increases as (normal load)$^{2/3}$.

Next we examine the stress distributions at the surface and within the two solids, for the Hertz pressure exerted between two frictionless elastic spheres in contact. The Cartesian components of the stress field are given by Hamilton and Goodman (1966). For pressure applied to a circular region, it is convenient to write expressions for the stress field in polar coordinates. The polar components of the stress field in the surface $z = 0$, inside the loaded circle ($r \leq a$) are (Johnson, 1985),

$$\frac{\sigma_r}{p_0} = \frac{1-2\nu}{3}\left(\frac{a^2}{r^2}\right)\left\{1-\left(1-\frac{r^2}{a^2}\right)^{3/2}\right\} - \left(1-\frac{r^2}{a^2}\right)^{1/2} \tag{3.2.6a}$$

$$\frac{\sigma_\theta}{p_0} = -\frac{1-2\nu}{3}\left(\frac{a^2}{r^2}\right)\left\{1-\left(1-\frac{r^2}{a^2}\right)^{3/2}\right\} - 2\nu\left(1-\frac{r^2}{a^2}\right)^{1/2} \tag{3.2.6b}$$

$$\frac{\sigma_z}{p_0} = -\left(1-\frac{r^2}{a^2}\right)^{1/2} \tag{3.2.6c}$$

and outside the circle

$$\frac{\sigma_r}{p_0} = -\frac{\sigma_\theta}{p_0} = \frac{(1-2\nu)a^2}{3r^2} \tag{3.2.7}$$

They are all compressive except at the very edge of contact where the radial stress is tensile having a maximum value of $(1-2\nu)p_0/3$ at the edge of the circle at $r = a$. This is the maximum tensile stress occurring anywhere in the contact and it is held responsible for the ring cracks which are observed to form when brittle materials such as glass are pressed into contact (Lawn, 1993). At the center the radial stress is compressive and of value $(1+2\nu)p_0/2$. Thus, for an incompressible material ($\nu = 0.5$) the stress at the origin is hydrostatic. Outside the contact area, the radial and hoop (circumferential) stresses are of equal magnitude and are tensile and compressive, respectively.

The stresses on the z axis may be calculated by considering a ring of concentrated force at radius r:

$$\frac{\sigma_r}{p_0} = \frac{\sigma_\theta}{p_0} = -(1+\nu)\left\{1-\left(\frac{z}{a}\right)\tan^{-1}\left(\frac{a}{z}\right)\right\} + \frac{1}{2}\left(1+\frac{z^2}{a^2}\right)^{-1} \tag{3.2.8a}$$

$$\frac{\sigma_z}{p_0} = -\left(1+\frac{z^2}{a^2}\right)^{-1} \tag{3.2.8b}$$

Expressions for σ_x and σ_y on the z axis are the same as those for σ_r and σ_θ (Hamilton and Goodman, 1966). The negative sign represents the compressive stresses. The stresses at other points throughout the solid have been calculated by Huber (1904) and Morton and Close (1922). Stress contours at various angles with respect to the z axis are presented by Davies (1949).

The stress distributions within the two solids with $\nu = 0.30$ are shown in Figure 3.2.2a. Contact pressure distribution in the contact plane is elliptical. Contours of principal shear stress [$=\frac{1}{2}$ (principal stress difference)] are shown in Figure 3.2.2a which compares well with the photo-elastic fringes presented by Johnson (1985). Along the z axis, σ_r, σ_θ and σ_z are principal stresses. (σ_r, σ_θ on the z axis are identical to σ_x and σ_y). The principal shear stress, $\tau_1 = \frac{1}{2}|\sigma_z - \sigma_r|$ on the z axis, is also plotted in Figure 3.2.2b. The principal shear stress τ_1 has a maximum value which lies below the surface. For the Hertz pressure distribution it has a maximum value of 0.31 p_0 which lies below the surface at a depth of 0.48 a (for $\nu = 0.30$). This is the maximum shear stress in the field, exceeding the shear stress at the origin,

$$= \frac{1}{2}\,|\sigma_z - \sigma_r| = 0.10 p_0$$

and also the shear stress in the surface at the edges of the contact

$$= \frac{1}{2}\,|\sigma_r - \sigma_\theta| = 0.13 p_0$$

Hence plastic yielding would be expected to initiate beneath the surface. We will discuss this issue in detail later. Note that for the wedge or cone, the maximum shear stress lies adjacent to the apex whereas for curved bodies, plastic enclave lies beneath the contact surface (Tabor, 1951, 1970).

The maximum shear stress $\tau_{max} = \frac{1}{2}|\sigma_z - \sigma_r|_{max}$ on the z axis and its location (z/a) for different values of ν can be calculated using Equation 3.2.8. Poisson's ratio plays a rather insignificant role relative to maximum shear stress and its location (Bhushan, 1996a).

3.2.1.2 Limit of Elastic Deformation

As the normal load between two contacting bodies is applied, they initially deform elastically according to their Young's moduli of elasticity. As the load is increased, one of the two bodies with lower hardness may start to deform plastically. As the normal load is further increased, the plastic zone grows until the entire material surrounding the contact has gone through plastic deformation. Metals, alloys and some nonmetals and brittle materials deform predominantly by "plastic shear" or "slip" in which one plane of atoms slides over the next adjacent plane. The load at which the plastic flow or plastic yield begins in the complex stress field of two contacting solids is related to the yield point of the softer material in a simple tension or pure shear test through an appropriate yield criterion.

Two of the yield criteria most commonly employed for most ductile materials as well as sometimes for brittle materials are described here (Hill, 1950). In Tresca's maximum shear

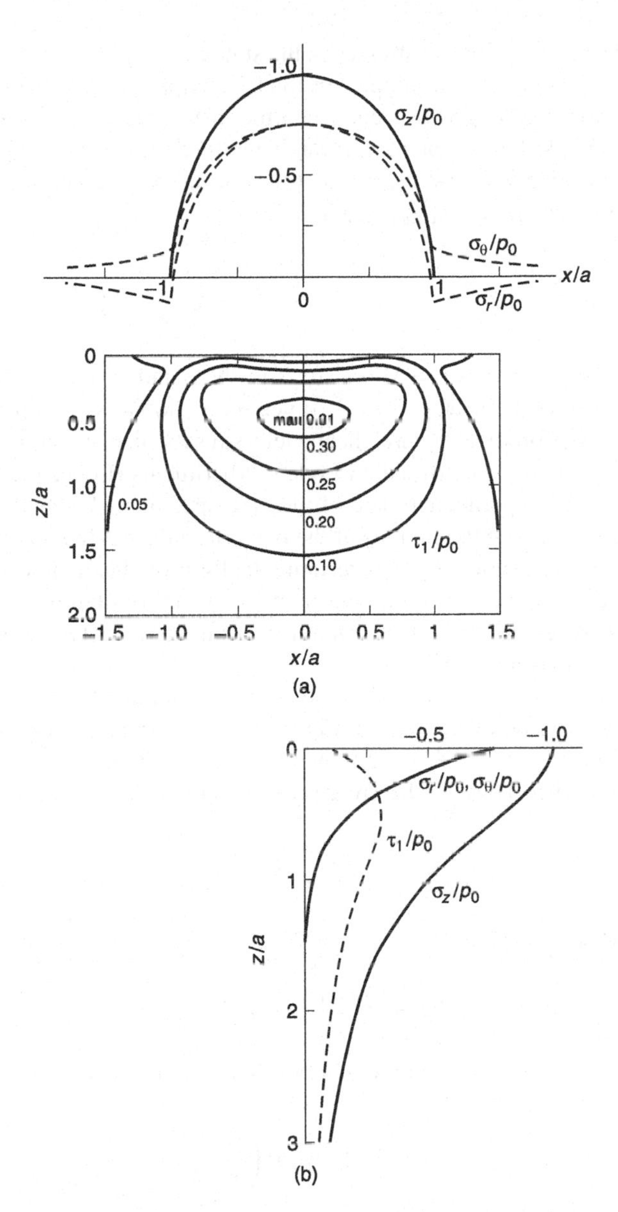

Figure 3.2.2 (a) Stress distributions at the surface and contours of principal shear stress in the subsurface (b) stresses along the z axis of symmetry caused by Hertz pressure acting on a circular area of radius a.

stress criterion, the yielding will occur when the maximum shear stress (half the difference between the maximum and minimum principal stresses) reaches the yield stress in the pure shear or half of yield stress in simple tension,

$$Max\left\{\frac{1}{2}\left|\sigma_1-\sigma_2\right|,\frac{1}{2}\left|\sigma_2-\sigma_3\right|,\frac{1}{2}\left|\sigma_3-\sigma_1\right|\right\}=k=\frac{Y}{2} \tag{3.2.9}$$

Here σ_1, σ_2 and σ_3 are the principal stresses in the state of complex stress. The yield point in pure shear k is half the yield stress in simple tension (or compression) Y. In the von Mises shear strain energy criterion, yielding will occur when the distortion energy equals the distortion energy at yield in simple tension or pure shear. Therefore yielding occurs when the square root of the second invariant of the stress deviator tensor (S_{ij}) reaches the yield stress in simple shear or $(1/\sqrt{3})$ of yield stress in simple tension,

$$J_2 = \frac{1}{2} S_{ij} S_{ij} \equiv \frac{1}{6} \left\{ (\sigma_1 - \sigma_2)^2 + (\sigma_2 - \sigma_3)^2 + (\sigma_3 - \sigma_1)^2 \right\} = k^2 = \frac{Y^2}{3} \tag{3.2.10}$$

$\sqrt{J_2}$ and $\sqrt{3J_2}$ are referred to as von Mises stress in shear and in tension, respectively. Note that the yield stress in pure shear is $(1/\sqrt{3})$ times the yield stress in simple tension. Thus the von Mises criterion predicts a pure shear yield stress which is about 15% higher than predicted by the Tresca criterion. Based on Lode's experiments (Lode, 1926), the von Mises criterion usually fits the experimental data of metallic specimens better than other theories. However, the difference in the predictions of the two criteria is not large. Tresca's criterion is employed for its algebraic simplicity to determine the limit of elastic deformation. However, this criterion often does not permit continuous mathematical formulation of the resulting yield surface, while von Mises criterion does. Therefore, von Mises criterion is employed more often than Tresca's in plasticity analyses.

In the case of axisymmetric contact of two spheres, maximum shear stress occurs beneath the surface on the axis of symmetry, z axis (Figure 3.2.1). Along this axis, σ_r, σ_θ, and σ_z are principal stresses and $\sigma_r = \sigma_\theta$. We have shown that for $\nu = 0.3$, the value of $\frac{1}{2}|\sigma_z - \sigma_r|$ is $0.31 p_0$ at a depth of 0.48 a. Thus, by the Tresca criterion, the value of p_0 for yield is given by

$$(p_0)_y = \frac{3}{2} (p_m)_y = 3.2\, k = 1.60\, Y \tag{3.2.11}$$

while by the von Mises criterion

$$(p_0)_y = 2.8\, k = 1.40\, Y \tag{3.2.12}$$

The load to initiate yield W_y is given by Equations 3.2.3b and 3.2.11,

$$W_y = 21.17\, R^2\, Y \left(\frac{Y}{E*} \right)^2 \tag{3.2.13}$$

The maximum normal approach before the onset of plastic deformation is given by Equations 3.2.2b, 3.2.3b and 3.2.11 or 3.2.12,

$$\delta_y = 6.32\, R \left(\frac{Y}{E*} \right)^2 \tag{3.2.14}$$

Note that yielding would occur in one of the two solids with a lower yield stress or hardness. Further note that to carry a high load (high interference) without yielding it is desirable to choose a material with a high yield strength or hardness and with a low elastic modulus.

Table 3.2.1 Stress and Deformation Formulas for Normal Contact of Elastic Solids (Hertz Contact).

Parameter	Circular contact (Diameter = $2a$, Load = W)	Line contact (Width = $2a$, Load = W'/unit length along y axis)
Semi-contact radius or width	$a = \left(\frac{3WR}{4E^*}\right)^{1/3}$	$a = 2\left(\frac{W'R}{\pi E^*}\right)^{1/2}$
Normal approach	$\delta = \frac{a^2}{R} = \left(\frac{9W^2}{16RE^{*2}}\right)^{1/3}$	$\delta = \frac{2W'}{\pi}\left\{\frac{1-\nu_1^2}{E_1}\left(\ln\left(\frac{4R_1}{a}\right)-\frac{1}{2}\right)+\frac{1-\nu_2^2}{E_2}\left(\ln\left(\frac{4R_2}{a}\right)-\frac{1}{2}\right)\right\}$
Contact Pressure	$p = p_0\left\{1-\left(\frac{r}{a}\right)^2\right\}^{1/2}$ $p_0 = \frac{3}{2}p_m = \frac{3W}{2\pi a^2}$ $= \left(\frac{6WE^{*2}}{\pi^3 R^2}\right)^{1/3}$	$p = p_0\left\{1-\left(\frac{x}{a}\right)^2\right\}^{1/2}$ $p_0 = \frac{4}{\pi}p_m = \frac{4W'}{\pi a}$ $= \left(\frac{W'E^*}{\pi R}\right)^{1/2}$
Maximum tensile stress	$(1-2\nu)\,p_0/3$ at $r = a$ (on the contact plane, $z = 0$)	Zero
Maximum shear stress	$0.31\,p_0$ at $r = 0$ and $z = 0.48a$ for $\nu = 0.3$	$0.30\,p_0$ at $x = 0$ and $z = 0.78a$ for all ν
Limit of elastic deformation	$(p_0)_y = 1.60Y = 3.2k$, Tresca criterion $= 1.60Y = 2.8k$, von Mises criterion	$(p_0)_y = 1.67Y = 3.3k$ Tresca criterion $= 1.79Y = 3.1k$, von Mises criterion ($\nu = 0.3$) (von Mises depends on ν)

Composite curvature, $\frac{1}{R} = \frac{1}{R_1} + \frac{1}{R_2}$ where R_1 and R_2 are the principal radii of curvature of the two bodies (convex positive).

Composite modulus $\frac{1}{E^*} = \frac{1-\nu_1^2}{E_1} + \frac{1-\nu_2^2}{E_2}$ where E and ν are Young's modulus and Poisson's ratio, respectively.

We summarize the results of elastic contact and onset of yielding results in Table 3.2.1. For completeness, we also include the result of two-dimensional contact of cylindrical bodies with their axes lying parallel to each other. For this case, the contact region is a long strip of width $2a$ lying parallel to the axes of the cylinders.

Most practical contact applications have to withstand many repeated passes of the load. If, in the first pass, the elastic limit is exceeded some plastic deformation will take place and thereby introduce residual stress. Generally, the residual stresses will increase the load required to initiate yielding in the second pass. After repeated loading (process of "shakedown"), the load required to initiate yielding would reach a steady value, higher than that after first loading (Johnson, 1985).

Example Problem 3.2.1

A ceramic ball with a radius of 5 mm is pressed into a hemispherical recess of 10 mm in radius in a steel plate. (a) What normal load is necessary to initiate yield in the steel plate; (b) what is the radius of the contact; and (c) at what depth does yield first occur? The given parameters are: $E_{ceramic} = 450$ GPa, $E_{steel} = 200$ GPa, $\nu_{ceramic} = 0.3$, $\nu_{steel} = 0.3$, $H_{ceramic} =$ 20 GPa *and* $H_{steel} = 5$ GPa. Assume that $H \sim 2.8\ Y$.

Solution

The composite modulus is given by

$$\frac{1}{E^*} = \frac{1-\nu_1^2}{E_1} + \frac{1-\nu_2^2}{E_2}$$
$$= \frac{1-0.3^2}{450} + \frac{1-0.3^2}{200}\ \text{GPa}^{-1}$$

Therefore

$$E^* = 152.2\ \text{GPa}$$

The composite radius is given by

$$\frac{1}{R} = \frac{1}{R_1} + \frac{1}{R_2}$$
$$= \frac{1}{5} - \frac{1}{10}\ \text{mm}^{-1}$$

Therefore,

$$R = 10\,\text{mm}$$

(a) Yield will occur when

$$W_y = 21.17\ R^2 Y \left(\frac{Y}{E^*}\right)^2$$

and

$$H \sim 2.8\ Y$$
$$W_y = \frac{21.17(10^{-2})^2(5\times 10^9/2.8)^3}{(152.2\times 10^9)^2}$$
$$= 515\ \text{N}$$

(b)

$$
\begin{aligned}
a &= \left(\frac{3WR}{4E^*}\right)^{1/3} \\
&= \left(\frac{3 \times 515 \times 10 \times 10^{-3}}{4 \times 152.2 \times 10^9}\right)^{1/3} \\
&= 0.29 \text{ mm}
\end{aligned}
$$

(c)
Yield occurs at a depth of 0.48 a:

$$= 0.48 \times 0.29 \text{ mm} = 0.14 \text{ mm}$$

3.2.1.3 Elastic-Plastic Contact of Frictionless Solids

During contact of two elastic-plastic bodies at small loads, the surface is deformed elastically with the maximum shear stress τ_{max} occurring at subsurface, some distance below the center of the contact region. At some critical load, τ_{max} exceeds the critical shear stress of the solid and a small amount of plastic flow occurs within the larger elastic hinterland. As the load is increased, the indentation grows in size, the plastic zone grows and the contact pressure increases until, eventually, the plastic zone reaches the surface and completely embraces the region around the indenter, Figure 3.2.3. Thus deformation grows from purely elastic to elastic-plastic (contained) followed by fully plastic (uncontained), common for most engineering material combinations. Plastic deformation will be initiated into one of the two solids but as the plastic deformation of one body proceeds, the mean contact pressure increases and as soon as it exceeds 1.1 Y of the mating solid, it begins to deform plastically as well. Consequently both solids will be permanently deformed.

The deformation depends on the nature of solids such as elastic (Figure 3.2.4a), rigid-perfectly plastic (Figure 3.2.4b), elastic-perfectly plastic (Figure 3.2.4c), elastic-plastic with strain hardening (Figure 3.2.4d), elastic-brittle, viscoelastic and viscoplastic solids.

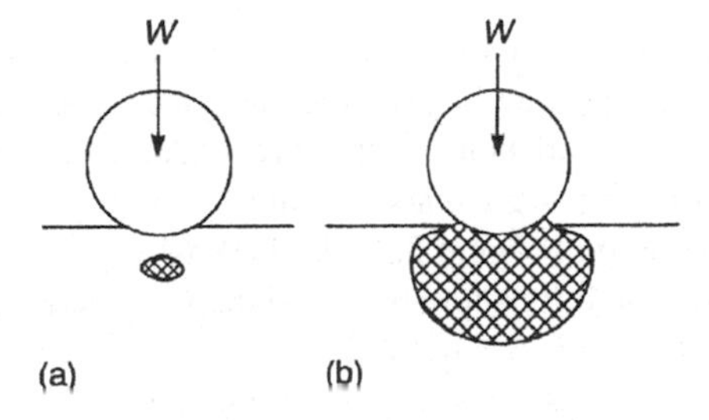

Figure 3.2.3 Indentation of an elastic-perfectly plastic solid by a spherical indenter; (a) onset of plasticity below the surface at an indentation pressure $p_m \sim 1.1Y$, and (b) at a higher load, full plasticity is reached and the plastic flow extends to the free surface (at this stage $p_m \sim 2.8Y$).

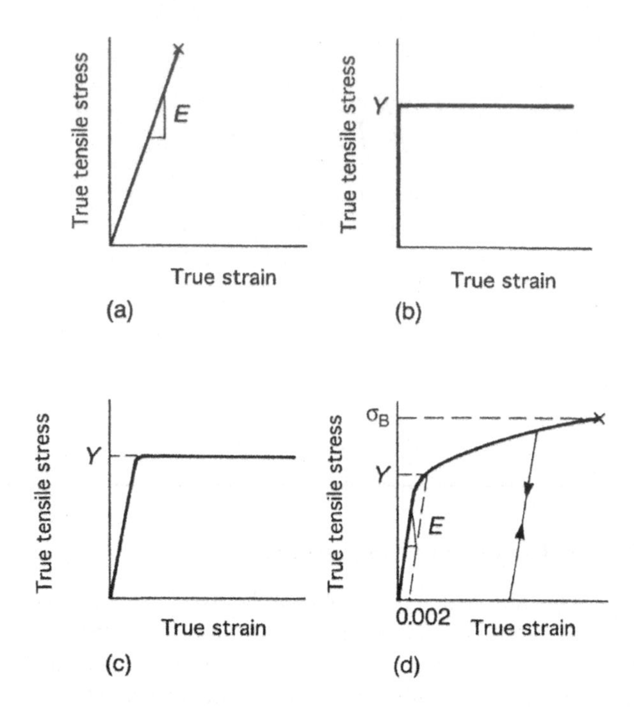

Figure 3.2.4 Schematic of stress-strain curve for (a) elastic, (b) rigid-perfectly plastic, (c) elastic-perfectly plastic solids and (d) real solids. E is the Young's modulus of elasticity, Y is the yield stress and σ_B is the breaking strength.

We consider an elastic-perfectly plastic uniaxial specimen with stress-strain curve as shown in Figure 3.2.4c; first it deforms elastically according to its Young's modulus and then, at tensile stress Y, it yields plastically at a constant yield or flow stress. At small loads, the specimen deforms elastically; then, at higher loads, the critical stress is first exceeded at a region below the center of the contact zone [Figure 3.2.3a]. This corresponds to the onset of plastic deformation, and it occurs for a mean contact pressure (Equation 3.2.11),

$$p_m = 1.07\,Y \sim 1.1\,Y \tag{3.2.15}$$

As the load W is increased further, the indentation becomes larger, and the plastic zone grows until the whole of the material surrounding the indenter undergoes plastic deformation.

An analytical expression for indentation pressure under conditions of full plasticity for a spherical indenter deforming a *rigid plastic* material (with no elastic deformation) was obtained using the slip-line field of Figure 3.2.5 by Ishlinsky (1944); it satisfied the plasticity equations and the boundary conditions for stress and displacement (Hill, 1950). For this case, the pressure over the indenter surface is not uniform over the contact region but is somewhat higher in the center than at the edges (Tabor, 1951). For this case, the mean contact pressure was obtained,

$$p_m = H = 2.8\,Y \sim 3\,Y. \tag{3.2.16a}$$

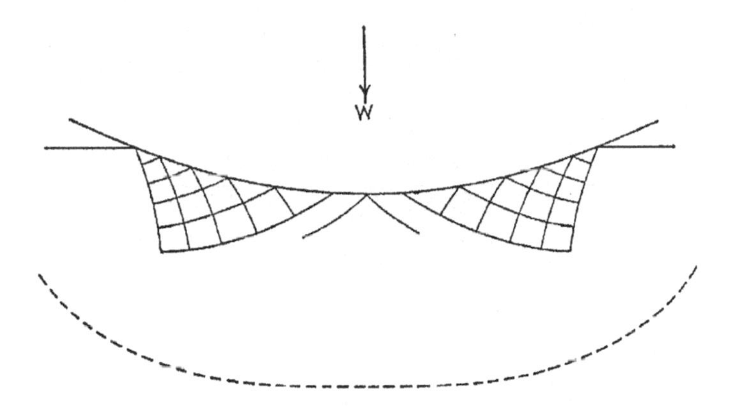

Figure 3.2.5 Part of the slip-line field obtained by Ishlinsky for a spherical indenter deforming a rigid plastic metal. The broken line is an approximate representation of the elastic-plastic boundary.

Based on a number of numerical analyses and experimental measurements of indentation of half-space of elastic-plastic materials with and without strain hardening by spheres and cones, Francis (1977) and Johnson (1985) have given a relationship between hardness (H) and the flow stress in simple compression (Y),

$$H \sim 2.8\, Y \tag{3.2.16b}$$

Next we study the deformation pattern of elastic-perfectly plastic materials. An approach is based on an early observation by R. Hill and others that the subsurface displacements produced by any blunt indenter (cone, pyramid, or sphere) are approximately radial from the point of first contact, with roughly hemispherical contours of equal strain (Hill, 1950; Tabor, 1970), Figure 3.2.6a. In this simplified model of elastic-plastic indentation, the contact surface of the indenter is encased in a hemispherical "core" of radius a. Within the core, there is assumed to be a hydrostatic component of stress p_m. (Of course the stress in the material immediately below an indenter is not purely hydrostatic.) This material under hydrostatic pressure could not yield plastically. Outside the hydrostatic core, plastic flow spreads into the surrounding material, the plastic strains gradually diminishing until they match the elastic strains in the hinterland at some radius c; this marks the elastic-plastic boundary. Clearly, in this model, the behavior depends little on the shape of the indenter itself. Outside the core it is assumed that the stresses and displacements have radial symmetry and are the same as an infinite elastic-perfectly plastic body which contains a spherical cavity under a pressure p_m. The elastic-plastic boundary lies at a radius c, where $c > a$.

Based on Hill (1950), the stresses in the plastic zone $a \le r \le c$ are given by

$$\frac{\sigma_r}{Y} = -2\ln\left(\frac{c}{r}\right) - \frac{2}{3} \tag{3.2.17a}$$

$$\frac{\sigma_\theta}{Y} = -2\ln\left(\frac{c}{r}\right) + \frac{1}{3} \tag{3.2.17b}$$

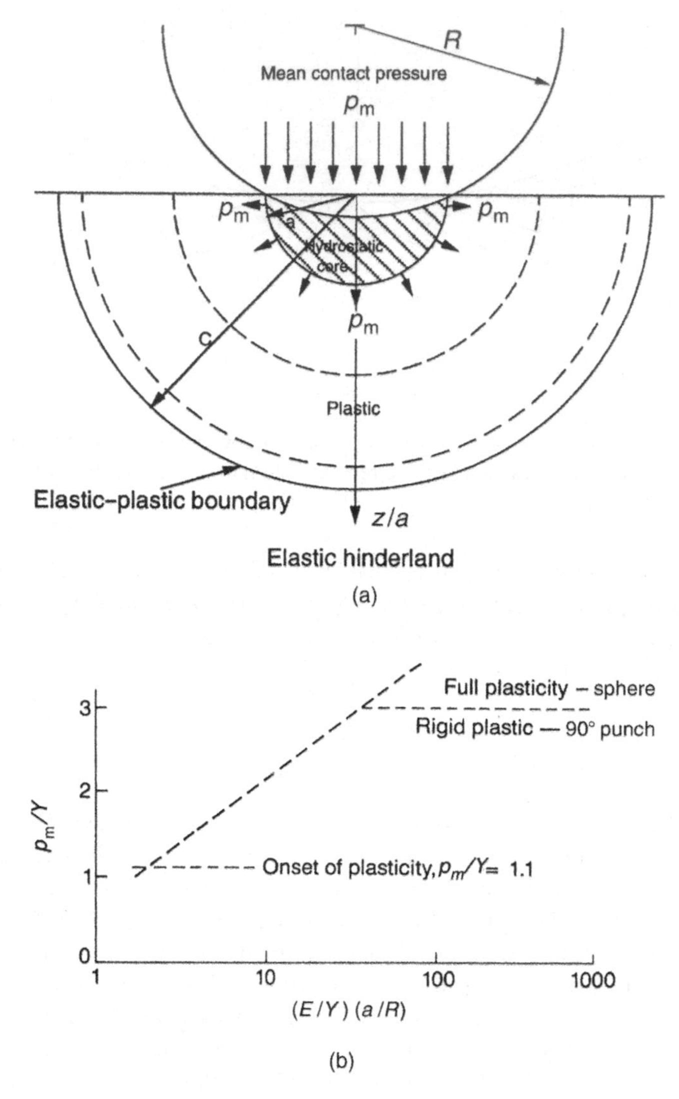

Figure 3.2.6 Cavity model of an elastic-plastic indentation by a sphere of radius R (a) hydrostatic core, plastic zone and elastic-plastic boundaries and (b) variation of indentation pressure p_m with (E/Y) (a/R). Reproduced with permission from Tabor, D. (1986), "Indentation Hardness and its Measurement: Some Cautionary Comments," in *Microindentation Techniques in Materials Science and Engineering* (P. J. Blau and B. R. Lawn, eds), 129–159, ASTM, Philadelphia. Copyright 1986 ASTM International.

In the elastic zone $r \geq c$

$$\frac{\sigma_r}{Y} = -\left(\frac{2}{3}\right)\left(\frac{c}{r}\right)^3, \quad \frac{\sigma_\theta}{Y} = \left(\frac{1}{3}\right)\left(\frac{c}{r}\right)^3 \tag{3.2.17c}$$

At the boundary of the core, the core pressure is given by

$$\frac{p_m}{Y} = \left(-\frac{\sigma_r}{Y}\right)_{r=a} = 2\ \ln\left(\frac{c}{a}\right) + \frac{2}{3} \tag{3.2.17d}$$

This implies that the elastic-plastic boundary coincides with the core boundary itself ($c = a$) at $p_m = (2/3)Y$, and below this contact pressure the analysis fails and no plastic flow can occur. However, there is no case of plastic deformation occurring in any system for an indentation pressure less than Y. On the other hand, in the region where $p_m \sim 3Y$, Equation 3.2.17d shows that $c = 3.2a$, and we must assume that the elastic yielding of the hinterland no longer influences the plastic flow of the material. The contact pressure now corresponds to the classical theory for a rigid plastic solid (Tabor, 1970). However, there is nothing in the expanding cavity model to indicate that the deformation pressure has an upper limit of 3 Y.

For an incompressible material indented by a spherical indenter of radius R, the pressure p_m in the core is given from Equation 3.2.17d (Hill, 1950),

$$\frac{p_m}{Y} = \frac{2}{3}\left\{1 + \ln\left[\frac{1}{3}\left(\frac{E}{Y}\right)\left(\frac{a}{R}\right)\right]\right\} \tag{3.2.18}$$

Equation 3.2.18 is plotted in Figure 3.2.6b and shows how the mean pressure for a sphere increases from $p_m \sim 1.1\ Y$ to $\sim 3\ Y$ (full plasticity) as the size of indentation (a/R) for a spherical indenter increases. Full plasticity is reached for a value of the horizontal abscissa (radius of the indentation) about 10 times greater than that at which the onset of plasticity occurs. Based on experiments with a steel ball sphere pressed against a work-hardened steel flat, Tabor (1970) found a straight line relationship between p_m/Y and log W. He reported that the condition of full plasticity is reached at load of about 300 times that at which onset of plastic deformation occurs.

Matthews (1980) has considered work-hardening materials which strain harden according to a power law of index n. Results for an elastic-perfectly plastic solid just presented, may be applied as a good approximation to a work-hardening solid if Y is replaced by a representative flow stress, measured in simple compression at a representative strain ε_R,

$$\varepsilon_R \sim 0.28\,(1+n)^{-n}(a/R) \tag{3.2.19}$$

Matthews (1980) explained experimental observations of piling-up and sinking-in during indentation made by Norbury and Samuel (1928). They found that piling-up around the indenter is observed in materials which exhibit little work hardening and sinking-in is observed in materials which exhibit strong work hardening.

3.2.2 Single Asperity Contact of Layered Solids in Frictionless and Frictional Contacts

3.2.2.1 Elastic Contact

Stress and deformation analyses for the cases of rigid and elastic cylindrical and spherical indenters contacting a two-dimensional elastic half-space bonded to one or two elastic layers

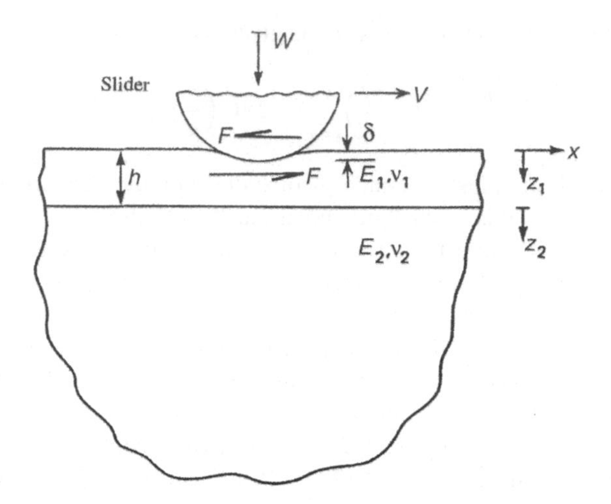

Figure 3.2.7 Schematic of spherical rigid body in a sliding contact with a layered elastic half-space.

in frictionless and frictional contacts have been performed by several investigators (Bhushan, 1996a). Here we consider a generalized case of a spherical body in static and sliding contacts with a homogenous (Hamilton and Goodman, 1966) and layered elastic half-space with a layer of thickness h (O'Sullivan and King, 1988), Figure 3.2.7.

Relative sliding consists of one body sliding past another at a relative peripheral velocity of the surfaces at their point of contact. In Figure 3.2.7, a slider moves from left to right over a fixed layered flat surface. We regard the point of initial contact as a fixed origin and assume the upper surface sliding through the contact region from left to right parallel to the x axis with a steady linear velocity *V*. (This sliding situation is equivalent to the lower body sliding over the fixed upper body, through the contact region from right to left parallel to the x axis). A normal load W is applied which gives rise to an area of contact for the elastic deformation of frictionless contacting bodies. In a frictionless contact, the contact stresses and deformations are unaffected by the sliding motion. Sliding motion or any tendency to slide introduces a tangential force (or traction) referred to as friction force F, active on both surfaces in a direction opposite to the sliding direction. During steady sliding motion, the friction force *F* represents the "kinetic" or "dynamic" friction between the surfaces. In the case of two bodies with no relative velocity but tending to slide (incipient sliding), the friction force arises from "static" friction (Chapter 4). The static friction force is greater than or equal to the kinetic friction force. From Amontons' law the friction force is proportional to the normal force (Chapter 4), $F = \mu W$, where μ is a constant known as the coefficient of friction. The tangential force at the contact surface affects the stress distributions and size and shape of the contact area. If the two solids sliding past each other are homogeneous and have the same elastic constants, any tangential force transmitted between them gives rise to equal and opposite normal displacements of any point on the interface. Thus, the warping of one surface conforms exactly with that of the other and does not disturb the distribution of normal pressure. The shape and size of the contact area are then fixed by the profiles of the two surfaces and the normal load, and are independent of the tangential force. With solids of different elastic properties (E, ν), this is no longer the case and the tangential forces do interact with the normal pressure. The influence of a difference in elastic constants has been analyzed by Bufler (1959). He has shown that the contact area and contact pressure distribution are no longer symmetrically placed; their center is displaced from the axis of symmetry and the contact pressure no longer has circular distribution. These differences are the function of differences in the elastic constants and coefficient of friction. Johnson (1985) has shown that the effect of tangential force on the normal pressure and the contact area is generally small, particularly when the coefficient of friction is less than 1. Therefore, the stresses and deformation due to the normal and tangential forces are generally assumed to be independent of each other and they are superimposed to find the resultant stresses.

Neglecting any interaction between normal pressure and tangential force arising from a difference in elastic constants of the two solids, the normal pressure distribution and the circular contact area can be obtained from Hertz theory. Assuming Amontons' law of friction, the tangential force per unit area acting on the surface can be obtained using Equation 3.2.3,

$$q(r) = \pm \frac{3\,\mu W}{2\pi a^2} \left\{1 - (r/a)^2\right\}^{1/2} \tag{3.2.20}$$

acting parallel to the x axis everywhere in the contact area. The positive sign is associated with a negative velocity of the lower body as shown in Figure 3.2.7. We now calculate the stress component in the solids produced by the surface forces. Explicit equations for calculating the stress components at any point in the solid for homogeneous lower body have been given by Hamilton (1983) and Sackfield and Hills (1983).

In the following results, Poisson's ratio was taken as 0.3 for both layer and substrate; Poisson's ratio has little effect on stresses. Various stress profiles on and beneath the contacting interface for homogeneous ($E_1 = E_2$) and layered ($E_1 \neq E_2$) elastic solids in static and sliding contacts are presented in Figures 3.2.8, 3.2.9, and 3.2.10. Figure 3.2.8 presents the pressure profile beneath the indenter at various E_1/E_2. The pressure is normalized by p_0 which is the maximum pressure under the center of the indenter for a homogeneous medium (E_1/E_2) when the radius of contact $a_0 = h$. Note that the so called Hertzian pressure distribution is elliptical with a maximum at the center of contact. The radius of the circular contact zone decreases and the maximum pressure increases with an increase in the value of E_1/E_2. Figure 3.2.9 presents the three nonzero stress components σ_z, σ_x and τ_{xz} as a function of depth in the layer and substrate under the center of the indenter ($x = y = 0$) for a coefficient of friction $\mu = 0.25$ and for various values of E_1/E_2. Note that σ_x drops off rapidly through the body (along z axis) for (at $z/a_0 \sim 1.3$ for homogeneous solids) whereas the σ_z drops off slowly (at $z/a_0 > 3$ for homogeneous and nonhomogenous solids). Maximum values of these stresses increase with an increase in friction and E_1/E_2. The stress component σ_x is tensile at the interface for a stiffer layer, which is significant for cracks at the base of the layer and orthogonal to the interface. The shear stress component is also aggravated by the stiffer layer; however, it decays rapidly into the depth. High interfacial shear stress adversely affects the adhesion of the layer to the substrate. For layers that are more compliant than the substrate, both the maximum value of normal stresses and the interfacial shear stress are reduced. With brittle materials, the appearance of tensile stresses is more important for yield than the value of the maximum shear stress.

We have seen in Figure 3.2.2 that one of the principal stresses in the surface is tensile near the edges of the contact. The effect of tangential force acting on the surface is to add tension

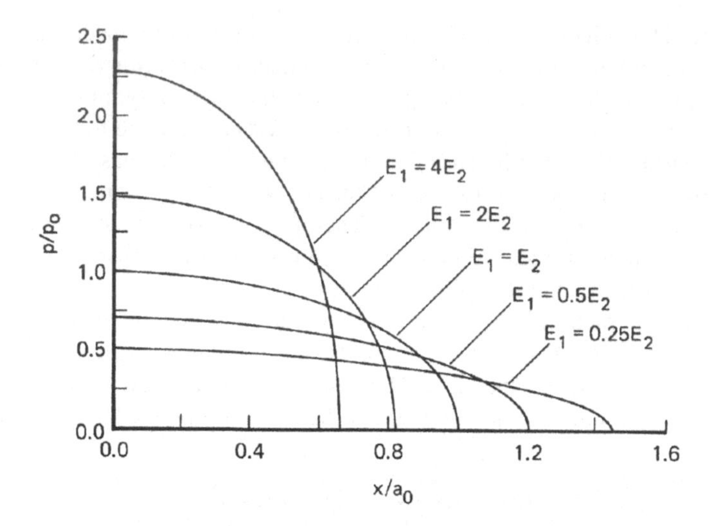

Figure 3.2.8 Normal contact pressure profile beneath the rigid spherical indenter along the x axis for a layered elastic half-space with different values of E_1/E_2 and $\mu = 0$ and when the radius of contact (at E_1/E_2) $a_0 = h$. Reproduced with permission from O'Sullivan, T.C. and King. R.B. (1988), "Sliding Contact Stress Field Due to a Spherical Indenter on a Layered Elastic Half-Space," *ASME J. Trib.* **110**, 235–240. Copyright 1988. ASME.

on one side of the contact and to subtract from it at the other. The tangential force in the lower body results in maximum compressive stresses along the x axis at the leading edge of the contact ($x = a$) and a maximum tensile at the trailing edge ($x = -a$). These stresses are superimposed over the stresses along the x axis as a result of normal pressure. Figure 3.2.10 shows the normal stress profile on the contact plane ($y = 0, z = 0$) along the x axis in the lower body. Note that for a homogenous solid, the normal stress (in all radial directions) is tensile outside the loaded circle. It reaches its maximum value at the edge of the circular contact. This is the maximum tensile stress occurring anywhere and it is held responsible for the surface ring cracks which are observed to form when brittle materials such as glass are pressed into contact with a blunt indenter (Lawn, 1993). Normal stresses are compressive inside the contact. A stiffer layer and friction both increase the maximum tensile and compressive stresses on the surface. As the coefficient of friction increases, σ_x becomes unsymmetrical, compressive at the leading edge of the contact area ($x = a$) and a maximum tensile at the trailing edge ($x = -a$). A stiffer layer and friction can thus degrade the brittle failure characteristics of a layered medium, whereas a more compliant layer can be beneficial. Based on the normal and shear stresses at the interface and maximum tensile normal stresses at the surface, layers that are more compliant (e.g., solid lubricants – Ag, MoS_2, graphite) than the substrate for multilayered bodies are preferred. However, high wear resistance may require harder (which are generally stiffer) layers. Low values of coefficient of friction are also preferable.

The results presented so far are for the case of $a_0 = h$ with various values of E_1/E_2. For values of $a_0 << h$, the maximum stresses in the layer are similar to the homogeneous case with modulus E_1. Similarly, for $a_0 >> h$, the stress field is dominated by the substrate. In the transition zone ($\sim 0.5h < a_0 < 6h$), the stress field depends strongly on the value of E_1/E_2.

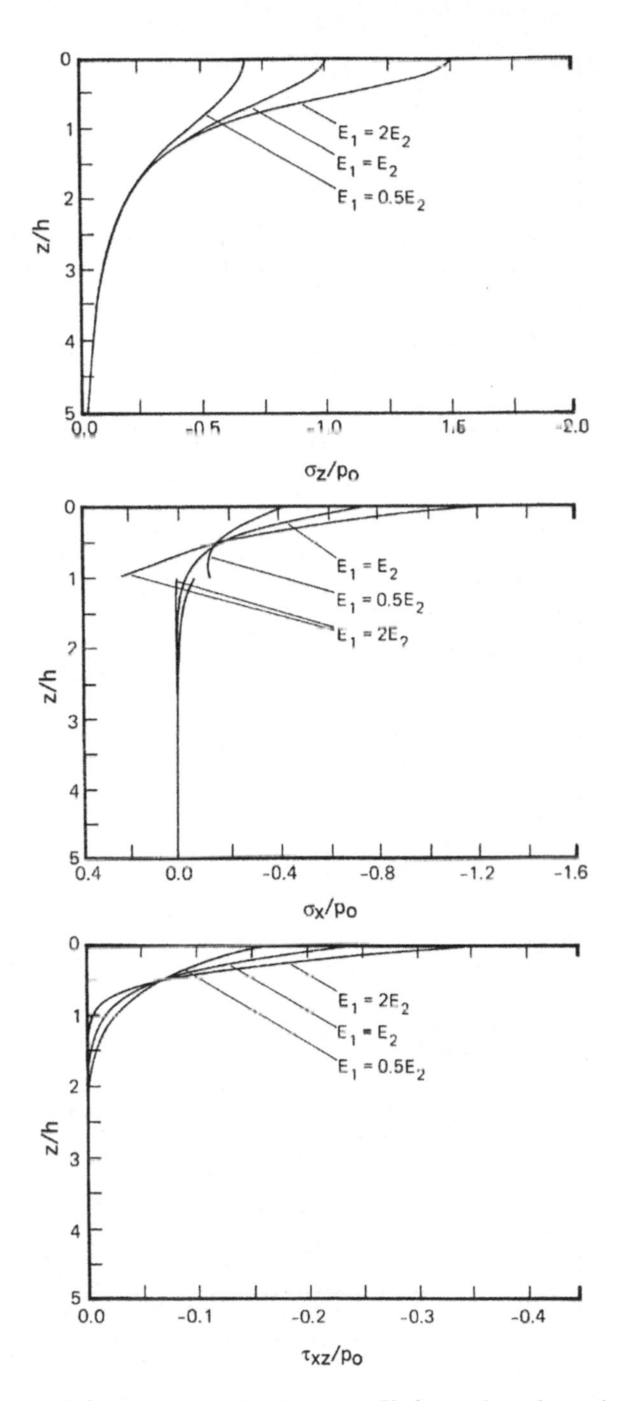

Figure 3.2.9 Normal and shear stresses under the center of indenter along the z axis for a rigid spherical indenter acting on a layered elastic half-space with different values of E_1/E_2 when $\mu = 0.25$ and when the radius of contact (at E_1/E_2) $a_0 = h$. Reproduced with permission from O'Sullivan, T.C. and King. R.B. (1988), "Sliding Contact Stress Field Due to a Spherical Indenter on a Layered Elastic Half-Space," *ASME J. Trib.* **110**, 235–240. Copyright 1988. ASME.

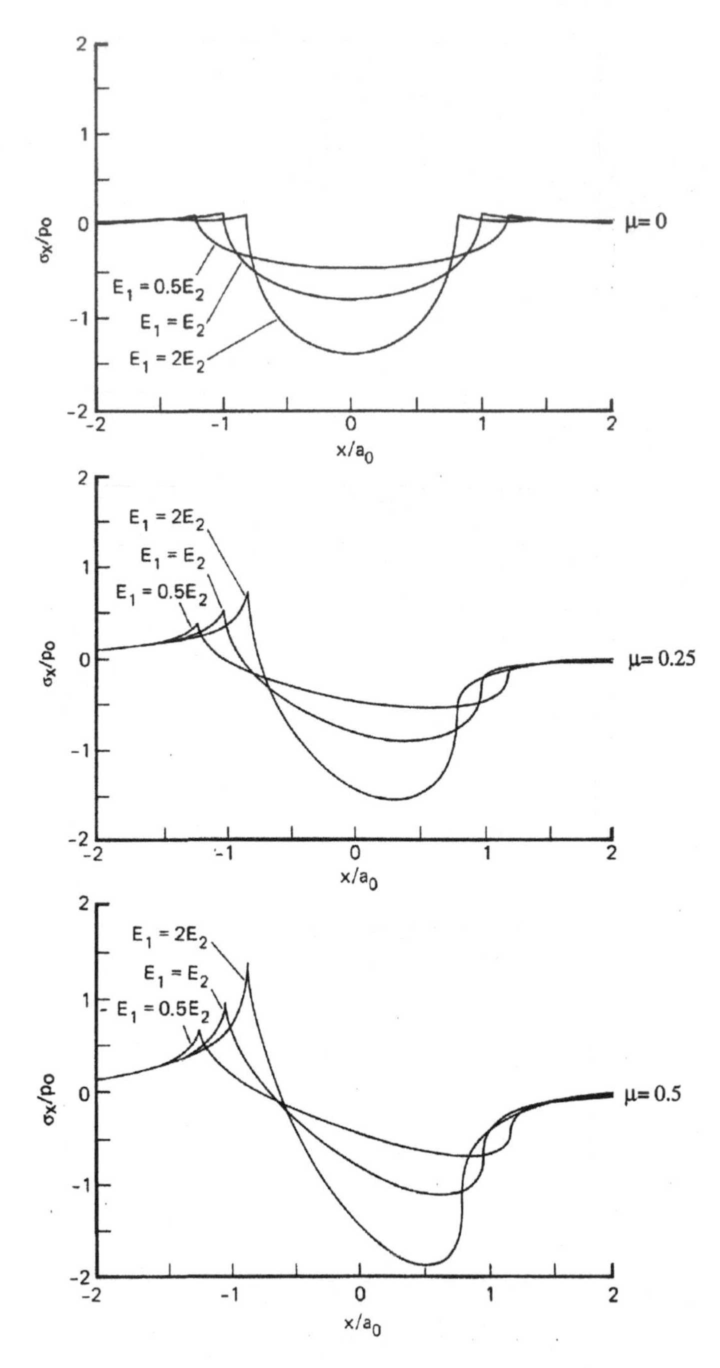

Figure 3.2.10 Variation of normal stress on the surface beneath the rigid spherical indenter along the x axis for an indenter acting on a layered elastic half-space with different values of E_1/E_2 and μ and when the radius of contact (at E_1/E_2) $a_0 = h$. Reproduced with permission from O'Sullivan, T.C. and King. R.B. (1988), "Sliding Contact Stress Field Due to a Spherical Indenter on a Layered Elastic Half-Space," *ASME J. Trib.* **110**, 235–240. Copyright 1988. ASME.

Stress distributions in Figures 3.2.8, 3.2.9, and 3.2.10 have been presented for the lower body. In the case of a homogeneous lower body, stress distribution for the upper body will be that of the lower body with x and z replaced with $-x$ and $-z$, respectively. For example, in the upper body the maximum shear stress location at $\mu = 0.5$ will be at $x/a < 0$ and the maximum tensile stress location will be at the leading edge ($x/a = 1.0$).

3.2.2.2 Limits of Elastic Deformation

As stated earlier, the onset of plastic yield or flow in a static or sliding contact will be governed by Tresca or von Mises yield criteria, Equations 3.2.9 and 3.2.10. According to the von Mises yield criterion, contact pressure p_0 at which yielding in the layer (in the case of a layered solid) and substrate is found by equating the square root of the invariant of the stress deviator tensor ($J_2^{1/2}$) (von Mises stress) to the yield stress in simple shear k. For static contact, the maximum Hertz contact pressure to initiate yield is 2.8 k. Contour plots of $(3J_2)^{1/2}/p_0$ in the lower body are shown in Figure 3.2.11 for $\mu = 0$, 0.25 and 0.50 for various values of E_1/E_2. For $\mu = 0$, the figure shows symmetrical contours and the $J_2^{1/2}$ (or maximum shear stress) occurs beneath the surface on the axis of symmetry, z axis (Figure 3.2.2). As the μ is increased, the region of maximum von Mises stress moves from a subsurface location towards the surface and becomes more *intense*; yield occurs at the surface when μ exceeds about 0.3. With respect to the center of contact, the maximum von Mises stress location moves in the direction of the friction force acting on the body (or in the opposite direction to the sliding velocity of the body). Contour plots of $E_1/E_2 = 2$ show higher von Mises stresses with significant discontinuities occurring at the interface. The presence of a layer increases the von Mises stress in the body. For a more compliant layer case of $E_1/E_2 = 0.5$, the von Mises stresses are lower than that in the case of a nonlayered homogeneous body and furthermore only a mild discontinuity occurs at the interface.

For a homogeneous solid, the maximum Hertz pressure at which yield will occur as a function of coefficient of friction, according to von Mises yield criteria, is presented in Figure 3.2.12. When the elastic limit occurs subsurface, it is not possible to write down an explicit form for the elastic limit, as the precise location where the maximum state of stress occurs needs to be located numerically. However, when elastic limit is surface controlled, the expression for elastic limit ($J_2^{1/2} = k$) is given as (Hamilton, 1983),

$$\frac{k}{(p_0)_y} = \frac{1}{\sqrt{3}}\left[\frac{(1-2\nu)^2}{3} + \frac{(1-2\nu)(2-\nu)\mu\pi}{4} + \frac{(16-4\nu+7\nu^2)\mu^2\pi^2}{64}\right]^{1/2} \quad (3.2.21)$$

3.2.2.3 Elastic-Plastic Contact

Rigorous elastic-plastic analyses of indentation of layered solids against a conical indenter, an axisymmetric punch of arbitrary profile and a rigid spherical indenter have been conducted (Bhushan, 1996a). Based on the finite element analysis of a rigid sphere against an elastic-perfectly plastic layered medium in a frictionless contact, evolution of the normalized von Mises stress, $(3J_2)^{1/2}/p_0$ for $E_1/E_2 = 2$ and the ratio of the layer thickness to the sphere radius of 0.02 is shown in Figure 3.2.13 (Kral *et al.*, 1995a, b). Loads and distances are normalized

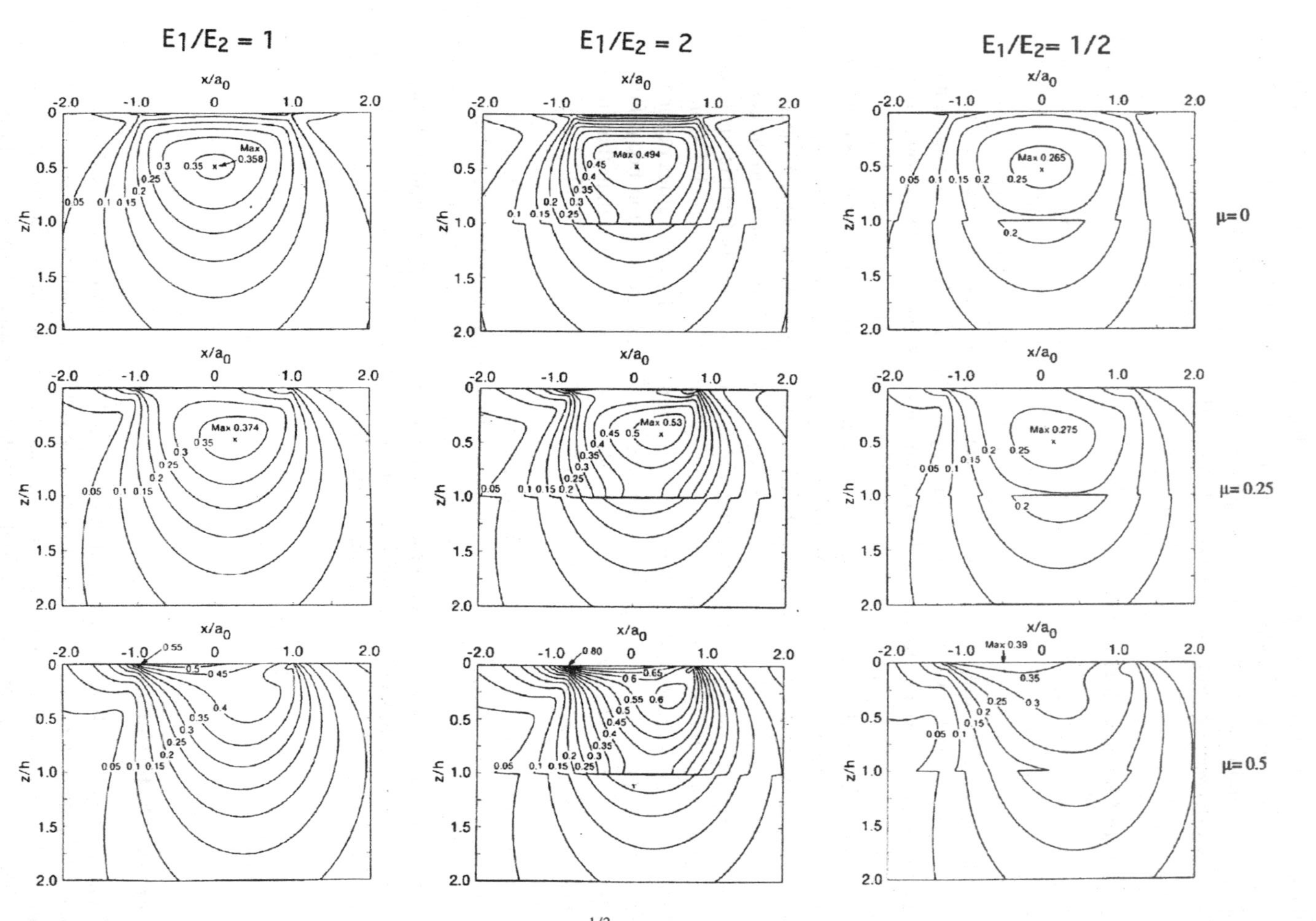

Figure 3.2.11 Contours of constant normalized von Mises stresses $J_2^{1/2}/p_0$, on the surface beneath the rigid spherical indenter for an indenter acting on a layered elastic half-space at $E_1 = E_2$, $E_1 = 2E_2$, and $E_1 = 0.5E_2$, and at $\mu = 0$, 0.25 and 0.5. Reproduced with permission from O'Sullivan, T.C. and King. R.B. (1988), "Sliding Contact Stress Field Due to a Spherical Indenter on a Layered Elastic Half-Space," *ASME J. Trib.* **110**, 235–240. Copyright 1988. ASME.

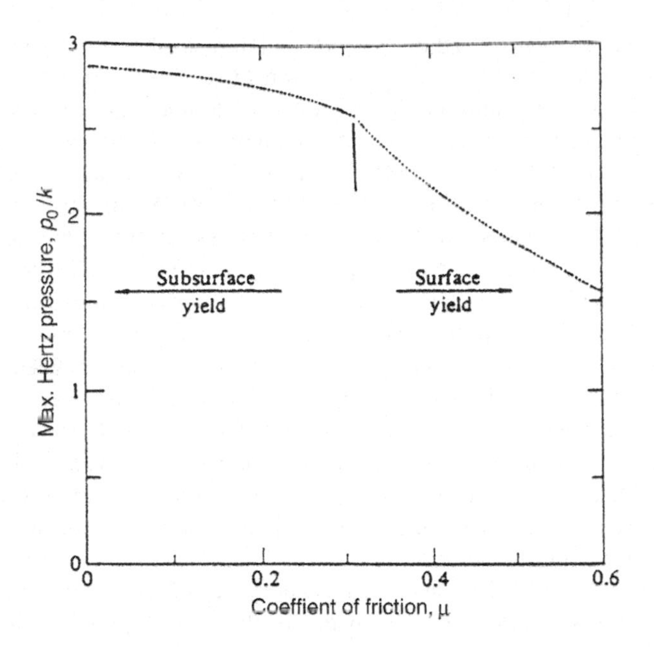

Figure 3.2.12 Effect of coefficient of friction on the maximum Hertz contact pressure for yield (von Mises) for a rigid spherical indenter acting on a homogeneous elastic-perfectly plastic half-space.

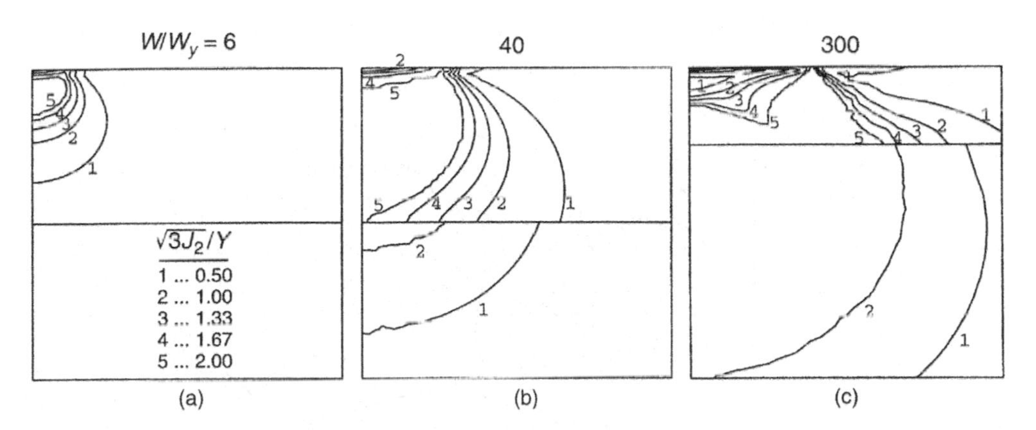

Figure 3.2.13 Contours of normalized von Mises stress $(3J_2)^{1/2}/p_0$ on the surface beneath the rigid spherical indenter for an indenter acting on a layered elastic-perfectly plastic half-space at $E_1 = 2E_2$. Parts (a) and (b) show the region $0 \leq r/a_y \leq 12$, $0 \leq z/a_y \leq 12$ and (c) shows the region $0 \leq r/a_y \leq 24$, $0 \leq z/a_y \leq 24$. Contour numbers 5 and 2 represent yielding in the layer $[(3J_2)^{1/2}/Y = 2]$ and substrate $[(3J_2)^{1/2}/Y = 1]$, respectively. Reproduced with permission from Kral, E.R., Komvopoulos, K., and Bogy, D.B. (1995b), "Finite Element Analysis of Repeated Indentation of an Elastic-Plastic Layered Medium by a Rigid Sphere, Part II: "Subsurface Results," *ASME J. App. Mech.* **62**, 29–42. Copyright 1995 ASME.

by the load W_y and the contact radius a_y, corresponding to the initial yield condition of a homogeneous substrate with $E/Y = 685\ Y$. The von Mises stress exhibits a discontinuity at the layer interface due to the different layer and substrate material properties. Figure 3.2.13 shows that yielding commences in the layer at a depth of about one-half the contact radius. As the load increases from $W/W_y = 6$, the plastic zone in the layer enlarges, eventually reaching both the surface and the interface for a load of $W/W_y = 40$. At this load, substrate yielding is also encountered. In all cases, substrate yielding first occurs at the interface on the axis of symmetry. As the load is further increased, the yielding region in the substrate continues to expand downward and along the interface, assuming an approximately elliptical shape. The yielding region in the layer forms a nose reaching to the surface moving outward with the contact edge. A stiffer and harder layer increases the load for inception of yielding in the substrate than that the load for inception of yielding in a homogeneous half-space with substrate properties. However, as reported earlier, a stiffer layer results in large tensile stresses on the layer surface and shear stresses at the layer-substrate interface which may result in formation of cracks and debonding of the interface. If minimization of plastic deformation is required for low wear, a stiffer layer should be used and its thickness must exceed that of the depth which undergoes plastic deformation.

Based on Figures 3.2.10, 3.2.11, and 3.2.13, schematics of the plastic zone and maximum tensile and shear stress locations are schematically shown in Figure 3.2.14. Note that with brittle materials, the appearance of tensile stresses is more important than the value of the maximum shear responsible for yield.

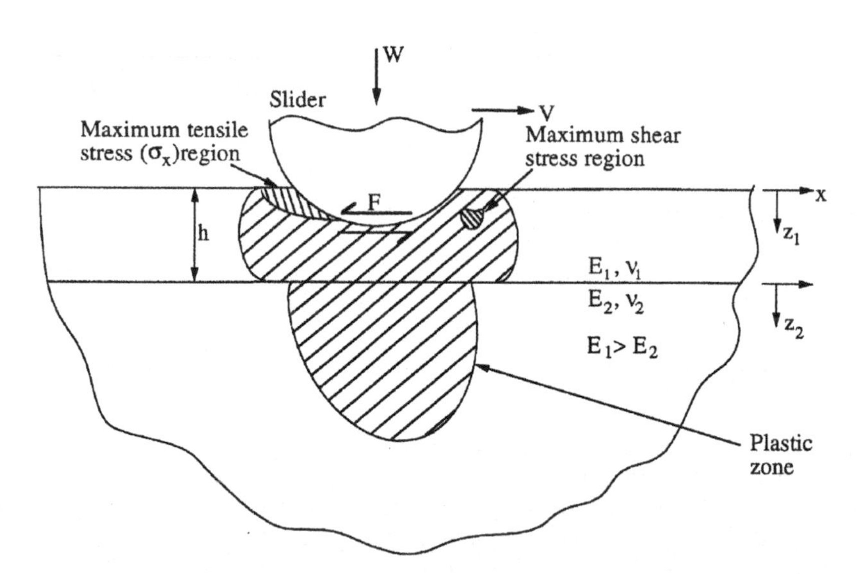

Figure 3.2.14 Schematic of plastic zone and maximum tensile and maximum shear stress locations in the lower body for the case of slider in a sliding contact with a layered solid. Note that stress distributions for the upper body will be that of the lower body with x and z replaced with −x and −z, respectively.

3.2.2.4 Effective Hardness and Young's Modulus of a Layered Medium

Effective hardness and Young's modulus of a layered medium is a function of indentation depth, the layer thickness, and elastic-plastic behavior of the layer and the substrate. For a thin layer, effective hardness and Young's modulus are mostly influenced by the hardness and Young's modulus of an underlying substrate whereas, for a thick layer, it is influenced by that of the layer. Bhushan and Venkatesan (2005) carried out elastic and plastic deformation analyses associated with indentation by one or multiple conical indenters and spherical indenters (representing asperities) on a layered medium, using a 3-D contact model, to be presented later. The indenters were considered to be perfectly rigid and the substrate and layer materials were assumed to be elastic-perfectly plastic. The effect of the elastic and plastic properties of both the layer and substrate on the hardness and Young's modulus of the layer/substrate composite were studied by determining the average pressure under the indenter as a function of the indentation depth. They developed empirical equations for layer/substrate combinations for which the substrate is either harder or softer than the layer for calculations of effective hardness and is whether stiffer or more compliant than the layer for calculations of effective Young's modulus.

As an example, for the case of a conical indenter indenting with elastic-plastic deformation of a soft layer on a harder substrate and of a hard layer on a softer substrate, the effective hardness can be given as

$$\frac{H}{H_s} = 1 + \left(\frac{H_f}{H_s} - 1\right)\exp\left[-\left(\frac{h_c}{h}\right)^{1.8}\left(\frac{E_f}{E_s}\right)^{-0.9}\left(\frac{H_f}{H_s}\right)^{1.0}\right], \quad H_f < H_s$$

$$\frac{H}{H_s} = 1 + \left(\frac{H_f}{H_s} - 1\right)\exp\left[-\left(\frac{h_c}{h}\right)^{1.1}\left(\frac{E_f}{E_s}\right)^{-0.5}\left(\frac{H_f}{H_s}\right)^{0.1}\right], \quad H_f > H_s \qquad (3.2.22)$$

where E_f and E_s are the Young's moduli, and H_f and H_s are the hardnesses of the layer and substrate, respectively. H is the effective hardness, h_c is the contact indentation depth, and h is the layer thickness.

For the case of a conical indenter indenting with only elastic deformation for a compliant layer on a more rigid substrate and on a rigid layer on a more compliant substrate, the effective Young's modulus can be given as

$$\frac{E_{eff}}{E_s} = 1 + \left(\frac{E_f}{E_s} - 1\right)\exp\left[-\left(\frac{h_c}{h}\right)^{0.9}\left(\frac{E_f}{E_s}\right)^{0.3}\right], \quad E_f < E_s$$

$$\frac{E_{eff}}{E_s} = 1 + \left(\frac{E_f}{E_s} - 1\right)\exp\left[-\left(\frac{h_c}{h}\right)^{0.5}\left(\frac{E_f}{E_s}\right)^{0.2}\right], \quad E_f > E_s \qquad (3.2.23)$$

Effective hardness and modulus results were found to depend only very weakly on Poisson's ratio(ν), and for this reason, this factor was not considered in the analysis. Figure 3.2.15 shows the effective hardness results as a function of (h_c/h) for different ratios of E_f/E_s and H_f/H_s. We note that hardness is generally less dependent on the substrate for indentation depths less than

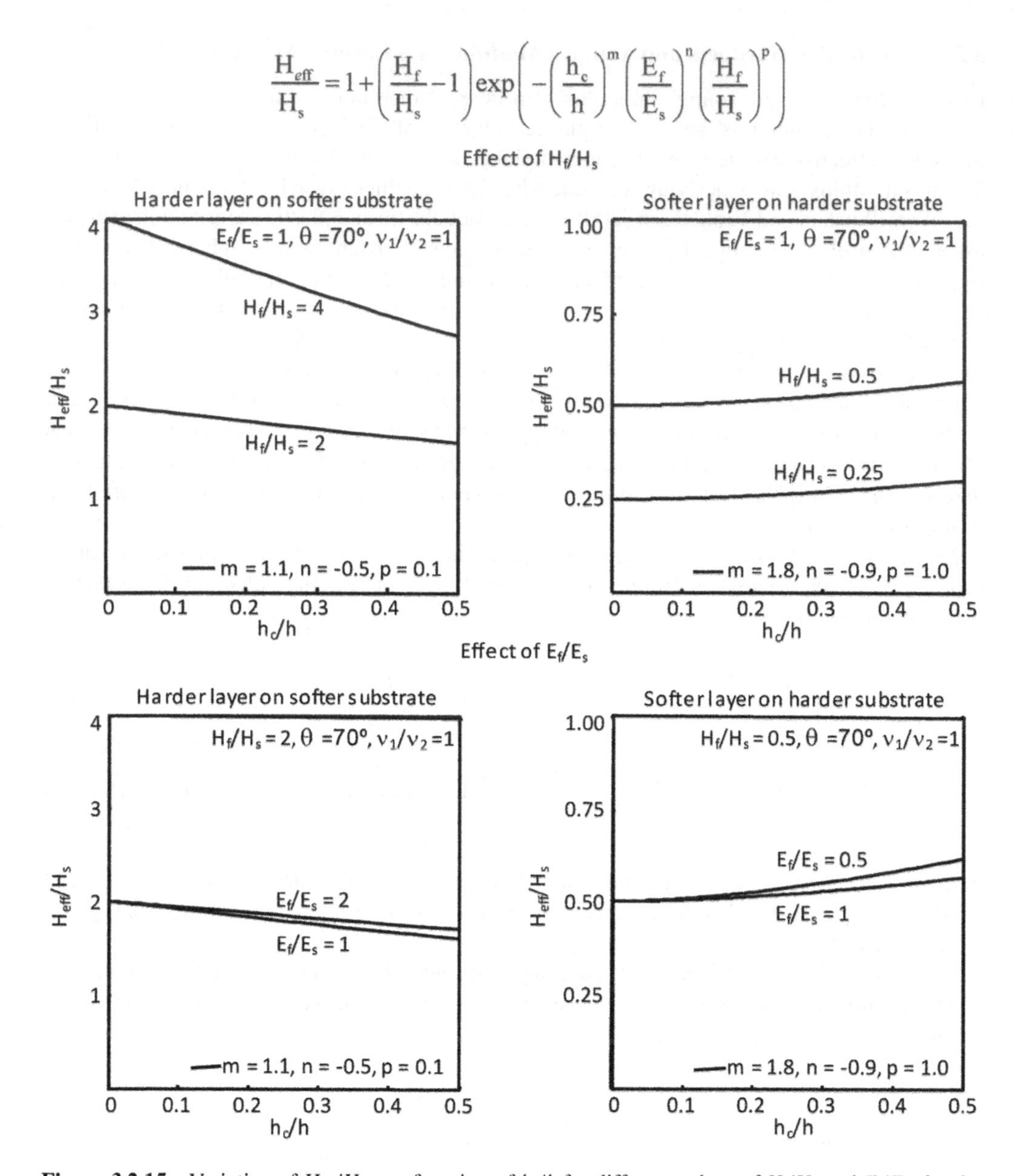

Figure 3.2.15 Variation of H_{eff}/H_s as a function of h_c/h for different values of H_f/H_s and E_f/E_s, for the contact of a conical indenter (representing an asperity) on a layered flat surface. The vertical range on the right hand column is expanded by a factor of four. Reproduced with permission from Bhushan, B. and Venkatesan, S. (2005), "Effective Mechanical Properties of Layered Rough Surfaces," *Thin Solis Films* **473**, 278–295. Copyright 2005. Elsevier.

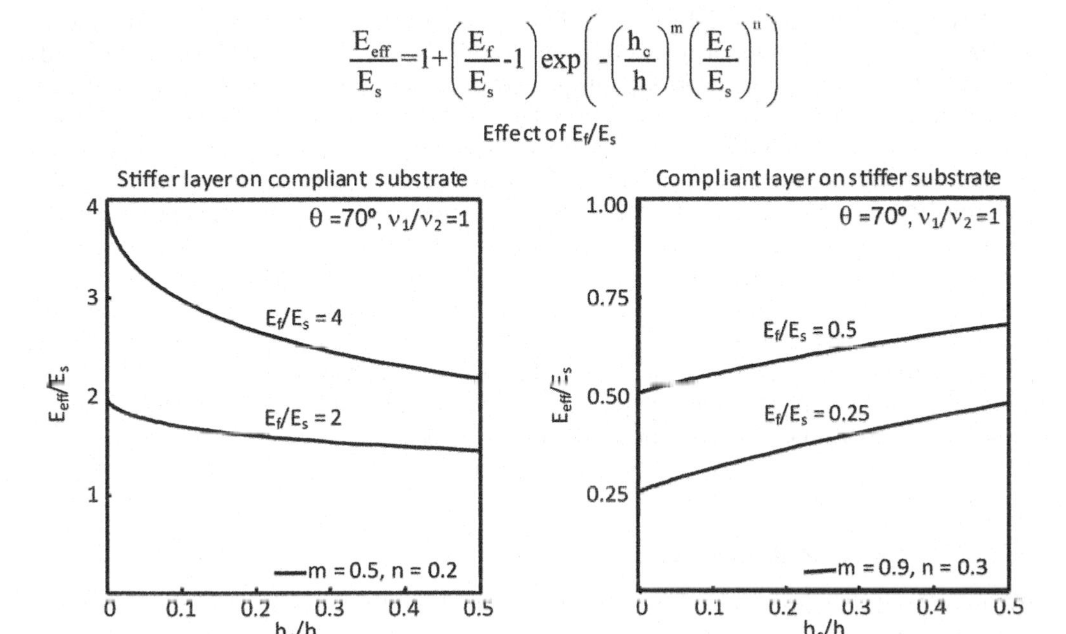

Figure 3.2.16 Variation of E_{eff}/E_s as a function of h_c/h, for different values of E_f/E_s, for the contact of a conical indenter (representing an asperity) on a layered flat surface. The vertical range on the right hand column is expanded by a factor of four. Reproduced with permission from Bhushan, B. and Venkatesan, S. (2005), "Effective Mechanical Properties of Layered Rough Surfaces," *Thin Solis Films* **473**, 278–295. Copyright 2005. Elsevier.

about 0.2 of the layer thickness, after which the hardness increases/decreases more rapidly because of the presence of the substrate. Figure 3.2.16 shows the effective modulus results as a function of (h_c/h) for different ratios of E_f/E_s.

3.2.3 Multiple Asperity Dry Contacts

Modeling of the contact of rough surfaces has been treated by several investigators using a number of approaches since the middle of the 1960s. The difficulty in the development of a theoretical model is that the surface is a random process and may be anisotropic so that stochastic models must be used. Due to the multiscale nature of surfaces, as reported in Chapter 2, the surface roughness parameters depend strongly on the resolution of the measuring instrument or any other form of filter, hence are not unique for a surface. Therefore, predictions of the contact models based on conventional roughness parameters may not be unique to a pair of rough surfaces. Roughness of engineering surfaces can be characterized by fractal geometry. A fractal theory of elastic and plastic contact between two rough surfaces, which uses scale-independent fractal parameters for surface characterization, can be used, but it is valid only for fractal surfaces.

Though the statistical models just mentioned can predict important trends on the effect of surface properties on the real area of contact, their usefulness is very limited because of over-simplified assumptions about asperity geometry and height distributions, the difficulty in determination of statistical roughness parameters, and the neglect of interactions between the adjacent asperities. With the advent of computer technology, a measured surface profile can be digitized and used for computer simulation. Digital maps of pairs of different surfaces can be brought together to simulate contact inside the computer and contours of contacts can be predicted. In computer simulations, the resulting contour maps can be analyzed to give contact parameters for various interplanar separations of the rough surfaces. Most of these analyses do not require assumptions of surface isotropy, asperity shape, and distribution of asperity heights, slopes and curvatures. However, one still has to select scan size and lateral resolution of the instrument relevant for the interface problem on hand (Chapter 2).

In this section, we first present a simple analysis of identical asperities, followed by statistical analysis, fractal analysis and numerical 3-D contact models.

3.2.3.1 Analysis of Identical Asperities

We first consider the contact between a smooth plane and a nominally flat surface covered with a number of spherical asperities with the same radius and the same height z, relative to the reference plane, Figure 3.2.17 (Bhushan and Tian, 1995; Chilamakuri and Bhushan, 1997). As the surfaces are loaded together, the total displacement (normal approach) δ is equal to $(z - d)$, where d is the current separation of the smooth surface and the reference plane of the rough surface. Each asperity is deformed equally, and carries the same normal load, W_i, so that for N asperities the total load W will be equal to NW_i. For each asperity, the load W_i and the area of contact A_i are known from the Hertz analysis presented in Section 3.2.1.1. Thus, if R is the radius of all identical asperities,

$$W_i = \frac{4E^*}{3} R^{1/2} \delta^{3/2} \tag{3.2.24a}$$

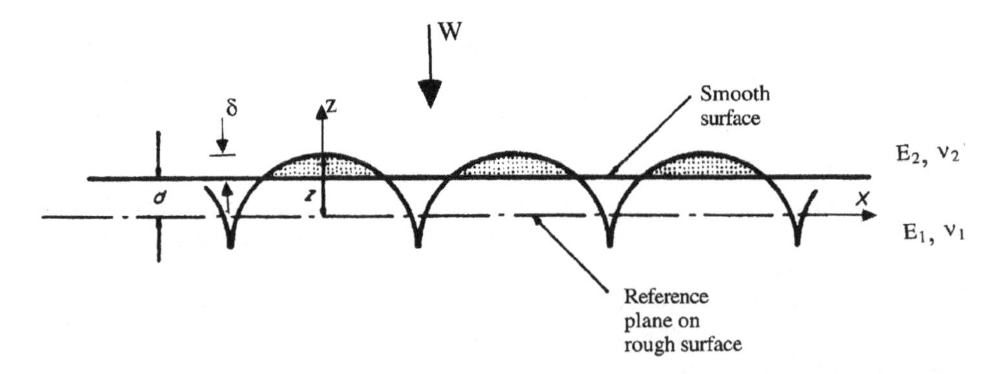

Figure 3.2.17 Schematic of contact of a regular patterned rough surface against a smooth plane surface.

and

$$A_i = \pi R\delta \tag{3.2.24b}$$

where E^* is the composite or effective modulus. The total load is

$$W = \frac{4}{3\pi^{3/2}} \frac{NE^* A_i^{3/2}}{R} \tag{3.2.25}$$

Load is related to the total real area of contact, A $(=NA_i)$ from Equations 3.2.24 and 3.2.25 as

$$W = \frac{4E^* A^{3/2}}{3\pi^{3/2} N^{1/2} R} \tag{3.2.26}$$

Equation 3.2.24b indicates that the surface just outside the footprint is displaced in such a way that the real area of contact is exactly half of the area $2\pi R\delta$ which would be obtained by plastic flattening of the spheres. Equation 3.2.26 shows that, for this particular model, the real area of contact is proportional to the two-thirds power of the applied normal load when the deformation is elastic.

The contact area is circular, having a radius a, and the contact pressure is elliptical with a maximum pressure at the center of the contact. The mean (p_m) and maximum contact pressure (p_0) are

$$p_m = \frac{2}{3} p_0 = \frac{W_i}{A_i} = \frac{4E^* \delta^{1/2}}{3\pi R^{1/2}} = \left(\frac{16WE^{*2}}{9\pi^3 NR^2}\right)^{1/3} \tag{3.2.27}$$

Propensity for yielding is governed by the Tresca maximum shear stress criterion or the von Mises shear strain energy criterion. From Equations 3.2.15 and 3.2.16, yielding is initiated when $p_m \sim 1.1\ Y \sim H/3$. Then from Equation 3.2.27, the critical load beyond which plastic deformation occurs is given by

$$\frac{W_{crit}}{N} \sim \frac{\pi^3 R^2}{48E^{*2}}(H^3) \tag{3.2.28}$$

where H is the hardness of the softer material. It is a general practice to introduce a factor of safety to account for the fluctuation of the hardness measurement and all the uncertainties involved in design. The factor includes any dynamic effect during asperity contacts. The value of a factor of safety is normally chosen between 2 and 3.

If the load exceeds the critical load, the softer material of contacting bodies deforms plastically. If the material deforms plastically at the interface, at full plasticity each asperity contact can be thought of as going through the indentation process. For elastic-perfectly plastic material (with no work hardening), the flow pressure under full plasticity is found to be almost independent of the load. In this case,

$$\left(\frac{W_i}{A_i}\right)_p = \left(\frac{W}{A}\right)_p = H \tag{3.2.29}$$

thus the real area of contact is proportional to the load.

3.2.3.2 Statistical Analysis of Contacts

If the two rough surfaces, both nominally flat, come into contact until their reference planes (taken to pass through the *mean* of the peak height distribution) are separated by a distance d, then there will be contact at those asperities whose total heights, $(z_1 + z_2)$, are greater than d. The contacts can be either elastic or plastic; viscous effects are normally neglected. Archard (1957) introduced the first statistical model for a multiple asperity-contact condition. The classical statistical model for a combination of elastic and elastic-plastic contacts between a rough surface and a smooth surface is that of Greenwood and Williamson (1966) (G&W). They assumed that: (1) the rough surface is covered with a large number of asperities, which, at least near their summit, are spherical; (2) asperity summits have the constant radius on each surface, R_{p1} and R_{p2} and composite radius (R_p) could be assigned to the rough surface; (3) that their heights vary randomly; and (4) that most engineering surfaces have a Gaussian distribution of peak heights. Many surfaces follow a Gaussian distribution. The assumption of peak radii being constant is clearly not valid (Chapter 2).

Greenwood and Tripp (1970–1971) have treated the contact of two rough surfaces instead of the one rough surface against a flat surface treated by G&W. For the case of two rough surfaces in contact with the pairs of asperities not aligned and the usual contact will be between the shoulders of the two hills, they found that, for the Gaussian peak-height distribution, the specification of asperity shape and the locations of asperities on one or both surfaces are unimportant. Therefore, although the asperity tips are assumed to be spherical for numerical simplicity, this will not affect contact area calculations. Also, they showed that the contact of two rough surfaces could be reduced to an equivalent, single, rough surface with a plane. O'Callaghan and Cameron (1976) and Francis (1977) also considered a case in which both surfaces can be rough and asperities need not contact at their tops. They concluded that the contact of two rough surfaces is negligibly different from the contact of a smooth and an equivalent rough surface.

The equivalent rough surface is defined as one whose asperity peak curvature, $1/R_p$, is the sum of the curvatures of two random rough surfaces (Chapter 2),

$$1/R_p = 1/R_{p1} + 1/R_{p2} \tag{3.2.30}$$

and by elementary statistics, if the peak-height distributions of two rough surfaces follow independent random distribution (not necessarily Gaussian) with standard deviations of σ_{p1} and σ_{p2}, the distribution of the equivalent rough surface has a standard deviation σ_p (Chapter 2),

$$\sigma_p = \left(\sigma_{p1}^2 + \sigma_{p2}^2\right)^{1/2} \tag{3.2.31}$$

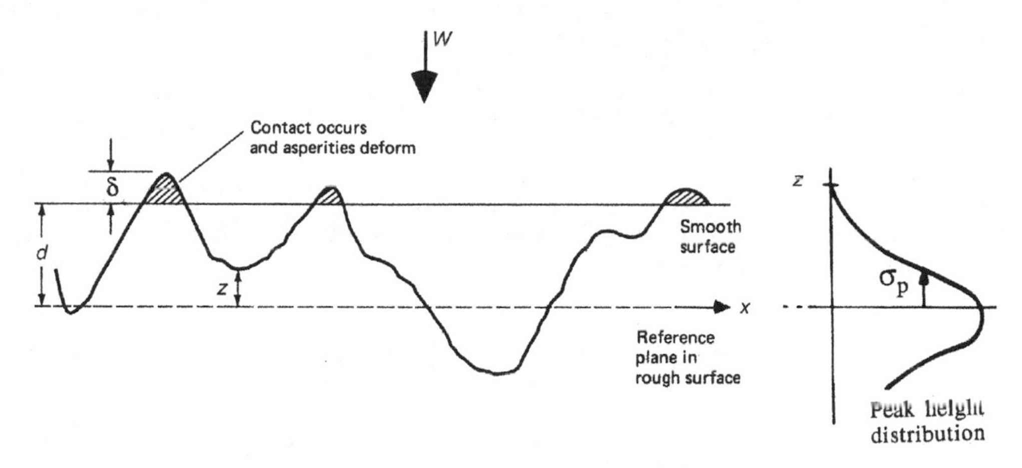

Figure 3.2.18 Schematic representation of the contact between a rough surface and a smooth surface.

Equations (3.2.31) and (3.2.32) are valid when the two surfaces are independent, as is likely when two surfaces are prepared separately. However, when the surfaces have slid together, this assumption may be violated. If so, the expression for σ_p must be modified by a covariance term.

Elastic Contact

For elastic contacts in static conditions or in dynamic conditions with no tangential stresses present at the contact, using G&W's assumptions, we calculate the apparent pressure, p_a, mean real pressure, p_r, (elastic) real area of contact, A_{re}, number of contact spots, n, and mean asperity real area of contact as a function of separation, d.

Based on G&W analysis, we consider the contact between a plane and a nominally flat surface covered with a large number of spherically tipped asperities of the same radius R_p and with their peak heights having a probability density function of $p(z)$, Figure 3.2.18.

If the two surfaces come together until their reference planes are separated by a distance d, then there will be contact at any asperity whose height was originally greater than d. Thus, the probability of making contact at any given asperity of height z is

$$P\,(z > d) = \int_d^\infty p\,(z)\,dz \tag{3.2.32a}$$

and if there are N asperities in all, the expected number of contacts will be

$$n = N \int_d^\infty p\,(z)\,dz \tag{3.2.32b}$$

Also, since $\delta = z - d$, the total (elastic) real area of contact is

$$A_{re} = \pi N R_p \int_d^\infty (z - d)\,p\,(z)\,dz \tag{3.2.32c}$$

Similarly, we find the expected total load is

$$W = p_r A_{re} = p_a A_a = \left(\frac{4}{3}\right) NE^* R_p^{1/2} \int_d^\infty (z-d)^{3/2} p(z)\, dz \tag{3.2.32d}$$

where p_r and p_a are real pressure and apparent pressure, respectively, and A_a is the apparent area. It is convenient to introduce nondimensional variables. The relationships in the form of dimensionless variables are presented here:

$$\frac{p_a}{(\eta R_p \sigma_p)\, E^* (\sigma_p/R_p)^{1/2}} = \left(\frac{4}{3}\right) F_{3/2}(D) \tag{3.2.33}$$

$$\frac{p_r}{E^* (\sigma_p/R_p)^{1/2}} = \left(\frac{4}{3\pi}\right) F_{3/2}(D)/F_1(D) \tag{3.2.34}$$

$$A_{re} E^* (\sigma_p/R_p)^{1/2} / p_a A_a = \left(\frac{3\pi}{4}\right) F_1(D)/F_{3/2}(D) \tag{3.2.35}$$

$$nR_p \sigma_p E^* (\sigma_p/R_p)^{1/2} / p_a A_a = F_0(D) / \left(\frac{4}{3}\right) F_{3/2}(D) \tag{3.2.36}$$

$$(A_{re}/n)\, R_p \sigma_p = \pi F_1(D)/F_0(D) \tag{3.2.37}$$

where D, the dimensionless separation, is d/σ_p; η is the density of asperity summits per unit area (N/A_a) on a surface with smaller density; and $F_m(D)$ is a parabolic cylinder function given by

$$F_m(D) = \int_D^\infty (s-D)^m p^*(s)\, ds \tag{3.2.38}$$

where $p^*(s)$ is the standardized peak-height probability density function in which the height distribution has been scaled to make its standard deviation unity. For the case of peak-height distribution following a Gaussian-height distribution (Chapter 2),

$$\begin{aligned} F_m(D) &= \left[\frac{1}{(2\pi)^{1/2}}\right] \int_D^\infty (s-D)^m \exp\left(-s^2/2\right) ds \\ &= \left[\frac{m!}{(2\pi)^{1/2}}\right] \left[\exp(-D^2/4)\right] U\left(m+\frac{1}{2}, D\right) \text{ for } m=0 \end{aligned} \tag{3.2.39}$$

The values of U are listed in Abramowitz and Stegun (1965). A short table of functions $F_m(D)$ is also given by Greenwood and Tripp (1970–1971).

Note that Equations 3.2.33–3.2.37 hold for all surface distributions. However, the assumption of two rough surfaces being the same as one equivalent rough surface with a plane is valid only for surfaces having a Gaussian distribution.

A simple relationship exists for an exponential height distribution,

$$p^*(s) = \exp(-s) \tag{3.2.40}$$

For this case, functions $F_m(D)$ are just $m!\exp(-D)$, and we have

$$p_r = \left(\frac{4}{3\pi}\right)(1.5!)E^*\left(\frac{\sigma_p}{R_p}\right)^{1/2} \tag{3.2.41a}$$

$$A_{re} = \left(\frac{3\pi}{4}\right)\frac{1}{(1.5!)}\frac{p_a A_a}{E^*\left(\sigma_p/R_p\right)^{1/2}} \tag{3.2.41b}$$

$$n = \frac{3}{4}\frac{1}{(1.5!)}\frac{p_a A_a}{R_p\sigma_p E^*\left(\sigma_p/R_p\right)^{1/2}} \tag{3.2.41c}$$

and

$$\frac{A_{re}}{n} = \pi R_p\sigma_p \tag{3.2.41d}$$

We note that the real area of contact and the number of contacts are both proportional to load, even though the asperities are deforming elastically. The real contact pressure and mean asperity real area of contact is independent of load. For other distributions, such a simple relationship will not apply.

For a Gaussian distribution, D vs. p_a is obtained from Equation 3.2.33, and is plotted in Figure 3.2.19. Then, with the help of this relationship and Equations 3.2.34 to 3.2.35, the relationships between p_a and p_r, A_{re}, n, and A_{re}/n are obtained in the dimensionless form in

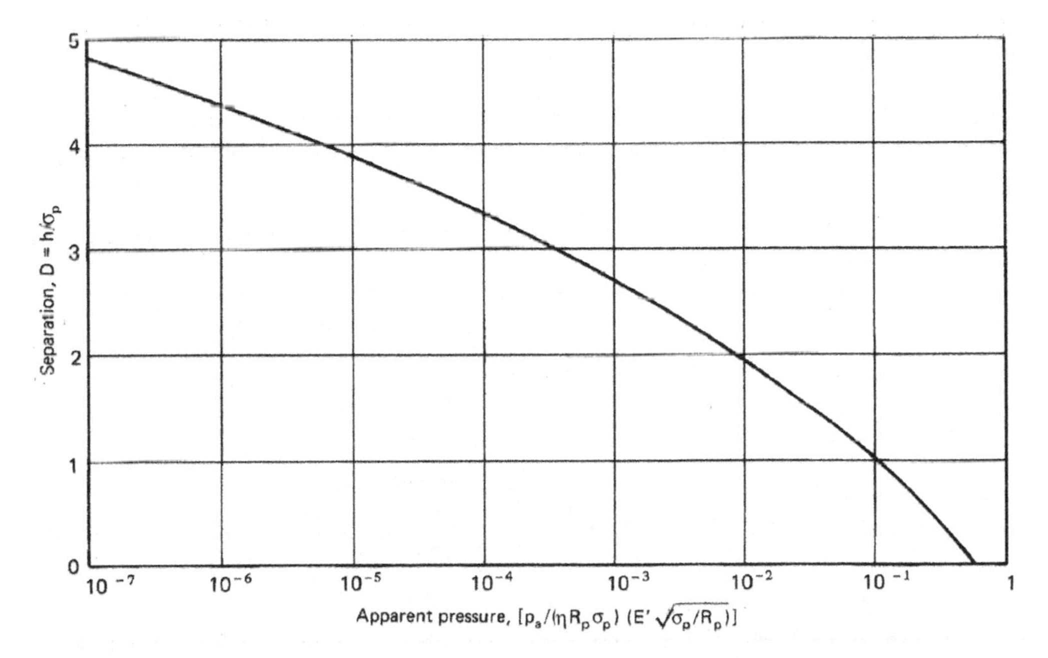

Figure 3.2.19 Relationship between separation and apparent pressure. Reproduced with permission from Bhushan, B. (1984), "Analysis of the Real Area of Contact Between a Polymeric Magnetic Medium and a Rigid Surface," *ASME J. Trib.* **106**, 26–34. Copyright 1984. ASME.

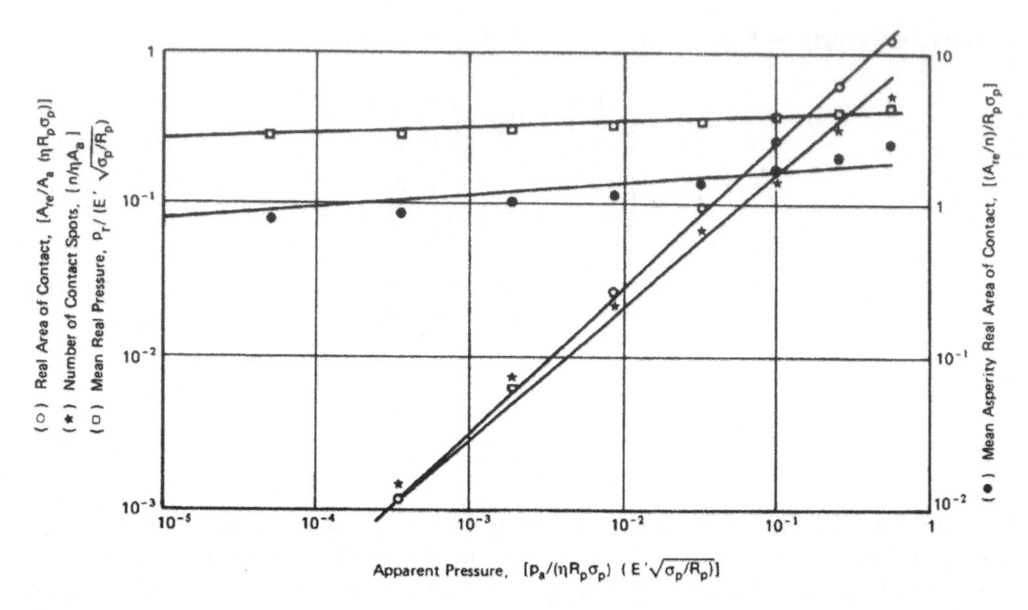

Figure 3.2.20 Relationship between mean real pressure, real area of contact, number of contact spots, and mean asperity real area of contact with apparent pressure. Reproduced with permission from Bhushan, B. (1984), "Analysis of the Real Area of Contact Between a Polymeric Magnetic Medium and a Rigid Surface," *ASME J. Trib.* **106**, 26–34. Copyright 1984. ASME.

Figure 3.2.20 . Next, the data are fitted to a power form using the least-squares fit. Approximate direct relationships of D, p_r, A_{re}, n, and A_{re}/n with p_a are presented in Table 3.2.2 (Bhushan, 1984). From equations in Table 3.2.2, note that D is a very weak function of p_a; p_r and A_{re}/n are practically independent of p_a; and A_{re} and n are approximately proportional to p_a. The increase in the load creates new contact areas proportionately which is responsible for the real

Table 3.2.2 Interplanar separation, mean real pressure, real area of contact, number of contact spots, and mean asperity real area of contact for elastic contacts.

Interplanar separation	$D = 1.40\left[\log(0.57/P_a)\right]^{0.65}$
Mean real pressure	$\dfrac{p_r}{E^*\left(\sigma_p/R_p\right)^{1/2}} = 0.42P_a^{0.04} \sim 0.32$
Real area of contact	$\dfrac{A_{re}}{A_a\left(\eta R_p\sigma_p\right)} = 2.40P_a^{0.96} \sim 3.20P_a$
Number of contact spots	$\dfrac{n}{\eta A_a} = 1.21P_a^{0.88} \sim 2.64P_a = 0.5\,(D \to 0)$
Mean asperity real area of contact	$\dfrac{A_{re/n}}{R_p\sigma_p} = 2.00P_a^{0.08} \sim 1.21$

$P_a = p_a/(\eta R_p\sigma_p)E^*(\sigma_p/R_p)^{1/2} \le 0.57$
Source: Bhushan, 1984.

area of contact being proportional to the load. An important relationship for the real area of contact in the elastic regime is listed here (Bhushan, 1984):

$$A_{re} \sim \frac{3.2\, p_a\, A_a}{E^* \left(\sigma_p / R_p\right)^{1/2}} \tag{3.2.42}$$

This contact model is defined by three parameters: σ_p, R_p and η, where η is the summit density. As described in Chapter 2, Whitehouse and Archard (1970) regarded the profile of a random surface as a random signal represented by a height distribution and an autocorrelation function. They showed that all features of a surface with Gaussian distribution of heights and an exponential autocorrelation function could be represented by two parameters: σ and β^*, where σ is the standard deviation of surface heights and β^* is the correlation length at which autocorrelation function $C(\tau) = 0.1$ (Chapter 2). For this surface, Onions and Archard (1973) expressed the peak heights, curvatures and asperity density in terms of σ and β^*, where σ is the standard deviation of surface heights and β^* is the correlation length (Chapter 2). In their contact model, they did not assume that peak heights follow a Gaussian distribution but follow a distribution derived from an assumed Gaussian distribution of surface heights. Second, peak radii are not constant, and have a distribution which is dependent upon the height. They reported that distribution of peak heights is not quite Gaussian and the peak curvature of the higher peaks tended to have higher values than those at lower levels. Based on their contact model,

$$A_{re} \propto \frac{p_a A_a}{E^* \left(\sigma / \beta^*\right)} \tag{3.2.43}$$

As reported in Chapter 2, for two random surfaces (1 and 2) in contact, σ and β^* of an equivalent surface are summed as

$$\begin{aligned} \sigma^2 &= \sigma_1^2 + \sigma_2^2 \\ \frac{1}{\beta^*} &= \frac{1}{\beta_1^*} + \frac{1}{\beta_2^*} \end{aligned} \tag{3.2.44}$$

Nayak (1973) and Bush *et al.* (1975) carried out the contact analysis by modeling the rough surfaces as isotropic, Gaussian surfaces in terms of spectral moments. Bush *et al.* (1979), Gibson (1982) and McCool (1986a) used Nayak microgeometry assumptions to develop an elastic contact model for anisotropic surfaces. The asperities were represented as elliptical paraboloids with random principal axis orientation and aspect ratio of the grains. They developed expressions for contact area in terms of five surface parameters – m_0, and m_2 and m_4 along the grains and across the grains.

Based on the contact analysis of non-Gaussian surfaces with skewness and kurtosis (Chapter 2), Figure 3.2.21 shows the effect of skewness and kurtosis on the fractional real area of contact at two nominal pressures (Kotwal and Bhushan, 1996; Chilamakuri and Bhushan, 1998; Bhushan, 1998, 1999). A positive skewness between 0 and 0.2 at low pressure and about 0.2 at high pressure results in lowest real area of contact. Real area of contact decreases with an increase in kurtosis. Note that kurtosis has more effect than skewness.

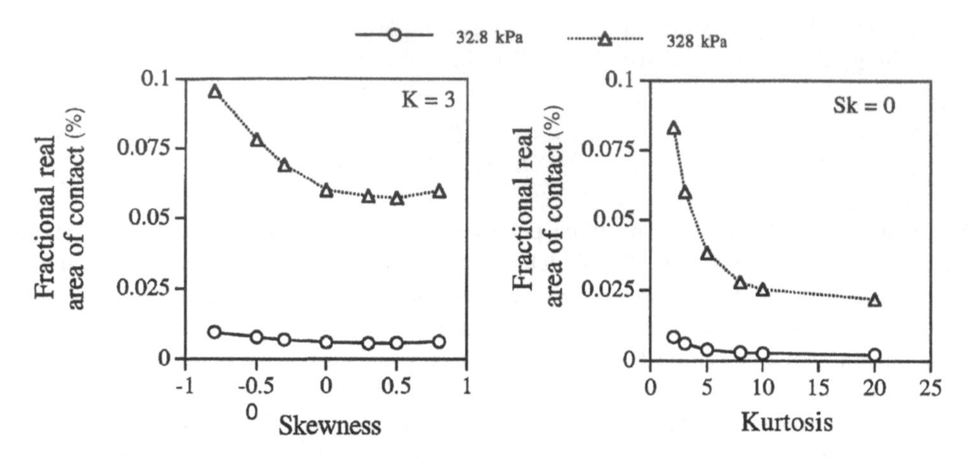

Figure 3.2.21 Effect of skewness and kurtosis on the fractional real area of contact at applied pressures of 32.8 and 328 kPa ($E^* = 100$ GPa, $\eta = 500{,}000/\text{mm}^2$, $A_a = 0.915\ \text{mm}^2$, $\sigma_p = 1$ nm, $R_p = 10\ \mu\text{m}$). Reproduced with permission from Kotwal, C.A. and Bhushan, B. (1996), "Contact Analysis of Non-Gaussian Surfaces for Minimum Static and Kinetic Friction and Wear," *Tribol. Trans.* **39**, 890–898. Copyright 1996. Taylor and Francis.

Limit of Elastic Deformation

For a random surface with asperities with Gaussian height distribution and constant radii, the normal approach in terms of p_m from Equation 3.2.27 is

$$\delta = \frac{9\pi^2}{16} p_m^2 \frac{R_p}{E^{*2}} \tag{3.2.45}$$

From Equations 3.2.15, 3.2.16 and 3.2.45, the critical value of δ for the asperity necessary to initiate subsurface plastic flow is

$$\delta_p \sim R_p \left(\frac{H}{E^*}\right)^2 \tag{3.2.46}$$

A contact will become plastic if the height of an asperity z is greater than $d + \delta_p$. Therefore, from Equation 3.2.32c, the plastic component of the real area of contact A_{rp} is given as

$$\begin{aligned} A_{rp}/A_a &= \pi\eta R_p\sigma_p \int_{D+\delta_p^*}^{\infty} (s-D)p(s)\,ds \\ &= \pi\eta R_p\sigma_p \left[\int_{D+\delta_p^*}^{\infty} \left(s-D-\delta_p^*\right) p(s)\,ds + \delta_p^* \int_{D+\delta_p^*}^{\infty} p(s)\,ds\right] \\ &= \pi\eta R_p\sigma_p \left[F_1\left(D+\delta_p^*\right) + \delta_p^* F_0\left(D+\delta_p^*\right)\right] \end{aligned} \tag{3.2.47}$$

where $\delta_p^* = \delta_p/\sigma_p$. Also, the elastic component of the real area of contact from Equation 3.2.32c is

$$A_{re}/A_a = \pi \eta R_p \sigma_p F_1(D) \tag{3.2.48}$$

Therefore, from Equations 3.2.47 and 3.2.48 the plastic to elastic real area of contact is given as

$$A_{re}/A_a = \left[F_1\left(D+\delta_p^*\right) + \delta_p^* F_0\left(D+\delta_p^*\right)\right]/F_1(D) \tag{3.2.49}$$

We define a plasticity index ψ as the square root of the inverse of δ_p normalized with σ_p as,

$$\Psi = \left(\frac{\sigma_p}{\delta_p}\right)^{1/2} = \left(\frac{E^*}{H}\right)\left(\frac{\sigma_p}{R_p}\right)^{1/2} \tag{3.2.50}$$

This index is indicative of the degree of plasticity. Note that plastic deformation would occur in the solid with a lower hardness.

Using Equation 3.2.49, we calculate A_{rp}/A_{re} as a function of ψ for different separation values of D. The corresponding apparent pressures for different values of D were calculated using Figure 3.2.19. The results are plotted in Figure 3.2.22. Assuming $A_{rp}/A_{re} = 0.02$ as the criterion for the onset of a significant degree of plasticity, it was found that if $\psi < 0.6$, the deformation is largely elastic and if $\psi > 1$, surface deformation is largely plastic. Note that $\psi > 1$, the plastic flow will occur even at trivial normal loads. Note that the probability of plastic flow is virtually independent of the load and solely a function of the plasticity index as long as the asperities continue to deform independently. The index depends on both the mechanical properties and the surface roughness of the contacting surfaces. Slight modifications to the elastic-plastic models have been presented in several papers (e.g., see Francis, 1977; Bhushan, 1984; McCool, 1986b; Chang *et al.*, 1987; Thomas, 1999).

Based on the G&W approach, Onions and Archard (1973) also defined a plasticity index based on σ and β^* as,

$$\psi = \frac{E^*}{H}\left(\frac{\sigma}{\beta^*}\right) \tag{3.2.51}$$

They found that if $\psi > 0.45$, plastic flow occurs even at trivial loads, and if $\psi < 0.25$ plastic flow is most unlikely.

One of the benefits of the Onions and Archard analysis is that the plasticity index and contact parameters are expressed in terms of surface parameters, which are most easily measured. σ and R_p can be expressed in terms of σ and β^* (Chapter 2).

In a plastic contact, each contact can be visualized as a small hardness indentation, and the mean contact pressure will be equal to the hardness and independent of the load and the contact geometry. Therefore, real area of contact, A_{rp} is inversely proportional to the hardness,

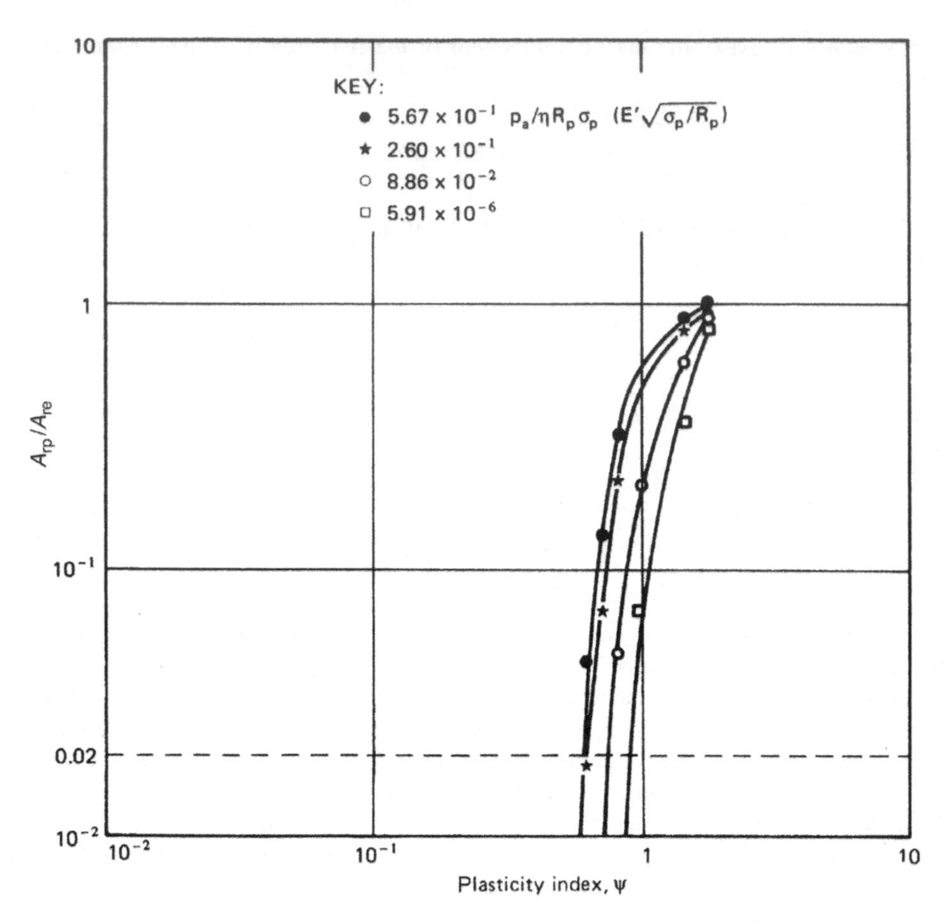

Figure 3.2.22 Influence of plasticity indices upon the proportion A_{rp}/A_{re} of the contact area that involves plastic flow. Reproduced with permission from Bhushan, B. (1984), "Analysis of the Real Area of Contact Between a Polymeric Magnetic Medium and a Rigid Surface," *ASME J. Trib.* **106**, 26–34. Copyright 1984. ASME.

proportional to the normal load, and independent of the apparent area (Bowden and Tabor, 1950, 1964):

$$A_{rp} = \frac{p_a A_a}{H} \tag{3.2.52}$$

where H is the hardness of the sliding surface layer of the softer material. If the asperities are plastically deformed, the details of the surface texture seem relatively unimportant, because the total real area of contact and the contact pressure do not depend upon surface texture.

Figure 3.2.23 shows a scanning electron micrograph (SEM) of a section taken normal to the sliding surface. It is clear that the surface has been severely strained, by the friction process,

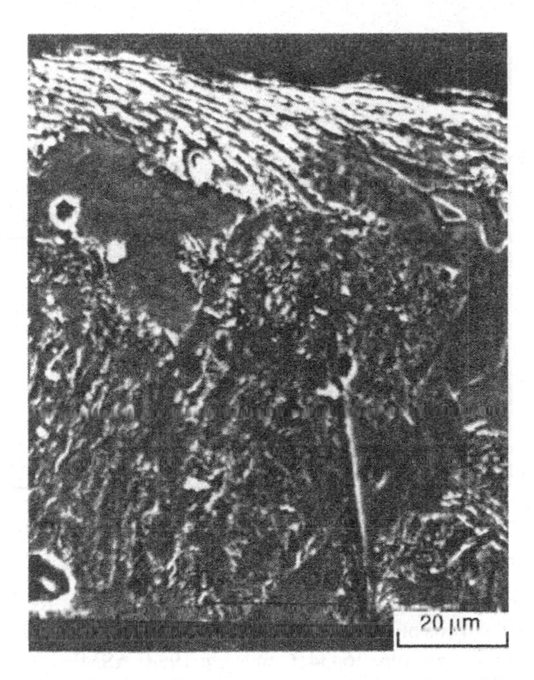

Figure 3.2.23 SEM micrograph of surface layer of AISI 8620 after sliding on AISI 4140 at 1 m/s and 415 kPa. Reproduced with permission from Cook, N.H. and Bhushan, B. (1973), "Sliding Surface Interface Temperatures," *ASME J. Lub. Tech.* **95**, 59–64. Copyright 1973. ASME.

to a depth of 5–10 μm. This depth is on the order of a typical contact diameter. Therefore, most of the plastic behavior associated with asperity contact formation must occur within this hardened layer. Thus, in contact analyses, one must use the surface hardness H_s rather than the bulk hardness H. Unfortunately, H_s is not easily measured. Cook and Bhushan (1973) estimated H_s by making a large number of microhardness tests on worn metal surfaces. The ratio H_s/H ranged from 1.4 to 2.1, and if anything tended to be on the low side. These authors have suggested that H_s/H for metals after sliding can be assumed to be 2.

It was presented earlier that in a sliding contact with friction present at the interface, maximum shear stress is larger and occurs nearer the surface. Therefore, the contacts become plastic at lower values of ψ. In addition, in a multilayered solid for a fixed value of substrate Young's modulus, the stresses increase for larger values of the overcoat Young's modulus, implying that the contact becomes plastic at lower values of ψ and vice versa. If the contact radius is much greater than the overcoat thickness, the effect of overcoat Young's modulus is negligible and the yielding is dominated by the modulus and hardness of the substrate (Bhushan and Doerner, 1989). Further, during sliding, polishing of the asperities generally occurs, which results in a smoother surface and an increase in the real area of contact (Bhushan, 1996b). During sliding, instantaneous roughness should be used.

For calculations of the real area of contact, E', Y, and H should be measured at a strain rate corresponding to the loading and unloading of the asperity contacts. During sliding, the asperities are loaded and unloaded periodically in a time corresponding to that taken

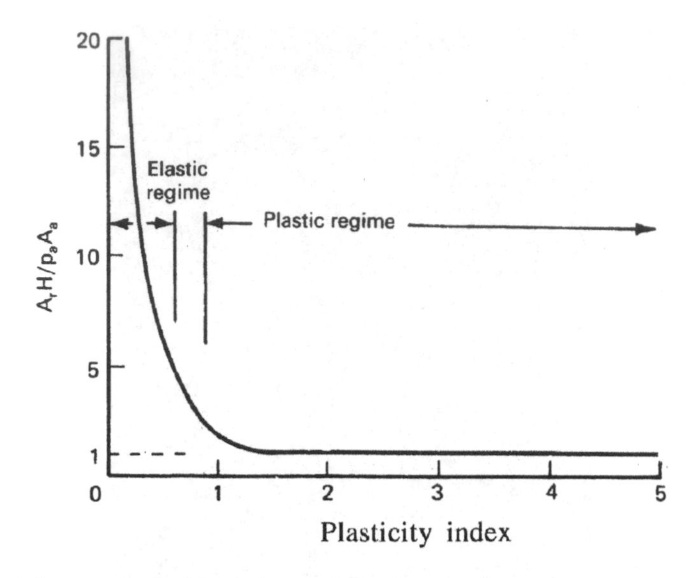

Figure 3.2.24 Influence of plasticity index on the real area of contact. Reproduced with permission from Bhushan, B. (1984), "Analysis of the Real Area of Contact Between a Polymeric Magnetic Medium and a Rigid Surface," *ASME J. Trib.* **106**, 26–34. Copyright 1984. ASME.

for a moving asperity to traverse its contact diameter. Therefore, the strain rate involved in the loading cycle (which determines the area of contact) can be estimated as the sliding velocity divided by the diameter of an asperity contact. Interface temperature rise caused by frictional heating during sliding would affect the mechanical properties and should be taken into account.

Finally, in the case of materials that creep to a marked extent such as polymers, the real area of contact will increase with time of application of the load (Bhushan, 1985b, 1996b).

Optimization of Mechanical Properties and Surface Roughness Parameters

To minimize friction and wear in a machine for given operating conditions, the fraction of real area of contact to apparent area of contact and the real contact pressure should be as low as possible. The real area of contact versus the plasticity indices given from Equations 3.2.43 or 3.2.44 and 3.2.51 are plotted in Figure 3.2.24. An examination of this figure shows that the plastic contact results in a minimum contact area. However, repeated plastic contact would lead to an undesirable permanent deformation and smoothening resulting in elastic contacts (and higher real area of contact). Wear is more probable when asperities touch plastically than in pure elastic contacts. Therefore, it is desirable to design machine components in the elastic-contact regime and ψ close to the elastic contact limit ($\psi \sim 0.6$) or $E^*(\sigma_p/R_p)^{1/2}$ to be as high as possible. Intuitively, we can explain the fact that $E^*(\sigma_p/R_p)^{1/2}$ has to be higher for lower real of contact by the following observation. The asperities with high E and low R_p produce high contact stresses and result in lower A_{re} for a given load. In addition, high σ_p allows contact with fewer asperities and again produces high contact stresses and results in lower A_{re} for a given load.

In principle, one can produce a well-defined (rather than statistical) roughness with few tall asperities of the same height and a small radius of curvature on a very smooth surface to minimize contact. Selection of number of asperities and radii of curvature of summits should be made such that the mean real pressure is just below the yield strength of the softer material (Bhushan and Tian, 1995; Chilamakuri and Bhushan, 1997). An advantage of this approach is that a well-defined localized roughness is produced and unwanted roughness is eliminated which results in better mechanical durability. However, creation of localized roughness may be an impractical approach because mechanical properties of the asperity tips and load variations in an application may be uncontrolled which makes it difficult to select optimum shape and number of asperities.

Typical Calculations of Contact Statistics

Surface roughness, Young's modulus, and microhardness are required for calculations of the contact statistics. Examples of roughness and mechanical property data of mating surfaces and contact statistics for magnetic head-polymeric tape and magnetic head-thin-film disk interfaces are presented in Table 3.2.3. Roughness measurements were made using a noncontact optical profiler (NOP) with a lateral resolution of 1 μm and atomic force microscope (AFM) with a lateral resolution of about 10 nm. The surface topography statistics calculated for the AFM data show significant differences from those calculated using the NOP data for the same samples. The average summit radius (R_p) for the AFM data is two to four orders of magnitude smaller than that for the NOP data. A surface with asperities having small radii of curvature will result in high contact stresses leading to plastic flow.

We note that the plasticity index (ψ) for head-tape and head-disk interfaces calculated using the AFM data suggests that all contacts are plastic, while ψ calculated with NOP data suggests that all contacts are elastic. It appears that as the two surfaces touch, the nanoasperities (detected by AFM) are the first to come into contact, Figure 3.2.25. As the load is applied, the small asperities are plastically deformed and the contact area increases. When the load is increased, the nanoasperities in the contact zone merge and the load is supported by elastic deformation of the larger scale asperities or microasperities (detected by NOP). The fractional contact area of the nanoasperities is small. The contact analysis using AFM data predicts that the contact area of individual asperities is a few square microns. However, before the contact size can become so large, these nanoasperities are completely crushed and become part of large (micro) asperities which are subsequently deformed elastically. Therefore, contact area statistics predicted using NOP data is believed to be more representative for contact area calculations. However, since nanoasperities go through plastic deformation which is undesirable, nanoroughness obtained using AFM needs to be measured and is of interest (Bhushan and Blackman, 1991; Poon and Bhushan, 1996b).

The depth of penetration during plastic deformation of the nanoasperities is small compared to that in the elastic deformation of the microasperities. (The elastic penetration depth is $\sim A_{re}/\pi n R_p \sim 1 - 2$ nm). Therefore, plastic deformation of nanoasperities is much more dependent on the near surface properties.

In the case of very hard vs. very hard and rough surfaces, ψ based on NOP and AFM data is generally greater than 1, which suggests that deformation of asperities is primarily plastic. Also note that the real area of contact of interfaces involving hard and rough surfaces is low as compared to soft vs. soft and/or smooth vs. smooth surfaces.

Table 3.2.3 Real area of contact statistics calculations for a magnetic polymeric tape *a* against a Ni-Zn ferrite head and a magnetic thin-film rigid disk against a Al_2O_3-TiC head slider.

	E	H	σ (nm)		σ_p (nm)		$1/R_p$ (1/mm)		η (1/mm^2)		ψ		$A_r/A_a p_a$ (1/GPa)	$n/A_a p_a$ (1/mN)	A_r/n (μm^2)	p_r (GPa)
Component Designation	(GPa)	(GPa)	NOP	AFM	NOP	AFM	NOP	AFM	NOP	AFM	NOP	AFM	NOP	NOP	NOP	NOP
Head-tape interface																
Polymeric tape – A	1.75	0.25	19.5	36.3	19.0	45.4	2.20	1.4×10^5	5.7×10^3	8.0×10^6	0.05	50.17	241.6	9.6	25.3	3.23×10^{-3}
Mn-Zn ferrite head	122	6.9	2.15	3.61	2.51	5.47	0.23	7.8×10^3	1.2×10^3	6.2×10^6						
Head-disk interface																
Magnetic disk	113	6.0	7.33	6.33	7	6.7	3.90	3.0×10^3	732	9.1×10^6	0.12	3.4	3.3	2.5	1.7	0.24
Al_2O_3-TiC head slider	450	22.6	1.63	1.55	2	1.4	0.53	1.2×10^3	2.4×10^3	13.3×10^6						

NOP - Noncontact optical profiler; AFM - Atomic force microscope.

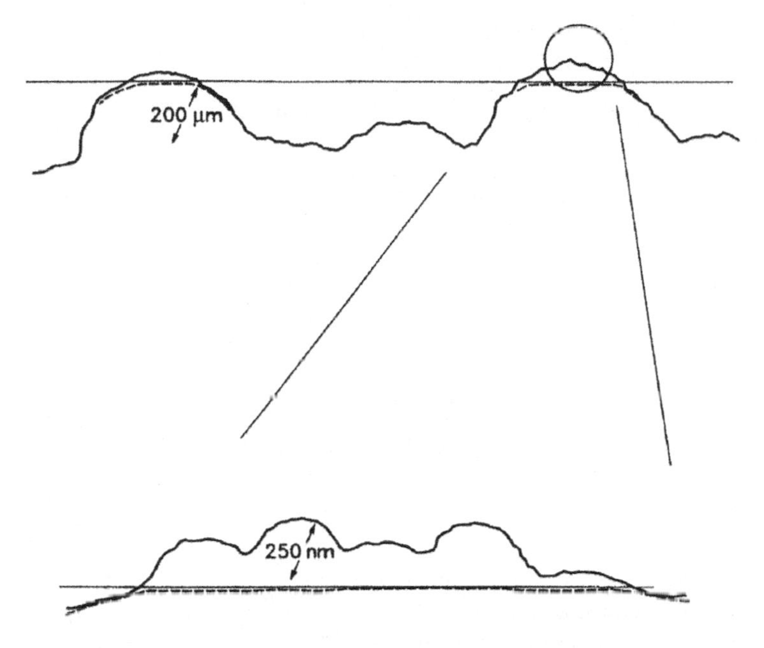

Figure 3.2.25 Schematic of local asperity deformation during contact of a rough surface; upper profile measured by an optical profiler and lower profile measured by AFM. Typical dimensions are shown for a polished, magnetic thin-film rigid disk against a flat magnetic slider surface. Reproduced with permission from Bhushan, B. and Blackman, G.S. (1991), "Atomic Force Microscopy of Magnetic Rigid Disks and Sliders and its Applications to Tribology," *ASME J. Trib.* **113**, 452–457. Copyright 1991. ASME.

Example Problem 3.2.2

Two nominally flat steel surfaces are lapped to give standard deviation of peak heights of 0.2 µm and mean peak radius of 5 µm for each of the two surfaces. When the surfaces are placed in contact, (a) would you expect the asperity deformation to be predominantly elastic, plastic or mixed elastic-plastic? (b) Does your answer depend on the normal load? (c) What is the real area of contact for a normal load of 100 N? The given parameters are: $E_{steel} = 200$ GPa, $\nu_{steel} = 0.3$ *and* $H = 8$ GPa.

Solution

Given

$$E_1 = E_2 = 200 \text{ GPa}, \ \nu_1 = \nu_2 = 0.3 \ and \ H = 8 \text{ GPa}$$

$$\frac{1}{E^*} = \frac{1-\nu_1^2}{E_1} + \frac{1-\nu_2^2}{E_2} = \frac{2(1-0.3^2)}{200} \text{ GPa}^{-1}$$

or

$$E^* = 109.9 \text{ GPa}$$

Roughness parameters of an equivalent surface are,

$$\sigma_p = \sqrt{\sigma_{p1}^2 + \sigma_{p2}^2} = \sqrt{2}\,(0.2)\ \mu\text{m} = 0.2828\ \mu\text{m}$$

$$\frac{1}{R_p} = \frac{1}{R_{p1}} + \frac{1}{R_{p2}} = \frac{2}{5}\mu\text{m}^{-1} = 0.4\ \mu\text{m}^{-1}$$

(a) We now calculate

$$\begin{aligned}\psi &= \frac{E^*}{H}\left(\frac{\sigma_p}{R_p}\right)^{1/2}\\ &= \frac{109.9}{8}(0.2828 \times 0.4)^{1/2}\\ &= 4.62\end{aligned}$$

Since $\psi > 1$, deformation is predominately plastic.

(b) The result does not depend upon load.

(c) Real area of contact in plastic contact regime,

$$\begin{aligned}A_{rp} &= \frac{W}{H}\\ &= \frac{100\ \text{N}}{8 \times 10^9\ \text{Pa}}\\ &= 1.25 \times 10^{-8}\ \text{m}^2\end{aligned}$$

3.2.3.3 Fractal Analysis of Contacts

Due to the multiscale nature of surfaces, it is found that the surface roughness parameters depend strongly on the resolution of the roughness measuring instrument or any other form of filter, hence are not unique for a surface. Therefore, the predictions of the contact models based on conventional roughness parameters may not be unique to a pair of rough surfaces. However, if a rough surface is characterized in a way such that the structural information of roughness at all scales is retained (possible with modern roughness measurement tools – atomic force microscope), then it will be more logical to use such a characterization in a contact theory. In order to develop such a contact theory, it is first necessary to quantify the multiscale nature of surface roughness.

A unique property of rough surfaces is that if a surface is repeatedly magnified, increasing details of roughness are observed right down to nanoscale. In addition, the roughness at all magnifications appears quite similar in structure. Such a behavior can be characterized by fractal geometry (Chapter 2). The main conclusions from these studies are that a fractal characterization of surface roughness is *scale-independent* and provides information of the roughness structure at all the length scales that exhibit the fractal behavior.

A fractal model of elastic-plastic contacts (Majumdar and Bhushan, 1991) has been developed to predict whether contacts experience elastic or plastic deformation, and to predict the real area of contact and statistical distribution of contact points. Based on the fractal model of elastic-plastic contact, whether contacts go through elastic or plastic deformation is determined by a critical area which is a function of fractal parameters (D, C), hardness, and modulus of elasticity of the mating surfaces. If the contact spot is smaller than the critical area, it goes through plastic deformation and large spots go through elastic deformation. The critical contact area of inception of plastic deformation for a magnetic thin-film disk ($\sigma = 7$ nm, $H/E^* = 0.06$) was reported by Majumdar and Bhushan (1991) to be about 10^{-27}m^2, so small that all contact spots can be assumed to be elastic at moderate loads.

Majumdar and Bhushan (1991) and Bhushan and Majumdar (1992) have reported relationships for cumulative size distribution of the contact spots, portions of the real area of contact in elastic and plastic deformation modes, and the load–area relationships.

3.2.3.4 Numerical Three-Dimensional Contact Models and Computer Simulations

Numerical techniques are used to provide a deterministic solution to stresses and areas for the approach of two three-dimensional (3-D) real rough surfaces, Figure 3.2.26 (Bhushan, 1998, 2013). In one of the techniques the complex stress-deformation analytical expressions are converted to a system of linear algebraic equations, which are generally solved, based on numerical methods, with a computer. This technique takes full account of the interaction of deformation from all contact points and predicts contact geometry of real surfaces under loading. It provides useful information on the contact pressure, number of contacts, their sizes and distributions, and the spacing between contacts. For contacts with a moderate number of contact points, both elastic and elastic-plastic analyses of two rough surfaces can be carried out (Tian and Bhushan, 1996). As shown in Figure 3.2.26, the real area of contact between the two bodies occurs at the tips of highest asperities. The contact area is a small fraction of the surface areas of the contacting bodies; therefore, we can assume that the asperity contacts of each body occurs on an elastic half-space. Another assumption used in this study is that the area of individual contact is much smaller than the radii of curvature of contacting asperities. This allows the use of linear theory of elasticity as well as the approximation of plane surface around the real contact area.

Computer programs have been developed to perform contact analysis. In a program, based on a minimum potential energy theory, first the three-dimensional surface profiles of the two surfaces are read in. Contacting surfaces are discretized into small elements corresponding to the different points of surface heights. For a given rigid body approach (or load) between two rough surfaces, the total surface displacement at the surface of real contact is equal to the interference of two contacting bodies. The elements with finite displacement are included in the formulation. A so-called influence matrix ($\underset{\sim}{C}$) is constructed to relate contact pressure ($\underset{\sim}{p}$) to the given displacements ($\underset{\sim}{u}$) dependent on the location of the pressure element and contact points and an expression for the total complementary potential energy involving displacement and pressure is developed. The minimum value of the total complementary potential energy is obtained using a direct quadratic mathematical programming technique. The real area of

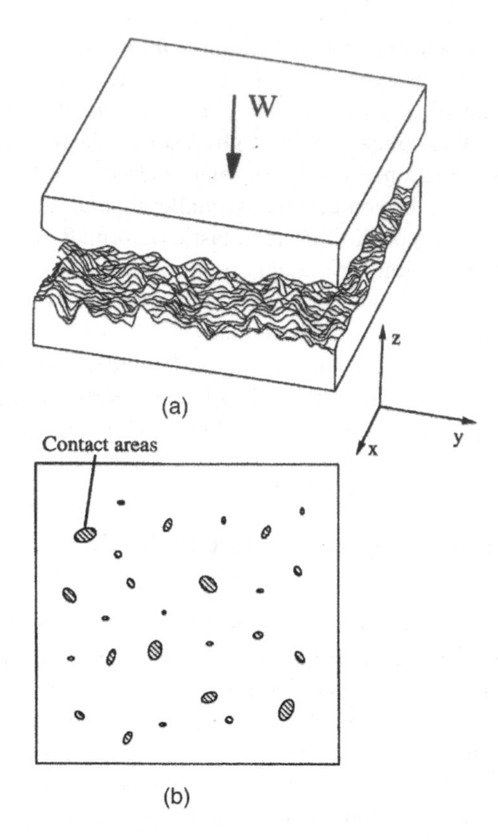

Figure 3.2.26 Schematic of (a) two 3-D rough surfaces in contact, and (b) corresponding contact areas.

contact and contact pressure distribution are those that minimize the total complementary potential energy. The analysis solves for contacts with positive pressure. Figure 3.2.27 shows the flow chart of the main program.

Subsurface stress fields and the effect of friction on stresses have been analyzed by Yu and Bhushan (1996). For friction effects, the contact pressure and contact locations (contact points) are considered not to be altered by the presence of surface friction. This assumption of no effect on contact pressure is strictly true only when the two bodies are smooth or have the same elastic contacts. In the case of a rough-on-rough contact, asperity interactions may invalidate this assumption, although for the case of a smooth-on-rough contact with approximately equal moduli, it is considered a reasonable assumption. For the case of friction present at the interface, the tangential force at each contact is equal to its contact pressure multiplied by the coefficient of friction.

A three-dimensional model to analyze contact of two-layered nominally flat surfaces has been developed by Bhushan and Peng (2002). They extended analysis to elastic and elastic-plastic solids and included the effect of friction. In addition, they presented surface and subsurface stress distributions for layers of varying thicknesses and elastic moduli.

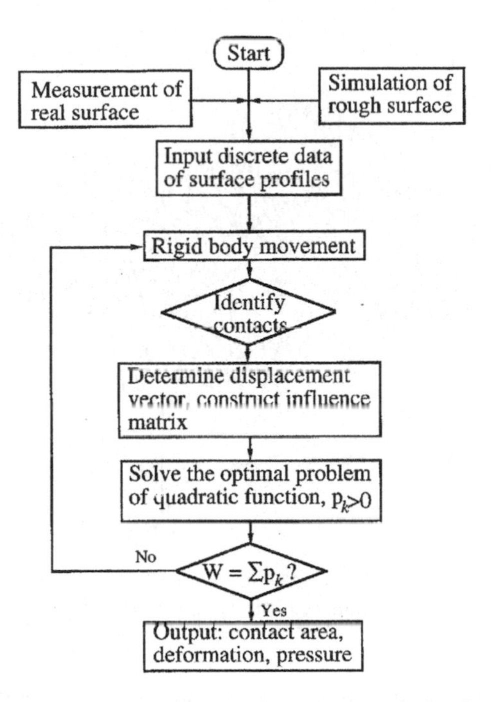

Figure 3.2.27 Flow chart of the computer program for contact analysis of two rough surfaces. Reproduced with permission from Tian, X. and Bhushan, B. (1996), "A Numerical Three-Dimensional Model for the Contact of Rough Surfaces by Variational Principle," *ASME J. Trib.* **118**, 33–42. Copyright 1996. ASME.

Simulation of Two Nominally Flat Surfaces

Three-dimensional Gaussian and non-Gaussian rough surfaces with given standard deviation of surface heights and correlation lengths were generated using a digital filter technique combined with a fast Fourier transform (FFT) (Hu and Tonder, 1992). A typical computer-generated three-dimensional Gaussian rough surface map shown in Figure 3.2.28, has σ of 1 nm and β^* is 0.5 μm; and the total scan size is 20 × 20 μm^2. The heights of the surface map consist of 256×256 data points. A numerical model based on the variational approach is used to predict the real area of contact and contact stresses between a rigid flat surface and a computer-generated or measured rough surface (Bhushan and Chilamakuri, 1996; Poon and Bhushan 1996a, 1996b; Tian and Bhushan, 1996; Yu and Bhushan 1996; Chilamakuri and Bhushan, 1998). Yu and Bhushan (1996) calculated surface and subsurface stresses. Figure 3.2.29 shows the contact pressure map at a nominal pressure of 32.8 kPa, typical of a magnetic head-disk interface. Contact pressure at the asperities is high as compared to the rest of the contact region. Under the applied load, a small number of contact spots (6) are obtained and the deformation is elastic for ceramic to ceramic contacts. Contours of von Mises stresses on the surface and subsurface (y = 11 μm) on and close to the maximum von Mises stress plane for frictionless and frictional contacts are shown in Figure 3.2.30. Maximum von Mises stress at both frictionless and frictional contacts occurs very close to the surface. The tensile principal stress distributions at and near the location of maximum von Mises stress are shown in

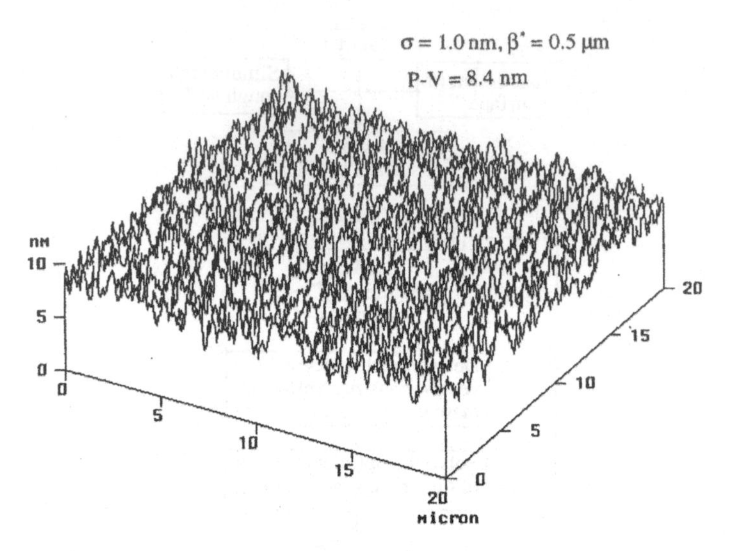

Figure 3.2.28 Computer generated surface profile: standard deviation of surface heights (σ) = 1 nm, correlation length (β^*) = 0.5 μm, and total scan size = 20 × 20 μm^2.

Figure 3.2.31. A pulse of high stress appears on the surface ($z = 0$) at the location of maximum $\sqrt{J_2}$, $x = 16.5$ μm and $y = 11$ μm. The friction effect is to increases the magnitude of the maximum stress but does not change the distribution.

Figure 3.2.32 shows the computed results of the real area of contact under various applied nominal pressures. As expected, real area of contact and number of contact points increase with nominal pressure. Bearing area of the contact surface corresponding to the same geometrical

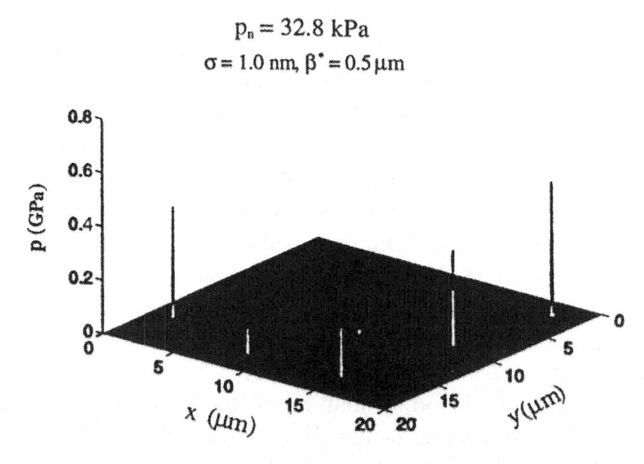

Figure 3.2.29 Contact pressure map of a computer generated rough surface (σ = 1 nm and β^* = 0.5 μm) on a rigid smooth flat surface (E^* = 100 GPa) at a nominal pressure of 32.8 kPa. Reproduced with permission from Yu, M.H. and Bhushan, B. (1996) "Contact Analysis of Three-Dimensional Rough Surfaces Under Frictionless and Frictional Contact," *Wear* **200**, 265–280. Copyright 1996. Elsevier.

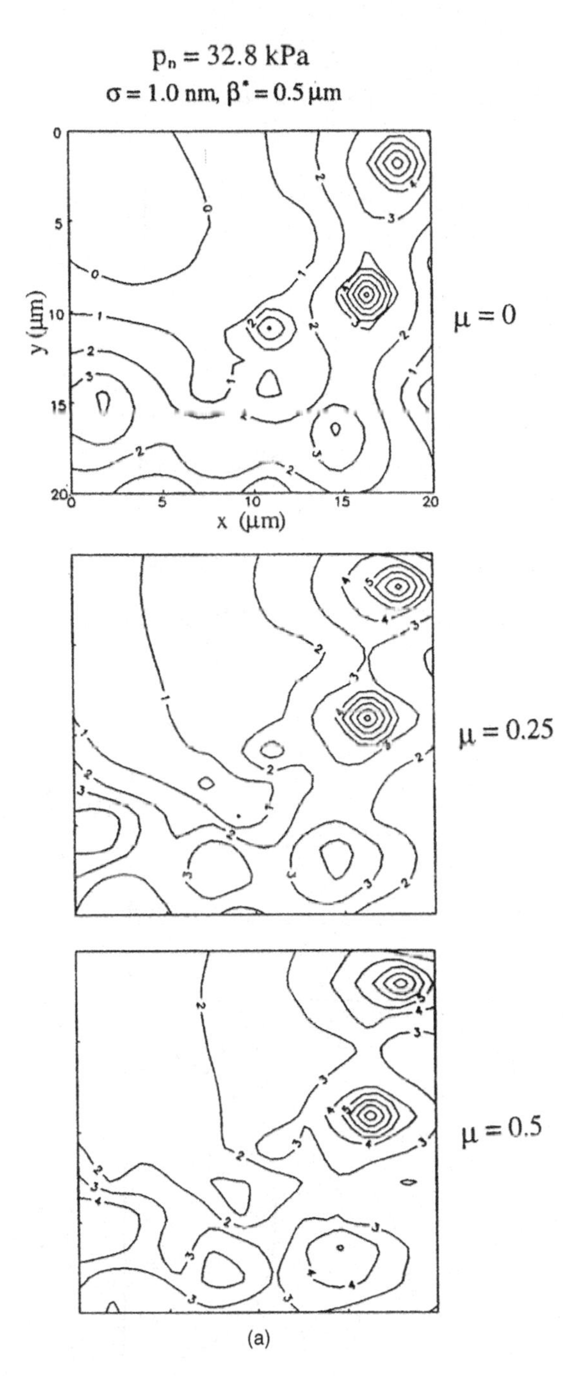

Figure 3.2.30 Contours of von Mises stresses (a) on the surface and (b) and (c) in the subsurface (y = 11 μm) at $\mu = 0$, 0.25 and 0.50, for the case of a rough surface (σ = 1 nm, β^* = 0.5 μm) on a rigid smooth flat surface (E^* = 100 GPa) at a nominal pressure of 32.8 kPa. The contour levels are natural log values of the calculated stresses expressed in kPa. Reproduced with permission from Yu, M.H. and Bhushan, B. (1996) "Contact Analysis of Three-Dimensional Rough Surfaces Under Frictionless and Frictional Contact," *Wear* **200**, 265–280. Copyright 1996. Elsevier. (*Continued*)

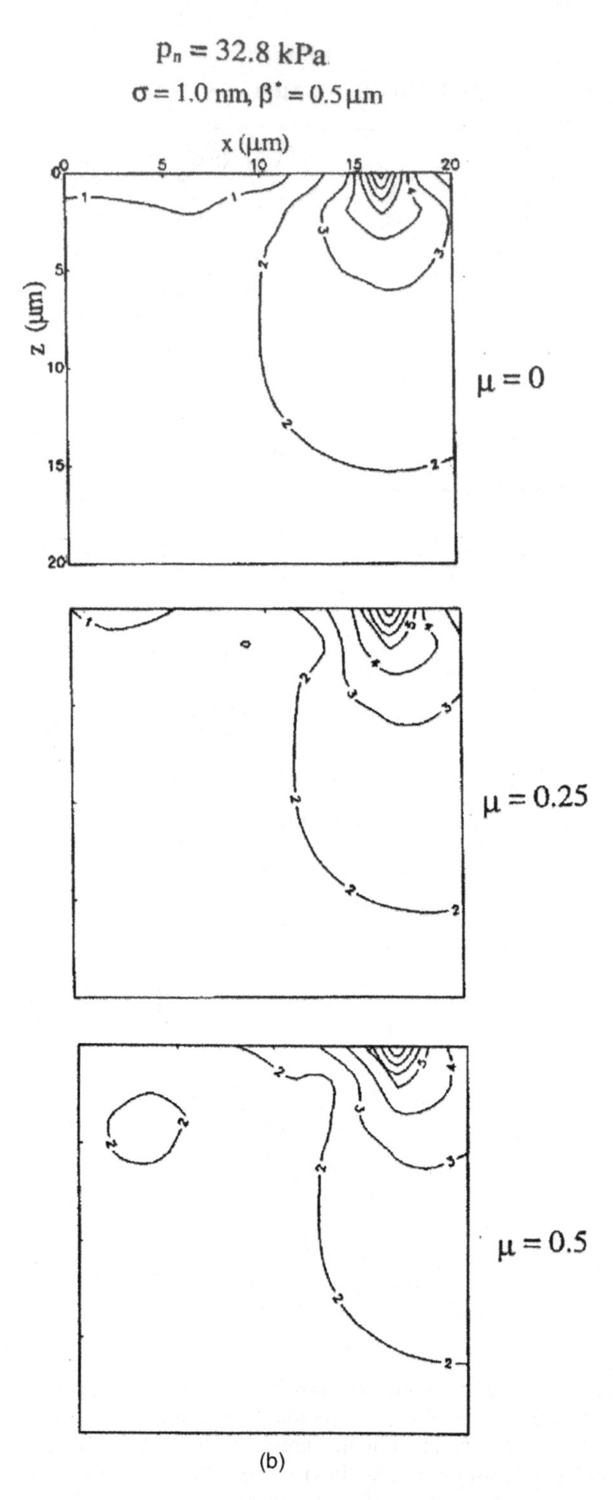

(b)

Figure 3.2.30 (*Continued*)

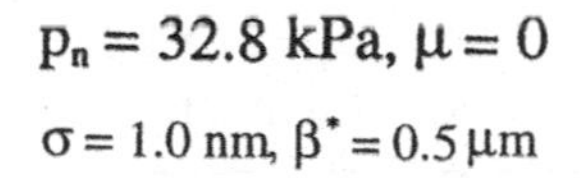

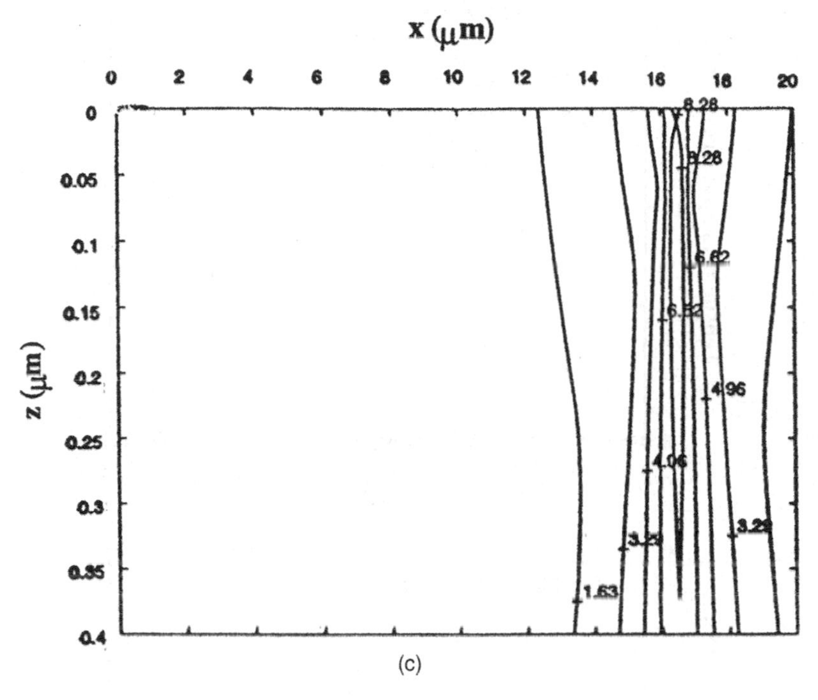

(c)

Figure 3.2.30 *(Continued)*

interference is given by slicing off the surface for a given interference. Tian and Bhushan (1996) showed that bearing area greatly overestimates the real area of contact because the elastic deformation is not included in bearing area calculations. The relationship between the maximum contact pressure and real area of contact and applied nominal pressure (load) for two rough surfaces is shown in Figure 3.2.33. It can be seen that the real area of contact is linearly proportional to the applied pressure when the pressure is relatively small. This relationship between the applied normal pressure and the real area of contact is consistent with what is commonly observed in engineering practice, i.e., the friction force is linearly proportional to the applied pressure. Figure 3.2.33 also shows the maximum contact pressure under different applied loads. We note that, contrary to common belief, contact pressure increases with applied nominal pressure. Figure 3.2.34 shows the contact pressure maps and contours of von Mises stresses at surface and subsurface ($y = 6.5$ μm) for two surfaces with $\sigma = 3$ nm and $\beta^* = 0.1$ μm, at a nominal pressure of 32.8 MPa (1000 × nominal pressure used in previous figures). Real area of contact is low and contact pressure and von Mises stresses are high for the surface with $\sigma = 3.0$ nm as compared to the other surface with $\sigma = 1.0$ nm. For the surface with $\sigma = 3.0$ nm, contact occurs at 1570 contact points, among them contact pressure at the surface and von Mises stress very close to the surface is very high at six points, which

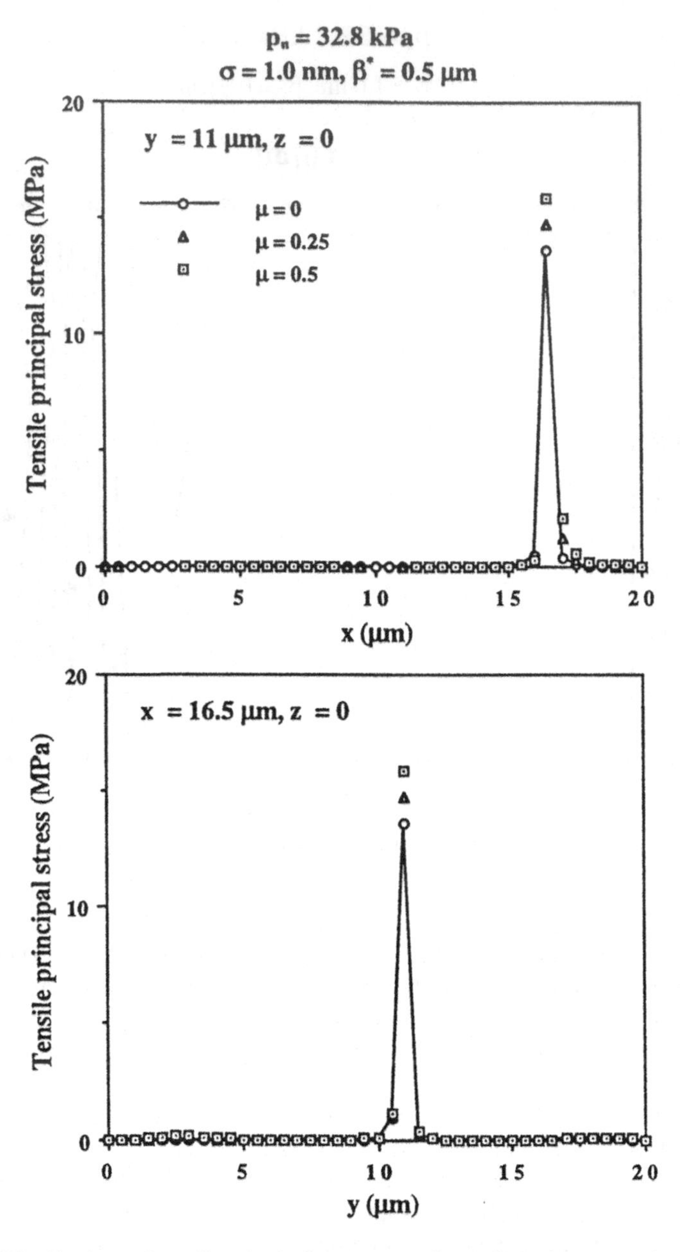

Figure 3.2.31 Distributions of tensile principal stresses on the surface along x- and y-axes at and near the location of maximum von Mises stress ($x = 16.5$ μm, $y = 11$ μm) for the case of a rough surface ($\sigma = 1$ nm, $\beta^* = 0.5$ μm) on a rigid smooth flat surface ($E^* = 100$ GPa) at a nominal pressure of 32.8 kPa. Reproduced with permission from Yu, M.H. and Bhushan, B. (1996) "Contact Analysis of Three-Dimensional Rough Surfaces Under Frictionless and Frictional Contact," *Wear* **200**, 265–280. Copyright 1996. Elsevier.

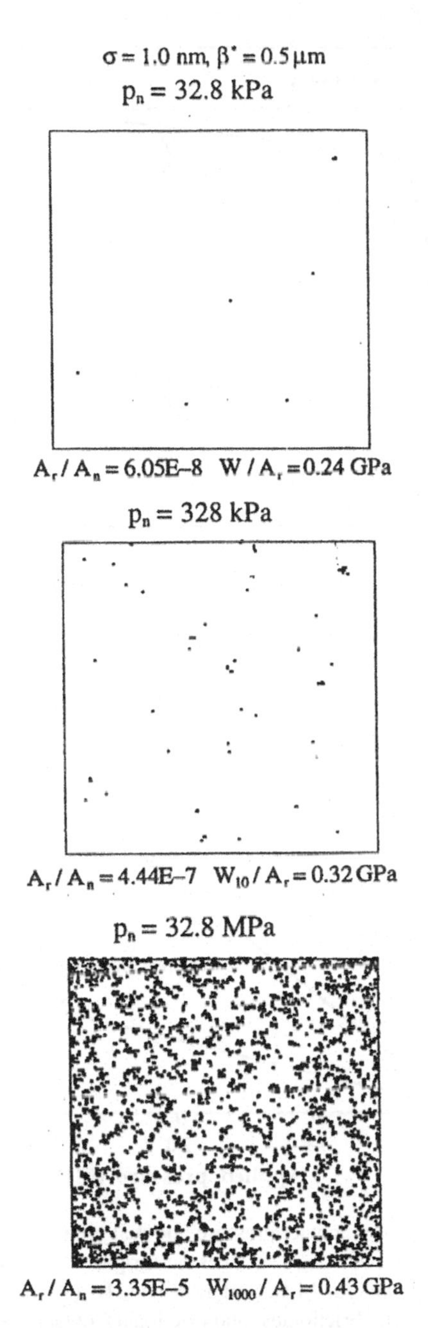

Figure 3.2.32 Two-dimensional images of real area of contact between a computer generated rough surface ($\sigma = 1$ nm, $\beta^* = 0.5$ μm) on a rigid smooth flat surface ($E^* = 100$ GPa) at three different nominal pressures. Reproduced with permission from Yu, M.H. and Bhushan, B. (1996) "Contact Analysis of Three-Dimensional Rough Surfaces Under Frictionless and Frictional Contact," *Wear* **200**, 265–280. Copyright 1996. Elsevier.

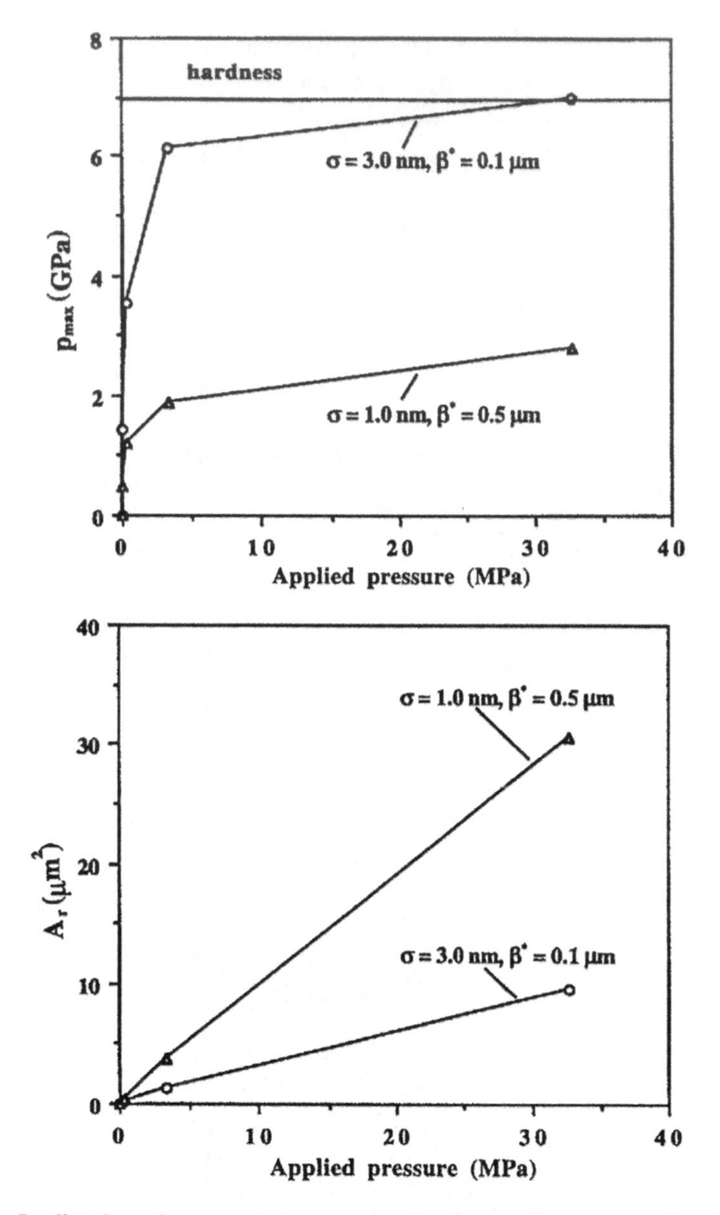

Figure 3.2.33 Predicted maximum contact pressure and real area of contact as a function of applied nominal pressure for the case of two rough surfaces on a rigid smooth flat surface ($E^* = 100$ GPa). Reproduced with permission from Yu, M.H. and Bhushan, B. (1996) "Contact Analysis of Three-Dimensional Rough Surfaces Under Frictionless and Frictional Contact," *Wear* **200**, 265–280. Copyright 1996. Elsevier.

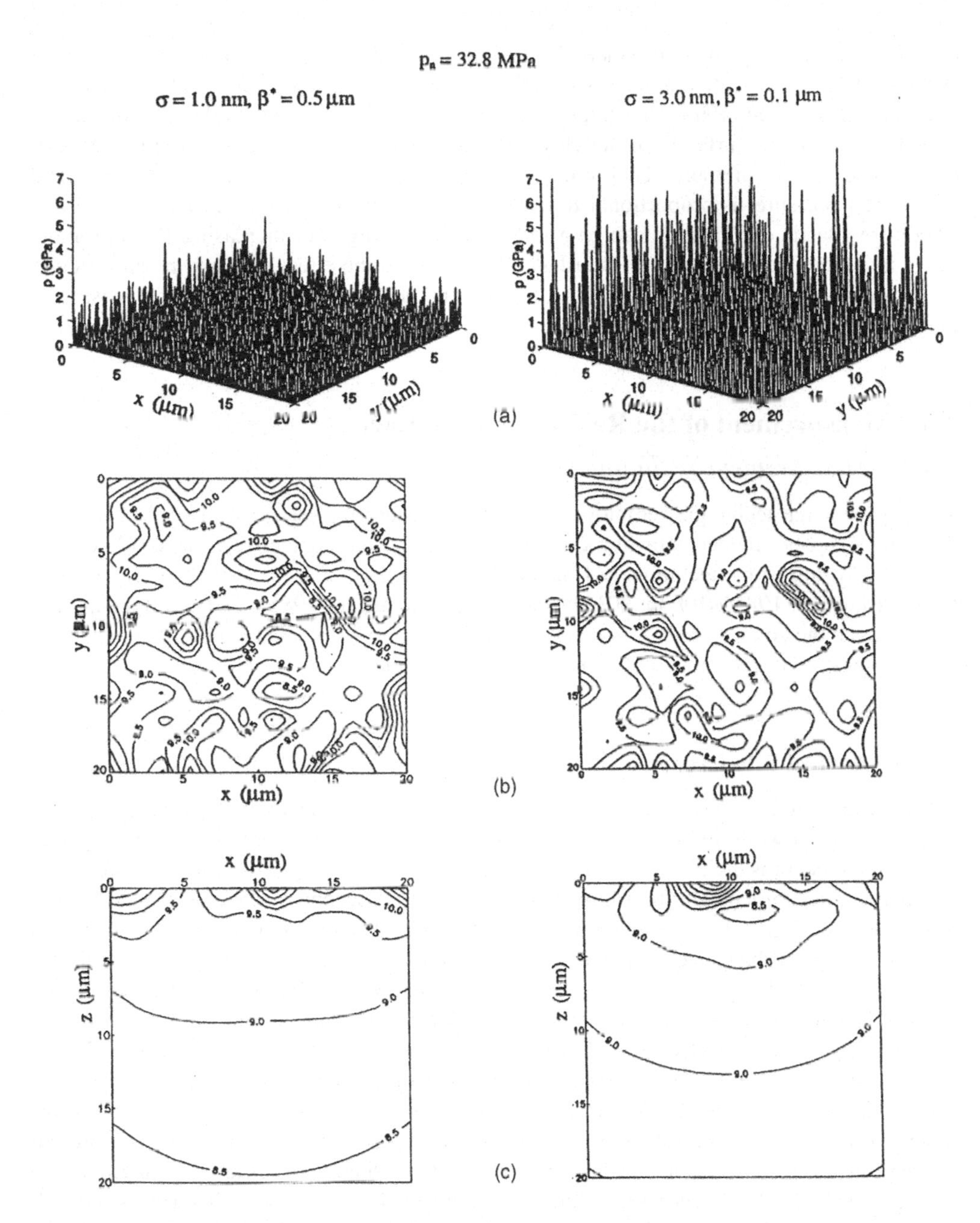

Figure 3.2.34 (a) Contact pressure maps and (b) contours of von Mises stresses on the surface and (c) in the subsurface at y = 6.5 μm at a nominal pressure of 32.8 MPa for two rough surfaces on a rigid smooth flat surface (E^* = 100 GPa). Reproduced with permission from Yu, M.H. and Bhushan, B. (1996) "Contact Analysis of Three-Dimensional Rough Surfaces Under Frictionless and Frictional Contact," *Wear* **200**, 265–280. Copyright 1996. Elsevier.

experience plastic deformation. Since the maximum von Mises stress occurs very close to the surface, plastic deformation and consequently wear will occur on and close to the surface.

Poon and Bhushan (1996c) developed computer-generated Gaussian surfaces with various σ and β^*. For these surfaces, in the elastic contact regime, they calculated the real area of contact and contact pressure. They found that contact area is inversely proportional to, and the contact pressure, is proportional to σ/β^*, as proposed by Onions and Archard (1973). Therefore, plastic deformation can be avoided by decreasing σ or increasing β^* to promote better interface durability. Yu and Bhushan (1996) studied the effect of σ and β^* and friction on subsurface stresses. Increase in σ and friction and decrease in β^* generally increased the magnitude of stresses.

3.3 Measurement of the Real Area of Contact

3.3.1 Measurement Techniques

The experimental techniques employed to date to measure the real area of contact in static conditions can be divided into five categories: (1) electrical-contact resistance; (2) optical methods; (3) ultrasonic technique; (4) neutrographic method; and (5) paints and radioactive traces (Bhushan, 1985a, 1996b, 2013). According to Bhushan (1985a, 2013), all techniques overestimate the contact area by as much as 400%. The most suitable technique for measurement of static contacts is the optical interference technique.

3.3.1.1 Optical Interference

If two partially reflecting surfaces, at least one of them being optically transparent, are placed in contact and a white light or a monochromatic light is projected through the transparent member, the beam is divided at the partially reflected surface and part of the light is reflected and part is transmitted. The transmitted light travels to a second surface and some of it is reflected back (Figure 3.3.1). Reflection from a rare to a dense medium causes a phase change of π but no phase change accompanies reflection from a dense to a rare medium. This is true in dielectric reflections; however, for metallic reflections (for example, metallic-film coated glass slides), the phase change is complex. If a glass slide (with no metallic coatings) is placed on a partially reflective dielectric surface in air, no phase change will occur at the air-dielectric surface. Hence, destructive interferences (black areas) for normal incidence in air will occur at zero thickness and at spacings of $m\lambda/2$, where λ is the wavelength and m is called the fringe order and takes on integer values of 0, 1, 2, 3, ... Zeroth orders of dark fringes represent the dark areas. In order to avoid the higher-order dark fringes, the peak-to-valley distance should be less than $\lambda/2$. Therefore, techniques work well only for smooth surfaces (Howell and Mazur, 1953; Bailey and Courtney-Pratt, 1955). For two-beam interference, the intensity distribution follows a sinusoidal distribution, and it is difficult to decide where the area of contact really ends. Some overestimation of the contact area is unavoidable. If the reflectivity of both surfaces is very high (for instance, metallized glass plates), higher-order reflected waves become significant and generate a larger number of multiple, internally reflected rays. Fringe visibility or contrast is very high in the case of multiple-beam interferences. A white light with a monochromatic filter is desirable.

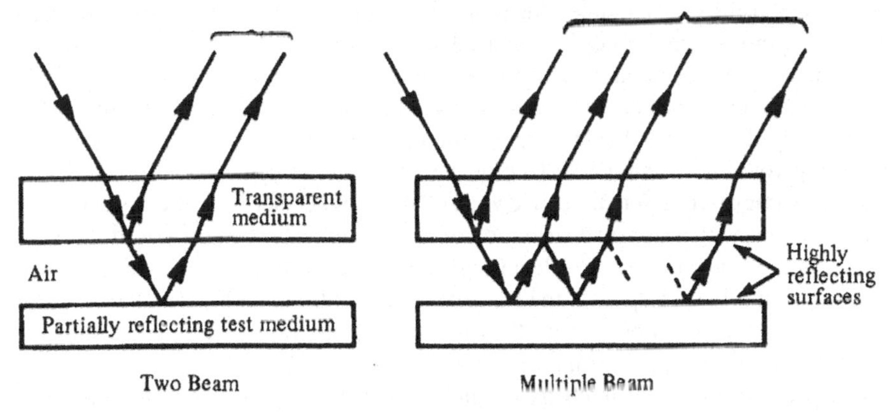

Figure 3.3.1 Schematic representation of the principles of optical interference techniques used for the measurement of the real area of contact.

3.3.2 *Typical Measurements*

Most measurements of the real area of contact have been made on interfaces involving magnetic tapes and rigid disks, using a two-beam optical interference technique with a white light source (Bhushan, 1985b, 1996b; Bhushan and Dugger, 1990; Bhushan and Lowry, 1995). Since the contact sizes in a softer tape are larger than a rigid disk, it is easier to make contact area measurements in a tape against a glass slide, with high accuracy. A schematic of the experimental apparatus for flat thin-film disks is shown in Figure 3.3.2. It consists of a thin plano-convex lens held around its circumference with the convex surface placed in contact

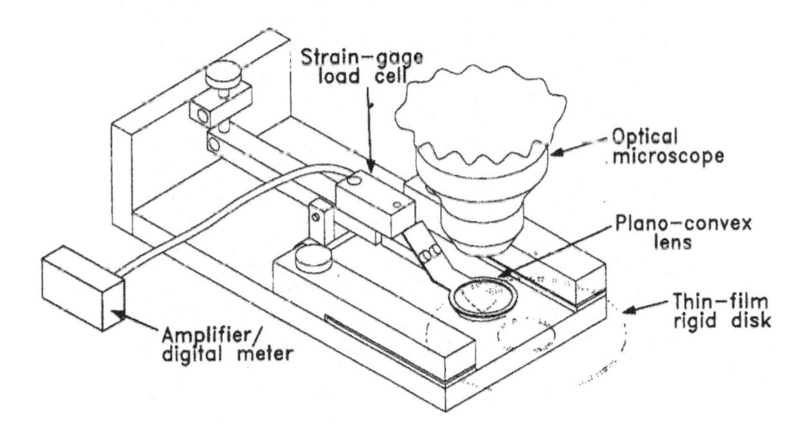

Figure 3.3.2 Schematic diagram of the experimental set-up for measuring the real area of contact between a flat disk and a convex lens, by the optical interference technique. Reproduced with permission from Bhushan, B. and Dugger, M.T. (1990), "Real Contact Area Measurements on Magnetic Rigid Disks," *Wear* **137**, 41–50. Copyright 1990. Elsevier.

with the disk and the contact zone viewed from the planar side of the lens using an optical microscope. A large-focal-length lens is used so that the normal pressure variation near the center of the contact zone could be neglected within the field of view (less than 10%). The lens is mounted on a suspension which is connected to the end of a rigid beam through a strain gauge load cell. The lens is loaded against the disk by rotating the threaded knob, causing the beam to pivot and the curved surface of the lens to be forced. The entire device rests on the stage of a microscope for direct viewing of the contact between the disk and the glass lens. Filtered light with a narrow distribution of wavelengths can be used for illumination, and destructive interference of beams reflected off areas of lens-disk contact and areas separated by an air gap result in dark spots at the areas of contact. Photographs are taken of the contact areas at a magnification ranging from 200x to 600x. Photographs are scanned into an image analysis program using a video frame grabber.

Because the distribution of intensity around the contacts is sinusoidal, the image analysis program is used to apply a gray-scale threshold to the image to eliminate all but the darkest regions. These are highlighted in the filtered image as light regions. Threshold values are generally kept constant for a given set of experiments. The image analysis program is used to obtain the total real area of contact as well as contact size distribution for each load. The lateral resolution (pixel size) of the measurements of images taken at a magnification of 600x is about 0.25 μm. Contact size distribution is obtained by assuming that a contact is circular and the diameter values are fitted to a log normal distribution.

A typical photograph showing the contact areas is shown in Figure 3.3.3. Only the darkest areas are read. Table 3.3.1 and Figure 3.3.4 show the various contact parameters for the thin-film rigid disk in contact with a plano-convex lens at four different loads. Statistical analysis was used with roughness data obtained with an optical profiler to predict contact

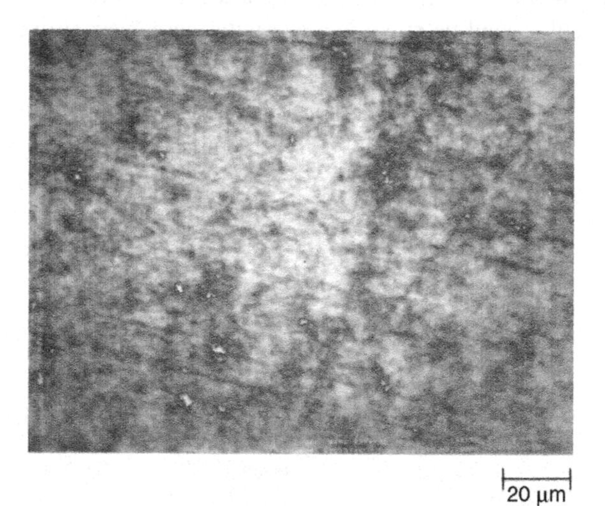

Figure 3.3.3 Photograph of the asperity contact areas of a thin-film disk in contact with a plano-convex lens at an applied load 500 mN (9.27 MPa at the contact center). Light regions correspond to contact spots. Reproduced with permission from Bhushan, B. and Dugger, M.T. (1990), "Real Contact Area Measurements on Magnetic Rigid Disks," *Wear* **137**, 41–50. Copyright 1990. Elsevier.

Table 3.3.1 Measured and predicted* values of the real area of contact of a thin-film rigid disk against a glass lens.

Normal load (mN)	Normal pressure (MPa)	A_r/A_a (%)		d_r (μm)		n/A_a (mm^{-2})	
		Meas.	Pred.	Meas.	Pred.	Meas.	Pred.
10	2.52	0.35	2.42	1.76	1.18	1376	22176
100	5.42	1.41	5.20	2.06	1.18	4196	47696
500	9.27	1.97	8.90	2.11	1.18	4458	81576
2000	13.72	2.74	13.13	2.04	1.18	4515	129540

*Parameters for the interface: $E^* = 110$ GPa, $\sigma_p = 2$ nm, $1/R_p = 2.24\ mm^{-1}$
Source: Bhushan and Dugger, 1990.

parameters and the predicted values are included in Table 3.3.1. Note that, as expected, new contacts are formed as the pressure is increased at low loads, however, the number of contacts remains roughly constant at high loads (Figure 3.3.4). Diameters, as measured and calculated from the microcontact analysis (Bhushan and Doerner, 1989), have a weak dependence on the normal load and are in fair agreement. Calculated values are slightly lower because, as we discussed earlier, the measurement technique overestimates the contact sizes. The measured values of number of contacts and total contact areas are about an order of magnitude smaller

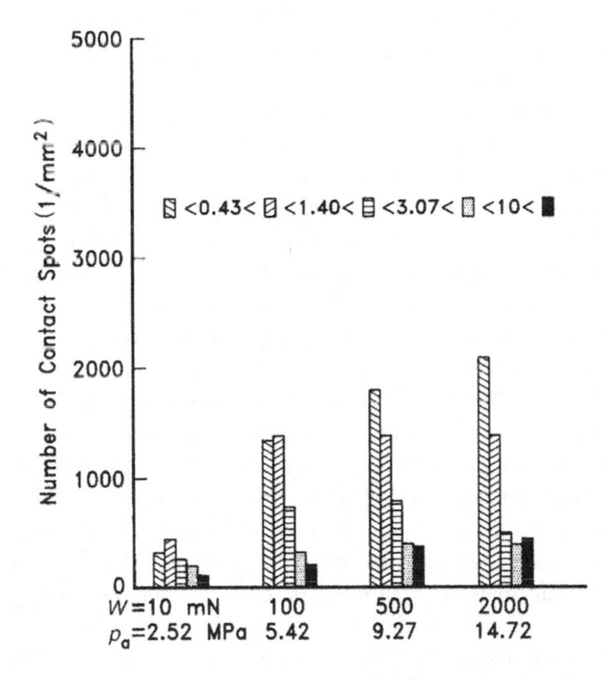

Figure 3.3.4 Histograms of asperity contact diameters at different pressures. Reproduced with permission from Bhushan, B. and Dugger, M.T. (1990), "Real Contact Area Measurements on Magnetic Rigid Disks," *Wear* **137**, 41–50. Copyright 1990. Elsevier.

than the predicted values. Since the lateral resolution of the image analysis technique is about 0.25 μm, any contact spots below this size are not detected; thus the number of contacts and the total contact area are underestimated. Observed trends of an increase in the measured values of number of contacts and total contact area with an increase in the normal load are as predicted.

Uppal *et al.* (1972) measured areas of metallic contacts using the Nomarski interference technique. They reported that the smallest contact spots were approximately circular and 3–4 μm in diameter, but became elliptical as they grew in size. For low loads, the real area of contact was proportional to applied load, and at high loads (0.025 times the hardness of the softer material), the real area of contact becomes proportional to $(\text{load})^{2/3}$, which resembles the relationship for a simple asperity Hertzian contact. They also reported that at low loads, the growth in total real area of contact occurred predominantly by emergence of first contact spots, however, at high loads, the growth occurred predominantly by increases in the areas of the individual contact spots. Similar observations have been made by Woo and Thomas (1980) based on a review of published data.

3.4 Closure

When the two nominally flat surfaces are placed in contact, surface roughness causes contact to occur at discrete contact spots. The sum of the areas of all the contact spots constitutes the real (true) area of contact. Deformation occurs in the region of the contact spots, establishing stresses that oppose the applied load. Relative sliding introduces a tangential force (referred to as the friction force) at the contact interface. The mode of surface deformation is either elastic, elastic-plastic, viscoelastic or viscoplastic.

If the shape of the asperity tips is assumed to be spherical, then a single asperity contact problem reduces to two spheres or one sphere of composite radius against a flat surface of composite mechanical properties in contact. Stresses and deformations of this problem in the elastic deformation regime can be obtained using the Hertz analysis. For this case, the contact area is proportional to the normal load raised to the exponent 2/3 and the contact pressure distribution is elliptical with the maximum pressure at the contact center; the average pressure is two-thirds of the maximum pressure. Radial, hoop, and normal stresses along the axis of symmetry are compressive inside the loaded circle; however, the radial stress is tensile just outside the loaded circle with a maximum at the edges of the circle which increases with an increase of the normal load. This tensile radial stress makes brittle material susceptible to ring cracks. For brittle materials, tensile stress may be more important than maximum shear stress at the subsurface. The maximum shear stress occurs at the axis of symmetry below the surface at about half of the contact radius. Based on the von Mises or Tresca yield criterion, plastic deformation is initiated at the maximum shear stress location. As the normal load is further increased, the plastic zone grows until the whole of the material surrounding the contact has gone through plastic deformation. A thin elastic core remains over the plastic zone below the central region of the contact interface prior to fully plastic deformation. The elastic core diminishes and eventually disappears as the load increases. The depth of the plastic zone is about twice the contact radius at the load.

Relative sliding introduces a frictional (tangential) force at the interface. As the coefficient of friction increases, the shear stress magnitude increases and the maximum no longer occurs on the axis of symmetry; it occurs near the surface and at the surface if the coefficient of friction exceeds 0.3. The contact pressure for which first yield occurs, decreases with an increase in the coefficient of friction. As the coefficient of friction increases for elastic deformation, the radially symmetrical tensile radial stress, which has a maximum occurring at the edge of contact at zero friction, becomes unsymmetrical, with compressive stress at the leading edge of the contact in the stationary body and greatly intensified tensile stress at the trailing edge whose magnitude increases with an increase in the coefficient of friction.

Elastic analysis of contact of the layered elastic half-space shows that the value of the layer Young's modulus relative to the substrate has a strong effect on the potential for yielding in both the layer and substrate as well as on the interfacial shear stresses. The von Mises stresses show significant discontinuities at the interface, and the maximum value for the sliding case moves closer to the surface as in the nonlayered case, with the layered case with a stiffer layer having a higher value. The maximum tensile stress on the surface depends strongly on both the coefficient of friction and the value of E_1 relative to E_2 in addition to the normal load. In general, for a fixed value of the substrate Young's modulus, the stresses increase for larger values of the layer Young's modulus. The stiffer layer further increases both the maximum tensile and compressive stresses on the surface. A stiffer layer can thus degrade the brittle failure characteristics of a layered medium, whereas a more compliant layer reduces the maximum tensile and compressive stresses and this can be beneficial. A more compliant layer further reduces the interfacial shear stresses. However, the elastic-plastic analyses of layered solids show that a stiffer and harder layer increases the load for inception of yielding for the substrate and reduces its plastic deformation, which is desirable for low wear. Thus, stiffer and harder layers may be more desirable for low wear in spite of higher surface tensile stresses and interfacial shear stresses. In order to make best use of stiffer layer, the layer thickness must exceed that of depth which undergoes plastic deformation. Optimization of the thickness and mechanical properties of the layer with respect to that of the substrate is necessary.

A rough surface is a random process and stochastic models are used to characterize and model the contact of two rough surfaces with significant and unrealistic assumptions. Nevertheless the model allows identification of important roughness parameters and mechanical properties which govern contact mechanics. Numerical models are more commonly used which can analyze contact of two rough surfaces with measured three-dimensional roughness maps and mechanical properties. Numerical models do not require any assumptions of surface isotropy, asperity shape, and distribution of asperity heights, slopes and curvatures. Further these can handle a large number of contact points (tens of thousands). These models are useful to study the effect of roughness distribution and mechanical properties on the real area of contact and surface and subsurface stresses.

In the elastic-contact situation of a simple asperity contact or if the number of contacts remains constant, the real area of contact is proportional to $(\text{load})^{2/3}$, whereas for the plastic-contact situation, it is proportional to load. For the multiple asperity-contact condition in two contacting rough surfaces, the real area of contact is proportional to load for both elastic and plastic contacts. An increase in normal load results in an increase in the number of contacts, which is responsible for an increase in the real area of contact. Whether contacts are elastic or plastic depends primarily on the mechanical properties and surface roughness and not load. In the case in which most of deformation at the asperities is elastic, fine asperities will go through plastic deformation. Contact sizes typically range from submicron to several microns with smaller ones for harder and rougher surfaces and lighter loads. The number of contacts can typically range from a few hundred to a few hundreds of thousands with a smaller number for harder and rougher surfaces and lighter loads. Contact pressures are typically two to six orders of magnitude larger than nominal pressure and real areas of contact are two to six orders of magnitude smaller than the apparent areas of contact.

A number of experimental techniques have been used to measure real area of contact. All techniques overestimate contact area by as much as 400%. The optical interference technique is the best choice with a lateral resolution of about 0.25 μm. The optical techniques can only be used for surfaces with good reflectivity. The technique provides contact-size distribution and the total area of contact in static conditions. Good repeatability of the measurement can be obtained by taking data of a large sample and at higher normal pressures.

Problems

3.1 A ceramic ball with a diameter of 5 mm is pressed into a steel ball of 10 mm in diameter under a normal load of 100 N. Calculate: (a) the minimum hardness of the steel ball for the contact to remain fully elastic; and (b) the radius of the contact zone. The given parameters are: $E_{ceramic} = 450$ GPa, $E_{steel} = 200$ GPa, $\nu_{ceramic} = 0.3$, $\nu_{steel} = 0.3$, and $H_{ceramic} > H_{steel}$.

3.2 Two nominally flat steel surfaces are lapped to give standard deviation of surface heights of 0.15 μm and correlation lengths of 1 μm. When the surfaces are placed in contact: (a) would you expect the asperity deformation to be elastic, predominantly plastic or elastic-plastic? (b) What is the real area of contact for a normal load of 100 N? Given parameters are: composite modulus $E^* = 110$ GPa and $H = 8$ GPa.

3.3 A polymer disk is placed in contact with a ceramic disk. Surface roughness of the disks was measured using a noncontact optical profiler (NOP) and atomic force microscope (AFM). Mechanical properties and surface roughness parameters of the two disks are given in the Table P3.3.

Calculate the plasticity indices using the NOP and AFM data. Why does one get different indices when using roughness obtained with different instruments? Calculate the real area of contact per unit load, number of contacts per unit load, mean asperity real area of contact, and contact pressure using the NOP data.

3.4 Two nominally flat steel surfaces are lapped to give standard deviation of peak heights of 0.2 μm and mean peak radius of 5 μm for each of the two surfaces. Calculate (a) the minimum value of hardness at which the deformation of the asperities is predominantly elastic, and (b) the maximum value of hardness at which the deformation of the asperities is predominantly plastic, given that $E^* = 110$ GPa.

Table P3.3

Component	E (GPa)	ν	H (GPa)	σ (nm)		σ_p (nm)		$1/R_p$ (1/mm)		η (1/mm^2)	
				NOP	AFM	NOP	AFM	NOP	AFM	NOP	AFM
Polymer disk	9.4	0.5	0.53	9.39	13.6	9.0	10.5	4.79	6.0×10^3	5.9×10^3	2.4×10^6
Ceramic disk	450	0.3	22.6	1.63	1.55	2.0	1.4	0.53	1.2×10^3	2.4×10^3	13.3×10^6

3.5 A rough surface with σ = 100 nm and β^* = 10 μm, slides upon a flat surface with E^* = 80 GPa and H_s = 4 GPa. The measured coefficient of friction μ is 0.1. (a) Calculate plasticity index for σ = 100 nm, 10 nm, and 1 μm. (b) Calculate the coefficient of friction if the rough surface is manufactured with σ = 10 nm. (c) Will coefficient of friction increase or decrease if the same surface is manufactured with σ = 1 μm? (Explain why.)

References

Abramowitz, M. and Stegun, I.A. (1965), *Handbook of Mathematical Functions*, pp. 685–720, Dover, New York.

Archard, J.F. (1957), "Elastic Deformation and the Laws of Friction," *Proc. Roy. Soc. Lond. A* **243**, 190–205.

Bailey, A.I. and Courtney-Pratt, J.S. (1955), "The Area of Real Contact and the Shear Strength of Monomolecular Layers of a Boundary Lubricant," *Proc. Roy. Soc. Lond. A* **227**, 500–515.

Bhushan, B. (1984), "Analysis of the Real Area of Contact Between a Polymeric Magnetic Medium and a Rigid Surface," *ASME J. Trib.* **106**, 26–33.

Bhushan, B. (1985a), "The Real Area of Contact in Polymeric Magnetic Media – I: Critical Assessment of Experimental Techniques," *ASLE Trans.* **28**, 75–86.

Bhushan, B. (1985b), "The Real Area of Contact in Polymeric Magnetic Media – II: Experimental Data and Analysis," *ASLE Trans.* **28**, 181–197.

Bhushan, B. (1996a), "Contact Mechanics of Rough Surfaces in Tribology: Single Asperity Contact," *Appl. Mech. Rev.* **49**, 275–298.

Bhushan, B. (1996b), *Tribology and Mechanics of Magnetic Storage Devices*, Second edition, Springer-Verlag, New York.

Bhushan, B. (1998), "Contact Mechanics of Rough Surfaces in Tribology: Multiple Asperity Contact," *Trib. Lett.* **4**, 1–35.

Bhushan, B. (1999), "Surfaces Having Optimized Skewness and Kurtosis Parameters for Reduced Static and Kinetic Friction," US Patent 6,007,896, Dec.28.

Bhushan, B. (2013), *Principles and Applications of Tribology*, Second edition, Wiley, New York.

Bhushan, B. and Blackman, G.S. (1991), "Atomic Force Microscopy of Magnetic Rigid Disks and Sliders and its Applications to Tribology," *ASME J. Trib.* **113**, 452–457.

Bhushan, B. and Chilamakuri, S. (1996), "Non-Gaussian Surface Roughness Distribution of Magnetic Media for Minimum Friction/Stiction," *J. Appl. Phys.* **79**, 5794–5996.

Bhushan, B. and Doerner, M.F. (1989), "Role of Mechanical Properties and Surface Texture in the Real Area of Contact of Magnetic Rigid Disks," *ASME J. Trib.* **111**, 452–458.

Bhushan, B. and Dugger, M.T. (1990), "Real Contact Area Measurements on Magnetic Rigid Disks," *Wear* **137**, 41–50.

Bhushan, B. and Lowry, J.A. (1995), "Friction and Wear Studies of Various Head Materials and Magnetic Tapes in a Linear Mode Accelerated Test Using a New Nano-Scratch Wear Measurement Technique," *Wear* **190**, 1–15.

Bhushan, B. and Majumdar, A. (1992), "Elastic-Plastic Contact Model of Bifractal Surfaces," *Wear* **153**, 53–63.

Bhushan, B. and Peng, W. (2002), "Contact Mechanics of Multilayered Rough Surfaces," *Appl. Mech. Rev.* **55**, 435–480.

Bhushan, B. and Tian, X. (1995), "Contact Analysis of Regular Patterned Rough Surfaces in Magnetic Recording," *ASME J. Electronic Packaging* **117**, 26–33.

Bhushan, B. and Venkatesan, S. (2005), "Effective Mechanical Properties of Layered Rough Surfaces," *Thin Solis Films* **473**, 278–295.

Bowden, F.P. and Tabor, D. (1950), *The Friction and Lubrication of Solids, Part I*, Clarendon Press, Oxford, UK.

Bowden, F.P. and Tabor, D. (1964), *The Friction and Lubrication of Solids, Part II*, Clarendon Press, Oxford, UK.

Bufler, H. (1959), "Zur Theorie der Rollenden Reibung," *Ing. Arch.* **27**, 137.

Bush, A.W., Gibson, R.D., and Thomas, T.R. (1975), "The Elastic Contact of a Rough Surface," *Wear* **35**, 87–111.

Bush, A.W., Gibson, R.D., and Keogh, G.P. (1979), "Strongly Aniostropic Rough Surfaces," *ASME J. Lub. Tech.* **101**, 15–20.

Chang, W.R., Etsion, I., and Bogy, D.B. (1987), "An Elastic-Plastic Model for the Contact of Rough Surfaces," *ASME J. Trib.* **109**, 257–263.

Chilamakuri, S.K. and Bhushan, B. (1997), "Optimization of Asperities for Laser-Textured Magnetic Disk Surfaces," *Tribol. Trans.* **40**, 303–311.

Chilamakuri, S.K. and Bhushan, B. (1998), "Contact Analysis of Non-Gaussian Random Surfaces," *Proc. Instn Mech. Engrs, Part J: J. Eng. Tribol.* **212**, 19–32.

Cook, N.H. and Bhushan, B. (1973), "Sliding Surface Interface Temperatures," *ASME J. Lub. Tech.* **95**, 59–63.

Davies, R.M. (1949), "Determination of Static and Dynamic Yield Stresses Using a Steel Ball," *Proc. Roy. Soc. A* **197**, 416–432.

Francis, H.A. (1977), "Application of Spherical Indentation Mechanics to Reversible and Irreversible Contact Between Rough Surfaces," *Wear* **45**, 221–269.

Gibson, R.D. (1982), "The Surface as a Random Process," in *Rough Surfaces* (T.R. Thomas ed.), Longman, London.

Greenwood, J.A. and Tripp. J.H. (1970–1971), "The Contact of Two Nominally Flat Rough Surfaces," *Proc. Instn Mech. Engrs* **185**, 625–633.

Greenwood, J.A. and Williamson, J.B.P. (1966), "Contact of Nominally Flat Surfaces," *Proc. Roy. Soc. Lond. A* **295**, 300–319.

Hamilton, G. M. (1983), "Explicit Equations for the Stresses Beneath a Sliding Spherical Contact," *Proc. Instn Mech. Engrs* **197C**, 53.

Hamilton, G.M. and Goodman, L.E. (1966), "Stress Field Created by a Circular Sliding Contact," *ASME J. App. Mech.* **33**, 371–376.

Hertz, H. (1882), "Uber die Beruhrung fester Elastische Korper und Uber die Harte (On the Contact of Rigid Elastic Solids and on Hardness)," Verhandlungen des Vereins zur Beforderung des Gewerbefleisses, Leipzig, Nov. 1882. (For English translation see Miscellaneous Papers by H. Hertz, Eds. Jones and Schott, MacMillan, London, 1896.)

Hill, R. (1950), *The Mathematical Theory of Plasticity*, Oxford University Press, London.

Hill, R., Storakers, B., and Zdunek, A.B. (1989), "A Theoretical Study of the Brinell Hardness Test," *Proc. Roy Soc. Lond. A* **423**, 301–330.

Howell, H.G. and Mazur, J. (1953), "Amontons' Law and Fibre Friciton," *J. Tex. Inst.* **44**, 159–169.

Hu, Y.Z. and Tonder, K. (1992), "Simulation of 3-D Random Surface by 2-D Digital Filter and Fourier Analysis," *Int. J. of Mach. Tool Manufact.* **32**, 82–90.

Huber, M. T. (1904), "Zür Theorie der Beruhrung fester Elastischer Korper," *Ann. der Phys.* **14**, 153–163.

Ishlinsky, A.J. (1944), *J. Appl. Math. Mech.* (USSR) **8**, 233.

Johnson, K.L. (1985), *Contact Mechanics*, Cambridge University Press, Cambridge, UK.

Kotwal, C.A. and Bhushan, B. (1996), "Contact Analysis of Non-Gaussian Surfaces for Minimum Static and Kinetic Friction and Wear," *Tribol. Trans.* **39**, 890–898.

Kral, E.R., Komvopoulos, K., and Bogy, D.B. (1995a), "Finite Element Analysis of Repeated Indentation of an Elastic-Plastic Layered Medium by a Rigid Sphere, Part I: Surface Results," *ASME J. App. Mech.* **62**, 20–28.

Kral, E.R., Komvopoulos, K., and Bogy, D.B. (1995b), "Finite Element Analysis of Repeated Indentation of an Elastic-Plastic Layered Medium by a Rigid Sphere, Part II: "Subsurface Results," *ASME J. App. Mech.* **62**, 29–42.

Lawn, B. (1993), *Fracture of Brittle Solids*, Second edition, Cambridge University Press, Cambridge, UK.

Lode, W. (1926), "Versuche ueber den Einfluss der mittleren Hauptspannung auf das Fliessen der Metalle Eisen Kupfer und Nickel," *Z. Physik* **36**, 913–939.

Majumdar, A. and Bhushan, B. (1991), "Fractal Model of Elastic-Plastic Contact Between Rough Surfaces," *ASME J. Trib.* **113**, 1–11.

Matthews, J.R. (1980), "Indentation Hardness and Hot Pressing," *Acta Met.* **28**, 311–318.

McCool, J.I. (1986a), "Predicting Microfracture in Ceramics via a Microcontact Model," *ASME J. Trib.* **108**, 380–386.

McCool, J.I. (1986b), "Comparison of Models for the Contact of Rough Surfaces," *Wear* **107**, 37–60.

Morton, W.B. and Close, L.J. (1922), "Notes on Hertz's Theory of Contact Problems," *Philos. Mag.* **43**(254), 320–329.

Nayak, P.R. (1973), "Random Process Model of Rough Surfaces in Plastic Contact," *Wear* **26**, 305–333.

Norbury, A.L. and Samuel, T. (1928), "Recovery and Sinking-In or Piling-Up of Material in the Brinell Test," *J. Iron and Steel Inst.* **117**, 673.

O'Callaghan, M. and Cameron, M.A. (1976), "Static Contact Under Load Between Nominally Flat Surfaces in Which Deformation is Purely Elastic," *Wear* **36**, 76–97.

Onions, R.A. and Archard, J.F. (1973), "The Contact of Surfaces Having a Random Structure," *J. Phys. D.: Appl. Phys.* **6**, 289–303.

O'Sullivan, T.C. and King. R.B. (1988), "Sliding Contact Stress Field Due to a Spherical Indenter on a Layered Elastic Half-Space," *ASME J. Trib.* **110**, 235–240.

Poon, C.Y. and Bhushan, B. (1996a), "Rough Surface Contact Analysis and its Relation to Plastic Deformation at the Head Disk Interface," *J. Appl. Phys.* **79**, 5799–5801.

Poon, C.Y. and Bhushan, B. (1996b), "Nano-Asperity Contact Analysis and Surface Optimization for Magnetic Head Slider/Disk Contact," *Wear* **202**, 83–98.

Poon, C.Y. and Bhushan, B. (1996c), "Numerical Contact and Stiction Analyses of Gaussian Isotropic Surfaces for Magnetic Head Slider/Disk Contact," *Wear* **202**, 68–82.

Sackfield, A. and Hills, D.A. (1983), "A Note on the Hertz Contact Problem: Correlation of Standard Formulae," *J. Strain Analysis* **18**, 195.

Tabor, D. (1951), *The Hardness of Metals*, Clarendon Press, Oxford, UK.

Tabor, D. (1970), "The Hardness of Solids," *Proc. Inst. Phys., F. Phys. in Technology.* **1**, 145–179.

Tabor, D. (1986), "Indentation Hardness and its Measurement: Some Cautionary Comments," in *Microindentation Techniques in Materials Science and Engineering* (P. J. Blau and B. R. Lawn, eds), pp. 129–159, ASTM, Philadelphia.

Thomas, T.R. (1999), *Rough Surfaces*, Second edition, Longman, London, UK.

Tian, X. and Bhushan, B. (1996), "A Numerical Three-Dimensional Model for the Contact of Rough Surfaces by Variational Principle," *ASME J. Trib.* **118**, 33–42.

Uppal, A.H., Probert, S.D., and Thomas, T.R. (1972), "The Real Area of Contact Between a Rough and a Flat Surface," *Wear* **22**, 163–183.

Whitehouse, D.J. and Archard, J.F. (1970), "The Properties of Random Surface of Significance in Their Contact," *Proc. Roy. Soc. Lond. A* **316**, 97–121.

Woo, K.L. and Thomas, T.Y. (1980), "Contact of Rough Surfaces: A Review of Experimental Work," *Wear* **58**, 331–340.

Yu, M.H. and Bhushan, B. (1996) "Contact Analysis of Three-Dimensional Rough Surfaces Under Frictionless and Frictional Contact," *Wear* **200**, 265–280.

Further Reading

Bhushan, B. (1996a), "Contact Mechanics of Rough Surfaces in Tribology: Single Asperity Contact," *Appl. Mech. Rev.* **49**, 275–298.

Bhushan, B. (1996b), *Tribology and Mechanics of Magnetic Storage Devices*, Second edition, Springer-Verlag, New York.

Bhushan, B. (1998a), "Contact Mechanics of Rough Surfaces in Tribology: Multiple Asperity Contact," *Trib. Lett.* **4**, 1–35.

Bhushan, B. (1998b), "Method of Texturing a Magnetic Recording Medium and Optimum Skewness and Kurtosis to Reduce Friction with a Magnetic Head," US Patent 5, 737, 229, April 7.

Bhushan, B. (1999), "Surface Having Optimized Skewness and Kurtosis Parameters for Reduced Static and Kinetic Friction," U.S. Patent 6, 007, 896, Dec. 28.

Bhushan, B. (2013), *Principles and Applications of Tribology*, Second edition, Wiley, New York.

Bhushan, B. and Peng, W. (2002), "Contact Mechanics of Multilayered Rough Surfaces," *Appl. Mech. Rev.* **55**, 435–480.

Bowden, F.P. and Tabor, D. (1950), *The Friction and Lubrication of Solids, Part I*, Clarendon Press, Oxford, UK.

Bowden, F.P. and Tabor, D. (1964), *The Friction and Lubrication of Solids, Part II*, Clarendon Press, Oxford, UK.

Goryacheva, I.G. (1998), *Contact Mechanics in Tribology*, Kluwer, Dordrecht, Netherlands.

Hills, D.A., Nowell, D., and Sackfield, A. (1993), *Mechanics of Elastic Contacts*, Butterworth-Heineman, Oxford, UK.

Hohm, R. (1967), *Electric Contacts Handbook*, Fourth edition, Springer-Verlag, New York.

Johnson, K.L. (1985), *Contact Mechanics*, Cambridge University Press, Cambridge, UK.

Ling, F.F. (1973), *Surface Mechanics*, Wiley, New York.

Popov, V.L. (2010), *Contact Mechanics and Friction*, Springer-Verlag, Berlin.

Timoshenko, S.P. and Goodier, J.N. (1970), *Theory of Elasticity*, Third edition, McGraw-Hill, New York.

4

Adhesion

4.1 Introduction

When two solid surfaces are brought into contact, adhesion or bonding across the interface can occur which requires a finite normal force, called adhesive force, to pull the two solids apart. A distinction must be made between adhesion and cohesion. Cohesion represents the atomic bonding forces associated within a material; that is, cohesion represents the forces that exist in the bulk of the material bonding one atom to another or one molecule to another. Thus, for example, if one cleaves a crystalline material in the bulk and generates two new surfaces, the bonds that are fractured are the cohesive bonds. When, however, two dissimilar (or even identical) materials are brought into solid-state contact with an interface, the bonding of the surface of one solid to that of another results in the formation of adhesive bonds. This is generally called adhesion as opposed to cohesion.

Again, adhesion is the phenomenon that occurs when two surfaces are pressed together, either under a pure normal force (load) or under combined normal and shear forces. A normal tensile force must be exerted to separate the surfaces, Figure 4.1.1. The ratio of the normal tensile force W' required for separation (normally referred to as adhesive force) to the normal compressive force W initially applied, is often referred to as the coefficient of adhesion, μ',

$$\mu' = \frac{W'}{W} \tag{4.1.1}$$

W' typically increases linearly with an increase of W and μ' generally increases with duration of static contact and separation rate.

Adhesion occurs both in solid–solid contacts and the two solids interposed with liquids or tacky solids. If two solid surfaces are clean and all of the chemical films and adsorbates are removed, strong adhesion or bonding of one solid to another generally occurs. Surface contaminants or thin films in many cases reduce adhesion; however, in some cases, the opposite may be true. With well-lubricated surfaces, weak adhesion is generally observed.

Adhesion can be either desirable or undesirable. Strong adhesion is required to bond the two surfaces together. In many engineering applications such as sliding and rotating machinery,

Introduction to Tribology, Second Edition. Bharat Bhushan.
© 2013 John Wiley & Sons, Ltd. Published 2013 by John Wiley & Sons, Ltd.

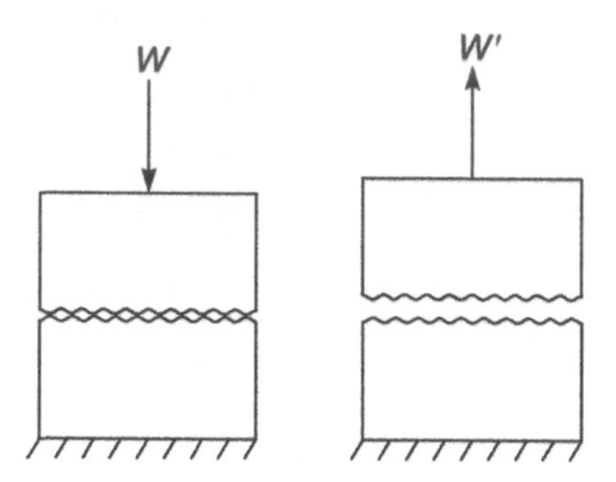

Figure 4.1.1 Schematic illustration of normal pull of two solid bodies; W is the compressure normal force (load) applied for a certain duration and W' is the tensile normal force needed to separate surfaces.

however, adhesion is undesirable. Adhesion results in friction and wear. In some applications, controlled adhesion is required.

4.2 Solid–Solid Contact

Proximity of the asperities results in an adhesive joint caused by interatomic attractions. In a broad sense, adhesion is considered to be either physical or chemical in nature (Bikerman, 1961; Zisman, 1963; Houwink and Salomon, 1967; Mahanty and Ninham, 1976; Derjaguin *et al.*, 1978; Buckley, 1981; Anonymous, 1986; Israelachvili, 1992; Bhushan, 1996, 2003; Maugis, 2000). A chemical interaction involves covalent bonds, ionic or electrostatic bonds, and metallic bonds; and physical interaction involves the hydrogen bonds and van der Waals bonds as a result of intermolecular forces (secondary forces of attraction). Hydrogen and van der Waals bonds are much weaker than that in the molecules that undergo chemical interaction because in secondary bonds, there is no electron exchange. The van der Waals forces are always present when two asperities are in close proximity. For two solid surfaces in contact, the interfacial bond may be stronger than the cohesive bond in the cohesively weaker of the two materials. In that case, on separation of the two solids, this results in the transfer of the cohesively weaker material to the cohesively stronger. In the example shown in Figure 4.2.1, gold contacted a single-crystal silicon surface and during separation, gold transferred to the silicon surface. Adhesion is a function of material pair and interface conditions such as crystal structure, crystallographic orientation, solubility of one material into another, chemical activity and separation of charges, surface cleanliness, normal load, temperature, duration of contact (rest time or dwell time), and separation rate (e.g., Sikorski, 1963; Buckley, 1981).

For clean surfaces, free from oxide and other surface films and from adsorbed gases, significant adhesion is observed between metal surfaces; such conditions can be achieved under ultra-high vacuum. Surface films, such as physisorbed, chemisorbed and chemically reacted films, and contaminants in the environment, generally decrease the adhesion of two reactive surfaces (Coffin, 1956; Bowden and Rowe, 1956; Johnson and Keller, 1967; Buckley, 1981). When exposed to ambient air, even noble metals adsorb oxygen and water vapor; this film may not be more than a few molecules thick. Small amounts of contaminants may be much more effective in reducing the adhesion of some metals than of others. For example, a

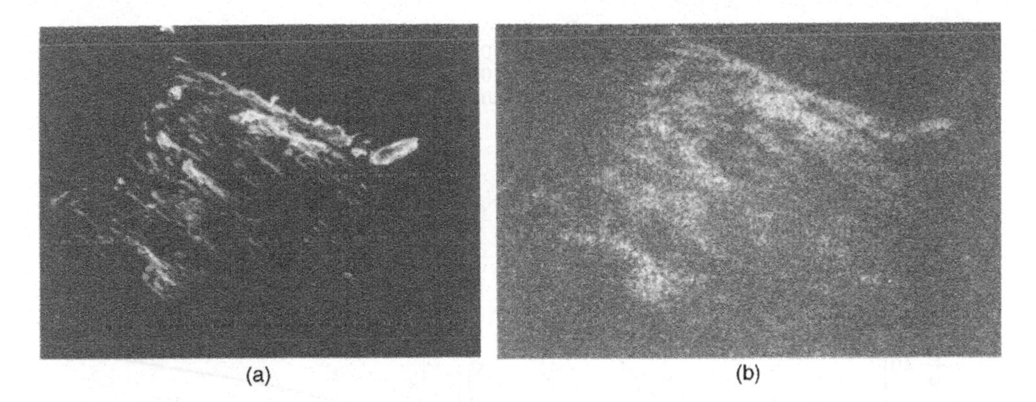

Figure 4.2.1 Silicon (111) surface after adhesive contact with gold (300 mN, 23°C, 10^{-8} Pa) showing (a) SEM micrograph of the transfer and (b) X-ray map for the transferred gold. Reproduced with permission from Buckley, D.H. (1981), *Surface Effects in Adhesion, Friction, Wear and Lubrication*, Elsevier, Amsterdam.Copyright 1981. Elsevier.

very small amount of oxygen (perhaps enough to give a monolayer) can produce a marked reduction in the adhesion of iron, whereas far more oxygen is required to produce a comparable reduction in the adhesion of copper.

Temperature affects the adhesive strength of a contact. At high temperatures, softening of surfaces result in greater flow, ductility and a larger real area of contact which results in stronger adhesion. High temperatures can also result in diffusion across the interface. In a metal–metal contact, high temperature may result in increased solubility, and in a polymer–polymer contact, interdiffusion strengthens the contact, which results in stronger adhesion.

If two surfaces are placed together, because of surface roughness, the real area of contact is usually very much smaller than the geometrical area. Adhesion is affected by the real area of contact, which is a function of normal load, surface roughness and mechanical properties (see Chapter 3). Adhesion force generally increases linearly with an increase in the normal load, Figure 4.2.2a (McFarlane and Tabor, 1950). Materials with higher roughness, modulus of elasticity and/or hardness and lack of ductility exhibit lower real area of contact, which leads to lower adhesion. Any viscoelastic or viscoplastic deformation (creep) under load would increase the real area of contact as a function of duration of contact leading to an increase in adhesion, Figure 4.2.2b (McFarlane and Tabor, 1950; Moore and Tabor, 1952). The real area of contact can also increase as a result of interatomic attraction (van der Waals or VDW forces) in the case of a soft solid, such as elastomer, that is in contact with a hard surface, both being smooth so that the asperity separation is on the order of molecular levels (1–10 nm) (Bhushan *et al.*, 1984). Contact first occurs at the tip of the asperities, as given by the analysis presented in Chapter 3. These are then drawn closer as a result of the van der Waals forces, with a normal pressure on the order of 1 atm, when asperity contacts are separated by 1–10 nm. This process goes on and may result in a very large contact area at no normal loads (Figure 4.2.3). This mechanism is also partially responsible for the behavior of thin polymer films, such as clingfilm wrap. Of course, this mechanism would be inoperative for hard material pairs and/or rough surfaces.

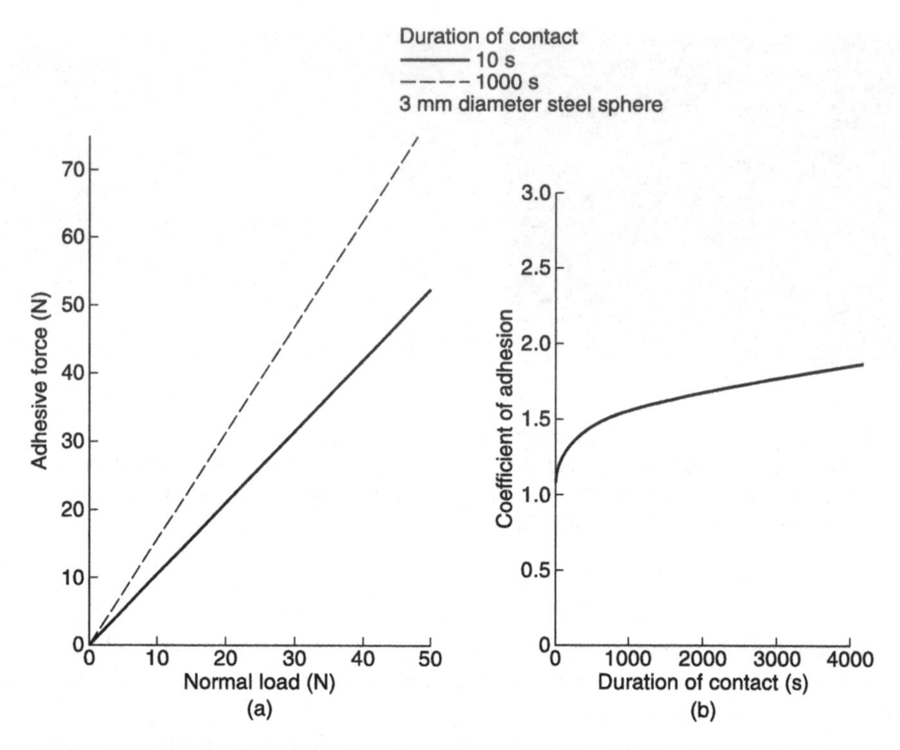

Figure 4.2.2 (a) Adhesive force as a function of normal load; and (b) coefficient of adhesion as a function of duration of contact for a clean steel sphere on indium. *Source*: McFarlane, J.S. and Tabor, D. (1950), "Adhesion of Solids and the Effects of Surface Films," *Proc. R. Soc. Lond. A* **202**, 224–243, by permission of the Royal Society.

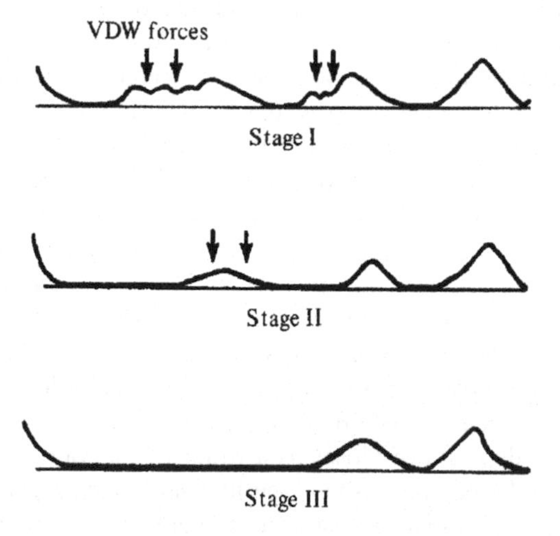

Figure 4.2.3 Diagram indicating how the real area of contact between a smooth elastomer and a smooth hard surface grows to a larger fraction of the geometric area.

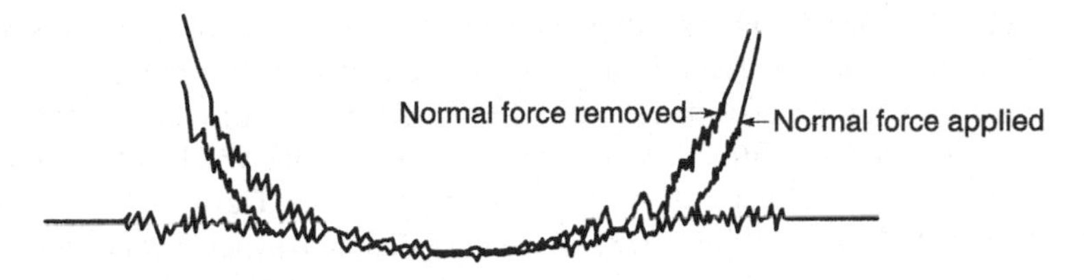

Figure 4.2.4 Schematic showing a sphere on a nominally flat surface with normal force applied and the force removed.

Another consideration in the real area of the contact is elastic recovery. When a normal force is decreased from two surfaces in intimate contact, contact is partially peeled apart by elastic forces in a process known as elastic recovery, Figure 4.2.4 (Bowden and Rowe, 1956). A lower elastic modulus would result in less elastic recovery and vice versa. Ductility also plays a role: the greater the ductility, the greater the elongation of the contacts and, therefore, less elastic recovery. Therefore, elasticity and ductility affect the real area over which adhesion occurs and influence adhesion and friction. Elastic recovery, to a large extent, is responsible for lower adhesion of clean interfaces than the theoretical values.

Adhesive forces significantly increase if a shear displacement (force) is added in addition to the normal load. When a tangential force is applied to the loaded metallic specimens, there is a growth in the real area of contact by plastic flow under the influence of combined normal and tangential stresses (see Chapter 5) and any relative sliding tends to produce penetration of surface layers that otherwise prevent metal-to-metal contact (Sikorski, 1963; Bowden and Rowe, 1956). Even hard metals subjected to sliding or twisting after being pressed can exhibit high adhesion.

Now, we discuss various surface interactions which are responsible for solid–solid adhesion.

4.2.1 Covalent Bond

A covalent bond consists of a pair of electrons (of opposite magnetic spins) shared between two atoms. When covalent solids are brought into intimate contact, one might expect the bonding across the interface to be similar to the bonding within the solid. However, there is some evidence that the bonds on the free surface are relaxed and that a finite amount of energy is required to activate them. Most covalent solids have a high elastic modulus and are generally extremely hard. Consequently it is often difficult to obtain large areas of contact even if appreciable joining loads are employed. However, molecularly smooth surfaces can result in high real area of contact, leading to high adhesion.

4.2.2 Ionic or Electrostatic Bond

Ionic bonds are formed whenever one or more electrons are transferred from one atom to another. Transfer of electrons results in the formation of negative and positive ions. Coulombic attraction of unlike ions results in the formation of ionic bonds (Callister, 2007; Hein and Arena, 2010). Metals, which have relatively little attraction for their valence electrons, tend to form ionic bonds when they combine with nonmetals. When the separation equals the atomic spacing, the bond resembles that within the bulk of the material. If a polymer (insulator) is brought into contact with a metal, there is a far larger separation of charge at the interface. This produces an electrostatic attraction in addition to the van der Waals interaction between the bodies (Johnsen and Rahbek, 1923; Skinner *et al.*, 1953; Davies, 1973; Wahlin and Backstrom, 1974; Derjaguin *et al.*, 1978). Based on detailed experiments with polymers, Derjaguin *et al.* (1978) stated that practically the whole of the adhesion is electrostatic in origin. These nonequilibrium charges will decay with time and do not result in permanent adhesion.

Transfer of charge occurs by contact and separation of two surfaces. Certain material combinations, generally nonconductive materials, become electrically charged, by friction, being rubbed. This effect is commonly referred to as the "triboelectric effect," and is a common source of static charge generation. Being electrically charged, either negatively or positively, upon contact with an uncharged object or one of opposite polarity, there may be a discharge of static electricity, a spark. These nonequilibrium static charges will decay with time and do not result in permanent adhesion.

4.2.3 Metallic Bond

The valence electrons of metals are not bound to any particular atom in the solid and are free to drift throughout the entire metal, referred to as delocalized electrons. They form a sea of electrons or an electron cloud. The remaining nonvalence electrons and atomic nuclei form ion cores which possess a net positive charge, equal in magnitude to the total valence electron charge per atom. The free electrons shield the positive ion cores from mutually repulsive electrostatic forces. The metal can be viewed as containing a periodic structure of positive ions surrounded by a sea of valence electrons (negative). The attraction between the two provided the metallic bond (Callister, 2007; Hein and Arena, 2010).

Broadly speaking, most clean metals stick strongly to one another. For separations greater than, say, 2 nm, they are attracted by van der Waals forces, which increase as the separation decreases. At a small separation, the metallic bond begins to develop. When the surfaces are at an atomic distance apart, the full metallic bond is generally formed and the short-range repulsive forces also come into operation to provide final equilibrium between the two bodies. If clean identical metals (e.g., gold) are pressed together with a force to produce plastic deformation at the contact region, one would expect the interfacial strength comparable with that of bulk metal so that the force required to pull two surfaces apart should be large; it is always appreciably less. The effect of released elastic stresses, surface roughness and degree of cleanliness are some of the reasons for adhesive strength being lower than expected. The ductility of the metals is important, particularly if the loading is sufficient to produce plastic deformation. Adhesion of ductile materials such as indium, lead, copper and gold is generally stronger than for less ductile metals, for example, the hexagonal metals with a small number of slip systems and ceramics.

The self-adhesion of a wide range of metals seems to fall into fairly well-defined groups, depending on structure. For example, hexagonal metals form a self-consistent, poorly adhering group; cobalt (hcp) exhibits markedly low adhesive forces when brought in contact with itself.

Table 4.2.1 Some properties of various metals and force of adhesion of these metals to (011) iron. Applied normal force = 200 μN, diameter of contacting flat = 3 mm, temperature = 20°C, ambient press. = 10^{-8} Pa, contact duration = 10 s (*Source*: Buckley, 1981).

Metals	Cohesive energy (kJ/g atom)	Free surface energy (mJ/m^2)	Atomic size (nm)	Solubility in iron (at %)	Adhesive force to iron (μN*)
Clean					
Iron	405	1800	0.286	100	>4000
Cobalt	426	1800	0.250	35	1200
Gold	366	1200	0.288	<1.5	500
Copper	338	1300	0.255	<0.25	1300
Aluminum	323	1000	0.280	22	2500
Lead	197	500	0.349	Insoluble	1400
With H_2S Adsorption					
Iron	–	–	–	–	100

*10 μN = 1 dyne

In general, similar metal pairs with non-hexagonal structures are metallurgically compatible and exhibit high adhesion and must be avoided, particularly iron against iron.

The orientation at the surface influences adhesive behavior. Contact of similar planes exhibit higher adhesive bonding forces than dissimilar crystallographic planes of the same metal in contact with itself. The lowest adhesion force is found on the close-packed, high atomic density and low free surface energy planes (to be discussed later). The polycrystalline form of a metal in contact with itself exhibit higher adhesive forces than single crystals in contact with themselves; this reflects the influence of grain boundary energies.

In the case of dissimilar metals, the mutually solubility of metals would affect adhesion; mutually insoluble metals would generally show poor adhesion (Keller, 1963, 1972; Rabinowicz, 1995). However, if the surfaces are thoroughly clean, regardless of mutual solubility, the adhesion would be strong. In general, but not always, transfer occurs from the softer metal to the harder metal. With some alloys, preferential segregation of one of the constituents could occur at the free surface.

Table 4.2.1 presents adhesion data for various metal–metal pairs. A clean iron surface against another iron is high. Surface film by adsorption of H_2S decreases the adhesive force dramatically. Cohesion or self-adhesion gives much stronger forces than does the adhesion of any other metal to iron. An increase in solubility does not always result in an increase in the adhesive forces. The other parameters that correlate with the observed adhesive force are the cohesive energy and free surface energy of the metals. This is not surprising, since both the cohesive and surface energies are measures of the strength of interatomic forces. Lead is insoluble, but being soft results in a large real area of contact responsible for high adhesion. Aluminum, being soft, also results in a large real area of contact and high adhesion. These observations demonstrate the importance of ductility. Strong adhesion of transition metal aluminum to iron has also been related to the nature of the d valence bond character or the chemical activity.

4.2.4 Hydrogen Bond

Hydrogen can exist both as a positively charged and as a negatively charged ion. The positive hydrogen ion, or proton, results from the removal of the only electron. The negative ion, on the other hand, is formed by the imperfect shielding of the positively charged nucleus by the single electron in the neutral atom. This imperfect shielding will result in a constantly shifting dipole that has a weak tendency to acquire another electron by purely ionic attraction. This property of the hydrogen atom enables it to bridge two negative ions in what is known as a hydrogen bond (Bhushan, 1996, 2003). It plays an important role in adhesion with polymers if there are certain polar atoms present capable of producing hydrogen bonding. Hydrogen bonds or hydrogen bridges are the strongest secondary forces of attraction.

Hydrophilic silica surfaces in microelectromechanical systems (MEMS) contain adsorbed water layers. When two of these hydrated surfaces are brought into close contact, hydrogen bonds may form between oxygen and the hydrogen atoms of the absorbed water layers. Hydrogen bonds are productively used in wafer bonding.

4.2.5 van der Waals Bond

The three types of bonding mentioned so far are all relatively strong primary bonds. Weaker, secondary bonds, which also result in interatomic attraction, are van der Waals forces. These act between molecules or within molecules with atoms between which chemical bonds have not formed. With polar molecules they arise from dipole–dipole interactions. With nonpolar molecules, they arise from the interaction of fluctuating dipoles in the individual atoms (London forces). Existence of van der Waals (VDW) forces between macroscopic bodies, such as crossed mica cylinders, has been measured by several investigators (Derjaguin *et al.*, 1987; Israelachvili, 1992). The effect of surface roughness on VDW forces has been studied by Meradudin and Mazur (1980). Based on calculations, they found that surface roughness increases the magnitude of the van der Waals force over its value when the two surfaces are smooth.

Assuming that the contact region can be modeled with two parallel plates, equations can be used to calculate attractive forces over the contact region. Figure 4.2.5 shows the calculated values of VDW and the electrostatic attractive forces per unit area exerted on the mica plates as a function of separation. Because of the $1/x^3$ dependence of VDW forces, they are only of significance in the region of true contact, for center-to-center separations of 0.6 nm (R_0 or twice the typical lattice spacing) to 20 nm. VDW forces are smaller than electrostatic forces.

4.2.6 Free Surface Energy Theory of Adhesion

A detailed calculation of van der Waals forces is difficult. A simpler approach is to use the concept of free surface energy. If one cleaves a crystalline solid along its cleavage plane, two highly chemically active surfaces are generated. The cleavage process causes the fracture of cohesive bonds across the cleavage interface, and these fractured bonds leave the surface in a highly energetic state. Thus, the energy that normally would be associated with bonding to other atoms (like other atoms in the bulk solid) is now available at the atoms on the surface. This energy required to create new surface, expressed over an area consisting of many atoms

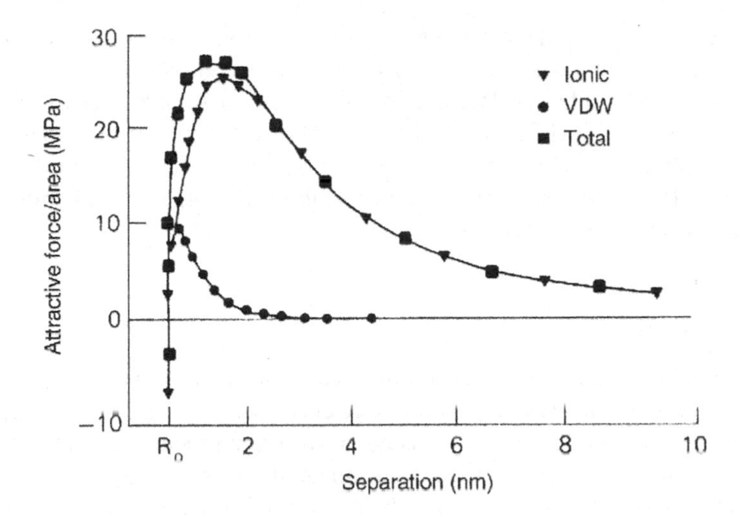

Figure 4.2.5 Ionic and VDW forces per unit area and total attractive force per unit area as a function of separation between two plane, parallel mica sheets at any point. Reproduced with permission from Bailey, A.I., and Daniels, H. (1972), "Interaction Forces Between Mica Sheets at Small Separations," *Nature Phys. Sci.* **240**, 62–63. Copyright 1972. Nature Publishing Group.

in the surface lattice, is referred to as free surface energy. It is a function of the material as well as the surface orientation.

Because the atoms at the surface have this unused energy, they can interact with each other, with other atoms from the bulk, and with species from the environment. Free surface energy influences adhesive bonds for solids in contact and, hence, friction and wear. In addition, it determines the nature of the interaction of lubricants with solids. When a bond is formed between two materials (having free surface energies per unit area in air $(\gamma_{SA})_1$ and $(\gamma_{SA})_2$ or simply γ_1 and γ_2) in contact, the surface energy of the interface per unit area changes to γ_{12}. Based on early work by Bradley (1932) and Bailey (1961), work of adhesion or the energy of adhesion per unit area is defined as:

$$W_{ad} = \Delta\gamma = \gamma_1 + \gamma_2 - \gamma_{12} \tag{4.2.1}$$

$\Delta\gamma$ is equal to a reduction in the surface energy of the system per unit area (always negative), in mJ/m^2, erg/cm^2, dynes/cm or mN/m (1 mJ/m^2 = 1 erg/cm^2 = 1 dyne/cm = 1 mN/m). Thus, $\Delta\gamma$ represents the energy that must be applied to separate a unit area of the interface or to create new surfaces. For two similar materials, $\Delta\gamma$ becomes the work of cohesion, equal to 2γ ($\gamma_{12} = 0$). This important thermodynamic relation (Equation 4.2.1) is valid for both solid and liquid interfaces. γ is generally called free surface energy for solids and surface tension for liquids. McFarlane and Tabor (1950) and Sikorski (1963) have reported a good correlation between the coefficient of adhesion and W_{ad}/H_s for metal–metal pairs where H_s is the hardness of the softer metal. The exception was the hcp metals pair which exhibited low values of coefficients of adhesion.

The higher the surface energy of a solid surface, the stronger the bonds it will form with a mating material. One obvious suggestion from the surface energy theory of adhesion is to select materials that have a low surface energy and low $\Delta\gamma$. Use of lubricants at the interface reduces the surface energy. The surface energy of solid surfaces typically ranges from a few hundred to a few thousand mJ/m^2, whereas for most liquids it is few tens of mJ/m^2. Nonpolar lubricants have a lower surface energy than polar lubricants. Organic contaminants can also reduce the surface energy considerably.

4.2.6.1 Contact Analysis

We consider an elastic sphere in contact with a hard flat surface under zero external load, Figure 4.2.6a. Because of a decrease in the surface energy during contact, an attractive molecular force between the surfaces exists. This attractive force produces a finite contact radius such that there is an energy balance between the released surface energy and the stored elastic energy around the interface, Figure 4.2.6b. The loss in free surface energy E_s is given by

$$E_s = -\pi a^2 \Delta\gamma \tag{4.2.2}$$

The force F_s associated with this energy change is

$$F_s = -dE_s/d\delta \tag{4.2.3}$$

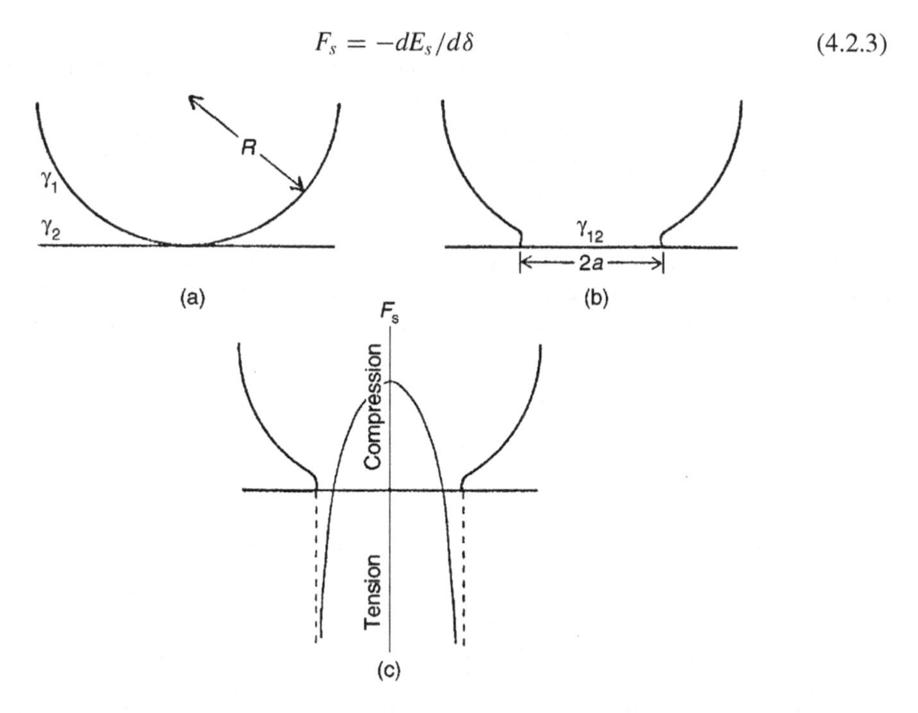

Figure 4.2.6 Contact between elastic sphere and hard flat surface with no applied force, (a) in the absence of attractive forces between the two bodies, (b) in the presence of attractive forces, surfaces are drawn together to make contact over a circle of radius a, and (c) pressure distribution in the presence of attractive forces.

where δ is the normal movement of the bodies, given by the Hertz equations (Chapter 3) as $\delta = a^2/R$. Combining Equations (4.2.2) and (4.2.3) with the Hertz equations, we get

$$F_s = \pi R \Delta\gamma \tag{4.2.4a}$$

From Hertz analysis the contact radius at no externally applied force (Chapter 3),

$$a = \left(\frac{3F_s R}{4E^*}\right)^{1/3} \tag{4.2.4b}$$

where R is the composite radius and E^* is the composite modulus (Chapter 3).

This theory is approximate since contact stresses, even in the enlarged area are assumed to be Hertzian. However, when spherical surfaces are maintained in contact over an enlarged area by surface forces, the stresses between the surfaces are tensile at the edge of the contact area (peripheral region) and only remain compressive in the center, Figure 4.2.6c (Johnson *et al.*, 1971). Since the applied force is zero, the integrated compressive force must equal the integrated tensile force. Furthermore, in the case of a sphere with relatively low elastic modulus, the deformed profile of the sphere outside the contact area is also changed. A rigorous determination of the contact equilibrium between elastic spheres under surface forces involves computation of the total energy in the system as a function of contact radius (Johnson *et al.*, 1971). Based on the modified Hertz analysis, referred to as JKR analysis, expressions for a tensile force F_s required to pull surfaces apart and the residual contact radius a when the external load is reduced to zero, are

$$F_s = \frac{3}{2}\pi R \Delta\gamma \tag{4.2.5a}$$

and

$$a = \left(\frac{9\pi \Delta\gamma R^2}{2E^*}\right)^{1/3} \tag{4.2.5b}$$

Note that F_s is independent of elastic modulus. The value of F_s is the same whether the surfaces are initially pressed together with an external force or not. As a result of surface forces, contact size is larger than the Hertzian value without adhesion and will be finite for zero external force.

If we pull the surfaces apart, the smallest force will begin to produce separation at the periphery of the contact region (where the forces are already tensile); the separating force will rapidly increase until a critical value is reached at which the rate of release of stored elastic energy just exceeds the rate of increase of surface energy arising from creation of free surface at the interface. The surfaces will then pull apart. The analysis predicts that at zero applied force, the contact area and attractive force between the surfaces should be finite, and they decrease as the applied force is made negative until a point is eventually reached at which the surfaces separate.

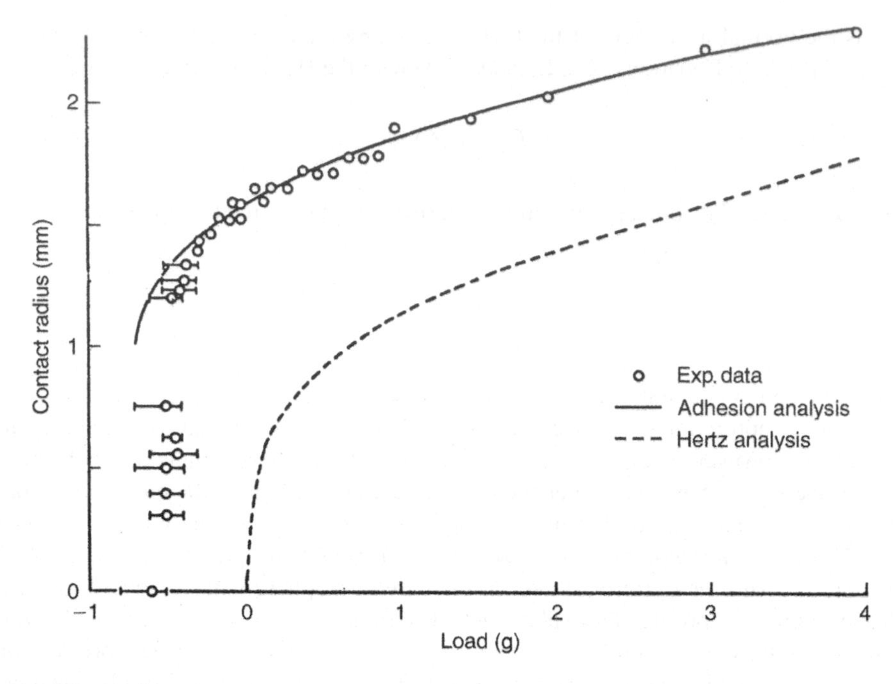

Figure 4.2.7 Radius of contact zone formed between a rubber sphere (22 mm radius) and a rubber flat as the initial joining load of 4 g is gradually reduced and then made negative. *Source*: Johnson, K.L., Kendall, K., and Roberts, A.D. (1971), "Surface Energy and the Contact of Elastic Solids," *Proc. Roy. Soc. Lond. A* **324**, 301–313, by permission of the Royal Society.

Experimental data of the contact zone formed between a rubber sphere and a rubber flat as the initial joining load of 4 g is gradually reduced and then made negative are shown in Figure 4.2.7. The contact radius remains finite until at a critical tensile force of about −0.75g, it suddenly falls to zero as the surfaces pull apart. Assuming a surface energy of rubber of about 34 mJ/m^2 for each rubber surface, agreement between the theory and data is very good. Hertz analysis does not predict the expected behavior.

Another analysis was developed by Derjaguin *et al.* (1975) (DMT analysis) for a sphere with high elastic modulus whose profile does not change outside the contact area. The contact region is under compression with the Hertzian distribution of stresses. For negligible elastic deformation of the sphere on a rigid surface,

$$F_s = 2\pi R \Delta\gamma \tag{4.2.6}$$

This equation is similar to Equation (4.2.5a) but has a coefficient of 2 instead of 3/2. This equation is the same as that derived by Bradley (1932). The interaction of the surfaces was assumed to be governed by a Lennard-Jones potential by Muller *et al.* (1980, 1983) which corresponds to an attractive pressure as a function of distance between the two surfaces and energy of adhesion $\Delta\gamma$. It is known that the surface forces are of reversible nature in equilibrium.

These analyses are recognized to apply to the opposite end of a spectrum of a non-dimensional parameter (Tabor, 1977):

$$\theta = \left[\frac{R(\Delta\gamma)^2}{E^{*2} z_0^3} \right]^{1/3} \tag{4.2.7}$$

where z_0 is the equilibrium spacing between two half-spaces made up of the Lennard-Jones 6–12 particles and modeled as a continuum. The parameter θ is a measure of the magnitude of the elastic deformation compared with the range of surface forces. For small θ (say less than 0.1) elastic deformation is negligible (hard solids) and the DMT analysis provides a good approximation; for large θ (greater than 5 say), elastic deformation is large (soft solids) and the JKR theory is good. A useful analysis of the intermediate range has been developed by Maugis (1992, 2000).

The aforementioned analyses include two simplifying assumptions. First, that the surfaces are so smooth that they make molecular contact over the whole of the region. If the surfaces are initially of optical quality and if the modulus of the rubber is very low, small protrusions are easily squeezed down to a common level and this assumption becomes reasonably valid. This is probably one of the reasons why very soft rubbers generally appear to be tacky. If the surfaces are rough and/or hard, true molecular contact will occur over a smaller area within the macroscopic region. Second, it is assumed that the deforming solid is ideally elastic.

We now extend the analysis of a sphere against a flat rough surface in elastic contact. In an interaction between elastic solids, elastic energy is stored in the asperities as they deform to bring surfaces into intimate contact. If this elastic energy is significant compared to the released surface energy ($\Delta\gamma$), the reduction in free energy is small and the resulting adhesion is small and vice versa. Fuller and Tabor (1975) modeled the asperity contacts of two rough surfaces following Greenwood and Williamson's approach described in Chapter 3. Their analysis predicts that the adhesion expressed as a fraction of maximum value (relative pull-off or adhesive force) depends upon a single parameter, called the adhesion parameter α, which is defined as:

$$\alpha = \left(\frac{4\sigma_p}{3} \right) \left[\frac{E^*}{\pi R_p^{1/2} \Delta\gamma} \right]^{2/3} \tag{4.2.8a}$$

where σ_p is the composite standard deviation of the summit heights, and R_p is the composite of mean radii of curvature of the summits of the two interacting surfaces (Chapter 3). The physical significance of the parameter α can be seen by considering

$$\alpha^{3/2} = \frac{1}{\pi} \left(\frac{4}{3} \right)^{3/2} \frac{E^* \sigma_p^{3/2} R_p^{1/2}}{R_p \Delta\gamma} \tag{4.2.8b}$$

We note that the denominator of Equation (4.2.8b) is a measure of the adhesive force experienced by spheres of radius R_p and the numerator of Equation 4.2.8b is a measure of the elastic force needed to push spheres of radius R_p to a depth of σ_p into an elastic solid of modulus E^*. Clearly, the adhesion parameter represents the statistical average of a competition between the compressive forces exerted by the higher asperities that are trying to separate the surfaces and

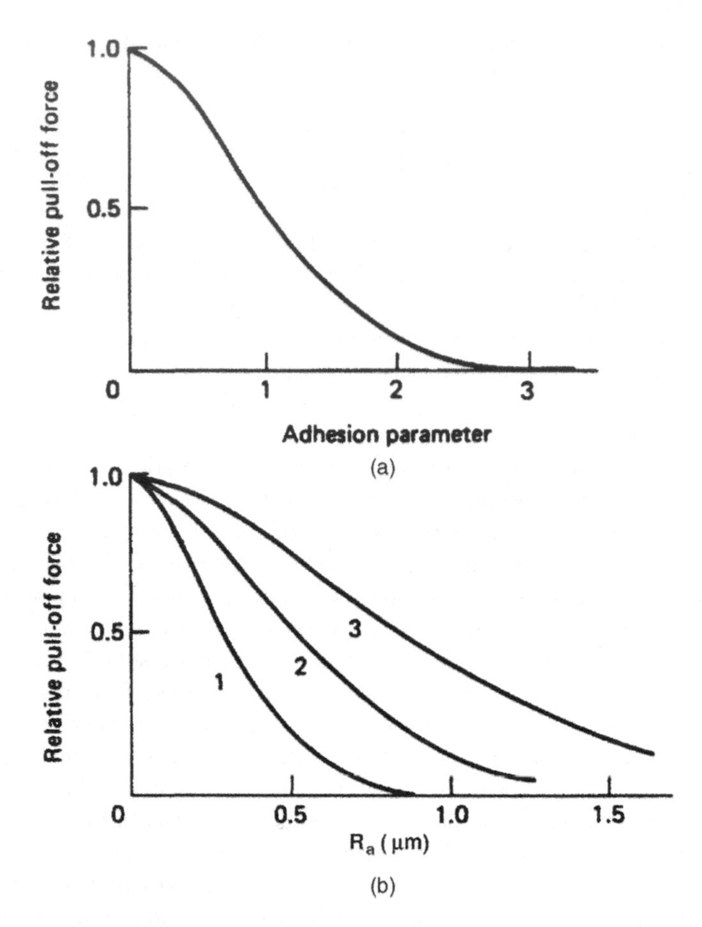

Figure 4.2.8 (a) Predicted relative pull-off force as a function of the adhesion parameter and (b) relative pull-off force for smooth rubber spheres in contact with a flat Perspex surface as a function of the R_a roughness of the Perspex for three moduli of the rubber; curve 1, 2.4 MPa; curve 2, 0.68 MPa; curve 3, 0.22 MPa. The pull-off force of smoothest surface was a few mN. *Source*: Fuller, K.N.G., and Tabor, D. (1975), "The Effect of Surface Roughness on the Adhesion of Elastic Solids," *Proc. Roy. Soc. Lond. A* **345**, 327–342, by permission of the Royal Society.

the adhesive forces between the lower asperities that are trying to hold the surfaces together. The relative pull-off (adhesive) force is virtually independent of the initial applied load, and is a function solely of the adhesion parameter, as shown in Figure 4.2.8a. When the adhesion parameter is small (less than 1) the adhesive factor dominates and the adhesion is high, and it is small if the adhesion parameter is large (2 or greater).

Relative pull-off forces measured between optically smooth rubber spheres of various moduli and a hard flat surface of Perspex of various roughnesses are shown in Figure 4.2.8b. The data show that an increase in surface roughness that is small compared with the overall deformation occurring at the interface can produce an extremely large reduction in adhesion and the effect is more marked for rubbers of higher modulus. An increase in the modulus or a decrease in the released surface energy also decreases the adhesion. On the other hand, the curvature of the sphere (over the range examined) had little influence. These results are consistent with the predictions of the analytical model (Figure 4.2.8a).

For smooth and clean surfaces, the attractive forces can be on the order of several grams. In normal circumstances, the adhesion observed between hard solids when placed in contact is very small. This may be due either to surface films of low surface energy and/or surface roughness.

Example Problem 4.2.1

Two mica spheres of 20 mm radius come into contact. Calculate the adhesive force. The free surface energy for mica per surface is 300 mJ/m^2 (= mN/m). Assume the surface energy of the interface to be equal to zero.

Solution

Based on JKR analysis,

$$F_s = \frac{3}{2}\pi R \Delta\gamma$$

$$\frac{1}{R} = \frac{1}{20} + \frac{1}{20} = \frac{1}{10}\,\mathrm{mm}^{-1}$$

$$\text{or} \quad R = 10\,\mathrm{mm}$$

$$\Delta\gamma = \gamma_1 + \gamma_2$$

$$= 600\,\mathrm{mN\,m}^{-1}$$

$$\text{and} \quad F_s = \frac{3}{2}\pi x 10^{-2} x\, 600\,\mathrm{mN}$$

$$= 28.3\,\mathrm{mN}$$

Based on DMT analysis,

$$F_s = 2\pi R \Delta\gamma$$

$$= 37.7\,\mathrm{mN}$$

4.2.7 Polymer Adhesion

Polymeric solids are used in many industrial applications where inherently low adhesion, friction and wear are desired. Interaction of polymeric solids primarily results in van der Waals attraction (Kaelble, 1971; Lee, 1974; Buckley, 1981). There are other factors involved with polymers. First, these materials are easily deformed by comparison with the other hard solids. With soft rubbers, for example, large areas of intimate contact can easily be established; consequently, although the interfacial forces themselves are weak, it is not difficult to obtain relatively high adhesive strengths. A similar factor probably accounts for the strong adhesion between sheets of thin polymeric films. Furthermore, being highly elastic solids, they can stretch appreciably under the influence of released elastic stresses without rupturing. Second, interdiffusion of polymeric chains across the interface may occur. This will greatly increase the adhesive strength, since valence bonds, as distinct from van der Waals bonds, will be established (Voyutski, 1963). Third, for dissimilar materials, charge separation may lead to an appreciable electrostatic component (Johnsen and Rahbek, 1923; Skinner *et al.*, 1953; Davies, 1973; Wahlin and Backstrom, 1974; Derjaguin *et al.*, 1978).

Experiments on tungsten against polytetrafluoroethylene (PTFE) and polyimide have shown that the polymer is transferred to the clean metal surface on simple touch contact (Buckley, 1981). The bonding is believed to be chemical in nature, and the formation of metal to carbon, nitrogen, or oxygen bonds (organometallics) takes place. Organometallics form covalent bonds with an ionic nature and have high bond strengths.

4.3 Liquid-Mediated Contact

Generally, any liquid that wets or has a small contact angle on (hydrophilic) surfaces will condense from vapor on surfaces as bulk liquid and in the form of an annular-shaped capillary condensate in the contact zone, Figure 4.3.1. The liquid film may also be deliberately applied for lubrication or other purposes. Adhesive bridges or menisci form around the contacting and near-contacting asperities due to surface energy effects in the presence of a thin liquid film. The presence of the liquid films of the capillary condensates or the pre-existing film of the liquid can significantly increase the adhesion between solid bodies (Adamson, 1990; Israelachvili, 1992; Bhushan, 1996; Cai and Bhushan, 2008a).

When separation of two surfaces is required, the viscosity of the liquid causes an additional attractive force, a rate-dependent viscous force, during separation. Thus liquid-mediated adhesive forces (F_{ad}) can be divided into two components: meniscus force (F_m) due to surface tension and a rate-dependent viscous force (F_v). These forces increase for smaller gaps and smoother surfaces so that the adhesion of ultraflat surfaces can be extremely strong. Thus

$$F_{ad} = F_m(t) + F_v(t) \tag{4.3.1}$$

The viscous component of the adhesive force is significant for more viscous liquids (dynamic viscosity $\sim$ 1 Pa s), but it can dominate for liquids of modest viscosity at high shear rates.

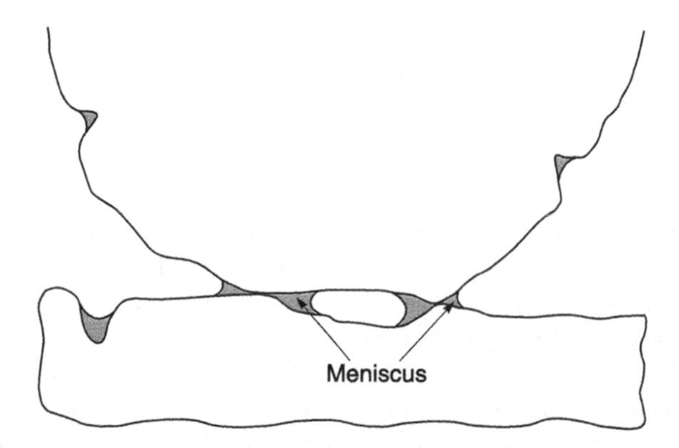

Figure 4.3.1 Condensation from liquid vapor on the surfaces at the interface.

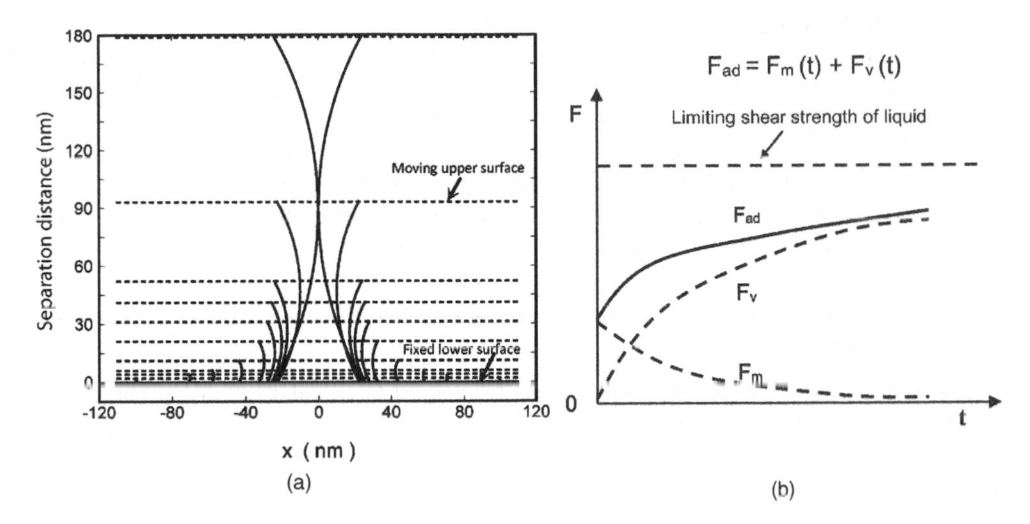

Figure 4.3.2 (a) Meniscus curvature as a function of separation distance when separating two parallel flat surfaces in the nominal direction, and (b) schematic of meniscus and viscous forces contribution to the total adhesive force during separation of two hydrophilic surfaces. Reproduced with permission from Cai, S. and Bhushan, B. (2008a), "Meniscus and Viscous Forces During Separation of Hydrophilic and Hydrophobic Surfaces with Liquid-Mediated Contacts," *Mater. Sci. Eng. R* **61**, 78–106. Copyright 2008. Elsevier.

During separation of two surfaces from liquid mediated contacts, an external force larger than the meniscus force is needed to initiate the process. After the initial motion, both meniscus and viscous forces operate inside the meniscus. During separation, the meniscus curvature decreases with separation, Figure 4.3.2a (Cai and Bhushan, 2008a). The meniscus force decreases with the separation distance because of the decrease in the meniscus area, whereas the viscous force increases with the separation distance, Figure 4.3.2b. Either the meniscus or the viscous force can be a dominant one during the separation process. It is well known that viscosity starts to drop above a certain shear stress and the liquid becomes plastic and can only support a certain value of stress, known as limiting shear strength at higher shear rates (Bhushan, 1996). This would limit the maximum value of viscous force.

Cai and Bhushan (2008a) carried out a separation analysis of both hydrophilic and hydrophobic surfaces with symmetric and asymmetric contact angles during normal and tangential separation. In this section, we present meniscus force analyses in static contact configuration and viscous analysis during normal and tangential separation.

4.3.1 Idealized Geometries

4.3.1.1 Kelvin Equation

For an incompressible liquid in equilibrium with its vapor in capillary condensation, the meniscus curvature $(1/r_1 + 1/r_2)$ is related to the relative vapor pressure (p/p_s) based on thermodynamic law, by the well-known Kelvin equation (Thomson, 1870),

$$r_K = \left(\frac{1}{r_1} + \frac{1}{r_2}\right)^{-1} = \frac{\gamma V}{RT\ell n\,(p/p_s)} \tag{4.3.2}$$

where r_K is the Kelvin radius, $1/r_1$ and $1/r_2$ are the meniscus curvatures along the two mutually orthogonal planes (sign is negative for concave shaped menisci), V is the molar volume of the liquid ($= 1.804 \times 10^{-5}$ m^3/mol at 20 °C), γ (also referred to as γ_{LA} in Chapter 2) is the surface tension of the liquid in air (= 73 mN/m for water; $\gamma V/RT$, = 0.54 nm for water at 20°C), R is the gas constant (= 8.31 J/mol K), T is the absolute temperature, and p/p_s is the relative vapor pressure or relative humidity (RH) for water in fraction (p is the pressure over the curved surface and p_s is the saturated vapor pressure at temperature T). For any capillary condensate, the water menisci must have $r_K < 0$ (concave) since $p < p_s$.

Example Problem 4.3.1

For a spherical concave water meniscus($r_1 = r_2 = r$) at 20°C, calculate meniscus curvature, r for p/p_s equal to 1 (100% RH), 0.9, 0.5 and 0.1.

Solution

$$r = \frac{2\gamma V}{RT}\left[\frac{1}{\ell n\,(p/p_s)}\right]$$
$$= \frac{1.08}{\ell n\,(p/p_s)}\ \text{nm}$$

For

$$\begin{aligned} p/p_s &= 1, & r &= \infty \\ p/p_s &= 0.9, & r &= -10.3\ \text{nm} \\ p/p_s &= 0.5, & r &= -1.56\ \text{nm} \\ p/p_s &= 0.1, & r &= -0.47\ \text{nm} \end{aligned}$$

4.3.1.2 Laplace–Young Equation

For a liquid introduced between two surfaces, menisci may be formed, Figure 4.3.1. In general, it is necessary to invoke two radii of curvature to describe a curved meniscus surface; these are equal for spherical menisci and are infinite for planar menisci. Surface tension results in a pressure difference across any meniscus surface because of Young and Laplace, sometimes credited to only Laplace, which is referred to as capillary pressure or Laplace pressure. If the surface is in mechanical equilibrium, the Laplace pressure in the liquid is given by the so-called Laplace–Young or simply Laplace equation (Adamson, 1990)

$$\Delta p = p_L = \frac{\gamma}{r_K} \tag{4.3.3}$$

The Laplace pressure acts on the projected meniscus area Ω, therefore the Laplace force is

$$F_L = \iint_{\Omega} \Delta p d\Omega \tag{4.3.4}$$

where γ is the surface tension of the liquid. Δp can be negative or positive depending upon whether the surface is hydrophilic or hydrophobic. If the liquid wets the (hydrophilic) surface ($0 \leq \theta < 90^\circ$, where θ is the contact angle between the liquid and the surface), the liquid surface is thereby constrained to lie parallel with the surface, and the complete liquid surface must therefore be concave in shape. The pressure inside the liquid in a concave meniscus ($r_K < 0$) is lower than that outside the liquid, which results in an intrinsic attractive force. If the surface is hydrophobic ($90^\circ < \theta \leq 180^\circ$), the liquid surface will be convex in shape. The pressure inside the meniscus ($r_K > 0$) is higher than outside the liquid, which results in a repulsive force.

4.3.1.3 Meniscus Forces

The total meniscus force due to the formation of a meniscus can be obtained by the Laplace force and the resolved surface tension around the circumference of the interface (Orr *et al.*, 1975; Fortes, 1982).

We study the effect of a liquid that wets, on the adhesion force between a macroscopic sphere and a flat surface and between two flat surfaces in a static contact configuration (Israelachvili, 1992; Cai and Bhushan, 2008a). In the former case, either a sphere can be in contact with a surface with a meniscus (Figure 4.3.3a), can be close to a surface with a separation and with a meniscus (Figure 4.3.3b), or can be close to a surface in the presence of a continuous film and meniscus formed on one of the surfaces (Figure 4.3.3c).

Sphere-on-Flat

We first consider the case of a sphere in contact with a flat surface with a meniscus (Figure 4.3.3). If a liquid is introduced at the point of contact, the surface tension results in a pressure difference across a meniscus surface, Equation 4.3.3. If $|r_2| \gg |r_1|$ (note that this condition is always satisfied for contacting asperities at the contact interface of rough surfaces where the asperity height is several orders of magnitude smaller than the asperity radius), then Equation 4.3.3 becomes

$$p_L \sim \frac{\gamma}{r_1} \tag{4.3.5a}$$

If the amount of liquid is small, the filling angle ϕ is small, and the top and bottom of liquid surfaces can be assumed to be parallel, then the meniscus height s, in terms of r_1, is given as

$$s = r_1(\cos\theta_1 + \cos\theta_2) \tag{4.3.5b}$$

where θ_1 and θ_2 are the contact angles between the liquid and the top and bottom surfaces.

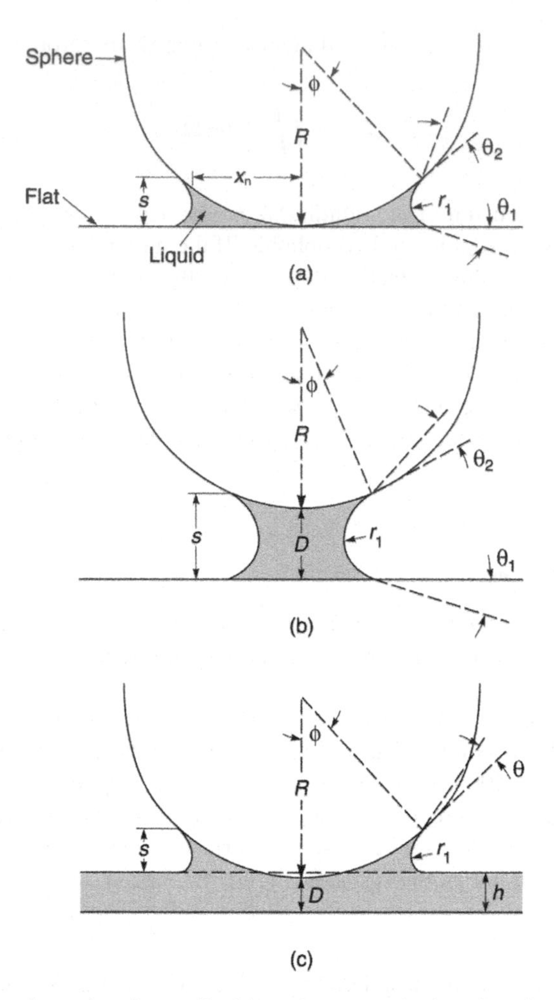

Figure 4.3.3 Meniscus formation from a liquid condensate at the interface for (a) a sphere in contact with a flat surface, (b) a sphere close to a flat surface, and (c) a sphere close to a flat surface with a continuous film.

The projected area of a circular meniscus formed between a sphere of radius R against a flat surface with a neck radius x_n, is

$$A_m = \pi x_n^2 \sim 2\pi Rs \tag{4.3.6}$$

For two spheres, R is replaced by $(1/R_1 + 1/R_2)^{-1}$ where R_1 and R_2 are the radii of two spheres. The attractive Laplace force F_L is a product of the Laplace pressure and the projected meniscus area (Equation 4.3.4). From Equations 4.3.5 and 4.3.6,

$$F_L \sim 2\pi R\gamma(\cos\theta_1 + \cos\theta_2) \tag{4.3.7a}$$

$$\sim 4\pi R\gamma \cos\theta(if\theta_1 = \theta_2) \tag{4.3.7b}$$

Note that F_L is independent of the amount of liquid at the interface, since the parameter r (or s) does not appear in its expression. However, the full meniscus force is realized only provided the film thickness exceeds the combined roughnesses of the contacting surfaces.

Another component of the adhesive force arises from the resolved surface tension around the circumference. The normal component of the surface tension force is (Orr *et al.*, 1975).

$$F_T = 2\pi R\gamma \sin\phi \sin(\phi + \theta) \tag{4.3.8}$$

F_T component is always small for small ϕ compared to the Laplace pressure contribution except for large θ close to 90° (when cos $\theta \sim 0$) as well as for large ϕ. The angle ϕ is generally small in asperity contacts. However, menisci formed around fine particles interposed between two surfaces can result in high ϕ (Patton and Bhushan, 1997). For most cases with small ϕ, the meniscus force,

$$\begin{aligned} F_m &= F_L + F_T \sim F_L \\ &= 4\pi R\gamma \cos\theta \end{aligned} \tag{4.3.9}$$

Equation 4.3.9 has been experimentally verified by McFarlane and Tabor (1950) and others. Israelachvili (1992) has reported that Laplace force expression is valid for water meniscus radii down to 2 nm.

Yet another adhesive force must be included in the preceding analysis. This arises from the direct solid–solid contact inside the liquid annulus, Figure 4.3.3. This force F_s is given by either Equation 4.2.6a or Equation 4.2.7. As an example based on DMT analysis (Equation 4.2.7), for two identical solids of free surface energies in liquid γ_{sL} ($\Delta\gamma = 2\gamma_{sL}$) using Equation 4.3.7b, the total meniscus force is

$$F_m = 4\pi R(\gamma \cos\theta + \gamma_{sL}) \tag{4.3.10}$$

For the case of a sphere close to a flat surface with a separation D with a meniscus shown in Figure 4.3.3b

$$A_m = \pi x_n^2 \sim 2\pi R(s - D) \tag{4.3.11a}$$

From Equations 4.3.3, 4.3.4, 4.3.5b and 4.3.11a, we get

$$F_L = \frac{2\pi R\gamma(\cos\theta_1 + \cos\theta_2)}{(1 + D/(s - D))} \tag{4.3.11b}$$

Maximum attraction occurs at $D = 0$ which is the same as Equation 4.3.7a.

For the case of a sphere close to a flat surface in the presence of a continuous liquid film of thickness h with a meniscus formed on the sphere (Figure 4.3.3c) (Gao *et al.*, 1995)

$$F_L = 2\pi R\gamma(1 + \cos\theta) \tag{4.3.12}$$

where θ is the contact angle between liquid and the sphere. The contact angle with the lower liquid film is zero and Equation 4.3.12 can be obtained from Equation 4.3.7a by substituting

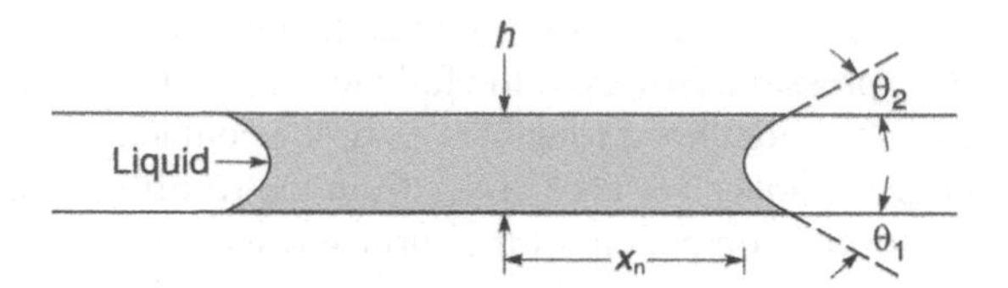

Figure 4.3.4 Meniscus formation from a liquid condensate between two flat surfaces.

$\theta_2 = \theta$ and $\theta_1 = 0$. Note that for a finite value of film thickness, F_L is independent of the film thickness.

Flat-on-Flat

For the case of two parallel flat surfaces ($R \to \infty$) separated by a liquid film of thickness h, $s = h$ and for projected area of the meniscus A_m comprising the liquid film (Figure 4.3.4), F_L based on Equations 4.3.4 and 4.3.5 is

$$F_L \sim \frac{A_m \gamma (\cos\theta_1 + \cos\theta_2)}{h} \tag{4.3.13a}$$

Meniscus area can be less than or equal to the interfacial area of any shape. For a circular meniscus of neck radius x_n, $A_m = \pi x_n^2$, and

$$F_L = \frac{\pi x_n^2 \gamma (\cos\theta_1 + \cos\theta_2)}{h} \tag{4.3.13b}$$

The normal component of the surface tension force can be expressed as (Fortes, 1982; Carter, 1988)

$$F_T = 2\pi \gamma x_n \sin\theta_{1,2} \tag{4.3.14}$$

where $\theta_{1,2}$ corresponds to the contact angle θ_1, or θ_2 depending on the surface being pulled. An interface can have asymmetric contact angles (θ_1, θ_2). This component is significant for a large meniscus.

An example of the effect of water vapor (relative humidity) on the adhesive force for a hemispherically ended pin of Ni-Zn ferrite in contact with a flat of Ni-Zn ferrite is shown in Figure 4.3.5. Note that the adhesive force remained low below about 60% RH; it increased greatly with increasing relative humidity above 60%. The adhesion at saturation is 30 times or more greater than that below 80% RH. The change in the adhesive force of contacts was reversible on humidifying and dehumidifying. Adhesion was independent of the normal load (in the range studied). Adhesive force measured in a saturated atmosphere of 1.35 mN can be predicted using meniscus analysis of sphere-flat contact. This concludes that an increase in adhesion of ferrite against itself at increasing humidity primarily arises from the meniscus (surface tension) effects of a thin film of water adsorbed in the interface.

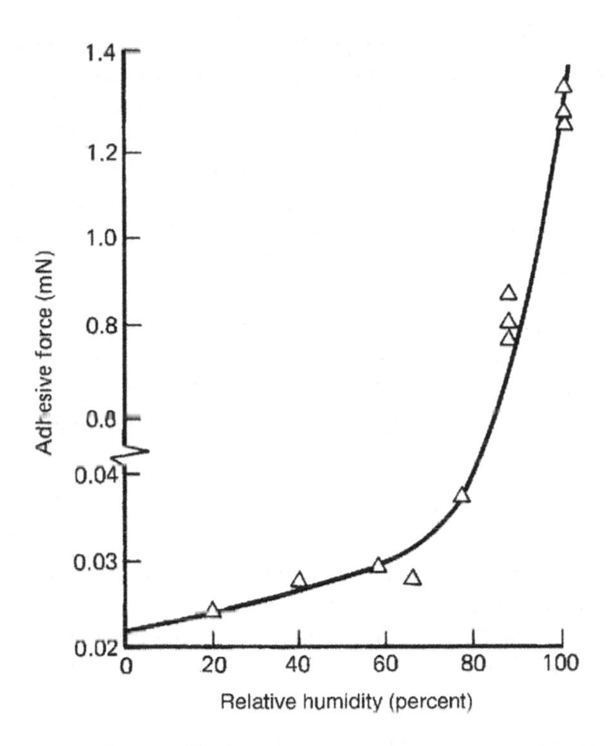

Figure 4.3.5 Effect of humidity on adhesion of a hemispherically ended pin of 2 mm radius of Ni-Zn ferrite in contact with a flat of Ni-Zn ferrite in nitrogen atmosphere in the load range of 0.67 to 0.87 mN. Reproduced with permission from Miyoshi, K., Buckley, D.H., Kusaka, T., Maeda, C., and Bhushan, B. (1988), "Effect of Water Vapor on Adhesion of Ceramic Oxide in Contact with Polymeric Magnetic Medium and Itself," in *Tribology and Mechanics of Magnetic Storage Systems*. (B. Bhushan and N.S. Eiss, eds), pp. 12–16, SP-25, ASLE, Park Ridge, IL. Copyright 1988. Springer.

4.3.1.4 Viscous Forces

Based on experimental evidence, the viscous component of the adhesive force for a liquid-mediated contact is given by (McFarlane and Tabor, 1950)

$$F_v = \frac{\beta\eta}{t_s} \tag{4.3.15}$$

where β is a proportionality constant (dimension of length^2), η is the dynamic viscosity of the liquid, and t_s is the time to separate (unstick) the two surfaces. We note that t_s is inversely related to acceleration or velocity of the interface during start-up. We further note that the fluid quantity has a weak dependence on the viscous force.

Normal separation of meniscus bridges takes place when two surfaces are pulled apart along an axis orthogonal to surfaces. Tangential separation takes place when two surfaces are slid with respect to each other in the tangential directions as encountered in sliding applications.

Viscous force occurs due to the viscosity of the liquid when separating two bodies within a short time. One may ignore viscous force for an infinitely long separation time t_s. However, an infinitely long separation time is not practically feasible. Thus, characterization of the relevant viscous force is needed in order to properly estimate the total force needed to separate two surfaces from a liquid-mediated contact. Matthewson (1988) and Bhushan (1996, 1999) presented viscous force analysis based on the critical viscous impulse. In the analysis presented by Cai and Bhushan (2007a, 2008a, b, c), Reynolds' lubrication theory is assumed to be feasible and is systematically applied to the process of separation. The results based on their analysis follow. It should be noted that meniscus necking occurs during separation and x_n decreases until it becomes zero at break (Figure 4.3.2).

Flat-on-Flat During Normal Separation

To separate two smooth flat surfaces for a liquid with kinematic viscosity η, the equation for the viscous force for separation of two flat surfaces was derived by Cai and Bhushan (2007a) by using the Reynolds' lubrication equation with a cylindrical coordinate system

$$\frac{\partial}{\partial r}\left(rh^3\frac{\partial p}{\partial r}\right) = 12\eta r\frac{dh}{dt} \tag{4.3.16}$$

where h is the separation distance and r is an arbitrary distance in the central plane of the meniscus in the direction of separation where separation occurs. Integrating the equation above with r and applying the boundary condition, $p(x_{ni}) = p$, the pressure difference at arbitrary radius r within a meniscus can be obtained,

$$\Delta p = \frac{3\eta}{h^3}\left(r^2 - x_{ni}^2\right)\frac{dh}{dt} \tag{4.3.17}$$

Subscript i represents the separation time step.

The pressure is maximum at the center of a meniscus, and it is equal to ambient pressure at the boundary. An average pressure difference is one half of the maximum pressure difference at the center of a meniscus

$$\Delta p_{avg} = -\frac{3\eta}{2h^3}x_{ni}^2\frac{dh}{dt} \tag{4.3.18}$$

The viscous force can be calculated by multiplying the average pressure difference based on the above equation with the meniscus area in the central plane in the direction of separation. The viscous force at a given separation distance can be expressed as

$$F_{V\perp} = \int_0^{x_{ni}} 2\pi\,\Delta p_{avg} r dr = -\frac{3\pi\eta}{2h^3}x_{ni}^4\frac{dh}{dt} \tag{4.3.19}$$

By integrating the above equation during the separation until break, one obtains the viscous force at the break point

$$F_{V\perp} = \frac{3\pi\eta x_{ni}^4}{4t_s}\left(\frac{1}{h_s^2} - \frac{1}{h_0^2}\right) \tag{4.3.20a}$$

$$\sim -\frac{3\pi\eta x_{ni}^4}{4t_s h_0^2}\ (\text{for } h_s \sim \infty) \tag{4.3.20b}$$

Equation 4.3.20a gives the expression for the total viscous force from the time step i to the separation point. Therefore, the total viscous force at the separation ($x_{ni} = x_{n0}$, initial meniscus neck radius)

$$= -\frac{3\pi\eta x_{n0}^2}{4t_s h_0^2} \tag{4.3.20c}$$

where t_s is the time to separate two bodies, h_0 is the initial meniscus depth, and h_s is the distance at the break point corresponding to a zero meniscus neck radius. The negative sign represents the attractive force which needs to be overcome during separation. One may take $h_s = \infty$ when separation occurs, however, this may lead to an over estimation of the real viscous force since a meniscus bridge may break very quickly when it is small and the meniscus radius is comparable to its height.

Sphere-on-Flat Surface with a Separation D_0 During Normal Separation

Similar to the approach in the previous section, for the calculation of viscous forces during separation of a sphere close to a flat surface with a separation D_0, h in the Reynolds equation Equation (4.3.16) is replaced with $H(r)$ (Cai and Bhushan, 2007a)

$$\frac{\partial}{\partial r}\left\{r[H(r)]^3\frac{\partial p}{\partial r}\right\} = 12\eta r\dot{D} \tag{4.3.21}$$

where $\dot{D}$ is the separation speed, and $H(r)$ is the shape of the upper boundary at radius r within x_{ni}, $H(x_{ni}) = x_{ni}^2/(2R) + D$. At the outside boundary x_{ni}, $p(x_{ni}) = p$. Integrating Equation (4.3.21) and applying this boundary condition, the pressure difference Δp at an arbitrary radius r within a meniscus is obtained

$$\Delta p = -3\eta R\dot{D}\left[\frac{1}{H^2(r)} - \frac{1}{H^2(x_{ni})}\right] \tag{4.3.22}$$

The viscous force at a given separation distance can be found by substituting the expression for $H(r)$ and $H(x_{ni})$ and integrating Δp over the meniscus area

$$F_{V\perp} = \int_0^{x_{ni}} 2\pi\,\Delta p\, r dr = -6\pi\eta R^2\left[1 - \frac{D}{H(x_{ni})}\right]^2 \frac{1}{D}\dot{D} \tag{4.3.23}$$

$H(x_{ni})$ changes with separation and needs to be calculated instantaneously. For $R >> x_{ni}$, the volume of the meniscus is

$$V = \int_0^{x_{ni}} 2\pi\, rH(r)dr = \pi R\left[H^2(x_{ni}) - D^2\right] \tag{4.3.24}$$

The conservation of volume leads to $V_m\,(i) = V_m\,(0)$ (the meniscus volume at the separation step i equals the initial volume), thus, the $H(x_{ni})$at a given separation distance can be found

$$H^2(x_{ni}) = H^2(x_{n0}) - D_0^2 + D^2 \tag{4.3.25}$$

where x_{n0} and D_0 are initial meniscus radius and gap, respectively. Substituting Equation (4.3.25) into Equation (4.3.23) and integrating the equation over time, the viscous force at a given separation distance can be obtained

$$F_{V\perp} = -\frac{1}{t_s}\int_{D_0}^{D_s} 6\pi\eta R^2 \left[1 - \frac{D}{\sqrt{H^2(x_{n0}) - D_0^2 + D^2}}\right]^2 \frac{1}{D}dD \tag{4.3.26}$$

where D_s is the distance when separation occurs. Separation occurs when a meniscus neck radius equals zero. Further integrating Equation (4.3.26) during the separation until break, one obtains the viscous force at the break point

$$F_{V\perp} = -\frac{6\pi\eta R^2}{t_s}\ln\frac{D_s\left[D_0 + H(x_{n0})\right]^2\sqrt{H^2(x_{n0}) - D_0^2 + D_s^2}}{D_0 H(x_{n0})\left[D_s + \sqrt{H^2(x_{n0}) - D_0^2 + D_s^2}\right]^2} \tag{4.3.27}$$

When D_s approaches infinity

$$F_{V\perp} \sim -\frac{6\pi\eta R^2}{t_s}\ln\left[\frac{(D_0 + H\,(x_{n0}))^2}{4\,D_0\,H\,(x_{n0})}\right] \text{(for } Ds \sim \infty) \tag{4.3.28a}$$

$$= -\frac{6\pi\eta R^2}{t_s}\ln\left[\frac{\left(4\,RD_0 + x_{n0}^2\right)^2}{8\,RD_0\left(x_{n0}^2 + 2\,RD_0\right)}\right] \tag{4.3.28b}$$

Flat-on-Flat During Tangential Separation

Cai and Bhushan (2008b) calculated the viscous forces during the tangential separation of two flat surfaces and a sphere on a flat surface. They used a couette flow model to derive the equations. They reported the viscous force during the tangential separation of two flat surfaces at the break point as

$$F_{v\parallel} = \frac{8\eta x'^3_n}{3t_s h_0} \tag{4.3.29}$$

where x_n' is the radius of the outermost solid–liquid circular interface.

Sphere-on-Flat with a Separation D_0 During Tangential Separation

For the tangential separation of a sphere and a flat surface, the viscous force is given as

$$F_{v\parallel} = \frac{8\eta[2R(s - D_0)]^{3/2}}{3t_s s} \tag{4.3.30}$$

Table 4.3.1 summarizes the equations for meniscus forces for static cases and viscous forces during normal and tangential separation.

Table 4.3.1 A summary of equations for meniscus and viscous forces for various cases

Force		Flat-on-flat	Sphere-on-flat
Static meniscus force		$F_m = \frac{\pi x_r^2 \gamma (\cos\theta_1 + \cos\theta_2)}{h} + 2\pi \gamma x_n \sin\theta_{1,2}$ (1 and 2 for lower and upper surface, respectively)	$F_m = 2\pi R\gamma(\cos\theta_1 + \cos\theta_2) + 2\pi R\gamma \sin\varphi \sin(\varphi + \theta_2)$ (sphere in contact with flat) $\sim 2\pi R\gamma(\cos\theta_1 + \cos\theta_2)$ (for small ϕ) $\sim \frac{2\pi R\gamma(\cos\theta_1 + \cos\theta_2)}{1 + D/(s - D)}$ (sphere close to flat and for small ϕ) $\sim 2\pi R\gamma(1 + \cos\theta)$ (sphere close to a flat with a continuous liquid film and for small ϕ)
Viscous force	Normal separation	$F_{V\perp} \sim -\frac{3\pi \eta x_{n0}^4}{4t_s h_0^2}$ (for $h_s \sim \infty$)	$F_{V\perp} \sim -\frac{6\pi \eta R^2}{t_s} \ln\left[\frac{(4RD_0 + x_{n0}^2)^2}{8RD_0(x_{n0}^2 + 2RD_0)}\right]$ (for $D_s \sim \infty$)
	Tangential separation	$F_{v\parallel} = \frac{8\eta x'^3_n}{3t_s h_0}$	$F_{v\parallel} = \frac{8\eta [2R(s - D_0)]^{3/2}}{3t_s s}$

Source: Cai and Bhushan, 2008a.

Division of Menisci

Cai and Bhushan (2007b) considered division of a big meniscus bridge into N number of meniscus bridges with equal areas. They reported that the total meniscus force of N menisci increases and total viscous forces decreases as

$$(F_m)_{total} = \sqrt{N}(F_m)_{individual} \tag{4.3.31a}$$

and

$$(F_v)_{total} = (F_v)_{individual}/N \tag{4.3.31b}$$

Example Problem 4.3.2

A drop of water ($\gamma = 73$ dyne/cm, $\theta = 60°$) and perfluoropolyether lubricant ($\gamma = 25$ dyne/cm, $\theta = 10°$) are introduced into the contact region of a 10 mm radius sphere touching a flat plate, calculate the meniscus forces (1 dyne/cm = 1 mN/m).

Solution

For small ϕ, $F_T = 0$, and meniscus force is,

$$F_m = 4\pi R\gamma(\cos\theta)$$

For a drop of water

$$\begin{aligned} F_m &= 4\pi x 0.01 x 73(\cos 60) \\ &= 4.6\,\text{mN} \end{aligned}$$

For a drop of lubricant

$$\begin{aligned} F_m &= 4\pi x 0.01 x 25(\cos 10) \\ &= 3.1\,\text{mN} \end{aligned}$$

Example Problem 4.3.3

A 10 nm thick film of water ($\gamma = 73$ dyne/cm, $\theta = 60°$, $\eta = 1$ mPa s) and 10 nm thick film of perfluoropolyether lubricant ($\gamma = 25$ dyne/cm, $\theta = 10°$, $\eta = 150$ mPa s) are placed between two flat surfaces with a circular geometry of 1 mm radius. Calculate the meniscus forces. If the two surfaces are separated in a normal direction in 10 s, calculate the viscous forces at the separation.

Solution

For small ϕ

$$F_m = \frac{2\pi x_n^2 \gamma(\cos\theta)}{h}$$

For water film

$$F_m = \frac{2\pi(0.001)^2 x 73(\cos 60)}{10^{-8}}\ \text{mN}$$
$$= 22.9\ \text{N}$$

For lubricant film

$$F_m = \frac{2\pi(0.001)^2 x 25(\cos 10)}{10^{-8}}\ \text{mN}$$
$$= 14.4\ \text{N}$$

The viscous force at normal separation is

$$F_{v\perp} = \frac{3\pi\eta x_{n0}^4}{4t_s h_0^2}$$

For water film

$$F_{v\perp} = \frac{3\pi\, x 1 x 10^{-3} x (0.001)^4}{4x10x(10^{-8})^2}\ \text{N}$$
$$= 2.4\ \text{N}$$

For lubricant film

$$F_{v\perp} = \frac{3\pi x 150x10^{-3}(0.001)^4}{4x10x(10^{-8})^2}\ \text{N}$$
$$= 3534\ \text{N}$$

4.3.1.5 Kinetic Meniscus Analysis

So far, we have discussed meniscus forces at equilibrium. When a body first comes in static contact (or rest) on another body, in the presence of a liquid film, the interface is not in equilibrium. The flow of liquid results in an increase in the wetted meniscus area which causes an increase in the meniscus force until it reaches equilibrium (Chilamakuri and Bhushan, 1999; Bhushan, 2013). This explains the experimentally observed increase in adhesive force with rest time in a liquid-mediated contact (Bhushan and Dugger, 1990).

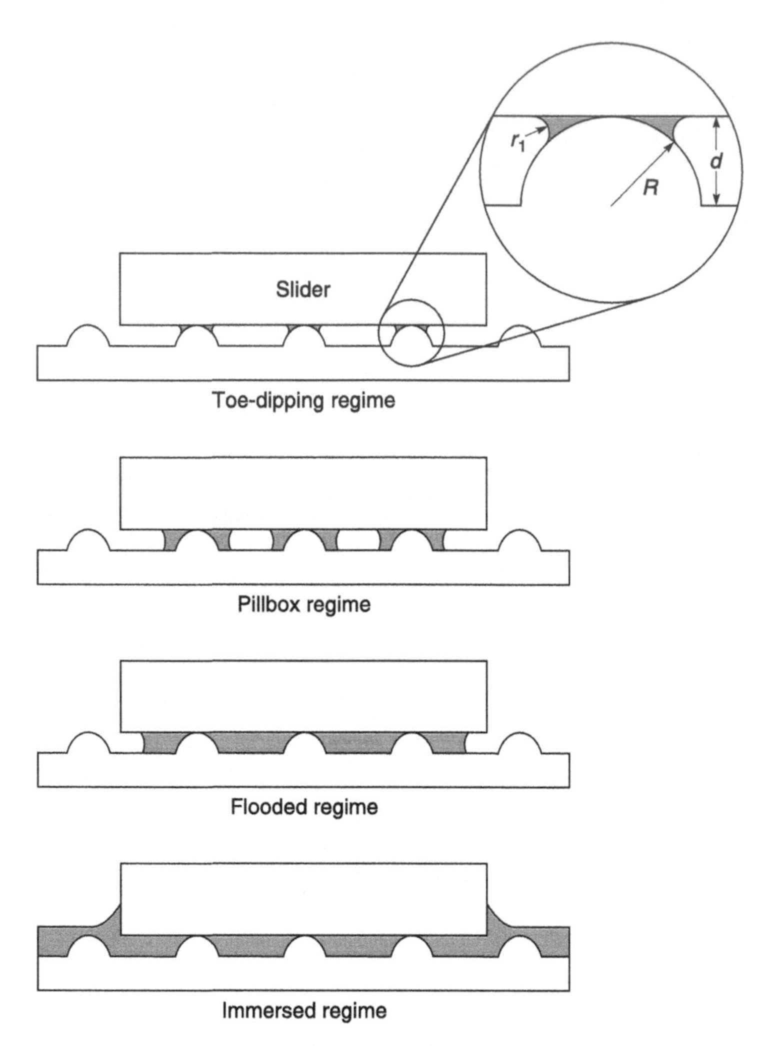

Figure 4.3.6 Regimes of different liquid levels at the interface with a smooth slider surface in contact with a rough surface.

4.3.2 Multiple-Asperity Contacts

Consider a smooth surface on a rough surface. Figure 4.3.6 shows a model of the contact region with different levels of fills of the interface dependent upon the mean interplanar separation and the liquid levels. Four distinct regimes are shown (Bhushan *et al.*, 1984; Matthewson and Mamin, 1988; Bhushan, 1996). In the first three regimes, menisci are formed which contribute to meniscus forces. The first and third are the extreme regimes in which either a small quantity of liquid bridges the surfaces around the tips of contacting asperities (the "toe-dipping" regime) or the liquid bridges the entire surface (the "flooded" regime) and in the second regime ("pillbox" regime), the liquid bridges the surface around one or more asperities to a large fraction of the apparent area. The flooded regime has the potential of generating very high adhesive forces. In the fourth regime (the "immersed" regime), the interface is immersed in the liquid and thus meniscus forces do not exist. Only viscous forces are present.

For a sufficiently thin liquid film, $r_1 > d/2$ (d = interplanar separation), the contacting surfaces will be in the toe-dipping regime. For a sufficiently thick film so that the equilibrium Kelvin radius is greater than half the interplanar separation d, the menisci will form pillbox-shaped cylindrical menisci with a capillary radius $r_1 < d/2$ around the contacting asperities. These pillbox menisci, which initially have an attractive Laplace pressure higher than the disjoining pressure in the lubricant film, grow by draining the surrounding lubricant film until it is thin enough to have a disjoining pressure equal to the Laplace pressure, $P_L = \frac{2\gamma}{d}$. The pillbox regime, however, is thermodynamically unstable, as the liquid film away from the interface has its original thickness and low disjoining pressure. Consequently, the high attractive Laplace pressure of the pillboxes will slowly pull in liquid from the film on the surface surrounding the contact regions, until the interface first becomes flooded, then immersed, and the appropriate equilibrium meniscus radius can form along the sides of the body.

Note that in the toe-dipping regime, the meniscus force is independent of the apparent area and proportional to the normal load (i.e. the number of asperity contacts). However, the flooded regime shows the opposite tendencies. The pillbox regime is intermediate and can exhibit either behavior at the extremes. Meniscus force generally decreases with an increase in roughness σ.

Example Problem 4.3.4

Calculate the meniscus forces at a magnetic head-disk interface with 1% of the area flooded with a perfluoropolyether lubricant of $\gamma = 25$ dynes/cm, $\theta = 10°$, and $\eta = 150$ mPa s. The interplanar separation is 20 nm, the apparent area of contact is 1 mm^2.

Solution

The interface is in a toe-dipping regime and

$$d = h = 20 \text{ nm}$$

$$F_m = \frac{2A_m\gamma\cos\theta}{h}$$

$$= \frac{2x10^{-2}x25\cos 10}{20}\text{N}$$

$$= 24.6\,\text{mN}$$

4.3.2.1 Statistical Analysis of Contacts

A schematic of a random rough surface in contact with a smooth surface with a continuous liquid film on the smooth surface is shown in Figure 4.3.7. Note that both contacting and

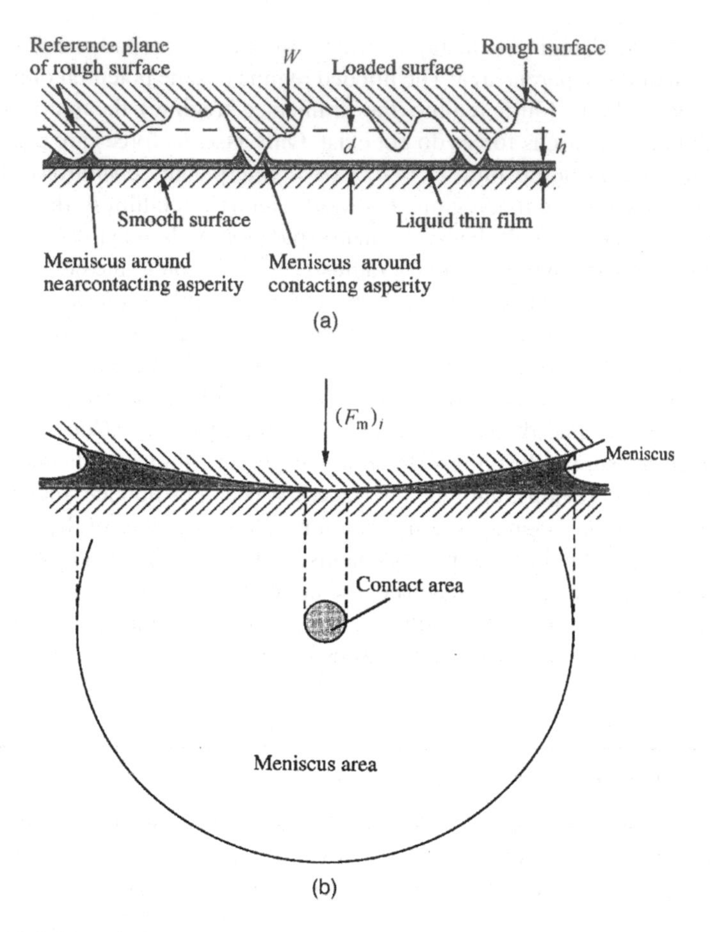

Figure 4.3.7 (a) Schematic for a rough surface in contact with a flat surface with a liquid film, and (b) schematic of contact area and meniscus area in a contacting asperity.

near-contacting asperities wetted by the liquid film contribute to the total meniscus force. A statistical approach, described in Chapter 3, is used to model the contact. The peak heights are assumed to follow a Gaussian distribution function and peak radii are assumed to be constant. In general, given the peak-height distribution function $p(z)$, the mean peak radius (R_p), the thickness of liquid film (h), the liquid surface tension (γ_ℓ), and the contact angle for the liquid in contact with the rough surface(θ), the total meniscus force (F_m) at the sliding interface is obtained by summing up the meniscus forces from all individual contacting and non-contacting asperities that form menisci over the nominal contact area $[(F_m)_i]$ shown in Figure 4.3.7 (Gao *et al.*, 1995):

$$F_m = \int_{d-h}^{\infty} (F_m)_i \, Np(z) \, dz = 2\pi R_p \gamma (1 + \cos\theta) N \int_{d-h}^{\infty} p(z) \, dz \tag{4.3.32}$$

where N is the total number of peaks in the nominal contact area. The interplanar separation, d, is determined from (see Chapter 3),

$$W + F_m = \frac{4}{3} E^* R_p^{1/2} N \int_d^{\infty} (z-d)^{3/2} p(z)\, dz \tag{4.3.33}$$

An iterative numerical approach is used to solve Equations 4.3.32 and 4.3.33.

It is evident that the maximum meniscus force can be obtained by setting h very large so that the integral in Equation 4.3.33 approaches its maximum value of unity. Therefore the maximum possible meniscus force is

$$F_{\max} = 2\pi R_p \gamma (1 + \cos\theta) N \tag{4.3.34}$$

regardless of the distribution function of peak heights. Conversely, when the film thickness h is very small, i.e., less than a molecular layer thick, F_m is zero since no meniscus can be formed and the problem reduces to dry contact.

The meniscus force increases as a function of liquid film thickness (h). For a given film thickness, the meniscus force decreases with an increase in the standard deviation of peak heights (σ_p) and it increases with an increase of radii of peaks (R_p) and number of peaks (N), Figure 4.3.8.

It has been reported that non-Gaussian surfaces with a range of positive skewness (between 0.3 and 0.7) and a high kurtosis (greater than 5) exhibit low real area of contact and meniscus forces and these surfaces are somewhat insensitive to liquid film as far as the magnitude of the meniscus force is concerned (Kotwal and Bhushan, 1996). Further discussion will be presented in the next section.

4.3.2.2 Numerical Three-Dimensional Contact Models

In a numerical model, the meniscus forces as a result of multi-asperity contacts with a pre-existing liquid film during contact of two rough surfaces are calculated. The meniscus force due to the Laplace pressure, P_L, is given by (Tian and Bhushan, 1996):

$$F_m = \int\!\!\int_{\Omega} p_L(x, y) d\Omega = \gamma \int\!\!\int_{\Omega} \frac{1}{r_1} d\Omega \tag{4.3.35}$$

where r_1 is the meniscus radius and Ω is the projected area of meniscus enclave which intersects the upper contacting asperity at a mean meniscus height. For multiple isolated menisci scattered over the whole contact interface, Ω should be the sum of the projected area of each meniscus enclave. To solve Equation 4.3.35, we need to know both the meniscus radius at different locations (or mean meniscus height) and the projected area of the meniscus enclave. These parameters are a function of the shape and the size of the meniscus (Bhushan, 2013). Using the numerical wet model developed by Tian and Bhushan (1996), Poon and Bhushan (1996) and Cai and Bhushan (2007c) carried out a wet analysis of two contacting rough surfaces with a liquid film sandwiched in between. (Also see Bhushan and Cai, 2008.) The elastic-plastic

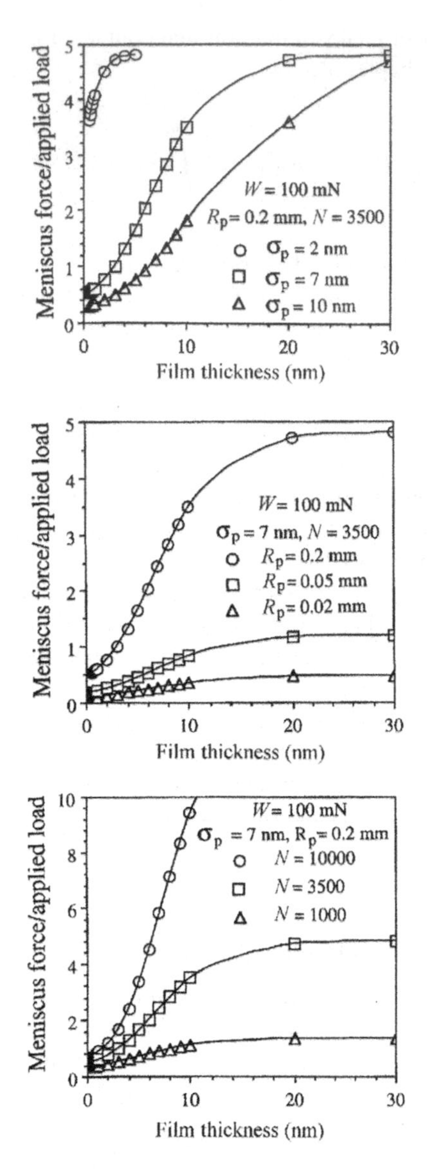

Figure 4.3.8 Ratio of the meniscus force to applied load (F_m / W) as a function of water film thickness at different σ_p, R_p and N for an interface. Reproduced with permission from Gao, C., Tian, X. and Bhushan, B. (1995), "A Meniscus Model for Optimization of Texturing and Liquid Lubrication of Magnetic Thin Film Rigid Disks," *Tribol. Trans.* **38**, 201–212. Copyright 1995 Taylor and Francis.

dry contact of rough surfaces (Chapter 3) was first analyzed. In the next step, a liquid film of known mean thickness was introduced over the deformed rough surfaces. Wetted areas were determined by selecting the areas where asperities of both contacting surfaces touch the liquid. The total projected meniscus area was determined by selecting those areas of islands of cross-cut area at a given mean meniscus height which overlap the wetted area. The meniscus

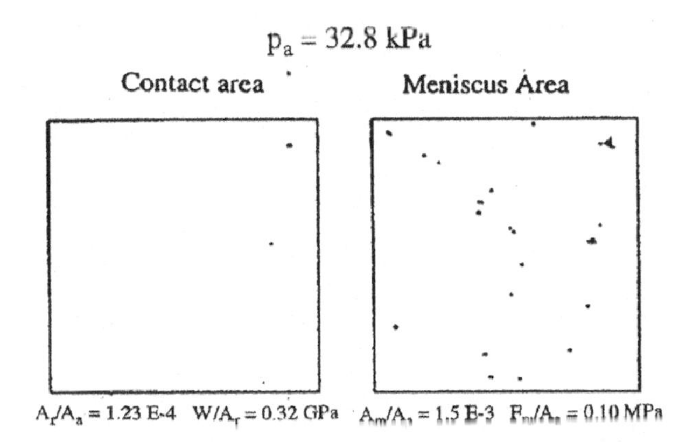

Figure 4.3.9 Contact area and meniscus area for the case of computer generated rough surface ($\sigma =$ 1 nm, $\beta^* = 0.5\mu$ m) in contact with a smooth surface with a composite elastic modulus of 100 GPa and a nominal pressure (p_a) of 32.8 kPa, in the presence of water film ($\gamma = 73$ dynes/cm, $\theta = 60^\circ$) thickness of 1 nm and meniscus height of 1 nm. Reproduced with permission from Poon, C.Y. and Bhushan, B. (1996), "Numerical Contact and Stiction Analyses of Gaussian Isotropic Surfaces for Magnetic Head Slider/Disk Contact," *Wear* **202**, 68–82. Copyright 1996. Elsevier.

force was then calculated using Equation 4.3.12. Figure 4.3.9 shows the representative contact area and the meniscus area maps for a computer-generated rough surface in contact with a smooth surface in the presence of a water film. As expected, the meniscus area is larger than the contact area and the meniscus force is three times that of the normal force. The effect of relative humidity on a glass ceramic disk substrate in contact with a smooth surface at various relative humidity is shown in Figure 4.3.10. The effect of the liquid film's thickness and the interface roughness on the meniscus force for computer-generated rough surfaces in contact with a smooth surface is shown in Figure 4.3.11. An increase in either relative humidity or

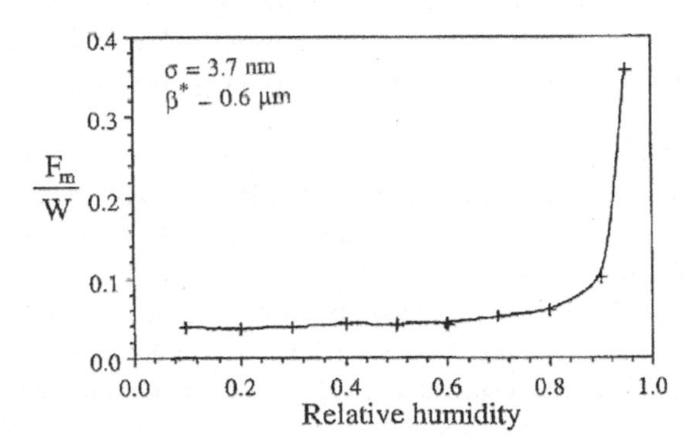

Figure 4.3.10 The effect of relative humidity on the relative meniscus force for a glass ceramic disk substrate in contact with a smooth surface. Reproduced with permission from Tian, X. and Bhushan, B. (1996), "The Micro-Meniscus Effect of a Thin Liquid Film on the Static Friction of Rough Surface Contact," *J. Phys. D: Appl. Phys.* **29**, 163–178. Copyright 1996. IOP Science.

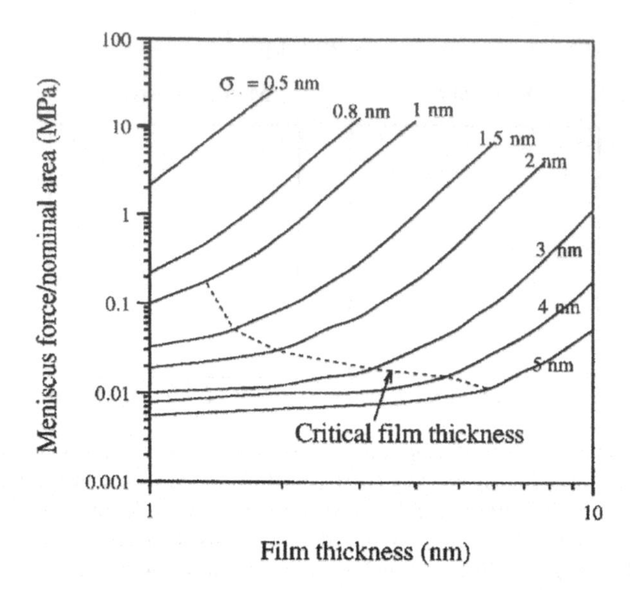

Figure 4.3.11 The effect of water film thickness and surface roughness on the relative meniscus force for computer generated Gaussian surfaces (correlation distance $\beta^* = 0.5\mu m$) in contact with a smooth surface. The dotted line defines the critical film thickness for different σ. Reproduced with permission from Poon, C.Y. and Bhushan, B. (1996), "Numerical Contact and Stiction Analyses of Gaussian Isotropic Surfaces for Magnetic Head Slider/Disk Contact," *Wear* **202**, 68–82. Copyright 1996. Elsevier.

liquid film thickness increases the liquid present at the interface. The thicker a liquid film, the more asperities touch the liquid surface and menisci form on the larger number of asperities. In addition, with a thicker film, a larger volume of liquid is present around the asperities resulting in a greater amount of meniscus volume accumulated at the contact interface and greater meniscus height. These effects lead to larger meniscus forces. There is a critical film thickness for a surface with a given roughness, above which the meniscus force increases rapidly. The critical film thickness is on the order of three-quarters of the liquid film thickness. The trends predicted by the numerical model are in agreement with experimental observations (Bhushan, 1996).

It was reported in Chapter 3 that selected non-Gaussian surfaces exhibit low real area of contact. Here we use the three-dimensional contact model to study the effect of skewness and kurtosis on a real area of contact and meniscus forces (Bhushan, 1998, 1999; Chilamakuri and Bhushan, 1998). Figure 4.3.12a shows the effect of skewness and kurtosis on the fractional real area of contact (A_r/A_a, where A_a is the apparent area) and the relative meniscus force (F_m/W) at different nominal pressures. A positive skewness between 0 and 0.2 at low pressure and about 0.2 at higher pressures results in the lowest real area of contact and meniscus force. Contact area and meniscus force decrease with an increase in the kurtosis. Fewer peaks present on a surface with positive skewness or high kurtosis can explain the trends. Figure 4.3.12b shows the variation of relative meniscus force with the h/σ ratio for different skewness and kurtosis values. Note that sensitivity of h/σ to meniscus force decreases at a range of positive skewness of 0 to 0.2 and kurtosis values of about five or larger are optimum.

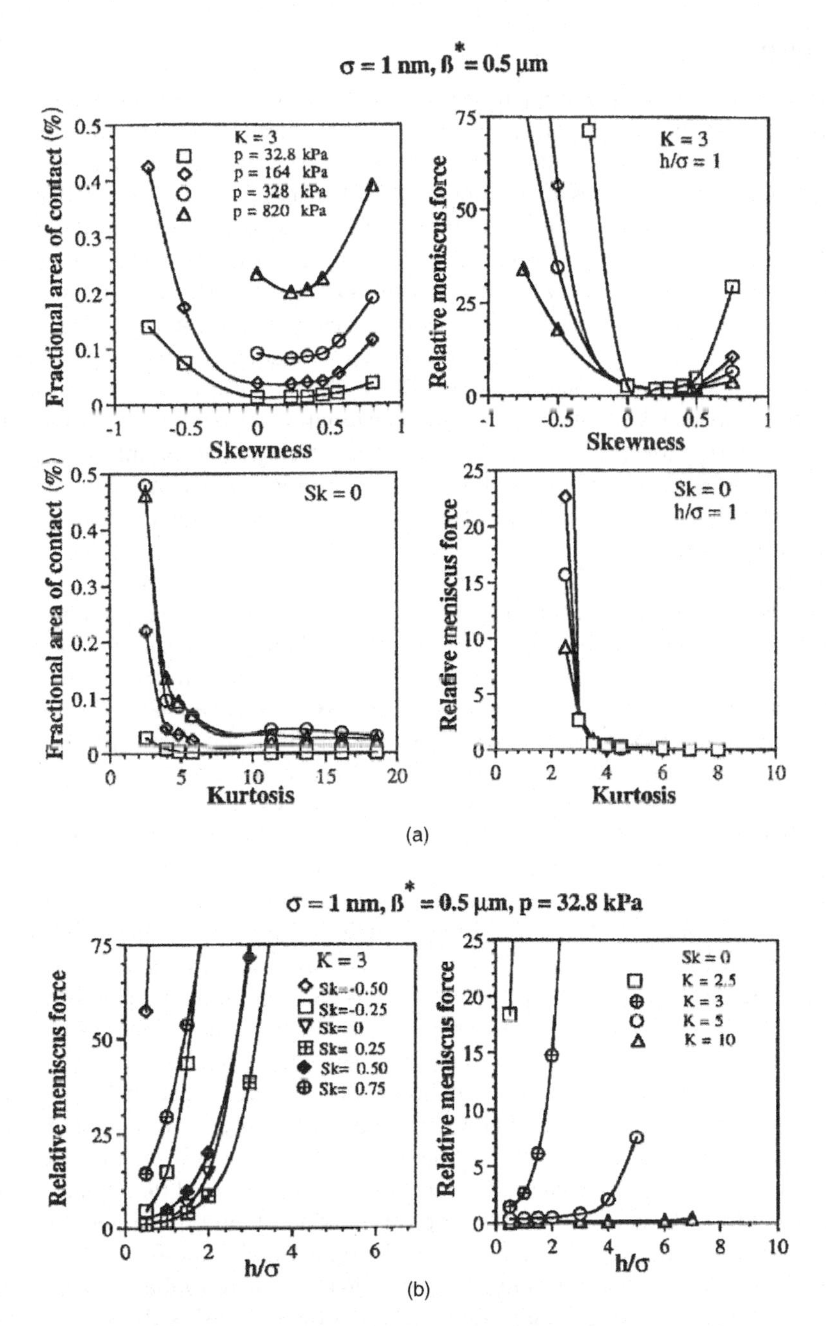

Figure 4.3.12 (a) Fractional real area of contact and relative meniscus force as a function of skewness and kurtosis at various nominal pressures, and (b) relative meniscus force as a function of h/σ for different skewness and kurtosis values, for an interface in the presence of perfluoropolyether liquid film ($\gamma = 25$ dynes/cm, $\theta = 10°$). Reproduced with permission from Chilamakuri, S.K. and Bhushan, B. (1998), "Contact Analysis of Non-Gaussian Random Surfaces," *Proc. Instn Mech. Engrs, Part J: J. Eng. Tribol.* **212**, 19–32.Copyright 1998 Sage Publications.

4.4 Closure

Adhesion between solids arises from the interatomic forces exerted across the interface. These forces may be strictly surface forces in the sense that they derive from the surface atoms themselves. Valence bonds provide surface forces. Surface charges provide surface forces; these occur when ionic surfaces are in contact with other ionic solids. They will also occur if an electrically charged layer is formed at the interface, e.g., during sliding (the triboelectric effect). Metallic bonds can form primarily in metal–metal pairs. All solids will, in addition, experience adhesion due to van der Waals interactions between atoms below the surface layers. Adhesion interactions may often be calculated in terms of free surface energies. The energy required to create new surface, expressed over an area consisting of many atoms in the surface lattice, is referred to as the free surface energy. The higher the surface energy of a solid surface, the stronger the bonds it will form with a mating material. One obvious suggestion is to select materials that have a low surface energy. The use of lubricants at the interface reduces the surface energy. Materials with low work of adhesion result in low adhesion, where work of adhesion represents the energy that must be applied to separate a unit area of the interface or to create new surfaces.

Broadly speaking, clean surfaces will adhere to most other clean surfaces. The real strength of hard solids is far lower than the theoretical strength because of the presence of surface films, roughness and lack of ductility leading to a low real area of contact (as compared to the apparent area of contact) as well as peeling apart of the contact due to elastic recovery during unloading. In general, highly elastic solids, such as polymers, adhere strongly if the surfaces are fairly smooth, in spite of the fact that the interfacial forces are relatively weak. Since the materials are soft and deformable they easily offer a large area of contact, and they can stretch appreciably under the influence of released elastic stresses without rupturing.

Liquids that have a small contact angle or wet such as water, will spontaneously condense from vapor as bulk liquid onto surfaces. The presence of the liquid films of the capillary condensates or the pre-existing film of the liquid can significantly increase the adhesion between solid bodies. Liquid-mediated adhesive forces include meniscus force due to surface tension and a rate-dependent viscous force. A wetting liquid between and around two contacting bodies results in the formation of curved (concave) menisci (liquid bridges). The Kelvin equation shows that the menisci should be concave shaped for condensed water. The attractive meniscus force occurs because the negative Laplace pressure inside the curved (concave) meniscus arises as a result of surface tension. The product of this pressure difference and the immersed surface area is the attractive meniscus force. In the early stages of meniscus formation, the meniscus force increases as a result of the flow of liquid into the low pressure region created inside the curved (concave) liquid–air interface, causing them to grow until the Laplace pressure is sufficiently reduced to match the disjoining pressure of the liquid remaining on the surface outside that contact. The disjoining pressure can be thought of as the force per unit area that the molecules on the surface of a liquid film experience relative to that experienced by the molecules on the surface of the bulk liquid. These attractive forces decrease rapidly with increasing liquid film thickness in a manner consistent with a strong van der Waals attraction. The increase in the wetted meniscus area causes an increase in the meniscus force, until it reaches equilibrium. The rates of increase of meniscus force and equilibrium time increase with the decreasing viscosity of the liquid. The equilibrium meniscus force increases with an increase in the surface tension of the liquid. The viscous component of the liquid-mediated adhesive force increases with the liquid viscosity and decreases with the time to separate the two surfaces. In the contact of two rough surfaces, the meniscus force increases with an increase in relative humidity and/or liquid film thickness and decrease of surface roughness of the interface. Selected non-Gaussian surfaces exhibit low meniscus forces.

During separation of two surfaces from liquid mediated contacts, an external force larger than the meniscus force is needed to initiate the process. After the initial motion, both meniscus and viscous forces operate inside the meniscus. During separation, meniscus curvature decreases with separation. The meniscus force decreases with the separation distance because of the decrease in the meniscus area, whereas the viscous force increases with the separation distance. Either the meniscus or the viscous force can be dominant during the separation process.

Problems

4.1 For a 10 nm thick liquid film of water ($\gamma_\ell = 73$ dynes/cm, $\theta = 60°$) between a 10 mm radius sphere and a flat surface, calculate the adhesive force. What is the adhesive force for a water film of 20 nm thickness?

4.2 For a 10 nm thick film of water ($\gamma_\ell = 73$ dynes/cm, $\theta = 60°$) of a projected area of 10 mm^2 between two flat circular surfaces of 10 mm radius, calculate the adhesive force. What is the effect of shape of the meniscus area?

4.3 Calculate the meniscus and viscous forces at a magnetic head-disk interface with 1% of the area flooded with water with $\gamma_\ell = 73$ dynes/cm, $\theta = 60°$, and $\eta_\ell = 1$ mPa s. The interplanar separation is 20 nm, the apparent area of contact is 1 mm^2 and the interface is being pulled apart at a constant rate of acceleration of 1 N/s^2.

4.4 A liquid with $\theta = 60°$ and $\gamma = 70$ mN/m forms a meniscus between a spherical asperity of radius R of 1 μm and a flat surface. Calculate the meniscus force.

References

Adamson, A.W. (1990), *Physical Chemistry of Surfaces*, Fifth edition, Wiley, New York.

Anonymous (1986) "Panel Report on Interfacial Bonding and Adhesion," *Mat. Sci. and Eng.* **83**, 169–234.

Bailey, A.I. (1961), "Friction and Adhesion of Clean and Contaminated Mica Surfaces," *J. Appl. Phys.* **32**, 1407–1412.

Bailey, A.I. and Daniels, H. (1972), "Interaction Forces Between Mica Sheets at Small Separations," *Nature Phys. Sci.* **240**, 62–63.

Bhushan, B. (1996), *Tribology and Mechanics of Magnetic Storage Devices*, Second edition, Springer-Verlag, New York.

Bhushan, B. (1998), "Method of Texturing a Magnetic Recording Medium for Optimum Skewness and Kurtosis to Reduce Friction with a Magnetic Head," US Patent No. 5,737,229, April 7.

Bhushan, B. (1999), "Surfaces Having Optimum Skewness and Kurtosis Parameter for Reduced Static and Kinetic Friction," US Patent No 6,007,896, Dec. 28.

Bhushan, B. (2003), "Adhesion and Stiction: Mechanisms, Measurement Techniques, and Methods for Reduction," (invited), *J. Vac. Sci. Technol. B* **21**, 2262–2296.

Bhushan, B. (2013), *Principles and Applications of Tribology*, Second edition, Wiley, New York.

Bhushan, B. and Cai, S. (2008), "Dry and Wet Contact Modeling of Multilayered Rough Solid Surfaces," *Appl. Mech. Rev.* **61**, #050803.

Bhushan, B. and Dugger, M.T. (1990), "Liquid-Mediated Adhesion at the Thin-Film Magnetic Disk/Slider Interface," *ASME J. Tribol.* **112**, 217–223.

Bhushan, B., Sharma, B.S., and Bradshaw, R.L. (1984), "Friction in Magnetic Tapes I: Assessment of Relevant Theory," *ASLE Trans.* **27**, 33–44.

Bikerman, J.J. (1961), *The Science of Adhesive Joints*, Academic, New York.

Bowden, F.P. and Rowe, G.W. (1956), "The Adhesion of Clean Metals," *Proc. Roy. Soc. A* **233**, 429–442.

Bradley, R.S. (1932), "The Cohesive Force Between Solid Surfaces and the Surface Energy of Solids," *Phil. Mag*. **13**, 853–862.

Buckley, D.H. (1981), *Surface Effects in Adhesion, Friction, Wear and Lubrication*, Elsevier, Amsterdam.

Cai, S. and Bhushan, B. (2007a), "Meniscus and Viscous Forces During Normal Separation of Liquid-Mediated Contacts," *Nanotechnology* **18**, #465704.

Cai, S. and Bhushan, B. (2007b), "Effects of Symmetric and Asymmetric Contact Angles and Division of Mensici on Meniscus and Viscous Forces During Separation," *Philos. Mag*. **87**, 5505–5522.

Cai, S. and Bhushan, B. (2007c), "Three-Dimensional Sliding Contact Analysis of Multilayered Solids with Rough Surfaces," *ASME J. Tribol*. **129**, 40–59.

Cai, S. and Bhushan, B. (2008a), "Meniscus and Viscous Forces During Separation of Hydrophilic and Hydrophobic Surfaces with Liquid-Mediated Contacts," *Mater. Sci. Eng. R* **61**, 78–106.

Cai, S. and Bhushan, B. (2008b), "Viscous Force During Tangential Separation of Meniscus Bridges," *Philos. Mag*. **88**, 449–461.

Cai, S. and Bhushan, B. (2008c), "Meniscus and Viscous Forces During Separation of Hydrophilic and Hydrophobic Smooth/Rough Surfaces with Symmetric and Asymmetric Contact Angles," *Phil. Trans. R. Soc. A*. **366**, 1627–1647.

Callister, W.D. (2007), *Materials Science and Engineering: An Introduction*, Seventh edition, Wiley, New York.

Carter, W.C. (1988), "The Force and Behavior of Fluids Constrained Solids," *Acta Metall*. **36**, 2283–2292.

Chilamakuri, S.K. and Bhushan, B. (1998), "Contact Analysis of Non-Gaussian Random Surfaces," *Proc. Instn Mech. Engrs, Part J: J. Eng. Tribol*. **212**, 19–32.

Chilamakuri, S.K. and Bhushan, B. (1999), "Comprehensive Kinetic Meniscus Model for Prediction of Long-Term Static Friction," *J. Appl. Phys*. **86**, 4649–4656.

Coffin, L.F. (1956), "A Study of the Sliding of Metals, With Particular Reference to Atmosphere," *Lub. Eng*. **12**, 50–59.

Davies, D.K. (1973), "Surface Charge and the Contact of Elastic Solids," *J. Phys. D: Appl. Phys*. **6**, 1017–1024.

Derjaguin, B.V., Muller, V.M., and Toporov, Y.P. (1975), "Effect of Contact Deformations on the Adhesion of Particles," *J. Colloid Interface Sci*. **53** 314–326.

Derjaguin, B.V., Krotova, N.A., and Smilga, V.P. (1978), *Adhesion of Solids (Translated from Russian by R.K. Johnston)*, Consultants Bureau, New York.

Derjaguin, B.V., Chugrev, N.V., and Muller, J.M. (1987), *Surface Forces*, Consultant Bureau, New York.

Fortes, M.A. (1982), "Axisymmetric Liquid Bridges between Parallel Plates," *J. Colloid Interf. Sci*. **88**, 338–352.

Fuller, K.N.G. and Tabor, D. (1975), "The Effect of Surface Roughness on the Adhesion of Elastic Solids," *Proc. Roy. Soc. Lond*. A **345**, 327–342.

Gao, C., Tian, X., and Bhushan, B. (1995), "A Meniscus Model for Optimization of Texturing and Liquid Lubrication of Magnetic Thin Film Rigid Disks," *Tribol. Trans*. **38**, 201–212.

Hein, M. and Arena, S. (2010), *Foundations of College Chemistry*, Thirteenth edition, Wiley, New York.

Houwink, R. and Salomon, G. (1967), *Adhesion and Adhesives*, Second edition, Elsevier, Amsterdam.

Israelachvili, J.N. (1992), *Intermolecular and Surface Forces*, Second edition, Academic, San Diego.

Johnsen, A. and Rahbek, K. (1923), "A Physical Phenomenon and its Applications to Telegraphy, Telephony, etc.," *J. Instn. Elec. Engrs*. **61**, 713–724.

Johnson, K.I. and Keller, D.V. (1967), "Effect of Contamination on the Adhesion of Metallic Couples in Ultra High Vacuum," *J. Appl. Phys*. **38**, 1896–1904.

Johnson, K.L., Kendall, K., and Roberts, A.D. (1971), "Surface Energy and the Contact of Elastic Solids," *Proc. Roy. Soc. Lond*. A **324**, 301–313.

Kaelble, D.H., ed. (1971), *Physical Chemistry of Adhesion*, pp. 22–83, Wiley Interscience, New York.

Keller, D.V. (1963), "Adhesion Between Solid Metals," *Wear*. **6**, 353–364.

Keller, D.V. (1972), "Recent Results in Particle Adhesion: UHV Measurements, Light Modulated Adhesion and the Effect of Adsorbates," *J. Adhesion*. **4**, 83–86.

Kotwal, C.A. and Bhushan, B. (1996), "Contact Analysis of Non-Gaussian Surfaces for Minimum Static and Kinetic Friction and Wear," *Trib. Trans*. **39**, 890–898.

Lee, L.H., ed. (1974), *Advances in Polymer Friction and Wear*, Vol. **5**A, Plenum, New York.

Mahanty, J. and Ninham, B.W. (1976), *Dispersion Forces*, Academic, New York.

Matthewson, M.J. (1988), "Adhesion of Spheres by Thin Liquid Films," *Phil. Mag*. A **57**, 207–216.

Matthewson, M.J. and Mamin, H.J. (1988), "Liquid-Mediated Adhesion of Ultra-Flat Solid Surfaces," *Proc. Mat. Res. Soc. Symp.* **119**, 87–92.
Maugis, D. (1992), "Adhesion of Spheres: The JKR-DMT Transition Using a Dugdale Model," *J. Colloid Interf. Sci.* **150**, 243–269.
Maugis, D. (2000), *Contact, Adhesion and Rupture of Elastic Solids*, Springer-Verlag, Berlin, Germany.
McFarlane, J.S. and Tabor, D. (1950), "Adhesion of Solids and the Effects of Surface Films," *Proc. R. Soc. Lond. A* **202**, 224–243.
Meradudin, A.A. and Mazur, P. (1980), "Effect of Surface Roughness on the van der Waals Forces Between Dielectric Bodies," *Phys. Rev.* **22**, 1684–1686.
Miyoshi, K., Buckley, D.H., Kusaka, T., Maeda, C., and Bhushan, B. (1988), "Effect of Water Vapor on Adhesion of Ceramic Oxide in Contact with Polymeric Magnetic Medium and Itself," in *Tribology and Mechanics of Magnetic Storage Systems*. (B. Bhushan and N.S. Eiss, eds), pp. 12–16, SP-25, ASLE, Park Ridge, IL.
Moore, A.C. and Tabor, D. (1952) "Some Mechanical and Adhesion Properties of Indium," *Br. J. Appl. Phys.* **3**, 299–301.
Muller, V.M., Yushchenko, V.S., and Derjaguin, B.V. (1980), "On the Influence of Molecular Forces on the Deformation of an Elastic Sphere and its Sticking to a Rigid Plane," *J. Colloid Interafce Sci.* **77**, 91–101.
Muller, V.M., Derjaguin, B.V., and Toporov, Y.P. (1983), "On Two Methods of Calculation of the Force of Sticking of an Elastic Sphere to a Rigid Plane," *Colloids and Surfaces* **7** 251–259.
Orr, F.M., Scriven, L.E., and Rivas, A.P. (1975), "Pendular Rings Between Solids: Meniscus Properties and Capillary Forces," *J. Fluid Mechanics* **67**, 723–742.
Patton, S.T. and Bhushan, B. (1997), "Environmental Effects on the Streaming Mode Performance of Metal Evaporated and Metal Particle Tapes," *IEEE Trans. Mag.* **33**, 2513–2530.
Poon, C.Y. and Bhushan, B. (1996), "Numerical Contact and Stiction Analyses of Gaussian Isotropic Surfaces for Magnetic Head Slider/Disk Contact," *Wear* **202**, 68–82.
Rabinowicz, E. (1995) *Friction and Wear of Material*, Second edition, Wiley, New York.
Sikorski, M. (1963), "Correlation of the Coefficient of Adhesion with Various Physical and Mechanical Properties of Metals," *Trans. ASME D.* **85**, 279–284.
Skinner, S.M., Savage, R.L., and Rutzler, J.E. (1953), "Electrical Phenomena in Adhesion. I. Electron Atmospheres in Dielectrics," *J. App. Phys.* **24**, 438–450.
Tabor, D. (1977), "Surface Forces and Surfaces Interactions," *J. Colloid Interface Sci.* **58**, 1–13.
Thomson, W. (1870), *Proc. R. Soc. Edinburgh* **1** 170–181.
Tian, X. and Bhushan, B. (1996), "The Micro-Meniscus Effect of a Thin Liquid Film on the Static Friction of Rough Surface Contact," *J. Phys. D: Appl. Phys.* **29**, 163–178.
Voyutski, S. S. (1963), *Autoadhesion and Adhesion of High Polymers*, Wiley, New York.
Wahlin, A. and Backstrom, G. (1974), "Sliding Electrification of Teflon by Metals," *J. Appl. Phys.* **45**, 2058–2064.
Zisman, W.A. (1963), "Adhesion", *Ind. Eng. Chem.* **55** (10), 19–38.

Further Reading

Adamson, A.W. (1990), *Physical Chemistry of Surfaces*, Fifth edition, Wiley, New York.
Anonymous (1986) "Panel Report on Interfacial Bonding and Adhesion," *Mat. Sci. and Eng.* **83**, 169–234.
Bhushan, B. (1996), *Tribology and Mechanics of Magnetic Storage Devices*, Second edition, Springer-Verlag, New York.
Bhushan, B. (2013), *Principles and Applications of Tribology*, Second edition, Wiley, New York.
Bikerman, J.J. (1961), *The Science of Adhesive Joints*, Academic, New York.
Buckley, D.H. (1981), *Surface Effects in Adhesion, Friction, Wear and Lubrication*, Elsevier, Amsterdam.
Cai, S. and Bhushan, B. (2008), "Meniscus and Viscous Forces During Separation of Hydrophilic and Hydrophobic Surfaces with Liquid-Mediated Contacts," *Mat. Sci. Eng. R* **61**, 78–106.
Derjaguin, B.V., Krotova, N.A., and Smilga, V.P. (1978), *Adhesion of Solids* (Translated from Russian by R.K. Johnston), Consultants Bureau, New York.
Derjaguin, B.V., Chugrev, N.V., and Muller, J.M. (1987), *Surface Forces*, Consultant Bureau, New York.
Houwink, R. and Salomon, G. (1967), *Adhesion and Adhesives*, Second edition, Elsevier, Amsterdam.
Israelachvili, J.N. (1992), *Intermolecular and Surface Forces*, Second edition, Academic, San Diego.

Kaelble, D.H., ed. (1971), *Physical Chemistry of Adhesion*, pp. 22–83, Wiley Interscience, New York.
Lee, L.H., ed. (1974), *Advances in Polymer Friction and Wear*, Vol. 5A, Plenum, New York.
Mahanty, J. and Ninham, B.W. (1976), *Dispersion Forces*, Academic, New York.
Maugis, D. (2000), *Contact, Adhesion and Rupture of Elastic Solids*, Springer-Verleg, Berlin, Germany.
Rabinowicz, E. (1995) *Friction and Wear of Material*, Second edition, Wiley, New York.
Ruths, M. and Israelachvili, J.N. (2011), "Surface Forces and Nanorheology of Molecularly Thin Films," in *Nanotribology and Nanomechanics II* (B. Bhushan, ed.), Third edition, pp. 107–202, Springer-Verlag, Heidelberg, Germany.
Voyutski, S. S. (1963), *Autoadhesion and Adhesion of High Polymers*, Wiley, New York.

5

Friction

5.1 Introduction

Friction is the resistance to motion during sliding or rolling, that is experienced when one solid body moves tangentially over another with which it is in contact, Figure 5.1.1. The resistive tangential force, which acts in a direction directly opposite to the direction of motion, is called the friction force. There are two main types of friction that are commonly encountered: dry friction and fluid friction. As its name suggests, dry friction, also called "Coulomb" friction, describes the tangential component of the contact force that exists when two dry surfaces move or tend to move relative to one another. Fluid friction describes the tangential component of the contact force that exists between adjacent layers in a fluid that are moving at different velocities relative to each other as in a liquid or gas between bearing surfaces. Fluid friction will be dealt with in a Chapter 8 on lubrication.

If the solid bodies are loaded together and a tangential force (F) is applied, then the value of the tangential force that is required to initiate motion is the static friction force, F_{static} or F_s. It may take a few milliseconds before relative motion is initiated at the interface. The tangential force required to maintain relative motion is known as the kinetic (or dynamic) friction force, $F_{kinetic}$ or F_k. The static friction force is either higher than or equal to the kinetic friction force, Figure 5.1.2.

Friction is not a material property, it is a system response. If two solid surfaces are clean without chemical films and adsorbates, high friction occurs. Surface contaminants or thin films affect friction. With well-lubricated surfaces, weak adhesion and friction are generally observed. However, a small quantity of liquid present at the interface results in liquid-mediated adhesion, which may result in high friction, especially between two smooth surfaces.

Friction forces can be either good or bad. Without friction it would be impossible to walk, use automobile tires on a roadway, or pick up objects. Even in some machine applications such as vehicle brakes and clutches and frictional transmission of power (such as belt drives), friction is maximized. However, in most other sliding and rotating components such as bearings and seals, friction is undesirable. Friction causes energy loss and wear of moving surfaces in contact. In these cases, friction is minimized.

Introduction to Tribology, Second Edition. Bharat Bhushan.
© 2013 John Wiley & Sons, Ltd. Published 2013 by John Wiley & Sons, Ltd.

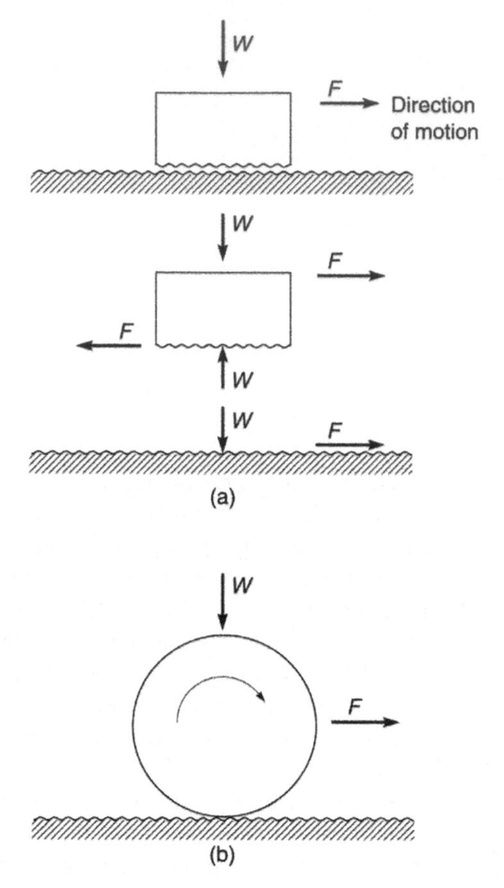

Figure 5.1.1 Schematic illustrations of (a) a body sliding on a surface with a free body diagram, and (b) a body rolling on a horizontal surface; W is the normal load (force) and F is the friction force.

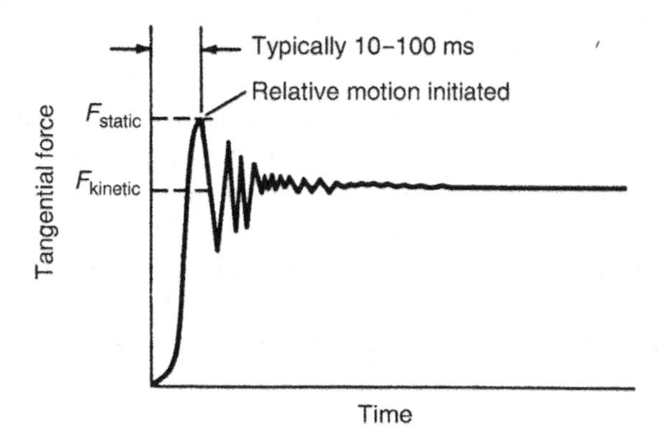

Figure 5.1.2 Tangential force as a function of time or displacement; F_{static} is the static friction force required to initiate motion and $F_{kinetic}$ is the kinetic friction force required to sustain motion.

In this chapter, we describe various mechanisms of friction in solid-solid and liquid-mediated contacts followed by representative data of friction of materials.

5.2 Solid–Solid Contact

5.2.1 Rules of Sliding Friction

Two basic rules of intrinsic (or conventional) friction are generally obeyed over a wide range of applications. These rules are often referred to as Amontons equations, after the French physicist Guillaume Amontons who rediscovered them in 1699 (Amontons, 1699); Leonardo da Vinci, however, was the first to describe them some 200 years earlier (Chapter 1). The first rule states that the friction force, F, is directly proportional to the nominal load, W, that is,

$$F = \mu W \tag{5.2.1}$$

where μ (also commonly labeled as f) is a proportionality constant known as the coefficient of static friction (μ_s) or kinetic friction (μ_k) which according to Equation 5.2.1 is independent of the normal load. Alternately, it is often convenient to express this rule in terms of constant angle of repose or frictional angle θ defined by

$$\mu_s = \tan\theta \tag{5.2.2}$$

In this equation, θ is the angle such that any body of any weight, placed on a plane inclined at an angle less than θ from the horizontal, will remain stationary, but if the inclination angle is increased to θ, the body will start to slide down, Figure 5.2.1. The coefficient of dry friction can vary over a wide range, from about 0.05 to a value as large as 10 or greater for soft and/or clean metals sliding against themselves in vacuum.

The second rule states that the friction force (or coefficient of friction) is independent of the apparent area of contact between the contacting bodies. Thus two bodies, regardless of their physical size, have the same coefficient of friction.

To these two rules, a third rule is sometimes added which is often attributed to Coulomb (1785). It states that the kinetic friction force (or coefficient of friction) is independent of the

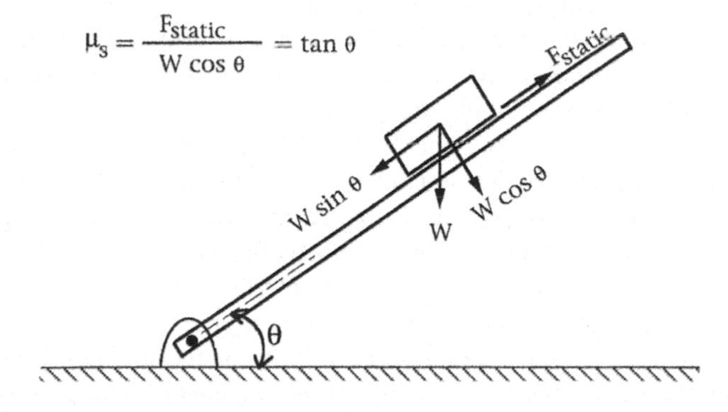

Figure 5.2.1 Force equilibrium diagram for a body on an inclined plane.

sliding velocity once motion starts. He also made a clear distinction between static friction and kinetic friction. These three rules are entirely empirical; situations in which these rules are not followed do not imply violation of more fundamental laws of nature.

The coefficient of friction as a function of load for a steel slider on unlubricated aluminum in air is shown in Figure 5.2.2a. The coefficient of friction remains essentially constant although the load is varied by a factor of 10^5. However, in the case of materials with surface films which are either deliberately applied or are produced by reaction with environment, the coefficient of friction may not remain constant as a function of load. For example, for copper sliding on copper in air, the coefficient of friction is low at low loads and a transition occurs to a higher value as the normal load is increased, Figure 5.2.2b. The factors responsible for low friction are (1) that copper readily oxidizes in air so that, at low loads, the oxide film effectively separates the two metal surfaces and there is little or no true metallic contact; and (2) that the oxide film has a low shear strength. At high loads, the film breaks down, resulting in intimate metallic contact, which is responsible for high friction and surface damage. This transition is common in other metals as well (Rabinowicz, 1995). In many metal pairs, in the high-load regime, the coefficient of friction decreases with load, Figure 5.2.2c. Increased surface roughening and a large quantity of wear debris are believed to be responsible for decrease in friction (Blau, 1992b; Bhushan, 1996).

The coefficient of friction may be very low for very smooth surfaces and/or at loads down to micro- to nanoNewton range (Bhushan and Kulkarni, 1996; Bhushan, 1999a, 2011). Figure 5.2.3 shows the coefficient of friction and wear depth as a function of load for a sharp diamond tip ($\sim$100 μm radius) sliding on three smooth materials. The coefficient of friction of Si(1 1 1) and SiO_2 coating starts to increase above some critical loads for which the contact stresses correspond to their hardnesses. Wear also starts to take place above the critical load. Very little plastic deformation and plowing contributions (to be discussed later) are responsible for low friction at loads below the critical load. In the case of diamond, transition does not occur in the load range because of its very high hardness. We will see later that in the elastic contact situation of a single-asperity contact or for a constant number of contacts, the coefficient of friction is proportional to $(load)^{-1/3}$.

The coefficient of friction of wooden sliders on an unlubricated steel surface as a function of apparent area of contact in air is shown in Figure 5.2.4. The apparent area of contact was varied by a factor of about 250 and the normal load was kept constant. The coefficient of friction remains essentially constant, which supports Amontons' second rule. The coefficient of friction may not remain constant for soft materials such as polymers and for very smooth and very clean surfaces (in which the real area of contact is effectively equal to the apparent area of contact). For example, the coefficient of friction of the automobile tire on the road surface increases with an increase of the tire width.

The third rule of friction, which states that friction is independent of velocity, is not generally valid. The coefficient of kinetic friction as a function of sliding velocity generally has a negative slope, Figure 5.2.5. Usually, the slope of a friction-velocity curve is small, that is the coefficient of friction changes a few percent for a change in velocity of an order of magnitude. High normal pressures and high sliding speeds can result in high interface (flash) temperatures which may form low shear strength surface films and in some cases, high temperatures may result in local melting and reduce the strength of materials. In addition, changes in the sliding velocity result in changes in the shear rate, which can influence the mechanical properties of the mating materials. The strength of many metals and nonmetals (especially polymers) is greater at

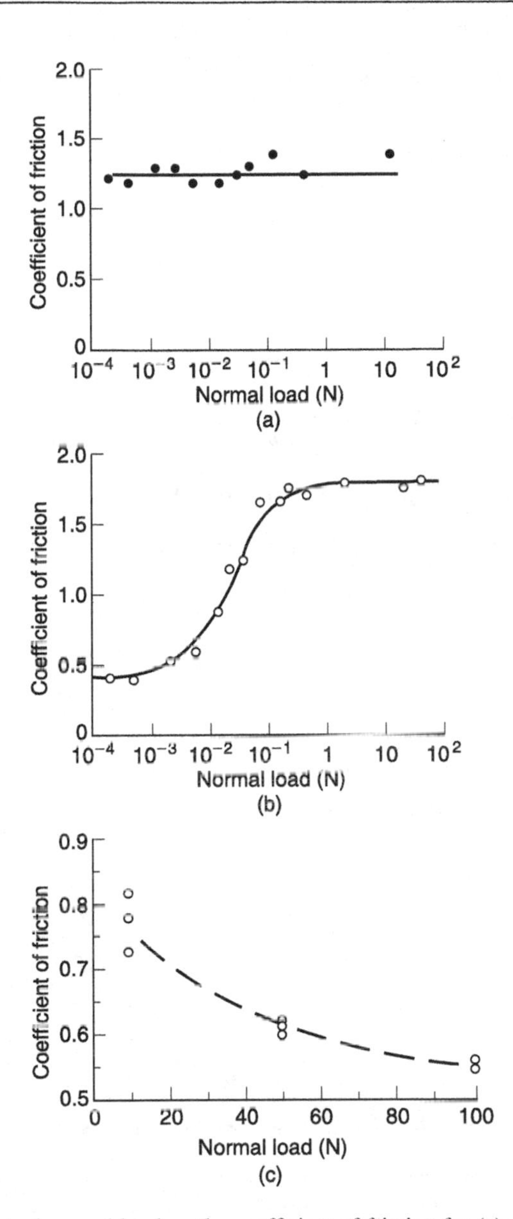

Figure 5.2.2 The effect of normal load on the coefficient of friction for (a) steel sliding on aluminum in air. (*Source*: Whitehead, J.R. (1950), "Surface Deformation and Friction of Metals at Light Loads," *Proc. Roy. Soc. Lond. A* **201**, 109–124, by permission of the Royal Society), (b) a copper on copper in air (*Source*: Whitehead, J.R. (1950), "Surface Deformation and Friction of Metals at Light Loads," *Proc. Roy. Soc. Lond. A* **201**, 109–124, by permission of the Royal Society), and (c) AISI 440C stainless steel on $Ni_3A\ell$ alloy in air. Reproduced with permission from Blau, P.J. (1992b), "Scale Effects in Sliding Friction: An Experimental Study," in *Fundamentals of Friction: Macroscopic and Microscopic Processes* (I.L. Singer and H.M. Pollock, eds.), pp. 523–534, Vol. E220, Kluwer Academic, Dordrecht, The Netherlands. Copyright 1992 Springer.

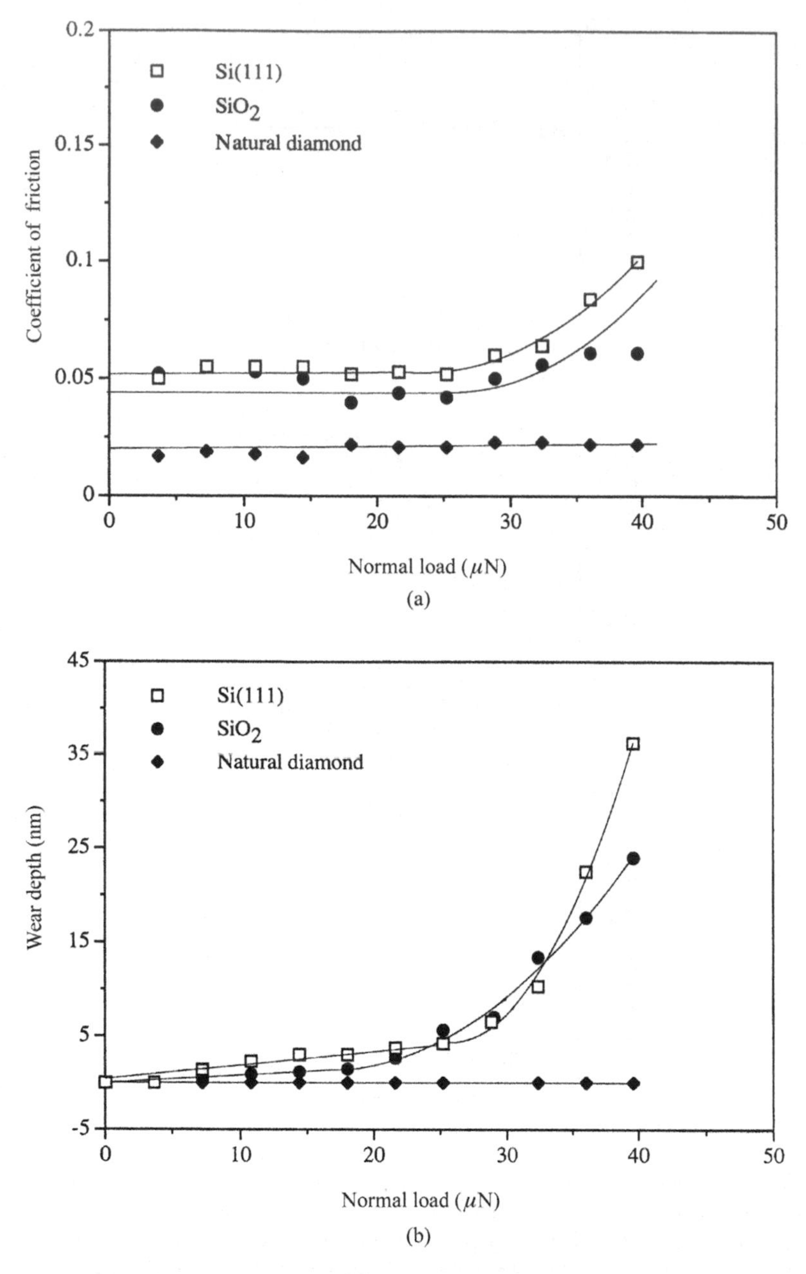

Figure 5.2.3 (a) Coefficient of friction and (b) corresponding wear depth as a function of normal load for a sharp diamond tip sliding on Si (1 1 1), SiO_2 coating and natural diamond in air at a sliding velocity of 4 μm/s using friction force microscopy. Reproduced with permission from Bhushan, B. and Kulkarni, A.V. (1996), "Effect of Normal Load on Microscale Friction Measurements," *Thin Solid Films* **278**, 49–56; 293, 333. Copyright 1996. Elsevier.

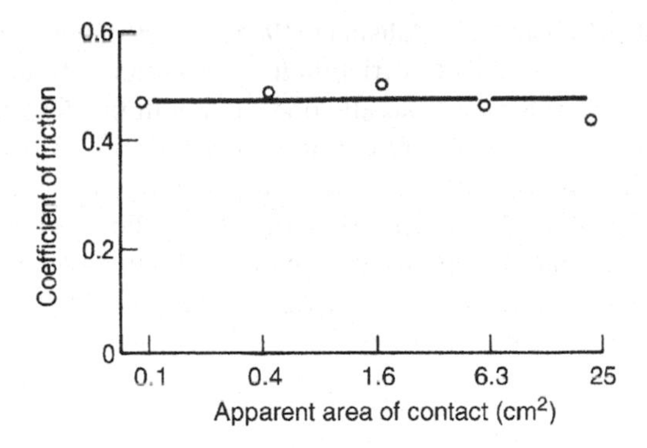

Figure 5.2.4 The effect of apparent area of contact on the coefficient of friction for wooden sliders on a steel surface in air at a normal load of 0.3 N. Reproduced with permission from Rabinowicz, E. (1995) *Friction and Wear of Material*, Second edition, Wiley, New York. Copyright 1995. Wiley.

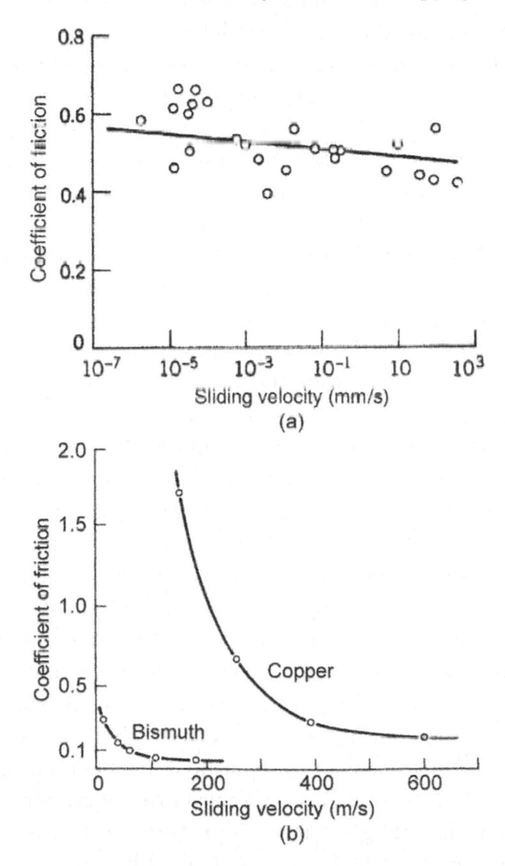

Figure 5.2.5 Coefficient of friction as a function of sliding velocity for (a) titanium sliding on titanium at a normal load of 3 N. Reproduced with permission from Rabinowicz, E. (1995) *Friction and Wear of Material*, Second edition, Wiley, New York. Copyright 1995. Wiley, and (b) pure bismuth and copper sliding on themselves. Reproduced with permission from Bowden, F.P. and Tabor, D. (1964), *The Friction and Lubrication of Solids, Part II*, Clarendon Press, Oxford. Copyright 1964 Oxford University Press.

higher shear strain rates (Bhushan and Jahsman, 1978a; 1978b), which results in a lower real area of contact and a lower coefficient of friction in a dry contact (Bhushan, 1981a).

In summary, the first two rules are generally obeyed to within a few percent in many cases. It should be emphasized that μ is strictly constant only for a given pair of sliding materials under a given set of operating conditions (temperature, humidity, normal pressure and sliding velocity). Many materials show dependence of normal load, sliding velocity and apparent area on the coefficients of static and kinetic friction in dry and lubricated contacts. In addition, μ is scale dependent (Bhushan and Nosonovsky, 2004; Bhushan, 2001b, 2011). Therefore, any reported values should be used with caution!

5.2.2 Basic Mechanisms of Sliding Friction

Amontons and Coulomb were the first to propose the mechanism of friction. Coulomb proposed that metallic friction can be attributed to the mechanical interaction of asperities of the contacting surfaces. In the so-called Coulomb model, the action of the wedge-shaped asperities causes the two surfaces to move apart as they slide from one position to another and then come close again. Work is done in raising the asperities from one position to another and most of the potential energy stored in this phase of the motion is recovered as surfaces move back. Only a small fraction of energy is dissipated in sliding down the asperities. Since *friction is a dissipative process*, the mechanical interaction theory was abandoned. A realistic friction theory should include mechanisms of energy dissipation.

Bowden and Tabor (1950) proposed that for two metals in sliding contact, high pressures developed at individual contact spots cause local welding and the contacts thus formed are sheared subsequently by relative sliding of the surfaces. Later, it was argued that asperities do not have to weld, but only the interfacial adhesion between asperities is sufficient to account for the friction of metals and ceramics (Bowden and Tabor, 1964, 1973). In addition to the frictional energy (or force) to overcome adhesion developed at the real areas of contact between the surfaces (asperity contacts), energy is required for micro-scale deformation of the contacting surfaces during relative motion. If the asperities of one surface (the harder of the two, if dissimilar) plow through the other via plastic deformation, energy is required for this macro-scale deformation (grooving or plowing). Macro-scale deformation can also occur through the particles trapped between the sliding surfaces. In viscoelastic materials (such as polymers), deformation force arises from elastic hysteresis losses. These theories, first advanced by Bowden and Tabor, are widely accepted theories for friction of metals and ceramics. The dominant mechanism of energy dissipation in metals and ceramics is plastic deformation. There is a little energy loss during the elastic deformation of interfaces; a loss of 0.1–10% (typically less than 1%) of the energy loss can occur by phonons. In engineering interfaces, even if deformation is primarily elastic, some plastic deformation also occurs. Regardless of the type of deformation, breaking of adhesive bonds during motion requires energy.

If we assume that there is negligible interaction between the adhesion and deformation processes during sliding, we may add them, and the total intrinsic frictional force (F_i) equals the force needed to shear adhered junctions (F_a) and the force needed to supply the energy of deformation (F_d). Therefore, we can write (see e.g., Bowden and Tabor, 1964)

$$F_i = F_a + F_d \tag{5.2.3}$$

or the coefficient of friction $\mu_i = \mu_a + \mu_d$. In polymers (especially elastomers) and rough surfaces in general, μ_d may be a significant fraction of μ_i.

The distinction between the adhesion and deformation theories is arbitrary, and the assumption of no interaction is too simplistic. In both cases, there is local deformation, and the magnitude of friction is influenced by the physical and chemical properties of the interacting surfaces, the load, the sliding velocity, the temperature, and so forth. There may be a continuous interplay between the two components.

For brittle materials, the fracture of adhesive contacts and brittle deformation of the materials need to be considered. An additional material property – fracture toughness – is important. Expressions for the friction of brittle materials based on fracture mechanics are available in the literature (Stolarski, 1990). The following analyses are applicable to ductile materials.

5.2.2.1 Adhesion

As described in Chapter 3, when two nominally flat surfaces are placed in contact under load, the contact takes place at the tips of the asperities, the load being supported by the deformation of contacting asperities, and discrete contact spots (junctions) are formed, Figure 5.2.6. The sum of the areas of all the contact spots constitutes the real (true) area of the contact (A_r) and for most materials under normal load, this will be only a small fraction of the apparent (nominal) area of contact (A_a). The proximity of the asperities results in adhesive contacts caused by either physical or chemical interaction. When these two surfaces move relative to each other, a lateral force is required to shear the adhesive bonds formed at the interface in the regions of real area of contact. Rupture occurs in the weakest regions, either at the interface or in one of the mating bodies. After shearing of the existing contacts, new contacts are formed. Because adhesion arises from molecular forces between the surfaces, the adhesive forces are

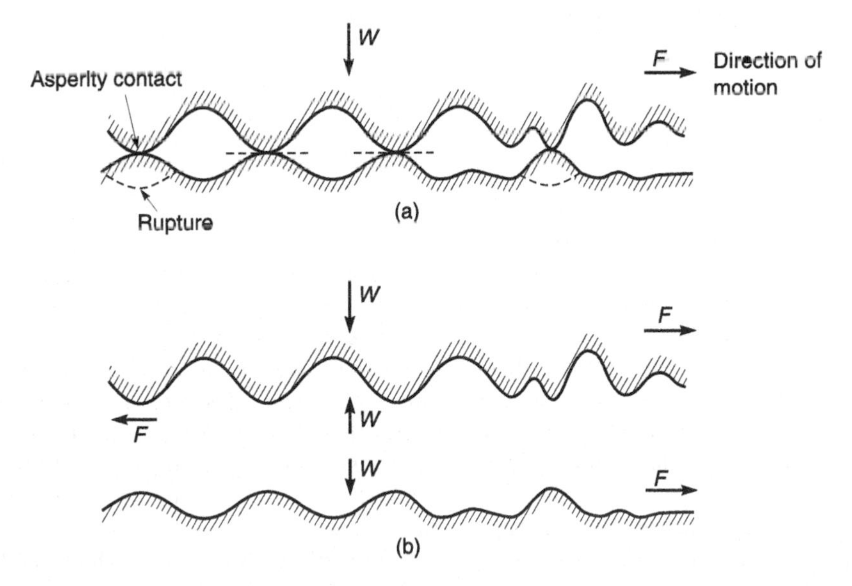

Figure 5.2.6 Schematic of (a) two rough surfaces in a sliding contact and (b) a corresponding free body diagram.

of the same nature as the forces existing between the molecules themselves. Consequently, the interface may be as strong as the bulk materials, and during sliding, the shearing process may actually tear out fragments of the materials. In that case, the friction force would depend on the bulk shear strength of the materials.

From the classical theory of adhesion to a very rough first approximation, the friction force (F_a) is defined as follows (Bowden and Tabor, 1950). For a dry contact,

$$F_a = A_r \tau_a \tag{5.2.4}$$

and for a contact with a partial liquid film,

$$F_a = A_r[\alpha \tau_a + (1 - \alpha)\tau_\ell] \tag{5.2.5a}$$

and

$$\tau_\ell = \frac{\eta_\ell V}{h} \tag{5.2.5b}$$

where τ_a and τ_ℓ are the average shear strengths of the dry contact and of the lubricant film, respectively; α is the fraction of the unlubricated area; η_ℓ is the dynamic (absolute) viscosity of the lubricant; V is the relative sliding velocity; and h is the liquid film thickness. A contribution to friction due to adhesion is always present at an interface. In boundary lubricated conditions and/or unlubricated interfaces exposed to humid environments, the presence of a liquid may result in the formation of menisci or adhesive bridges and the mensicus/viscous effects may become important, in some cases even dominating the overall friction force. These effects will be discussed in detail in a later section.

The coefficient of adhesional friction for a dry contact is

$$\mu_a = \frac{A_r \tau_a}{W} \tag{5.2.6a}$$

$$= \frac{\tau_a}{p_r} \tag{5.2.6b}$$

where p_r is the mean real pressure.

If shear (slip) occurs in one of the sliding bodies, the shear strength of the relevant body should be used. For a single crystal, the shear stress required to produce slip over the slip plane in the absence of dislocation is on the order of G/30, where G is the shear modulus of the material. If dislocations are present, shear strength would be on the order of thousand times less. Interfacial shear strength used for calculating friction should be measured at appropriate strain rates. From Bhushan (1981a), we assume that the depth of a shear zone (the transition distance between a moving surface and the surface where the rupture takes place) is equal to the linear dimension of a wear particle. If we assume that the average size of a wear particle is 1 μm for a sliding speed of 1 m/s, the shear-strain rate would be $1 \times 106\ s^{-1}$. The interfacial shear process is unique insofar as it corresponds to very high rates of strain which is different from the conventional bulk deformation process. Hence, the correlation between the interface shear process and those in bulk shear is highly approximate. Further note that the shear rates involved in shear are generally an order of magnitude or more, higher than that in the real area of contact analysis (Chapter 3).

The adhesion strength of the interface depends upon the mechanical properties and the physical and chemical interaction of the contacting bodies. The adhesion strength is reduced by reducing the surface interactions at the interface. For example, the presence of contaminants or deliberately applied fluid film (e.g., air, water or lubricant) would reduce the adhesion strength. Generally, most interfaces in a vacuum with intimate solid-solid contact would exhibit very high values of adhesion and consequently coefficient of friction. A few ppm of contaminants (air, water) may be sufficient to reduce friction dramatically. Thick films of liquids or gases would further reduce μ as it is much easier to shear into a fluid film than to shear a solid-solid contact.

The contacts can be either elastic or plastic, depending primarily on the surface roughness and the mechanical properties of the mating surfaces. We substitute expressions for A_r from Chapter 3 in the expression for coefficient of friction in Equation 5.2.6a. For elastic contacts

$$\mu_a \sim \frac{3.2\tau_a}{E^*(\sigma_p/R_p)^{1/2}} \tag{5.2.7a}$$

or

$$\propto \frac{\tau_a}{E^*(\sigma/\beta^*)} \tag{5.2.7b}$$

where E* is the composite or effective elastic modulus, σ_p *and* R_p are the composite standard deviation and composite radius of summits, σ is the composite standard deviation of surface heights and β^* is the composite correlation length. Note that μ_a is a strong function of surface roughness in the elastic contact regime.

In a single asperity contact or in a contact situation in which number of contacts remains constant, A_r is proportional to$(W)^{2/3}$, see Chapter 3. Therefore, for these situations,

$$\mu_a \propto W^{-1/3} \tag{5.2.8}$$

μ_a is not independent of load for these situations.

For plastic contacts,

$$\mu_a = \frac{\tau_a}{H} \tag{5.2.9}$$

where H is the hardness of the softer of the contacting materials. Note that μ_a is independent of the surface roughness unlike that in elastic contacts. Typical data for effect of roughness on μ for the elastic and plastic contact situations are presented in Figure 5.2.7. In the elastic contact situation of a thin-film disk against a ceramic slider in Figure 5.2.7a, μ decreases with an increase in roughness. In the plastic contact situation of copper against copper in Figure 5.2.7b, for moderate range of roughnesses, μ is virtually independent of roughness. It tends to be high at very low roughness because of the growth of real area of contact, it also tends to high at very high roughnesses because of mechanical interlocking (see p. 214).

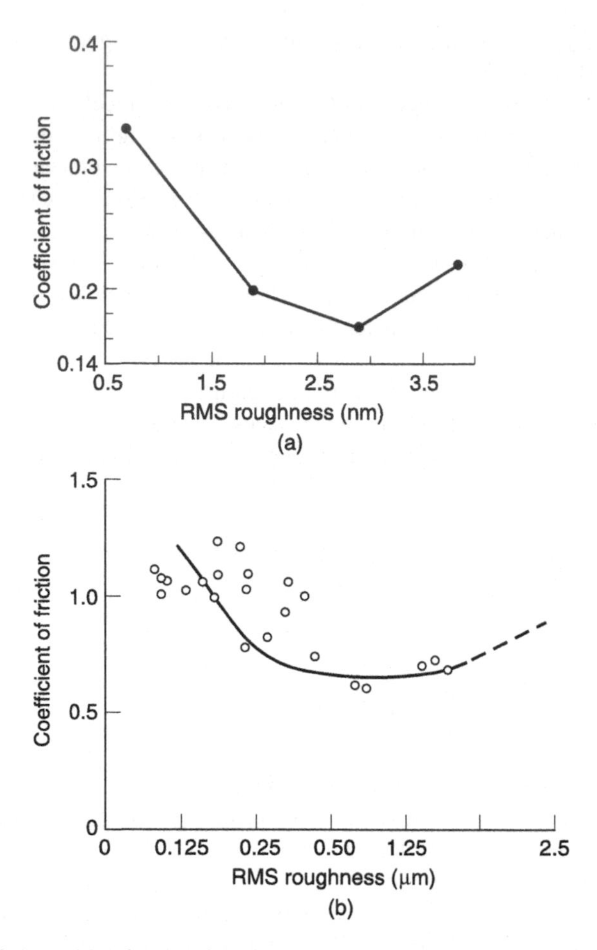

Figure 5.2.7 Coefficient of friction as a function of surface roughness for (a) a thin-film magnetic rigid disk with sputtered diamond like carbon overcoat against a Mn-Zn ferrite slider at 0.1 N load and 1 m/s. Reproduced with permission from Bhushan, B. (1996) *Tribology and Mechanics of Magnetic Storage Devices*, Second edition, Springer-Verlag, New York. Copyright 1996 Springer, and (b) copper against copper at 10 N and 0.1 mm/s. Reproduced with permission from Rabinowicz, E. (1995) *Friction and Wear of Material*, Second edition, Wiley, New York. Copyright 1995. Wiley.

Calculation of μ_a requires knowledge of τ_a. By using a limit analysis, it can be seen that the interfacial shear strength τ_a cannot substantially exceed the bulk shear strength k (the yield strength in shear) of the softer of the contacting materials for plastic contacts. If it did, each contact spot would shear within the softer material. For ductile metals (Chapter 3),

$$H \sim 5k$$

therefore,

$$\mu_a \leq 1/5 \tag{5.2.10}$$

The maximum value is independent of the metal pair. The predicted value is much smaller than the typical values observed under sliding, which typically range from 0.3 to greater than 1. The analysis so far includes an adhesional effect without other factors such as contact area growth and does not include other sources of friction such as deformation.

In most cases with plastic contacts, particularly in the case of ductile metals, there is a growth in contact area under the influence of combined normal and tangential stresses by as much as an order of magnitude, and this affects friction (Courtney-Pratt and Eisner, 1957). The growth in contact area occurs because plastic yielding of the contact is controlled by the combined effect of the normal (p) and tangential (or shear) stresses (τ) according to the form for asperities of general shape:

$$p^2 + \alpha\tau^2 = p_m^2 \quad (5.2.11)$$

where α is a constant, determined empirically, with a value of about 9 (McFarlane and Tabor, 1950) and p_m is the contact pressure for full plasticity (flow pressure) in normal compression, equal to the hardness H. In the case of plastic contacts, when the normal load is first applied, the local pressure at asperity tips rapidly approaches the mean contact pressure H under full plasticity, and at a particular location i, a contact area $(A_r)_i = W_i/H$. The subsequent application of shear stress causes the critical value of normal pressure p for plastic flow to decrease from the value required when only a normal load is applied. If the normal load remains constant then the maintenance of plasticity allows the real area of contact to grow. As a first approximation,

$$A_r = (A_r)_0 \left[1 + \alpha \left(\frac{F}{W}\right)^2\right]^{1/2} \quad (5.2.12)$$

where $(A_r)_0$ is the real area of contact without any shear stresses. Another factor that influences the area of contact is the interface temperature rise (discussed in the Chapter 6) caused by frictional heating. Under high load and speed conditions, this could have a substantial effect on the area of contact and consequently friction (Bhushan, 1996).

Rabinowicz (1995) has argued that the real area of contact is much larger than that given by deformation as a result of the applied load because of the work of adhesion (see Section 4.2). As the two surfaces come into contact, there is a decrease in the overall surface energy referred to as the work of adhesion (W_{ad}). For illustration, if a conical asperity with a roughness angle, or attack angle, of θ penetrates a half-space (Figure 5.2.8) by a distance dx, the work done by the normal load (W) is equal to the work done in deformation of the material and the change in the surface energy, given by

$$W\,dx = \pi r^2 p dx - (2\pi r)\, W_{ad} \frac{dx}{\sin\,\theta}$$

or

$$A_r = \pi r^2 = \frac{W}{p} + \frac{2\pi r}{\sin\,\theta}\frac{W_{ad}}{p} \quad (5.2.13a)$$

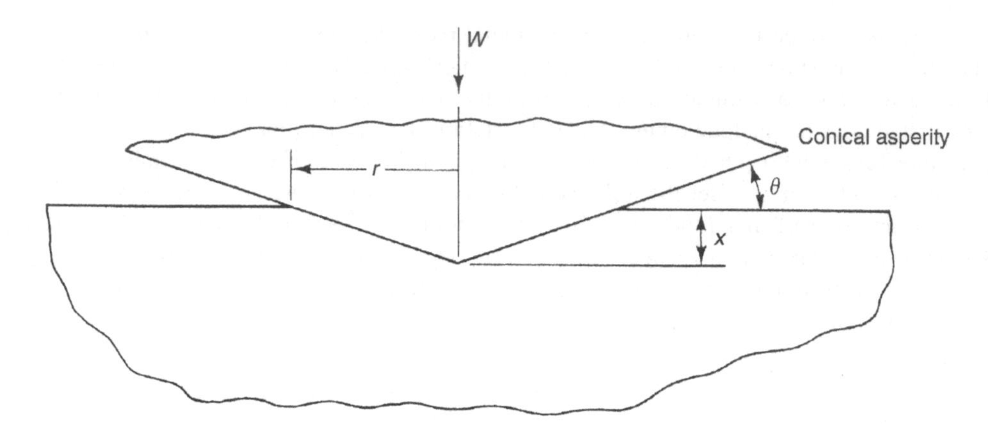

Figure 5.2.8 Indentation of a hard cone into a softer body. Reproduced with permission from Rabinowicz, E. (1995) *Friction and Wear of Material*, Second edition, Wiley, New York. Copyright 1995. Wiley.

where p is equal to H for plastic contacts. This equation shows that change in surface energy results in an increase in the real area of contact. For interfaces with high W_{ad}, the contribution of surface energy to the real area of contact can be large. We further note that the coefficient of friction is

$$\mu_a = \frac{\tau_a}{H}\left[\frac{1}{1 - 2W_{ad}/(rH\sin\theta)}\right] \tag{5.2.13b}$$

$$\sim \frac{\tau_a}{H}\left(1 + K\frac{W_{ad}}{H}\right) \tag{5.2.13c}$$

where K is a geometric factor (Suh and Sin, 1981). If we neglect the surface energy term ($W_{ad} = 0$), then Equation 5.2.13b reduces to Equation 5.2.9. In the presence of surface energy, μ is high when W_{ad}/H is large or roughness angle θ is small. Rabinowicz (1995) has shown that the friction is a function of a change in free surface energy for metals. The coefficient of friction generally increases with an increase of W_{ad}/H, whereas in the hexagonal metals, the friction is low and is constant. (Reasons for the low friction of hexagonal metals will be discussed later). Lee (1974) has shown a correlation between a change in free surface energy and coefficient of friction for polymers.

Adhesional Friction of Plastics

The shear strength of most solids is a function of the contact conditions such as mean contact pressure (real pressure). For plastics and some nonmetals,

$$\tau_a = \tau_0 + \alpha p_r \tag{5.2.14a}$$

and

$$\mu_a = \frac{\tau_0}{p_r} + \alpha \tag{5.2.14b}$$

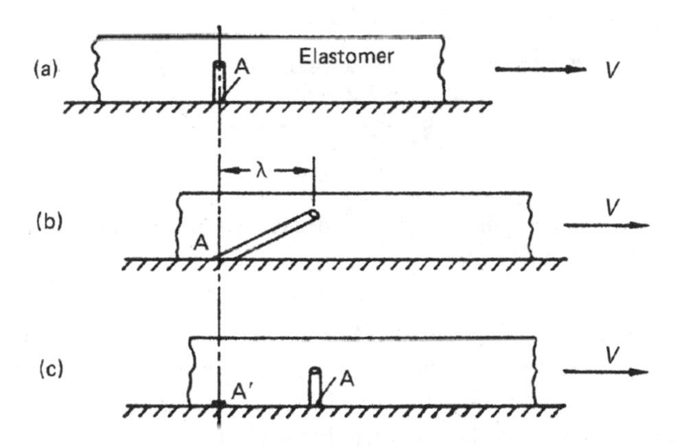

Figure 5.2.9 A simple mechanism of adhesion behavior between the elastomer and the hard surface. Reproduced with permission from Bulgin, D., Hubbard, G.D., and Walters, M.H. (1962), *Proc. 4th Rubber Technology Conf.* London, May, p.173, Institution of the Rubber Industry. Copyright Maney. http://www.maney.co.uk/

where τ_0 is the intrinsic characteristic shear strength and α is a pressure coefficient. For metals, the term τ_0/p_r is often as large as or larger than α. But for polymers and some nonmetals, it turns out that τ_0/p_r is small compared to α; for organic materials, α is on the order of 0.2. Consequently, to a first approximation, the coefficient of friction of these materials will be very close to α. This has been proven for plastic films by Briscoe and Tabor (1978). τ_a is also a function of sliding velocity and temperature. τ_a normally decreases with an increase in temperature as well as sliding velocity because of thermal heating effects.

Adhesional Friction of Elastomers

The classical adhesion theories of friction just described are generally accepted for all materials except fully viscoelastic materials-elastomers. Several molecular-kinetic and mechanical models have been proposed in the literature (Moore, 1972; Bartenev and Lavrentev, 1981; Bhushan *et al.*, 1984). In an adhesion model advanced by Bulgin *et al.* (1962), part of the physical model considers a simplified stick slip event on a molecular level, and part uses information from a mechanical model. Consider an elastomer sliding on a rigid surface and assume that adhesion takes place at a point A, Figure 5.2.9a. Let the adhesion persist for a time during which the system moves a distance and then release takes place. An associated strain develops in the material causing energy to be stored elastically in the element, Figure 5.2.9b. When the elastic stress exceeds the adhesive force, failure of the adhesive bonds take place at A and the element relaxes. Adhesion takes place at a new point A', and so one, Figure 5.2.9c. The coefficient of adhesional friction (μ_a) is given by

$$\mu_a = \frac{\pi}{2}\left(\frac{A_r}{W}\right)\tau_a \tan\delta \tag{5.2.15}$$

where $\tan\delta$ is the tangent modulus or damping factor.

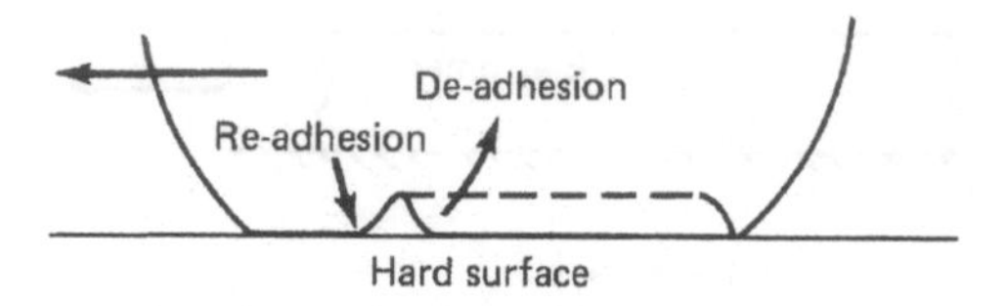

Figure 5.2.10 Schematic showing how a wave of detachment travels through the contact zone of a smooth rubber sliding over a smooth glass.

In the case of a smooth, hemispherical rubber slider moving over a clean, smooth, glass surface, waves of detachment have been reported to be generated, which traverse the contact area from the front (compression side) to the rear (tension side) at a very high speed (Schallamach, 1971). Adhesion appears to be complete between these waves which move folds in the rubber surface, probably produced by buckling attributed to tangential compressive stresses. The driving force for the waves of detachment is a tangential stress gradient. The motion of the rubber over the glass does not involve interfacial sliding but resembles the passage of a "ruck" through a carpet, or the motion of a caterpillar. There is continuous de-adhesion on one side of the ruck and re-adhesion on the other side as it passes through the contact zone, Figure 5.2.10. The energy for re-adhesion is much smaller than the energy required for de-adhesion. Frictional work is associated with the energy lost during the continuous de-adhesion and re-adhesion processes.

5.2.2.2 Deformation

Two types of interactions can occur during the sliding of two surfaces with respect to each other: the microscopic interaction where primarily plastic deformation and displacement of the interlocking surface asperities are required, and the more macroscopic interaction where the asperities of the harder material either plow grooves in the surface of the softer one via plastic deformation or result in fracture, tearing or fragmentation, Figure 5.2.11. Plowing of one or both surfaces can also occur by wear particles trapped between them, and truly macroscopic

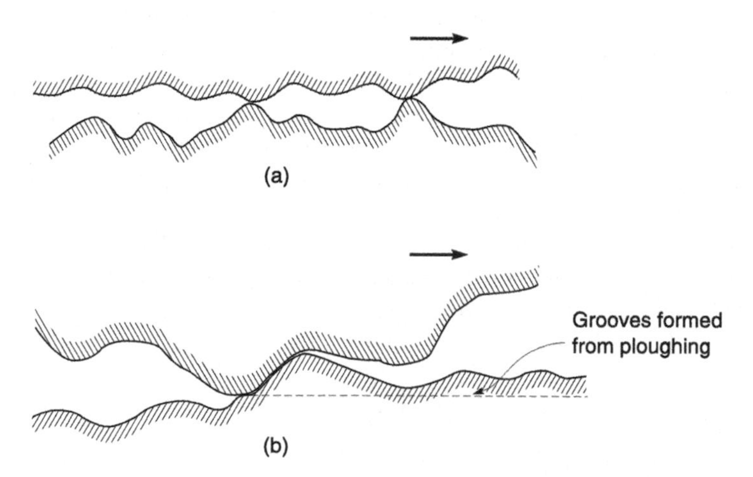

Figure 5.2.11 Schematic of interactions (a) asperity interaction, and (b) macroscopic interaction of two sliding surfaces.

plowing of the softer material by the harder, with the dimensions of the plowed groove being orders of magnitude greater than those of the asperities on either surface. Plowing deals with relatively large-volume deformations and small strains, whereas the shearing mechanism and local asperity interactions involve very thin, interfacial regions (a fraction of a nanometer thick) and large strains. During any relative motion, adhesion and asperity interactions are always present. The plowing contribution may or may not be significant; its magnitude depends on the surface roughnesses and relative hardnesses of the two surfaces, and on the size, shape and hardness of any wear debris and reaction products trapped between them. Before the onset of sliding between two surfaces, μ_d largely controls the coefficient of static friction.

Energy can be dissipated through the deformation of contacting bodies during sliding where no groove (macroscale deformation) is produced. The usual approach to the analysis of the micro-scale deformation of a single asperity is the slip-line field theory of a rigid perfectly material, similar to that used by Green for adhesional friction (Suh and Sin, 1981). However, in this analysis, μ_d does not depend on adhesion.

If one of the sliding surfaces is harder than the other, the asperities of the harder surface may penetrate and plow into the softer surface and produce grooves if shear strength is exceeded. Plowing into the softer surface may also occur as a result of impacted wear particles. In addition, interaction of two rather rough surfaces may result in mechanical interlocking on a micro- or macroscale. During sliding, interlocking would result in plowing of one of the surfaces. Because of the plowing displacement, a certain lateral (friction) force is required to maintain motion. Plowing not only increases the friction force, it creates wear particles, which in turn increase subsequent friction and wear.

In the case of metal and ceramic pairs with two rough surfaces and/or with trapped wear particles, the deformation term constitutes the force needed for plowing, grooving or cracking of surfaces, and it is generally dominant compared to the adhesion component. The dominant mechanism of energy dissipation in metals is plastic deformation (e.g., Rigney and Hirth, 1979). Rigid plastic materials can also be stressed beyond their yield point, and undergo plastic deformation, with no energy feedback. In the case of viscoelastic (rubbery) materials, the deformation term includes an energy loss caused by the delayed recovery of the material after indentation by a particular asperity, and gives rise to what is generally called the hysteresis friction.

We now calculate the plowing component of the friction force for four model rigid asperities or trapped wear particles – conical, spherical and cylindrical with two orientations (Bhushan, 1999a), Figure 5.2.12. First, we consider a circular cone of roughness angle, or attack angle, θ, pressed into a softer body, Figure 5.2.12a (Rabinowicz, 1995). During sliding only the front surface of the asperity is in contact with the softer body. Therefore, the load-support area (the horizontal projection of the asperity contact), A_ℓ, which supports the normal load is given by

$$A_\ell = \frac{1}{2}\pi r^2 \tag{5.2.16a}$$

The friction force is supported by the plowed (grooved) area (vertical projection of the asperity contact), A_p and

$$\begin{aligned} A_p &= \frac{1}{2}(2rd) \\ &= r^2 \tan\theta \end{aligned} \tag{5.2.16b}$$

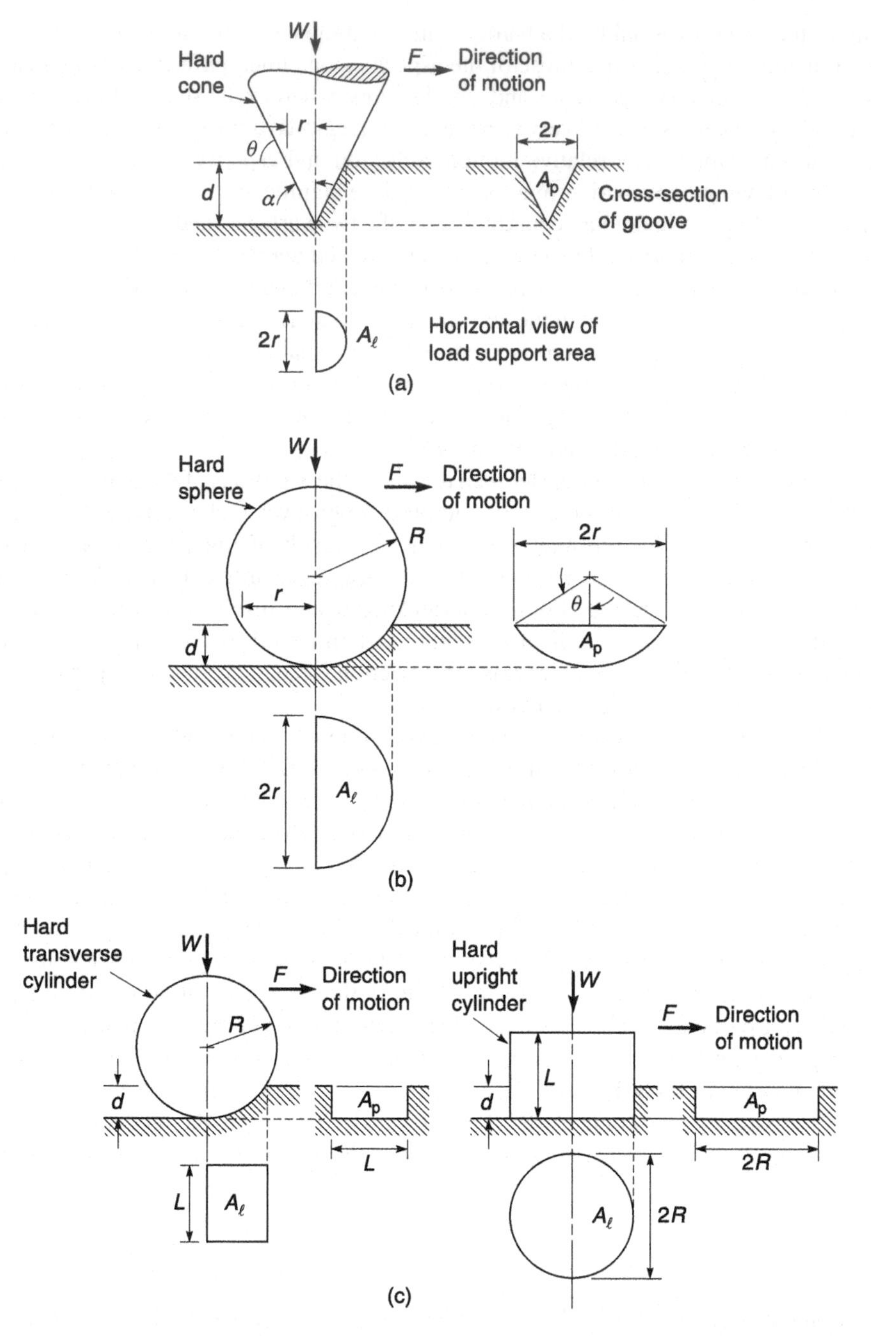

Figure 5.2.12 Schematics of (a) hard cone, (b) hard sphere, and (c) hard transverse and upright cylinders sliding on a softer material.

Assuming that the yielding of the body is isotropic and that its yield pressure is p, then

$$W = pA_\ell, \tag{5.2.17a}$$

$$F = pA_p \tag{5.2.17b}$$

and

$$\mu_p = \frac{F}{W} = \frac{A_p}{A_\ell} \tag{5.2.17c}$$

From Equations 5.2.16 and 5.2.17, we get

$$\mu_p = \frac{2\tan\theta}{\pi} \tag{5.2.18a}$$

This expression can be written in terms of the apex semi-angle of the cone $\alpha = 90^\circ - \theta$,

$$\mu_p = \frac{2\cot\alpha}{\pi} \tag{5.2.18b}$$

For most engineering surfaces, the angles of asperities with the horizontal surface (roughness angles) are very small and the plowing component of friction is correspondingly small. For example, for a conical asperity with a roughness angle of 5° on a very rough surface, the plowing component of friction is only 0.056. This is a low value because the piling up of the material ahead of the sliding asperity is neglected in the analysis. Abrasive materials and impacted wear particles may be very angular with large θ values which would result in large values of plowing component of friction.

Next we consider a spherical asperity of radius R in contact with a softer body, Figure 5.2.12b (Moore, 1972). The expression for μ_p is (Bhushan, 1999a)

$$\mu_p = \frac{A_p}{A_\ell} = \frac{4}{3\pi}\frac{r}{R} \tag{5.2.19}$$

For a relatively large width of the groove as compared to radius of sphere, an expression for μ_p is given by (Suh and Sin, 1981)

$$\mu_p = \frac{2}{\pi}\left\{\left(\frac{R}{r}\right)^2 \sin^{-1}\left(\frac{r}{R}\right) - \left[\left(\frac{R}{r}\right)^2 - 1\right]^{1/2}\right\} \tag{5.2.20}$$

μ_p increases rapidly with an increase of $\frac{r}{R}$, i.e., the coefficient of friction increases as the sphere digs deeper, Figure 5.2.13 (Suh and Sin, 1981).

Next we consider a cylinder placed in transverse and upright positions, Figure 5.2.12c (Moore, 1972). Expression for μ_p is (Bhushan, 2013)

$$\mu_p = \left[\frac{1}{2(R/d) - 1}\right]^{1/2} \tag{5.2.21}$$

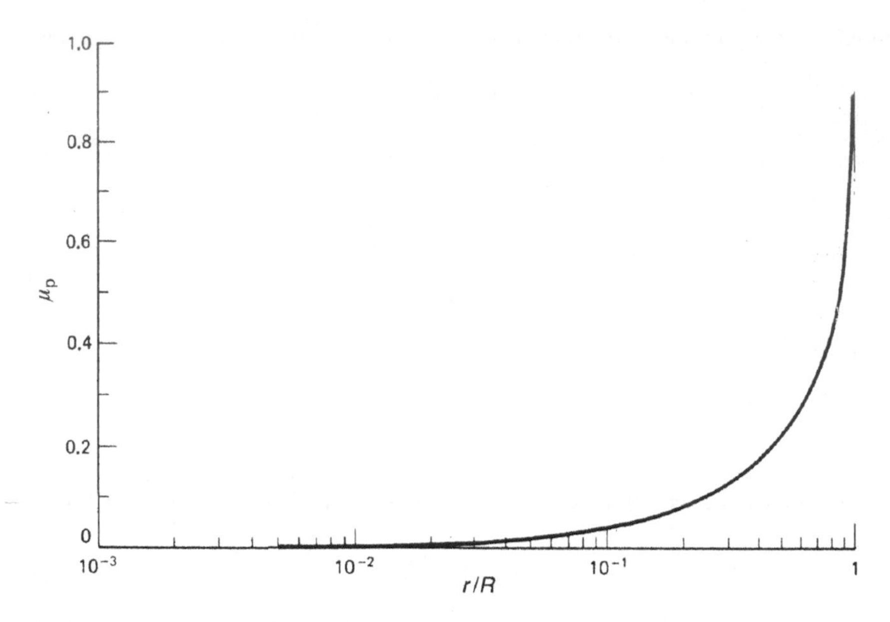

Figure 5.2.13 Coefficient of plowing friction as a function of the ratio of groove width to asperity diameter for a spherical asperity. Reproduced with permission from Suh, N.P. (1986), *Tribophysics*, Prentice-Hall, Englewood Cliffs, New Jersey. Copyright 1986. Prentice-Hall.

During sliding, the material is piled up ahead of the slider in the sliding /grooving path as a result of the accumulation of the material. In the calculations of model asperities, pile-up of material ahead of the slider has been neglected. However, the contribution by the pile-up material in some cases may be significant.

The deformation component of friction can be reduced by reducing the surface roughness, selecting materials of more or less equal hardness and by removing wear and contaminant particles from the interface. One of the ways to remove particles from the interface is to provide dimples (recesses) on the surface. Suh (1986) produced a modulated surface with a checkerboard pattern by etching away every other block (50 μm × 50 μm) of the checkerboard. The dimples created on the surface by the etching process provided a recessed space 50 μm deep in which wear particles would drop and be removed during sliding. The rise in friction of the modulated copper surface against the copper pin was negligible with modulated surfaces whereas the coefficient of friction of unmodulated copper surface against the copper pin increased by a factor of 3.

Hysteresis

A deformation (hysteresis) component of friction occurs in viscoelastic materials (such as polymers) in the so-called elastic limit, because of elastic hysteresis losses. For most metals, the fraction of energy lost in the elastic limit is less than 1%, but for viscoelastic materials such as polymers (especially elastomers), it may be large. During sliding, the material is first stressed and then the stress is released as sliding continues and the point of contact moves on, Figure 5.2.14. Note that tangential force produces an increase in deformation ahead of the

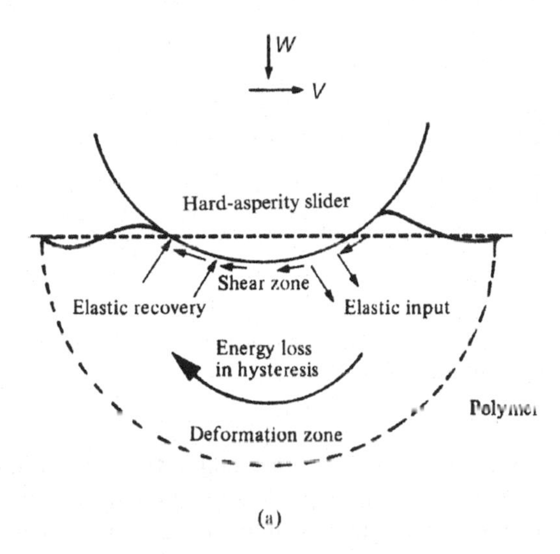

Figure 5.2.14 Sliding of a rough, hard sphere (representing a portion of an asperity) over a polymer.

indenter (Tanaka, 1961). Each time an element of volume is stressed, elastic energy is taken up by it. Most of the energy is later released as the stress is removed from the element of the body, but a small part is lost (in the form of heat) as a result of elastic hysteresis losses. Therefore, energy is fed ahead of an asperity, and some of the energy is restored at the rear of the asperity. If we recovered as much energy in the rear portion as we expended on the front portion, the net work required in sliding would be zero. However, polymers are not ideally elastic; in the course of deforming and relaxing polymers some energy is lost, referred to as hysteresis losses. The net loss of energy is related to the input energy and loss properties of the polymer at the particular temperature, pressure, and rate of deformation of the process.

Hegmon (1969) has proposed a relaxation theory that uses a Maxwell model of viscoelastic behavior to develop an equation that quantitatively defines μ_h for the hard mating-material asperities on a polymer surface in the absence of adhesional effects (for details, see Moore, 1972), as follows:

$$\mu_d = \mu_h = k_h \frac{p_a}{E'} \tan \delta \tag{5.2.22}$$

where k_h is a constant dependent on the shape of the asperity and contact length, and so forth, p_a is the apparent pressure in the whole contact area, and E' is the complex modulus and tan δ is the tangent modulus of the elastomer. E' and tan δ should be measured at the frequency of deformation. This theoretical relationship has been found to be in close agreement with experimental results on rubber (Ludema and Tabor, 1966; Moore, 1972).

5.2.2.3 Adhesion and Surface Roughness (Ratchet Mechanism)

If asperities of one surface are much smaller in lateral dimensions than that of the mating surface and contact stresses are lower than plastic flow stress, sharper asperities climb up and down over broader asperities without creating any interface damage. Energy (or force) is

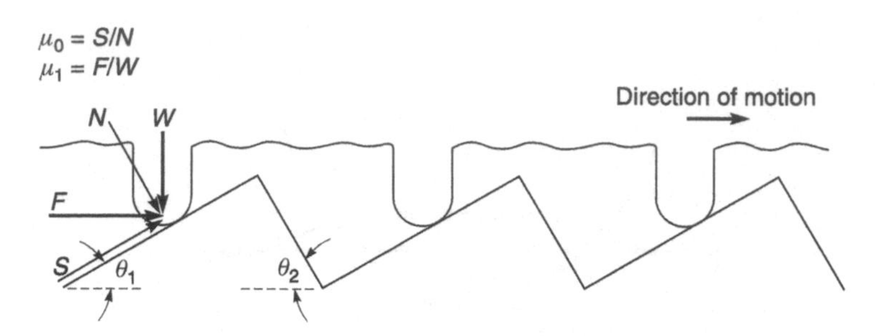

Figure 5.2.15 Schematic contact of two rough surfaces with asperities of one surface being much smaller in lateral size than that of the mating surface.

required to climb up the asperity of a given slope, and it decreases during climbing down. It is not a totally nondissipative process (a frictionless roller coaster) since the energy expended in ascending is higher than the energy in descending the asperity slopes. In sliding down the asperity, there may be impact and energy may be lost either by impact deformation or by the generation of phonons. It is believed that up to 10% of the energy used in ascending the asperities is lost during the descent (Samuels and Wilks, 1988). This dissipative mechanism is sometimes referred to as the ratchet (ride-over) mechanism. This mechanism resembles the Coulomb model.

Surface roughness can have an appreciable influence on friction if the adhesive friction is also present (Makinson, 1948; Tabor, 1979; Samuels and Wilks, 1988; Bhushan and Ruan, 1994; Bhushan, 1999a, 2011). Consider the contact of two rough surfaces with asperities of one surface much smaller in lateral size than that of the mating surface such that a small tip slides over an asperity making angles θ_1 *and* θ_2 with the horizontal plane, Figure 5.2.15. The normal force, W (normal to the general surface) applied by the small asperity to the mating surface is constant. The friction force F on the sample along the global horizontal axis varies as a function of the surface roughness. We assume that there is a true adhesive component of the coefficient of friction μ_0 such that if the local force normal to the asperity is N, the local friction force S on the asperity would be equal to $\mu_0 N$. In the presence of a symmetrical asperity ($\theta_1 = \theta_2 = \theta$), the local coefficient of friction μ_1 defined with respect to global horizontal and vertical axes in the ascending part is

$$\mu_1 = \frac{F}{W} = \frac{\mu_0 + \tan\theta}{1 - \mu_0 \tan\,\theta} \tag{5.2.23a}$$
$$\sim \mu_0 + \tan\theta,\ \textit{for small}\ \theta$$

where $\mu_0 = S/N$. It indicates that in the ascending part of the asperity one may simply add the friction force and the asperity slope to one another. Similarly, on the right-hand side (descending part) of the asperity,

$$\mu_2 = \frac{\mu_0 - \tan\theta}{1 + \mu_0 \tan\theta}$$
$$\sim \mu_0 - \tan\theta,\ \textit{for small}\ \theta \tag{5.2.23b}$$

For a symmetrical asperity ($\theta_1 = \theta_2$ in Figure 5.2.15), the average coefficient of friction experienced by a small asperity sliding across a larger asperity is

$$\mu_{av} = \frac{\mu_1 + \mu_2}{2}$$
$$\sim \mu_0 \left(1 + \tan^2 \theta\right) \quad (5.2.23c)$$

We thus see that for the case under consideration, surface roughness can influence friction. The roughness effects are important if one surface is compliant (such as fibers sliding on a hard surface) or one surface is much rougher than another surface with comparable hardness (Tabor, 1979), and in microscale friction measurements using a sharp tip on a rough surface (Bhushan, 1999a, 2011). This mechanism generally plays an insignificant role in engineering applications.

Example Problem 5.2.1

A hard ball is slid against a soft and flat surface at two different loads. At one load, the coefficient of friction is 0.20 and the groove width is 0.5 mm and at another load, the coefficient of friction is 0.25 and the groove width is 1 mm. Calculate the radius of the ball and the adhesive component of the coefficient of friction. Assume that the dominant sources of friction are adhesion and plowing and that these are additive.

Solution

$$\mu = \mu_a + \mu_p$$

For a ball on a flat surface,

$$\mu = \mu_a + \frac{4r}{3\pi R}$$

where 2r is the groove width and R is the radius of the ball. For the first load,

$$0.20 = \mu_a + \frac{4 \times 0.25}{3\pi R}$$

For the second load,

$$0.25 = \mu_a + \frac{4 \times 0.5}{3\pi R}$$

From Equations 5.2.24 and 5.2.25, we get

$$R = 2.1 \text{ mm}$$

and

$$\mu_a = 0.15$$

5.2.3 Other Mechanisms of Sliding Friction

5.2.3.1 Structural Effects

Hexagonal close-packed (HCP) metals exhibit low coefficient of friction (about 30%) less and much less wear (about a factor of ten less) than face-centered cubic (FCC) metals. One key factor that affects friction and wear is the number of slip planes. Hexagonal metals (c/a ~ 1.628 for HCP metals) have a limited number of slip planes. Five slip planes are required so that, as two rough surfaces deform, at each contact there is perfect conformance between one surface and the other. Accordingly, hexagonal metals like cobalt deform by slippage when pressed against each other, leaving many air gaps at each junction (see Figure 5.2.16). In contrast, cubic metals, which have 12 slip planes, have no such air gaps, and for this reason the contact is stronger, and the friction and wear correspondingly higher (Rabinowicz, 1995).

5.2.3.2 Grain Boundary Effects

Strained metal, that is, metal that contains a high concentration of dislocations, is chemically more active on the surface because the presence of defects increases the energy in the material. A grain boundary is a strained condition in that there are many dislocations present to help accommodate the misfit or mismatch in adjacent orientations, and there are rows of strained atoms that must help in accommodating the mismatch. Consequently, these regions are high-energy regions at the surface. The energy is greater at the boundary, and the boundary has its own characteristic energy that is separate and distinct from the energy of the grains on either side of the boundary.

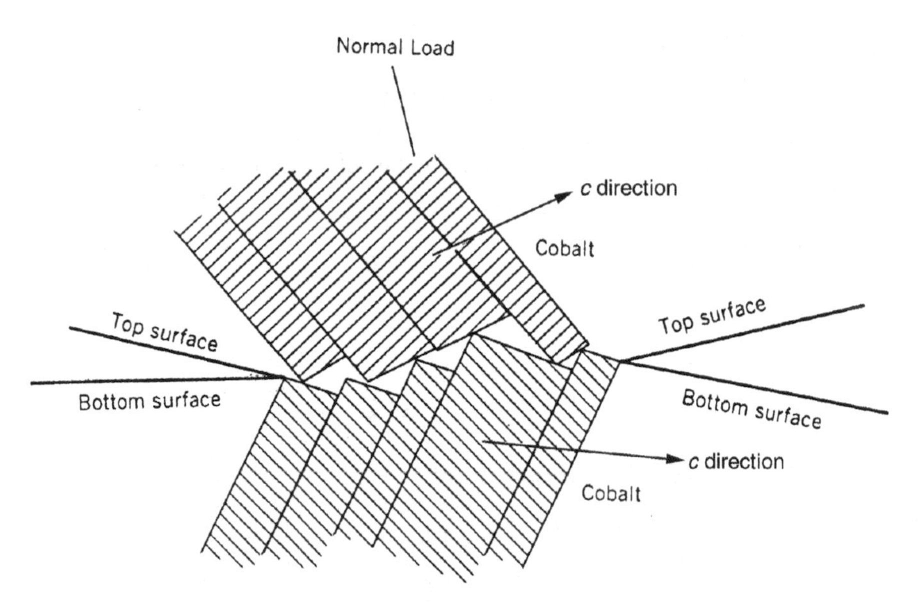

Figure 5.2.16 Schematic of plastic flow for hexagonal metals. Because of the limited number of slip planes, the two metals do not conform after slippage which reduces the degree of metal-metal contact. Reproduced with permission from Rabinowicz, E. (1995) *Friction and Wear of Material*, Second edition, Wiley, New York. Copyright 1995. Wiley.

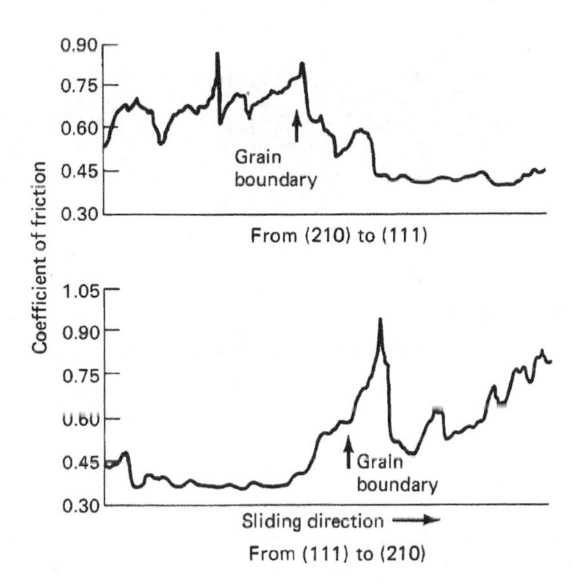

Figure 5.2.17 Coefficient of friction of polycrystalline copper slider across grain boundary on copper bicrystal, load = 1 N, sliding velocity = 0.023 mm/s. Reproduced with permission from Buckley, D.H. (1982), "Surface Films and Metallurgy Related to Lubrication and Wear," in *Progress in Surface Science* (S.G. Davison, ed.), Vol. 12, pp. 1–153, Pergamon, New York. Copyright 1982. Elsevier.

For polycrystalline materials, the presence of grain boundaries in the material influences adhesion and friction behavior, surface fracture, and wear. The near surface dislocations in the sliding process are blocked in their movement by a grain boundary, they accumulate at the grain boundary and produce strain hardening in the surficial layers. This strain hardening makes sliding more difficult and increases the friction force of materials in sliding contact (Anonymous, 1986; Buckley, 1981, 1982; Weick and Bhushan, 2000).

Sliding friction experiments have been conducted by Buckley (1982) across the surface of grain boundaries to measure the influence of the grain boundary on friction. Studies with a polycrystalline copper slider moving across a copper bicrystal [one grain the (111) and the other the (210) orientation] resulted in differences in friction not only on the surface of the grains but also in the grain boundary region as was observed with tungsten, Figure 5.2.17. In sliding from the (210) grain to the (111) grain, friction is higher on the (210) plane and in the grain boundary region than it is on the (111) plane. Grain boundary effects can be seen much more readily when sliding is initiated on the (111) surface. There is a pronounced increase in the friction for the slider-grain boundary interface. The grain boundary is atomically less dense than the grain surfaces on either side of that boundary.

Friction studies with polycrystalline tungsten (polytungsten) also indicate a grain boundary effect. Figure 5.2.18a is a magnified optical micrograph of the polytungsten sample with the 200 μm scan path superimposed, where friction measurements were obtained with a 20 μm radius diamond tip. These friction measurements are shown below the optical micrograph in Figure 5.2.18.b. Grain boundary locations are shown as dotted lines drawn between Figure 5.2.18a and Figure 5.2.18b. The data show that changes in friction indeed occur when

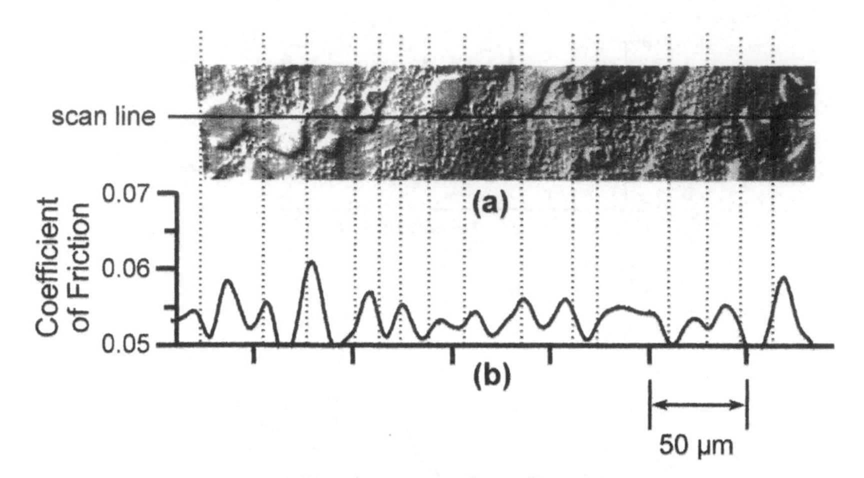

Figure 5.2.18 (a) Optical micrograph of polytungsten surface, and (b) coefficient of friction data acquired along the 200 μm scan path in (a) by sliding a 20-μm radius diamond tip, load = 1 g, sliding velocity = 0.3 mm/s. Reproduced with permission from Weick, B.L. and Bhushan, B. (2000), "Grain Boundary and Crystallographic Orientation Effects on Friction," *Tribol. Trans.* **43**, 33–38. Copyright 2000 Taylor and Francis.

grain boundaries are crossed. Although these changes appear to be mainly peaks in friction at the grain boundaries, there are some boundaries where decreases in friction appear to occur, and others where the friction appears to be highest within the grain itself. Since the grain boundaries are known to be high-energy sites in polycrystalline metals, one would expect an increase in the work of adhesion to be associated with passing over grain boundaries. This would in turn cause an increase or peak in friction. On the other hand, different crystallographic directions will have different hardness values. Directions with high atomic densities will have high hardness values and vice versa. Furthermore, as a slider moves out of a grain, across a boundary, and into another grain, the orientation of the crystallographic slip planes will change, and this could also be associated with changes in friction. As a result, the relative maxima and minima in friction may be located within the grain itself or along a grain boundary. Whether or not peaks in friction occur at the grain boundaries depends on both the increased work of adhesion at the grain boundaries due to the higher surface energies, and whether or not this increased work of adhesion exceeds the friction associated with the particular crystallographic direction in which the slider is traveling on either side of the grain boundary.

5.2.4 Friction Transitions During Sliding

During sliding, changes in the conditions of mating surfaces occur which affect friction and wear properties. After some period, the so-called "run-in," "break-in," or "wearing-in" period, the friction force generally stabilizes into what is called steady-state sliding. Typically after sliding for a period of time, friction increases again and reached another plateau as shown by the S-shaped curve in Figure 5.2.19a. This process can continue approaching more than two plateaus. After a useful interface lifetime, the interface fails and friction may become

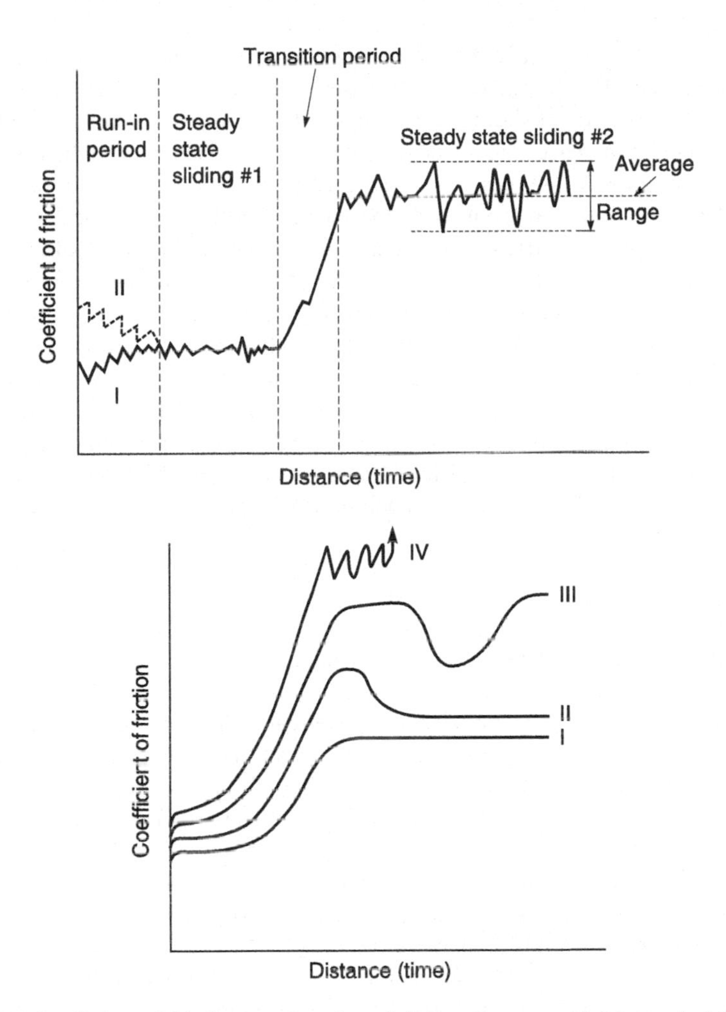

Figure 5.2.19 Coefficient of friction as a function of sliding distance with (a) a typical S-shaped curve showing run-in period, and (b) four hypothetical cases.

very high. During run-in, for example, high asperities may be knocked off, surfaces may mate better, initial surface films may be worn, new steady films may be formed, or structural changes may occur. These changes result in friction either going up or coming down from the initial value. The run-in period is critical for long interface life as lack of run-in can result in serious damage and early failure. After the first steady-state period, changes in the interface may further occur, such as roughening and trapped particles which lead to an increase in friction to another plateau, a steady-state period. The shape of friction curves can be affected by the interface materials as well as by the operating conditions.

Friction may increase in different patterns, such as: (1) the friction may remain at its initial value for some time and slowly increases to another steady-state value; (2) after being at initial value for some time, it may first increase to a high value then level off at a lower value (but higher than the initial value); or (3) it may increase to a high value, level off to this value, drop to a lower value and increase again to a high value; (IV) it may change in a nonrepeatable manner, Figure 5.2.19b. In all cases, the coefficient of friction would reach a high value after some period of sliding. Identical metals against each other exhibit the behavior shown in case (1); the increase is associated with plowing because of roughening and trapped wear particles. In smooth surfaces involving elastic deformation with dominant adhesive component of friction, the increase is associated with smoothing of the surfaces leading to a larger component of adhesive friction (Bhushan, 1996). The drop in the coefficient of friction in the case (2) is associated with smoothing of the two hard surfaces experiencing plastic deformation, which results in a drop in the plowing component of the friction. For elastic contacts where adhesive component is dominant, roughening and/or trapped wear particles reduce the real area of contact, which in turn reduces the adhesive component of the friction. The drop in the friction in case (3) in plastic contacts is associated with the ejection of wear particles, and a subsequent increase is associated with the generation and entrapment of wear particles. A significant increase in the friction to an unacceptably high value in a short period in case IV is associated with a poor material pair in which friction is contributed by all sources.

5.2.5 *Static Friction*

For two contacting surfaces, the friction force required to initiate motion is either more than or equal to the force needed to maintain the surfaces in the subsequent relative motion. In other words, the coefficient of static friction μ_s is greater than or equal to the coefficient of kinetic friction μ_k. The static friction is time-dependent. The length of the rest time (dwell time, time of sticking, or duration of contact) for which two solids are in contact affects adhesion consequently the coefficient of static friction. The coefficient of static friction can either decrease or increase with the rest time. During rest, if the contact becomes contaminated with lower shear-strength species, the coefficient of static friction will tend to decrease. On the other hand, if the contact is clean and a more tenacious interfacial bond develops, the coefficient of static friction tends to increase.

For freshly cleaved rock salt, the formation of surface films over time lowers the static friction (Kragelskii, 1965), Figure 5.2.20. On the other hand, the opposite trend is observed in the case of clean metallic and other surfaces. Sampson *et al.* (1943) have shown that the coefficient of static friction for small durations of stationary contact is equal to the kinetic friction. Coulomb (1785) reported that after four days of rest time, the coefficient of static friction for an oak slider on an iron bed grew by a factor of about 2.4. Dokos (1946) has reported that a plot of μ_s for steel on steel surfaces as a function of log t approximates to a straight line, Figure 5.2.21. Therefore, for small values of t, the slope of the μ_s versus time on a linear scale will be steep, and for large values it will be small. Other data show that μ_s approaches a maximum value after some time (Kragelskii, 1965).

The static friction of dry surfaces is believed to increase because of plastic flow and creep of interface materials and the degree of interaction of the atoms on the mating surfaces under load. There are two models for time-dependent static friction. The first is based on an exponential growth in friction with time (e.g., Ishlinskii and Kragelskii, 1944; Kato *et al.*, 1972). The coefficient of static friction at a time t,

$$\mu_s(t) = \mu_\infty - (\mu_\infty - \mu_0)\exp(-\alpha t_s) \tag{5.2.24a}$$

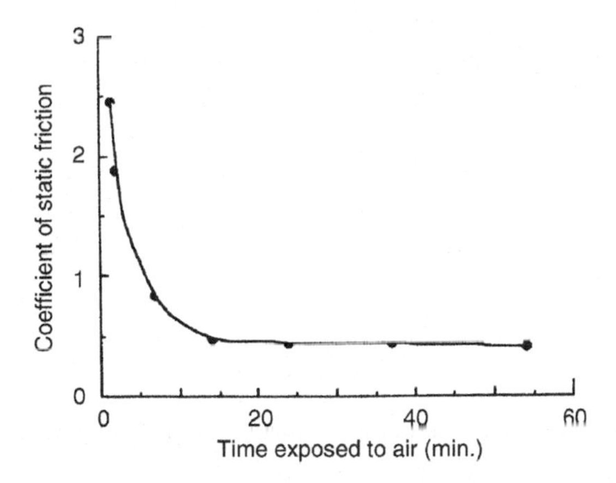

Figure 5.2.20 Coefficient of static friction as a function of time of exposure to air for cleaved salt surface. Reproduced with permission from Kragelskii, I.V. (1965), *Friction and Wear*, Butterworths, London, UK.

where μ_∞ is the limiting value of the coefficient of static friction at long times, μ_0 is the initial value of coefficient of static friction, t_s is the rest time, and α is a constant. This model suggests that static friction reaches a maximum value after a certain time, Figure 5.2.22. The second model is based on the power law (e.g., Rabinowicz, 1958; Brockley and Davis, 1968):

$$\mu_s(t) = \mu_0 + \alpha\ t_s^\beta \tag{5.2.24b}$$

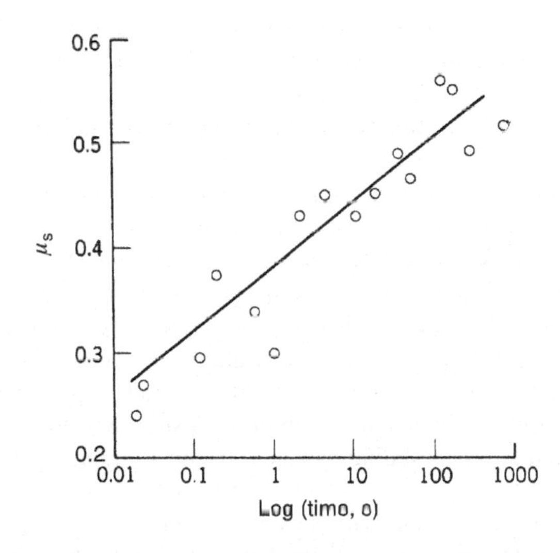

Figure 5.2.21 Coefficient of static friction as a function of rest time for steel on steel in air. Reproduced with permission from Dokos, S.J. (1946), "Sliding Friction Under Extreme Pressures," *J. Appl. Mech.* **13**, 148–156. Copyright 1946. ASME.

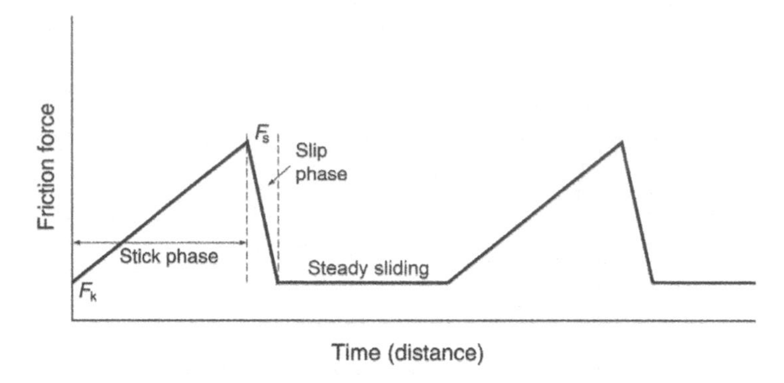

Figure 5.2.22 Friction force as a function of time or distance showing stick-slip behavior.

where α and β are two constants. In this model, the rate of increase in friction force decreases with time but it is not bounded asymptotically. μ_0 in either model is approximately equal to μ_k.

The increase in static friction with rest time is undesirable for many industrial applications requiring the intermittent operation of remotely controlled mechanisms, such as the antennas and other moving parts in Earth-observing satellites.

5.2.6 Stick-Slip

Sliding of one body over another under a steady pulling force proceeds sometimes at constant or nearly constant velocity, and on other occasions at velocities that fluctuate widely. If the friction force or sliding velocity does not remain constant as a function of distance or time and produces a form of oscillation, it may be based on a so-called "stick-slip" phenomenon, Figure 5.2.22. The term "stick-slip" was coined by Bowden and Leben (1939). During the stick phase, the friction force builds up to a certain value, and once a large enough force has been applied to overcome the static friction force, slip occurs at the interface. Usually, a saw tooth pattern in the friction force-time curve is observed during the stick-slip process. This classical stick-slip requires that the coefficient of static friction is markedly greater than the coefficient of kinetic friction. The stick-slip events can occur either repetitively or in a random manner (Bowden and Tabor, 1964; Armstrong-Helouvry, 1991; Rabinowicz, 1995). If the coefficient of kinetic friction varies with the sliding velocity such that slope of μ_k versus velocity curve is negative at a velocity, harmonic oscillations are sometimes observed, generally at high velocities. It should be noted that oscillatory variation in friction force with time does not necessarily always mean that the variation is caused by a stick-slip process.

The stick-slip process generally results in vibration, resulting in an audible squeal ($\sim$ 0.6–2 kHz) and chatter ($<$ 0.6 kHz) in sliding systems. In most sliding systems, the fluctuations of sliding velocity (or unstable motion) resulting from the stick-slip process, and associated squeal and chatter, are considered undesirable, and measures are normally taken to eliminate, or at any rate to reduce, the amplitude of the fluctuations (Bhushan, 1980a). The stick-slip process can be responsible for the squeal and chatter in bearings, the jerking of brakes, the chatter of windshield wipers on partly wet window glass, earthquakes, and inaccuracies in machining and positioning. Stick-slip can also be a source of pleasure, as in bowing stringed musical instruments.

Stick-slip occurs over a wide range of time scales. Oscillation frequencies can range from within the audible range ($\sim$ 2-5 kHz) to earthquakes with less than one slip in every couple of hundred years ($\sim 10^{-9}$ Hz). The elasticity of a mechanical system and frictional properties are critical in permitting stick-slip to occur.

5.2.6.1 Mechanisms of Stick-Slip

The classical form of stick-slip can arise whenever the coefficient of static friction is markedly greater than the coefficient of kinetic friction. To model the stick-slip behavior, we consider a typical sliding situation of a block of mass m with a normal force W acting on the block against a lower surface which is moving at a velocity V, as shown in Figure 5.2.23. The block is restrained by a spring element (of stiffness k) and a linear dashpot (of coefficient η) attached to a fixed support. The absolute displacement x of the block is measured from a reference line. If the coefficient of friction between the lower surface and the block is μ and it is sufficiently large at the equilibrium position, the block will stick to the lower surface and move along it with an absolute velocity of value $\dot{x} = V$. During the stick period the force relationship may be written as

$$\eta V + kx < \mu_s W \tag{5.2.25}$$

During the stick, the spring force increases with time at a rate kVt (or kx) as the slider will be displaced from point A to point B as indicated in Figure 5.2.24. Up to point B, the

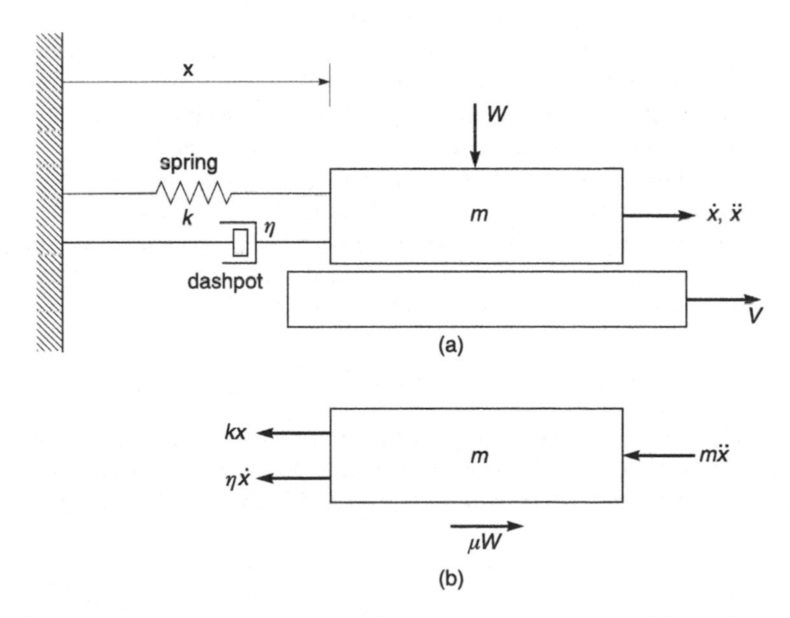

Figure 5.2.23 (a) A friction test apparatus of a rough horizontal body sliding relative to a block restrained by a spring element and a dashpot, and (b) a free body diagram of the block.

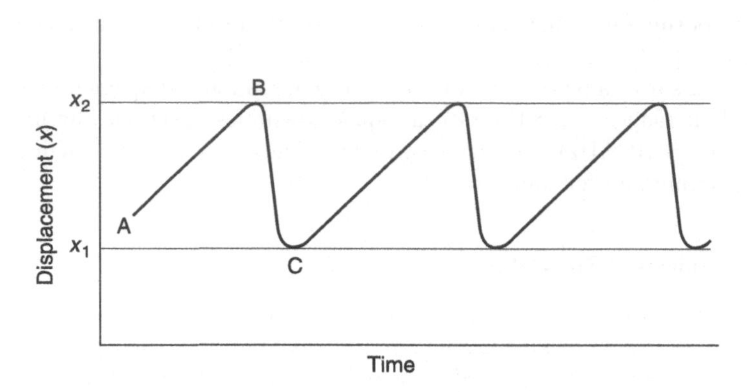

Figure 5.2.24 Displacement of block as a function of time during stick-slip behavior.

static friction force is capable of withstanding the combined restoring forces consisting of the constant damping force ηV and the increasing spring force kx. At point B, the restoring forces overcome the static friction force $\mu_s W$ and slip occurs to point C. As we have stated earlier, μ_s is a function of rest time. Therefore, time of stick during initial sliding will be longer for a longer rest time.

During the slip period, the motion of the block can be described by the equation

$$m\ddot{x} + \eta\dot{x} + kx = \mu_k W \tag{5.2.26}$$

Under equilibrium slip conditions, the block remains stationary and the lower surface moves at a constant velocity. During slip, the coefficient of kinetic friction, μ_k, may vary with velocity. It must, in fact, be lower than the static value if stick-slip vibrations are to occur. Harmonic oscillations are sometime observed at high sliding velocities. Their characteristic feature is that the movement of the block remains closely approximating to a simple harmonic oscillation. For this form of oscillation, the slope of μ_k versus velocity curves at the sliding velocity must be negative. At the instant considered, if the block is moving right with a velocity $\dot{x}$ and the bottom surface is moving at velocity V, the velocity of the bottom surface relative to the block is $V_r = V - \dot{x}$. Several empirical models have been proposed to relate μ_k to V_r, such as linear, exponential or polynomial in velocity. If μ_k is assumed to decrease linearly with an increase in V_r, i.e., if the μ_k versus V_r curve has a negative slope, then

$$\mu_k = \mu_k^0 - \alpha V_r \tag{5.2.27}$$

where μ_0 is the intercept of the tangent of the $\mu - V_r$ curve with the μ axis, which is approximately equal to μ_s, α is the slope of the $\mu_k - V_r$ curve at a given value of relative velocity, $d\mu_k/dV_r$ (with units of s/m). From Equations 5.2.26 and 5.2.27, we get

$$m\ddot{x} + (\eta - \alpha W)\dot{x} + kx = (\mu_k^0 - \alpha V)W \tag{5.2.28}$$

If the slope of the $\mu_k - V$curve is such that $\alpha > \eta/W$, the system has no damping and the negative damping coefficient in Equation 5.2.28 can be shown to give an exponentially increasing amplitude of vibration. Note that the negative damping coefficient feeds energy into the system and makes the stick-slip phenomenon possible. Their characteristic feature is that the block vibrates in a manner that closely approximates to a simple harmonic oscillation.

For the undamped oscillation case such that $\alpha = \eta/W$ and assuming that the block is stationary when first brought in contact with the moving surface (Barwell, 1956),

$$x = \left(\frac{W}{k}\right)\left(\mu_k^0 - \alpha V\right)\left[1 - \cos\left(\frac{k}{m}\right)_t^{1/2}\right] \tag{5.2.29}$$

The frequency of simple harmonic motion is $(k/m)^{1/2}$. This represents an undamped oscillation which does not tend to increase with time. As the damping is further increased ($\eta > \alpha W$), oscillations will diminish with time. These harmonic oscillations are observed mainly at high sliding velocities. For reference, the negative $\mu_k - V$ slopes observed at high speeds are connected with thermal softening, which produces a low-shear surface film on a harder substrate at high interface temperatures.

5.2.6.2 Prevention of Stick-Slip

We have just seen that the stick-slip process may occur when one or both of the following two features along with some elasticity of a mechanical system are present: the coefficient of static friction is greater than the coefficient of kinetic friction and/or the rate of change of the coefficient of kinetic friction as a function of velocity at the sliding velocity employed is negative. A number of methods are used to minimize or prevent the stick-slip process. One can design the mechanical system so that the amplitude of any such oscillations would be small; this can be done by reducing the system compliance (i.e. by making the spring stiff) and/or increasing both the system damping and inertias of sliding bodies. Another approach is to select the friction pair so that the difference between μ_s and μ_k is small; the practical value of doing this is by covering one of the moving surfaces by a boundary lubricant film (Bhushan, 1980a). A third approach is to select a friction pair that exhibits a positive $\mu_k - V$ characteristic at the sliding velocity employed; this may require the use of soft lubricant films which exhibit desirable properties. With most metals, μ_k decreases as velocity increases, and with softer metals and polymers μ_k may increase with increasing velocity up to some rather low velocity; at higher velocities, the friction then decreases. At high velocities, the μ_k of practically all materials decreases with increasing velocity.

Example Problem 5.2.2

In the sliding system shown in Figure 5.2.23, a block of mass m of 1 kg contacts a lower surface which is sliding with a velocity V of 10 m/s. The normal load W being applied at the interface is 10 N. The $\mu_k - V$ curve has a negative slope and can be expressed by the equation

$$\mu_k = 0.3 - 0.1V_r$$

where V_r is the relative velocity in m/s. If the system stiffness can be modeled with a spring constant of 10 N/mm, how much damping coefficient of the system is required to avoid stick-slip?

Solution

From Equation 5.2.28, the damping coefficient of the system required to avoid stick-slip is

$$\begin{aligned}\eta &> \alpha W \\ &= 0.1 \times 10\text{N s/m} \\ &= 1\text{N s/m}\end{aligned}$$

5.2.7 Rolling Friction

It is much easier to roll surfaces than to slide them. Rolling friction is the resistance to motion that takes place when a surface is rolled over another surface. The term rolling friction is usually restricted to bodies of near perfect (continuous) shapes with very small surface roughness. With hard materials, the coefficient of rolling friction between a cylindrical or spherical body against itself or a flat body generally is in the range of 5×10^{-3} *to* 10^{-5}. In comparison, the coefficient of sliding friction of dry bodies ranges typically from 0.1 to sometimes much greater than 1.

During rolling of two surfaces relative to each other, any relative motion can be regarded as a combination of rolling, sliding and spin (Johnson, 1985). Consider two nonconforming bodies which touch at a single point, O, Figure 5.2.25. Sliding (or slip) is the relative linear velocity between the two surfaces at the contact point O in the tangent plane, rolling is the relative angular velocity between the two bodies about an axis lying in the tangent plane, and spin is the relative angular velocity between the two surfaces about the common normal

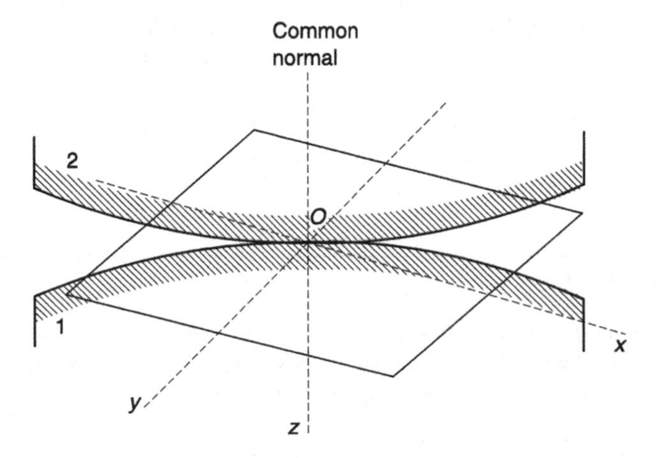

Figure 5.2.25 Two non-conforming bodies 1 and 2 in contact at point O.

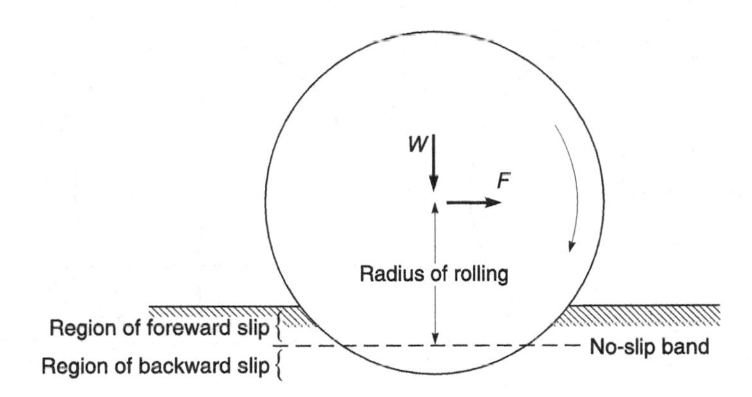

Figure 5.2.26 Schematic of a sphere rolling on a flat surface. Reproduced with permission from Rabinowicz, E. (1995) *Friction and Wear of Material*,Second edition, Wiley, New York. Copyright 1995. Wiley.

through O. We define free rolling as a rolling motion in which no tangential (friction) force or sliding (slip) can occur. Tractive rolling is the rolling motion in which the friction force or slip is nonzero; in the driving wheels of a train on the tracks or traction drives, large tangential forces are transmitted. The simplest form of free rolling occurs between two bodies which have the same elastic properties, are geometrically identical and experience little deformation in the contact region. In tractive rolling, the friction force must be less than or equal to μW, in the contact region where μ is coefficient of the sliding friction and W is the normal load. When friction force attains μW, local sliding (microslip) or gross sliding (in the entire contact) occurs.

5.2.7.1 Types of Slip

Three different cases of microslip are identified in the literature, discussion of which follows.

First, we consider a Hertzian contact of two nonconforming bodies having different elastic constants. If the two bodies roll freely together, the load that acts on each gives rise to unequal tangential displacements of the surfaces, leading to slip at the interface. This type of slip is called Reynolds slip (Reynolds, 1876).

Next, we consider tractive rolling of two nonconforming bodies having the same elastic properties, subjected to a tangential force which is less than required to cause gross sliding. It has been found that in the case of tractive rolling of elastically similar cylinders, the stick (no slip) region coincides with the leading edge of the rectangular contact area and the slip is confined to the trailing edge. As the tangential force increases, the slip zone extends forward until $F = \mu W$ holds across the whole area and gross sliding occurs. In contrast, in the two stationary cylinders subjected to a tangential force, there is a central stick area and two outer slip areas. This type of slip is called Carter–Poritsky–Foppel slip (Poritsky, 1950).

Now, we describe a third type of slip. If the contact of two nonconforming bodies, for example a sphere rolling on a flat surface, as shown in Figure 5.2.26, were a point, pure rolling conditions would prevail. However, in most cases, the contact region is elastically or plastically

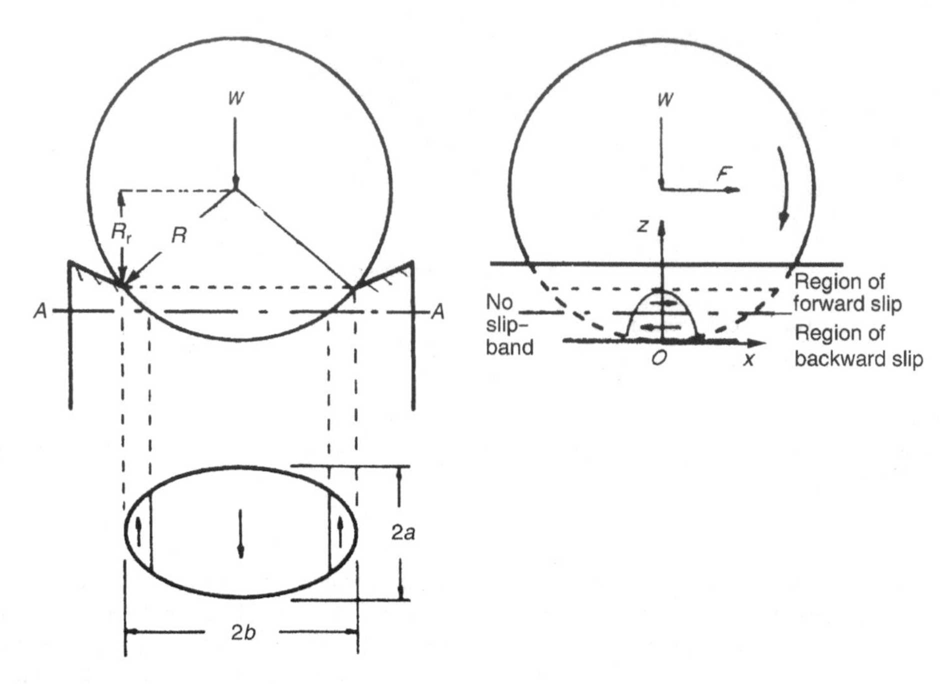

Figure 5.2.27 Schematic of a ball rolling on a grooved track and the slip pattern in the contact size. Reproduced with permission from Arnell, R.D., Davies, P.B., Halling, J., and Whomes, T.L. (1991), *Tribology-Principles and Design Applications*, Springer-Verlag, New York. Copyright 1991. Springer.

deformed, so the contact is made over an area and the points within the contact region lie in different planes. As a result, pure rolling takes place at a very small number of points but a combination of rolling with a small degree of sliding or slip takes place at all other points. To achieve this slipping, the sliding resistance at the interface has to be overcome and rolling friction must be present (Rabinowicz, 1995). In this case, the contact area lies substantially in a single plane. In many engineering applications, such as a ball rolling along a grooved track in ball bearings, the contact area lies in very different planes, Figure 5.2.27. The contact area is an ellipse. During rolling of the ball through one complete revolution along the grooved track, the center of the ball measures out its full diameter 2R along the track, while points on the ball at the edges of the contact zone measure out a smaller distance corresponding to radius R_r. These differences can be accommodated by the ball slipping in its track similar to the ball on a deformed track on a flat surface just described. The contact area in either case consists of three zones of slip - a single central zone of backward slip and two outer zones of forward slips, Figure 5.2.25 (Arnell *et al.*, 1991). In this situation, the axis AA is the instantaneous axis of rotation which passes through the two boundaries between the regions of forward and backward slip. This type of slip is known as Heathcote slip (Heathcote, 1921).

In rolling contact systems, such as rolling element bearings and gear teeth contacts, slip can occur by factors other than geometrical ones. In some cases, motion involves spin about the region of contact, such as in a thrust ball bearing (Johnson, 1985). In some cases, gross slippage may occur resulting in high friction.

5.2.7.2 Mechanisms of Rolling Friction

In the case of rolling contacts, rolling friction arises from the resistance to rolling and because of slip. The magnitude of the resistance to rolling is usually much less than that during slip. As discussed earlier for sliding, energy dissipation in rolling also occurs due to adhesive losses and deformation losses (plastic deformation or elastic hysteresis) during stress cycling of the contacting surfaces. The joining and separation of surface elements under rolling contacts are little different from those of sliding contacts. Owing to differences in kinematics, the surface elements approach and separate in a direction normal to the interface rather than in a tangential direction. Therefore, contact growth is unlikely in the main part of the contact area. Consequently, at the regions within the rolling contact interface where no relative motion in a tangential direction occurs, adhesive forces may be mainly of the van der Waals type. Short range forces such as strong metallic bonds may act only in microcontacts within the microslip area. If adhesive bonds are formed, they are separated at the trailing end of the rolling contact in tension rather than in shear, as in a sliding contact. Therefore, the adhesion component of friction may be only a small fraction of the friction resistance. Since the rolling friction in free rolling conditions is not affected much by the presence of lubricants, the contribution of adhesion to the rolling friction is small and it arises mainly from deformation losses. For rolling metals under specific situations, an adhesion component can nevertheless be the dominant factor in determining the order of coefficient of rolling friction for different metal pairs.

In elastic contact during free rolling, energy dissipation occurs because of elastic hysteresis. If the fraction of energy loss of the maximum elastic strain energy stored during the cycles is α (hysteresis loss factor), based on D. Tabor, the coefficient of rolling friction for a cylinder of radius R rolling freely on a plane (rectangular contact) is given as (Stolarski, 1990)

$$\mu_r = \frac{2\alpha a}{3\pi R} \tag{5.2.30a}$$

where a is the half width of the contact. For a sphere of radius R rolling freely on a plane (elliptical contact) it is

$$\mu_r = \frac{3\alpha a}{16R} \tag{5.2.30b}$$

where a is the half-width of the contact in the direction of rolling. Based on Hertz analysis, a is a function of the normal load and of the elastic properties of mating surfaces, thus μ_r is a function of these parameters as well.

In the case of tractive rolling, sliding resistance during slip arises from both adhesive and deformation losses. Slip velocities are generally on the order of 10% of the overall rolling velocity or less. The coefficient of rolling friction as a result of slip is given as

$$\mu_r = \frac{V_s}{V_r}\mu_k \tag{5.2.31}$$

where V_s and V_r are the slip velocity and rolling velocity, respectively and μ_k is the coefficient of kinetic friction in sliding.

Most rolling contacts are subjected to repeated stress cycles and the conventional yield criterion does not hold. During the first contact cycle, material at the subsurface or surface is plastically deformed at contact stresses above the first yield according to a yield criterion and residual compressive stresses are introduced. During subsequent rolling cycles, the material is subjected to the combined action of residual and contact stresses. Accumulation of residual stresses results in a decrease of the degree of plastic deformation with repeated cycles, and the deformation becomes fully elastic after a certain number of cycles. The process is referred to as *shakedown* and the maximum pressure for which it occurs is called the shakedown limit. However, if the contact stress exceeds a second critical value, which is higher than the value of the first yield, there is some plastic deformation on each stress cycle. For example, the ratio between the maximum shakedown pressure and the pressure for first yield for free rolling of an elastic sphere on an elastic-plastic half space is about 2.2 (Johnson, 1985).

Finally, we discuss other miscellaneous factors for friction losses. Lack of roundness and roughness of a rolling body or presence of contaminant particles at the interface contribute to the roughness component of friction. Asperities on the rolling surfaces may result in plastic deformation, leading to deformation losses. Therefore, the friction of smoother surfaces is less than that of rough surfaces. In the case of lubricated interfaces, viscous losses in the liquid lubricants would occur.

5.3 Liquid-Mediated Contact

With the presence of a thin liquid film with a small contact angle (wetting characteristics) such as a lubricant or adsorbed water layer at the contact interface, curved (concave) menisci form around contacting and noncontacting asperities due to the surface energy effects. The attractive meniscus force arises from the negative Laplace pressure inside the curved (concave) meniscus as a result of surface tension. The product of this pressure difference and the immersed surface area of the asperity is the attractive (adhesive) force and is referred to as the meniscus force. This intrinsic attractive force may result in high static friction, kinetic friction and wear. The problem of high static friction in liquid-mediated contacts is particularly important in an interface involving two very smooth surfaces such as in the computer data storage industry and in micro/nanodevices and is commonly referred to as "stiction" (Bhushan, 1996, 2010).

The total normal force on the wet interface is the externally applied normal force plus the intrinsic meniscus force. Therefore, during sliding, in the absence of any hydrodynamic effects, the force required to initiate or sustain sliding is equal to the sum of the intrinsic (true) friction force F_i and the stiction force F_s; the latter is a combination of the friction force due to the meniscus and viscous effects (Bhushan, 1996):

$$F = F_i + F_s = \mu_r(W + F_m) + F_{v||} \tag{5.3.1}$$

where μ_r is the true coefficient of friction in the absence of meniscus, and is smaller than the measured value of $\mu = F/W$. The sum of W and F_m is the total normal load. F_m is the meniscus force in the normal direction, and $F_{v||}$ is the viscous force in the sliding direction. The friction force ($\mu_r W$) depends on the material properties and surface topography, whereas F_m depends on the roughness parameters as well as the type of liquid and its film thickness.

$\mu_r F_m + F_{v||}$ is the friction force due to liquid-mediated adhesion. In a well-lubricated contact, the shear primarily occurs in the liquid film. The stress required to shear the liquid increases with an increase in the sliding velocity and the acceleration. Consequently, the coefficients of static and kinetic friction generally increase with the sliding speed or acceleration.

The coefficient of friction, μ, including the effect of the meniscus and viscous force, is given by

$$\mu = \frac{F}{W} = \mu_r \left(1 + \frac{F_m}{W}\right) + \frac{F_{v||}}{W} \tag{5.3.2}$$

F_m *and* $F_{v||}$ calculations can be made based on the analyses presented in Chapter 4 on adhesion. For static friction calculations at low velocities and accelerations, the viscous effect can be neglected.

For two surfaces in contact in the presence of liquid, coefficients of static and kinetic friction are a function of the amount of liquid present at the surface with respect to the interplanar separation (see Chapter 4). For rough surfaces with a composite roughness σ with a uniform liquid film of thickness h, to first-order, friction force is a function of h/σ (Gao and Bhushan, 1995; Bhushan, 1996), Figure 5.3.1. The coefficient of friction remains low below a certain values of h/σ and increases, in some cases rapidly, beyond this value. Larger values of h/σ correspond to a larger number of asperities wetted by the liquid film, resulting in a larger meniscus and viscous contributions. Below the critical value, much of the liquid remains in the valleys and does not readily form menisci. It appears that for low static and kinetic friction, h/σ should be less than or equal to about 0.5. Of course, if the contact is immersed in the liquid, menisci are not formed and shear occurs in the liquid film, leading to very low friction. Durability data in Figure 5.3.1 show that the durability increases with an increase in lubricant film thickness and with a decrease in the surface roughness.

In a humid environment, the amount of water present at the interface increases with an increase in the relative humidity. The adsorbed water film thickness on a diamond-like carbon-coated magnetic disk, for example, can be approximated as follows (Bhushan and Zhao, 1999),

$$h = h_1(RH) + h_2 \exp\left[\alpha\,(RH - 1)\right] \tag{5.3.3}$$

where $h_1 = 0.3$ nm, $h_2 = 0.5$ nm, $\alpha = 20$ and RH is the relative humidity fraction ranging from 0 to 1. In the data shown in Figure 5.3.2, the coefficient of static friction of the lubricated disk increases rapidly above a relative humidity (RH) of about 60%. This critical humidity is dependent upon the interface roughness. Trends observed in Figures 5.3.1 and 5.3.1 are consistent with those predicted by contact modeling in Chapter 4. The coefficient of friction of the unlubricated disk remains low at high humidities. It is the total liquid film thickness (including water and lubricant), which contributes to the meniscus effect; therefore, an unlubricated disk can sustain much more water condensation than a lubricated disk before friction increases significantly. For kinetic friction, little change is observed. The coefficient of kinetic friction of the unlubricated disk remains unchanged with humidity, whereas the kinetic friction of the lubricated disk increases slightly above 60% RH. The durability of a lubricated disk increases with an increase in the relative humidity but decreases at high humidities. The durability of an unlubricated disk increases with an increase in humidity. Condensed water acts as a lubricant

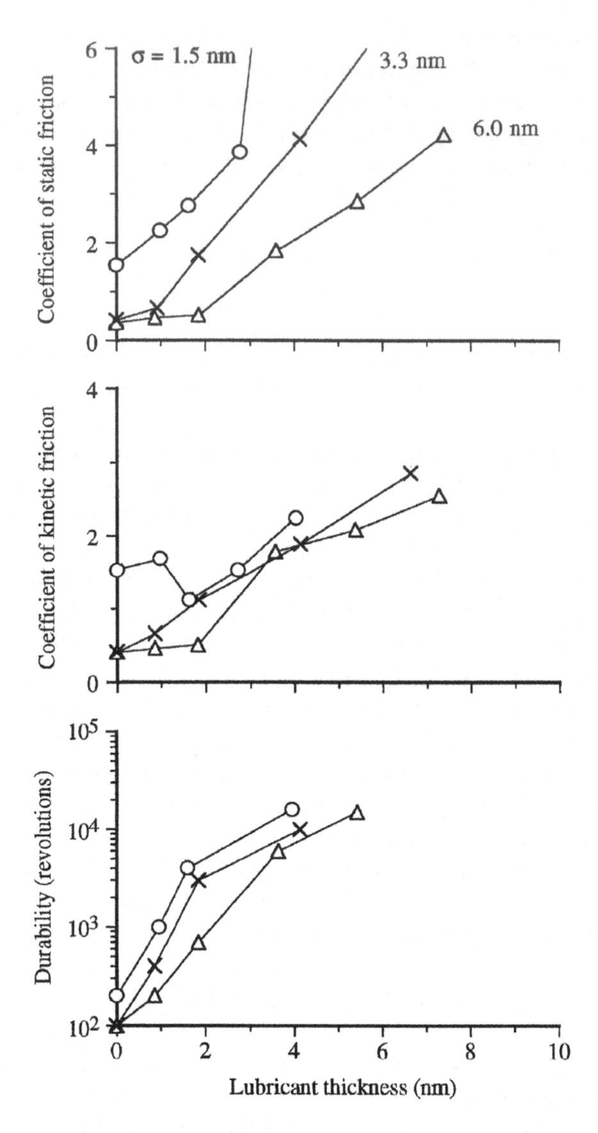

Figure 5.3.1 Coefficients of static friction (after a rest time of 100 s) and kinetic friction and durability as a function of lubricant film thickness (kinematic viscosity = 150 cSt) for a disk-head interface with lubricated thin-film magnetic disks with three roughnesses against a smooth slider.

and is responsible for an increase in the durability, whereas the drop in durability at high humidity in the case of lubricated disk occurs because of high static friction.

Static friction starts to increase in some cases, rapidly beyond a certain rest time and then levels off, Figure 5.3.3. The rest time required for increased static friction is again dependent upon the total liquid present at the interface; a lubricated disk requires less rest time than an unlubricated disk (Zhao and Bhushan, 1998).

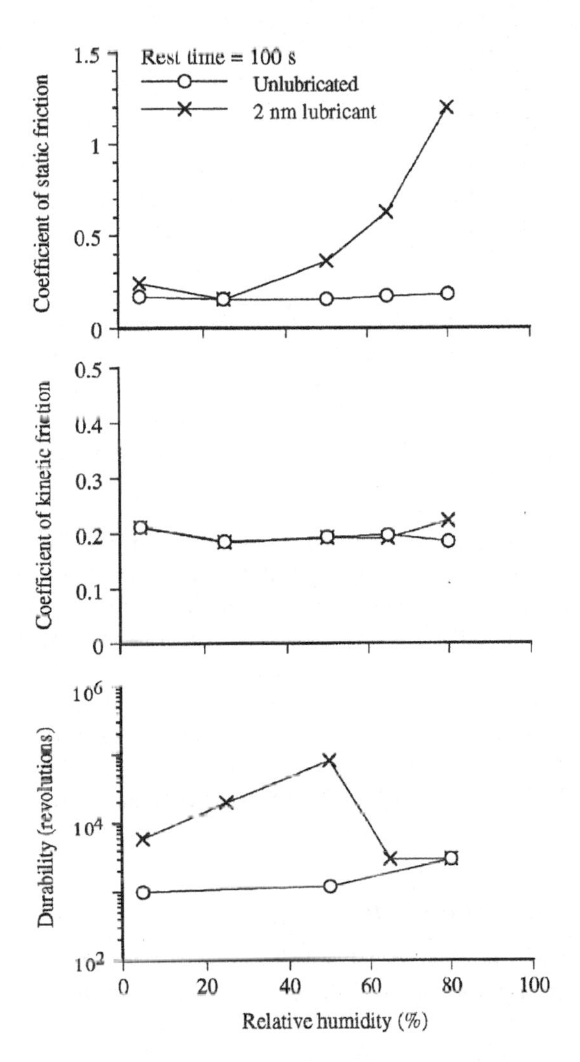

Figure 5.3.2 Coefficients of static friction (after a rest time of 100 s) and kinetic friction and durability as a function of relative humidity for unlubricated and lubricated disk-head interface.

Figure 5.3.4 shows the coefficient of static friction as a function of acceleration (Gao and Bhushan, 1995; Bhushan, 1996). Static friction increases with the acceleration because of the viscous effects as predicted from the analysis presented in Chapter 4.

5.4 Friction of Materials

The coefficient of friction of a material is dependent upon the counterface or mating material (or material pair), surface preparation and operating conditions. For example, material handling, such as the transfer of greasy materials from hands to the material surface, and the formation of chemically reacted products due to exposure to an environment, can change the surface

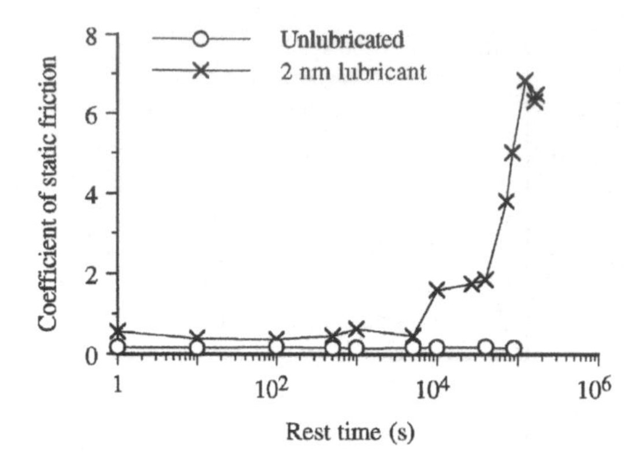

Figure 5.3.3 Coefficient of static friction as a function of rest time for unlubricated and lubricated disk-head interface.

chemistry, which may significantly affect the frictional properties. Therefore, the usefulness of the coefficient of friction values from any published literature lies more in their relative magnitudes than in their absolute values. A number of handbooks present typical friction values of a variety of material pairs (Neale, 1973; Avallone and Baumeister, 1987; Blau, 1992a, 1996; Bhushan and Gupta, 1997). Wood, leather and stones were among the earliest materials to be used for bearings and other structural components and their typical friction values are presented by Bhushan (1999a). Typical friction values of metals, alloys, ceramics, polymers and solid and liquid lubricants are presented in Tables 5.4.1 to 5.4.3. The coefficient of static friction values may be as much as 20-30% higher than that for kinetic friction values in some cases.

5.4.1 Friction of Metals and Alloys

As indicated in Chapter 4, the clean metal and alloy surfaces in contact exhibit high adhesion, and consequently high friction and wear. The coefficient of friction of contacting metallic

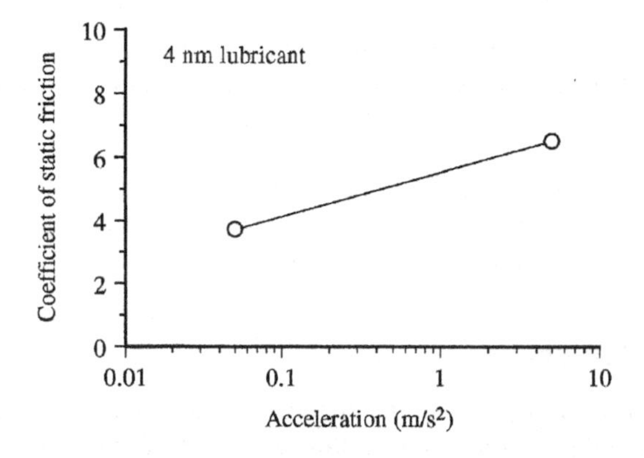

Figure 5.3.4 Coefficient of static friction (after a rest time of 100 s) as a function of acceleration of the disk during start-up of the magnetic disk drive.

Table 5.4.1 Typical values of the coefficient of kinetic friction of unlubricated metals and alloys sliding on themselves or on mild steel at room temperature in air.

Material	Coefficient of friction	
	Self-mated	On mild steel
Pure Metals		
Precious Metals		
Au, Pt	1–1.5	0.4–0.5
Ag	0.8–1.2	0.3–0.5
Soft Metals		
In, Pb, Sn	0.8–2	0.5–0.8
Metals		
Al	0.8–1.2	0.5–0.6
Co	0.5–0.6	0.4–0.5
Cr	0.5–0.6	0.4–0.5
Cu	0.8–1.2	0.6–0.7
Fe	0.8–1.5	0.8–1.5
Mg	0.5–0.6	0.4–0.6
Mo	0.5–0.6	0.4–0.6
Ni	0.7–0.9	0.6–0.9
Sn	0.8–1	0.6–0.8
Ti	0.5–0.6	0.4–0.6
W	0.7–0.9	—
Zn	0.5–0.6	—
Zr	0.7–0.8	—
Alloys		
Lead-based white metal or babbitt (Pb, Sn, Sb)	—	0.25–0.6
Leaded brass (Cu, Zn, Pb)	—	0.2–0.4
Leaded bronze (Cu, Sn, Pb)	—	0.2–0.4
Gray cast iron	0.8–1	0.3–0.5
Mild steel	0.7–0.9	—
Intermetallic Alloys		
Co-based alloy (Stellite, Tribaloys)	0.3–0.5	—
Ni-based alloys (Haynes alloy 40)	0.6–0.9	—

surfaces cleaned in a high vacuum, can be very high, typically 2 and much higher. Strong metallic bonds are formed across the interface and significant transfer of metal from one body to another, or as loose wear debris, occurs during sliding. The slightest contamination mitigates contact or forms chemical films which reduce adhesion resulting in reduction of the friction (Bowden and Tabor, 1950, 1964; Buckley, 1981).

Most metals oxidize in air to some extent and form oxide films, typically between 1 and 10 nm thick within a few minutes of exposure of an atomic clean surface. The oxide film acts as a low shear-strength film and in addition because of low ductility leads to low friction. The oxide film may effectively separate the two metallic surfaces. However, during sliding, these thin oxide films may be penetrated. Furthermore, the film is penetrated at higher loads, and

Table 5.4.2 Typical values of the coefficient of kinetic friction of unlubricated ceramics sliding on themselves at room temperature in air.

Material	Coefficient of friction
Al_2O_3	0.3–0.6
BN	0.25–0.5
Cr_2O_3	0.25–0.5
SiC	0.3–0.7
Si_3N_4	0.25–0.5
TiC	0.3–0.7
WC	0.3–0.7
TiN	0.25–0.5
Diamond	0.1–0.2

Table 5.4.3 Typical values of the coefficient of kinetic friction of polymers and solid and liquid lubricants on a hard surface at room temperature in air.

Material	Coefficient of friction
Polymers-Plastics	
Acetal	0.2–0.3
Polyamide (Nylon)	0.15–0.3
High-density polyethylene (HDPE)	0.15–0.3
Polyimide	0.2–0.4
Polyphenyl sulfide	0.15–0.3
PTFE (Teflon)	0.05–0.10
Polymers-Elastomers	
Natural and synthetic rubber	0.3–0.6
Butadiene-acrylonitrile rubber (Buna-N or Nitrile)	0.2–0.6
Styrene-butadiene rubber (SBR)	0.2–0.6
Silicone rubber	0.2–0.6
Solid Lubricants	
Layer-lattice solids	
MoS_2	0.05–0.10
Graphite	0.05–0.15
Graphite - fluoride	0.05–0.15
Nonlayer-lattice solids	
$CaF_2/CaF_2 - BaF_2$	0.2–0.3
Fullerenes (C_{60})	0.05–0.10
Liquid Lubricants	
Animal fats and vegetable oils	0.02–0.05
Petroleum based oils	0.02
Synthetic oils	0.02–0.03

transition occurs to high values of friction. Transitions of this kind are common in metals, although the change in friction value may not be as high as in copper. Some of the precious metals (such as Au and Pt) do not form oxide layers and exhibit high friction.

In the case of soft and ductile metals such as In, Pb and Sn, the contact area is large even at low loads but the shear strength of the contacts may be low. The coefficient of friction is generally high because of large contact areas and small elastic recovery. Hexagonal metals such as Co and Mg as well as other non-hexagonal metals such as Mo and Cr exhibit low friction. Chromium forms a tenacious oxide film which is responsible for low friction. Co, Mo and Cr are common alloying elements in steels to reduce friction, wear, and corrosion.

In general, the coefficient of friction for an alloy tends to be lower than that of its pure components. Binary alloys of cobalt and chromium with more than 10% Cr exhibit excellent resistance to oxidation and corrosion. Tungsten and molybdenum are added to increase their strength and to improve friction and wear properties. Haynes Stellites (Co-Cr-W-C alloys) and Tribaloys (Co-Cr-Mo-Si-C) are commonly used for tribological applications. Nickel-based alloys are poor in galling resistance (a severe form of adhesive wear) and are inferior to cobalt-based alloys.

Lead-based white metals (babbitts), brass and bronze and gray cast iron exhibit relatively low friction. All contain phases which form films of low shear strength. In the lead-based alloys, a thin film of lead is formed during sliding and in gray cast iron, the low shear strength film is provided by the graphite constituent. Thus, these alloys exhibit intrinsically low coefficients of friction in dry sliding against steel, which do not depend on the formation of a protective oxide layer. These alloys are commonly used as bearing and seal materials.

5.4.1.1 Effect of Operating Conditions

The coefficient of friction of metals and alloys is affected, in addition to surface cleanliness, by sliding velocity, contact pressure, temperature, gaseous environment, and relative humidity (Peterson *et al.*, 1960; Bowden and Tabor, 1964; Buckley, 1981; Hutchings, 1992; Rabinowicz, 1995; Blau, 1996; Bhushan, 2001a).

As stated earlier, oxide films are produced on metals and alloys (except noble metals) when exposed to air, which usually exhibit low friction, Figure 5.2.2b. These films are penetrated at high loads, resulting in an increase in friction. At very low loads in some material couples, deformation is primarily elastic and little plowing is responsible for low friction, Figure 5.2.3. In the high-load regime, the coefficient of friction of many metallic pairs starts to decrease with an increase in load, Figure 5.2.2c. Increased surface roughening and a large quantity of wear debris are believed to be responsible for a decrease in friction at higher loads. Thus, in general, the coefficient of friction of metallic pairs increases with an increase in load at low loads because of the oxide film breakdown and/or elastic deformation; it remains at a high value for a load range and begins to drop at high loads because of interfacial changes caused by wear, Figure 5.4.1.

Figure 5.2.5 shows a drop in the coefficient of friction as a function of sliding velocity. High sliding velocities and/or high contact pressures result in surface frictional heating. Surface heating may result in the formation of a low shear strength surface film and even local melting. On the other hand, interfacial softening may result in increased plowing in the softer material. Interplay among these factors makes it difficult to predict the effect of sliding velocity on friction.

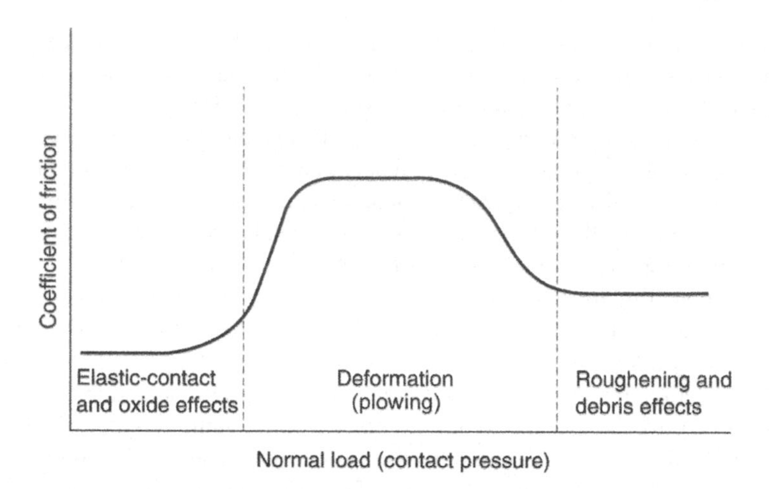

Figure 5.4.1 Schematic of effect of load on coefficient of friction for metallic pairs.

An increase in the temperature generally results in metal softening. An increase in temperature may result in solid-state phase transformation which may either improve or degrade mechanical properties. The most drastic effect occurs if a metal approaches its melting point and its strength drops rapidly, and thermal diffusion and creep phenomena become more important. The resulting increased adhesion at contacts and ductility lead to an increase in friction. High temperature also increases the rate of oxidation, which in many cases may result in low adhesion and low friction. Figure 5.4.2 shows the effect of temperature on cobalt on stainless steel. Cobalt exhibits a phase transformation at 417°C from a hexagonal close-packed structure with limited slip ductility to a cubic close-packed structure which is fully ductile, and this phase change is responsible for a peak in friction at about 500°C. The drop in friction above 550°C may be due to an increase in oxide thickness and also to a change in oxide species from CoO, which is a poor solid lubricant to Co_3O_4, which can be expected to give low friction.

Gaseous environment and relative humidity also affect the friction. For example, severe friction even seizure is experienced with most metallic pairs in a high vacuum.

5.4.2 *Friction of Ceramics*

Friction and wear data of ceramics in ambient and extreme environments can be found in various references (Bhushan and Sibley, 1982; Anonymous, 1987; Chandrasekar and Bhushan, 1990; Jahanmir, 1994; Bhushan and Gupta, 1997). Ceramics exhibit high mechanical strength, do not lose much mechanical strength or oxidize readily at elevated temperatures and are resistant to corrosive environments. Therefore, ceramic couples are commonly used in extreme environmental applications, such as high loads, high speeds, high temperatures, and corrosive environments. The mechanical behavior of ceramics differs from that of metals/alloys because of the different nature of the interatomic forces with covalent or ionic bonding in ceramics compared to that of metallic bonding in metals/alloys. Ceramic materials of either bond type show only limited plastic flow at room temperature and much less ductility than metals. Although adhesive forces, of covalent, ionic, or van der Waals origin, are present between

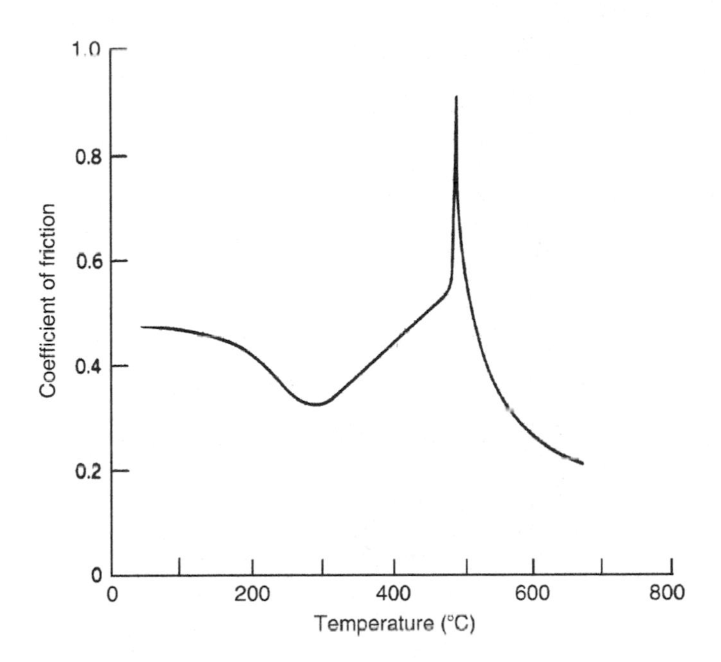

Figure 5.4.2 Coefficient of friction as a function of temperature for cobalt sliding on stainless steel at a normal load of 5 N and sliding velocity of 25 mm/s. Reproduced with permission from Rabinowicz, E. (1995) *Friction and Wear of Material*, Second edition, Wiley, New York. Copyright 1995. Wiley.

ceramic materials in contact, the low real area of contact results in relatively low values of the coefficient of friction comparable to metallic couples sliding in air in the presence of intact oxide films. Under clean environments, the coefficients of friction of ceramic pairs do not reach the very high values observed in clean metals, especially in ultra-high vacuum or in the absence of oxygen.

Fracture toughness of ceramics is an important property in the friction of ceramics (Ishigaki *et al.*, 1986; Stolarski, 1990). Figure 5.4.3 shows the coefficient of friction as a function of fracture toughness for a sharp diamond tip on silicon nitride disks produced with various hot pressing conditions. The coefficient of friction values of all ceramics decreases with an increase in fracture toughness. Fracture readily occurs in concentrated contacts, such as a hard sharp pin or stylus sliding against a flat. Energy dissipated during fracture at the sliding contact contributes to the friction.

Figure 5.4.4 shows the coefficient of friction as a function of a normal load for a spherical diamond rider sliding on single-crystal silicon carbide (0001) surface in air. The coefficient of friction increases with an increase in the normal load. The dominant source of friction is shear and plowing of silicon carbide with the diamond rider. Increased cracking at higher loads was believed to be responsible for higher wear as well as friction. Figure 5.4.5 shows the coefficient of friction as a function of sliding velocity at various temperatures (Skopp *et al.*, 1990). The data show that there is a minimum in the coefficient of friction at 1 m/s which shifts with increasing temperature. The maximum value of coefficient friction increases with temperature. The initial decrease in friction with velocity may be associated with an increase

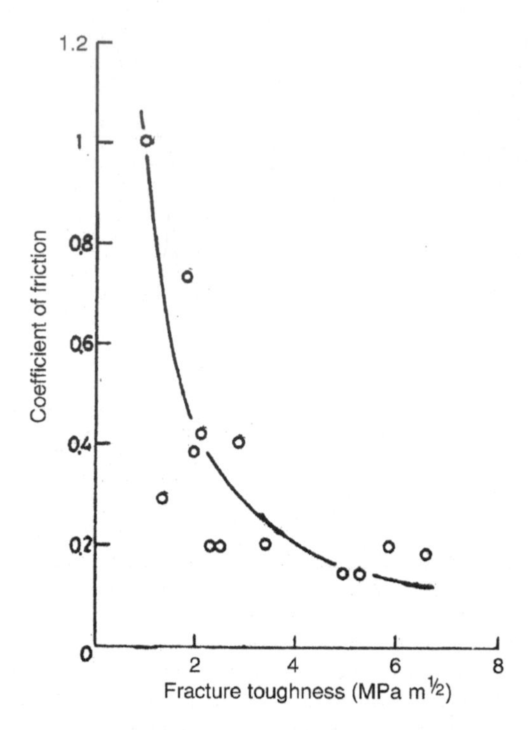

Figure 5.4.3 Coefficient of friction as a function of fracture toughness for a sharp diamond pin (5 μm radius) on disks made of silicon carbide, silicon nitride, alumina and zirconia oxide in air. Reproduced with permission from Ishigaki, H., Kawaguchi, I., Iwasa, M. and Toibana, Y. (1986), "Friction and Wear of Hot Pressed Silicon Nitride and Other Ceramics," *ASME J. Trib.* **108**, 514–521. Copyright 1986. ASME.

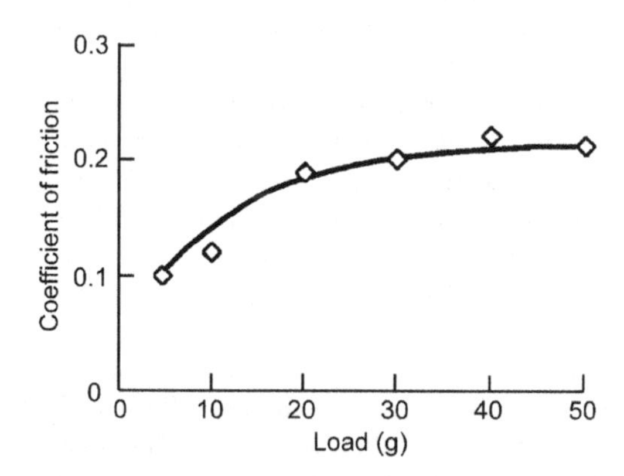

Figure 5.4.4 Coefficient of friction as a function of normal load for spherical diamond rider of 0.15 mm radius sliding on single-crystal silicon carbide (0001) in air. Reproduced with permission from Miyoshi, K. and Buckley, D.H. (1979), "Friction, Deformation and Fracture of Single-Crystal Silicon Carbide," *ASLE Trans.* **22**, 79–90. Copyright 1979. Taylor and Francis.

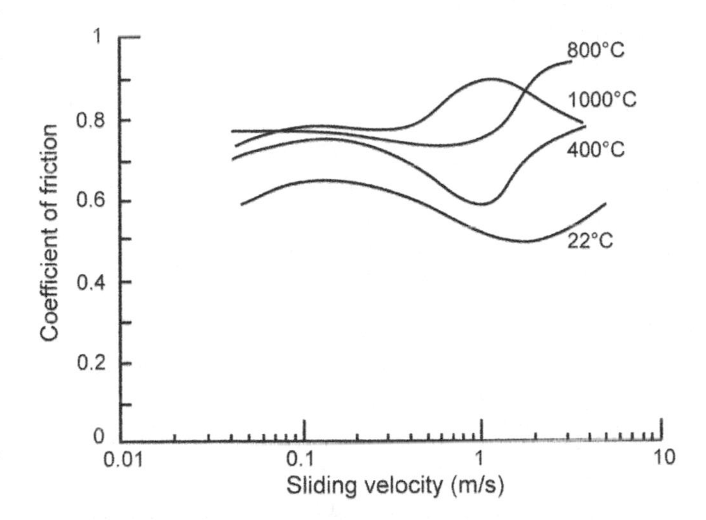

Figure 5.4.5 Coefficient of friction as a function of sliding velocity for silicon nitride sliding on silicon nitride at various temperatures in air. Reproduced with permission from Skopp, A., Woydt, M., and Habig, K.M. (1990), "Unlubricated Sliding Friction and Wear of Various Si3N4 Pairs Between 22° and 1000°C," *Tribol. Inter.* **23**, 189–199. Copyright 1990. Elsevier.

in the interface temperature which increases the formation of the tribochemical films on the sliding surface with lower shear strength. After a contact velocity, a high interface temperature may lead to material softening leading to a higher contact area and friction. Figure 5.4.6 shows the coefficient of friction as a function of temperature for alumina sliding on itself and alumina sliding on partially stabilized zirconia (PSZ) (Cox and Gee, 1997). The coefficient of friction is initially relatively low at temperatures below about 300°C, rises rapidly, and then either reduces at temperatures of 800°C, or continues to increase and reaches a plateau. The initial rise in friction can be due to the removal of absorbed water from the interface.

To summarize, the effects of normal load, sliding velocity, and temperature on the friction of ceramics can usually be interpreted in terms of changes in the tribochemical surface films and the extent of fracture in the contact region. Both load and sliding velocity affect the rate of frictional energy dissipation and hence the temperature at the interface. The environment, in general, plays a significant role in the friction and wear of ceramics and it will be described further in Chapter 7 on Wear.

Next, we discuss friction mechanisms of diamond and titanium nitride which are commonly used in tribological applications. These materials exhibit high friction when sliding against themselves in a vacuum. The coefficient of friction of diamond is 0.05–0.1 in air and is typically 0.1–0.2 for titanium nitride. The coefficient of friction of diamond does not change as a function of relative humidity and lubrication, Figure 5.4.7 (Tabor, 1979; Seal, 1981; Samuels and Wilks, 1988). Significant surface oxidation has been reported in titanium nitride and appears to be responsible for its low friction in air, but in diamond the surface modification is more likely to be due to adsorption of a gaseous species, rather than the formation of reaction products (Bowden and Young, 1951). Diamond has dangling carbon bonds on the surface and is reactive. Hydrogen adsorbs readily on the surface of diamond and adsorbed hydrogen from the environment forms a hydrocarbon layer which reduces friction (Pepper, 1982). In addition,

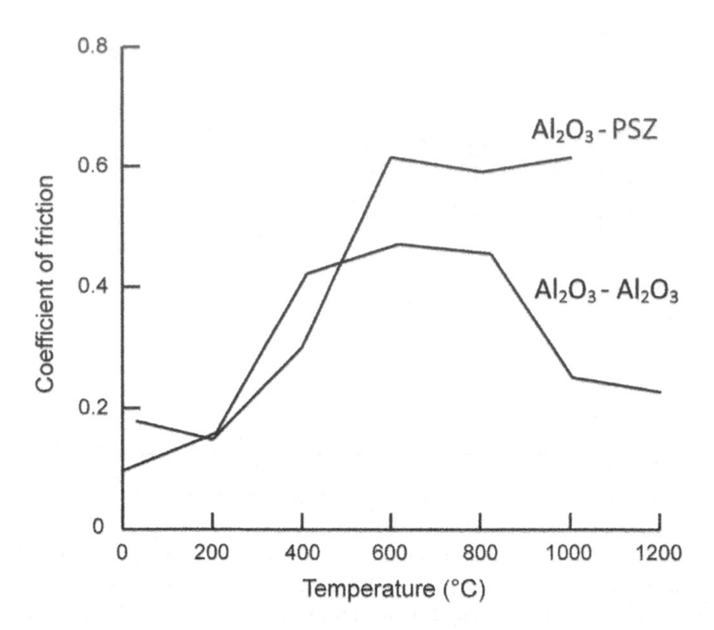

Figure 5.4.6 Coefficient of friction as a function of temperature for alumina sliding itself and alumina sliding on partially stabilized zirconia (PSZ) in air. Reproduced with permission from Cox, J.M. and Gee, M.G. (1997), "Hot Friction Testing of Ceramics," *Wear* **203–204**, 404–417. Copyright 1997. Elsevier.

a high thermal conductivity (the highest of any material) dissipates frictional energy, and is believed to result in low friction and wear.

Finally, a variety of ceramic coatings are used in a large number of industrial applications. One of the coatings of interest in applications requiring wear resistance is amorphous carbon, commonly referred to as diamondlike carbon (DLC) coating (Bhushan, 1996, 1999b, 2011; Bhushan and Gupta, 1997). Operating conditions including environment affect its friction and wear properties. As an example, Figure 5.4.8 shows the coefficient of friction as a function of sliding distance in vacuum, argon, oxidizing and humidity environment for a $A\ell_2 0_3$-$T_i C$ magnetic slider sliding against DLC-coated magnetic disk (Bhushan *et al.*, 1995). Wear lives are shortest in a high vacuum and the longest in atmospheres of mostly nitrogen and argon with the following order (from best to worst): argon or nitrogen, $Ar+H_2 0$, ambient, $Ar+0_2$, $Ar+H_2 0$ and vacuum. From this sequence, we can see that having oxygen and water in an operating environment worsens the wear performance of the coating (due to tribochemical oxidation), but having a vacuum is even worse (because of intimate contact).

5.4.3 Friction of Polymers

Polymers include plastics and elastomers. The coefficient of friction of selected polymers used for tribological applications, sliding against themselves or against metals or ceramics, ranges from 0.15 to 0.6 except for polytetrafluoroethylene (PTFE) which exhibits a very low coefficient of friction of about 0.05, comparable to that of conventional solid lubricants (Lancaster, 1972; Bhushan and Dashnaw, 1981; Bhushan and Wilcock, 1981; Bhushan and Winn, 1981;

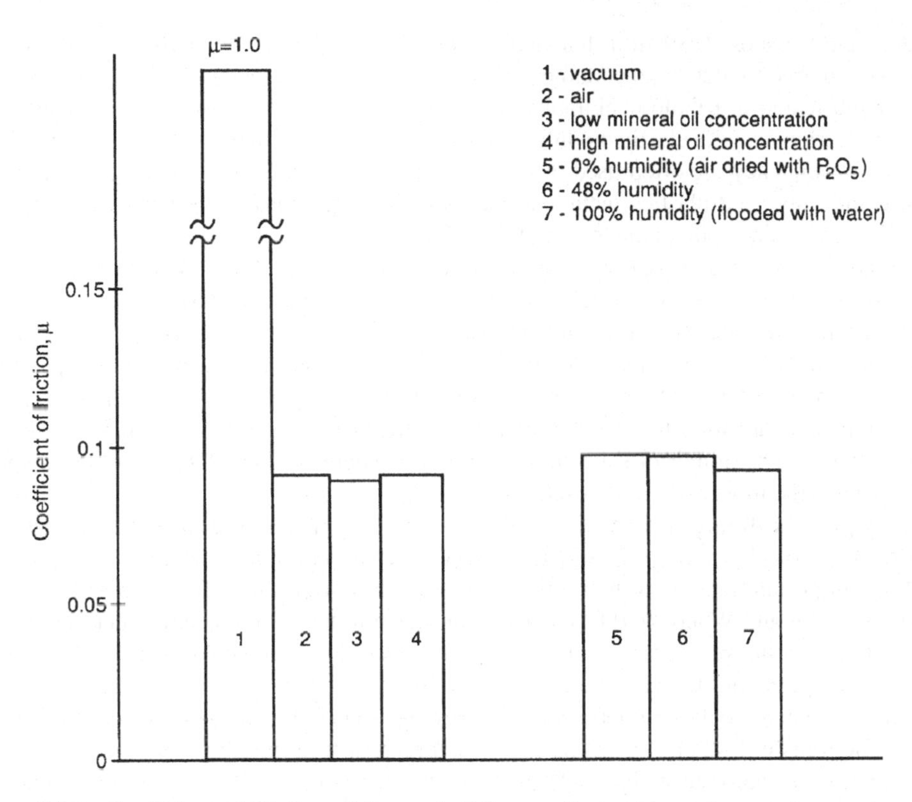

Figure 5.4.7 Coefficient of friction of diamond sliding on diamond in various environments under lightly loaded conditions.

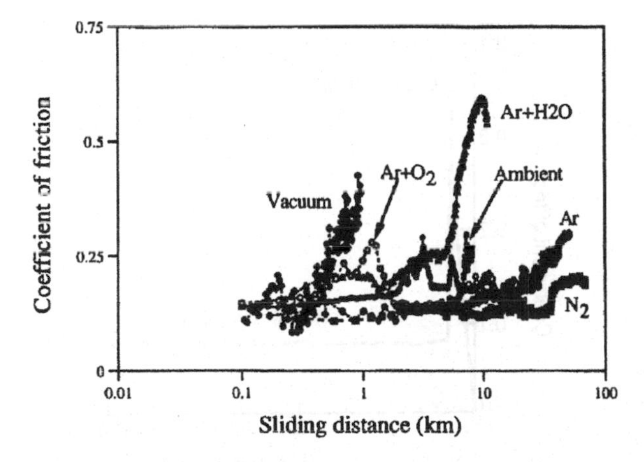

Figure 5.4.8 Wear performance for $A\ell_2O_3$-TiC magnetic slider sliding against DLC coated magnetic disk measured at a speed of 0.75 m/s and for a load of 10 g. Vacuum refers to 2×10^{-7} Torr. Reproduced with permission from Bhushan, B., Yang, L., Gao, C., Suri, S., Miller, R.A., and Marchon, B. (1995), "Friction and Wear Studies of Magnetic Thin-Film Rigid Disks with Glass-Ceramic, Glass and Aluminum-Magnesium Substrates," *Wear* **190**, 44–59. Copyright 1995. Elsevier.

Santner and Czichos, 1989; Bhushan and Gupta, 1997). Polymers generally exhibit low friction as compared to metal and ceramic couples but exhibit moderate wear. Polymers are often used unlubricated in tribological applications. Polymers are very compliant as compared to metals or ceramics, with elastic modulus values typically one tenth or even less. Their strength is much lower. They are often used in sliding applications against hard mating surfaces. The most commonly used plastics include acetal, polyamide (Nylon), high-density polyethylene (HDPE), polyimide, polyphenylene sulfide (PPS), and PTFE. A number of elastomers are also used in bearing and seal applications at loads and speeds lower than that for plastics. The most common ones are natural and synthetic rubber, butadiene-acrylonitrile (Buna-N or nitrile) rubber, styrene-butadiene rubber (SBR) and silicone rubber. These polymers are the family of self-lubricating solids. Polymers flow readily at modest pressures and temperatures. Polymers, in comparison to metals and ceramics, lack rigidity and strength. Therefore, polymer composites are used to provide a balance of mechanical strength and low friction and wear. Fillers can be liquids or solids in the form of powders or fibers. PTFE, carbon, graphite, and glass are the most commonly used fillers.

Many plastics sliding against hard mating surfaces (e.g. metals) result in the formation of transfer films of plastic onto the mating surface. The formation and behavior of the transfer films are important factors in the friction and wear of these plastics (Bhushan and Dashnaw, 1981; Bhushan and Wilcock, 1981). Once a transfer film has formed, subsequent interaction occurs between the plastic and a layer of similar material, irrespective of the substrate. On further sliding, the plastic may continue to wear by adding material to the transfer film, since the interfacial bond to the counterface is often stronger than that within the bulk of the polymer itself. The transfer film also wears through generation of wear particles and reaches a steady thickness in an interface with low friction and wear. Figure 5.4.9 shows the coefficient of friction of high-density polyethylene sliding on a glass surface. The coefficient of friction for initial sliding on a clean hard substrate is not particularly low and the transfer film is on the

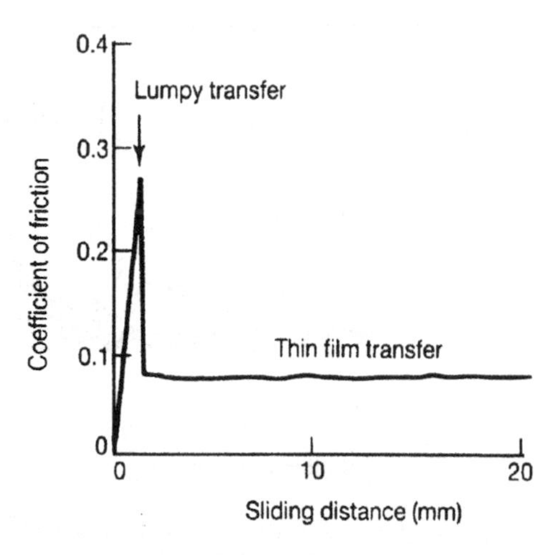

Figure 5.4.9 Coefficient of friction as a function of sliding distance for high density polyethylene (HDPE) sliding against glass. Reprinted from Pooley, C.M. and Tabor, D. (1972), "Friction and Molecular Structure: The Behavior of Some Thermoplastics," *Proc. Roy. Soc. Lond. A* **329**, 251–274, by permission of the Royal Society.

order of micrometers thick. As sliding progresses, the coefficient of friction drops to a much lower value; the transfer film becomes much thinner and contains molecular chains strongly oriented parallel to the sliding direction.

Asperity deformation with polymers is primarily elastic. In this respect, the friction of polymers differ from metals and ceramics. As shown in Chapter 3, the mechanical property ratio E/H along with surface roughness determines the extent of plasticity in the contact region. For metals and ceramics E/H is typically 100 or greater, whereas for polymers it is on the order of 10. Thus, the plasticity index for a polymer is on the order of one-tenth of that of a metal or a ceramic, consequently contact is primarily elastic except for very rough surfaces.

The forces of friction are mainly adhesion, deformation and elastic hysteresis. Adhesion responsible in polymers results from the weak bonding forces such as van der Waals forces and hydrogen bonding, which are also responsible for the cohesion between polymer chains themselves in the bulk of the material.

According to the analysis of adhesion, surface roughness and normal load affect coefficient of friction (Bhushan and Wilcock, 1981; Santner and Czichos, 1989; Bhushan and Gupta, 1997). On moderately rough surfaces as the load is increased, at some load, elastic deformation at the asperities is so great that the individual asperities on the contacting surfaces are totally deformed, and the contact region approximates to the contact of a large single asperity. In this case, $\mu \propto W^{-1/3}$, μ decreases with W after some load. For smooth surfaces, the coefficient of friction decreases with load to start with as the contact region starts out as approximately one giant asperity contact. The effect of normal load and surface roughness on friction is illustrated in Figure 5.4.10 which presents the coefficient of friction values for rough and smooth PMMA sliding on themselves as a function of load. In the case of smooth PMMA, the contact of friction decreases with an increase in normal load, whereas in the case of rough PMMA, the coefficient of friction remains constant at low loads (the case of multi-asperity contact) and starts to decrease with an increase in the normal load at high loads (single-asperity contact). For polymers under load, asperity deformation is generally large, leading to one giant asperity contact which results in a decrease of coefficient of friction with an increase in load (Steijn, 1967). As stated earlier, for polymers, $\mu = (\tau_0/p_r) + \alpha$. This also means that μ decreases with an increase in normal load. This relationship is generally followed (Bowers, 1971; Briscoe and Tabor, 1978).

Since polymers are viscoelastic materials, sliding velocity (loading time) has a significant effect on friction (Ludema and Tabor, 1966; Santner and Czichos, 1989; Bhushan, 1996). The coefficient of friction of butadiene acrylonitrile (Buna-N) rubber was measured at various sliding velocities and temperatures. For viscoelastic materials, deformation due to an increase in temperature is equivalent to decreasing sliding velocities, and vice versa. This equivalence of the time and temperature effects can be used to interpret the frictional behavior of certain polymeric materials, using the time–temperature superposition principle. By using this transformation, the coefficient of friction as a function of sliding velocity at 20°C was obtained as shown in Figure 5.4.11b; note a peak in the friction. These trends can be explained by the product of trends in the real area of contact (a function of the modulus of elasticity) and shear strength as a function of strain rates (rate of deformation) or sliding velocity, Figure 5.4.11a. Strain rates in shear processes involved in τ_a were 10^5 times as rapid as those involved in A_r (Ludema and Tabor, 1966). Experimental evidence on the frictional behavior of polymers other than rubber indicates that, in general, the superposition principle may not be applicable to most polymers.

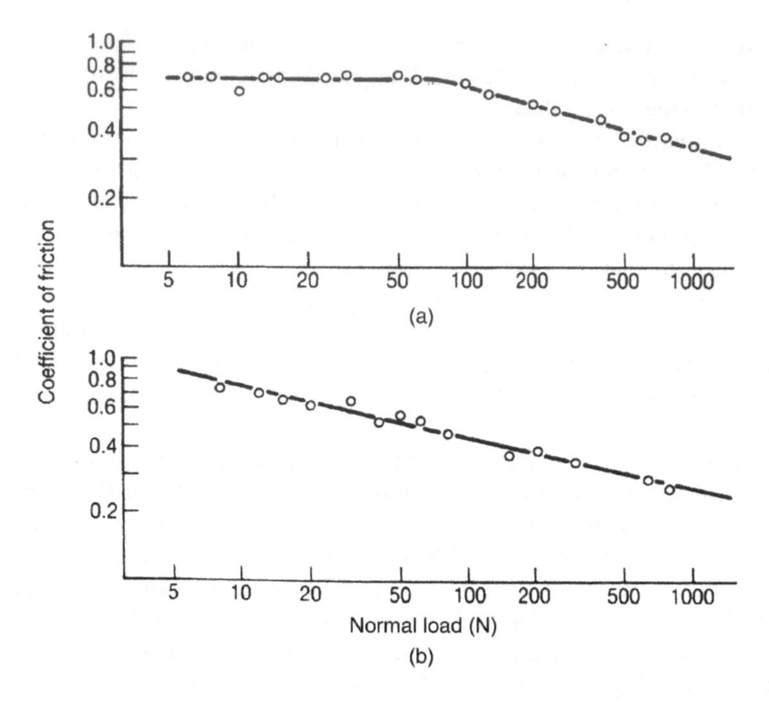

Figure 5.4.10 Coefficient of friction as a function of normal load for sliding of crossed cylinders of polymethylmethacrylate (PMMA) with two surface roughnesses, (a) lathe turned and (b) smooth polished. Reprinted from Archard, J.F. (1957), "Elastic Deformation and The Laws of Friction," *Proc. Roy. Soc. Lond. A* **243**, 190–205, by permission of the Royal Society.

PTFE (Teflon, a trade mark of DuPont Company) exhibits very low friction comparable to that of solid lubricants. PTFE is a fluorocarbon $(C_2F_4)_n$. It is a crystalline polymer with a melting point of 325°C. The molecular structure of PTFE is shown in Figure 5.4.12. Bunn and Howells (1954) have shown that the simple zigzag backbone of $-CF_2 - CF_2-$ groups is given a gentle twist of 180° over a distance corresponding to 13 CF_2 groups. The lateral packing of these rodlike molecules is hexagonal, with a lattice constant a of 0.562 nm. On a larger scale, the individual units in the structure of PTFE are believed to consist of thin, crystalline bands that are separated from each other by amorphous or disorder regions. It has been suggested that the smooth profile of rodlike molecules permits easy slippage of the molecules with respect to each other in planes parallel to the c axis. This probably accounts for the low coefficient of friction (0.05 and up) of PTFE and easy transfer of PTFE material onto the sliding partner (Steijn, 1966, 1967; Lancaster, 1973; Tanaka *et al.*, 1973). The fundamental reasons for the low coefficient of friction of PTFE are: the low adhesion of the PTFE surface (it has a slippery feel); the strong adhesions that are formed across the interface and then are sheared at quite low stresses; and the transfer of PTFE onto the bare surface, which results in subsequent PTFE-PTFE contact. The formation of a thin, coherent transfer film of PTFE on the mating surface seems indeed to be one of the essential prerequisites for easy sliding and low friction. Figure 5.4.13 illustrates the dependence of the coefficient of friction of PTFE on normal load, sliding velocity, and temperature.

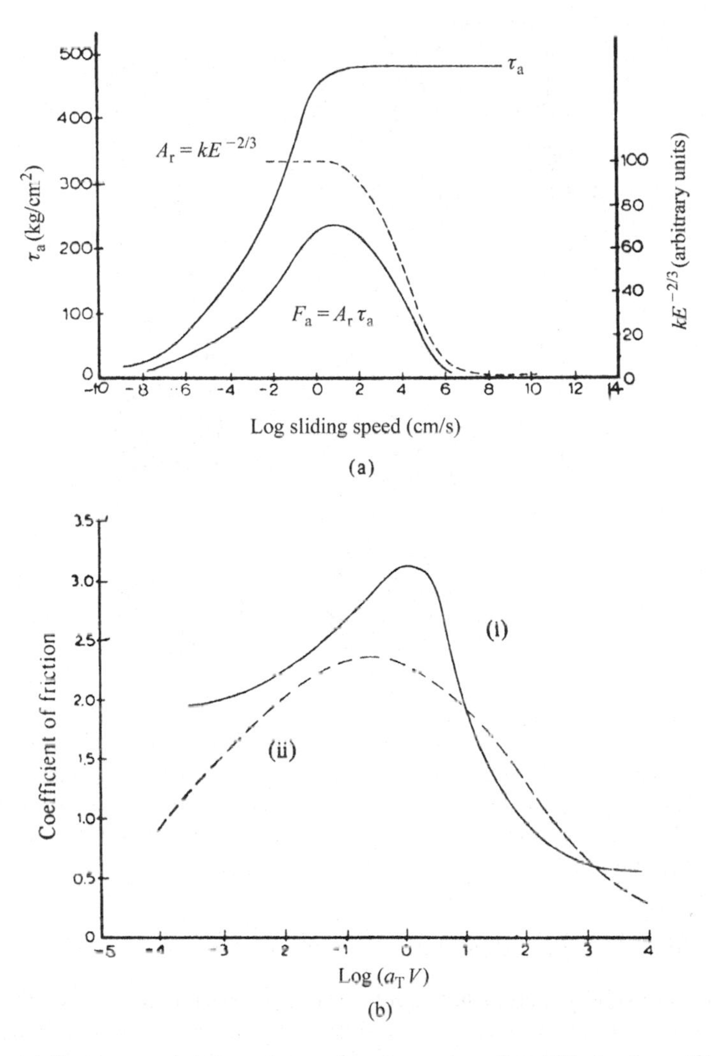

Figure 5.4.11 (a) Shear strength (τ_a), real area of contact and predicted friction force (F_a) as a function of sliding velocity. Reproduced with permission from Ludema, K.C. and Tabor, D. (1966), "The Friction and Viscoelastic Properties of Polymeric Solids," *Wear* **9**, 329–348. Copyright 1966. Elsevier; and (b) measured coefficient of friction as a function of sliding velocity for a mild steel hemisphere sliding on butadiene acrylonitrile rubber (i), plotted by means of Williams-Landel-Ferry (WLF) transformation at 20°C, where a_T is the shift factor (ii). *Source*: (i) Reproduced with permission from Ludema, K.C. and Tabor, D. (1966), "The Friction and Viscoelastic Properties of Polymeric Solids," *Wear* **9**, 329–348. Copyright 1966. Elsevier and (ii) Grosch, K.A. (1963), "The Relation Between the Friction and Viscoelastic Properties of Rubber," *Proc. Roy. Soc. Lond. A* **274**, 21–39, by permission of the Royal Society.

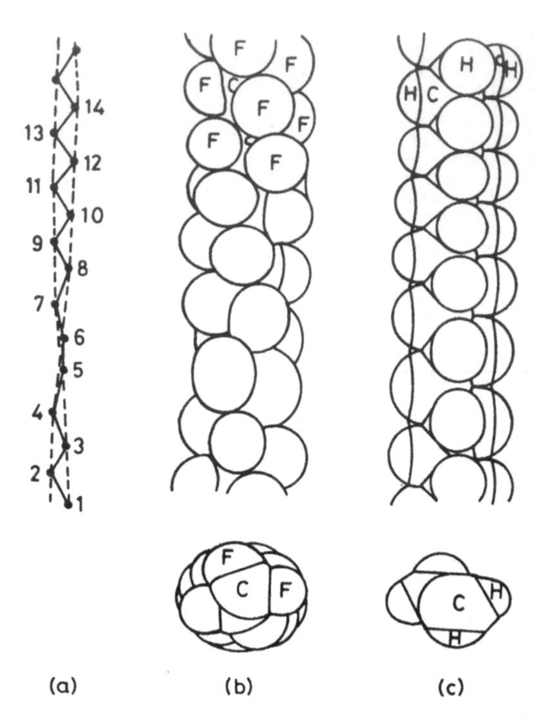

Figure 5.4.12 Molecular structure of PTFE: (a) twisted zigzag chain with 13 carbon atoms per 180° twist and repeat distance of 1.69 nm, (b) side and end views of a rodlike molecule of PTFE, and (c) ribbonlike hydrocarbon (e.g., polyethylene) molecule for comparison. Reproduced with permission from Bunn, C.W. and Howells, E.R. (1954), "Structure of Molecules and Crystals of Fluorocarbons," *Nature* **174**, 549–551. Copyright 1954 Nature Publishing Group.

5.4.4 Friction of Solid Lubricants

Solid lubricants are solid materials that exhibit very low friction and moderately low wear in sliding in the absence of an external supply of lubricant. The most commonly used solid lubricants are graphite and molybdenum disulfide as well as PTFE discussed earlier (Braithwaite, 1964, 1967; Anonymous, 1971, 1978, 1984; Clauss, 1972; Paxton, 1979; Iliuc, 1980; Bhushan and Gupta, 1997). A new form of carbon – fullerenes or Buckyballs (C_{60}) – is also proposed as a solid lubricant (Bhushan *et al.*, 1993; Gupta *et al.*, 1994). CaF_2 *and* $CaF_2 - BaF_2$ eutectic based coatings are also used for solid lubrication (Bhushan and Gupta, 1997).

Graphite, a planar molecule, has a hexagonal layered structure with a large number of parallel layers in the ABAB stacking sequence along the c axis, stacked 0.3354 nm apart, Figure 5.4.14. Within each layer (plane), atoms are arranged in hexagonal structure (benzene ring) with each carbon atom bonded (C–C distance = 0.1415 nm) to three other carbon atoms, arranged at the apexes of an equilateral triangle. The three hybridized valence electrons of carbon atoms create covalent (σ) bonds and the remaining unhybridized fourth electron creates π bonds between the two carbon atoms. The sheets of carbon atoms are attracted to each other only by the weak van der Waals forces. The graphite material is anisotropic. The existence of σ bonds explains the high electrical and thermal conductivity in the hexagonal plane—over

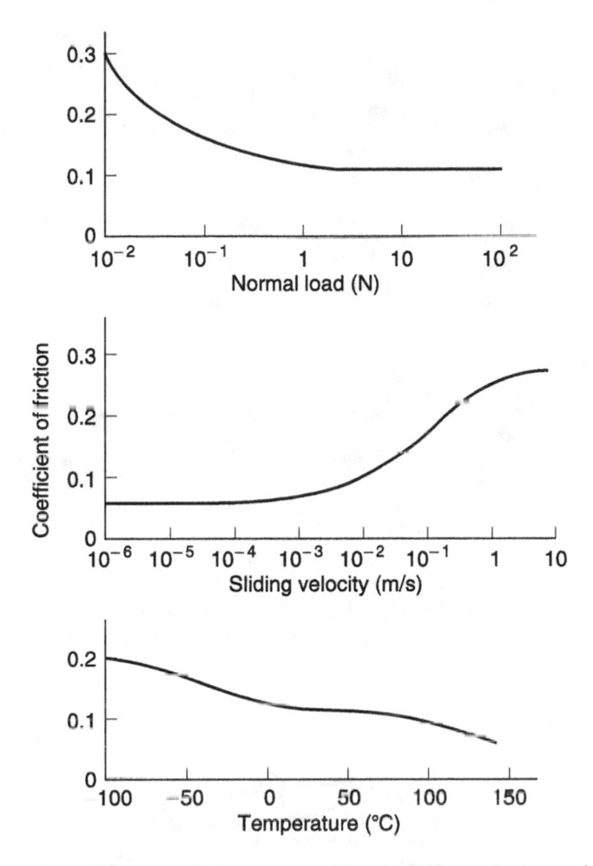

Figure 5.4.13 Effect of operating conditions – normal load, sliding velocity, and temperature – on the coefficient of friction of PTFE.

100 times that normal to the plane. They cleave (separate) easily, which accounts for the typical low friction of graphite.

Unlike two crystalline forms of carbon—graphite and diamond—which are infinite periodic network solids, this third form of carbon has a molecular form that is purely cage-like, nonplanar, and finite. Fullerene molecules take the form of hollow, geodesic domes commonly referred to as Buckyballs. All such domes are networks of pentagons and hexagons made from tetrahedral subunits with covalently bonded carbon atoms. Like other aromatic molecules, a carbon atom in this new geodesic form is bonded to only three other carbon atoms, being satisfied by a strong double bond, delocalized over the geodesic sphere. All single bonds are strong covalent (σ) bonds and the bonds associated with delocalized electrons in the double bond are the π bonds. Carbon–carbon bonds in the pentagon subunits are single bonds and alternate bonds in the hexagonal subunits are double bonds. To maintain the aromatic network, pentagonal subunits are not placed next to each other. An example of a most stable molecule, C_{60}, is shown in Figure 5.4.15.

The structure of MoS_2 is shown in Figure 5.4.16. MoS_2 has a hexagonal structure and consists of planes of molybdenum atoms alternating with planes of sulfur atoms in the sequence

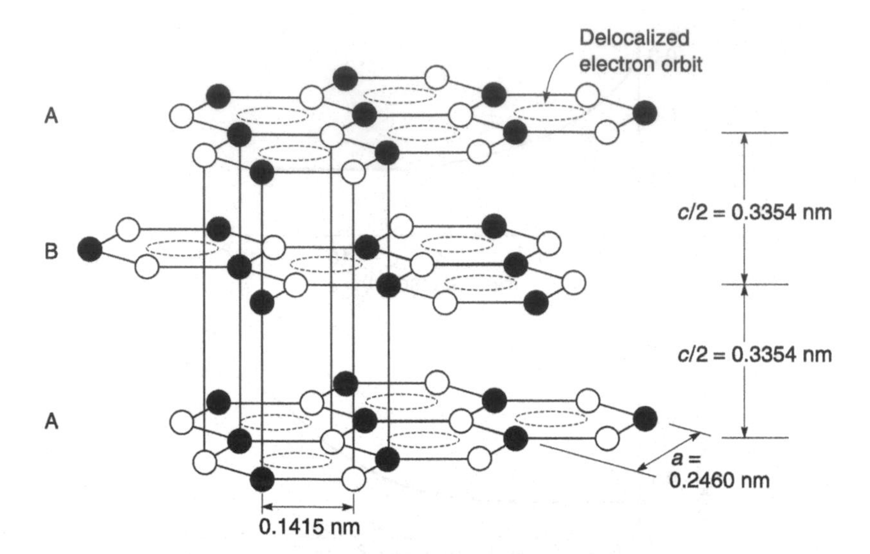

Figure 5.4.14 Three-dimensional representation of hexagonal layered structure of the graphite showing three staggered layers. There are two distinct types of carbon sites in graphite: solid and open circles. Solid circle atoms have neighbor atoms directly above and below in the adjacent layers and open circle atom to not have such neighbors.

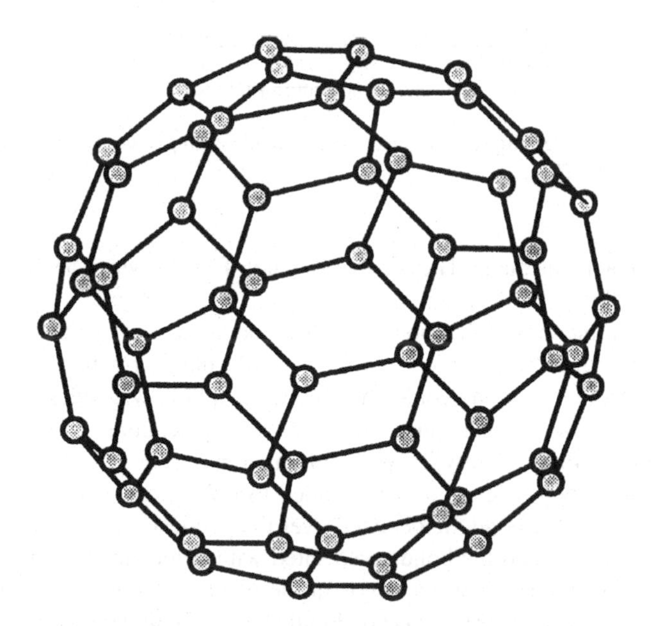

Figure 5.4.15 Structure of a most-stable fullerene-soccerball C_{60} (with 12 pentagons and 20 hexagons) with a cage diameter f 0.71 nm.

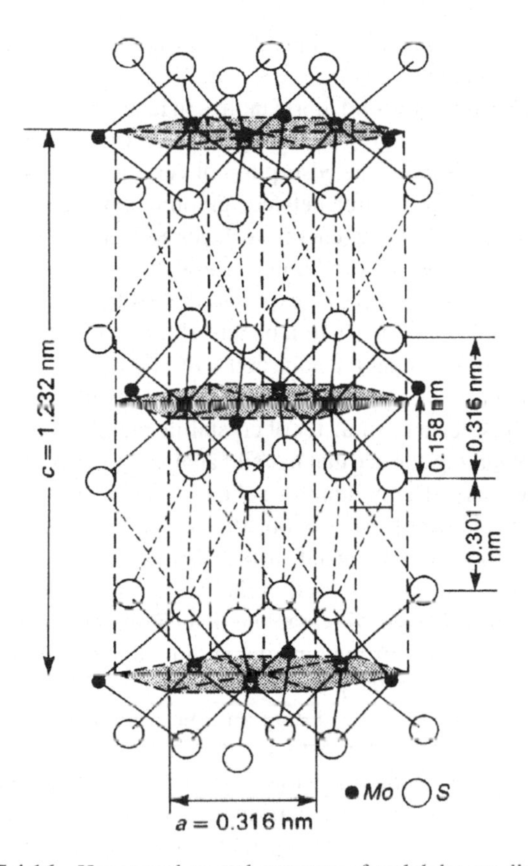

Figure 5.4.16 Hexagonal crystal structure of molybdenum disulfide.

S: Mo:S: S: Mo: S: ... The atomic arrangement in each layer is hexagonal, in which each atom of molybdenum is surrounded at equal distances by six atoms of sulfur placed at the corners of a trigonal prism. Each layer of S: Mo: S consists of two planes of sulfur atoms and an intermediate plane of the molybdenum atom. The distance between the planes of molybdenum and sulfur atoms is 0.158 nm and adjacent planes of sulfur atoms are 0.301 nm apart. The bonds between the atoms of molybdenum and sulfur are covalent (σ), while between the atoms of sulfur they are of the van der Waals type (π).

To summarize, graphite and MoS_2 have a lamellar structure or a hexagonal layer-lattice structure. Their crystal structure is such that layers or sheets exist within their crystal lattices, within which the atoms are tightly packed and bonding between the atoms is covalent and strong. These layers are separated by relatively large distances, and held together by weak van der Waals type bonding. The interplanar bond energy is about one-tenth to one-hundredth of that between atoms within the layers. Both materials are strongly anisotropic in their mechanical and other physical properties; in particular, they are much less resistant to shear deformation in the basal planes (i.e., parallel to the atomic planes) than in other directions. Graphite and MoS_2 may be visualized as infinite parallel layers of hexagonals stacked by a distance c apart. Under the action of a relatively small force, displacement of the layers with a high density of atoms occurs.

The key to graphite's value as a solid lubricant lies in its layer-lattice structure and its ability to form strong chemical bonds with gases such as water vapor (Savage, 1948; Bowden and Young, 1951; Bryant *et al.*, 1964). The adsorption of certain molecules is necessary to ensure a low coefficient of friction. The friction between graphite lamellae slid parallel to their planes appears always to be low; they are low-energy surfaces and show little adhesion. But the edges of the lamellae are high-energy sites, and bond strongly to other edge sites or to basal planes. In sliding, some edge sites will always be exposed, and so the friction of graphite in vacuum is high. Condensed vapors lower the friction by adsorbing selectively to the high-energy edge sites, saturating the bonds, reducing the adhesion with the mating surface, and thus lowering the friction. Only a small concentration of adsorbed molecules is needed to produce this effect. Further, sliding results in the transfer of crystallite platelets to the mating surface. Close examination of a metal mating surface that has slid against graphite in an ambient environment reveals a stained wear track. The buildup of solids in the wear track is generally 200–1000 nm thick (Buckley and Johnson, 1964; Paxton, 1979). The wear track is believed to be coated with graphite platelets bonded to the metal surface through oxide linkages. This transfer film plays an important role in controlling friction and wear rates. Coefficients of friction in the range of 0.05–0.15 can be achieved with graphite in an ambient environment. If the surface of graphite is examined by electron diffraction after sliding, it is found that the basal planes have become oriented nearly parallel to the plane of the interface, with a misalignment of the order of 5°. The coefficient of friction when sliding on a face perpendicular to the sheet direction compared to sliding parallel to sheet direction is three times or more higher.

The coefficient of friction of graphite varies significantly with the gaseous composition of the environment. Figure 5.4.17 shows the dependence of the coefficient of friction of graphite on the vapor pressure of various gases: nitrogen, hydrogen, water vapor, oxygen, and heptane. Any reactive gas results in reduction of friction. The quantity of water vapor or other reactive gases necessary to reduce the coefficient of friction is very small. Figure 5.4.18 shows the effect

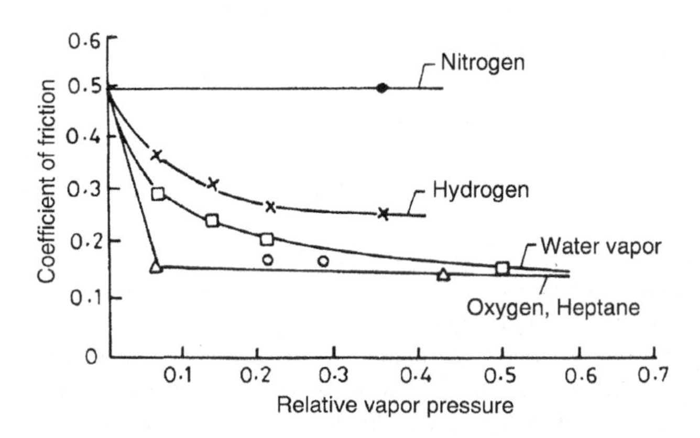

Figure 5.4.17 Effect of environment composition and its vapor pressure on coefficient of friction of graphite sliding against steel. Reproduced with permission from Rowe, G.W. (1960), "Some Observations on the Friction Behavior of Boron Nitride and of Graphite," *Wear* **3**, 274–285. Copyright 1960. Elsevier.

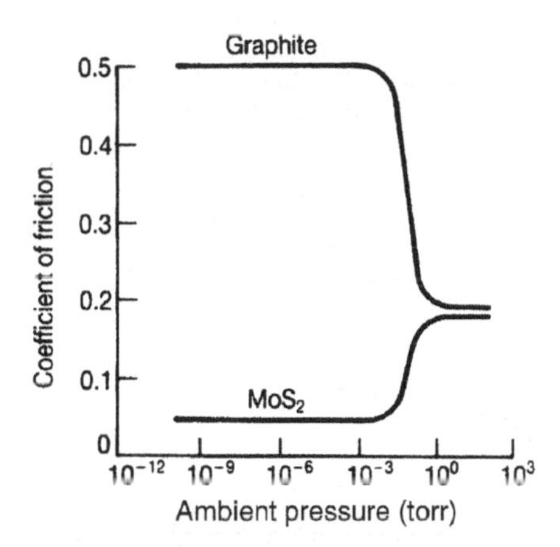

Figure 5.4.18 Coefficient of friction as a function of ambient pressure (partial pressure of air) for graphite and molybdenum disulphide. Reproduced with permission from Buckley, D.H. (1981), *Surface Effects in Adhesion, Friction, Wear and Lubrication*, Elsevier, Amsterdam.Copyright 1981. Elsevier.

of partial pressure of air (oxygen concentration) on the friction. In vacuum or dry nitrogen, the coefficient of friction is typically ten times greater than in air, and graphite under these conditions wears rapidly. A 1000-fold increase in the wear rate of graphite has been reported when a wear experiment with graphite sliding against a metal was conducted in a vacuum (13 μPa or 10^{-7}torr) (Buckley and Johnson, 1964). Concurrent with this increase in wear was the disappearance of graphite from the wear track. The necessity for adsorbed vapors to maintain low friction restricts the use of graphite to high-humidity environments. In air, the amount of physically adsorbed water may decrease at around 100°C to such an extent that low friction can no longer be maintained. Organic vapors are very effective substitutes for water and may be available as contaminants in the surrounding environment. Graphite also performs well under boundary lubrication conditions because of its good affinity for hydrocarbon lubricants. Graphite is used in powder form burnished to solid surfaces, mixed with resin binder then sprayed onto a surface, vacuum deposited on a surface, or as an additive to liquid lubricants, greases and solids. Graphite is widely used for many sliding applications, such as bearings, seals, and electrical contacts.

The use of graphite in ordinary air environments is limited by the onset of oxidation at about $430°C$. Some added inorganic compounds such as CdO, are able to extend the temperature range over which low friction occurs. Silver is added to improve friction properties (Bhushan, 1980b, 1981b, 1982).

The low friction of MoS_2 does not depend on adsorbed vapor and is, therefore, an *intrinsic* solid lubricant (Peterson and Johnson, 1953; Braithwaite, 1964, 1967, Winer, 1965; Clauss, 1972; Holinski and Gansheimer, 1972; Farr, 1975; Sliney, 1982; Bhushan and Gupta, 1997). Like graphite, MoS_2 forms an oriented film on a sliding surface, with the basal planes tending to be aligned parallel to the surface; bonding of the crystallites to the surface is probably aided by the internal polarization of the lamellar which results from their sandwich structure. A coefficient of friction ranging from 0.05 to 0.15 is typically found for sliding between basal planes; for edge-oriented crystallites sliding against basal planes, the coefficient is three times

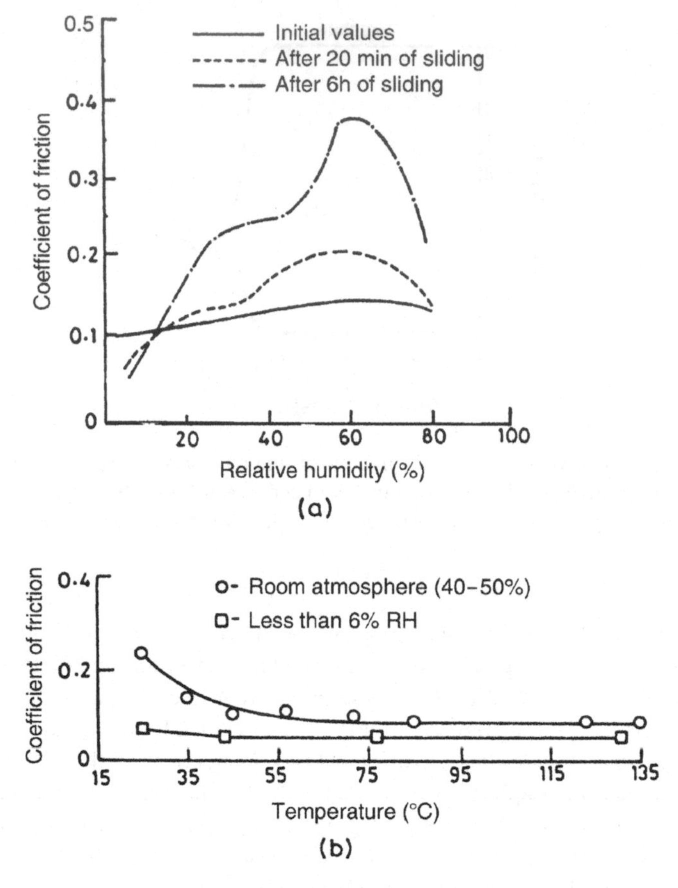

Figure 5.4.19 (a) Coefficient of friction as a function of relative humidity, and (b) temperature and relative humidity of steel specimens lubricated with powdered MoS_2 in air (*Source*: Peterson and Johnson, 1953).

or more higher. The low coefficient of friction of MoS_2 is believed to arise from the strong orientation of the films of MoS_2 produced by sliding, and the intrinsically low adhesion and shear strength between the basal planes of the MoS_2 structure. In contrast to graphite, the coefficient of friction of MoS_2 is lower in a vacuum than in air, Figure 5.4.18. Loss of adsorbed surface films, principally water vapor, actually slightly enhances the lubricating properties of MoS_2, and some of the other layer-lattice compounds. Peterson and Johnson (1953) reported that in a steel slider-steel disk interface lubricated with powdered MoS_2, humidity had a very definite effect on the friction of MoS_2, Figure 5.4.19a. The coefficient of friction increased with increasing relative humidity up to a relative humidity of approximately 60% and then dropped off again. The reason for this decrease at high humidity is not fully known, but it is believed that the MoS_2 coating was disrupted by moisture during the 30–60% humidity period responsible for the high friction. Peterson and Johnson (1953) further stated that an increase in temperature could drive volatile materials (most of which is adsorbed water) out of an MoS_2

film and restore lubrication at high humidity. This can be seen from Figure 5.4.19b where the coefficient of friction in the 40 to 50% humidity range was lowered significantly to values approaching those in dry air (less than 6 % relative humidity) as the temperature increased. The coefficient of friction in dry air is less than that in moist air, and it does not show as pronounced a drop with temperature. Peterson and Johnson also reported that as the relative humidity increased from 15 to 70%, the wear in a 6-hour period increased by 200% (along with an increase in friction from 0.1 to 0.38).

When used in air, MoS_2 not only is a poorer lubricant, but also is corrosive to the mating metallic surface in the presence of adsorbed water vapor. MoS_2 in the presence of water vapor can form a surface layer of oxysulfide and sulfuric acid according to the following reactions:

$$MoS_2 + H_2O \rightarrow MoOS_2 + H_2 \tag{5.4.1a}$$

$$2MoS_2 + 9O_2 + 4H_2O \rightarrow 2MoO_3 + 4H_2SO_4 \tag{5.4.1b}$$

Sulfuric acid is corrosive to steel surfaces (Bhushan, 1987).

The MoS_2 can be used as a lubricant from cryogenic temperatures up to a maximum temperature of about 315°C in air (or about 500°C in an inert environment, e.g., N_2 or vacuum). The maximum-use temperature in air atmosphere is limited by oxidation of MoS_2 to MoO_3,

$$2MoS_2 + 7O_2 \rightarrow 2MoO_3 + 4SO_2 \tag{5.4.2}$$

Figure 5.4.20 shows that MoS_2 lubricates at a higher temperature in a nonreactive argon atmosphere up to 500°C.

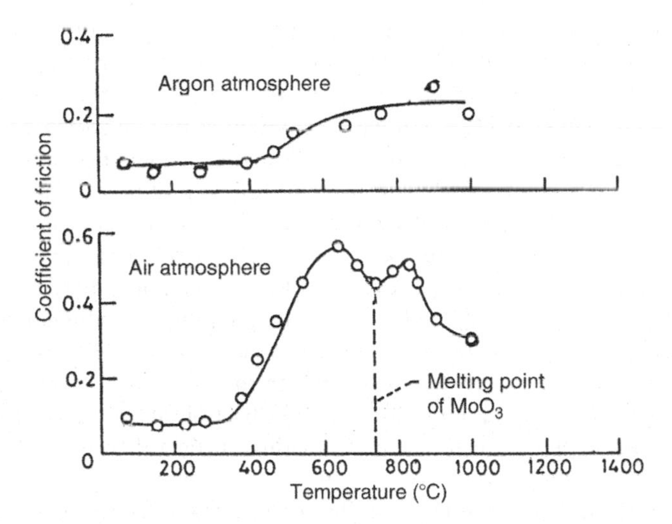

Figure 5.4.20 Coefficient of friction as a function of temperature for burnished MoS_2 disks against a pin in argon and air atmosphere. Reproduced from Sliney, H.E. (1982), "Solid Lubricant Materials for High Temperatures: A Review," *Tribol. Int.* **15**, 303–314. Copyright 1982,with permission from Elsevier.

The lubricating ability from cryogenic temperatures up to 300–400°C in air and even higher temperatures in nonoxidizing atmospheres makes it attractive for aerospace applications. Note that the low friction of graphite depends on the presence of water vapors and hydrocarbons, etc.; therefore, it is a poor lubricant in a vacuum or nonoxidizing atmosphere. However, graphite in air can be used to a higher temperature than can MoS_2. PTFE (discussed earlier) does not perform well under high loads because of its tendency to cold flow. Therefore, for high loads, MoS_2 and graphite are preferred.

MoS_2, like graphite, is used in powder form, as resin-bonded coatings, vacuum deposited coatings, or as an additive to lubricants, greases and solids. MoS_2 is widely used in industry and its sputtered coatings are used in bearings and other sliding applications, especially in non-oxidizing environments, for example, satellites, space shuttle, and other aerospace applications.

As stated earlier, the low friction of both graphite and MoS_2 is associated with their lamellar structures and weak interplanar bonding. Many other compounds with lamellar structures, like talc, tungsten disulphide, tungsten diselenide, graphite fluoride $\left[(CF_x)_n\right]$ and cadmium iodide, do show low friction and are potentially useful as solid lubricants (Bhushan and Gupta, 1997). All compounds with similar structures, such as mica, necessarily do not show low friction, and the low friction values cannot therefore be associated with these factors alone.

The C_{60} molecule has the highest possible symmetry and assumes the shape of soccer ball. At room temperature, fullerene molecules pack in an fcc lattice bonded with weak van der Waals attraction between other molecules (Bhushan, 1999a). Since the C_{60} molecules are very stable and do not require additional atoms to satisfy chemical bonding requirements, they are expected to have low adhesion to the mating surface and low surface energy. Since C_{60} molecules with a perfect spherical symmetry are weakly bonded to other molecules, during sliding, C_{60} clusters are detached readily, similar to other layer-lattice structures, and are transferred to the mating surface by mechanical compaction or are present as loose wear particles which may roll like tiny ball bearings in a sliding contact, resulting in low friction and wear. The wear particles are expected to be harder than as-deposited C_{60} molecules because of their phase transformation at the high-asperity contact pressures present in a sliding interface. The low surface energy, spherical shape of C_{60} molecules, weak intermolecular bonding, and high load-bearing capacity are some of the reasons for C_{60}-rich molecules to perform as good solid lubricants (Bhushan *et al.*, 1993). It has been reported that an increased amount of C_{70} and impurities in the fullerene films do not exhibit low friction (Gupta *et al.*, 1994).

The coefficient of friction of a sublimed C_{60}-rich (fullerene) film is compared with that of sputtered MoS_2 and dipped graphite films in Figure 5.4.21a. Measurements on fullerene films produced in separate depositions were made and a range of friction data is presented in the figure. The fullerene films exhibit low coefficient of friction comparable to that of MoS_2 and graphite films. Friction takes off above 10 m of sliding as fullerene film is not optimized for durability. The effect of temperature on friction and wear is shown in Figure 5.4.21b. There is a significant reduction in the coefficient of friction from 0.2 at 20°C to 0.08 at 100°C. High friction and wear were observed above 100°C. Low friction at 100°C is probably because of more facile transfer of C_{60} material to the mating surface than at 20°C. High friction at higher ambient temperatures $>$ 220°C may result from partial oxidation of C_{60} film and absence of transfer film. (C_{60} sublimes at 450°C in air.) The effect of relative humidity and environmental gas on friction and wear is shown in Figure 5.4.21c. The friction was low at a high humidity of 80%, but wear life was extremely short. The coefficient of friction in the dry nitrogen environment was low with a longer life. Absence of oxidation and formation of

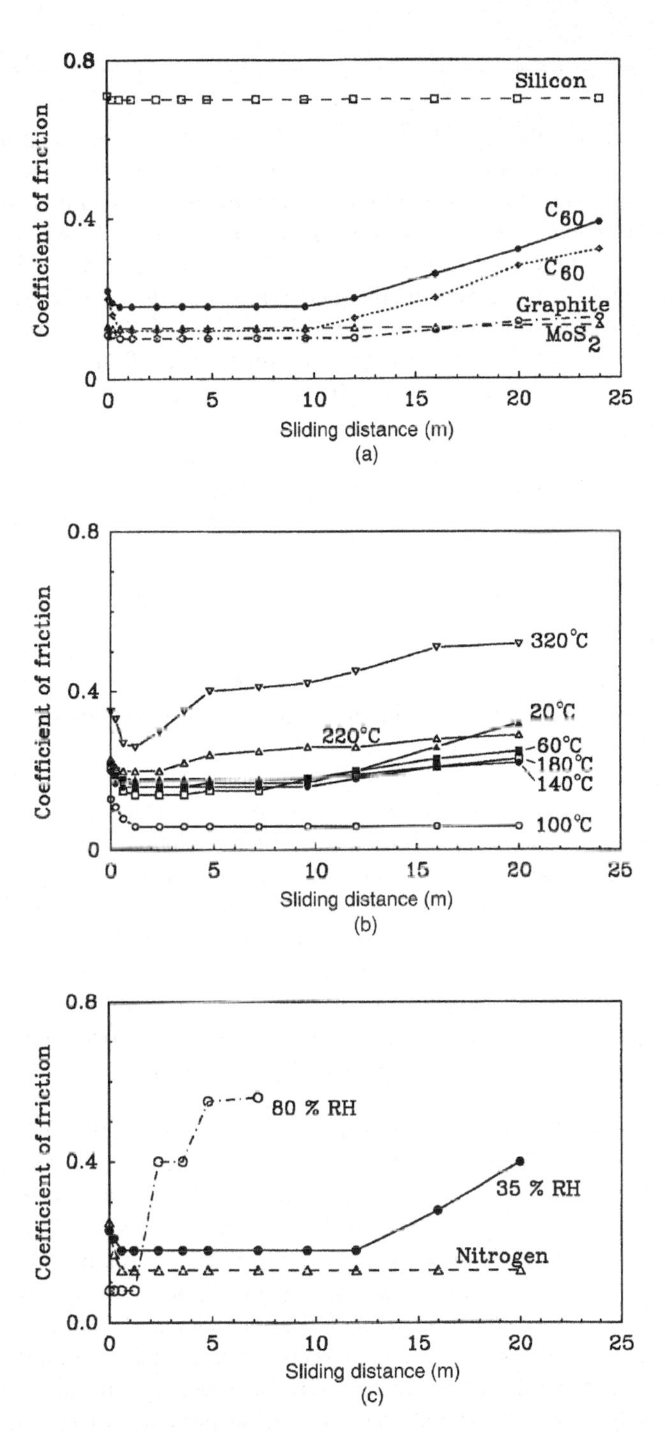

Figure 5.4.21 (a) Coefficient of friction as a function of sliding distance for two typical C_{60}-rich films, MoS_2 and graphite films, and silicon substrate (for comparisons) slid against 52100 steel ball at ambient atmosphere, and (b) the effect of temperature and (c) environment on the coefficient of friction of C_{60}-rich film. Reproduced with permission from Bhushan, B., Gupta, B.K., Van Cleef, G.W., Capp, C., and Coe, J.V. (1993), "Fullerene (C60) Films for Solid Lubrication," *Tribol. Trans.* **36**, 573–580. Fig. 6 p. 578, Figs. 10 and & 11 p. 579. Copyright 1993 Taylor and Francis.)

tenacious transfer film is believed to be responsible for low friction and wear in a dry nitrogen environment. Thus, C_{60}-rich films perform as solid lubricant in dry nitrogen environment and in an ambient up to about 100°C. Environmental envelopes of C_{60} similar to that of MoS_2. Gupta *et al.* (1994) have reported that ion implantation can improve the wear life of C_{60} films.

Conventional solid lubricants such as graphite, MoS_2, and graphite fluoride oxidize or dissociate above about 500°C. Several inorganic salts with low shear strength and film-forming ability are used as solid lubricants. CaF_2 *and* $CaF_2 - BaF_2$ are nonlayered inorganic compounds. These exhibit low shear strength and form a surface film of low shear strength at high ambient temperatures. Some CaF_2-based coatings can be used from room temperature to 900°C (Bhushan and Gupta, 1997).

Solid lubricant materials are used in the form of bulk solids or films as well as dry powders, solids impregnated with solid lubricants, and dispersions (suspensions in lubricating oils and greases).

5.5 Closure

Friction is the resistance to motion, during sliding or rolling, that is experienced whenever one solid body moves tangentially over another with which it is in contact. The coefficient of friction is defined as the ratio of friction force to the normal load. The tangential force that is required to initiate relative motion is known as the static friction force and the force to maintain relative motion is known as the kinetic (dynamic) friction force. The coefficient of static friction is either greater than or equal to the coefficient of kinetic friction. According to Amontons' rules of friction, the friction force is proportional to normal load and is independent of apparent area of contact. A third rule credited to Coulomb states that the coefficient of kinetic friction is independent of the sliding velocity. The first two rules are generally adhered to within a few percent in many macroscopic sliding conditions, but the third rule generally does not hold. In the elastic-contact situation of a single-asperity contact or for a constant number of contacts, the coefficient of friction is proportional to $(load)^{-1/3}$. Strictly, the coefficient of friction is not an inherent material property; it depends very much on the operating conditions and surface conditions.

Friction involves mechanisms of energy dissipation during relative motion. As the two engineering surfaces are brought into contact, contact occurs at the tips of asperities and the load is supported by the deformation of contacting asperities and the discrete contact spots are formed. The proximity of asperities results in adhesive contacts caused by either physical or chemical interactions. Friction arises due to adhesion and deformation. The adhesion term constitutes the force required to shear the adhesive bonds formed at the interface in the regions of real area of contact. In the case of metals, ceramics and other hard materials, the deformation term constitutes the force needed for deformation of asperities (micro-scale) and/or plowing, grooving or cracking of one surface by asperities of the harder, mating material (macro-scale). Plowing of one or both surfaces can also occur by wear particles trapped between them. In the case of viscoelastic (rubbery) materials, the deformation term includes energy loss as a result of elastic hysteresis losses, which gives rise to what is generally called the hysteresis friction. If we assume that there is negligible interaction between these two processes – adhesion and deformation – we may linearly add them. Since in contacts involving primarily elastic deformation, the energy dissipation is very small, the coefficient of friction is very low, as in ultra-smooth surfaces (particularly made of low modulus materials) under lightly loaded conditions. In these interfaces, the dominant contribution to friction is adhesive and the deformation component is small.

Adhesion is present in all contacts; the degree of adhesion is a function of the interface conditions. The adhesive component of friction can be reduced, for example, by reducing the real area of contact and the adhesion strength. Methods to decrease the real area of contact were described in Chapter 3. One method to reduce adhesive strength is to use films of low shear strength at the interface. Contaminants from the environment reduce the adhesion of metal–metal contacts resulting in friction lower than that of clean metals. The deformation component of the friction, for example, is a function of the relative hardnesses and surface roughnesses of the interface materials, and probability of wear particles trapped at the interface. The deformation component of friction can be reduced by reducing the interface roughness, by selecting materials of more or less equal hardness and by removing wear and contaminant particles from the interface. Reduction of plastic deformation at asperity contacts also reduces energy dissipation.

Static friction is time-dependent. There are two models for time dependence. The first model is based on an exponential growth in friction with rest time and the second is based on the power law. If the friction force or sliding velocity does not remain constant as a function of distance or time and produces a form of oscillation, it may be based on the stick-slip phemonenon. During the stick phase, the friction force builds up to a certain value and a large enough force has to be applied to overcome the static friction force, and slip occurs at the interface. Usually, a saw tooth pattern in the friction–force–time curve is observed during the stick-slip process. The stick-slip process can cause resonance of the mechanical system which shows up as a sinusoidal pattern in friction force. This classical stick-slip requires that the coefficient of static friction is greater than the coefficient of kinetic friction. If the coefficient of kinetic friction varies with sliding velocity such that the slope of the μ_k versus velocity curve is negative at a velocity, harmonic oscillations are sometimes observed, generally at high velocities. It should be noted that variation in friction force with time does not always mean that this variation is caused by the stick-slip process.

Rolling friction is much lower than sliding friction. Typical values of the coefficient of rolling friction of a hard cylindrical or spherical body against a cylindrical, spherical, or flat body are in the range of 5×10^{-3} *to* 10^{-5}, whereas typical values of the coefficient of sliding friction of dry bodies range typically from 0.1 to sometimes greater than 1. The term "rolling friction" is usually restricted to bodies of near perfect (continuous) shapes with very small roughness. In the case of rolling contacts, rolling friction arises from the resistance to rolling and because of sliding (slip). The joining and separation of surface elements under rolling contacts are different from those of sliding contacts. Owing to differences in kinematics, the surface elements approach and separate in a direction normal to the interface rather than in a tangential direction. The contributions of adhesion to the rolling friction are small, and arise mainly from deformation losses. In elastic contacts, energy dissipation occurs because of elastic hysteresis. Sliding resistance during slip arises from both adhesive and deformation losses. The magnitude of the resistance to rolling is usually much less that during slip.

In liquid-mediated contacts, the high coefficient of static friction, and in some cases kinetic friction, is a function of the meniscus and viscous contributions. Surface roughness, type of liquid and its film thickness, rest time, and start-up acceleration affect the static friction. Very high static friction can be reached in very smooth surfaces in the presence of some liquid.

Friction of a material is dependent on the mating material (or material pair), surface preparation and operating conditions. For example, material handling, such as the transfer of greasy material from hands to the material surface and the formation of chemically reacted products with the environment, significantly affects friction properties. Therefore, the usefulness of coefficient of friction values from any published literature lies more in their relative magnitudes than in their absolute values.

Clean metals and alloys exhibit high adhesion, and consequently high friction and wear. Any contamination mitigates contact, and chemically produced films, which reduce adhesion, result in the reduction of friction. In dry sliding, identical metals, particularly iron on iron, are metallurgically compatible and exhibit high friction, and identical pairs should be avoided. Soft and ductile metals such as In, Pb and Sn exhibit high friction. Hexagonal metals such as Co and Mg as well as some non-hexagonal metals such as Mo and Cr exhibit low friction. Lead-based white metals (babbitts), brass and bronze, and gray cast iron generally exhibit relatively low friction and wear, and are commonly used as dry and lubricated bearing and seal materials.

In dry sliding conditions, similar or dissimilar ceramic pairs are commonly used. In ceramics, fracture toughness is an important mechanical property which affects friction. An oxidizing and humid environment has a significant effect on friction; tribochemically produced surface films at high interface temperatures generally affect friction. Diamond against itself or other sliding materials generally exhibits very low friction, on the order of 0.1.

Polymers include plastics and elastomers. These generally exhibit low friction. Among polymers, PTFE exhibits the lowest friction, as low as 0.05. PTFE and other polymers flow at modest pressures and modest temperatures, and therefore polymer composites are commonly used. Solid lubricants, namely graphite and MoS_2, exhibit friction as low as 0.05, and are the most commonly used solid lubricants. MoS_2 performs best in friction and wear at low humidities and ultra-high vacuum, whereas graphite works best at high humidities. PTFE performs well in all environments. These lubricants - MoS_2 and graphite - are used in the form of powder, thin films or as an additive. Most soft-solid lubricants which lubricate effectively form a strongly adherent transfer film on the surface being lubricated so that, after a short running-in period during which this film is formed, the actual contact is between lubricant and lubricant.

Problems

5.1 Two hard conical sliders of semiangles 70° and 80° are slid against a lubricated metal surface. The ratio of the coefficient of friction obtained using the two sliders is 1.2. Calculate the adhesive component of the coefficient of friction. Assume that the dominant sources of friction are adhesion and plowing and that these are additive.

5.2 A flat body and a conical indenter are slid against a soft surface. Measured values of the coefficients of friction are 0.2 and 0.4, respectively. Calculate the coefficient of friction for a conical indenter having a semi-angle half of that used in the measurements. The cross-sectional area of the flat body is large enough for plowing contributions to be negligible.

5.3 A hard metal ball of 10 mm diameter slid across a soft metal surface, produces a groove of 2 mm width. For a measured coefficient of friction of 0.4, calculate the adhesive contribution to the coefficient of friction.

5.4 Coefficient of friction values for the following material pairs in a sliding contact are 0.02, 0.1, 0.1, 0.2, and 0.5. Enter appropriate values in the Table P5.4.1.

Table P5.4.1

Material pair	Coefficient of friction
Steel vs steel	?
Steel vs graphite	?
Alumina vs graphite	?
Steel vs brass	?
Steel vs brass in the presence of thick lubricant film	?

5.5 In an AFM measurement, the friction force of 2 mN is measured at a normal load of 10 μN and 4 mN for a normal load of 30 μN. Calculate the meniscus force and the coefficient of dry friction.

5.6 A rough surface with $\sigma = 1$ μm and $\beta^* = 30$ μm comes in contact with flat surface with $E^* = 300$ GPa and $H_s = 5$ GPa. The shear strength τ_a is 0.5 GPa. Determine whether the contact is predominantly elastic or plastic. Calculate the coefficient of friction.

5.7 For two bodies in plastic contact, the hardness of a softer material H is 2 GPa, and the shear strength τ_a is 0.1 GPa. Calculate the real area of plastic contact at a load W of 10 N and the coefficient of adhesional friction.

References

Amontons, G. (1699), "De la résistance causée dans les Machines," *Mémoires de l'Académie Royale* **A**, 257–282.

Anonymous (1971), *Proc 1st. Int. Conf. on Solid Lubrication*, SP-3, ASLE, Park Ridge, Illinois.

Anonymous (1978), *Proc 2nd Int. Conf. on Solid Lubrication*, SP-6, ASLE, Park Ridge, Illinois.

Anonymous (1984), *Proc 3rd Int. Conf. on Solid Lubrication*, SP-14, ASLE, Park Ridge, Illinois.

Anonymous (1986), "Panel Report on Interfacial Bonding and Adhesion," *Mat. Sci. and Eng.* **83**, 169–234.

Anonymous (1987), *Tribology of Ceramics*, Special Publications SP-23 and SP-24, STLE, Park Ridge, Illinois.

Archard, J.F. (1957), "Elastic Deformation and The Laws of Friction," *Proc. Roy. Soc. Lond. A* **243**, 190–205.

Armstrong-Helouvry, B. (1991), *Control of Machines with Friction*, Kluwer Academic Pub., Dordrecht, Netherlands.

Arnell, R.D., Davies, P.B., Halling, J. and Whomes, T.L. (1991), *Tribology-Principles and Design Applications*, Springer-Verlag, New York.

Avallone, E.A. and Baumeister, T. (1987), *Marks' Standard Handbook for Mechanical Engineers*, Ninth edition, Mc-Graw-Hill, New York.

Bartenev, G.M. and Lavrentev, V.V. (1981), *Friction and Wear of Polymers*, Elsevier, Amsterdam.

Barwell, F.T. (1956), *Lubrication of Bearings*, Butterworths, London, UK.

Bhushan, B. (1980a), "Stick-Slip Induced Noise Generation in Water-Lubricated Compliant Rubber Bearings," *ASME J. Lub. Tech.* **102**, 201–212.

Bhushan, B. (1980b), "High Temperature Low Friction Surface Coatings," US Patent No. 4, 227, 756, Oct. 14.

Bhushan, B. (1981a), "Effect of Shear-Strain Rate and Interface Temperature on Predictive Friction Models," *Proc. Seventh Leeds-Lyon Symposium on Tribology* (D. Dowson, C.M. Taylor, M. Godet and D. Berthe, eds.), pp. 39–44, IPC Business Press, Guildford, UK.

Bhushan, B. (1981b), "High Temperature Low Friction Surface Coatings and Methods of Application," US Patent No. 4, 253, 714, March 3.

Bhushan, B. (1982), "Development of CdO-Graphite-Ag Coatings for Gas Bearings to 427°C," *Wear* **75**, 333–356.

Bhushan, B. (1987), "Overview of Coating Materials, Surface Treatments and Screening Techniques for Tribological Applications Part I: Coating Materials and Surface Treatments," *Testing of Metallic and Inorganic Coatings* (W.B. Harding and G.A. DiBari, eds.), STP947, pp. 289–309, ASTM, Philadelphia.

Bhushan, B. (1996), *Tribology and Mechanics of Magnetic Storage Devices*, Second edition, Springer-Verlag, New York.

Bhushan, B. (1999a), *Handbook of Micro/Nanotribology*, Second edition, CRC Press, Boca Raton, Florida.

Bhushan, B. (1999b), "Chemical, Mechanical and Tribological Characterization of Ultra-Thin and Hard Amorphous Carbon Coatings as Thin as 3.5 nm: Recent Developments," *Diamond and Related Materials* **8**, 1985–2015.

Bhushan, B. (2001a), *Modern Tribology Handbook Vol. 1: Principles of Tribology*, CRC Press, Boca Rotan, Florida.

Bhushan, B. (2001b), *Fundamentals of Tribology and Bridging the Gap Between Macro- and Micro/Nanoscales*, NATO Science Series II–Vol. 10, Kluwer Academic Pub., Dordrecht, Netherlands.

Bhushan, B. (2010), *Springer Handbook of Nanotechnology*, third edition, Springer-Verlag, Heidelberg, Germany.

Bhushan, B. (2011), *Nanotribology and Nanomechanics I & II*, Third edition, Springer-Verlag, Heidelberg, Germany.

Bhushan, B. (2013), *Principles and Applications of Tribology*, Second edition, Wiley, New York.

Bhushan, B. and Dashnaw, F. (1981), "Material Study For Advanced Stern-Tube Bearings and Face Seals," *ASLE Trans.* **24**, 398–409.

Bhushan, B. and Gupta, B.K. (1997), *Handbook of Tribology – Materials, Coatings, and Surface Treatments*, McGraw-Hill, New York (1991) reprinted with corrections, Krieger, Malabar, Florida (1997).

Bhushan, B. and Jahsman, W.E. (1978a), "Propagation of Weak Waves in Elastic-Plastic and Elastic-Viscoplastic Solids With Interfaces," *Int. J. Solids and Struc.* **14**, 39–51.

Bhushan, B. and Jahsman, W.E. (1978b), "Measurement of Dynamic Material Behavior Under Nearly Uniaxial Strain Conditions," *Int. J. Solids and Struc.* **14**, 739–753.

Bhushan, B. and Kulkarni, A.V. (1996), "Effect of Normal Load on Microscale Friction Measurements," *Thin Solid Films* **278**, 49–56; **293**, 333.

Bhushan, B. and Nosonovsky, M. (2004), "Scale Effects in Dry and Wet Friction, Wear, and Interface Temperature," *Nanotechnology* **15**, 749–761.

Bhushan, B. and Ruan, J. (1994), "Atomic-Scale Friction Measurements Using Friction Force Microscopy: Part II – Application to Magnetic Media," *ASME J. Tribol.* **116**, 452–458.

Bhushan, B. and Sibley, L.B. (1982), "Silicon Nitride Rolling Bearings for Extreme Operating Conditions," *ASLE Trans.* **35**, 628–639.

Bhushan, B. and Wilcock, D.F. (1981), "Frictional Behavior of Polymeric Compositions in Dry Sliding," in *Proc. Seventh Leeds-Lyon Symp. on Tribology* (D. Dowson, C.M. Taylor, M. Godet and D. Berthe, eds.), pp. 103–113, IPC Business Press, Guildford, UK.

Bhushan, B. and Winn, L.W. (1981), "Material Study for Advanced Stern-Tube Lip Seals," *ASLE Trans.* **24**, 398–409.

Bhushan, B. and Zhao, Z. (1999), "Macro- and Microscale Tribological Studies of Molecularly-Thick Boundary Layers of Perfluoropolyether Lubricants for Magnetic Thin-Film Rigid Disks," *J. Info. Storage Proc. Syst.* **1**, 1–21.

Bhushan, B., Sharma, B.S., and Bradshaw, R.L. (1984) "Friction in Magnetic Tapes I: Assessment of Relevant Theory," *ASLE Trans.* **27**, 33–44.

Bhushan, B., Gupta, B.K., Van Cleef, G.W., Capp, C., and Coe, J.V. (1993), "Fullerene (C_{60})Films for Solid Lubrication," *Tribol. Trans.* **36**, 573–580.

Bhushan, B., Yang, L., Gao, C., Suri, S., Miller, R.A., and Marchon, B. (1995), "Friction and Wear studies of Magnetic Thin-Film Rigid Disks with Glass-Ceramic, Glass and Aluminum-Magnesium Substrates," *Wear* **190**, 44–59.

Blau, P.J. (1992a), *ASM Handbook: Vol. 18 Friction, Lubrication, and Wear Technology*, ASM International, Metals Park, Ohio.

Blau, P.J. (1992b), "Scale Effects in Sliding Friction: An Experimental Study," in *Fundamentals of Friction: Macroscopic and Microscopic Processes* (I.L. Singer and H.M. Pollock, eds.), pp. 523–534, Vol. **E220**, Kluwer Academic, Dordrecht, Netherlands.

Blau, P.J. (1996), *Friction Science and Technology*, Marcel Dekker, New York.

Bowden, F.P. and Leben, L. (1939), "The Nature of Sliding and the Analysis of Friction," *Proc. Roy. Soc. Lond. A* **169**, 371–379.

Bowden, F.P. and Tabor, D. (1950), *The Friction and Lubrication of Solids*, Part I, Clarendon Press, Oxford, UK.

Bowden, F.P. and Tabor, D. (1964), *The Friction and Lubrication of Solids*, Part II, Clarendon Press, Oxford, UK.

Bowden, F.P. and Tabor, D. (1973), *Friction: An Introduction to Tribology*, Doubleday and Company, Garden City, N.Y.; Reprinted Krieger Publishing Co., Malabar, Florida (1982).

Bowden, F.P. and Young, J.E. (1951), "Friction of Clean Metals and The Influence of Adsorbed Films," *Proc. Roy. Soc. Lond. A* **208**, 311–325.

Bowers, R.C. (1971), "Coefficient of Friction of High Polymers as a Function of Pressure," *J. Appl. Phys.* **42**, 4961–4970.

Braithwaite, E.R. (1964), *Solid Lubricants and Solid Surfaces*, Pergamon, Oxford, UK.

Braithwaite, E.R. (1967), *Lubrication and Lubricants*, Elsevier, Amsterdam.

Briscoe, B.J. and Tabor, D. (1978), "Shear Properties of Thin Polymeric Films," *J. Adhesion* **9**, 145–155.

Brockley, C. and Davis, H. (1968), "The Time Dependence of Static Friction," *ASME J. Lub. Tech.* **90**, 35–41.

Bryant, P.J., Gutshall, P.L., and Taylor, L.H. (1964), "A Study of Mechanisms of Graphitic Friction and Wear," *Wear* **7**, 118–126.

Buckley, D.H. (1981), *Surface Effects in Adhesion, Friction, Wear, and Lubrication*, Elsevier, Amsterdam.

Buckley, D.H. (1982), "Surface Films and Metallurgy Related to Lubrication and Wear," in *Progress in Surface Science* (S.G. Davison, ed.), Vol. **12**, pp. 1–153, Pergamon, New York.

Buckley, D.H. and Johnson, R.L. (1964), "Mechanism of Lubrication for Solid Carbon Materials in Vacuum to 10^{-9}mm of Mercury," *ASLE Trans.* **7**, 97.

Bulgin, D., Hubbard, G.D., and Walters, M.H. (1962), *Proc. 4th Rubber Technology Conf. London*, May, p. 173, Institution of the Rubber Industry.

Bunn, C.W. and Howells, E.R. (1954), "Structure of Molecules and Crystals of Fluorocarbons," *Nature* **174**, 549–551.

Chandrasekar, S. and Bhushan, B. (1990), "Friction and Wear of Ceramics for Magnetic Recording Applications-Part I: A Review," *ASME J. Trib.* **112**, 1–16.

Clauss, F.J. (1972), *Solid Lubricants and Self-Lubricating Solids*, Academic, New York.

Coulomb, C.A. (1785), "Theorie des Machines Simples, en ayant egard au Frottement de leurs Parties, et a la Roideur des Cordages," *Mem. Math. Phys.*, **X**, Paris, 161–342.

Courtney-Pratt, J.S. and Eisner, E. (1957), "The Effect of a Tangential Force on the Contact of Metallic Bodies," *Proc. Roy. Soc. Lond. A* **238**, 529–550.

Cox, J.M. and Gee, M.G. (1997), "Hot Friction Testing of Ceramics," *Wear* **203–204**, 404–417.

Dokos, S.J. (1946), "Sliding Friction Under Extreme Pressures," *J. Appl. Mech.* **13**, 148–156.

Farr, J.P.G. (1975), "Molybdenum Disulphide in Lubrication: A Review," *Wear* **35**, 1–22.

Gao, C. and Bhushan, B. (1995), "Tribological Performance of Magnetic Thin-Film Glass Disks: Its Relation to Surface Roughness and Lubricant Structure and its Thickness," *Wear* **190**, 60–75.

Grosch, K.A. (1963), "The Relation Between the Friction and Viscoelastic Properties of Rubber," *Proc. Roy. Soc. Lond. A* **274**, 21–39.

Gupta, B.K., Bhushan, B., Capp, C., and Coe, J.V. (1994), "Materials Characterization and Effect of Purity and Ion Implantation on The Friction and Wear of Sublimed Fullerene Films," *J. Mater. Res.* **9**, 2823–2838.

Heathcote, H.L. (1921), "The Ball Bearings," *Proc. Instn. Auto. Eng.* **15**, 1569.

Hegmon, R.R. (1969), "The Contribution of Deformation Losses to Rubber Friction," *Rubber Chem. and Technol.* **42**, 1122–1135.

Holinski, R. and Gansheimer, J. (1972), "A Study of the Lubricating Mechanism of Molybdenum Disulfide," *Wear* **19**, 329–342.

Hutchings, I.M. (1992), *Tribology: Friction and Wear of Engineering Materials*, CRC Press, Boca Raton, Florida.

Iliuc, I. (1980), *Tribology of Thin Layers*, Elsevier, Amsterdam.

Ishigaki, H., Kawaguchi, I., Iwasa, M., and Toibana, Y. (1986), "Friction and Wear of Hot Pressed Silicon Nitride and Other Ceramics," *ASME J. Tribol.* **108**, 514–521.

Ishlinskii, A. and Kragelskii, I. (1944), "On Stick-Slip in Sliding," *Sov. J. Tech. Phys.* **14**, 276–282.

Jahanmir, S. (ed.), (1994), *Friction and Wear of Ceramics*, Marcel Dekker, New York.

Johnson, K.L. (1985), *Contact Mechanics*, Clarendon Press, Oxford, UK.

Kato, S. Sato, N., and Matsubayashi, T. (1972), "Some Considerations on Characteristics of Static Friction of Machine Tool Sideways," *ASME J. Lub. Tech.* **94**, 234–247.

Kragelskii, I.V. (1965), *Friction and Wear*, Butterworths, London, UK.

Lancaster, J.K. (1972), "Friction and Wear," in *Polymer Science* (A.D. Jenkins, ed.), Vol. **2**, pp. 959–1046, North-Holland, Amsterdam.

Lancaster, J.K. (1973), "Dry Bearings: A Survey of Materials and Factors Affecting Their Performance," *Tribol. Int.* **6**, 219–251.

Lee, L.H. (1974), "Effect of Surface Energetics on Polymer Friction and Wear," in *Advances in Polymer Friction and Wear* (L.H. Lee, ed.), Vol. **5A**, pp. 31–68, Plenum, New York.

Ludema, K.C. and Tabor, D. (1966), "The Friction and Viscoelastic Properties of Polymeric Solids," *Wear* **9**, 329–348.

Makinson, K.R. (1948), "On the Cause of Frictional Difference of the Wool Fiber," *Trans. Faraday Soc.* **44**, 279–282.

McFarlane, J.S. and Tabor, D. (1950), "Adhesion of Solids and the Effects of Surface Films," *Proc. Roy. Soc. Lond. A* **202**, 224–243.

Miyoshi, K. and Buckley, D.H. (1979), "Friction, Deformation and Fracture of Single-Crystal Silicon Carbide," *ASLE Trans.* **22**, 79–90.

Moore, D.F. (1972), *The Friction and Lubrication of Elastomers*, Pergamon, Oxford, UK.

Neale, M.J. (ed.) (1973), *Tribology Handbook*, Newnes-Butterworths, London, UK.

Paxton, R.R. (1979), *Manufactured Carbon: A Self Lubricating Material for Mechanical Devices*, CRC, Boca Raton, Florida.

Pepper, S.V. (1982), "Effect of Electronic Structure of The Diamond Surfaces on the Strength of Diamond-Metal Interface," *J. Vac. Sci. Technol.* **20**, 643–647.

Peterson, M.B. and Johnson, R.L. (1953), "Friction and Wear Characteristics of Molybdenum Disulfide I. Effect of Moisture," Tech. Rep. TN-3055, NACA Washington, D.C.

Peterson, M.B., Florek, J.J., and Lee, R.E. (1960), "Sliding Characteristics of Metals at High Temperatures," *Trans. ASLE* **3**, 101–115.

Pooley, C.M. and Tabor, D. (1972), "Friction and Molecular Structure: The Behavior of Some Thermoplastics," *Proc. Roy. Soc. Lond. A* **329**, 251–274.

Poritsky, H. (1950), "Stresses and Deflections of Cylindrical Bodies in Contact With Application to Contact of Gears and of Locomotive Wheels," *ASME J. Appl. Mech.* **17**, 191–201.

Rabinowicz, E. (1958), "The Intrinsic Variables Affecting the Stick-Slip Process," *Proc. Phys. Soc. Lond.* **71**, 668–675.

Rabinowicz, E. (1995), *Friction and Wear of Materials*, Second edition, Wiley, New York.

Reynolds, O. (1876), "On Rolling Friction," *Philos. Trans. R. Soc. Lond.* **166**, 155–174.

Rigney, D.A. and Hirth, J.P. (1979), "Plastic Deformation and Sliding Friction of Metals," *Wear* **53**, 345–370.

Rowe, G.W. (1960), "Some Observations on the Friction Behavior of Boron Nitride and of Graphite," *Wear* **3**, 274–285.

Sampson, J.B., Morgan, F., Reed, D.W., and Muskat, M. (1943), "Friction Behavior During the Slip Portion of the Stick-Slip Process," *J. Appl. Phys.* **14**, 689–700.

Samuels, B. and Wilks, J. (1988), "The Friction of Diamond Sliding on Diamond," *J. Mater. Sci.* **23**, 2846–2864.

Santner, E. and Czichos, H. (1989), "Tribology of Polymers," *Tribol. Int.* **22**, 104–109.

Savage, R.H. (1948), "Graphite," *Lubrication* **19**, 1–10.

Schallamach, A. (1971), "How Does Rubber Slide," *Wear* **17**, 301–312.

Seal, M. (1981), "The Friction of Diamond," *Philos. Mag. A* **43**, 587–594.

Skopp, A., Woydt, M., and Habig, K.M. (1990), "Unlubricated Sliding Friction and Wear of Various Si_3N_4 Pairs Between 22° and 1000°C," *Tribol. Inter.* **23**, 189–199.

Sliney, H.E. (1982), "Solid Lubricant Materials for High Temperatures: A Review," *Tribol. Int.* **15**, 303–314.

Steijn, R.P. (1966), "The Effect of Time, Temperature, and Environment on the Sliding Behavior of Polytetrafluoropolyethene," *ASLE Trans.* **9**, 149–159.

Steijn, R.P. (1967), "Friction and Wear of Plastics," *Metals Eng. Q.* **7**, 371–383.

Stolarski, T.A. (1990), *Tribology in Machine Design*, Heinemann Newnes, Oxford, UK.

Suh, N.P. (1986), *Tribophysics*, Prentice-Hall, Englewood Cliffs, New Jersey.

Suh, N.P. and Sin, H.C. (1981), "The Genesis of Friction," *Wear* **69**, 91–114.

Tabor, D. (1979), "Adhesion and Friction," in *The Properties of Diamond* (J.E. Field, ed.), pp. 325–348, Academic, New York.

Tanaka, K. (1961), "Friction and Deformation of Polymers," *J. Phys. Soc. Jpn.* **16**, 2003–2016.

Tanaka, K., Uchiyama, Y., and Toyooka, S. (1973), "The Mechanism of Wear of Polytetrafluoropolyethene," *Wear* **23**, 153–172.

Weick, B.L. and Bhushan, B. (2000), "Grain Boundary and Crystallographic Orientation Effects on Friction," *Tribol. Trans.* **43**, 33–38.

Whitehead, J.R. (1950), "Surface Deformation and Friction of Metals at Light Loads," *Proc. Roy. Soc. Lond. A* **201**, 109–124.

Winer, W.O. (1965), "Molybdenum Disulfide as a Lubricant: A Review of the Fundamental Knowledge," *Wear* **10**, 422–452.

Zhao, Z. and Bhushan, B. (1998), "Effect of Lubricant Thickness and Viscosity and Rest Time on Long-Term Stiction in Magnetic Thin-Film Rigid Disks," *IEEE Trans. Magn.* **34**, 1708–1710.

Further Reading

Bartenev, G.M. and Lavrentev, V.V. (1981), *Friction and Wear of Polymers*, Elsevier, Amsterdam.
Bhushan, B. (1996), *Tribology and Mechanics of Magnetic Storage Devices*, Second edition, Springer-Verlag, New York.
Bhushan, B. (1999), *Principles and Applications of Tribology*, Wiley, New York.
Bhushan, B. (2001a), *Modern Tribology Handbook Vol. 1: Principles of Tribology*, CRC Press, Boca Rotan, Florida.
Bhushan, B. (2001b), *Fundamentals of Tribology and Bridging the Gap Between Macro- and Micro/Nanoscales*, NATO Science Series II-Vol. **10**, Kluwer Academic Pub., Dordrecht, Netherlands.
Bhushan, B. (2011), *Nanotribology and Nanomechanics I & II*, Third edition, Springer-Verlag, Heidelberg, Germany.
Bhushan, B. and Nosonovsky, M. (2004), "Scale Effects in Dry and Wet Friction, Wear, and Interface Temperature," *Nanotechnology* **15**, 749–761.
Blau, P.J. (1996), *Friction Science and Technology*, Marcel Dekker, New York.
Bowden, F.P. and Tabor, D. (1950), *The Friction and Lubrication of Solids*, Part I, Clarendon Press, Oxford, UK
Bowden, F.P. and Tabor, D. (1964), *The Friction and Lubrication of Solids*, Part II, Clarendon Press, Oxford, UK.
Bowden, F.P. and Tabor, D. (1973), *Friction: An Introduction to Tribology*, Doubleday and Company, Garden City, N.Y.; Reprinted Krieger Publishing Co., Malabar, Florida (1982).
Bruce, R.W. (2012), *Handbook of Lubrication and Tribology, Vol. II: Theory and Design*, Second edition, CRC Press, Boca Raton, Florida.
Buckley, D.H. (1981), *Surface Effects in Adhesion, Friction, Wear, and Lubrication*, Elsevier, Amsterdam.
Donnet, C. and Erdemir, A. (2008), *Tribology of Diamond-like Carbon Films*, Springer-Verlag, Berlin.
Kragelskii, I.V. (1965), *Friction and Wear*, Butterworths, London, UK.
Miyoshi, K. (2001), *Solid Lubrication-Fundamentals and Applications*, Marcel Dekker, New York.
Moore, D.F. (1972), *The Friction and Lubrication of Elastomers*, Pergamon, Oxford, UK.
Rabinowicz, E. (1995), *Friction and Wear of Materials*, Second edition, Wiley, New York.
Rigney, D.A. (ed.) (1981), *Fundamentals of Friction and Wear of Materials*, Am. Soc. Metals, Metals Park, Ohio.
Singer, I.L. and Pollock, H.M. (1992), *Fundamentals of Friction: Macroscopic and Microscopic Processes*, Vol. **E220**, Kluwer Academic Publisher, Dordrecht, Netherlands.

6

Interface Temperature of Sliding Surfaces

6.1 Introduction

The majority of surface effects are temperature-dependent. This is not surprising because on an atomic scale, mechanical, chemical, and electrical phenomena are generally dependent on the thermal energy available to assist or activate these phenomena. During sliding, the effect of operating conditions such as load and velocity on friction and wear are frequently manifestations of the effect of temperature rise on the variable under study. The mechanical properties (such as elastic modulus and hardness) and lubricating properties of many materials start to degrade with a rise in interface temperature which affects their tribological performance. Therefore, an estimate of the interface temperature rise is necessary for the design of an interface.

In any sliding operation, most of the frictional energy input is generally used up in plastic deformation which is directly converted to heat in the material close to interface. The plastic deformation results in increased atomic lattice vibrations, which are really sound waves and are called phonons. The sound energy is eventually transferred into heat. There is a little energy loss during the elastic deformation of interfaces; a loss of 0.1–10% (typically less than 1%) of the energy loss can occur by phonons. In viscoelastic deformation, elastic hysteresis losses result in heat. In the absence of lubricants, this heat is conducted into the two sliding members through contact spots.

Contact between two bodies can be approximated as a single contact or as multiple contacts, Figure 6.1.1. During a high contact-stress situation, the real area of contact (A_r) is close to the apparent area of contact (A_a) and essentially a single contact occurs during sliding. Contact of two very smooth surfaces, even at a relatively low load, may be approximated as a single contact. In most engineering contacts of interest, a low contact-stress situation exists. During a low contact-stress sliding situation, asperity interaction results in numerous, high, transient temperature flashes of as high as several hundred degrees Celsius over areas of a fraction of a micron to a few microns in diameter with a few nanoseconds to a few microseconds in duration. These temperature flashes shift from one place to another during sliding (Bhushan, 1971). Since the heat generated is dissipated over microcontacts, the temperature rise at

Introduction to Tribology, Second Edition. Bharat Bhushan.
© 2013 John Wiley & Sons, Ltd. Published 2013 by John Wiley & Sons, Ltd.

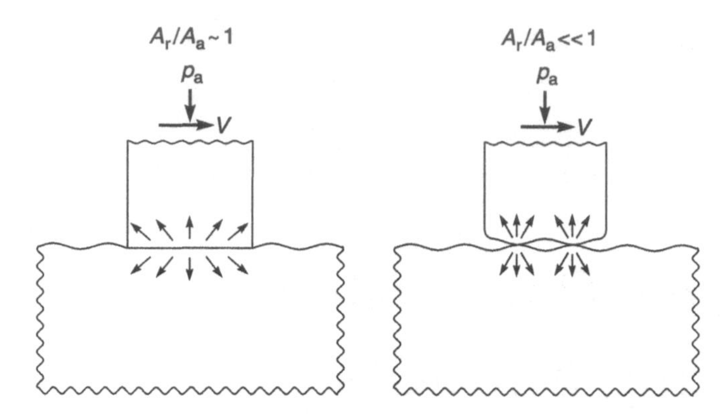

Figure 6.1.1 Schematics of two bodies in sliding contact.

asperity contacts can be very high (Griffioen *et al.*, 1986). If one of the sliding members is compliant, such as a polymer, the temperature rise is small, because a polymer would result in a very large real area of contact which reduces the heat generated per unit of apparent area (Gulino *et al.*, 1986; Bhushan, 1987b).

A number of attempts have been made to predict the transient temperature flashes of asperity contacts (such as Blok, 1937; Jaeger, 1942; Bowden and Tabor, 1950; Ling and Saibel, 1957; Archard, 1958; Holm, 1967; Bhushan and Cook, 1973; Cook and Bhushan, 1973; Bhushan, 1987a). The most complete solution of temperature rise as a result of moving sources of heat on a semi-infinite body was presented by Jaeger (1942). Based on a concept of partition of heat advanced by Blok (1937), the interface temperature can be calculated. This analysis is adequate for a high contact-stress situation where the real area of contact is approximately equal to the apparent area. In sliding contact of two surfaces under low contact-stress conditions, multiple contacts occur. A detailed thermal analysis to calculate the temperature rise during the life of an asperity contact and to analyze cumulative effect of multi-asperity contacts was developed by Bhushan (1971, 1987a). In all of the analyses, it is assumed that the heat transfer occurs by the conduction of heat in solids and convection and radiation effects are neglected. In this chapter, thermal analyses with empirical relationships to predict the interface temperature rise for various cases are presented, which can be conveniently used in the design of any sliding surface.

The temperature rise over large contact areas can be measured with some accuracy. Measurement of temperature rise over isolated micro-contacts is very difficult to measure. A number of techniques have been used with limited success to measure the transient temperature rise in a sliding contact (Kennedy, 1992; Bhushan, 1996). Therefore, the transient temperature rise is generally calculated rather than measured. A description of various measurement techniques and typical results of *in situ* measurements are presented in this chapter.

6.2 Thermal Analysis

There is no single thermal analysis that will reasonably represent all the conditions of sliding. As a result, we shall develop solutions that are valid over limited ranges of contact stress and velocity. We shall consider high contact-stress (individual contact) and low contact-stress (multiple-asperity contact) conditions.

6.2.1 Fundamental Heat Conduction Solutions

Thermal analyses in this section are based on the classical equations for heat conduction in a homogeneous isotropic solid (Carslaw and Jaeger, 1980),

$$\frac{\partial^2\theta}{\partial x^2}+\frac{\partial^2\theta}{\partial y^2}+\frac{\partial^2\theta}{\partial z^2}=\nabla^2\theta=\frac{1}{\kappa}\frac{\partial\theta}{\partial t} \tag{6.2.1}$$

where θ is the temperature rise (°C), x, y, z are Cartesian coordinates (mm), κ is thermal diffusivity (m^2/s) and t is time (s). And

$$\kappa=\frac{k}{\rho c_p} \tag{6.2.2}$$

where k is the thermal conductivity (W/m K) and ρc_p is the volumetric specific heat (J/m^3 K); specifically ρ is the mass density (kg/m^3) and c_p is the specific heat (J/g K). In Equation 6.2.1, it is assumed that thermal properties are constant (thermal isotropy) and heat is only introduced at the boundaries. It is known that thermal properties can, in fact, vary considerably with temperature.

In an infinite solid, if a quantity of heat Q (J or Ws) is instantaneously released at the origin (instantaneous point source) at $t=0$, the temperature rise at time t and distance r from the origin is

$$\theta(r,t)=\frac{Q}{8\rho c_p(\pi\kappa t)^{3/2}}\exp\left[-\frac{r^2}{4\kappa t}\right] \tag{6.2.3}$$

Or, if heat Q is released at x', y', z' at $t=0$, the temperature rise at x, y, z and time t is,

$$\theta(x,y,z,t)=\frac{Q}{8\rho c_p(\pi\kappa t)^{3/2}}\exp\left[-\frac{(x-x')^2+(y-y')^2+(z-z')^2}{4\kappa t}\right] \tag{6.2.4}$$

If heat per unit length Q' is instantaneously released uniformly along the y' axis (instantaneous line source) from $-\infty$ *to* ∞ and through the point (x', 0, z'), then the temperature rise at x, 0, z and time t is obtained by replacing Q in Equation 6.2.4 by $Q'dy'$, and integrating with respect to y' from $-\infty$ *to* ∞,

$$\theta(x,z,t)=\frac{Q'}{4\pi kt}\exp\left[-\frac{(x-x')^2+(z-z')^2}{4\kappa t}\right] \tag{6.2.5}$$

Solutions will be obtained by suitably integrating the equations presented here, with respect to time and space.

6.2.2 *High Contact-Stress Condition ($A_r/A_a \sim 1$) (Individual Contact)*

For materials sliding under high contact-stress conditions, where apparent contact stress (or pressure) p_a approaches the hardness of the softer material (H_s) or $A_r/A_a \sim 1$, frictional heating is assumed to be liberated uniformly over the contact area. A single contact can also occur during sliding of very smooth surfaces even at relatively low loads. High contact stresses are achieved in only a few deformation processes, such as metal cutting (Shaw, 2005). This case of frictional heat uniformly distributed over the contact area is also applicable to an *individual sliding asperity of constant size.*

We now calculate the temperature rise of point, band, and rectangular/square/circular sources of heat moving across the surface of a semi-infinite solid, basing the calculations on classical work by Jaeger (1942). It will be first assumed that heat only goes into the solid and with no loss of heat from the surface. The partition of heat takes place and heat goes into both solids, a procedure that will be described later. This procedure will be used to calculate the steady-state temperature rise with heat flowing in both bodies.

6.2.2.1 Point Source

Consider a point source of heat moving with a relative sliding velocity V across the surface of a semi-infinite solid ($z > 0$) in the $+x$ direction, as shown in Figure 6.2.1 with no loss of heat from the plane $z = 0$. If the source is at the origin now, $t = 0$; then at time t ago it was at $x = -Vt$ (considering time positive in the past). The strength of the heat source is $\dot{Q}$ (J/s or W). Thus, the temperature rise due to heat ($Q = \dot{Q}dt$) liberated at ($x = -Vt$) is, from Equation 6.2.4,

$$d\theta\,(x, y, z, 0) = \frac{2\dot{Q}dt}{8\rho c_p\,(\pi\kappa t)^{3/2}}\,\exp\left[-\frac{(x + Vt)^2 + y^2 + z^2}{4\kappa t}\right] \tag{6.2.6}$$

The factor of 2 is due to shifting from an infinite solid to a semi-infinite solid. Equation 6.2.6 is integrated over all past time to obtain the steady-state temperature as (Jaeger, 1942)

$$\theta = \frac{\dot{Q}}{4\pi rk}\,\exp\left[-\frac{V\,(r + x)}{2\kappa}\right]$$

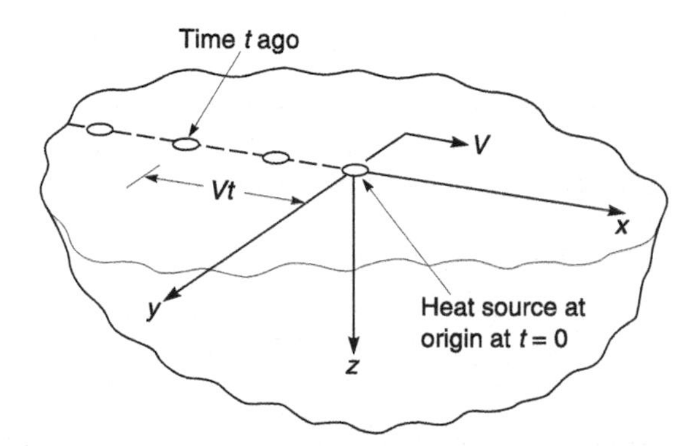

Figure 6.2.1 Schematic of a moving point heat source over a semi-infinite body.

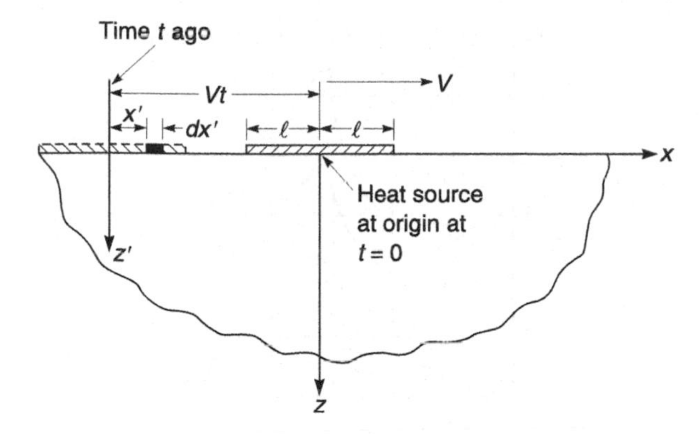

Figure 6.2.2 Schematic of a moving band heat source of length 2ℓ over a semi-infinite body.

where

$$r^2 = x^2 + y^2 + z^2 \tag{6.2.7}$$

6.2.2.2 Band Source

Next, consider a band heat source of length (2ℓ) in the x direction, and infinite in the y direction, as shown in Figure 6.2.2 The temperature rise due to the heat source of strength per unit area per unit time, q (W/mm^2), is calculated by integrating Equation 6.2.5 across the band length for the band in all prior locations. The temperature rise due to portion dx' of the band centered at $x = -Vt$ and $Q' = qdx'dt$ is

$$d\theta = \frac{2qdx'dt}{4\pi kt} \exp\left[-\frac{(x + Vt - x')^2 + z^2}{4\kappa t}\right] \tag{6.2.8}$$

Equation 6.2.8 must be integrated from $t = \infty$ to $t = 0$ and from $x' = -\ell$ to $x' - +\ell$. The solution is not in closed form, but can be evaluated numerically. It is convenient to express the results in terms of dimensionless length and three position parameters, called Peclet numbers:

$$L = \frac{V\ell}{\kappa}, \quad X = \frac{Vx}{\kappa}, \quad Y = \frac{Vy}{\kappa}, \quad Z = \frac{Vz}{\kappa} \tag{6.2.9}$$

Figure 6.2.3 shows a dimensionless temperature rise parameter $\frac{\pi\rho c_p V\theta}{2q}$ as functions of L and x/ℓ (Jaeger, 1942). Note that at low speed ($L < 0.4$), the temperature distribution is almost symmetrical about $x = 0$, which would be expected for a *stationary* source. At high speed $L > 20$, the temperature distribution is highly asymmetrical with the maximum rise at the trailing edge. Approximate equations for both maximum (θ_{max}) and mean ($\bar{\theta}$) temperature rise for two extreme cases of large (> 10) and small (< 0.5) values of L are given below. For the

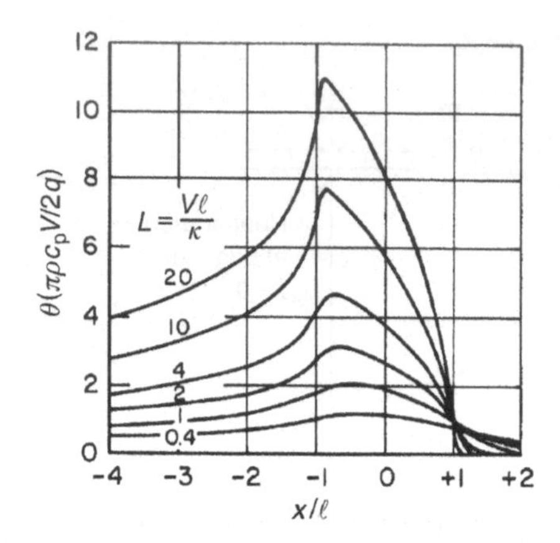

Figure 6.2.3 Steady-state dimensionless temperature rise caused by an infinitely long moving-band heat source of length 2ℓ as a function of sliding distance x and Peclet number L. The band source is transversely oriented to the direction of motion. Reproduced with permission from Jaeger, J.C. (1942), "Moving Sources of Heat and the Temperature at Sliding Contacts," *Proc. Roy. Soc. N.S.W.* **76**, 203–224. Copyright 1942. Royal Society of New South Wales.

high-speed case ($L > 10$)

$$\theta_{max} \sim 1.6 \frac{q\ell}{k} \left(\frac{V\ell}{\kappa} \right)^{-1/2} \tag{6.2.10a}$$

$$= 1.6 \frac{q}{\rho c_p V} \left(\frac{V\ell}{\kappa} \right)^{1/2} \tag{6.2.10b}$$

$$\bar{\theta} \sim \frac{2}{3} \theta_{max} \tag{6.2.10c}$$

For the low speed case ($L < 0.5$),

$$\theta_{max} \sim 0.64 \frac{q\ell}{k} \ell n \left(\frac{6.1\kappa}{V\ell} \right) \tag{6.2.11a}$$

$$\bar{\theta} \sim 0.64 \frac{q\ell}{k} \ell n \left(\frac{5.0\kappa}{V\ell} \right) \tag{6.2.11b}$$

6.2.2.3 Rectangular, Square, and Circular Sources

As calculated for the band source, Equation 6.2.4 can be used to calculate the temperature distribution due to a rectangular source ($2\ell \times 2b$) or any other shape. The results are similar to those for a band source in that at low values of L, the solution is essentially that of a stationary source, while at high values of L, the distribution becomes most non-uniform. In fact, at high speed (L > 10) solutions of rectangular, square, and circular heat sources are similar to those of a band source because side (heat) flow is negligible. Therefore, maximum and mean temperatures for a square source ($2\ell \times 2\ell$), for a rectangular source ($2\ell \times 2b$ *with* 2ℓ in the direction of sliding), or for a circular source of diameter, 2ℓ, are given by Equation 6.2.10.

At low speed, the maximum temperature rise for a circular heat source of diameter 2ℓ is the same as that for a stationary source and is given by (Jaeger, 1942)

$$\theta_{\max} \sim \frac{q\ell}{k} \tag{6.2.12a}$$

$$= \frac{q}{\rho c_p V}\left(\frac{V\ell}{\kappa}\right) \tag{6.2.12b}$$

At low speed, the maximum and mean temperature rise for a rectangular source ($2\ell \times 2b$) are the same as those for a stationary source and are given as (Loewen and Shaw, 1954),

$$\theta_{\max} = 0.64\frac{q\ell}{k}\left[\sinh^{-1}\left(\frac{b}{\ell}\right) + \left(\frac{b}{\ell}\right)\sinh^{-1}\left(\frac{\ell}{b}\right)\right] \tag{6.2.13a}$$

$$\bar{\theta} = 0.64\frac{q\ell}{k}\left\{\left(\frac{b}{\ell}\right)\sinh^{-1}\left(\frac{\ell}{b}\right) + \sinh^{-1}\left(\frac{b}{\ell}\right) + 0.33\left(\frac{b}{\ell}\right)^2 + 0.33\left(\frac{\ell}{b}\right) - 0.33\left[\frac{\ell}{b} + \frac{b}{\ell}\right]\left[1 + \left(\frac{b}{\ell}\right)^2\right]^{1/2}\right\} \tag{6.2.13b}$$

Equations 6.2.13a and 6.2.13b may be rewritten as

$$\theta_{\max} = A_m \frac{q\ell}{k} \tag{6.2.14a}$$

and

$$\bar{\theta} = \bar{A}\frac{q\ell}{k} \tag{6.2.14b}$$

$\bar{A}$ *and* A_m are the area factors (as a function of the aspect ratio of the surface area, b/ℓ) and are plotted in Figure 6.2.4 Note that for a square heat source ($b/\ell = 1$),

$$\bar{A} = 0.95 \tag{6.2.15a}$$

$$A_m = 1.12 \tag{6.2.15b}$$

For an individual contact for $A_r/A_a \sim 1$, the rate at which frictional heat is liberated at an interface is

$$q = \mu p_a V \tag{6.2.16}$$

where μ is the coefficient of friction.

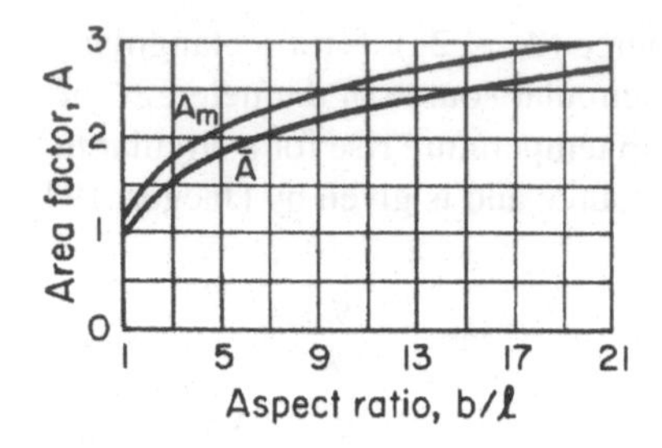

Figure 6.2.4 Area factors for a stationary rectangular heat source ($2\ell \times 2b$) as a function of the aspect ratio. Reproduced with permission from Loewen, E.G. and Shaw, M.C. (1954), "On the Analysis of Cutting Tool Temperatures," *Trans. ASME* **76**, 217–231. Copyright 1954. ASME.

6.2.2.4 Partition of Heat

So far, we have assumed that heat is going into only one surface and that the other surface is insulated. The partition of heat clearly takes place at the individual asperity contacts. If we have two materials, 1 and 2, a portion (r_1) of the heat (q) will go into material 1 and a portion (r_2) will go into material 2, and

$$r_1 + r_2 = 1 \tag{6.2.17a}$$

To determine the partition quantity, we assume that the interface temperature is the same on both contacting surfaces (Blok, 1937), Figure 6.2.5. Therefore, the interface temperature rise is

$$\bar{\theta} = \bar{\theta}_1 = \bar{\theta}_2 \tag{6.2.17b}$$

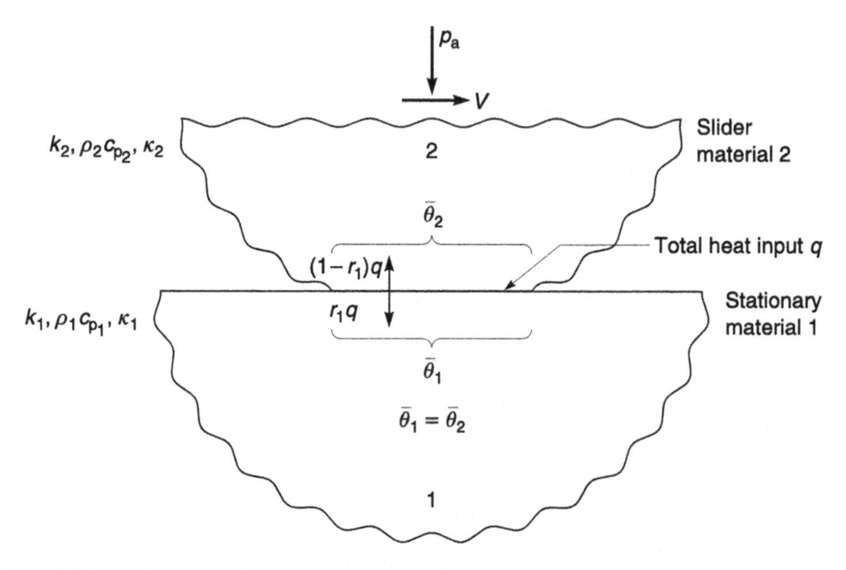

Figure 6.2.5 Schematic of an individual asperity contact between two semi-infinite bodies.

Note that frictional heat is generated in the mating materials close to the interface as a result of plastic deformation. If most heat is generated in one of the sliding bodies (e.g., in a soft material on a hard material), the assumption of the same temperature on both surfaces may not be very accurate.

We first consider the high-speed case. We use the subscript 2 for the slider and subscript 1 for the material on which the slider moves, see Figure 6.2.5. Assume that the portion $r_1 q$ of the total heat is conducted into material 1, and that $r_2 q = (1 - r_1)q$ goes into the slider. From Equation 6.2.10a, the mean temperature of material 1 is

$$\bar{\theta}_1 \sim \frac{r_1 q \ell}{k_1} \left(\frac{V\ell}{\kappa_1} \right)^{-1/2} \tag{6.2.18a}$$

The heat source is stationary relative to the slider; therefore the expression for mean temperature of material 2 is obtained from Equations 6.2.12–6.2.14. For example, if the heat source is rectangular, the mean temperature of material 2 is

$$\bar{\theta}_2 \sim \frac{(1 - r_1) q \ell}{k_2} \tag{6.2.18b}$$

If we let $\bar{\theta}_1 = \bar{\theta}_2$, then

$$r_1 = \left[1 + \left(\frac{k_2}{k_1} \right) \left(\frac{\kappa_1}{V\ell} \right)^{1/2} \right]^{-1} \tag{6.2.19}$$

Note that for $k_2 \ll k_1$ or $\frac{V\ell}{\kappa} \gg 1$, essentially all of the heat goes into material 1.

For a low-speed case for a heat source of any size,

$$r_1 = \left[1 + \left(\frac{k_2}{k_1} \right) \right]^{-1} \tag{6.2.20}$$

Equations 6.2.19 and 6.2.20 apply only to the cases where a single slider is identifiable such as is the case for high contact stress sliding or an individual asperity contact.

Example Problem 6.2.1

Calculate the temperature rise of a brass pin with 20 mm × 20 mm cross section sliding on a steel AISI 1095 disk at a relative velocity of 10 mm/s and a high stress of 1.8 GPa in the dry condition ($\mu = 0.22$). Assume that the entire surface of the brass pin contacts the steel surface. Mechanical and thermal properties of the two materials are as shown in Table 6.2.1.

Table 6.2.1 Mechanical and thermal properties of brass and steel AISI 1095.

Material	Bulk hardness (GPa)	Thermal diffusivity (mm^2/s)	Thermal conductivity (W/m K)
Brass	2.14	35.3	115
Steel AISI 1095	1.93	12.5	45

Solution

$$L = \frac{V\ell}{\kappa}$$

$$= \frac{10^{-2} \times 10^{-2}}{12.5 \times 10^{-6}} = 8 \qquad \text{for steel}$$

$$= \frac{10^{-2} \times 10^{-2}}{35.3 \times 10^{-6}} = 2.83 \qquad \text{for brass}$$

(low speed case)

From Equations 6.2.14, 6.2.15, 6.2.16 and 6.2.20,

$$\bar{\theta} = \frac{0.95\, \mu p_a V \ell}{k_1 + k_2}$$

$$\theta_{max} = \frac{1.12\, \mu p_a V \ell}{k_1 + k_2}$$

$$\bar{\theta} = \frac{0.95 \times 0.22 \times 1.8 \times 10^9 \times 10^{-2} \times 10^{-2}}{115 + 45}$$

$$= 235.1°\text{C}$$

$$\theta_{max} = 277.2°\text{C}$$

6.2.2.5 Transient Conditions

It is of interest to know how long it takes before the steady-state conditions are reached. Basic equations can be numerically solved for time duration less than infinite. Consider a square source ($2\ell \times 2\ell$) moving at velocity V on a semi-infinite body with no loss of heat from the plane $z = 0$. Figure 6.2.6 shows the dimensionless temperature at the center of a square source that has been moving for time t (Jaeger, 1942). If we plot the time required for the curves of Figure 6.2.6 to reach steady state, t_s, as a function of L, we find that (Bhushan, 1987a)

$$\frac{V^2 t_s}{\kappa} \sim 2.5 \frac{V\ell}{\kappa}$$

or

$$V t_s \sim 2.5\ell \qquad (6.2.21)$$

Now $V t_s$ is simply the distance slid during time t_s. Therefore, a square source reaches the steady state after moving a distance of only 1.25 slider lengths. For example, for contact size of 10 μm × 10 μm and at a sliding speed of 1 m/s, it takes about 12.5 μs for the temperature to reach the steady-state value.

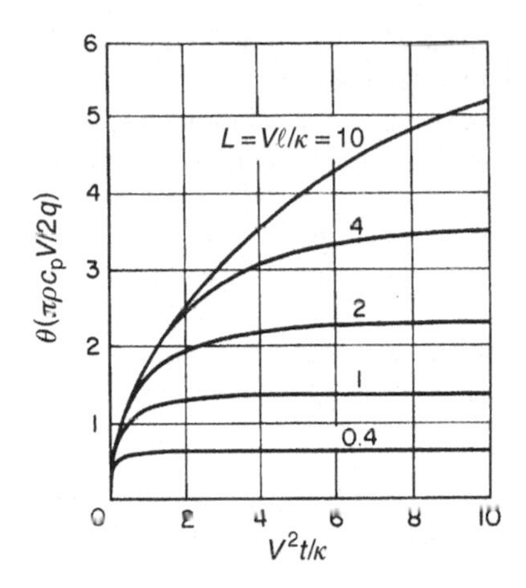

Figure 6.2.6 Dimensionless temperature rise at the center of a moving square source ($2\ell \times 2\ell$) as a function of time t and Peclet number L. Reproduced with permission from Jaeger, J.C. (1942), "Moving Sources of Heat and the Temperature at Sliding Contacts," *Proc. Roy. Soc. N.S.W.* **76**, 203–224. Copyright 1942. Royal Society of New South Wales.

6.2.2.6 Temperature Variation Perpendicular to the Sliding Surface

It is important to know the temperature gradient perpendicular to the sliding surface. We will discuss the analysis only for the band source because that is easier to calculate and it contains all the essential features; the temperatures for the square source for the same heat input arc, of course, lower. Figure 6.2.7 shows the variation of the dimensionless trailing edge temperature

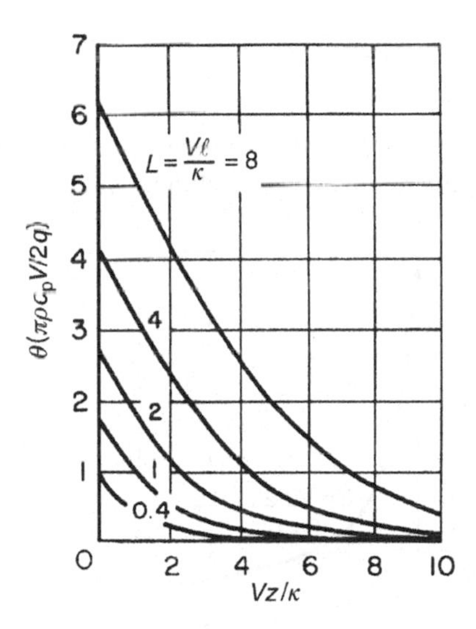

Figure 6.2.7 Dimensionless temperature rise at the trailing edge (maximum temperature location) of a moving-band source of length 2ℓ as a function of depth z and Peclet number L. Reproduced with permission from Jaeger, J.C. (1942), "Moving Sources of Heat and the Temperature at Sliding Contacts," *Proc. Roy. Soc. N.S.W.* **76**, 203–224. Copyright 1942. Royal Society of New South Wales.

(maximum temperature location) as a function of Z (=Vz/κ, where z is the depth) for various values of L (= $V\ell/\kappa$), for a band source of length 2ℓ.

For the low-speed case ($L < 0.5$), where the solution is essentially the same as for a stationary contact, the thermal gradient is based on the definition of thermal conductivity,

$$\frac{d\theta}{dz} = -\frac{q}{k} \tag{6.2.22}$$

The temperature beneath a stationary, circular heat source of diameter d on the surface of a semi-infinite body is given by (Carslaw and Jaeger, 1980)

$$\theta(z,t) = \left[\frac{2q\,(\kappa t)^{1/2}}{k}\right]\left\{ierfc\left[\frac{z}{2\,(\kappa t)^{1/2}}\right] - ierfc\left[\left(z^2 + \frac{d^2}{4}\right)^{1/2} \Big/ 2\,(\kappa t)^{1/2}\right]\right\} \tag{6.2.23a}$$

where *ierfc* is the integral of the complementary error function.

At steady-state conditions, Equation (6.2.23a) reduces to

$$\frac{\theta(z)}{\theta(0)} = -\frac{2z}{d} + \left[1 + \left(\frac{2z}{d}\right)^2\right]^{1/2} \tag{6.2.23b}$$

Equation 6.2.23b can be used to estimate the thermal penetration near a contact (see Figure 6.2.8). At $z/d = 1$ and 2, the temperature drops to 23% and 12% of the surface temperature, respectively.

6.2.3 *Low Contact-Stress Condition ($A_r/A_a \ll 1$) (Multiple Asperity Contact)*

In the previous section, we considered sliding at high contact stress where there is reasonably uniform contact and heat generation between sliding members. In this section we study

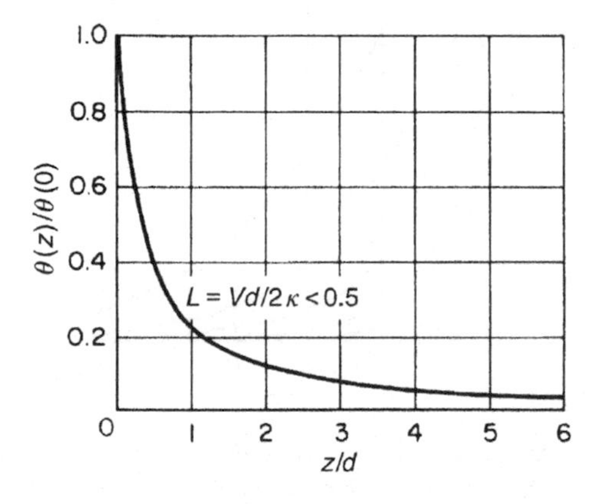

Figure 6.2.8 Steady-state temperature rise as a function of depth z for a stationary, circular asperity contact of diameter d.

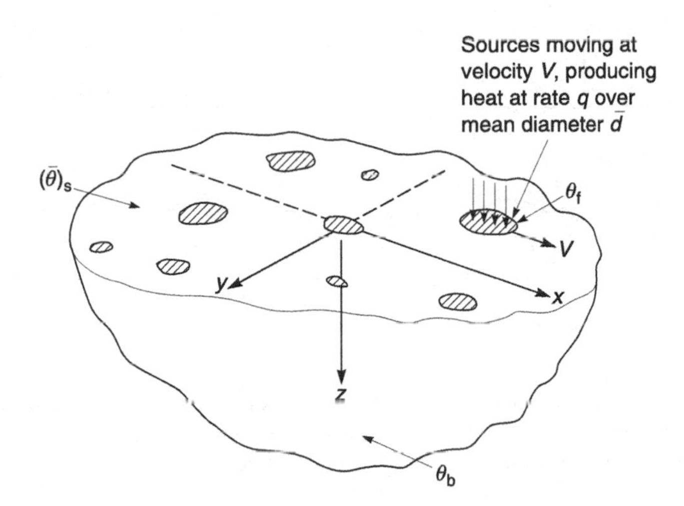

Figure 6.2.9 Schematic of asperity contacts of different sizes moving across a semi-infinite body; these contacts continually grow, shrink and disappear.

sliding at low contact stress ($A_r/A_a \ll 1$) where neither the contact nor the heat generation is uniform (Bhushan, 1971, 1987a; Bhushan and Cook, 1973; Cook and Bhushan, 1973). Consider relative sliding of two "rough" surfaces. At any instant, contact is between a number of pairs of contacting asperities. The load per asperity, size of contact, location of asperity and duration of contact vary from point to point and from time to time, Figure 6.2.9 Contacts grow, shrink, and disappear and this process goes on. Contacts last for a short time. As some contacts disappear, others are continuously formed.

There are three levels of temperature rise in multiple-asperity sliding contacts: maximum asperity contact temperature, average surface temperature, and bulk temperature. The highest contact temperatures occur at asperity contacts and last as long as the pair of asperities are in contact. The integrated (in space and time) average of the temperature of all contacts is referred to as the average temperature rise of the interface, or average surface temperature, $(\bar{\theta})_s$. This average temperature rise is less than the peak temperature rise at a contact, typically on the order of half of a peak asperity-contact temperature. The surface temperature outside the contacts rapidly decreases as the distance from the contacts increases. Frictional heating results in a modest temperature rise of the bulk (θ_b), generally less than about 100°C. The "flash temperature" θ_f is defined as the asperity-contact temperature above the bulk temperature (Blok, 1937). The high, transient temperature flashes of as high as several hundred or even over a thousand degrees Celsius occur over areas of a fraction of a micron to few microns in diameter with a few nanoseconds to a few microseconds in duration. These temperature flashes shift from one place to another during sliding.

The low contact-stress condition is divided into two cases:

1. Both surfaces are of more or less equal roughness. The contact is made between pairs to tips of asperities and each asperity acts as a slider for the other. The center of the contact moves at approximately half the sliding velocity with respect to each asperity (Figure 6.2.10a);

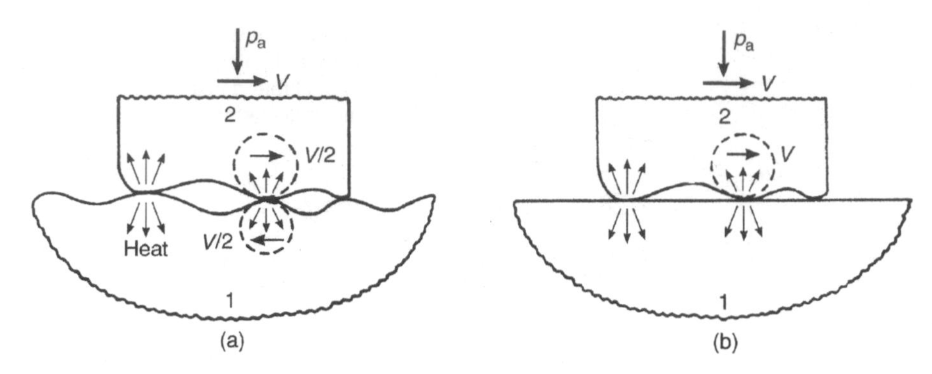

Figure 6.2.10 Schematics of two bodies in contact during sliding at a relative sliding velocity V and a mean normal stress p_a at low-stress conditions: (a) rough-rough surfaces, and (b) rough-smooth surfaces. Heat is dissipated at contacts resulting in high flash temperatures.

2. One surface is much rougher than the other so that the asperities of the rougher surface can be identified as the sliders and the smoother surface is assumed to be stationary (Figure 6.2.10b). Both cases are analyzed here.

We will first discuss the method for finding the temperature history of one asperity contact (flash temperature, θ_f) and the method for averaging that temperature history. We will then discuss the effect of other asperity contacts on an individual asperity temperature (interaction). Next, we will discuss the concept of partition of frictional heat to calculate the asperity temperature at the interface. Finally, we will discuss the method used to predict an average transient temperature of an interface by averaging the temperature of individual asperity contacts, ($\bar{\theta}_s$). Transient conditions and thermal gradients normal to the sliding surfaces have already been analyzed in the previous section.

6.2.3.1 Sliding of Equally Rough Surfaces

Independent (Flash) Temperature Rise (θ_f) of an Asperity Contact

For two rough surfaces sliding under low contact-stress conditions, neither the contact nor the heat generation is uniform. When one of the surfaces slides against another rough surface, contact is made between a finite number of pairs of contacting asperities. The sizes and lifetimes of asperity contacts are different; therefore, the heat produced by friction at one asperity contact is different from another. To calculate the temperature history of one asperity contact, we must know the time-dependent geometry of the contact, the energy dissipated at the contact (which depends on the coefficient of friction, the contact stress, and the sliding velocity), and the thermal properties of the materials involved.

The geometry of the asperity contact can be obtained from statistical methods. Briefly, surface-topography statistics are generated from digitized roughness data; an elastic/plastic analysis is used, along with the topography statistics, to find the statistical distribution of asperity-contact areas under load. This results in a model in which each surface consists of a series of spherically topped asperities having height and radius-of-curvature distributions, which are measured. The degree of interaction between the two surfaces at any time during sliding depends on the average contact stress (the normal load divided by the apparent area of contact). The problem reduces to a sphere of a certain radius sliding against another sphere of another radius, assuming the distance to the center of the two spheres is fixed. The contacts can be either elastic or plastic.

A contact is formed and destroyed as one sphere slides past the other at a given velocity. When one sphere comes into contact with the other, the real area of contact starts to grow; when one sphere is directly above the other, the area is at maximum; as one sphere moves away, that area starts to get smaller. The center of the contact moves at approximately half the relative sliding velocity with respect to each asperity, Figure 6.2.10a. The real area of contact is a source of frictional heat, and the heat intensity is assumed to be proportional to the real area.

If we plot a profile of most engineering surfaces on a 1:1 scale, the surface looks almost flat, and therefore, for heat transfer purposes, the individual asperities can be assumed to be semi-infinite solids. As we have seen earlier, the thermal gradients perpendicular to the sliding surface are so high that even most multi-layered bodies can be assumed to be infinitely thick.

The rate at which frictional heat is liberated at an interface is

$$q = \mu p_a V \left(\frac{A_a}{A_r}\right) \tag{6.2.24}$$

where q is the heat produced per unit real area per unit time (W/m^2), A_a is the apparent area of contact and A_r is the real area of contact (Chapter 3).

First, we calculate the temperature history during the life of a contact; then, we calculate the average temperature of that contact. The results of the analyses fall into two classes, depending on a dimensionless Peclet number $L(= V\ell/\kappa)$. If $L > 10$, it falls in the category of high speed; if $L < 0.5$, it falls in the category of low speed. For $0.5 < L < 10$, approximate transition curve can be used.

Note that the center of the contact (or a velocity of a particle on asperity 1 with respect to asperity 2) moves at $V/2$ relative to each asperity; therefore, V in the definition of L is $V/2$ and ℓ is the average contact radius equal to $3d_{\max}/8$ (see Equation 6.2.31b to be derived later), so that the expression for L can be rewritten as

$$L = \frac{3Vd_{\max}}{16\kappa} \tag{6.2.25}$$

where $d_{\max}$ is the maximum contact diameter.

(a) High-Speed Sliding ($3Vd_{max}/16\kappa > 10$)

In high-speed sliding, there is not enough time for the heat to flow to the sides, and the heat flow is assumed to be only in one direction perpendicular to the sliding surface. The manner in which a contact grows and then diminishes is shown schematically in Figure 6.2.11. For calculation purposes, we divide the life of a contact into 20 equal parts. At any time interval, each heat-source segment (Figure 6.2.11) is considered separately, and the temperature rise caused by these individual heat sources is calculated on the basis of the length of time for which each heat source is on. Knowing the temperature rise of the individual heat sources, we then

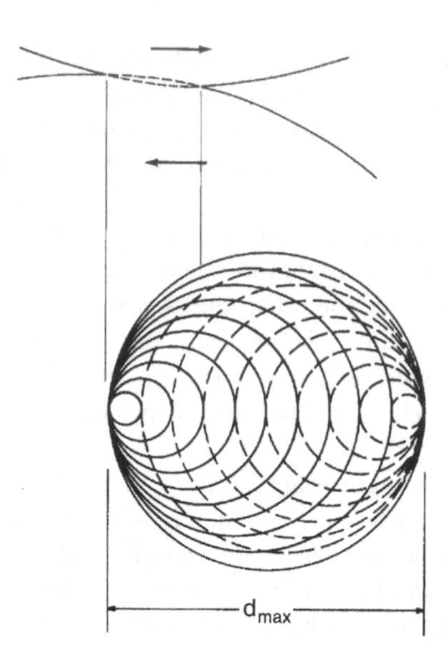

Figure 6.2.11 As one spherical asperity slides past the other spherical asperity, the circular asperity contact grows to d_{max} and then shrinks to zero in the high-speed sliding case. Dotted circles show the shrinking process.

calculate the area-weighted-average temperature (relevant for calculating the instantaneous asperity-contact temperature; see a later section on average surface temperature) at any instant during the life of the contact.

Based on computer runs of various interfaces, the manner in which the contact temperature varies over the contact life ($2t_{max}$) is shown in Figure 6.2.12. For average temperature

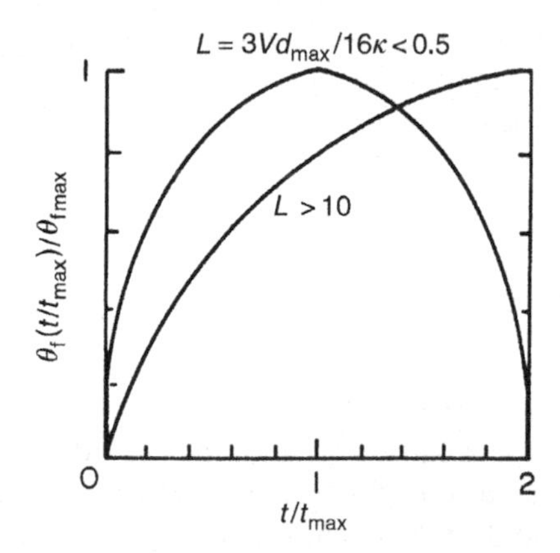

Figure 6.2.12 Independent (flash) asperity temperature rise as a function of time for the high-speed and low-speed sliding cases for an asperity contact during its life. Reproduced with permission from Bhushan, B. (1987a), "Magnetic Head-Media Interface Temperatures-Part 1 – Analysis," *ASME J. Trib.* **109**, 243–251. Copyright 1987. ASME.

calculations, we use the area-weighted average (see a later section on average surface temperature), and is given as (Bhushan, 1999)

$$\bar{\theta}_{fa} = 0.68\theta_{f\,\max} \tag{6.2.26}$$

The maximum contact temperature ($\theta_{f\,\max}$) is correlated to the maximum contact diameter and the thermal properties. Based on the number of computer runs for several materials with large variations in thermal and mechanical properties having different surface roughnesses at different interferences (which provide a large variation in contact diameter) and at different sliding velocities, we find that the following relationships holds:

$$\frac{\theta_{f\,\max}\rho c_p V}{q} = 0.95\left(\frac{Vd_{\max}}{\kappa}\right)^{1/2} \tag{6.2.27}$$

Equation 6.2.27 indicates that the flash temperature rise is proportional to the square root of the velocity. This equation has the same form except the constant as Equation 6.2.10 for a single asperity contact of uniform size.

(b) Low-Speed Sliding ($3Vd_{max}/16\kappa < 0.5$)

In low-speed sliding, the problem reduces almost to the case of a stationary heat source where the heat flow is three-dimensional. For calculation purposes, we assume the contacts to be square rather than circular - the difference in the temperature rise of a square and a circular source of the same diameter is less than 10% (see Section 6.2.2.3). Therefore, we can assume that the contact area grows as discussed previously. Based on the computer runs of various interfaces, and the manner in which the contact temperature varies over the contact life is shown in Figure 6.2.12. The area-weighted temperature is (Bhushan, 1999)

$$\bar{\theta}_{fa} = 0.85\theta_{f\,\max} \tag{6.2.28}$$

The maximum flash temperature is correlated to the maximum contact diameter and the thermal properties as follows:

$$\frac{(\theta_{f\,\max}\rho c_p V)}{q} = 0.33\left(\frac{Vd_{\max}}{\kappa}\right) \tag{6.2.29}$$

Equation 6.2.29 shows that the temperature is proportional to the speed. Again, this equation has the same form, except for the constant, as Equations 6.2.12–6.2.14 for a single asperity contact of uniform size.

(c) Study of the Transition Range

The maximum temperature rise for both high- and low-speed cases as a function of $Vd_{\max}/\kappa$ is plotted in Figure 6.2.13. The curves for the high- and low-speed cases were extrapolated into the range of $L\ (= 3Vd_{\max}/16\kappa)$ between 0.5 and 10.

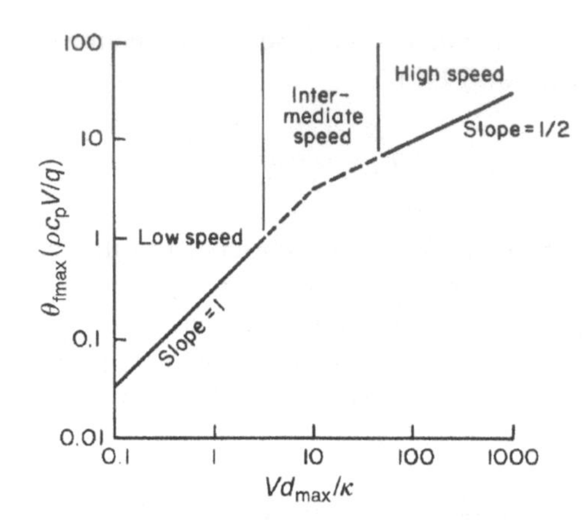

Figure 6.2.13 Independent (flash) asperity temperature rise (maximum) as a function of Vd_{max}/κ for low-speed and high-speed sliding cases. Reproduced with permission from Bhushan, B. (1987a), "Magnetic Head-Media Interface Temperatures-Part 1 – Analysis," *ASME J. Trib.* **109**, 243–251. Copyright 1987. ASME.

Steady-state Interaction Temperature Rise (θ_i)

The independent (flash) asperity temperature rise (θ_f) shows no explicit load dependency, but in practice, we observe that the normal stress affects the interface temperature. The reason is that the number of contacts increases with the load, and therefore the contacts become closer together. As the separation decreases, the thermal interaction between contacts increases; that is, the temperature rise at one contact produces a subsequent temperature rise at a neighboring contact. The cumulative effect of all neighboring asperities in contact at high stresses results in a fair contribution to the total temperature rise.

To develop a model to calculate the interaction temperature at the interface, let n square contacts ($d \times d$) be symmetrically arranged on a square slider $2\ell \times 2\ell$ with a mean contact spacing $\ell_i \left[= d\,(A_a/A_r)^{1/2}\right]$, as shown in Figure 6.2.14. To calculate the interaction

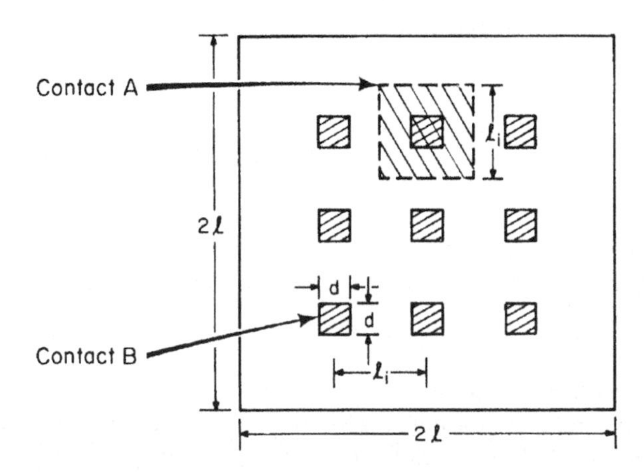

Figure 6.2.14 Model for thermal interaction between neighboring asperity contacts.

temperature, note that when we consider the heat source of diameter d or a square heat source $d \times d$), the important consideration at A is the total energy released, not the size of the spot. In fact, the energy released at A can be considered to be uniformly released over a much larger area (Figure 6.2.14). On average, then, we can let the total frictional energy be released uniformly over the total apparent area. The primary error in this analysis is that we count the effect of source A twice: once independently and once spread over $\ell i \times \ell i$ at its own location. We can subtract the rise caused by this latter term, which is negligible (Bhushan, 1971).

Such a calculation leads to a steady-state, average, surface-temperature rise, which is the ambient temperature of an asperity before it makes contact and undergoes the independent asperity temperature rise. The temperature can be calculated from the methods of Jaeger (1942) and Loewen and Shaw (1954), presented in section 6.2.2.3. The heat produced per unit apparent area per unit time, q', is given as

$$q' = \mu p_a V \tag{6.2.30}$$

At high speed, maximum ($\theta_{i\,\max}$) and mean ($\bar{\theta}_i$) interaction temperature rises for rectangular square, and circular heat sources are given by Equation 6.2.10 for a heat source of strength q'. At low speed, the interaction temperature rise $\bar{\theta}_i$ for circular, rectangular and square heat sources are given by Equations 6.2.12 to 6.2.14.

Partition of Heat

So far, we have assumed that heat is going into only one surface and that the other surface is insulated. As discussed earlier, the partition of heat takes place at the individual asperity contacts such that a portion (r_1) of the heat (q) will go into material 1 and a portion ($r_2 = 1 - r_1$) will go into material 2. To determine the partition quantity, we assume that the total asperity-contact temperature is the same on both contacting surfaces, Equation 6.2.17b.

Note that the individual slider is identifiable in the case of the interaction temperature; therefore, one of the sliders is stationary. Also, for the total interface temperature, the high-speed case is valid when $3Vd_{\max}/16\kappa > 10$, where the κ of the material with a higher value is used. Combining Equations 6.2.26, 6.2.27, 6.2.10, 6.2.14 and 6.2.15 for a rectangular slider ($2\ell \times 2b$ with 2ℓ in the direction of sliding), we get the average asperity-contact temperature for the high-speed case as

$$\bar{\theta} = r_1 \left[\frac{0.65\, \mu p_a \left(\frac{A_a}{A_r}\right) \left(\frac{Vd_{\max}}{\kappa_1}\right)^{1/2}}{\rho_1 c_{p1}} + \frac{\mu p_a \left(\frac{V\ell}{\kappa}\right)^{1/2}}{\rho_1 c_{p_1}} \right]$$

$$= (1 - r_1) \left[\frac{\mu p_a \left(\frac{A_a}{A_r}\right) \left(\frac{Vd_{\max}}{\kappa_2}\right)^{1/2}}{\rho_2 c_{p\,2}} + \frac{\bar{A} \mu p_a \left(\frac{V\ell}{\kappa_2}\right)}{\rho_2 c_{p\,2}} \right] \tag{6.2.31a}$$

or

$$r_1 = \left\{ \frac{1 + \left(\frac{k_2\rho_2 c_{p2}}{k_1\rho_1 c_{p1}}\right)^{1/2} \left[1 + 1.54\left(\frac{A_r}{A_a}\right)\left(\frac{\ell}{d_{\max}}\right)^{1/2}\right]}{\left[1 + 1.54\bar{A}\left(\frac{A_r}{A_a}\right)\left(\frac{\ell}{d_{\max}}\right)^{1/2}\left(\frac{V\ell}{\kappa_2}\right)^{1/2}\right]} \right\}^{-1} \tag{6.2.31b}$$

(for a square slider $\bar{A} = 0.95$).

Note that if the normal stress is very low, or $A_r/A_a \ll 1$,

$$r_1 \sim \left[1 + \left(\frac{k_2\rho_2 c_{p2}}{k_1\rho_1 c_{p1}}\right)^{1/2}\right]^{-1} \tag{6.2.31c}$$

For a total interface temperature, the low-speed case is defined when $V\ell/\kappa < 0.5$, where the κ of the material with lower value is used. Combining Equations 6.2.27, 6.2.28, 6.2.14 and 6.2.15 for a rectangular slider ($2\ell \times 2b$), we get the average asperity-contact temperature for the low speed case:

$$\bar{\theta} = r_1 \left[\frac{0.28\,\mu p_a \left(\frac{A_a}{A_r}\right)\left(\frac{V d_{\max}}{\kappa_1}\right)}{\rho_1 c_{p1}} + \frac{\bar{A}\mu p_a \left(\frac{V\ell}{\kappa_1}\right)}{\rho_1 c_{p1}} \right]$$

$$= (1 - r_1) \left[\frac{0.28\,\mu p_a \left(\frac{A_a}{A_r}\right)\left(\frac{V d_{\max}}{\kappa_2}\right)}{\rho_2 c_{p2}} + \frac{\bar{A}\mu p_a \left(\frac{V\ell}{\kappa_2}\right)}{\rho_2 c_{p2}} \right] \tag{6.2.32a}$$

or

$$r_1 = \left[1 + \frac{k_2}{k_1}\right]^{-1}$$

Therefore

$$\bar{\theta} = \left[\frac{\mu p_a V}{k_1 + k_2}\right]\left[0.28\left(\frac{A_a}{A_r}\right) d_{\max} + \bar{A}\ell\right] \tag{6.2.32b}$$

Similarly, maximum asperity-contact temperatures are given as

$$\theta_{\max} = r_{1m} \left[\frac{0.95\,\mu p_a \left(\frac{A_a}{A_r}\right)\left(\frac{V d_{\max}}{\kappa_1}\right)^{1/2}}{\rho_1 c_{p1}} + \frac{1.5\,\mu p_a \left(\frac{V\ell}{\kappa_1}\right)^{1/2}}{\rho_1 c_{p1}} \right]$$

for

$$3Vd_{\text{max}}/16\kappa > 10 \quad (6.2.33a)$$

and r_{1m} is the same as r_1 in Equation 6.2.31b, except that $\bar{A}$ is replaced by 0.68 A_m,

$$\theta_{\text{max}} = \left[\frac{\mu p_a V}{k_1 + k_2}\right]\left[0.33\left(\frac{A_r}{A_a}\right)d_{\text{max}} + A_m \ell\right] \quad \text{for } V\ell/\kappa < 0.5 \quad (6.2.33b)$$

(For a square slider, $A_m = 1.12$).

Average Surface Temperature Rise

We must ensure that the methods used to predict an average surface temperature $(\bar{\theta})_s$ from averaging the temperature of individual asperity contacts employ the same type of averaging used in the measurements. For the case where the interface temperature is measured through the thermal electromotive force (EMF) produced at the sliding surface, an analysis is derived that suggests how the various EMFs produced at the individual asperities should be averaged. In this analysis, we developed an electrically analogous model that represents the individual contact temperatures by voltage sources (from the Seebeck effect, the temperature rise is a function of EMF) and thermal constriction by electrical resistance. The model developed here is also valid for calculating the instantaneous (independent) asperity temperature from the temperature of asperity segments at a given time and for calculating the average independent asperity temperature.

The contact resistance of an asperity contact is given as (Holm, 1967)

$$R_c = \frac{\rho_1 + \rho_2}{2d} + \frac{\xi}{A} \quad (6.2.34a)$$

where ρ_1 and ρ_2 are the specific resistances of materials 1 and 2 (ohm cm), d is the contact diameter (μm), ξ is the tunnel resistivity of the contact (ohm cm^2), and Λ is true contact area $\left[= (\pi/4)d^2\right]$. From Bhushan (1971) we know that for many material pairs, for example, steel–bronze, the first term is two orders of magnitude lower than the second term; therefore,

$$R_c \sim \frac{\xi}{A} \quad (6.2.34b)$$

Thus the contact resistance, like the electrical resistance, is inversely proportional to the area. The tunnel resistivity depends on the material pair. For a steel–bronze pair with ξ being approximately 5.5×10^{-11} ohm m^2 and d being 10 μm, the contact resistance is on the order of 0.7 ohm. This order of magnitude of the contact resistance was also checked by measuring the total resistance of the contacting surfaces.

To estimate the resistance between the asperity contacts, an electrical analogous model of the asperity contacts was made on an electrically conducting teledeltos paper for an ℓ_i/d (distance between the asperity contacts divided by the diameter of the contact) of 7.5. We found that the resistance between the asperities of 10 μm in diameter separated by 75 μm is

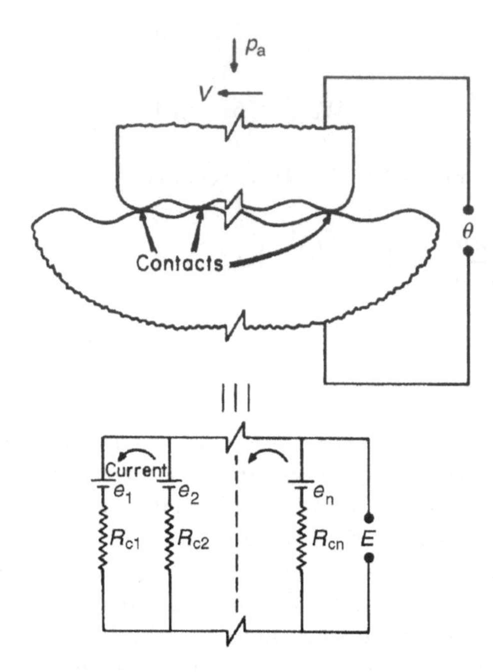

Figure 6.2.15 Electrically analogous model for averaging the temperature rise of individual asperity contacts in order to calculate average surface temperature. Reproduced with permission from Bhushan, B. (1987a), "Magnetic Head-Media Interface Temperatures-Part 1 – Analysis," *ASME J. Trib.* **109**, 243–251. Copyright 1987. ASME.

on the order of 10^{-3} ohm. Because the contact resistance is much bigger than the resistance between the asperity contacts, the latter can be neglected.

Assume n contacts have voltage sources $e_i (i = 1, \ldots, n)$ and contact resistances $R_{ci} (i = 1, \ldots, n)$ in series, respectively. The contacts are connected in parallel (Figure 6.2.15). Voltage E is the overall effect of these sources and is given as

$$E = \frac{\sum_{i=1}^{n} (e_i / R_{ci})}{\sum_{i=1}^{n} (1/R_{ci})} \tag{6.2.35}$$

Using Equations (6.2.34b) and (6.2.35), we obtain

$$E = \frac{\sum_{i=1}^{n} (e_i A_i)}{\sum_{i=1}^{n} A_i} \tag{6.2.36}$$

Therefore, the average temperature at the surface, $(\bar{\theta})_s$, is the area-weighted average of the temperature of all asperity contacts.

Equations 6.2.31–6.2.33 presented the relationships for the average and maximum asperity-contact temperature. The interaction temperature is essentially steady and does not need to be averaged. We must now use the surface's topographical data to calculate the area-weighted average temperature ($\bar{\theta}$) of all the contacting asperities, for which a computer program can be written. Alternatively, because the flash temperature is always directly related to the contact diameter, we can calculate the area-weighted average of the maximum contact diameters ($\bar{d}_{max}$) and use them in Equations 6.2.31–6.2.33 to calculate the average surface temperature directly.

6.2.3.2 Sliding of a Rough Surface on a Smooth Surface

Steady-State Independent (Flash) Temperature Rise (θ_f) of an Asperity Contact

Asperities in the rough-smooth surfaces form contacts and unlike rough rough surfaces, the size of the contacts does not change and the contacts can be continuous during sliding, unless the contact results in wear-debris generation (Figure 6.1.1). Asperities in rough–smooth surfaces can be identified as the sliders; the smoother surface is assumed to be stationary. Asperity contacts are assumed to be either square or circular and to move across a stationary semi-infinite body. The steady-state temperature rise for a circular asperity contact is given by Equations 6.2.10 and 6.2.12 by substituting ℓ by $\bar{d}/2$, where $\bar{d}$ is the area-weighted-average diameter.

Steady-State Interaction Temperature Rise (θ_i)

The analysis presented earlier for rough-rough surfaces is also applicable for rough–smooth surfaces.

Partition of Heat

The individual slider is identifiable for both independent asperity and interaction temperature rises. For the high-speed case($V\bar{d}/2\kappa > 10$), combining Equations 6.2.10, 6.2.12, and 6.2.14, we obtain, for a rectangular slider $2\ell \times 2b$ *with* 2ℓ in the direction of sliding, and an average asperity contact diameter $\bar{d}$,

$$\bar{\theta} = r_1 \left[\frac{\mu p_a \left(\dfrac{A_a}{A_r}\right)\left(\dfrac{V\bar{d}}{2\kappa_1}\right)^{1/2}}{\rho_1 c_{p1}} + \frac{\mu p_a \left(\dfrac{V\ell}{\kappa_1}\right)^{1/2}}{\rho_1 c_{p1}} \right]$$

$$= (1 - r_1) \left[\frac{\bar{A}\mu p_a \left(\dfrac{A_a}{A_r}\right)\left(\dfrac{V\bar{d}}{2\kappa_2}\right)}{\rho_2 c_{p2}} + \frac{\bar{A}\mu p_a \left(\dfrac{V\ell}{\kappa_2}\right)}{\rho_2 c_{p2}} \right] \quad (6.2.37a)$$

or

$$r_1 = \left\{ 1 + \left(\frac{k_2}{\bar{A}k_1}\right)\left(\frac{2\kappa_1}{V\bar{d}}\right)^{1/2} \left[1 + \left(\frac{A_r}{A_a}\right)\left(\frac{2\ell}{\bar{d}}\right)^{1/2}\right] \Big/ \left[1 + \left(\frac{A_r}{A_a}\right)\left(\frac{2\ell}{\bar{d}}\right)\right] \right\}^{-1} \quad (6.2.37b)$$

(for a square slider $\bar{A} = 0.95$).

Note that if the normal stress is very low, or $A_r/A_a << 1$,

$$r_1 \sim \left[1 + \left(\frac{k_2}{\bar{A}k_1}\right)\left(\frac{2\kappa_1}{V\bar{d}}\right)^{1/2}\right]^{-1} \tag{6.2.37c}$$

For the low-speed case ($V\ell/\kappa < 0.5$), we get the average temperature as

$$\bar{\theta} = r_1 \left[\frac{\bar{A}\mu p_a \left(\frac{A_a}{A_r}\right)\left(\frac{V\bar{d}}{2\kappa_1}\right)}{\rho_1 c_{p1}} + \frac{\bar{A}\mu p_a \left(\frac{V\ell}{\kappa_1}\right)}{\rho_1 c_{p1}}\right]$$

$$= (1 - r_1)\left[\frac{\bar{A}\mu p_a \left(\frac{A_a}{A_r}\right)\left(\frac{V\bar{d}}{2\kappa_2}\right)}{\rho_2 c_{p2}} + \frac{\bar{A}\mu p_a \left(\frac{V\ell}{\kappa_2}\right)}{\rho_2 c_{p2}}\right] \tag{6.2.38a}$$

or

$$r_1 = \left[1 + \frac{k_2}{k_1}\right]^{-1}$$

Therefore

$$\bar{\theta} = \left[\frac{\mu p_a V}{k_1 + k_2}\right]\left[0.5\bar{A}\left(\frac{A_a}{A_r}\right)\bar{d} + \bar{A}\ell\right] \tag{6.2.38b}$$

Similarly, maximum temperatures are given as

$$\theta_{\text{max}} = r_{1m}\left[\frac{1.5\,\mu p_a \left(\frac{A_a}{A_r}\right)\left(\frac{V\bar{d}}{2\kappa_1}\right)^{1/2}}{\rho_1 c_{p1}} + \frac{1.5\,\mu p_a \left(\frac{V\ell}{\kappa_1}\right)^{1/2}}{\rho_1 c_{pl}}\right] \quad \text{for } \frac{V\bar{d}}{2\kappa} > 10, \tag{6.2.39a}$$

and r_{1m} is the same as r_1 in Equation 6.2.37b, except that $\bar{A}$ is replaced by 0.67 A_m. And

$$\theta_{\text{max}} = \left[\frac{\mu p_a V}{k_1 + k_2}\right]\left[0.5A_m\left(\frac{A_a}{A_r}\right)\bar{d} + A_m\ell\right] \quad \text{for } \frac{V\ell}{\kappa} < 0.5. \tag{6.2.39b}$$

Note that for high-stress sliding discussed in Section 6.2.2, the temperature rise is given by Equations 6.2.37 and 6.2.38 with contribution by the q $\left[= \mu p_a \left(\frac{A_a}{A_r}\right) V\right]$ term removed.

Average Surface Temperature Rise Again

The analysis presented earlier for rough-rough surfaces is applicable to rough–smooth surfaces.

The analyses presented here have been used to predict temperature rises of various interfaces (Cook and Bhushan, 1973; Bhushan, 1987b, 1992).

Example Problem 6.2.2

Calculate the transient temperature rise at the interface of a rough brass pin with 20 mm × 20 mm cross section, sliding on a rough AISI 1095 steel disk at a relative speed of 10 mm/s and a nominal pressure of 415 kPa in dry ($\mu = 0.22$) and lubricated ($\mu = 0.05$) conditions. Assume that most contacts are plastic contacts. The area-weighted average of the maximum contact diameter of the contact is 25 μm. Mechanical and thermal properties of the two materials are as shown in Table 6.2.1.

Solution

This problem is the case of sliding of equally rough surfaces under low contact stresses. We first calculate the Peclet number (L) to determine if sliding should be considered as a high speed or low-speed case.

The Peclet number based on the maximum contact diameter is

$$\begin{aligned} L &= \frac{3Vd_{\max}}{16\kappa} \\ &= \frac{3 \times 10^{-2} \times 25 \times 10^{-6}}{16 \times 12.5 \times 10^{-6}} = 3.8 \times 10^{-3} \quad \text{for steel} \\ &= \frac{3 \times 10^{-2} \times 25 \times 10^{-6}}{16 \times 35.3 \times 10^{-6}} = 1.3 \times 10^{-4} \quad \text{for brass} \end{aligned}$$

(Low-speed case)

The Peclet number based on the slider length is

$$\begin{aligned} L &= \frac{V\ell}{\kappa} \\ &= \frac{10^{-2} \times 10^{-2}}{12.5 \times 10^{-6}} = 8 \quad \text{for steel} \\ &= \frac{10^{-2} \times 10^{-2}}{35.3 \times 10^{-6}} = 2.83 \quad \text{for brass} \end{aligned}$$

From Equations 6.2.32, for plastic contacts and the low-speed case, the average temperature rise is

$$\bar{\theta} = \frac{\mu V (0.28\ H_s d_{\max} + 0.95 p_a \ell)}{k_1 + k_2}$$

and

$$\theta_{\max} = \frac{\mu V (0.33\ H_s d_{\max} + 1.12 p_a \ell)}{k_1 + k_2}$$

Assuming $H_s = 2H$, for dry contact,

$$\bar{\theta} = \frac{0.22 \times 10^{-2}(0.28 \times 1.93 \times 10^9 \times 2 \times 25 \times 10^{-6} + 0.95 \times 415 \times 10^3 \times 10^{-2})}{115 + 45}$$

$$= 0.43^\circ\text{C}$$

$$\theta_{\text{max}} = \frac{0.22 \times 10^{-2}(0.33 \times 1.93 \times 10^9 \times 2 \times 25 \times 10^{-6} + 1.12 \times 415 \times 10^3 \times 10^{-2})}{115 + 45}$$

$$= 0.50^\circ\text{C}$$

For a lubricated contact

$$\bar{\theta} = 0.43 \frac{0.05}{0.22}$$

$$= 0.097^\circ\text{C}$$

$$\theta_{\text{max}} = 0.011^\circ\text{C}$$

Note that $\bar{\theta}$ and θ_{max} increase with an increase of V and p_a.

6.3 Interface Temperature Measurements

A number of techniques have been used to measure the transient interface temperature rise at sliding interfaces (Kennedy 1992; Bhushan, 1996, 2001). These include thermocouples, thermistors, magnetoresistive sensors, radiation detection techniques, metallographic techniques. and liquid crystals. Thermocouples are probably the most commonly used sensors to measure interface temperatures. These are simple to use but do not measure true flash temperature or surface temperature. Radiation techniques come close to the true measurement of flash temperatures. Metallographic and liquid crystals only give a crude estimate.

6.3.1 Thermocouple and Thin-Film Temperature Sensors

6.3.1.1 Embedded Thermocouples

A thermocouple involves wires of two dissimilar metals connected together at the two ends to give rise to a thermal electromotive force (EMF) potential (referred to as the Seebeck potential) which is a function of the difference in the temperature between the two junctions and is independent of the gradients in the wires. One junction is held at a known reference temperature (cold junction) and the temperature of the other measuring junction (hot junction) can be inferred by comparison of the measured total EMF with an empirically derived calibraton table (Reed, 1982).

In the thermocouple technique, a small hole is drilled through the stationary component of a sliding pair (e.g., Ling and Simkins, 1963). The hole may extend just beneath the sliding surface or to the surface, Figure 6.3.1. A small thermocouple is then inserted through the hole such that

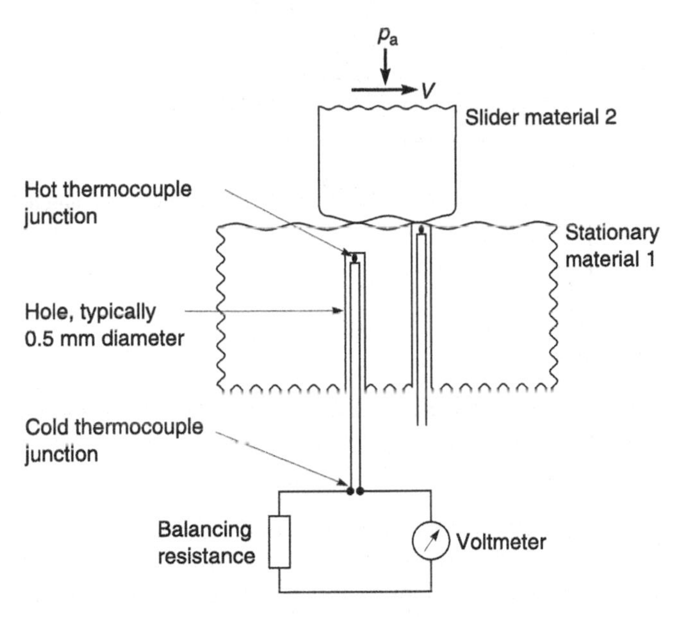

Figure 6.3.1 Schematic of two embedded thermocouples, one flushed with the sliding surface and another at subsurface, for measurement of surface and subsurface temperature; longitudinal axes of thermocouples can be in the plane or perpendicular to the plane of the paper.

its measuring junction rests either at or just beneath the sliding surface. An inorganic cement or an epoxy is used to bond it in the hole, which also insulates the thermocouple wires from the surrounding material. In some cases, a set of thermocouples may be embedded at varying distances from the sliding surface to measure thermal gradients. Thermocouples provide a good measure of transient changes in frictional heating as well as a relative measure of the coefficient of friction. These are commonly used to monitor frictional changes in sliding tests (Bhushan, 1971). However, because of the finite mass of the measuring junction and distance from contacts where heat is being generated, thermocouples cannot be used to measure true flash temperature or surface temperature peaks. The thermal gradient perpendicular to the sliding surface is very high; in metal cuttings it can be as high as 1000°C/mm. A thermocouple can extend to the sliding surface by placing it in a hole that extends to the surface and then grinding the thermocouple even with the surface. The temperatures measured by a thermocouple flush with the surface are altered by the thermal properties of the thermocouple, and the finite mass of the thermocouple junction measures average temperature of the entire mass and does not allow response to flash temperatures of short duration (Spurr, 1980).

6.3.1.2 Dynamic Thermocouples

Temperature rise in a metallic contact of dissimilar metals can be simply obtained from the measurement of the thermal EMF potential produced at the interface, Figure 6.3.2. A thermocouple junction is formed at the sliding interface by the contacting bodies themselves. Electrical connection with the moving component is typically made using a mercury cup or

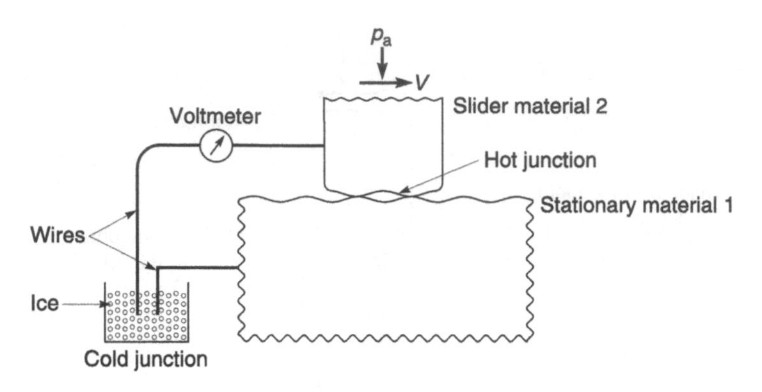

Figure 6.3.2 Schematic of dynamic thermocouple technique used for measurement of surface temperature of two dissimilar metals.

slip ring/brush arrangement. By calibrating the sliding pair against a standard thermocouple, the thermal EMF is converted into the temperature rise. Calibration is usually carried out by heating the sliding pair and a calibrated thermocouple (such as chromel alumel) in a molten lead bath and measuring the EMFs of the two pairs (Bhushan, 1971; Shaw, 2005). The dynamic thermocouple technique was originally developed to measure contact temperatures during metal-cutting operations (Shore, 1925; Chandrasekar *et al.*, 1990; Shaw, 2005). It is now used to measure the surface temperature of bearing interfaces involving dissimilar metals (such as steel sliding on graphite, brass or babbitt) (Bhushan, 1971; Bhushan and Cook, 1973, 1975; Cook and Bhushan, 1973).

The dynamic thermocouple technique gives a good measure of an average surface temperature rise. Since dynamic thermocouples have a very thin junction, consisting only of the contact zone, these can respond very rapidly to changes in surface temperatures. Dynamic thermocouples have been found to give higher values of measured temperatures and faster transient response than embedded thermocouples. However, this technique can only be used for metallic pairs made with dissimilar metals and requires electrical contact with a moving body as well.

6.3.1.3 Thin-Film Temperature Sensors

Thin-film microelectronic fabrication techniques involving vapor deposition are used to form thin-film temperature sensors on surfaces, with a small size to more accurately measure temperature rise on a small region and very small mass to realize rapid response time. One of the first vapor-deposited surface temperature sensors was a thermistor used to measure surface temperatures on gear teeth (Kannel and Bell, 1972). The sensor consisted of a thin strip of titanium coated onto an alumina insulator on the surface of one of a pair of meshing teeth, Figure 6.3.3. The resistance of the titanium strip is sensitive to temperature and pressure; therefore, measurement of any change in resistance allows the measurement of change in transient temperature and pressure. Since the strip has a finite length, it gives lower than the true temperature rise on a small point.

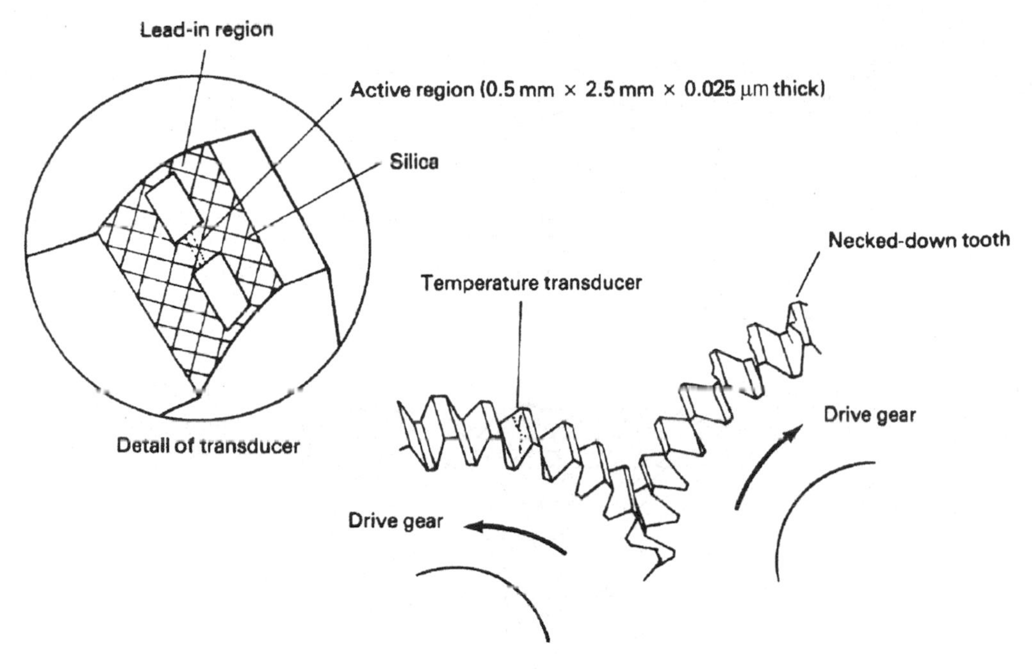

Figure 6.3.3 Schematic of a thermistor (thin-film temperature sensor) for measurement of surface temperature on gear teeth. Reproduced with permission from Kannel, J.W. and Bell, J.C. (1972), "A Method for Estimation of Temperatures in Lubricated Rolling-Sliding Gear on Bearing Elastohydrodynamic Contacts," *Proc. EHD Symposium*, pp. 118–130, Instn Mech. Engrs, London, UK. Copyright 1972. Council of the Institution of Mechanical Engineers.

Magnetoresistive (MR) sensors are used as read heads in magnetic recording (Bhushan, 1996). An MR sensor is a thin strip of a ferromagnetic alloy (for example, $Ni_{80}Fe_{20}$). In addition to variation of resistance as a function of the variation of the magnetic field, resistance of an MR strip is extremely sensitive to temperature. MR sensors have been used to measure interface flash temperature at magnetic head-medium interfaces.

Thin-film microfabrication techniques have also been used to produce thin-film thermocouples (Marshall *et al.*, 1966). These thermocouples use vapor-deposition techniques to produce thermocouple pairs from thin films of two different metals, such as nickel and copper, sandwiched between thin layers of hard, dielectric material such as alumina. A schematic cross-section of a thin-film thermocouple developed for measurement of sliding interface temperatures is shown in Figure 6.3.4 (Tian *et al.*, 1992). The dielectric layer beneath the thermocouple junction is necessary to electrically insulate the thermocouple sensor from the underlying metallic substrate. A protective top layer is needed above the junction to protect the soft metal films and connecting leads. A total thickness of the sensor as small as 1 μm, with a measuring junction 10–100 μm square and about 0.5 μm thick, can be achieved. A small mass results in a fast response and a small junction size results in a better measure of flash temperature than that by wire thermocouples.

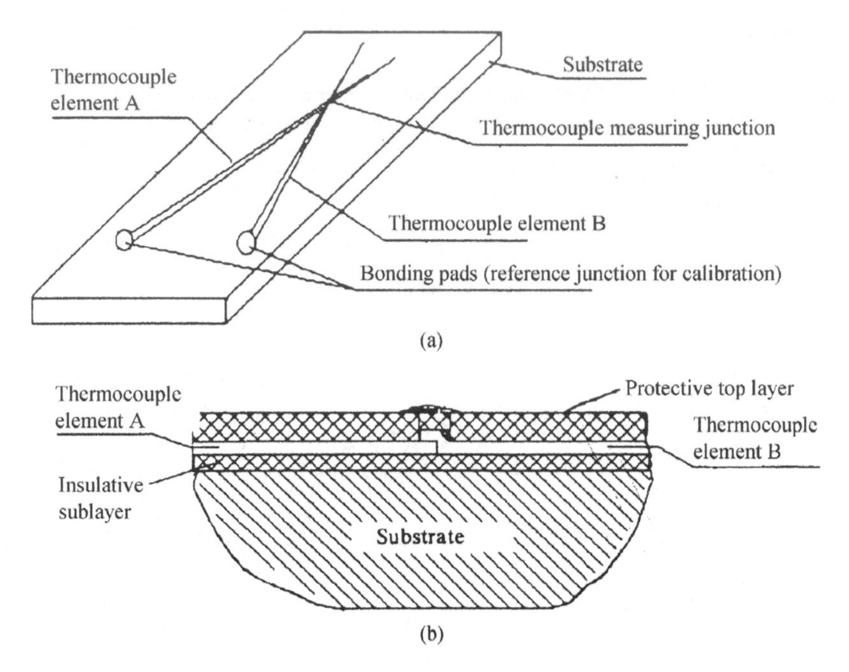

Figure 6.3.4 Schematics of (a) surface, and (b) cross section of a thin-film thermocouple. Reproduced with permission from Tian, X., Kennedy, F.E., Deacutis, J.J., and Henning, A.K. (1992), "The Development and Use of Thin-Film Thermocouples for Contact Temperature Mesurements," *Tribol. Trans.* **35**, 491–499. Copyright 1992. Taylor and Francis.

6.3.2 Radiation Detection Techniques

All materials emit radiation that depends upon the surface temperature and structure of the material, referred to as thermal radiation. Thermal radiation is distributed continuously over the entire electromagnetic spectrum but, when the body's temperature is between 10 and 5000K, it is concentrated in the infrared region. To determine the temperature, the radiative characteristics of the material, as well as the radiant heat transfer properties of the particular geometry configuration, must be known. Once these factors have been computed or measured, the temperature may be deduced from the output of the radiometer. Radiation detection techniques have been most successfully employed to measure the transient flash temperatures; however, these require that one of the bodies be transparent to the radiation to be detected. Sapphire is a useful material for these studies as it is essentially transparent near infrared regions [internal transmission of a 4 mm thick sapphire is 0.93, Union Carbide (1972)], and its mechanical and thermal properties are similar to those of steel.

6.3.2.1 Basics of Radiation

Any surface at a temperature above absolute zero radiates thermal energy, known as thermal radiation (Siegel and Howell, 1981). Emissive power radiated by the surface is a function of its temperature and is given by the Stefan-Boltzmann law,

$$\underline{\overline{\Phi}} = \varepsilon \sigma T^4 A \tag{6.3.1}$$

Table 6.3.1 Representative values of total emissivity of solid surfaces at 300K.

Material	Emissivity at 25°C
Aluminum foil	0.02
Copper, polished	0.03
Copper, oxidized	0.5
Iron, polished	0.08
Iron, oxidized	0.8
Carbon	0.8
Blackbody	1.0

(*Source*: Bedford, 1991).

where $\overline{\Phi}$ is the power (rate of energy), T is the absolute temperature, σ is the Stefan-Boltzmann constant, A is the area of the heat source, and ε is the total emissivity of the surface emitting the radiation. ε is dependent on the surface structure and the temperature. For typical emissivity values see Table 6.3.1. The radiation is composed of photons of many wavelengths. Monochromatic emissive power of a black body in a vacuum at a wavelength λ is given by Planck's law:

$$W_{b,\lambda} = \frac{2\pi hc^2\lambda^{-5}}{\exp(ch/k\lambda T) - 1} \tag{6.3.2}$$

where $W_{b,\lambda}$ is the power emitted per unit area at wavelength λ, c is the velocity of light in vacuum, and h and k are the Planck and Boltzmann constants, respectively. Integration of Equation 6.3.2 over all wavelengths leads to Equation 6.3.1 for the case of a black body ($\varepsilon = 1$). Planck's law (Equation 6.3.2) is plotted in Figure 6.3.5 for selected temperatures (Bedford, 1991). Note that $W_{b,\lambda}$ is very low at small and long wavelengths, so most emissive power is found at wavelengths in the range $1\ \mu m < \lambda < 10\ \mu m$. For this reason, most successful attempts at measuring temperature by the detection of thermal radiation have concentrated on the infrared region of the spectrum (wavelengths of 0.75–500 μm). If the surface temperature T is high enough, radiation in the visible part of the spectrum (400–750 nm) can also be detected. Several different radiation measurement techniques have been successfully used to measure surface temperatures, including photography, pyrometry, thermal imaging and photon detection (Kennedy, 1992). In the following, we describe infrared detection and photon collection techniques; the former technique is a widely used temperature measurement technique.

6.3.2.2 Infrared Detection

A photoelectric detector integrates Equation 6.3.2, multiplied by the emissivity, over all wavelengths within its spectral range and over the surface area viewed by the detector, to obtain the equivalent of Equation 6.3.1. In an infrared (IR) radiometric microscope used for local surface temperature measurements, the detector is equipped with optics to limit the field of view to a small spot size in order to permit a small spatial resolution. A high-resolution IR microscope (Barnes RM2A) used by Winer and coworkers (e.g., Nagaraj *et al.*, 1978; Gulino *et al.*, 1986) consists of a liquid-nitrogen-cooled indium antiminide detector with reflective

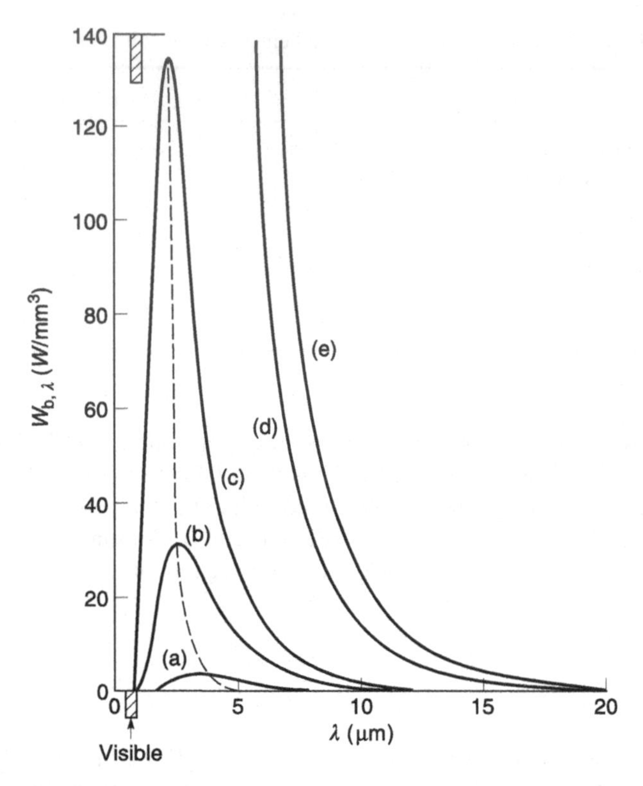

Figure 6.3.5 Spectral radiant emittance of a blackbody (Equation 7.3.1) at various temperatures: (a) 800 K, (b) 1200 K, (c) 1600 K, (d) 6000 K, and (e) 10,000 K. Reproduced with permission from Bedford, R.E. (1991), "Blackbody Radiation" in *Encyclopedia of Physics* (R.G. Lerner and G.L. Trigg, eds.), Second edition, pp. 83–84, Addison-Wesley, Reading, Mass. Copyright 1991. Wiley.

optics which permits a spatial resolution of 38 μm. The detector is basically a photometer measuring radiant energy arriving from an object in the wavelength range of 1.8–5.5 μm. The electrical output signal of the detector is an integrated value of radiation received over this range. The IR microscope measures the difference in the radiant energy arriving from an object and the infrared energy received in an immediately preceding time interval. In the DC mode, it is produced by chopping the radiation arriving from the target area. In the AC mode it arises from thermal fluctuations. The time resolution of the instrument depends on the DC or AC mode of operation with a minimum time constant of 8 μs for the AC mode, used for rapidly varying temperatures.

IR microscopes have been effectively used to measure surface temperatures where the detector is focused on a spot emerging from the contact zone, or through the sapphire onto the contact zone between the sapphire and the sample. Figure 6.3.6 shows the rays diagram to calculate radiation contributions of a sample in contact with the sapphire block. Radiation can enter the objective from three sources, neglecting double reflections (Gulino *et al.*, 1986): (1) radiation from the environment reflected from the top surface of the sapphire; (2) radiation passing through the top surface and then reflected from the bottom surface of the sapphire; and

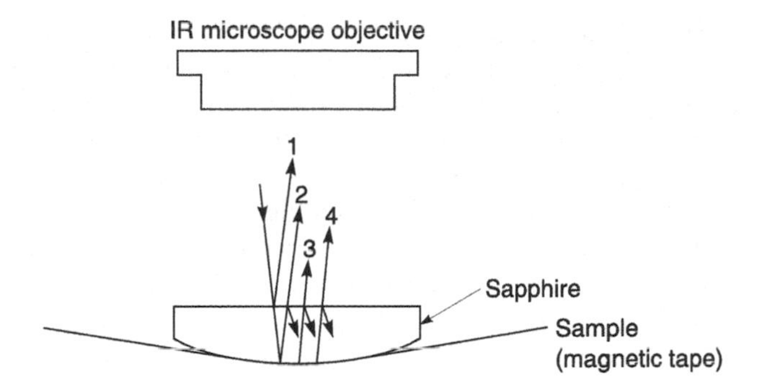

Figure 6.3.6 Radiative contributions into IR microscope objective from a sample (magnetic tape)-sapphire interface.

(3) radiation emitted by the sample – the radiation emitted by the sapphire is negligible [ε for 3.2 mm thick sapphire is 0.11, Union Carbide (1972)]. The reflectivity of sapphire is known [0.063 for a 4-mm thick sapphire plate, Union Carbide (1972)]. In order to calculate the surface temperature of the sample, its radiative properties (emissivity, reflectivity and transmittance) needs to be measured. After suitable calibrations, surface temperature of the sample can be obtained (Nagaraj *et al.*, 1978; Gulino *et al.*, 1986).

The output signal corresponds to a temperature derived from a weighted average of intensities over the target area. The target area at focus is the minimum spot size. The minimum spot size for a specific objective with air as the transmitting medium supplied by the manufacturer must be adjusted due to the presence of the sapphire within the optical path which increases the spot diameter. In the experiments reported by Gulino *et al.* (1986), the spot diameter increased by about four times, to a value of 120 μm for the 15x objective. Both the minimum spot size and time constant are larger than the typical contact size and duration of flash temperature for most engineering interfaces, and therefore the peak flash temperature cannot be accurately measured. The maximum temperature rise of a ball in a sliding contact with the sapphire disk in the presence of lubrication has been reported on the order of 100–200°C (Nagaraj *et al.*, 1978).

By limiting the field of view to a single spot, the IR microscope can miss many contact events occurring at other spots within the area of contact. To overcome this limitation, a scanning type IR camera (or microimager) (such as AGA Thermovision 750, Lindingo, Sweden) is used (Meinders *et al.*, 1983; Griffioen *et al.*, 1986; Gulino *et al.*, 1986). The camera is an optical scanning device which has a detector similar to that of IR microscope, but the detector is optically scanned over the contact surface in either line scan or area scan mode. During scanning, radiant thermal energy is converted into an optical pattern visible on a video display screen. In the line scan mode, a fixed line, perhaps several millimeters in length, is continuously scanned at a maximum scan rate of 2500 lines/s with 100 lines comprising one frame. The minimum spot size of these devices is very large, on the order of 1.5 mm, and the time taken to complete a scan is in milliseconds, much larger than the duration of flash temperature. However, it scans a large region and captures many flash events. In spite of this fact, the scanning type instruments underestimate the flash temperature rise. Griffioen *et al.* (1986) have reported a temperature rise of more than 1000°C of a silicon nitride pin in sliding contact with a sapphire disk at a velocity of 1.5 m/s and a normal load of 8.9 N. The temperature rise of an elastomer in contact with the sapphire block is on the order of 10°C (Brink, 1973; Meinders *et al.*, 1983; Gulino *et al.*, 1986). As stated earlier, the large real area of contact in elastomeric contact reduces the energy dissipated per unit area, resulting in low surface temperatures.

The peak flash temperature rise can be obtained from the spectral distribution of the radiation emitted by using two separate detectors (Bair *et al.*, 1991). The emitted radiation is split between the two detectors and a different band-pass filter is placed in front of each detector to measure the radiated power in two different wavelength ranges. Both measured values are a function of two variables: hot-spot area and temperature. The ratio of detected power at the two wavelengths can be used to determine the maximum temperature within the field of view. The hot-spot area can also be determined once its temperature has been calculated. The optical setup for this method is shown in Figure 6.3.7, and a typical result is shown in Figure 6.3.8 for

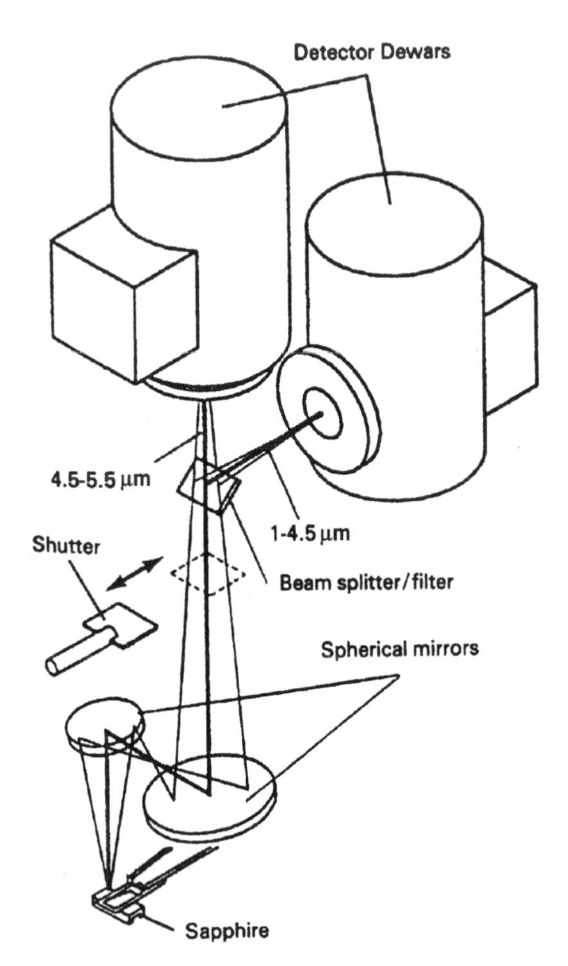

Figure 6.3.7 Optical arrangement for an IR measurement system with two detectors equipped with different band-pass filters. Reproduced with permission from Bair, S., Green, I., and Bhushan, B. (1991), "Measurements of Asperity Temperatures of a Read/Write Head Slider Bearing in Hard Magnetic Recording Disks," *ASME J. Trib.* **113**, 32–37. Copyright 1991. ASME.

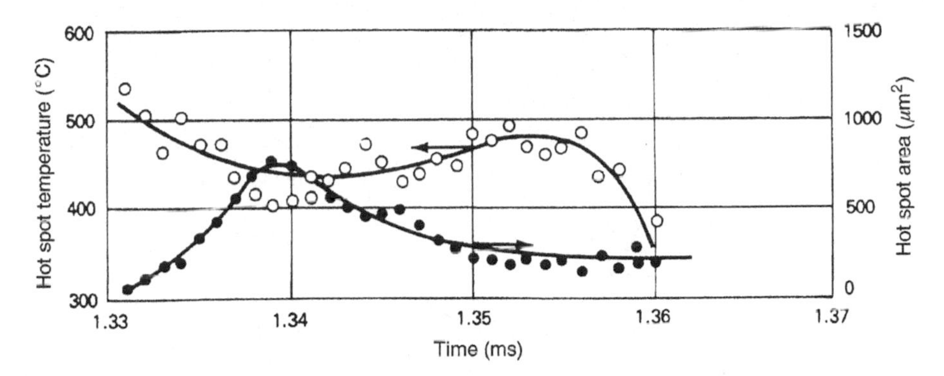

Figure 6.3.8 Hot-spot temperature and area measured using an IR measurement system with two detectors, for a sapphire slider flying on a thin-film magnetic disk at a normal load of 0.1 N after a short overload (1.8 N for 0.3 s). Reproduced with permission from Bair, S., Green, I., and Bhushan, B. (1991), "Measurements of Asperity Temperatures of a Read/Write Head Slider Bearing in Hard Magnetic Recording Disks," *ASME J. Trib.* **113**, 32–37. Copyright 1991. ASME.

a sapphire slider against a thin-film magnetic disk. This method appears to hold much promise for future development. The field of view can be made large enough so that any temperature flashes from the area of interest (larger than the resolution limit) can be detected. Uncertainty about the emissivity of contacting interfaces during the sliding process can lead to inaccurate temperature determination with any of the IR techniques. The total emissivity of a metallic (or nonmetallic) surface can vary considerably owing to oxidation, wear, or other changes in surface characteristics. These difficulties are overcome in the two detector measurement techniques, because the calculated temperature is independent of the emissivity, provided the spectral distribution of emissivity remains unchanged.

6.3.2.3 Photon Collection

As an alternative to IR detectors, which generally respond too slowly to accurately measure flash temperatures in rapidly moving hot spots, a surface temperature measurement method has recently been developed that uses a photomultiplier to collect photons emitted by a hot contact spot (Suzuki and Kennedy, 1991). Use of a photomultiplier with sensitivity to wavelengths in the 500–900 nm range allows detection of photons emitted from a hot contact between a moving surface and a transparent (sapphire) slider. Surface temperatures generally need to be at least 400–500°C in order to generate photons with enough energy to be detected by the photomultiplier, but the response time of the photomultiplier is very rapid (less than 30 ns). Therefore, the technique can be used to detect flash temperatures of very short duration (2 μs or less). However, it is not too useful for measuring mean surface temperatures, which are generally lower than 500°C. A further restriction is that the sliding test must be run in complete darkness to eliminate noise, which can dominate the output signal. This method is also subject to a limitation similar to that of most IR detectors; that is, the area of the spot emitting the photons must be known in order to accurately determine the temperature of the spot. Calibration of the relationship between output voltage and temperature can be accomplished using reference surfaces of the contacting materials at known temperatures (Suzuki and Kennedy, 1991).

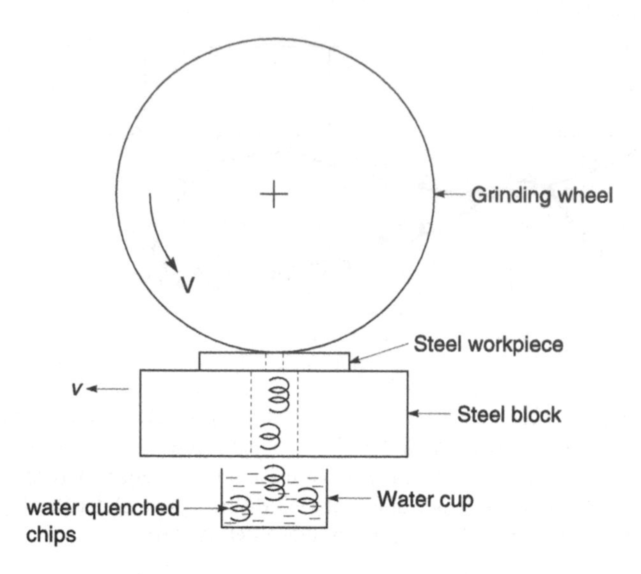

Figure 6.3.9 Schematic of a grinding wheel in contact with a steel workpiece, the water cup underneath the workpiece is used to collect chips (water quenched) as produced during grinding.

6.3.3 Metallographic Techniques

Surface and near surface heating result in microstructural changes of many materials. These changes can be detected by using an optical or scanning electron microscopic examination of the cross sections of the sliding body, in a plane perpendicular to the sliding direction (Wright and Trent, 1973). For some materials, microhardness measurements can provide a measure of the surface temperature rise. This technique can be used for materials which go through known changes in the microstructure or microhardness at the temperatures expected in sliding. This technique gives only a crude estimate of the temperature rise. A major limitation of this technique is that structural and hardness changes, in addition to being due to temperature rise, may occur because of plastic deformation in surface and near-surface regions.

In metal cutting, the flash temperatures during contact can be estimated from the microstructural changes in the hot metal chips after water quenching. In a method shown in Figure 6.3.9, a hole is drilled through the steel workpiece. A water cup is placed under the hole. The steel chips produced during grinding are collected in the water cup. Water quenching of the hot chips result in microstructural changes which are a function of the chip temperature. The technique is simple and can again be used to provide only a crude estimate of the interface temperature.

6.3.4 Liquid Crystals

Cholesteric liquid crystals are commonly used in surface thermography (Fergason, 1968; Gray, 1978). These exhibit dramatic changes in color with very small changes in temperature. In practice, a surface of the body to be tested is merely coated with a specific liquid crystal system and any change in the body temperature induces a change in colors in the liquid crystal material. This pattern is reversible, it will reappear over and over as the body is cycled back and forth through a particular temperature range.

Cholesteric liquid crystals flow like a liquid and simultaneously exhibit the optical properties of a crystal by scattering light selectively. When white light shines on liquid crystals from several directions at once, a different wavelength is reflected at each angle and the resulting mix of different colors is seen. A change in temperature or other environmental effect causes a shift in molecular structure and thus a different color at the same angle.

Since the colors scattered by liquid crystals are unique for a specific temperature, the quantitative measurement of temperature is possible to an accuracy of about 0.1°C. By selection of a particular liquid crystal system having the desired sensitivity, the entire visible spectrum, from red to violet, can be traversed in the 1 – 50°C range, beginning at essentially any temperature from −20 to 250°C (Liquid Crystal Industries Inc., Turtle Creek, PA). For example, a liquid crystal material may appear dark red at 30°C under incandescent light, but as the temperature is raised to 31°C, the color will shift toward yellow, then green, then blue and finally violet. The liquid crystal is colorless above and below its operating range, which in this example is 30–31°C.

Liquid crystals can be sprayed or applied by brush. To increase the brilliance of colors, heat is applied to realign their molecular structure. After the test is completed, these can be removed with petroleum ether.

Liquid crystals can be used to measure bulk temperature of the bodies rather than flash temperatures.

6.4 Closure

Thermal analyses for the interface temperatures of sliding interfaces under high-stress ($A_r/A_a \sim 1$) and low-stress ($A_r/A_a \ll 1$) conditions have been presented. Almost all of the frictional energy input is directly converted to heat in the material close to the interface. The partition of heat is used to calculate the interface temperature. Two extremes, the high-speed and the low-speed cases, have been considered separately. The temperature rise in high-speed sliding in all cases is proportional to the square root of the speed, and in low-speed sliding it is proportional to the speed. The temperature rise of a contact reaches a steady state value after a moving distance of only 1.25 times slider length. Temperature gradients perpendicular to the sliding surface are very large.

In the low-stress condition, the heat is dissipated at microscopic contacts leading to high temperatures at asperity contact. There are three levels of temperature rise in multiple-asperity sliding contacts: maximum asperity-contact (flash) temperature above the bulk temperature; average surface temperature; and bulk temperature. The flash temperature is the highest contact temperature. Flash temperature rise in either high contact-stress conditions or in ceramic couples can be as high as a few hundred degrees Celsius, and interface with compliant materials exhibit low temperatures on the order of tens of degrees Celsius. The integrated (in space and time) average of the temperature of all contacts is referred to as average surface temperature; this is typically on the order of half of a peak asperity-contact (flash) temperature. The bulk temperature is usually small, generally less than about 100°C.

Analysis for the low-stress condition is further subdivided into two cases: (1) when both surfaces are of more or less equal roughness and (2) when one surface is much rougher than the other. In the case of two surfaces of more or less equal roughness, the contact size and its duration are very small. Contacts continue to be made and broken and the hot spots on the surfaces shift their location. The temperature of an asperity contact increases during the entire life of the contact in high-speed sliding, and the temperature rise peaks at the half-life of the contact and is symmetrical in low-speed sliding. The detailed analysis provides an empirical relationship for predicting the independent asperity-contact temperature. The temperature rise of an asperity contact must include the steady-state interaction temperature, which is a significant portion of the total temperature at high normal stresses or in the case of a larger real area of contact. For completeness, note that the mean contact diameter is roughly three-quarters the maximum diameter in an asperity contact of two surfaces of more or less equal roughness.

If one surface is much rougher than the other, the contact size does not change and the contact may be continuous during sliding unless it generates wear debris. The total flash temperature rise in this case is slightly higher than that in the previous case for identical parameters. In sliding of a very soft surface on a hard surface of more or less equal roughness, an assumption of a rough, hard surface on a smooth, soft surface is appropriate because the asperities of a soft surface deform much more readily that those of a hard surface.

We found that the average transient surface temperature is the area-weighted average temperature of all asperity contacts.

A number of techniques have been used to measure the transient interface temperature rise in a sliding contact. Surface temperature rise in a metallic contact of dissimilar metals can be obtained from the measurement of the thermoelectric voltage produced at the interface. By calibrating the sliding pair against a standard thermocouple, the thermoelectric voltage is converted into the temperature rise. This technique is simple and works well to measure the average surface temperature, and is commonly used. In the case of metallic interfaces with identical metals pairs or nonmetallic interfaces, embedded thermocouple(s) located at the subsurface or surface are extensively used to measure the transient temperatures. One cannot obtain the peaks of transient temperature flashes from the measurements by thermocouples located at the subsurface because the thermal gradient perpendicular to the surface in a sliding contact is very high. The temperatures measured by a thermocouple flushed with the surface are altered by the thermal properties of the thermocouple. Measured temperature is lower than the true flash temperature because it averages over its cross-sectional area, which is generally much larger than the size of the asperity contact and the measuring junction has a finite mass which does not allow a response to flash temperatures of short duration. Thin-film thermocouples, thermistors and MR sensors are used to reduce the spatial size and mass effects. In spite of their limitations, embedded thermocouples are most commonly used because of ease of application.

Infrared radiation measurement techniques have been employed to measure flash temperatures more accurately; however, these require that one of the surfaces be transparent to infrared. In order to accurately measure the transient temperature flashes, an infrared microscope is needed with a spot size on the order of a micrometer, comparable to an asperity size, and a detector response on the order of a microsecond, comparable to an asperity duration. These measurements become difficult with current technology. Moreover, because asperity contacts move continuously, a scanning-type measurement tool is preferred; however, these tools have a larger spot size and slower detector response than needed. Therefore, measured contact temperatures by IR devices may be less than actual flash temperatures if the size of the hot spot is smaller than the spot size of the detector and is very short-lived. The peak flash temperature rise, however, can be obtained from the spectral radiation emitted by using two separate IR detectors. This method appears to hold promise for future development. Photon detection techniques can be used to measure the mean surface temperatures of very short duration (2 μs or less); however, these are not too useful for measurements of temperatures lower than 500°C; a further restriction is that the sliding test must be run in complete darkness.

Finally, metallographic techniques can be used to get a crude estimate of the surface temperatures. Liquid crystals can be used to measure bulk temperature.

Problems

6.1 Calculate the transient mean temperature rise at the interface of a rough polymeric magnetic tape (12.7 mm wide on a rough Ni-Zn ferrite head at a relative speed of 2 m/s and a nominal pressure of 14 kPa with a coefficient of friction of 1. Most contacts are elastic with a relatively large real area of contact of $A_{re}/A_a \sim 0.003$ and $d_{max} =$ 8 μm. The thermal diffusivities of the tape coating and ferrite are 0.27 and 2.31 mm^2/s, respectively. The thermal conductivities of the tape and ferrite are 0.41 and 8.69 W/m°K, respectively.

6.2 A rectangular block (10 mm × 20 mm) of brass (k = 117 W/m°K) and babbitt (k = 24 W/m°K) slide on an AISI 1095 steel disk (k = 44.8 W/m°K). Interface conditions are such that L < 0.5 (low-speed case). Calculate the ratio of the heat flowing into the steel disk when sliding against brass to that against the babbitt block.

6.3 The flash temperature rises of the following interfaces are measured to be 200°C, 500°C, and 1000°C. Enter appropriate values in the Table P.6.3.

Table P.6.3

Material pair	Flash temperature rise (°C)
Steel vs. steel in vacuum	?
Steel vs. steel in ambient	?
Steel vs. copper in ambient	?

6.4 Calculate the average temperature rise of a brass pin with a cross section 2 cm × 2 cm, sliding upon a steel disk at a relative velocity V of 10 m/s under the normal load W of 100 N. Assume that the entire surface of the pin contacts with the disk, the coefficient of friction $\mu = 0.15$, and the thermal conductivity for brass $k_b = 115$ W/mK and for steel $k_s = 45$ W/mK. Use $\theta = 0.95\frac{q\ell}{k_b+k_s}$ where θ is the average temperature rise, $2\ell \times 2\ell$ is the cross sectional area of the pin, and q is the rate at which the heat is generated (equal to the friction energy per unit time per unit area).

References

Archard, J.C. (1958), "The Temperature of Rubbing Surfaces," *Wear* **2**, 438–455.

Bair, S., Green, I., and Bhushan, B. (1991), "Measurements of Asperity Temperatures of a Read/Write Head Slider Bearing in Hard Magnetic Recording Disks," *ASME J. Trib.* **113**, 32–37.

Bedford, R.E. (1991), "Blackbody Radiation" in *Encyclopedia of Physics* (R.G. Lerner and G.L. Trigg, eds), Second edition, pp. 83–84, Addison-Wesley, Reading, MA.

Bhushan, B. (1971), *Temperature and Friction of Sliding Surfaces*, MS Thesis, MIT, Cambridge, MA.

Bhushan, B. (1987a), "Magnetic Head-Media Interface Temperatures – Part 1: Analysis," *ASME J. Trib.* **109**, 243–251.

Bhushan, B. (1987b), "Magnetic Head-Media Interface Temperatures – Part 2: Application to Magnetic Tapes," *ASME J. Trib.* **109**, 252–256.

Bhushan, B. (1992), "Magnetic Head-Media Interface Temperatures – Part 3: Application to Rigid Disks," *ASME J. Trib.* **114**, 420–430.

Bhushan, B. (1996), *Tribology and Mechanics of Magnetic Storage Devices*, Second edition, Springer-Verlag, New York.

Bhushan, B. (1999), *Principles and Applications of Tribology*, Wiley, New York.

Bhushan, B. (2001), *Modern Tribology Handbook, Vol. 1: Principles of Tribology*, CRC Press, Boca Raton, Florida.

Bhushan, B. and Cook, N.H. (1973), "Temperatures in Sliding," *ASME J. Lub. Tech.* **95**, 535–536.

Bhushan, B. and Cook, N.H. (1975), "On the Correlation Between Friction Coefficients and Adhesion Stresses," *ASME J. Eng. Mat. and Tech.* **97**, 285–287.

Blok, H. (1937), "Theoretical Study of Temperature Rise at Surface of Actual Contact Under Oiliness Lubricating Conditions," *Gen. Disn. Lubn. Inst. Mech. Eng.* **2**, 222–235.

Bowden, F.P. and Tabor, D. (1950), *The Friction and Lubrication of Solids*, Part I, Clarendon Press, Oxford, UK.

Brink, R.V. (1973), "The Heat Load on an Oil Seal," Paper C1, in *Proc. 6th Int. Conf. on Fluid Sealing*, BHRA Fluid Eng., Cranfield, Bedford, UK.

Carslaw, H.S and Jaeger, J.C. (1980), *Conduction of Heat in Solids*, Second edition, Oxford University Press, UK.

Chandrasekar, S., Farris, T.N., and Bhushan, B. (1990), "Grinding Temperatures for Magnetic Ceramics and Steel," *ASME J. Tribol.* **112**, 535–541.

Cook, N.H. and Bhushan, B. (1973), "Sliding Surface Interface Temperatures," *ASME J. Lub. Tech.* **95**, 59–64.

Fergason, J.L. (1968), "Liquid Crystals in Nondestructive Testing," *Appl. Opt.* **7**, 1729–1737.

Gray, G.W. (1978), *Advance in Liquid Crystal Materials for Applications*, BDH Chemicals Ltd. Pool, Dorset, UK.

Griffioen, J.A., Bair, S., and Winer, W.O. (1986), "Infrared Surface Temperature Measurements in a Sliding Ceramic-Ceramic Contact," in *Mechanisms and Surface Distress: Proc. Twelfth Leeds-Lyon Symp. on Trib.* (D. Dowson, C.M. Taylor, M. Godet, and D. Berthe, eds), pp. 238–245, Butterworths, Guildford, U.K.

Gulino, R., Bair, S., Winer, W.O., and Bhushan, B. (1986), "Temperature Measurement of Microscopic Areas within a Simulated Head/Tape Interface Using Infrared Radiometric Technique," *ASME J. Tribol.* **108**, 29–34.

Holm, R. (1967), *Electrical Contacts: Theory and Application*, Springer-Verlag, Berlin, Germany.

Jaeger, J.C. (1942), "Moving Sources of Heat and the Temperature at Sliding Contacts," *Proc. Roy. Soc. N.S.W.* **76**, 203–224.

Kannel, J.W. and Bell, J.C. (1972), "A Method for Estimation of Temperatures in Lubricated Rolling-Sliding Gear on Bearing Elastohydrodynamic Contacts," *Proc. EHD Symposium*, pp. 118–130, Instn Mech. Engrs, London, UK.

Kennedy, F.E. (1992), "Surface Temperature Measurement," in *Friction, Lubrication and Wear Technology* (P.J. Blau, ed.), pp. 438–444, Vol. 18, ASM Handbook, ASM International, Metals Park, Ohio.

Ling, F.F. and Saibel, E. (1957), "Thermal Aspects of Galling of Dry Metallic Surfaces in Sliding Contact," *Wear* **1**, 80–91.

Ling, F.F. and Simkins, T.E. (1963), "Measurement of Pointwise Juncture Condition of Temperature at the Interface of Two Bodies in Sliding Contact," *ASME J. Basic Eng.* **85**, 481–486.

Loewen, E.G. and Shaw, M.C. (1954), "On the Analysis of Cutting Tool Temperatures," *Trans. ASME* **76**, 217–231.

Marshall, R., Atlas, L., and Putner, T. (1966), "The Preparation and Performance of Thin Thermocouples," *J. Sci. Instrum.* **43**, 144.

Meinders, M.A., Wilcock, D.F., and Winer, W.O. (1983), "Infrared temperature Measurements of a Reciprocating Seal Test," *Tribology of Reciprocating Engines: Proc. Ninth Leeds-Lyon Symp. on Trib.* (D. Dowson, C.M. Taylor, M. Godet and D. Berthe, eds.), pp. 321–328, Westbury House, Butterworths, Guildford, UK.

Nagaraj, H.S., Sanborn, D.M., and Winer, W.O. (1978), "Direct Surface Temperature Measurement by Infrared Radiation in Elastohydrodynamic Contacts and the Correlation with the Blok Flash Temperature Theory," *Wear* **49**, 43–59.

Reed, R.P. (1982), "Thermoelectric Thermometry: A Function Model," in *Temperature: Its Measurement and Control in Science and Industry* (J.F. Schooley, ed.), Vol. 5, pp. 915–922, American Institute of Physics, Woodbury, New York.

Shaw, M.C. (2005), *Metal Cutting Principles*, Second edition, Oxford University Press, Oxford, UK.

Shore, H. (1925), "Thermoelectric Measurement of Cutting Tool Temperatures," *J. Wash. Acad. Sci.* **15**, 85–88.

Siegel, R. and Howell, J.R. (1981), *Thermal Radiation Heat Transfer*, Second edition, Hemisphere Publishing Co., New York.

Spurr, R.T. (1980), "Temperatures Reached by Sliding Thermocouples," *Wear* **61**, 175–182.

Suzuki, S. and Kennedy, F.E. (1991), "The Detection of Flash Temperatures in a Sliding Contact by the Method of Tribo-Induced Thermoluminescence," *ASME J. Trib.* **113**, 120–127.

Tian, X., Kennedy, F.E., Deacutis, J.J., and Henning, A.K. (1992), "The Development and Use of Thin-Film Thermocouples for Contact Temperature Mesurements," *Tribol. Trans.* **35**, 491–499.

Union Carbide (1972), "Technical Bulletin: Optical Properties and Applications of Linde Cz Sapphire," F-CPD 72950, San Diego, California.

Wright, P.K. and Trent, E.M. (1973), "Metallographic Methods of Determining Temperature Gradients in Cutting Tools," *J. Iron Steel Inst.* **211**, 364–388.

7

Wear

7.1 Introduction

Wear is the surface damage or removal of material from one or both of two solid surfaces in a sliding, rolling, or impact motion relative to one another. In most cases, wear occurs through surface interactions at asperities. During relative motion, first, material on the contacting surface may be displaced so that properties of the solid body, at least at or near the surface, are altered, but little or no material is actually lost. Later, material may be removed from a surface and may result in the transfer to the mating surface or may break loose as a wear particle. In the case of transfer from one surface to another, the net volume or mass loss of the interface is zero, although one of the surfaces is worn (with a net volume or mass loss). Wear damage precedes the actual loss of material, and it may also occur independently. The definition of wear is generally based on the loss of material, but it should be emphasized that damage due to material displacement on a given body (observed using microscopy), with no net change in weight or volume, also constitutes wear.

Wear, as friction, is not a material property, it is a system response. Operating conditions affect interface wear. Erroneously it is sometimes assumed that high-friction interfaces exhibit high wear rates. This is not necessarily true. For example, interfaces with solid lubricants and polymers exhibit relatively low friction and relatively high wear, whereas ceramics exhibit moderate friction but extremely low wear.

Wear can be either good or bad. Examples of productive wear are writing with a pencil, machining, polishing, and shaving, which require controlled wear. Wear is undesirable in almost all machine applications such as bearings, seals, gears and cams. Components may need replacement after a relatively small amount of material has been removed or if the surface is unduly roughened. In well-designed tribological systems, the removal of material is a very slow process but it is very steady and continuous. The generation and circulation of wear debris, particularly in machine applications where the clearances are small relative to the wear particle size, may be more of a problem than the actual amount of wear.

In this chapter, we describe various mechanisms of wear and types of particles present in wear debris, followed by representative data of wear of materials.

Introduction to Tribology, Second Edition. Bharat Bhushan.
© 2013 John Wiley & Sons, Ltd. Published 2013 by John Wiley & Sons, Ltd.

7.2 Types of Wear Mechanism

Wear occurs by mechanical and/or chemical means and is generally accelerated by frictional heating (or thermal means). Wear includes six principal, quite distinct phenomena that have only one thing in common: the removal of solid material from rubbing surfaces (Burwell, 1957/1958; Kragelski, 1965; Engel, 1976; Eyre, 1976; Rigney and Glaeser, 1978; Scott, 1979; Peterson and Winer, 1980; Suh and Saka, 1980; Buckley, 1981; Rigney, 1981; Bhushan *et al.*, 1985a, 1985b; Loomis, 1985; Suh, 1986; Zum Gahr, 1987; Blau, 1992; Hutchings, 1992; Bayer, 1994; Rabinowicz, 1995; Bhushan, 1996, 2001a, 2001b, 2011; Shipley and Becker, 2002). These are: (1) adhesive; (2) abrasive; (3) fatigue; (4) impact by erosion and percussion; (5) chemical (or corrosive); and (6) electrical-arc-induced wear. Other commonly encountered wear types are fretting and fretting corrosion. These are not distinct mechanisms, but rather combinations of the adhesive, corrosive, and abrasive forms of wear. According to some estimates, two-thirds of all wear encountered in industrial situations occurs because of adhesive- and abrasive-wear mechanisms. Wear by all mechanisms, except by fatigue mechanism, occurs by the gradual removal of material.

Of the aforementioned wear mechanisms, one or more may be operating in one particular piece of machinery. In many cases, wear is initiated by one mechanism and it may proceed by other wear mechanisms, thereby complicating failure analysis. Failed components are generally examined to determine the type of wear mechanism(s) responsible for eventual failure. Microscopy and a variety of surface analytical techniques are generally used in failure analyses.

7.2.1 Adhesive Wear

Adhesive wear occurs when two nominally flat solid bodies are in sliding contact, whether lubricated or not. Adhesion (or bonding) occurs at the asperity contacts at the interface, and these contacts are sheared by sliding, which may result in the detachment of a fragment from one surface and attachment to the other surface. As the sliding continues, the transferred fragments may come off the surface on which they are transferred and be transferred back to the original surface, or else form loose wear particles. Some are fractured by a fatigue process during repeated loading and unloading action resulting in formation of loose particles.

Several mechanisms have been proposed for the detachment of a fragment of a material. In an early theory of sliding wear (still well recognized), it was suggested that shearing can occur at the original interface or in the weakest region in one of the two bodies (Archard, 1953), Figure 7.2.1. In most cases, interfacial adhesion strength is expected to be small as compared to the breaking strength of surrounding local regions; thus, the break during shearing occurs at the interface (path 1) in most of the contacts and no wear occurs in that sliding cycle. In a small fraction of contacts, break may occur in one of the two bodies (path 2) and a small

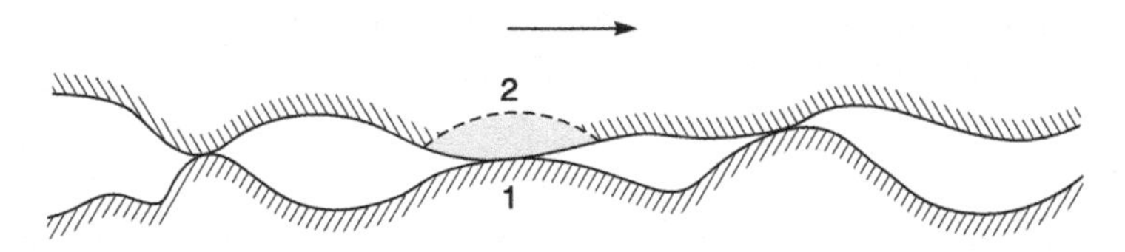

Figure 7.2.1 Schematic showing two possibilities of break (1 and 2) during shearing of an interface.

fragment (the shaded region in Figure 7.2.1) may become attached to the other surface. These transfer fragments are irregular and blocky shaped. In another mechanism, plastic shearing of successive layers of an asperity contact result in detachment of a wear fragment. According to this theory, plastic shearing of successive layers based on a slip line field occurs in conjunction with the propagation of a shear crack, along which the fragment detaches, Figure 7.2.2 (Kayaba

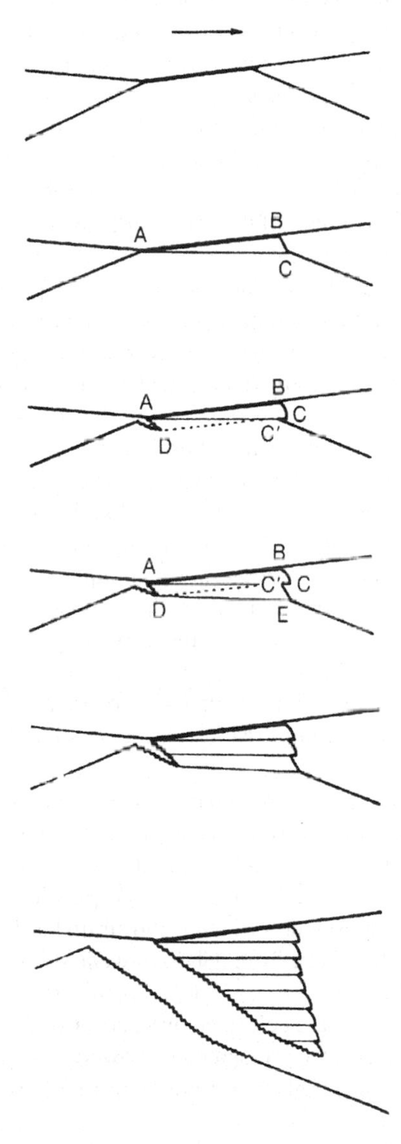

Figure 7.2.2 Schematic showing detachment of fragment of a material from plastic shearing of successive layers of an asperity contact. Reproduced with permission from Kayaba, T. and Kato, K. (1981), "Adhesive Transfer of the Slip-Tongue and the Wedge," *ASLE Trans.* **24**, 164–174. Copyright 1981. Taylor and Francis.

and Kato, 1981). This process results in thin wedge-shaped transfer fragments. The fragment is detached from one surface and transferred to the mating surface because of adhesion. Further sliding causes more fragments to be formed by either of the two mechanisms. These remain adhering to a surface, transfer to the mating surface, or to another previously attached fragment; in the latter case a larger agglomerate becomes detached as a large loose wear particle. These particles may be of roughly equal size in each dimension.

Although the adherence of fragments presupposes a strong bond between the fragments and the surface onto which they are transferred, the formation of the final loose particle implies a weak bond. The formation of a loose particle often results from chemical changes in the fragment. The fragments have a large surface area and tend to oxidize readily, which reduces the adhesive strength, and they readily break loose. A second mechanism responsible for the formation of loose particles involves the residual elastic energy of adherent fragments. When sandwiched between two surfaces, the fragment is heavily stressed. As the other surface moves on, only residual elastic stresses remain. If the elastic energy is larger than the adhesive energy, a fragment breaks loose as a wear particle.

In material combinations with dissimilar materials, wear particles of both materials are formed, although more wear particles of the softer material are formed, and are usually larger than that of the harder counterpart. Because of defects and cracks within the harder material, there are local regions of low strength. If the local regions of low strength of the harder material coincide with local regions of high strength of the softer material at a strong contact, the fragment of the harder material is formed. Formation of fragments of the harder material may also be produced by detachment of the material transferred by adhesion to the harder surface by a fatigue process as a result of a number of loading and unloading cycles.

The transfer of material from one surface to another has been studied by several investigators. In the early 1950s, an autoradiography technique was used in which one sliding material was made radioactive and the transfer of the radioactive material to the mating surface during sliding was demonstrated by placing a photographic film in contact with the mating surface after rubbing, and later developing the film to obtain an autoradiograph of any transferred material (Rabinowicz and Tabor, 1951; Rabinowicz, 1953; Kerridge and Lancaster, 1956; Bhushan *et al.*, 1986). Black impressions on the developed film are produced by each fragment.

A scanning electron microscope (SEM) micrograph of a stainless steel shaft surface after adhesive wear by sliding in a stainless steel journal bearing under unlubricated conditions is shown in Figure 7.2.3 (Bhushan *et al.*, 1985b). Evidence of adhesive debris pullout can be clearly seen. During sliding, surface asperities undergo plastic deformation and/or fracture. The subsurface also undergoes plastic deformation and strain hardening. The SEM micrograph of the cross section of the shaft surface from adhesive wear shows visible plastic deformation with a 25 μm thick layer, Figure 7.2.4. (A copper plate was applied before sectioning to protect the worn surface.) Selected area electron diffraction studies in a transmission electron microscope of the cross sections of worn samples showed that extensive structural changes had occurred. We believe that material close to the worn surface (~15 μm thick) may have recrystallized from an instantaneous surface temperature rise during sliding. Microhardness measurements of the cross-section of worn samples showed that a 10–80% increase of hardness in the worn layer had occurred (Cook and Bhushan, 1973; Bhushan *et al.*, 1985b).

Severe types of adhesive wear are often called galling, scuffing, welding or smearing, although these terms are sometimes used loosely to describe other types of wear.

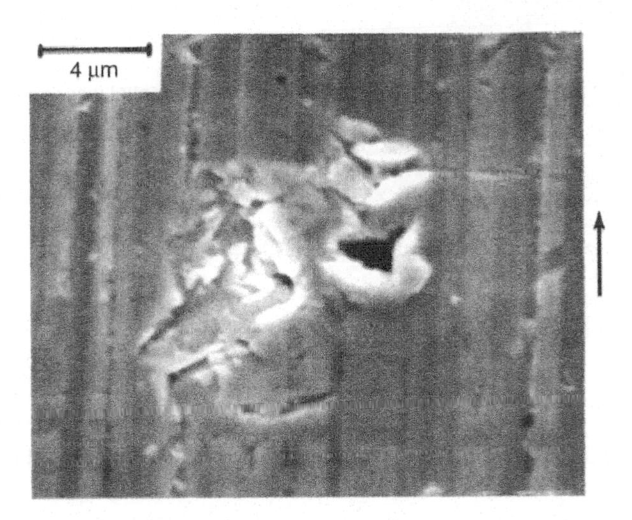

Figure 7.2.3 SEM micrograph of 303 stainless steel shaft surface after adhesive wear under unlubricated conditions. Sliding direction is along the vertical axis.

7.2.1.1 Quantitative Equations

Based on experimental data of various unlubricated material pairs, the vast majority being metallic, it is possible to write the rules of adhesive wear as follows. The amount of wear is generally proportional to the applied load W and sliding distance x and generally inversely

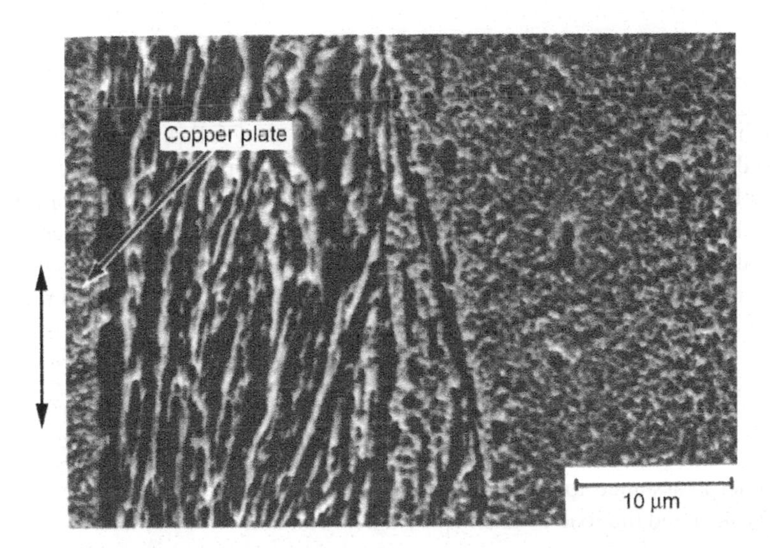

Figure 7.2.4 SEM micrograph of cross section of 303 stainless steel shaft after adhesive wear. Sliding direction is along the vertical axis.

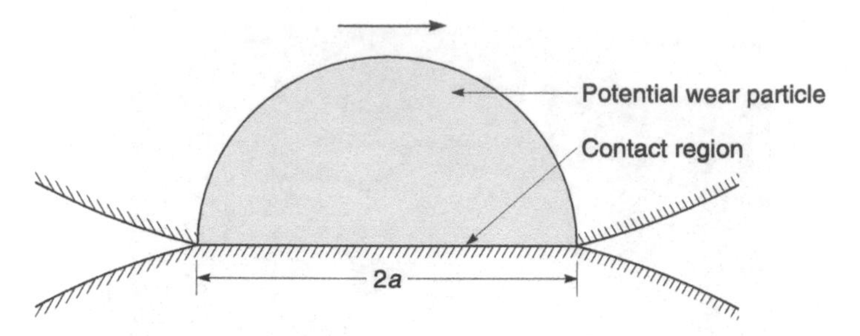

Figure 7.2.5 Schematic of a hypothetical model of generation of a hemispherical wear particle during a sliding contact.

proportional to the hardness H of the surface being worn away. That is, the volume of wear being worn away (Holm, 1946) is

$$v = \frac{kWx}{H} \tag{7.2.1}$$

where k is a nondimensional wear coefficient dependent on the materials in contact and their cleanliness.

Archard (1953) presented the theoretical basis for the expression in Equation 7.2.1. Consider two surfaces in a sliding contact under applied load W. Assume that during an asperity interaction, the asperities deform plastically under the applied load and that at each unit event there is a definite probability that a wear particle will be produced. Further assume that contact is made up of asperities with an average radius of a, Figure 7.2.5. If the material has yielded under the maximum normal load dW, supported by an asperity,

$$dW = \pi a^2 H \tag{7.2.2a}$$

where H is the mean contact pressure under the condition of full plasticity, flow pressure, or hardness of the softer material. We now assume that this asperity contact results in a worn particle of volume dv. The dimension of this worn particle will be directly proportional to the contact size. Physical examination of the wear particles shows that particles are generally of roughly equal lengths in three dimensions rather than, say, layers. Thus, dv is expected to be proportional to a^3. If a particle is assumed to be hemispherical in shape with radius equal to the contact radius, then,

$$dv = \frac{2}{3}\pi\, a^3 \tag{7.2.2b}$$

Finally, contact is assumed to remain in existence for a sliding distance dx equal to $2a$ after which it is broken and the load is taken up by a new contact,

$$dx = 2a \tag{7.2.2c}$$

From Equations 7.2.2a, 7.2.2b and 7.2.2c,

$$\frac{dv}{dx} = \frac{1}{3}\frac{dW}{H} \tag{7.2.3a}$$

If only a fraction α (= $3k$) of all encounters produce wear particles, then the volume of wear by all asperities is

$$v\,\alpha\,\frac{1}{3}\frac{Wx}{H} = \frac{kWx}{H} \quad \text{(plastic contacts)} \tag{7.2.3b}$$

Equation 7.2.3b is identical to Equation 7.2.1 and is commonly referred to as Archard's equation of adhesive wear. Equation 7.2.3b is generally considered to give the amount of wear removed from the softer of the two surfaces. This equation can also be used to calculate the amount of wear of the harder surface by using its hardness. Therefore, in the calculation of k of either surface in Equation 7.2.1 or 7.2.3b, the hardness of the surface which is wearing away should be used. The term k is usually interpreted as the probability that transfer of a material fragment occurs or a wear particle is formed to a given asperity encounter. The value of k ranges typically from 10^{-8} to 10^{-4} for mild wear and from 10^{-4} to 10^{-2} for severe wear for most material combinations, dependent on the operating conditions.

Archard's analysis suggests that there should be two simple rules of wear, i.e., that the wear rate is independent of the apparent area and is directly proportional to the applied load. These rules are analogous to Amontons' equations of friction discussed in Chapter 5. Further, the wear rate is constant with sliding distance (or time) and independent of sliding velocity.

Equation 7.2.1 suggests that the probability of decohesion of a certain volume of material and/or formation of a wear particle (worn volume) increases with each asperity interaction, i.e., an increase in the real area of contact, A_r ($A_r = W/H$ for plastic contacts) and the sliding distance. For elastic contacts which occur in interfaces with one of the materials with a low modulus of elasticity or with very smooth surfaces (such as in magnetic recording interfaces, Bhushan, 1996), Equation 7.2.1 can be rewritten as (Bhushan's equation of adhesive wear) (Bhushan, 1996)

$$v = \frac{k'Wx}{E^*\left(\sigma_p/R_p\right)^{1/2}} \quad \text{(elastic contact)} \tag{7.2.4a}$$

or

$$v = \frac{k'Wx}{E^*\left(\sigma/\beta^*\right)} \tag{7.2.4b}$$

where E^* is the composite or effective modulus of elasticity, σ_p and $1/R_p$ are the composite standard deviation and composite mean curvature of the summits of the mating surfaces, respectively, and σ and β^* are the composite standard deviation of surface heights and correlation length, respectively (Chapter 3), and k' is a nondimensional wear coefficient. In an elastic contact, though the normal stresses remain compressive throughout the entire contact (Chapter 5), strong adhesion of some contacts can lead to generation of wear particles. Repeated elastic contacts can also fail by surface/subsurface fatigue. In addition, in all contacts, contact first occurs on the nanoasperities which always deform by plastic deformation regardless of the deformation on the microscale (Chapter 3), and the plastic contacts are specially detrimental from the wear standpoint.

For a designer who is interested in the rate of wear depth, Equations 7.2.1 and 7.2.4a can be rewritten as

$$\dot{d} = \frac{kpV}{H} \quad \text{(plastic contacts)} \tag{7.2.5a}$$

and

$$\dot{d} = \frac{kpV}{E^* \left(\sigma_p / R_p\right)^{1/2}} \quad \text{(elastic contacts)} \tag{7.2.5b}$$

where $\dot{d}$ is the rate of wear depth (d/t) (mm/s) (where t is the sliding time or duration), p is the apparent normal pressure ($= W/A_a$, where A_a is the apparent area) and V is the sliding velocity. Note that the wear rate is proportional to the pV factor or the life of an interface is inversely proportional to the pV factor. The pV factor is generally used in the selection of materials for dry bearings, see discussion later.

As discussed in Chapter 5, flow pressure or yield pressure under combined normal and shear stresses, p_m, is lower than that under a static normal load p_m ($= H$)

$$p_m = \frac{H}{\left(1 + \alpha\ \mu^2\right)^{1/2}} \tag{7.2.6}$$

where α is a constant (about 9) and μ is the coefficient of friction. This expression for the hardness may be used in Archard's wear equation.

Rabinowicz (1995) has suggested that average diameter of a loose wear particle,

$$d = 60{,}000\, \frac{W_{ad}}{H} \tag{7.2.7}$$

where W_{ad} is the work of adhesion, as described in Chapter 4. The size of the particles in metallic contact typically ranges from submicrons to tens of microns.

The adhesive wear mechanism may be the only mechanism in which there may be some correspondence between the coefficient of friction and the wear rate for metals and nonmetals since the same adhesion factors affect friction and wear.

Example Problem 7.2.1

The flat face of a brass annulus having an outside diameter of 20 mm and an inside diameter of 10 mm is placed on a flat carbon steel plate under a normal load of 10 N and rotates about its axis at 100 rpm for 100 h. As a result of wear during the test, the mass losses of the brass and steel are 20 mg and 1 mg, respectively. Calculate the wear coefficients and wear depths for the bronze and the steel. (Hardness of steel = 2.5 GPa, density of steel = 7.8 Mg/m^3, hardness of brass = 0.8 GPa, and density of brass = 7.5 Mg/m^3.)

Solution

In a brass–steel contact at the test load, the load is expected to be supported by plastic contacts. Therefore, wear coefficients are given by Equation 7.2.3.

Given

$$W = 10\,\text{N}$$
$$\text{Mass loss of brass } (\text{m}_\text{b}) = 20\,\text{mg}$$
$$\text{Mass loss of steel } (\text{m}_\text{s}) = 1\,\text{mg}$$
$$\text{Rotational speed} = 100\,\text{rpm}$$
$$\text{Test duration} = 100\,\text{h}$$

Now,

$$v_b = \frac{2 \times 10^{-2}}{8.5 \times 10^6}\,\text{m}^3 = 2.35 \times 10^{-9}\,\text{m}^3$$

$$v_s = \frac{10^{-3}}{7.8 \times 10^6}\,\text{m}^3 = 1.28 \times 10^{-10}\,\text{m}^3$$

$$\text{Average contact diameter} = \frac{10 + 20}{2} = 15\,\text{mm}$$

and average sliding distance,

$$\begin{aligned} x &= \pi \times 15 \times 10^{-3} \times 100 \times 100 \times 60\,\text{m} \\ &= 2.82 \times 10^4\,\text{m} \end{aligned}$$

$$k_b = \frac{v_b H_b}{Wx} = \frac{2.35 \times 10^{-9} \times 0.8 \times 10^9}{10 \times 2.82 \times 10^4} = 6.65 \times 10^{-5}$$

$$k_s = \frac{v_s H_s}{Wx} = \frac{1.29 \times 10^{-10} \times 2.5 \times 10^9}{10 \times 2.82 \times 10^4} = 1.14 \times 10^{-5}$$

$$d_b = \frac{v_b}{A_b} = \frac{2.5 \times 10^{-9}}{\pi\left(10^2 - 5^2\right) \times 10^{-6}}\,\text{m} = 10.6\,\mu\text{m}$$

$$d_s = \frac{v_s}{A_s} = \frac{1.28 \times 10^{-10}}{\pi\left(10^2 - 5^2\right) \times 10^{-6}}\,\text{m} = 0.54\,\mu\text{m}$$

7.2.1.2 Experimental Evidence

Adhesive wear equations, Equations 7.2.1 and 7.2.4, imply that if k is a constant for a given sliding system, then the volume of worn material should be inversely proportional to H or E^* and interface roughness and proportional to the normal load (pressure) and the sliding distance.

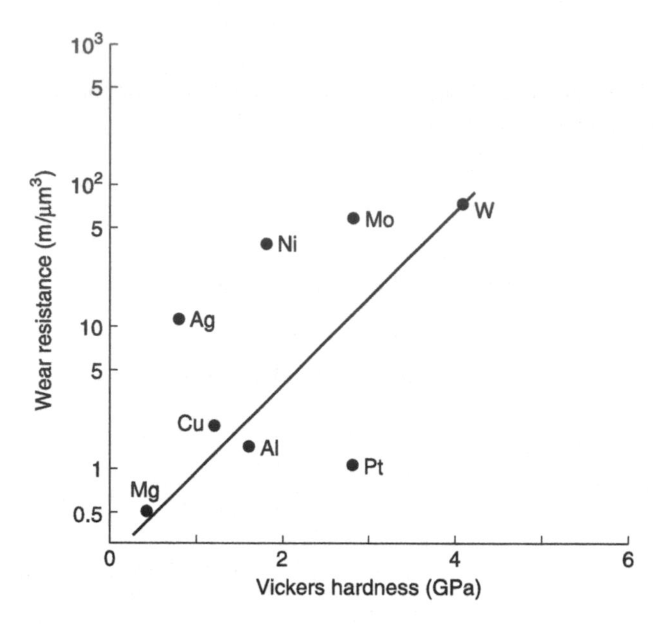

Figure 7.2.6 Wear resistance of self-mated pure metals under unlubricated conditions as a function of Vickers hardness. Reproduced with permission from Zum Gahr, K.H. (1987), *Microstructure and Wear of Materials*, Elsevier, Amsterdam. Copyright 1987. Elsevier.

For a given material combination with primarily plastic contacts, the wear rate generally decreases with an increase in hardness, Figure 7.2.6; more data will be presented in the next section on abrasive wear. However, in the case of extremely hard and/or brittle materials, the fracture toughness generally affects the wear rate, but is not included in the wear equations. For a material combination with primarily elastic contacts, the wear rate generally decreases with an increase in the modulus of elasticity, Figure 7.2.7. (Contacts in magnetic head-medium interfaces are primarily elastic, Bhushan, 1996.) For a material combination with primarily elastic contacts, the wear rate, in adhesive wear mode, should decrease with an increase in surface roughness. However, if the wear occurs by other wear modes, such as abrasive wear, the wear rate may increase with an increase in the surface roughness.

In many material combinations, wear rate increases linearly with the load (pressure) over a limited range; wear rate may either increase or decrease abruptly at some critical loads (Archard and Hirst, 1956). It is the apparent pressure which determines the critical value of the load. This can be explained by the breaking or formation of oxide layers as a function of pressure or frictional heating. For example, the wear rate of a brass pin sliding on a tool steel ring increases linearly with load (according to Archard's wear equation), however, the wear rate of the ferritic stainless steel pin increases linearly at low loads and increases rapidly above a critical load, Figure 7.2.8. Tests conducted with a steel cone sliding on a steel plate show that the wear coefficient increases rapidly above an apparent pressure equal to one-third of the indentation hardness, Figure 7.2.9 (Burwell, 1957/1958). It is generally observed that at these higher loads, large-scale welding and seizure occur.

Experimental data suggest that the wear volume increases with the sliding distance or sliding time at a constant velocity (Archard and Hirst, 1956). At the start of sliding, during

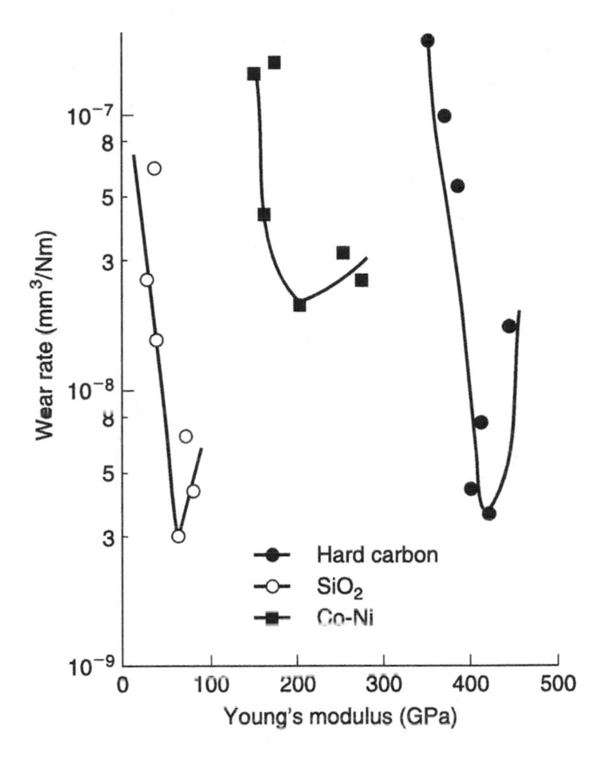

Figure 7.2.7 Wear rate of ceramic thin films as a function of their Young's modulus of elasticity sliding against magnetic rigid disk heads of $A\ell_2O_3 - TiC$, in the elastic-contact regime. Reproduced with permission from Tsukamoto, Y., Yamaguchi, H., and Yanagisawa, M. (1988), "Mechanical Properties and Wear Characteristics of Various Thin Films for Rigid Magnetic Disks," *IEEE Trans. Magn. MAG-24*, 2644–2646. Copyright 1988 IEEE.

the so-called running-in period, the wear rate may be either higher or lower, followed by steady-state wear rate until failure of the interface. Figure 7.2.10 shows the wear data obtained from pin-on-ring tests for a wide range of material combinations under unlubricated conditions in air. In each case, the steady state wear rate (wear volume per unit distance) is essentially constant for each material combination.

Next, wear rate should be independent of the sliding velocity according to wear Equations 7.2.1 and 7.2.4. For many sliding combinations, this assumption holds for a range of values of sliding velocity. However, sharp transitions in wear rate are seen at critical sliding velocities and apparent pressures, which are described using wear maps, to be discussed later.

7.2.1.3 Role of Metallurgical Compatibility

Rabinowicz (1980, 1995) has argued that the tendency of the sliding metals to adhere strongly to each other is indicated by their metallurgical compatibility, which is the degree of solid solubility when the two metals are melted together. The increasing degree of incompatibility reduces wear, leading to lower value of the wear coefficients. This is also true for the coefficient

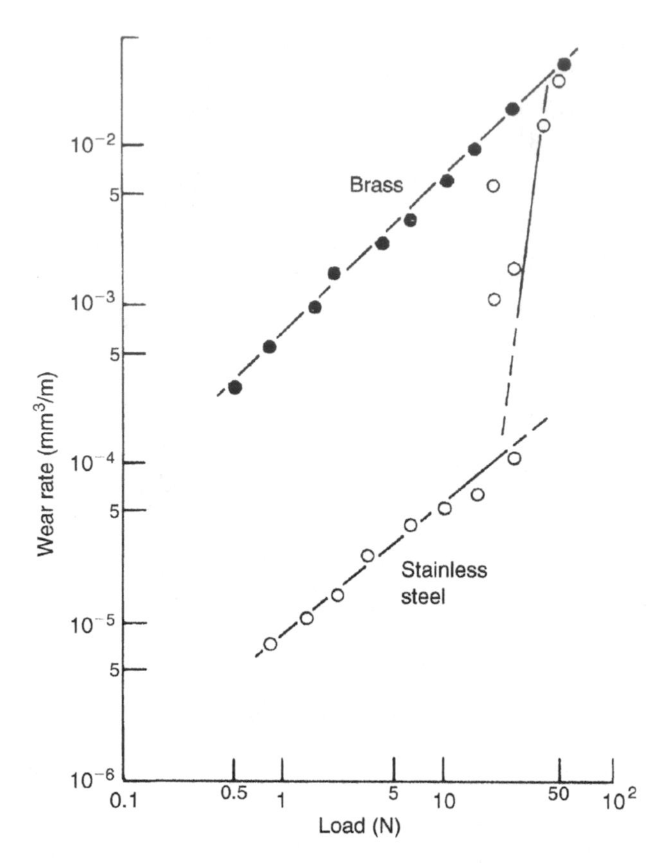

Figure 7.2.8 Wear rate as a function of load (logarithmic scales) for brass and ferritic stainless steel pins sliding against tool steel counterfaces in unlubricated pin-on-ring tests. Reproduced with permission from Archard, J.F. and Hirst, W. (1956), "The Wear of Metals Under Unlubricated Conditions," *Proc. R. Soc. Lond. A* **236**, 397–410, by permission of the Royal Society.

of friction. Table 7.2.1 shows typical values of wear coefficients of metal on metal and nonmetal on metals with different degrees of lubrication at the sliding interface. Both degrees of metallurgical compatibility and lubricant significantly affect wear. The wear coefficient varies by up to two orders of magnitude depending on the degree of compatibility and by up to three orders of magnitude depending on the extent of lubrication at the sliding interface. It is clear that identical metal pairs must be avoided for low wear and friction.

7.2.1.4 Structural Effects

Hexagonal close packed (HCP) metals exhibit lower wear (an order of magnitude less) and friction than cubic metals (Rabinowicz, 1995). A material pair involving two hexagonal metals behaves the same way as a pair with only one hexagonal metal. As stated in Chapter 5, hexagonal metals have a limited number of slip planes, responsible for low wear and friction.

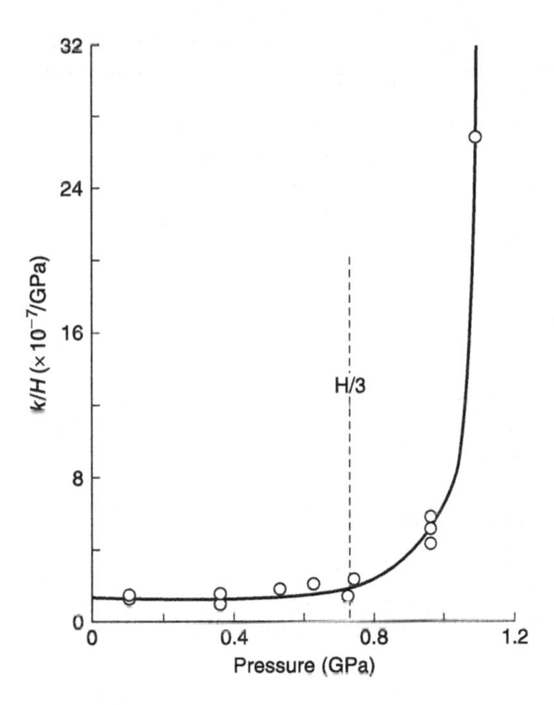

Figure 7.2.9 Wear coefficient/hardness ratio as a function of the average pressure for SAE 1095 steel having hardness of 223 Brinell against a 120° conical slider. Reproduced with permission from Burwell, J.T. (1957/1958), "Survey of Possible Wear Mechanisms," *Wear* **1**, 119–141. Copyright 1957/1958. Elsevier.

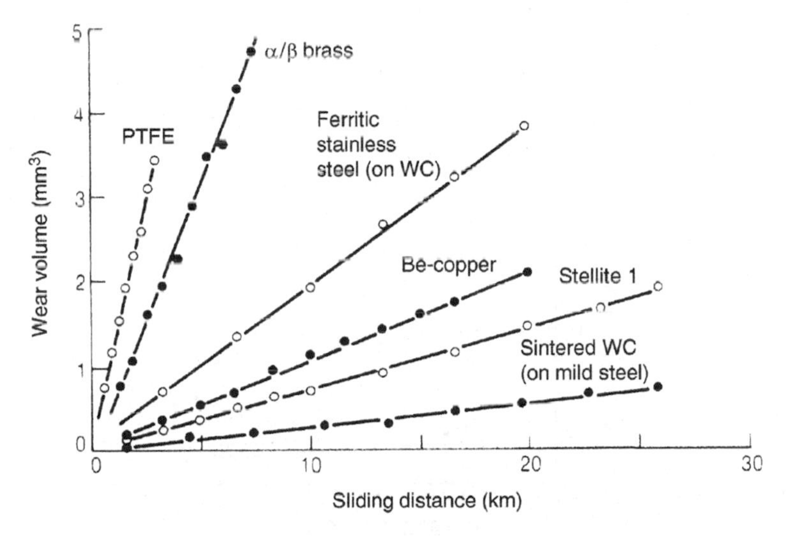

Figure 7.2.10 Wear volume removed from the specimen pin sliding against a tool steel ring (unless otherwise indicated) as a function of total sliding distance from unlubricated pin-on-ring tests on the materials indicated. Reproduced with permission from Archard, J.F. and Hirst, W. (1956), "The Wear of Metals Under Unlubricated Conditions," *Proc. R. Soc. Lond. A* **236**, 397–410, by permission of the Royal Society.

Table 7.2.1 Typical values of wear coefficients (k) for metal on metal (both with non-hexagonal structure) and nonmetal on metal (both with non-hexagonal structure) combinations under different degrees of lubrication.

Condition	Metal on metal k ($\times 10^{-6}$)		Nonmetal on metal k ($\times 10^{-6}$)
	Like	Unlike*	
Clean (Unlubricated)	1500	15–500	1.5
Poorly lubricated	300	3–100	1.5
Average lubrication	30	0.3–10	0.3
Excellent lubrication	1	0.03–0.3	0.03

* The values are dependent upon the metallurgical compatibility with increasing degree of incompatibility corresponding to lower wear.

7.2.1.5 Grain Boundary Effects

As stated in Chapter 5, grain boundary regions are high-energy regions at the surface. For polycrystalline materials, the presence of grain boundaries in the materials influences adhesion, friction, surface fracture, and wear. Sliding friction experiments conducted by Kehr *et al.* (1975) show that wear rate of Ni-Zn ferrite sliding against two magnetic tapes decreases with an increase in the grain size, Figure 7.2.11. Bhushan (1996) also reported that single-crystal Mn-Zn ferrites generally have lower wear rate (by about 10–25%) than the polycrystalline materials.

These observations suggest that polycrystalline materials with high grain boundary densities (finer grains) would exhibit higher wear rates than those with lower grain boundary densities (coarser grains) or single-crystalline materials.

7.2.2 Abrasive Wear (by Plastic Deformation and Fracture)

Abrasive wear occurs when asperities of a rough, hard surface or hard particles slide on a softer surface and damage the interface by plastic deformation or fracture. In the case of

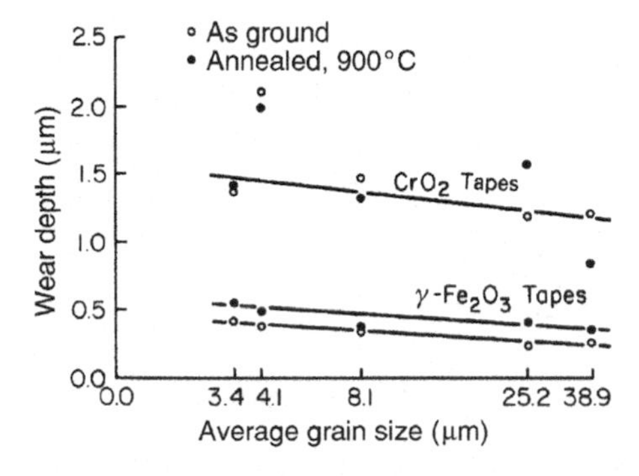

Figure 7.2.11 Wear depth of an Ni-Zn ferrite rod in a sliding contact with $\gamma - Fe_2O_3$ and CrO_2 tapes as a function of ferrite grain size. Reproduced with permission from Kehr, W.D., Meldrum, C.B., and Thornley, R.F.M. (1975), "The Influence of Grain Size on the Wear of Nickel-Zinc Ferrite by Flexible Media," *Wear* **31**, 109–117. Copyright 1975. Elsevier.

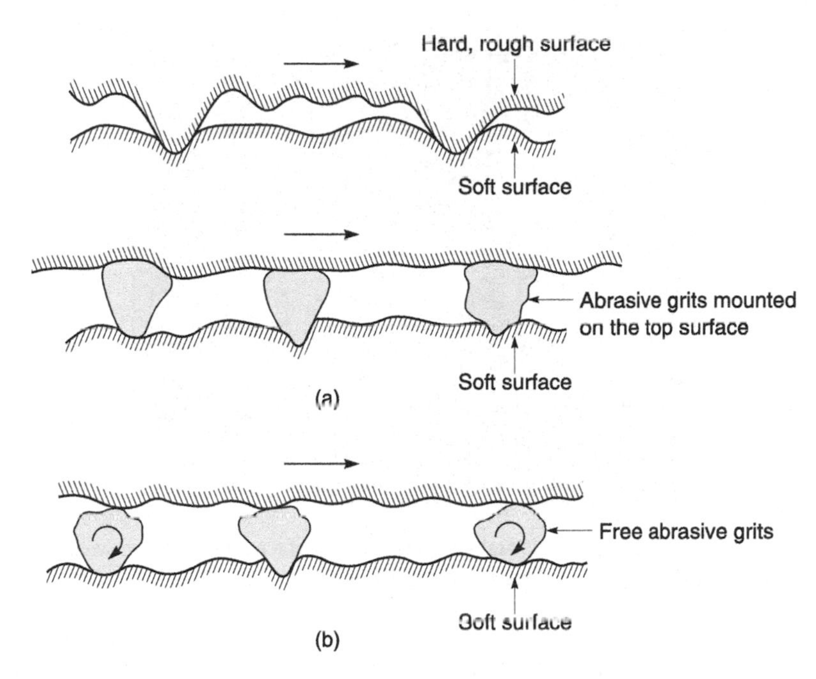

Figure 7.2.12 Schematics of (a) a rough, hard surface or a surface mounted with abrasive grits sliding on a softer surface, and (b) free abrasive grits caught between the surfaces with at least one of the surfaces softer than the abrasive grits.

ductile materials with high fracture toughness (e.g., metals and alloys), hard asperities or hard particles result in the plastic flow of the softer material. Most metallic and ceramic surfaces during sliding show clear evidence of plastic flow, even some for ceramic brittle materials. Contacting asperities of metals deform plastically even at the lightest loads. In the case of brittle materials with low fracture toughness, wear occurs by brittle fracture. In these cases, the worn zone consists of significant cracking.

There are two general situations for abrasive wear, Figure 7.2.12. In the first case, the hard surface is the harder of two rubbing surfaces (two-body abrasion), for example, in mechanical operations, such as grinding, cutting and machining; and in the second case, the hard surface is a third body, generally a small particle of abrasive, caught between the two other surfaces and sufficiently harder, that it is able to abrade either one or both of the mating surfaces (three-body abrasion), for example, in free-abrasive lapping and polishing. In many cases, the wear mechanism at the start is adhesive, which generates wear particles that get trapped at the interface, resulting in a three-body abrasive wear (Bhushan *et al.*, 1985b).

In most abrasive wear situations, scratching (of mostly the softer surface) is observed as a series of grooves parallel to the direction of sliding (plowing). A scanning electron micrograph of a stainless steel surface after abrasive wear by sliding against a stainless steel journal surface in the presence of alumina particles under unlubricated conditions is shown in Figure 7.2.13. Scratching in the sliding direction can be seen. An SEM examination of the cross-section of a sample from abrasive wear showed some subsurface plastic deformation, but not as much as

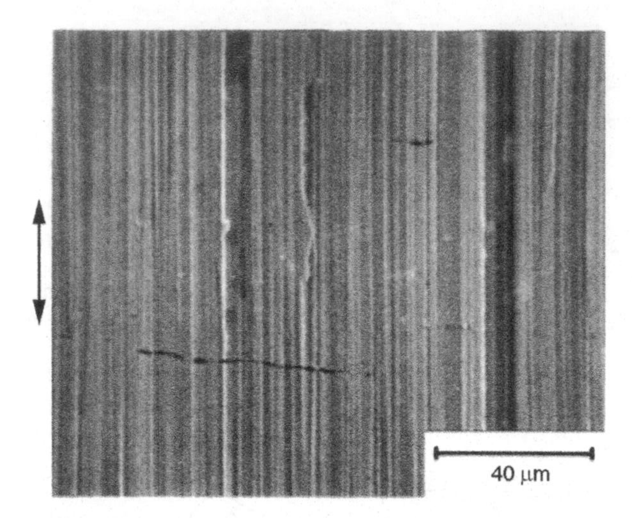

Figure 7.2.13 SEM micrograph of 303 stainless steel shaft surface after abrasive wear under unlubricated conditions. Sliding direction is along the vertical axis.

in adhesive wear (Bhushan *et al.*, 1985b). However, a 10–80% increase in microhardness of the worn surfaces was observed.

Other terms for abrasive wear also loosely used are scratching, scoring or gouging, depending on the degree of severity.

7.2.2.1 Abrasive Wear by Plastic Deformation

Material removal from a surface via plastic deformation during abrasion can occur by several deformation modes which include plowing, wedge formation and cutting, Figure 7.2.14. Plowing results in a series of grooves as a result of the plastic flow of the softer material. In the plowing (also called ridge formation) process, material is displaced from a groove to the sides without the removal of material, Figure 7.2.14a. However, after the surface has been plowed several times, material removal can occur by a low-cycle fatigue mechanism. When plowing occurs, ridges form along the sides of the plowed grooves regardless of whether or not wear particles are formed. These ridges become flattened, and eventually fracture after repeated loading and unloading cycles, Figure 7.2.15 (Suh, 1986). The plowing process also causes subsurface plastic deformation and may contribute to the nucleation of surface and subsurface cracks (Bhushan, 1999). Further loading and unloading (low-cycle, high-stress fatigue) cause these cracks and pre-existing voids and cracks to propagate (in the case of subsurface cracks to propagate parallel to the surface at some depth) and join neighboring cracks which eventually shear to the surface leading to thin wear platelets, Figure 7.2.15. In very soft metals, such as indium and lead, the amount of wear debris produced is small and the deformed material is displaced along the sides of the groove. This plowing wear process should not be confused with rolling contact fatigue (to be described later) which develops macroscopic pits and results due to the initiation of subsurface, high-cycle, low-stress fatigue cracks at the level at which Hertzian elastic stresses are a maximum.

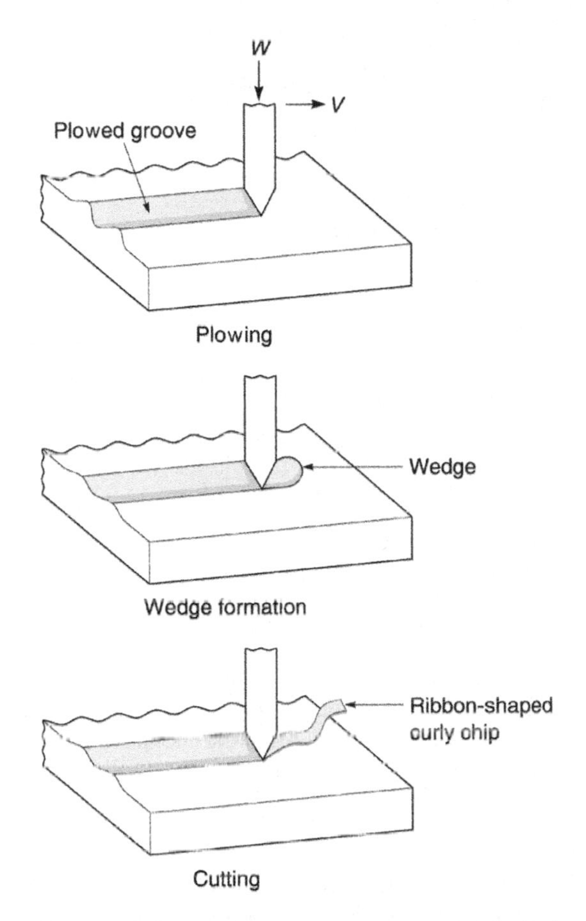

Figure 7.2.14 Schematics of abrasive wear processes as a result of plastic deformation by three deformation modes.

In the wedge formation type of abrasive wear, an abrasive tip plows a groove and develops a wedge on its front. It generally occurs when the ratio of shear strength of the interface relative to the shear strength of the bulk is high (about 0.5–1). In this situation, only some of the material displaced from the groove is displaced to the sides and the remaining material shows up as a wedge, Figure 7.2.14b.

In the cutting form of abrasive wear, an abrasive tip with large attack angle plows a groove and removes the material in the form of discontinuous or ribbon-shaped debris particles similar to that produced in a metal cutting operation, Figure 7.2.14c. This process results in generally significant removal of material and the displaced material relative to the size of the groove is very little.

The controlling factors for the three modes of deformation are the attack angle (Figure 7.2.17, to be presented later) or degree of penetration, and the interfacial shear strength of the interface. In the case of a sharp abrasive tip, there is a critical angle for which there is a transition from plowing and wedge formation to cutting. This critical angle depends on the material being

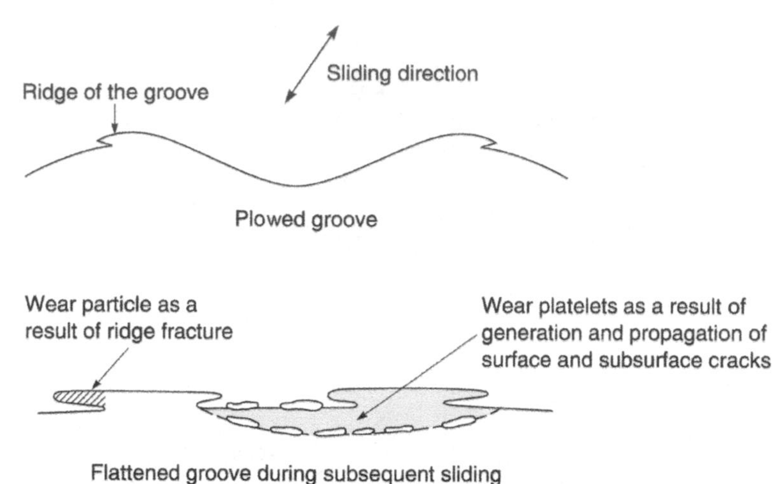

Figure 7.2.15 Schematics of plowed groove and formation of wear particle due to plowing as a result of fracture of flattened ridge and propagation of surface and subsurface cracks.

abraded. The degree of penetration is critical in the transition from plowing and wedge formation to cutting as the coefficient of friction increases with an increase in the degree of penetration (Hokkirigawa and Kato, 1988). For ductile metals, the mechanisms of plowing, wedge formation, and cutting have been observed, Figure 7.2.16.

Quantitative Equation (Plowing)

To obtain a quantitative expression for abrasive wear for plastic contacts, we consider a simplified model, in which one surface consists of an array of hard conical asperities sliding on a softer and flat surface and plows a groove of uniform depth (Rabinowicz, 1995). Figure 7.2.17 shows a single conical asperity, with a roughness angle (or attack angle) of θ (apex semi-angle

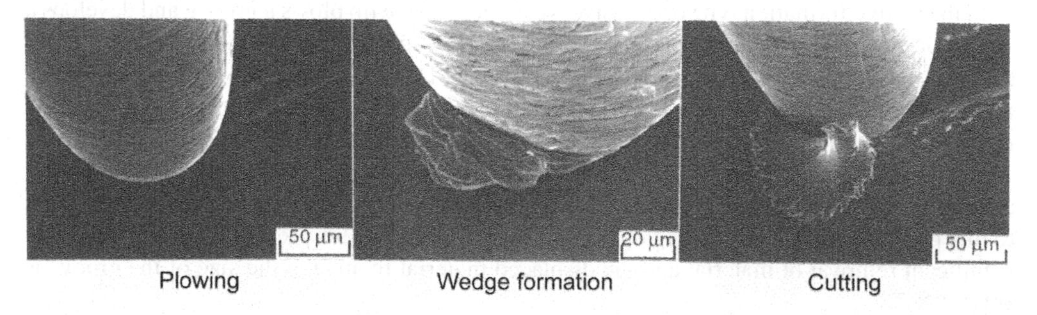

Figure 7.2.16 SEM micrographs observed of wear process during wear of unlubricated brass by a steel pin. Reproduced with permission from Hokkirigawa, K. and Kato, K. (1988), "An Experimental and Theoretical Investigation of Ploughing, Cutting and Wedge Formation During Abrasive Wear," *Tribol. Inter.* **21**, 51–57. Copyright 1988. Elsevier.

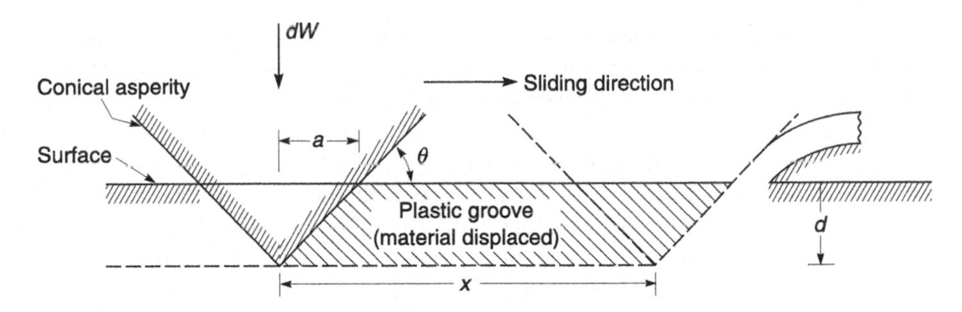

Figure 7.2.17 A hard conical asperity in sliding contact with a softer surface in an abrasive wear mode.

of asperities 90° – θ), creating a track through the softer surface with a depth of d and width of $2a$. We assume that the material has yielded under the normal load dW; therefore

$$dW = \frac{1}{2}\pi a^2 H \tag{7.2.8}$$

where H is the hardness of the softer surface. The volume displaced in a distance x is,

$$dv = a^2 x(\tan\theta) \tag{7.2.9}$$

From Equations 7.2.8 and 7.2.9, we get

$$dv = \frac{2dWx(\tan\theta)}{\pi H} \tag{7.2.10a}$$

The total volume of material displaced by all asperities is,

$$v = \frac{2Wx\overline{\tan\theta}}{\pi H} \tag{7.2.10b}$$

where $\overline{\tan\theta}$ is a weighted average of the tan θ values of all the individual conical asperities, called the roughness factor.

The derivation of Equation 7.2.10 is based on an extremely simple model. For example, the distribution of asperity heights and shapes and any material build-up ahead of the asperities are not taken into account. An equation of the form similar to Archard's equation for adhesive wear is found to cover a wide range of abrasive situations, and is

$$v = \frac{k_{abr}Wx}{H} \tag{7.2.11}$$

where k_{abr} is a nondimensional wear coefficient that includes the geometry of the asperities ($\overline{\tan\theta}$ for a simple case of conical asperities) and the probability that given asperities cut (remove) rather than plows. Thus, the roughness effect on the volume of wear is very distinct. The value of k_{abr} typically ranges from 10^{-6} to 10^{-1}. The rate of abrasive wear is frequently very large – two to three orders of magnitude larger than the adhesive wear.

Note that in the elastic contact regime, the real area of contact and consequently the coefficient of friction decreases with an increase in surface roughness, whereas the abrasive wear rate in the plastic contact regime increases with roughness.

The wear equation for two-body abrasive wear is also valid for three-body abrasive wear. However, k_{abr} is lower, by about one order of magnitude, because many of the particles tend to roll rather than slide (Rabinowicz *et al.*, 1961). It seems that the abrasive grains spend about 90% of the time rolling, and the remaining time sliding and abrading the surfaces, Figure 7.2.12. (The coefficient of friction during three-body abrasion is generally less than that in two-body abrasion, by as much as a factor of two.) In some cases, such as in free abrasive polishing, a surface can elastically deform sufficiently to allow the particles to pass through, which minimizes the damage. In this instance, Young's modulus has a direct bearing on abrasive wear.

During wear, some blunting of the hard asperities or abrasive particles occurs, thus reducing the wear rate. However, a brittle abrasive particle can fracture which would result in resharpening of the edges of the particle and an increase in wear rate.

Example Problem 7.2.2

A hard steel surface consisting of an array of conical asperities of an average semi-angle of 60° slides on a soft lead surface ($H = 75$ MPa) under a load of 10 N. Calculate the volume of lead displaced in unit slid distance. Given that the volume of lead material removed is 10^{-6} m^3 for a sliding distance of 1 km, calculate the wear coefficient of lead.

Solution

Given,

$$\text{Roughness angle, } \theta = 30^\circ$$

$$W = 10 \text{ N}$$

$$v_{lead} = 10^{-6} \text{ m}^3$$

$$H_{lead} = 75 \text{ MPa}$$

The volume of material displaced by all asperities in unit slid distance is

$$\frac{2W\overline{\tan\theta}}{\pi H} = \frac{2 \times 10 \times \tan 30}{\pi \times 75 \times 10^6} \text{ m}^3/\text{m}$$
$$= 4.9 \times 10^{-8} \text{ m}^3/\text{m}$$

The wear coefficient of lead material,

$$k_{abr} = \frac{v_{lead} H_{lead}}{Wx} = \frac{10^{-6} \times 75 \times 10^6}{10 \times 10^3}$$
$$= 7.5 \times 10^{-3}$$

Experimental Evidence

There is significant experimental evidence that the wear rate in two-body abrasion is generally inversely proportional to the hardness and proportional to the normal load and sliding distance for many pure metals; alloys often exhibit more complex behavior (Kruschov, 1957, 1974; Goddard and Wilman, 1962; Mulhcarn and Samuels, 1962; Misra and Finnie, 1981). Hardness is an important parameter for abrasive wear resistance. Wear resistance (proportional to 1/wear rate) of annealed pure metals is generally directly proportional to their hardness but is more complex for alloys, Figure 7.2.18 (Kruschov, 1957, 1974; Kruschov and Babichev, 1958). These authors reported that prior work hardening of the pure metals and alloys had no effect on the wear rate. Cold working of the 0.4% carbon steel resulted in a significant increase in bulk hardness but had no effect on its wear resistance. These and other experiments show that a metal surface strain hardens by plastic flow during abrasion to a maximum value, and it is this value of hardness which is important for abrasion resistance. Also note that if a material is hardened, it generally becomes more brittle. Brittle materials can produce larger particles, resulting in high wear rates. During three-body abrasion with alumina particles, the wear resistance of metals is also found to be proportional to the hardness of the workpiece (Rabinowicz *et al.*, 1961).

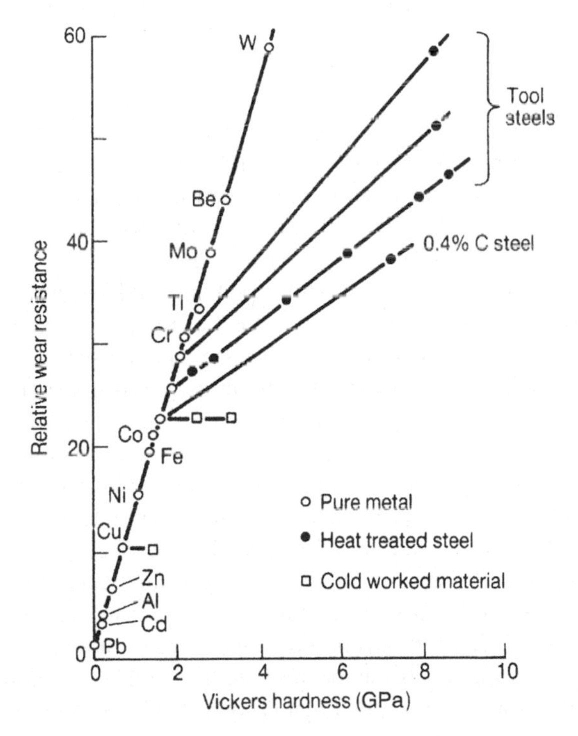

Figure 7.2.18 Relative wear resistance of pure metals and heat treated and cold worked steels as a function of hardness in two-body abrasion. Reproduced with permission from Kruschov, M.M. (1957), "Resistance of Metals to Wear by Abrasion, as Related to Hardness," *Proc. Conf. Lubrication and Wear*, pp. 655–659, Instn Mech. Engrs, London, UK. Copyright 1957. Institution of Mechanical Engineers.

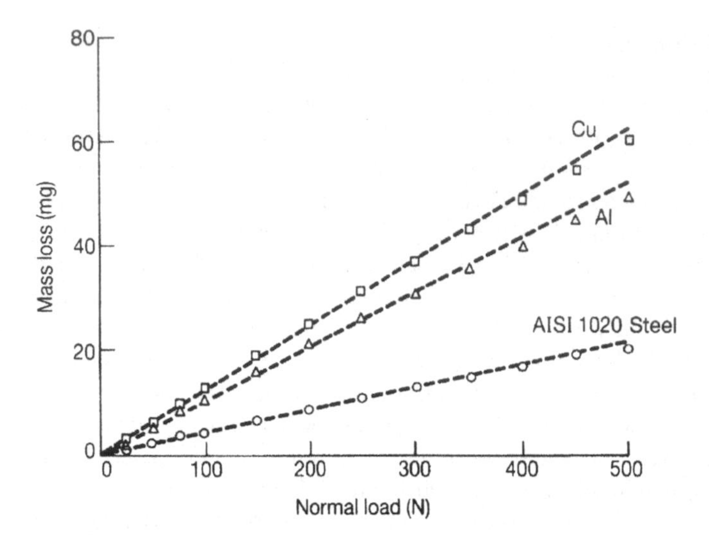

Figure 7.2.19 Mass loss of three ductile metals as a function of applied normal load subjected to two-body abrasion by 115 μm SiC abrasive paper. Reproduced with permission from Misra, A. and Finnie, I. (1981), "Some Observations on Two-Body Abrasive Wear," *Wear* **68**, 41–56. Copyright 1981. Elsevier.

The volume of wear generally increases linearly with an increase in applied normal load; often the linear relationship is not maintained at high loads. For example, in Figure 7.2.19, mass loss for three ductile metals increases with applied load when abraded against an abrasive paper. Bhushan (1985, 1996) has also reported that the wear rate increases with applied load for abrasive magnetic tapes sliding against ceramic heads.

The wear rate changes as a function of the sliding velocity and the particle size of an abrasive paper or roughness of the abrading surface. The effect of the particle size on the wear rate in two-body and three-body abrasion is discussed by Xie and Bhushan (1996a). Figure 7.2.20 shows the dependence of the sliding velocity and the abrasive grit size for copper being abraded against an abrasive wear. Note that the wear rate increases by a few percent for an increase in the sliding velocity by three orders of magnitude, which suggests that the wear rate is not very sensitive to the sliding velocity, as expected. An increase in the wear resistance with sliding velocity is due presumably to the increase in the strain rate which increases the yield stress of the material. At very high sliding velocities, high interface temperatures as a result of frictional heating result in a decrease in the yield stress of the material being abraded, which counteracts the effect due to the increased strain rate. Wear rate increases with an increase in grit size up to about 100 μm and beyond this size the wear rate becomes less sensitive to the particle size. In explaining this behavior, we note that in the wear equation, although the abrasive particle size does not enter explicitly, it is conceivable that the roughness factor tan θ may be size-dependent. With large abrasive particles, the shape of the abrasive particles does not depend on the particle size. The surface roughness of the abrasive surface with the same type and density of abrasive particles has an effect on the wear rate. Wear rate increases with an increase in the surface roughness of the abrasive tape with CrO_2 magnetic particles sliding (partially flying) on an Ni-Zn ferrite head, Figure 7.2.21.

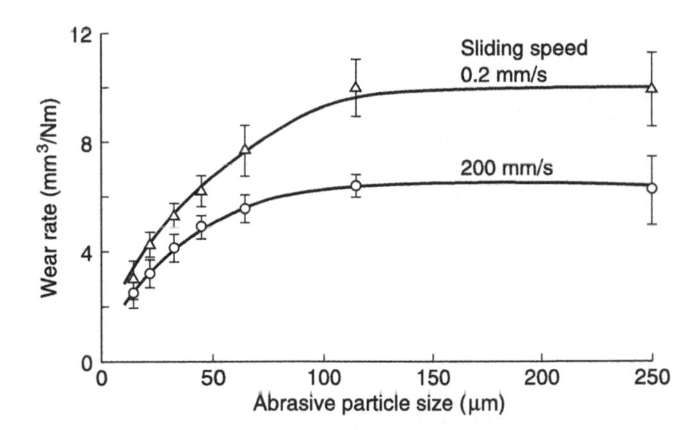

Figure 7.2.20 Wear rate of copper, subjected to two-body abrasion by SiC abrasive paper, as a function of abrasive particle size at two different sliding velocities. Reproduced with permission from Misra, A. and Finnie, I. (1981), "Some Observations on Two-Body Abrasive Wear," *Wear* **68**, 41–56. Copyright 1981. Elsevier.

The dependence of the abrasive wear rate as a function of the sliding distance is more complex. If wear takes place with fresh abrasive paper (in two-body wear) or fresh abrasive particles (in three-body wear), wear continues at a steady rate (Rabinowicz, 1995; Bhushan, 1985, 1996). However, if a limited amount of abrasive is used as the sliding continues, the

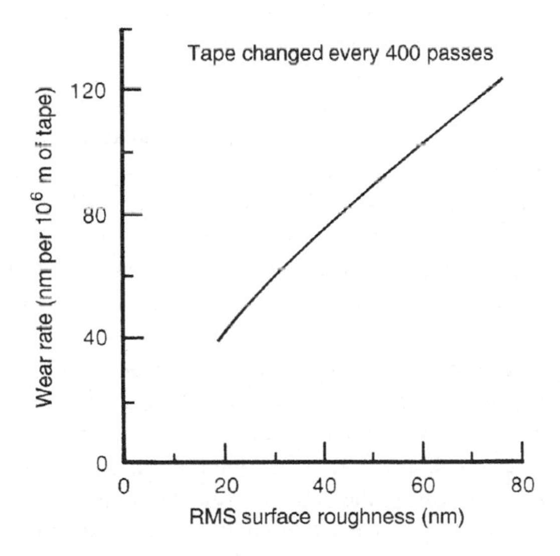

Figure 7.2.21 Wear rate of Ni-Zn ferrite as a function of RMS surface roughness of a magnetic tape with CrO_2 magnetic particles. Reproduced with permission from Bhushan, B. (1985), "Assessment of Accelerated Head-Wear Test Methods and Wear Mechanisms," in Tribology and Mechanics of Magnetic Storage Systems, Vol. 2 (B. Bhushan and N.S. Eiss, eds), pp. 101–111, special publication SP-19, ASLE, Park Ridge, Illinois. Copyright 1985. ASLE. (*Source*: Bhushan, 1985).

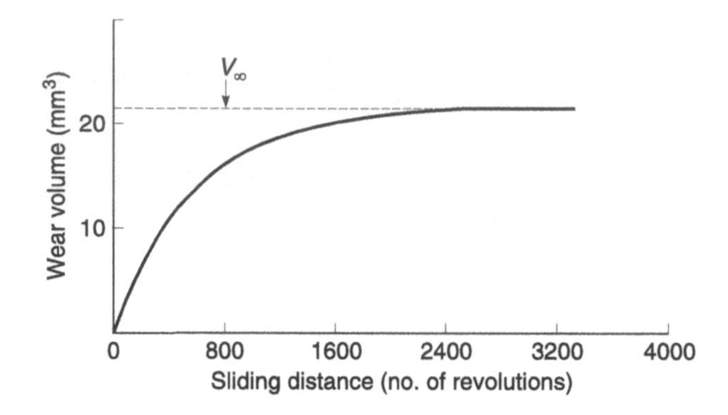

Figure 7.2.22 Wear volume of steel as a function of sliding distance subjected to two-body abrasion by 220 grade silicon carbide paper. Reproduced with permission from Mulhearn, T.O. and Samuels, L.E. (1962), "The Abrasion of Metals: A Model of the Process," *Wear* **5**, 478—498. Copyright 1962. Elsevier.

wear rate generally decreases as a function of time. For example, in Figure 7.2.22 (Mulhearn and Samuels, 1962), the wear rate decreases as a function of the sliding distance when steel is abraded on silicon carbide abrasive paper. Mulhearn and Samuels (1962) reported that the data fit the following form:

$$v = v_\infty \left[1 - \exp\left(-\alpha x\right)\right] \tag{7.2.12}$$

where v_∞ is the total volume of metal removed if the sliding is continued indefinitely and α is a constant. Similar results for abrasive magnetic tapes sliding against ceramic heads or another abrasive tape have been reported by Bhushan (1985, 1996).

A decrease in the wear rate as a function of sliding distance is believed to occur as a result of blunting of the abrasive surfaces in two-body wear or abrasive particles in three-body wear, Figure 7.2.23a. In addition, clogging of the abrasive surface by abraded debris occurs during wear, Figure 7.2.23b (Rabinowicz, 1995). If at any instance, the wear debris is larger than abrasive particles, it may leave the material being abraded above the level of the abrasive grains and result in no additional wear. One can see that abrasive action should cease much more rapidly in wear with fine grades of abrasive paper than with coarse grades.

Effect of Relative Hardness of Abrasive Medium to Workpiece

In two-body (Aleinikov, 1957; Richardson, 1968) and three-body (Rabinowicz, 1977, 1983) abrasive situations, if the abrading medium is softer than a workpiece, the wear coefficient does not remain constant. It is known that when the hardness ratio of the workpiece to the abrasive particles is less than unity, the wear coefficient remains approximately constant; however, if the ratio is equal to or greater than unity, the wear coefficient decreases rapidly with an increase in the hardness ratio, Figure 7.2.24 (plotted by Rabinowicz, 1983 based on data by Aleinikov, 1957 and Richardson, 1968). When the hardness of the workpiece is of the same order of magnitude as the hardness of the abrasive particles, the wear of the workpiece is not rapid, since deformation occurs both in the abrasive particles and the workpiece, and wear generally occurs in both. When the workpiece is significantly harder than the abrasive

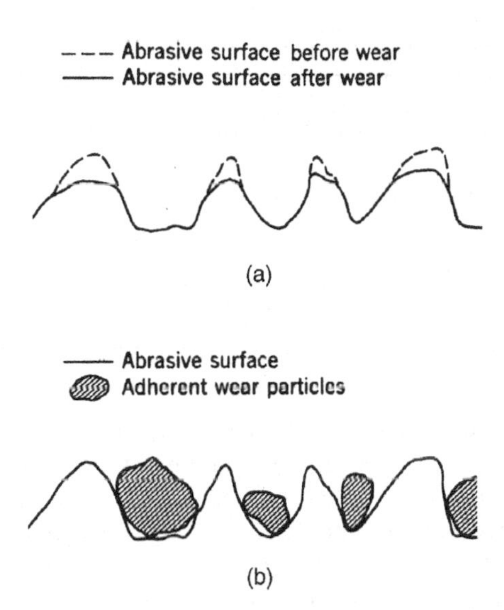

Figure 7.2.23 Schematics (a) of an abrasive surface before and after wear, showing blunting, and (b) of an abrasive surface clogged by wear debris. Reproduced with permission from Rabinowicz, E. (1995) *Friction and Wear of Material*, Second edition, Wiley, New York. Copyright 1995. Wiley.

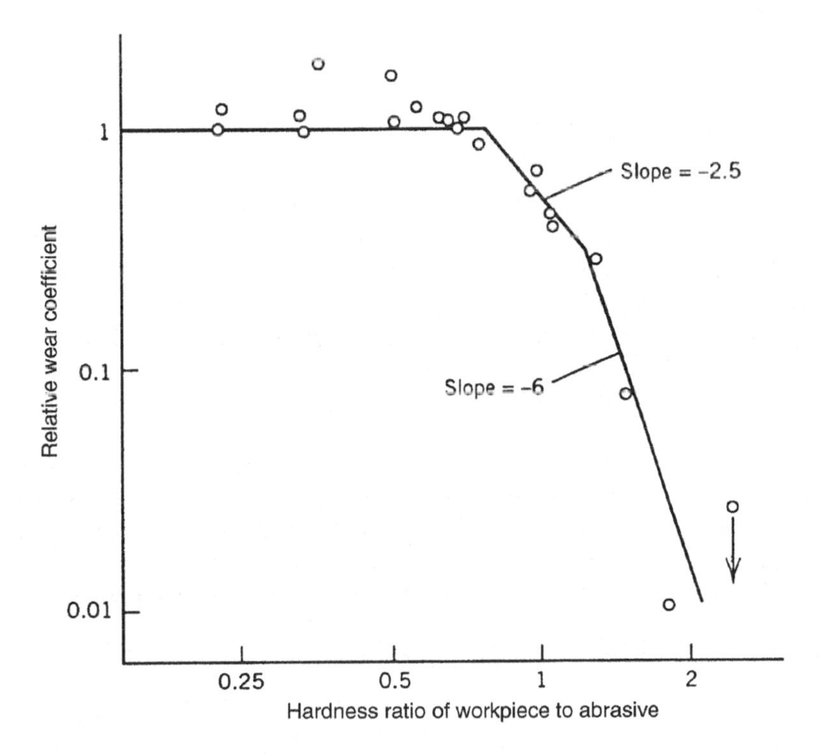

Figure 7.2.24 Relative abrasive wear coefficient of materials, covering a range of hardnesses, as a function of workpiece hardness to abrasive hardness quotient.

Table 7.2.2 Typical hardnesses of commonly used abrasives.

Abrasive material	Hardness (GPa)
Diamond	80
Cubic boron nitride	40
Silicon carbide	25
Alumina	21
Quartz, silica	8
Magnesium oxide	8

particles, negligible deformation, and consequently, wear occur. Accompanying the change in hardness ratio from less than unity to greater than unity, there is a significant change in the roughness of the worn surface because the wear mechanism has changed. The wear coefficients from the wear tests with workpieces with varying hardnesses against an abrasive can be used to estimate the hardness of the abrasive (Rabinowicz, 1977; Xie and Bhushan, 1996b). Next, we note that if the abrasive wear is required, the abrasive material must be harder that the surface to be abraded; it does not have to be much harder. The desired criteria for high wear are hardness and sharpness. Thus it is advantageous if the abrasive is brittle so that it results in sharp corners when it is subjected to high stresses; many nonmetals meet this criterion. Hardnesses of commonly used abrasives are listed in Table 7.2.2.

7.2.2.2 Abrasive Wear by Fracture

Quantitative Equation

To obtain a quantitative expression for abrasive wear of brittle solids by brittle fracture, we consider an asperity with sharp geometry on a flat surface of a brittle solid, Figure 7.2.25 (Evans and Marshall, 1981). At low loads, a sharp asperity contact will cause only plastic deformation and wear occurs by plastic deformation. Above a threshold load, brittle fracture occurs, and wear occurs by lateral cracking at a sharply increased rate. The threshold load is proportional to $(K_c/H)^3\ K_c$ (Lawn and Marshall, 1979). The H/K_c is known as the index of brittleness, where H is hardness (resistance to deformation) and K_c is fracture toughness (resistance to fracture).

Lateral cracks in amorphous materials develop from the residual stresses associated with the deformed material (Lawn, 1993). The maximum extension of the crack is thus realized when the penetrating asperity is removed. As a sharp asperity slides over the surface, lateral cracks grow upward to the free surface from the base of the subsurface-deformed region and material is removed as platelets from the region bounded by the lateral cracks and the free surface.

The lateral crack length c for a sliding asperity contact is given by (Evans and Marshall, 1981)

$$c = \alpha_1 \left[\frac{(E/H)^{3/5}}{K_c^{1/2} H^{1/8}} \right] W^{5/8} \tag{7.2.13}$$

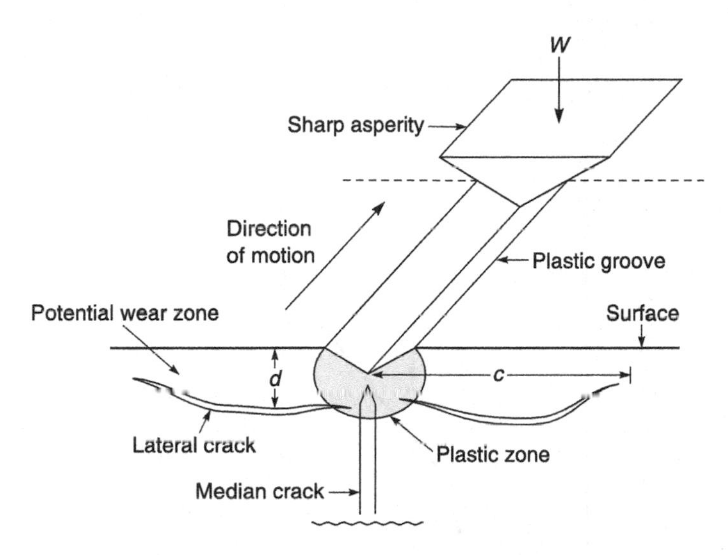

Figure 7.2.25 Schematic of the mechanism of wear by a sharp asperity sliding on the flat surface of a brittle material by lateral fracture. Reproduced by permission from Evans, A.G. and Marshall, D.B. (1981), "Wear Mechanisms in Ceramics," in *Fundamentals of Friction and Wear of Materials* (D.A. Rigney, ed), pp. 439–452, Amer. Soc. Metals, Metals Park, Ohio. Copyright 1981. ASM International.

where α_1 is a material-independent constant that depends on the asperity shape. The depth, d, of the lateral crack is given by (Evans and Marshall, 1981)

$$d = \alpha_2 \left(\frac{E}{H}\right)^{2/5} \left(\frac{W}{H}\right)^{1/2} \tag{7.2.14}$$

where α_2 is another material-independent constant. The maximum volume of material removed per asperity encounter per unit sliding distance is 2 dc. If N asperities contact the surface with each carrying the load W, then from Equations 7.2.13 and 7.2.14, the volume of wear per unit sliding distance of the interface is given by (Evans and Marshall, 1981),

$$v = \alpha_3 N \frac{(E/H)\, W^{9/8}}{K_c^{1/2} H^{5/8}} \tag{7.2.15}$$

where α_3 is a material-independent constant. The ratio (E/H) does not vary by much for different hard brittle solids. Therefore, wear rate is inversely proportional to the (fracture toughness)$^{1/2}$ and (hardness)$^{5/8}$. Wear rate is proportional to (normal load)$^{9/8}$ which implies that wear rate by lateral fracture increases more rapidly than linearly with the applied normal load as in plastic deformation. This implies that the wear coefficient in the wear equation is not independent of load. Based on the method of calculation of c and d for the wear model, variations to Equation 7.2.15 have been reported in the literature.

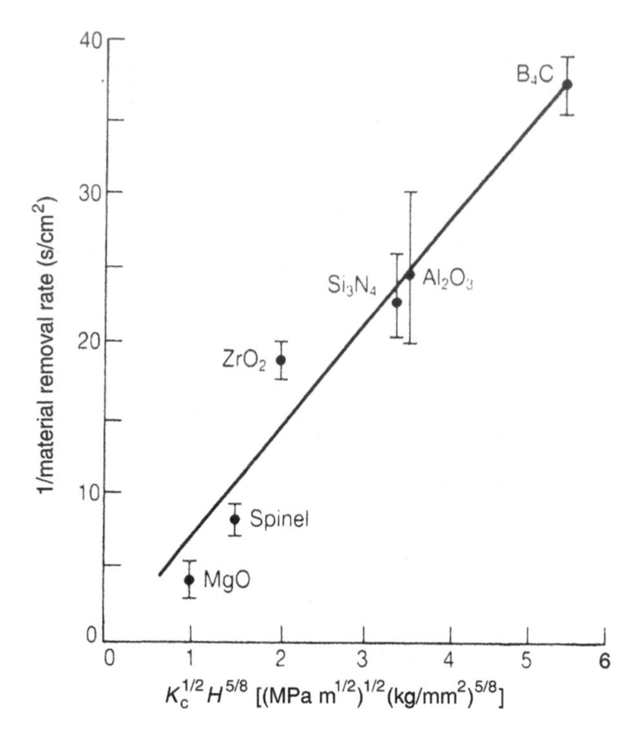

Figure 7.2.26 Correlation between the reciprocal of wear rate at constant load with the material property quantity $K_c^{1/2} H^{5/8}$ for several ceramic materials. Reproduced by permission from Evans, A.G. and Marshall, D.B. (1981), "Wear Mechanisms in Ceramics," in *Fundamentals of Friction and Wear of Materials* (D.A. Rigney, ed), pp. 439–452, Amer. Soc. Metals, Metals Park, Ohio. Copyright 1981. ASM International.

Experimental Evidence

There is some experimental evidence that wear rate in two-body abrasion is inversely proportional to $(Kc^{1/2} H^{5/8})$. Figure 7.2.26 shows a good correlation in grinding experiments between wear rate at constant loads and material properties for various ceramic material combinations as predicted by the lateral fracture model. For additional wear data, for example, see Yamamoto *et al.* (1994).

7.2.3 Fatigue Wear

Subsurface and surface fatigues are observed during repeated rolling (negligible friction) and sliding, respectively. The repeated loading and unloading cycles to which the materials are exposed may induce the formation of subsurface or surface cracks, which eventually, after a critical number of cycles, will result in the breakup of the surface with the formation of large fragments, leaving large pits in the surface, also known as pitting. Prior to this critical point (which may be hundreds, thousands, or even millions of cycles), negligible wear takes place, which is in marked contrast to the wear caused by an adhesive or abrasive mechanism, where wear causes a gradual deterioration from the start of running. Therefore, the amount of material removed by fatigue wear is not a useful parameter. Much more relevant is the useful life in terms of the number of revolutions or time before fatigue failure occurs.

Chemically enhanced crack growth (most common in ceramics) is commonly referred to as static fatigue. In the presence of tensile stresses and water vapor at the crack tip in many ceramics, a chemically induced rupture of the crack-tip bonds occurs rapidly, which increases the crack velocity. Chemically enhanced deformation and fracture result in an increased wear of surface layers in static and dynamic (rolling and sliding) conditions.

7.2.3.1 Rolling Contact Fatigue

Adhesive and abrasive wear mechanisms are operative during direct physical contact between two surfaces moving relative to each other. If the two surfaces are separated by a fluid film (and abrasive particles are excluded), these wear mechanisms do not operate. However, in an interface with nonconforming contact, the contact stresses are very high and the fatigue mechanism can be operative. In these cases, although direct contact does not occur, the mating surfaces experience large stresses, transmitted through the lubricating film during the rolling motion. Well-designed rolling element bearings usually fail by subsurface fatigue.

From a Hertz elastic stress analysis, the maximum compressive stresses occur at the surface, but the maximum shear stresses occur some distance below the surface, Figure 7.2.27. As rolling proceeds, the directions of the shear stresses for any element change sign. Time to fatigue failure is dependent on the amplitude of the reversed shear stresses, the interface lubrication conditions, and the fatigue properties of the rolling materials (Lundberg and Palmgren,

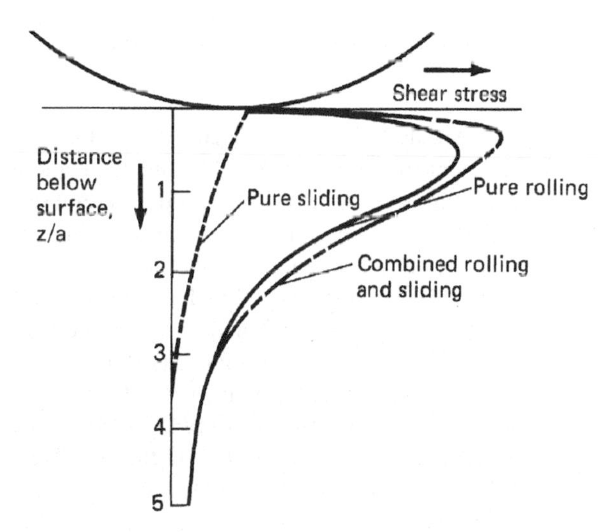

Figure 7.2.27 Variation of principal shear stress at various depths directly below the point of contact of two hard surfaces in pure rolling ($\mu = 0$), pure sliding (μ, high value), and combined contact (μ, moderate value). The z is the distance below the surface in the vertical direction and a is half of the Hertzian diameter.

Figure 7.2.28 Spalling of a 52100 ball bearing race from subsurface fatigue. Reproduced with permission from Tallian, T.E., Baile, G.H., Dalal, H., and Gustafsson, O.G. (1974), *Rolling Bearing Damage*, SKF Industries Inc., King of Prussia, Pennsylvania. Copyright 1974. SKF USA Inc.

1947, 1951). When a fatigue crack does develop, it occurs below the surface, until a region of metal is separated to some extent from the base metal by the crack and ultimately becomes detached and spalls out. By the time cracks grow large enough to emerge at the surface and produce wear particles, these particles may become large spalls or flakes. A typical example of spalling of a ball bearing race due to subsurface fatigue is shown in Figure 7.2.28. Since materials in rolling contact applications are often surface hardened, the surface could be brittle. Hence, cracks may also initiate at the surface as a result of tensile stresses and lead to surface fatigue.

The earliest method of rating rolling element bearing is based on the stochastic life prediction methods based on the Weibull distribution (Weibull, 1951; Bhushan, 1999). The Weibull life model was developed by Lundberg and Palmgren (1947, 1951). This method was later standardized by ISO in 1962 (281) and is recommended by the Antifriction Bearing Manufacturers Association (AFBMA). It is still widely used. The life of rolling element bearings (normally referred to as L_{10} *or* B_{10}) in millions of revolutions for 90% of the bearing population is determined from

$$L_{10} = (C/W)^p \tag{7.2.16a}$$

where

$$p = 3 \text{ for ball bearings (point contacts)}$$
$$= \frac{10}{3} \text{ for roller bearings (line contacts).}$$

Here, C is the bearing's basic load capacity and W is the equivalent radial or thrust load for the radial or thrust bearing, respectively. The basic load capacity of a bearing is the load that 90% of the bearings can endure for 1 million revolutions under given running conditions.

The AFBMA method for determining bearing load rating as published in bearing manufacturer's catalogs is based on bearing tests conducted in the 1940s. There have been significant improvements in materials and processing. The bearing life also depends on the lubrication and operating conditions. In other words, the bearing life depends on the bearing geometry (size and accuracy of manufacture), the physical properties (such as the modulus of elasticity and fatigue strength), the metallurgy of bearing materials, and the lubrication conditions (such as viscosity, speed of rotation, and surface roughness). It is assumed that various bearing design factors, as a first approximation, are multiplicative. Then the expected bearing life for given operating conditions, L_A, can be related to the calculated rating life L_{10} by using the various life adjustment factors (Bamberger *et al.*, 1971):

$$L_A = (D)(E)(F)(G)(H)\, L_{10} \tag{7.2.16b}$$

where D is the bearing material factor, E is the bearing processing factor, F is the lubrication factor, G is the speed effect factor and H is the misalignment factor. Typical values of life adjustment factors are presented by Bamberger *et al.* (1971), Harris (1991) and Zaretsky (1992) and these can be less than or greater than 1.

Because of the continued improvement in bearing materials and better understanding of bearing behavior, in many cases, bearings manufactured from clean, homogeneous steel, virtually infinite life can be obtained. However, Lundberg-Palmgren theory, based on probability of survival from subsurface-initiated fatigue, predicts a finite life. Ioannides and Harris (1985) developed a modified theory in which they introduced the fatigue limit as the lower limit of the fatigue behavior, i.e., they assumed that no failure can occur if the stress in a volume element is less than or equal to the endurance limit of the material. This modification is able to predict infinite L_{10} life.

Based on statistical methods (e.g., Johnson, 1964; Nelson, 1982), the life of a system with multiple bearings of lives $L_1, L_2, \ldots$ for a given probability of survival is

$$L = \left[\left(\frac{1}{L_1}\right)^{\beta} + \left(\frac{1}{L_2}\right)^{\beta} + \cdots\right]^{1/\beta} \tag{7.2.17}$$

where β is the Weibull slope for a Weibull distribution. β is about 1.5 for rolling element bearings. The probability of the survival of multiple bearings with probabilities $1–P_i$ in fraction ($i = 1, \ldots n$) for n bearings is $(1 - P_1)\,(1 - P_2), \ldots (1 - P_n)$. For n bearings with identical $(1 - P)$, the probability of survival of the system is $(1 - P)^n$.

Example Problem 7.2.3

The basic load capacity of a radial ball bearing is 8 kN. Calculate its life based upon a 90% probability of survival for the bearing operating at 600 rpm and at radial loads of 6 kN and 12 kN.

Solution

For a ball bearing,

$$L_{10} = (C/W)^3$$

$$\text{For } W = 600 \text{ N}, \ L_{10} = \left(\frac{8}{6}\right)^3 \times 10^6 \text{ rev.}$$

$$= 2.37 \times 10^6 \text{ rev.}$$

$$\text{Bearing life in hours} = 2.37 \times 10^6/(600 \times 60)\ h = 65.8 \text{ h}$$

$$\text{For } W = 1200 \text{ N}, \ \text{bearing life in hours} = 65.8\ (1/2)3 = 8.3 \text{ h}$$

Example Problem 7.2.4

A ball bearing spindle with two radially loaded ball bearings and belt driven at 600 rpm, drives two pumps. The radial load on each bearing is 10 kN. Given that the basic load capacity of each bearing is 25 kN, calculate the L_{10} life of the system.

Solution

$$\begin{aligned} L_{10} &= (C/W)^3 \times 10^6 \text{ rev.} \\ &= (25/10)^3 \times 10^6 \text{ rev.} \\ &= 15.63 \times 10^6 \text{ rev.} \\ &= \frac{15.63 \times 10^6}{600 \times 60} = 434 \text{ h} \end{aligned}$$

Since the spindle has two bearings, the system probability of survival for a life of 434 h is 0.9^2 or 81%. Assuming $\beta = 1.5$, the life of the system at a 90% probability of survival, from Equation 7.2.17, is

$$L_{10} \text{ of the system} = \frac{434}{2^{1/1.5}} \text{h} = 273 \text{ h}$$

Rolling/Sliding Contact Fatigue

Rolling contact is frequently accompanied by slip or sliding. The complex motions in most rolling contact situations produce at least a small fraction of slip or sliding (on the order of 1 to 10%) such as in rolling bearings, hypoid gear teeth, can roller followers, and wheel-rail contacts. The friction stresses due to sliding cause the maximum shear stresses to be nearer the surface, Figure 7.2.27, and the failure occurs by near surface fatigue. (See Hertz analysis of an elastic sphere acting on an elastic semi-infinite solid presented in Chapter 3.) Slip can result in severe adhesive wear (scuffing) damage to the mating surfaces. Proper lubrication is important to minimize the deleterious effects of slip in these rolling-contact situations.

Sliding Contact Fatigue

We have seen that, when sliding surfaces make contact via asperities, wear can take place by adhesion and abrasion. However, it is conceivable that asperities can make contact without adhering or abrading and can pass each other, leaving one or both asperities plastically deformed from the contact stresses. As the surface and subsurface deformation continues, cracks are nucleated at and below the surface. Once the cracks are present (either by crack nucleation or from pre-existing voids or cracks), further loading and deformation cause cracks to extend and propagate. After a critical number of contacts, an asperity fails due to fatigue, producing a wear fragment. In a sliding contact, friction is generally high compared to a rolling contact, and the maximum shear stress occurs at the surface (Figure 7.2.27), which leads to surface fatigue. This may be the situation in a boundary lubrication system in which one or several monolayers of lubricant or absorbed surface layers at the interface separate the asperities but contact stresses are still experienced by the asperities.

7.2.3.2 Static Fatigue

Static fatigue results from a stress-dependent chemical reaction between water vapor and the surface of the ceramic. The rate of reaction depends on the state of stress at the surface and the environment. The stress is greatest at the roots (or tips) of small cracks in the material, and consequently the reaction proceeds at its greatest rate from these roots. The small cracks gradually lengthen and failure occurs when the cracks are long enough to satisfy the Griffith failure criteria for fracture (Wiederhorn, 1967). Thus, two stages of crack growth can be visualized: (1) slow crack motion occurs because of chemical attack at the crack tip; and (2) a catastrophic stage of crack motion is initiated when the crack is long enough to satisfy the Griffith criteria. The time to failure is the time required for the crack to grow from the subcritical to critical Griffith size (Wiederhorn, 1967, 1969; Westwood, 1977; Lawn, 1993). This moisture-assisted crack propagation and fracture is called static fatigue.

The stresses (residual stresses produced during machining and stresses introduced during static or dynamic contact) at the crack tip control the rate of crack growth. The kinetics of fluid flow from the environment also control the rate of rupture of the crack-tip bonds. For a start, the gas molecules do not have direct access to the crack-tip bonds. The mean free path for intermolecular collisions at ambient conditions is typically 1 μm, which will clearly exceed the crack-wall separation some distance behind the tip of a brittle crack. Thus, as the gas molecules migrate along the crack interface, a point will be reached where collisions with the walls become more frequent than with other gas molecules. The gas then enters a zone of

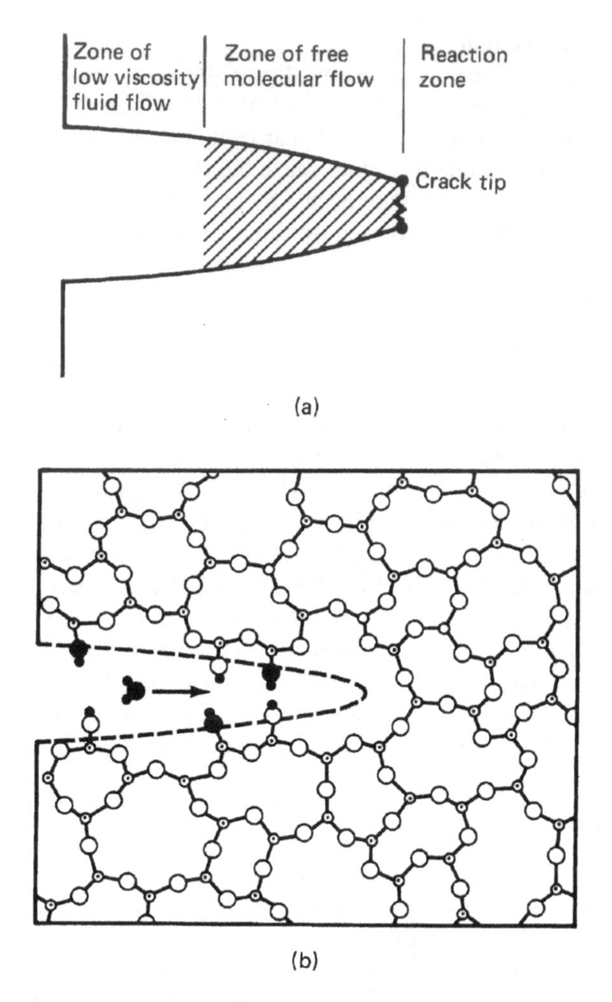

Figure 7.2.29 (a) Gaseous flow along interface to tip of brittle crack. Flow changes from that of a low-viscosity fluid to that of a dilute gas as the crack-wall separation diminishes below the intermolecular mean free path. (b) Two-dimensional representation of water-induced bond rupture in silica glass. Large circles, oxygen; intermediate circles, silicon; small circles, hydrogen. Solid circles denote species originally from the environment. Reproduced with permission from Lawn, B.R. (1993), *Fracture of Brittle Solids*, Second edition, Cambridge University Press, Cambridge. Copyright 1993. Cambridge University Press.

free molecular flow within which diffused molecular scattering at the walls may considerably attenuate the flow rate (Figure 7.2.29a).

One of the best studied examples of chemically enhanced crack growth is that of silica glass in the presence of a water environment. The basic crack-tip reaction is as follows:

$$(H - O - H) + (-Si - O - Si-) \rightarrow (-Si - OH\ HO - Si-) \tag{7.2.18}$$

That is, an incident water molecule hydrolyzes a siloxane bridging bond at the crack tip to form two terminal silane groups. A two-dimensional representation of the process is given in Figure 7.2.29b.

The dependence of crack velocity on applied stress and water vapor for soda-lime glass (amorphous structure) is shown in Figure 7.2.30a. A similar effect is also seen for a single-crystal sapphire in water vapor, Figure 7.2.30b. As the level of applied stress is raised, the crack velocity shows an initial, rapid increase, followed by a rather abrupt saturation. Humidity also plays a significant role; crack velocity changes by three orders of magnitude (0.1–100 μm/s) when the relative humidity is changed by three orders of magnitude (0.1–100% RH) for the same nominal stress. We note that the moisture-assisted fracture occurs for both crystalline and amorphous ceramics. For a given material and the environment, wear due to static fatigue can be reduced with lower residual stresses and a lesser degree of microcracking of the surface (Bhushan, 1996).

Wear of metals and ceramics in three-body abrasion with ceramic abrasive particles (Larsen-Basse, 1975) and of ceramics in two-body abrasion (Larsen-Basse and Sokoloski, 1975; Wallbridge *et al.*, 1983; Takadoum, 1993; Bhushan, 1996; Bhushan and Khatavkar, 1996) is reported to increase sharply with an increase in relative humidity of the environment above the ambient conditions while there is little effect at low humidities for example, see Figure 7.2.31.

Moisture-assisted fracture of the abrasive particles in three-body abrasion, at high humidities, brings more and sharper cutting asperities into contact with the abrading surface. However, moisture-assisted fracture of one of the sliding ceramic bodies in two-body abrasion produces sharp particles; some of which become trapped at the interface and result in high wear rates from three-body abrasion. Even in the absence of relative sliding, loose particles can be generated from ceramic bodies with high density of cracks placed in tension and exposed to high humidities. In an experiment, an Mn-Zn ferrite (MnO-ZnO-Fe_2O_3) ceramic rod under a static tensile stress was exposed to close to 100% RH for seven days. Debris particles could be collected in a dish placed under the rod, Figure 7.2.32. In a sliding interface, some of these particles would be trapped at the interface resulting in high wear rates.

7.2.4 Impact Wear

Two broad types of wear phenomena belong under this heading: erosive and percussive wear. Erosion can occur by jets and streams of solid particles, liquid droplets, and implosion of bubbles formed in the fluid. Percussion occurs from repetitive solid body impacts. Repeated impacts result in progressive loss of solid material.

7.2.4.1 Erosion

Solid Particle Erosion

Solid particle erosion occurs by impingement of solid particles, Figure 7.2.33. It is a form of abrasion that is generally treated rather differently because the contact stress arises from the kinetic energy of particles flowing in an air or liquid stream as it encounters a surface. The particle velocity and impact angle combined with the size of the abrasive give a measure of the kinetic energy of the impinging particles, that is, of the square of the velocity. Wear debris formed in erosion occurs as a result of repeated impacts.

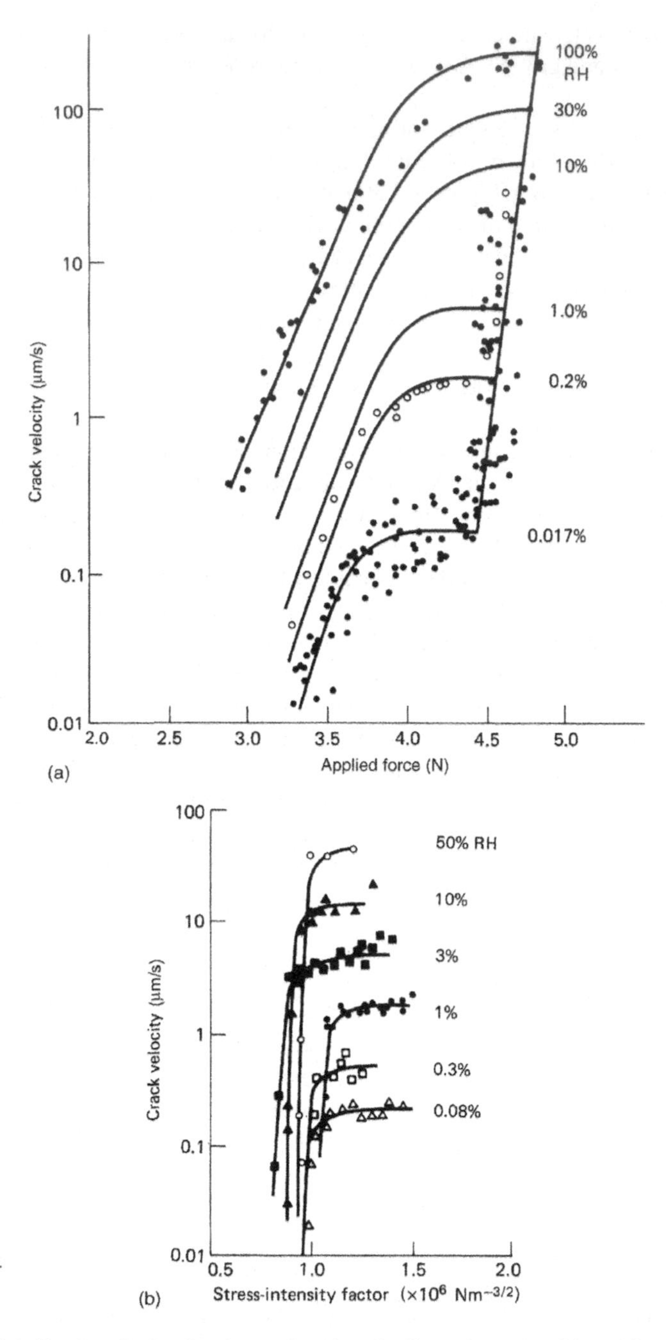

Figure 7.2.30 (a) Crack velocity for (amorphous) soda-lime glass tested in moistened nitrogen gas (relative humidities indicated) at room temperature. Some data points are included to demonstrate the scatter of the data between runs. Reproduced with permission from Wiederhorn, S.M. (1967), "Influence of Water Vapor on Crack Propagation in Soda-Lime Glass," *J. Amer. Cer. Soc.* **50**, 407–414. Fig. 3 p. 409. Copyright 1967. Wiley, (b) Crack velocity for single-crystal sapphire tested in moistened nitrogen gas (relative humidities indicated) at room temperature. (*Source*: Wiederhorn, 1969).

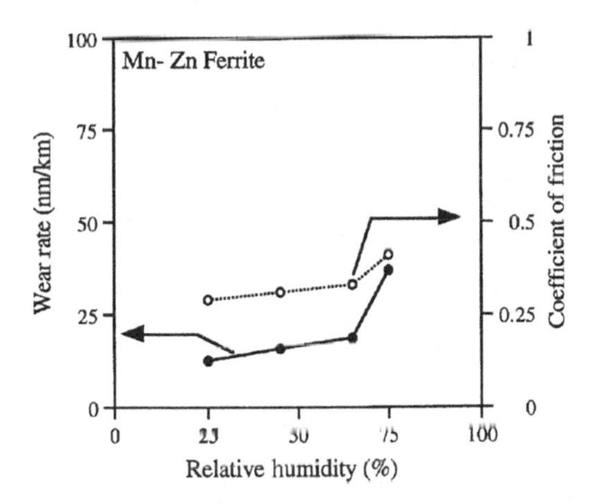

Figure 7.2.31 Wear rate and coefficient of friction as a function of relative humidity for Mn-Zn ferrite sliding against a magnetic tape with CrO_2 magnetic particles. Reproduced with permission from Bhushan, B. and Khatavkar, D.V. (1996), "Role of Water Vapor on the Wear of Mn-Zn Ferrite Heads Sliding Against Magnetic Tapes," *Wear* **202**, 30–34.

As in the case of abrasive wear, erosive wear occurs by plastic deformation and/or brittle fracture, dependent upon material being eroded away and upon operating parameters. Wear-rate dependence on the impact angle for ductile and brittle materials is different, as shown in Figure 7.2.34 (Bitter, 1963). Ductile materials will undergo wear by a process of plastic

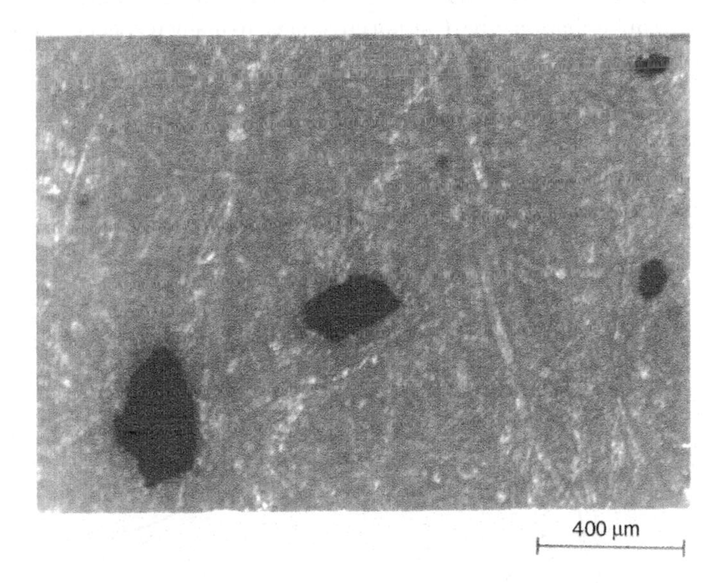

Figure 7.2.32 Optical micrographs of particles shed by stressed Mn-Zn ferrite rod placed at 22°C and 100% RH for seven days. Reproduced with permission from Bhushan, B. and Khatavkar, D.V. (1996), "Role of Water Vapor on the Wear of Mn-Zn Ferrite Heads Sliding Against Magnetic Tapes," *Wear* **202**, 30–34.

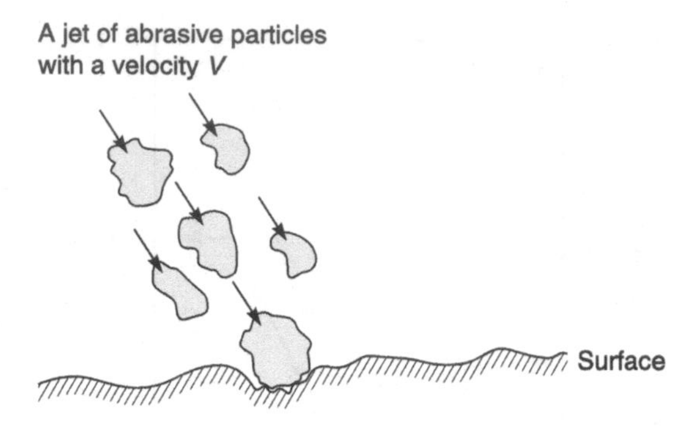

Figure 7.2.33 Schematic of a jet of abrasive particles hitting a surface at a high velocity.

deformation in which the material is removed by the displacing or cutting action of the eroded particle. In a brittle material, on the other hand, material will be removed by the formation and intersection of cracks that radiate out from the point of impact of the eroded particle (Finnie, 1960). The shape of the abrasive particles affects that pattern of plastic deformation around each indentation, consequently the proportion of the material displaced from each impact. In the case of brittle materials, the degree and severity of cracking will be affected by the shape of the abrasive particles. Sharper particles would lead to more localized deformation and consequently wear, as compared to the more rounded particles.

Two basic erosion mechanisms have been observed for erosion of ductile materials (Bellman and Levy, 1981; Soderberg *et al.*, 1983): cutting erosion and deformation (plowing) erosion. In cutting erosion, the detachment of crater lips occurs by one or several impacts of the micromachining, plowing or lip formation type. Cutting erosion is, in many respects, similar to abrasive

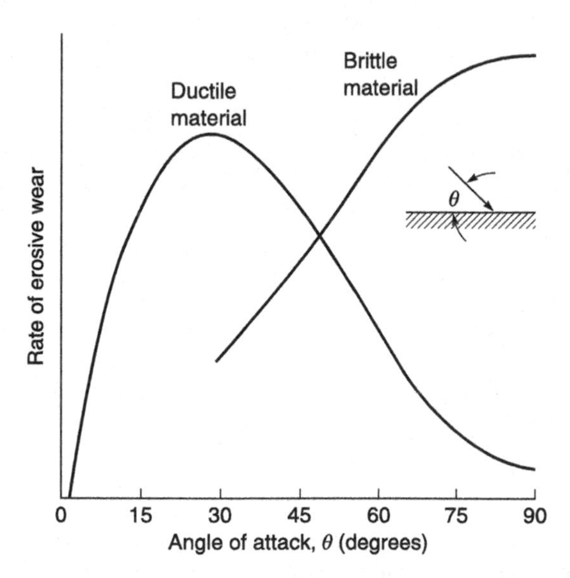

Figure 7.2.34 Rate of erosive wear as a function of angle of attack (with respect to the material plane) of impinging particles.

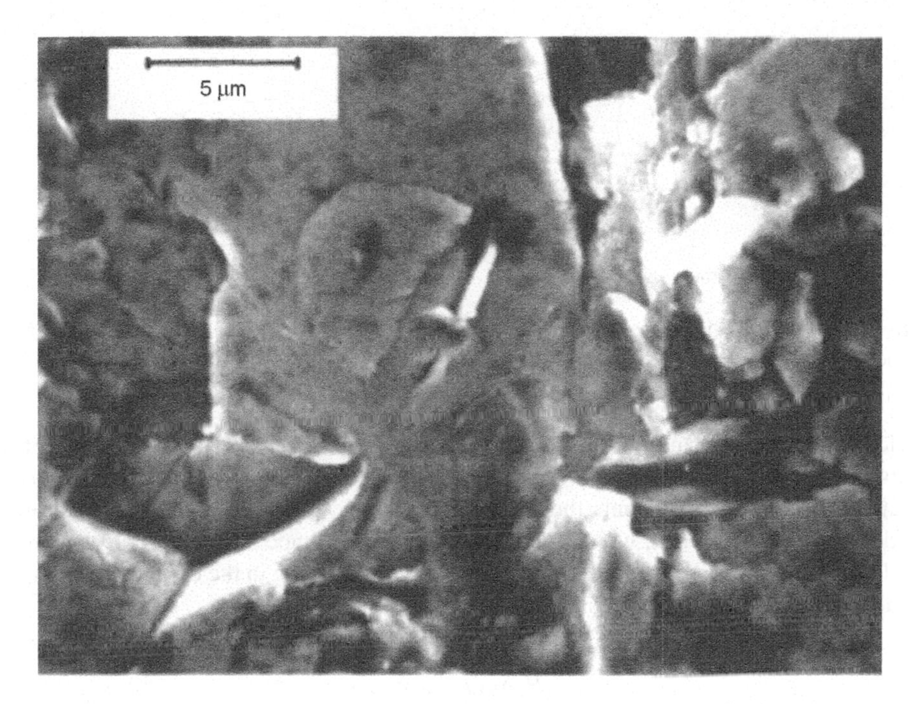

Figure 7.2.35 SEM micrograph of 303 stainless steel surface after solid particle erosion.

wear. In deformation erosion, the detachment of the material occurs by surface fragmentation due to several impacts of the indentation type. The effect of a single-indentation-type crater is to raise small lips of material around the impact type. The effect of successive impacts is to flatten and to strain further the lips, creating thin platelets of highly stressed metal that are finally knocked off the surface by succeeding particles. The relative importance of the two erosion mechanisms at multiple impacts is strongly dependent on the angle of impingement. Cutting erosion and deformation erosion dominate at grazing and normal incidence, respectively (Hutchings and Winter, 1974; Hutchings *et al.*, 1976). Surface hardness and ductility are the most important properties for cutting and deformation erosion resistance, respectively. An SEM micrograph of 303 stainless steel (microhardness of 320 kg/mm^2) eroded by sand blasting is shown in Figure 7.2.35. An SEM of the cross-section of worn samples showed the presence of a very thin layer ($\sim$2 μm) of visible plastic deformation. From Figure 7.2.35, it appears that erosion results primarily from a deformation erosion mechanism.

Solid particle erosion is a problem in machinery such as ingested sand particles in gas turbine blades, helicopter and airplane propellers, the windshields of airplanes, the nozzles for sand blasters, coal turbines, hydraulic turbines and the centrifugal pumps used for coal slurry pipelines. It has useful application in processes such as sand blasting, abrasive deburring, and erosive drilling of hard materials.

Quantitative Equation

We first consider erosion, involving plastic deformation, by a single hard particle striking a softer surface at normal incidence, Figure 7.2.36. Based on Hutchings (1992), assume that the

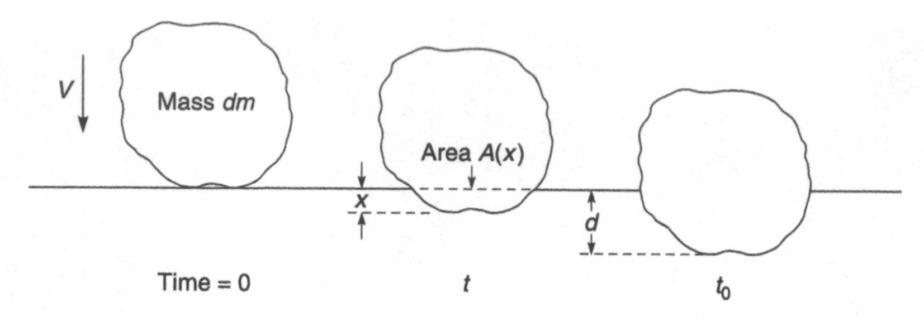

Figure 7.2.36 Schematic of erosion by a single hard particle striking a softer surface at normal incidence.

particle does not deform and the deformation of the surface is perfectly plastic with a constant indentation pressure (hardness), H. At time t after initial contact, the particle, of mass dm with an initial velocity V, indents the surface to a depth x such that cross-sectional area of the indent impression is $A(x)$, which is dependent upon the shape of the particle. The upward force decelerating the particle is due to the plastic pressure acting in the contact area $A(x)$. The equation of motion of the particle is given as

$$-H\,A(x) = dm\frac{d^2x}{dt^2} \tag{7.2.19}$$

If the particle comes to rest at a depth d after time t_0, the work done by the retarding force is equal to the initial kinetic energy of the particle

$$\int_0^d H\,A(x)dx = \frac{1}{2}dmV^2 \tag{7.2.20}$$

or

$$dv = \frac{dm\;V^2}{2H} \tag{7.2.21a}$$

where dv is the volume of material displaced from the indentation. If there are particles of a total mass m, then

$$v = \frac{mV^2}{2H} \tag{7.2.22b}$$

All of the displaced material does not end up as wear debris. If k is the proportion of the displaced material result as wear debris, then

$$v = \frac{kmV^2}{2H} \tag{7.2.23a}$$

Summation of Equation 7.2.23a, over many impacts gives the volume of wear for a period over which erosion takes place. The erosion wear equation is normally written in terms of dimensionless erosion ratio (E), the mass of material removed divided by the mass of erosive particles striking the surface. Equation 7.2.23a can be rewritten as

$$E = \frac{k\rho V^2}{2H} \tag{7.2.23b}$$

where ρ is the density of the material being eroded.

Compared with abrasive wear equations, the volume of erosive wear is inversely proportional to the hardness as in abrasive equations. The normal load in abrasive wear is replaced by mV^2 in erosive wear (Hutchings, 1992). The derivation of Equation 7.2.23 is based on an extremely simple model. It does not include the effect of the impact angle and the shape and size of particles. k depends on the impact angle and the shape and size of particles. The value of k typically ranges from 10^{-5} to 10^{-1}.

The erosive wear rate by brittle fracture depends, in addition, on the fracture toughness of the material being eroded (Hutchings, 1992).

Experimental Evidence

The erosion of pure metals shows strong sensitivity to particle impact velocity, and the hardness of metals. The erosion is related to the particle velocity by $E \propto V^n$, where n ranges between 2.3 and 3. For erosive wear data on copper as a function of impact velocity at two impact angles, see Figure 7.2.37. For pure annealed metals, erosion decreases with an increase in the hardness, where as observed for abrasive wear, dependence for work hardened metals is not linear. A better correlation is found between the erosion resistance and the hardness of a surface after work hardening by erosion.

Liquid Impingement Erosion

When small drops of liquid strike the surface of a solid at high speeds (as low as 300 m/s), very high pressures are experienced, exceeding the yield strength of most materials. Thus, plastic deformation or fracture can result from a single impact, and repeated impact leads to pitting and erosive wear. In many cases, the probable impact velocities and impact angles are such that pure liquid impingement erosion is an unlikely mechanism; an erosion-corrosion mechanism usually does more damage (Preece, 1979). The damage by this process is important in the so-called moisture erosion of low-pressure steam turbine blades operating with wet steam, rain erosion of aircraft or missile surfaces and helicopter rotors, nuclear power plant pipes, and heat exchangers.

Based on Haymann (1992), the high-velocity impact of a liquid drop against a solid surface produces high contact pressure in the impact region followed by liquid jetting flow along the surface, radiating out from the impact area. In ductile materials, a single intense impact may produce a central depression, with a ring of plastic deformation around it where the jetting-out flow may remove the material by a tearing action. In brittle materials, circumferential cracks may form around the impact site caused by tensile stress waves propagating outward along the surface. In subsequent impacts, material can spall off the inside surface due to the compressive stress wave from the impact reflecting there as a tensile wave.

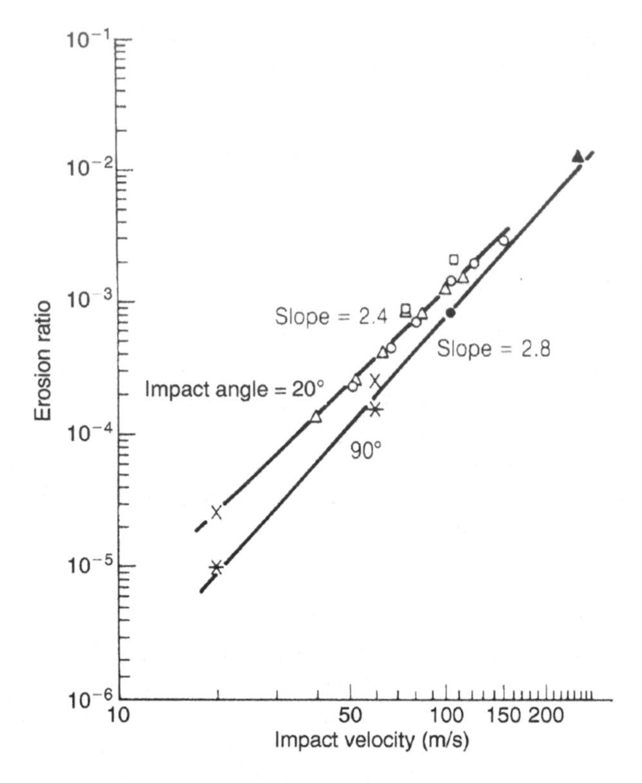

Figure 7.2.37 Erosion ratio of copper as a function of impact velocity for two impact angles. Reproduced with permission from Ives, L.K. and Ruff, A.W. (1979), in *Erosion: Prevention and Useful Applications* (W.F. Adler, ed), pp. 5–35, Special Tech. Pub. ASTM, Philadelphia. Copyright 1979. ASTM International.

Cavitation Erosion

Cavitation is defined as the repeated nucleation, growth, and violent collapse of cavities or bubbles in a liquid. Cavitation erosion arises when a solid and fluid are in relative motion, and bubbles formed in the fluid become unstable and implode against the surface of the solid. When bubbles collapse that are in contact with or very close to a solid surface, they will collapse asymmetrically, forming a microjet of liquid directed toward the solid. The solid material will absorb the impact energy as elastic deformation, plastic deformation or fracture. The latter two processes may cause localized deformation and/or erosion of the solid surface (Preece, 1979). Damage by this process is found in components such as ships' propellers and centrifugal pumps.

All liquids contain gaseous, liquid and solid impurities, which act as nucleation sites for the bubbles or vapor-filled voids. When a liquid is subjected to sufficiently high tensile stresses, bubbles are formed at weak regions within the liquid. Subsequently if this liquid is subjected to compressive stresses, i.e. to higher hydrostatic pressures, these bubbles will collapse. In practice, cavitation can occur in any liquid in which the pressure fluctuates either because of flow patterns or vibration in the system. If, at some location during liquid flow, the local pressure falls below the vapor pressure of the liquid, then cavities may be nucleated, grow to a stable size and be transported downstream with the flow. When they reach the high-pressure region, they become unstable and collapse (Hansson and Hansson, 1992). The stability of a bubble is dependent on the difference in pressure between the inside and outside of the bubble and the surface energy of the bubble. The damage created is a function of the pressures produced and the energy released by collapse of the bubble. Thus, reduction of surface tension of the liquid reduces damage, as does an increase in vapor pressure.

Materials that are resistant to fatigue wear, namely, hard but not brittle materials, are also resistant to cavitation. Resistance to corrosive attack by the liquid, however, is an additional requirement for cavitation resistance.

7.2.4.2 Percussion

Percussion is a repetitive solid body impact, such as experienced by print hammers in high-speed electromechanical applications and high asperities of the surfaces in a gas bearing. In most practical machine applications, the impact is associated with sliding; that is, the relative approach of the contacting surfaces has both normal and tangential components known as compound impact (Engel, 1976). Percussive wear occurs by hybrid wear mechanisms which combines several of the following mechanisms: adhesive, abrasive, surface fatigue, fracture, and tribochemical wear.

To model compound impact, Figure 7.2.38a shows a slug of radius R normally directed at a speed V against a tangentially moving platen with a speed u (Engel, 1976). The slug of mass m may be idealized as a point mass, supported by a tangential spring with a stiffness of k. After a time t_s (called the slipping time), the slug comes up to the horizontal speed u of the platen. They then travel together in the horizontal direction for the rest of the impact duration t_i, after which the two bodies separate. The Hertz impact force during the impact duration is shown in Figure 7.2.38b. To simplify contact analysis, we can assume the Hertz impact force, $F(t)$, to have a sinusoidal distribution,

$$F(t) = F_0 \sin\left(\frac{\pi t}{t_i}\right) \tag{7.2.24}$$

The approximation retains the peak force F_0 while raising the rest of the force-time curve from the bell shaped one, typical of Hertz impact, to sinusoidal. The peak force F_0 depends on geometrical, material, and normal impact parameters. For an elastic contact of a slug on an infinitely massive flat platten with an infinitely high modulus of elasticity (Figure 7.2.38a),

$$F_0\bigg|_{elastic} = \left(\frac{5}{3}mV^2\right)^{3/5}\left[\frac{4E}{3\left(1-\nu^2\right)}R^{1/2}\right]^{2/5} \tag{7.2.25}$$

We now write the equation of tangential motion for an infinitely massive platen,

$$m\dot{x} = \mu\int_0^t F(t)\,dt,\ 0 \le t \le t_s \tag{7.2.26a}$$

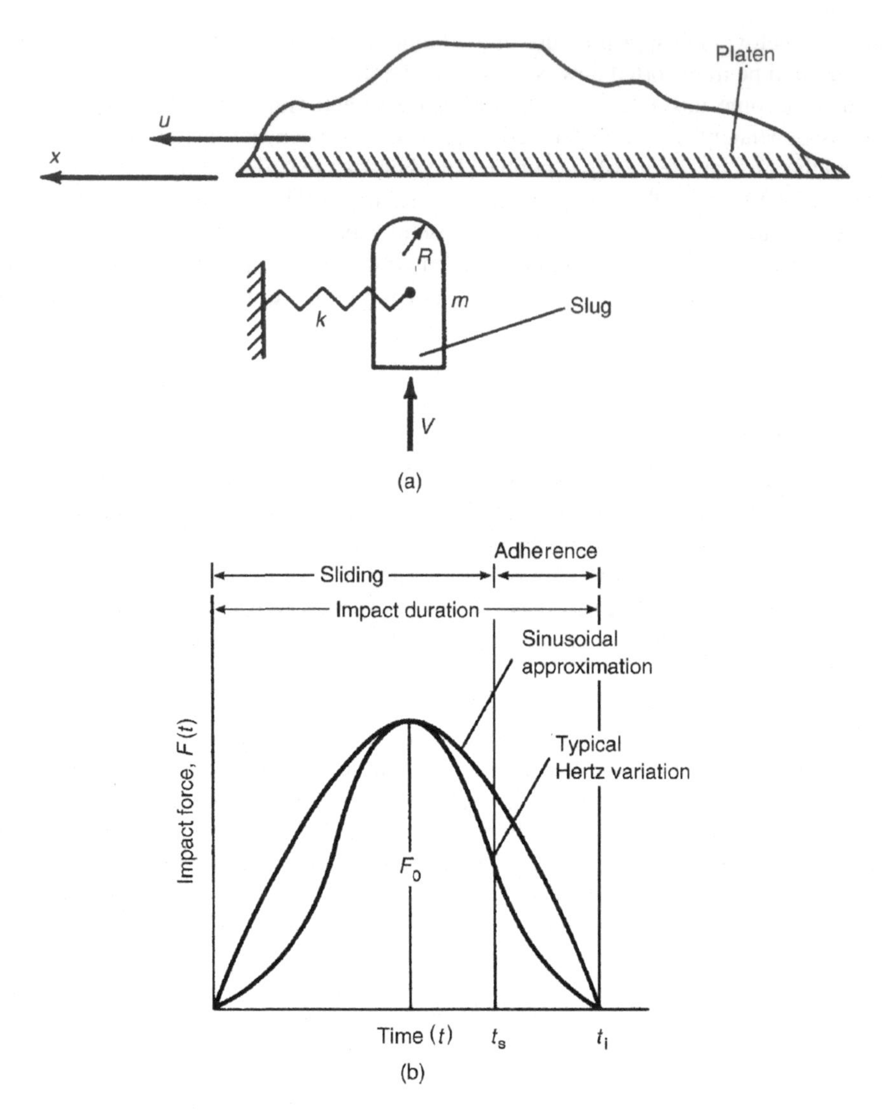

Figure 7.2.38 Schematic of (a) impact of a slug on a tangentially moving platen, and (b) impact force cycle.

where μ is the coefficient of friction. Using Equations 7.2.24 and 7.2.26, we calculate the slipping time, t_s

$$t_s \sim \frac{t_i}{\pi} \cos^{-1}(1 - S) \tag{7.2.27a}$$

where

$$S = \frac{\pi m u}{\mu F_0 t_i} \tag{7.2.27b}$$

S is called the slip factor. If $S = 0$, normal impact occurs. For larger S, slipping persists for a longer time during impact (compound impact). If $S \geq 2$, slipping persists during the entire contact time (t_i).

The impact wear is proportional to the slip factor because wear primarily occurs during the portion of the impact spent in relative sliding. Normal impact on a harder substrate can produce fracture, and repeated impacts can give rise to a subsurface fatigue wear mechanism. An impact with associated sliding (compound impact) gives rise to surface fatigue and/or adhesive/abrasive wear; specific wear mechanisms depend on the geometrical, material, and operative conditions. For materials with high toughness, the contribution due to surface fatigue is negligible.

We apply the impact wear analysis to a print head striking on a paper covering the platen (Engel, 1976). Abrasive wear of the print head occurs during the slipping time. The abraded volume v with respect to the sliding distance x of the print head relative to the paper is

$$\frac{dv(t)}{dx(t)} = \frac{kF(t)}{H} \tag{7.2.28}$$

where k is the abrasive wear coefficient and H is the hardness of the print head.

Therefore, the total wear volume per impact cycle is

$$v = \int_0^{t_s} \frac{k}{H} F(t) dx(t) = \frac{kut_i F_0 S}{2\pi H}, \; 0 \leq S \leq 2 \tag{7.2.29a}$$

$$= \frac{kmV^2}{2\,\mu H} = \frac{2kut_i F_0}{\pi H}\left(1 - \frac{1}{S}\right), \; S \geq 2 \tag{7.2.29b}$$

Once the wear volume per cycle has been determined, the total wear after N cycles can be predicted by multiplying the unit wear by N cycles.

7.2.5 Chemical (Corrosive) Wear

Chemical or corrosive wear occurs when sliding takes place in a corrosive environment. In air, the most dominant corrosive medium is oxygen. Therefore chemical wear in air is generally called oxidative wear. In the absence of sliding, the chemical products of the corrosion (e.g., oxides) would form a film typically less than a micrometer thick on the surfaces, which would tend to slow down or even arrest the corrosion, but the sliding action wears the chemical film away, so that the chemical attack can continue. Thus, chemical wear requires both chemical reaction (corrosion) and rubbing. Machinery operating in an industrial environment or near the coast generally produces chemical products (i.e., it corrodes) more rapidly than when operating in a clean environment. Chemical wear is important in a number of industries, such as mining, mineral processing, chemical processing, and slurry handling.

Corrosion can occur because of the chemical or electrochemical interaction of the interface with the environment. Chemical corrosion occurs in a highly corrosive environment and in high-temperature and high-humidity environments. Electrochemical corrosion is a chemical reaction accompanied by the passage of an electric current, and for this to occur, a potential difference must exist between two regions. The region at low potential is known as an anode

Figure 7.2.39 SEM micrograph of 52100 quenched and tempered roller bearing after corrosive wear. Reproduced with permission from Tallian, T.E., Baile, G.H., Dalal, H., and Gustafsson, O.G. (1974), *Rolling Bearing Damage*, SKF Industries Inc., King of Prussia, Pennsylvania. Copyright 1974. SKF USA Inc.

and the region at high potential is known as a cathode. If there is a current flow between the anode and cathode through an electrolyte (any conductive medium), at the anode the metal dissolves in the form of ions and liberates electrons. The electrons migrate through the metal to the cathode and reduce either ions or oxygen. Thus, electrochemical corrosion is equivalent to a short-connected battery with partial anodic and partial cathodic reactions occurring on the two sliding members (commonly referred to as galvanic corrosion) or in a sliding member on two regions atomic distances away. These regions may shift to different locations (Wagner and Traud, 1938). Electrochemical corrosion is influenced by the relative electropotential. Electrochemical corrosion may accelerate in a corrosive environment because corrosive fluids may provide a conductive medium necessary for electrochemical corrosion to occur on the rubbing surfaces. The most common liquid environments are aqueous, and here small amounts of dissolved gases, commonly oxygen or carbon dioxide, influence corrosion.

A typical example of a corroded roller subsequent to running in a bearing is shown in Figure 7.2.39. The corrosion left a multitude of dark-bottomed pits, the surroundings of which are polished by running. The condition subsequently creates extensive surface-originated spallings from a multitude of initiated points.

7.2.5.1 Tribochemical Wear

Friction modifies the kinetics of chemical reactions of sliding bodies with each other, and with the gaseous or liquid environment, to the extent that reactions which occur at high temperatures occur at moderate, even ambient, temperatures during sliding. Chemistry dealing with this modification of chemical reaction by friction or mechanical energy is referred to as tribochemistry, and the wear controlled by this reaction is referred to as tribochemical wear (Heinicke, 1984; Fischer, 1988). The most obvious mechanism by which friction increases the rate of chemical reaction (tribochemistry) is frictional heat produced at contacting asperities. Besides the friction heat, other mechanisms are: the removal of product scale resulting in fresh surfaces; accelerated diffusion and direct mechanochemical excitation of surface bonds. The tribochemical reactions result in oxidative wear of metals, the tribochemical wear of ceramics, formation of friction polymer films on surface sliding in the presence of organics, and the dissolution of silicon nitride in water during sliding without fracture. Oxidative wear of metals and tribochemical wear of ceramics will be described later in Section 7.4.

Applications of tribochemistry include formation of friction polymer films for low friction and wear and tribochemical polishing without fracture (also called chemomechanical polishing or CMP).

The formation of organic films on sliding surfaces occurs as a result of the repolymerization of organics (in the form of solid, liquid, or vapors) in the sliding contact due to a local increase in surface temperature in the regions of greatest contact and possibly due to the added catalytic action of certain freshly exposed surfaces. These films are known as friction polymers or tribopolymers (Chaikin, 1967; Furey, 1973; Lauer and Jones, 1986; Bhushan and Hahn, 1995). The chemical structure of the friction polymers must be different from the organic source from which they are formed. In many cases, these films reduce friction and wear of sliding surfaces operating in boundary or hydrodynamic lubrication regimes. These protective polymeric films may be intentionally produced by adding additives to the lubricating oil. The process of forming friction polymers and degrading them is a dynamic one, that is, initially formed films will degrade and must be replenished. Degraded friction polymers can produce high friction and wear. Also the thickness of the friction polymer films may become so large that they begin to delaminate and generate wear debris. A second form of tribochemistry is the tribochemical polishing in ceramics in liquids. Friction in water enhances the tribochemical reaction. For example, the tribochemical dissolution of silicon nitride in water occurs at contacting asperities that are removed, and it results in extremely smooth surfaces (Fischer, 1988). If the material removal is purely tribochemical, no microfracture or plastic deformation takes place. Therefore, the polished surface is free of microcracks and other defects.

7.2.6 *Electrical-Arc-Induced Wear*

When a high potential is present over a thin air film in a sliding process, a dielectric breakdown results that leads to arcing. During arcing, a relatively high-power density (on the order of 1 kW/mm^2) occurs over a very short period of time (on the order of 100 μs). The heat-affected zone is usually very shallow (on the order of 50 μm). Heating is caused by the Joule effect due to the high power density and by ion bombardment from the plasma above the surface. This heating results in considerable melting and subsequent resolidification, corrosion, hardness changes, and other phase changes, and even in the direct ablation of material (Guile and Juttner, 1980; Bhushan and Davis, 1983). Arcing causes large craters, and any sliding or oscillation after an arc either shears or fractures the lips, leading to three-body abrasion, corrosion, surface fatigue, and fretting (Figure 7.2.40). Arcing can thus initiate several modes of wear resulting in catastrophic failures in electrical machinery (Bhushan and Davis, 1983).

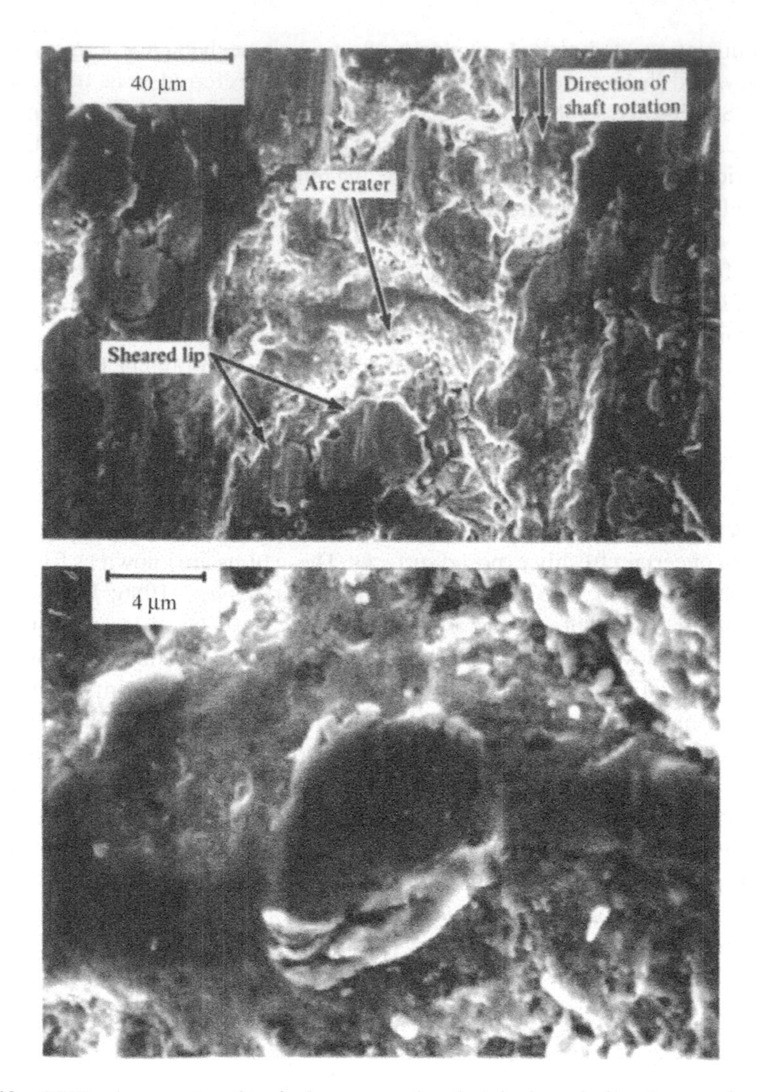

Figure 7.2.40 SEM micrographs of typical worn area by electrical-arc-induced wear of a 303 stainless steel surface. Arc craters and sheared lips can be seen.

When contacts are rubbing, as in the case of a copper commutator or copper slip rings against graphite-based brushes (Johnson and Moberly, 1978), the sparking damage on the copper commutator or slip ring can cause excessive wear of the brush by abrasion. In certain applications of bearings in electrical machinery, there is the possibility that an electric current will pass through a bearing. When the current is broken at the contact surfaces between rolling elements and raceways, inner-race and shaft, or outer-race and housing, arcing results. Both surfaces should be in the path of least resistance to a potential difference. Electrical-arc-induced wear has been productively used as a method of metal removal in electrodischarge machining.

Methods to minimize electrical-arc-induced wear are as follows: (1) to eliminate the gap between the two surfaces with a potential difference; (2) to provide an insulator of adequate dielectric strength (e.g., an elastomer or Al_2O_3 coating) between the two surfaces; (3) to provide a low impedance connection between the two surfaces to eliminate the potential difference; or (4) to have one of the surfaces not ground. Bearing manufacturers recommend that bearings should be press-fitted to the shaft and conducting grease should be used to eliminate arcing, for example in the case of rolling-element bearings, shaft and the inner race, the inner and outer races, and rolling elements.

7.2.7 Fretting and Fretting Corrosion

Fretting occurs where low-amplitude oscillatory motion in the tangential direction (ranging from a few tens of nanometers to few tens of microns) takes place between contacting surfaces, which are nominally at rest (Anonymous, 1955; Hurricks, 1970; Waterhouse, 1981, 1992). This is a common occurrence, since most machinery is subjected to vibration, both in transit and in operation. Examples of vulnerable components are shrink fits, bolted parts, and splines. The contacts between hubs, shrink- and press-fits, and bearing housings on loaded rotating shafts or axles are particularly prone to fretting damage. Flexible couplings and splines, particularly where they form a connection between two shafts and are designed to accommodate some misalignment, can suffer fretting wear.

Basically, fretting is a form of adhesive or abrasive wear, where the normal load causes adhesion between asperities and oscillatory movement causes ruptures, resulting in wear debris. Most commonly, fretting is combined with corrosion, in which case the wear mode is known as fretting corrosion. For example, in the case of steel particles, the freshly worn nascent surfaces oxidize (corrode) to Fe_2O_3, and the characteristic fine reddish-brown powder is produced, known as cocoa. These oxide particles are abrasive. Because of the close fit of the surfaces and the oscillatory small amplitude motion (on the order of a few tens of microns), the surfaces are never brought out of contact, and therefore, there is little opportunity for the products of the action to escape. Further oscillatory motion causes abrasive wear and oxidation, and so on. Therefore the amount of wear per unit sliding distance due to fretting may be larger than that from adhesive and abrasive wear. The oscillatory movement is usually the result of external vibration, but in many cases it is the consequence of one of the members of the contact being subjected to a cyclic stress (i.e., fatigue), which results in early initiation of fatigue cracks and results in a usually a more damaging aspect of fretting, known as fretting fatigue.

Surfaces subjected to fretting wear have a characteristic appearance with red-brown patches on ferrous metals and adjacent areas that are highly polished because of the lapping quality of the hard iron-oxide debris. Figure 7.2.41 shows the SEM micrograph of the 303 stainless steel shaft after it underwent fretting corrosion.

A rapid increase in wear rate occurs with slip amplitude over an amplitude range, Figure 7.2.42. For a given slip amplitude, the amount of wear per unit of sliding distance per unit of applied normal load linearly increases with the number of oscillating cycles up to an amplitude of about 100 μm. Above this amplitude, the wear rate per unit sliding distance becomes constant, identical with unidirectional or reciprocating sliding wear rates. This then gives a possible upper limit for the slip amplitude for the case of true fretting. At small amplitudes, characteristic of fretting, the relative velocities are much lower, even at high frequencies,

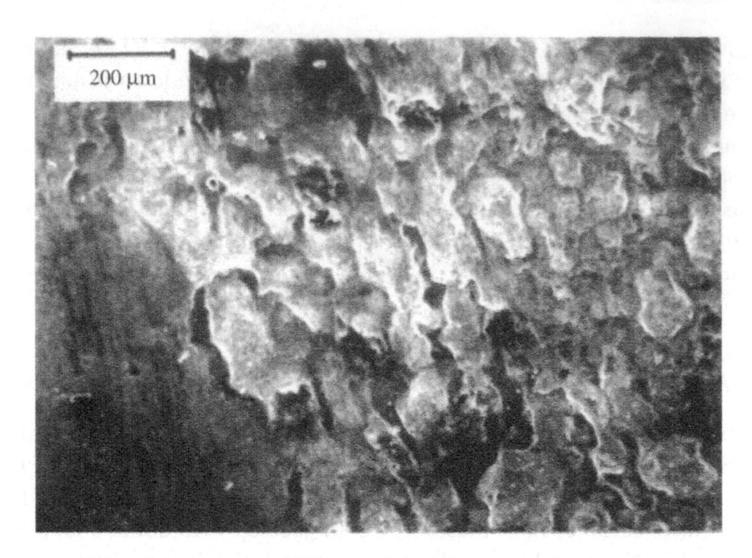

Figure 7.2.41 SEM micrographs of 303 stainless steel shaft surface after fretting corrosion.

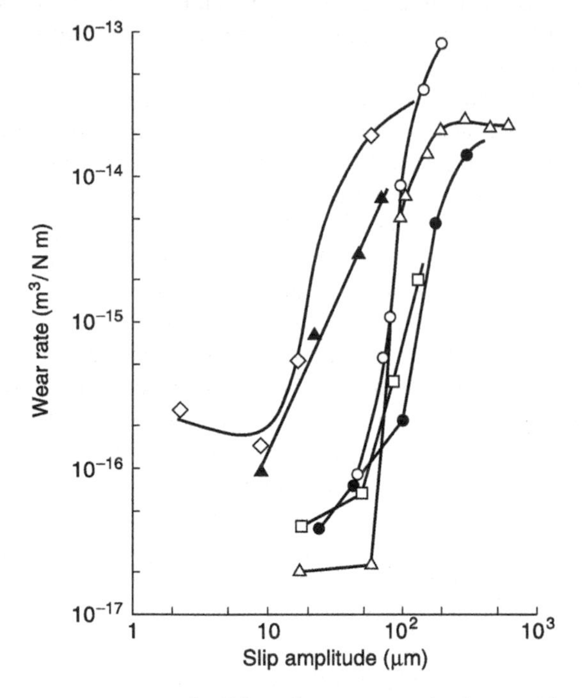

Figure 7.2.42 Volume wear rate per unit sliding distance per unit of normal load as a function of slip amplitude for mild steel against itself. Each curve is the result of a separate investigation. Reproduced with permission from Waterhouse, R.B. (1992), "Fretting Wear," in *ASM Handbook, Vol. 18: Friction, Lubrication and Wear Technology*, pp. 242–256, ASM International, Metals Park, Ohio. Copyright 1992. ASM International.

compared with conditions in typical unidirectional sliding. The fretting wear rate is directly proportional to the normal load for a given slip amplitude. In a partial slip situation, the frequency of oscillation has little effect on the wear rate per unit distance in the low-frequency range, whereas the increase in the strain rate at high frequencies leads to increased fatigue damage and increased corrosion due to rise in temperature. However, in the total-slip situation, there is little effect of the frequency (Waterhouse, 1992).

There are various design changes which can be carried out to minimize fretting wear. The machinery should be designed to reduce oscillatory movement, reduce stresses or eliminate two-piece design altogether.

7.3 Types of Particles Present in Wear Debris

The size and shape of debris may change during sliding in dry and lubricated systems; therefore, the condition of a system can be monitored by debris sampling and maintenance can be scheduled, known as condition-based maintenance. Mild wear is characterized by finely divided wear debris (typically 0.01–1 μm in particle size). The worn surface is relatively smooth. Severe wear, in contrast, results in much larger particles, typically on the order of 20–200 μm in size, which may be visible even with the naked eye; the worn surface is rough. Particles are collected during sliding for analysis. Particles are collected from dry interfaces by sucking air from the sliding interface on a filter paper. Particles in a sample of lubricant from an oil-lubricated system are recovered by filtration, centrifuging, or magnetically (for magnetic particles). The technique commonly used for ferrous metals, known as ferrography, uses a magnetic field to sort particles by the size and shape of the magnetic particles (Scott and Westcott, 1977). The size, shape, structural, and chemical details of particles are analyzed using various techniques including optical microscopy, scanning electron microscopy (SEM), transmission and scanning transmission electron microscopy (TEM/STEM), energy dispersive and wavelength dispersive spectroscopy (EDS and WDS), Auger electron spectroscopy (AES), X-ray photoelectron spectroscopy (XPS), X-ray and electron diffraction. Size analysis of airborne particles is also carried out using particle counters, generally based on a light-scattering method.

Particles can be classified based on the wear mechanism or their morphology. Particles collected from a wear test may not be in the same state in which these were first produced because of changes in subsequent sliding. Since it is difficult to identify the exact possible wear mechanism, particles are generally classified based on their morphology, and their description follows (Scott, 1975; Scott and Westcott, 1977; Ruff *et al.*, 1981; Samuels *et al.*, 1981; Hokkirigawa and Kato, 1988; Rigney, 1992; Glaeser, 2001).

7.3.1 Plate-Shaped Particles

Thin, plate-shaped or flake-type wear particles with an aspect ratio of 2–10, are commonly found in wear debris from dry and lubricated interfaces. These particles are produced as a result of plowing followed by repeated loading and unloading fatigue, as a result of nucleation and propagation of subsurface cracks or plastic shear in the asperity contacts, Figures 7.3.1 and 7.3.2. Note that particles produced in rolling-contact fatigue are large and their formation develops macroscopic pits.

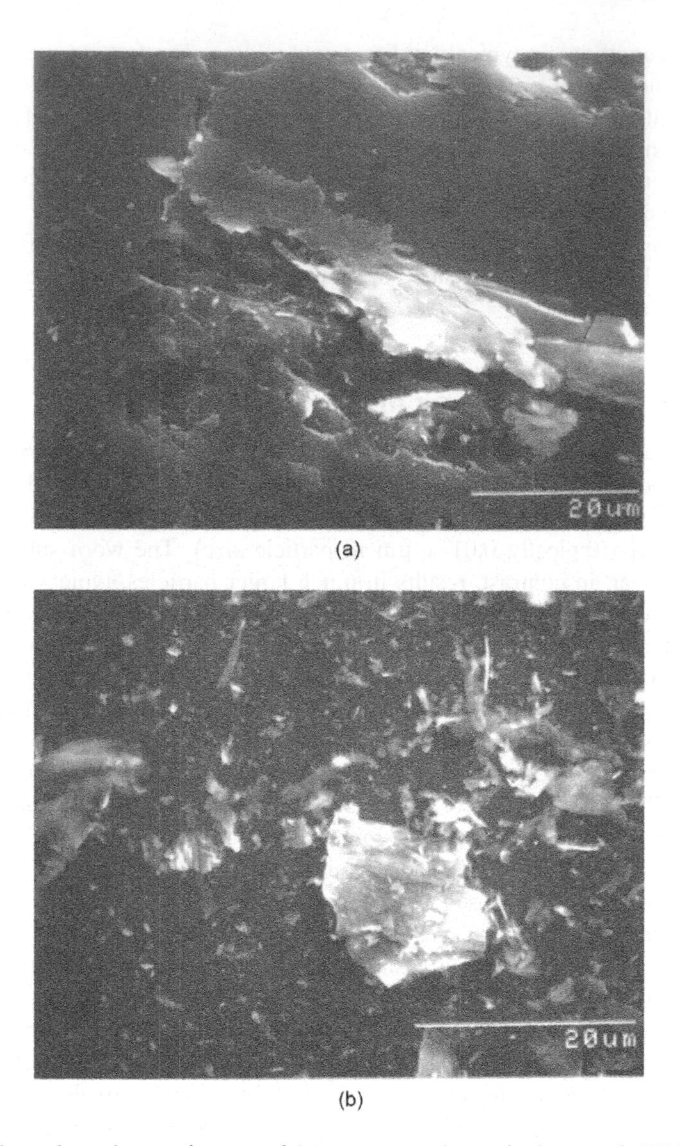

Figure 7.3.1 Secondary electron images of (a) a wear track on single-crystal Si(111) after sliding against a diamond pin at a normal load of 0.5 N and a sliding velocity of 25 mm/s in vacuum, and (b) a flake-type debris particle. Reproduced with permission from Rigney, D.A. (1992), "The Role of Characterization in Understanding Debris Generation" in *Wear Particles* (D. Dowson, C.M. Taylor, T.H.C. Childs, M. Godet, and G. Dalmaz, eds), pp.405–412, Elsevier Science Publishers, Amsterdam. Copyright 1992. Elsevier.

7.3.2 Ribbon-Shaped Particles

Ribbon-shaped or cutting-type particles are frequently found with aspect ratios, on the order of ten or more; and usually are curved and even curly. These are produced as a result of plastic deformation. They have all the characteristics of machining chips: as a result, the ribbon-shaped

Figure 7.3.2 SEM micrograph of flake-type debris particle generated by sliding a Cu-Be block on a M2 tool steel ring at a normal load of 133 N, a sliding velocity of 50 mm/s and a sliding distance of 360 m in dry argon. Reproduced with permission from Rigney, D.A. (1988), "Sliding Wear of Metals," *Ann. Rev. Mater. Sci.* **18**, 141–163. Copyright 1988. Annual Reviews.

particles are referred to as microcutting chips or cutting chips. They are generally produced during run-in, as a result of detachment of fin-like ridges generally present at the edges of the abrasion grooves in machined (e.g., ground) surfaces. These particles are produced with sharp asperities or abrasive particles digging into the mating surface with material flowing up the front face of the asperity or abrasive particles and being detached from the wearing surface in the form of a chip, Figure 7.3.3. Typically, changes in chemical composition are small.

7.3.3 Spherical Particles

Spherical particles are not common. Wear particles of various shapes may not escape from the interface to become loose debris. Some of them remain trapped and are processed further as in the spherical shape. Spherical particles have been observed in sliding (Rigney, 1992), fretting and rolling contact fatigue (Smith, 1980; Samuels *et al.*, 1981). Spherical particles with 1–5 μm in diameter are reported to be associated with rolling-contact fatigue just prior to fatigue failure, Figure 7.3.4.

7.3.4 Irregularly Shaped Particles

The majority of particles have an irregular morphology. Wear debris produced by detachment of the transferred fragment in adhesive wear and brittle fracture are irregularly shaped, Figure 7.3.5.

Figure 7.3.3 SEM micrograph of ribbon-shaped or cutting-type debris particle generated by sliding a 304 stainless steel block on an M2 tool steel ring at a normal load of 67 N and sliding velocity of 50 mm/s in dry argon. Reproduced with permission from Rigney, D.A. (1992), "The Role of Characterization in Understanding Debris Generation" in *Wear Particles* (D. Dowson, C.M. Taylor, T.H.C. Childs, M. Godet, and G. Dalmaz, eds), pp.405–412, Elsevier Science Publishers, Amsterdam. Copyright 1992. Elsevier.

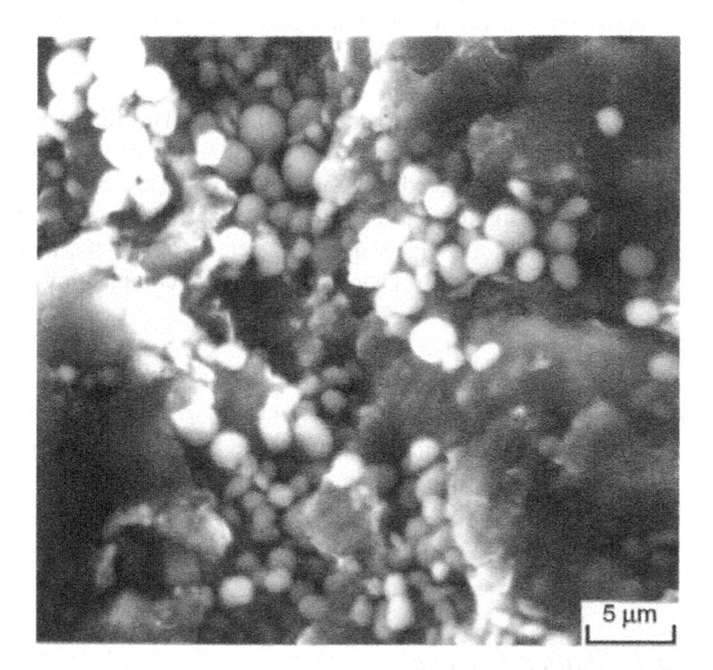

Figure 7.3.4 SEM micrograph of spherical particles present on the surface of a crack produced as a result of rolling contact fatigue of steel surfaces. Reproduced with permission from Smith, R.A. (1980), "Interfaces of Wear and Fatigue," in *Fundamentals of Tribology* (N.P. Suh and N. Saka, eds), Figure 5, © 1980 Massachusetts Institute of Technology, by permission of The MIT Press.

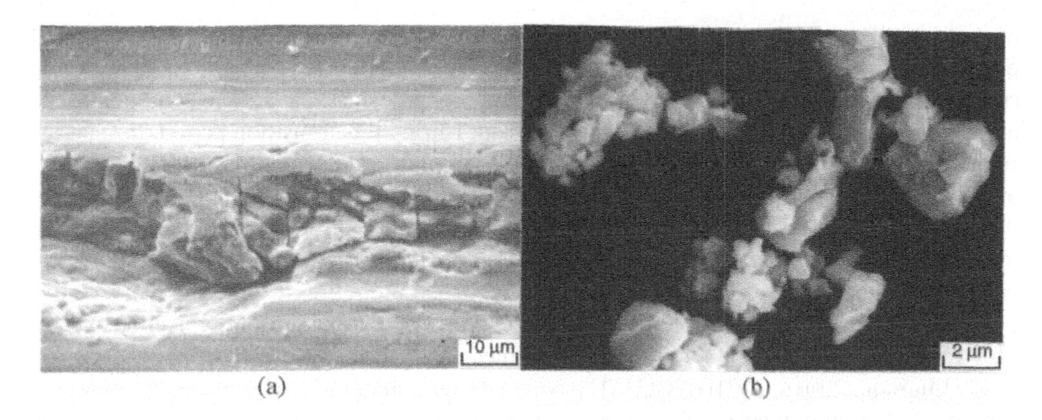

Figure 7.3.5 SEM micrographs of (a) a wear surface of an austenitic steel showing irregular fragmentation of the surface, and (b) typical debris particles produced from multiple fracturing in a brittle layer produced on the surface during sliding. Reproduced with permission from Samuels, L.E., Doyle, E.D., and Turley, D.M. (1981), "Sliding Wear Mechanisms," in *Fundamentals of Friction and Wear of Materials* (D.A. Rigney, ed), pp. 13–41, Amer. Soc. Metals, Metals Park, Ohio. Copyright 1981. ASM International.

7.4 Wear of Materials

Wear process is generally quantified by wear rate. Wear rate is defined as the volume or mass of material removed per unit time or per unit sliding distance. Other forms could be dimensionless, such as the depth of material per unit sliding distance, or the volume removed per apparent area of contact and per unit sliding distance. Wear rate is generally not constant. In general, wear rate is a complex function of time. Wear rate may start low and later rise, or vice versa, Figure 7.4.1. After a certain duration, the wear rate remains constant for a period and

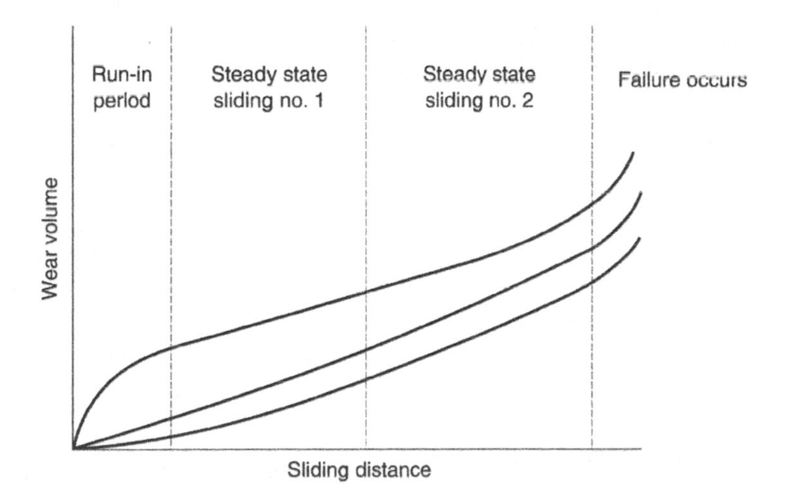

Figure 7.4.1 Three hypothetical cases of wear volume as a function of sliding distance showing run-in, steady-state and failure regions.

may change if transition from one mechanism to another occurs during a wear test. The initial period during which wear rate changes is known as the run-in or break-in period. Wear during run-in depends on the initial material structure and properties and on surface conditions such as surface finish and the nature of any films present. During this transition period, the surface roughness is modified to a steady-state condition by plastic deformation. Initial conditions affect the damage during the transition period and its duration.

The wear rate, like friction, of a material is dependent upon the counterface or mating material (or material pair), surface preparation, and operating conditions. The usefulness of wear coefficients or wear data presented in the published literature lies more in their relative magnitudes than in their absolute values. A number of handbooks present typical wear rates of a variety of material pairs (Peterson and Winer, 1980; Blau, 1992; Bhushan and Gupta, 1997, Bhushan, 2001a, 2001b, 2011). The wear rate for a material pair is normally presented in terms of the nondimensional wear coefficient (k). Wear rate is also presented in terms of a wear factor defined as the wear volume per unit applied normal load and per unit sliding distance (mm^3/Nm). Typical ranges of the coefficient of friction and the wear coefficients of metals, alloys, ceramics, polymeric and solid lubricant pairs are presented in Table 7.4.1. Values of selected pairs are presented in Table 7.4.2.

A self-mated steel pair exhibits high friction and wear. Pairss of dissimilar metals exhibit moderate friction and wear. These are generally used in lubricated applications. Ceramic versus metal or versus another ceramic or versus itself exhibits moderate friction but extremely low wear. Self-mated ceramic pairs as opposed to self-mated metal pairs are desirable as they are

Table 7.4.1 Typical ranges of friction and wear coefficients of various material pairs.

Material combination	Typical range of coefficient of friction	Typical range of wear coefficient	Comments
Self-mated metals pairs	0.5– > 1 (high)	5×10^{-3} (high)	Undersirable in dry contacts
Dissimilar metals pairs Alloy-alloy pairs	0.3–0.9 (moderate) 0.2–0.6 (moderate)	10^{-4}–10^{-3} (moderate) 10^{-6}–10^{-3} (moderate)	Easy to fabricate, low cost
Ceramic-metal pairs Ceramic-ceramic pairs	0.25–0.8 (moderate) 0.25–0.7 (moderate)	10^{-7}–10^{-4} (very low) 10^{-7}–10^{-5} (very low)	For high temperature applications, for lowest wear requirements, self-mated ceramic pairs desirable, good for unlubricated conditions
Polymer-hard surface pairs	0.05*–0.6 (low)	10^{-6}–10^{-3} (low)	For corrosive environment and low friction applications, good for low loads, good for unlubricated conditions
Solid lubricant-hard surface pairs	0.05–0.15 (very low)	10^{-4}–10^{-3} (low)	For lowest friction requirement

*PTFE

Table 7.4.2 Coefficient of friction and wear coefficients of softer material for various material pairs in the unlubricated sliding at a normal load of 3.9 N and sliding speed of 1.8 m/s (Archard, 1980). The stated value of the hardness is that of the softer (wearing) material in each example.

Materials		Vickers microhardness	Coefficient of	
Wearing surface	Counter surface	(kg/mm^2)	friction	Wear coefficient (k)
60/40 leaded brass	Tool steel	95	0.24	6×10^{-4}
Mild steel	Mild steel	186	0.62	7×10^{-3}
Ferritic stainless steel	Tool steel	250	0.53	1.7×10^{-5}
Stellite	Tool steel	690	0.60	5.5×10^{-5}
Tungsten carbide	Tungsten carbide	1300	0.35	1×10^{-6}
PTFE	Tool steel	5	0.18	2.4×10^{-5}

not abusive to the mating surface. Since they exhibit very low wear, these are used in both unlubricated and lubricated conditions. Polymers and solid lubricants against hard surfaces exhibit very low friction but not very low wear.

Metal pairs are most commonly used because of the ease of machinability and low cost. Ceramics are used because they are somewhat inert, strong, and can be used at high temperatures. Polymers are inexpensive and ideal in corrosive environments. In the case of polymeric materials and solid lubricants, their mechanical properties degrade and in some cases oxidize at temperatures somewhat higher than ambient, making them unusable at elevated temperatures. The temperature rise occurs as a result of friction heating, which is a function of a product of pressure and velocity known as PV limit. These materials are classified based on the PV limit. Since polymeric materials and solid lubricants exhibit a low coefficient of friction and wear whether self-mated or sliding against other materials, they are commonly used in unlubricated applications. These are commonly used against harder mating materials.

In journal-bearing applications with soft liners, embeddability and conformability are important considerations. If particles are longer than the thinnest region of the oil film, the particle may reside at the interface and may result in significant abrasive wear. One of the ways to minimize the damage is to select a hard journal shaft and a soft bearing alloy or a polymer such that particles are embedded in the bearing material. The ability to embed the abrasives in this way is referred to as "embeddability." Further in the applications, the use of bearings with significant misalignment, resulting in high loads, can lead to significant damage. Side loads from the misalignment can be accommodated without severe damage by the use of soft bearing alloys or polymers. The soft materials can deform plastically and accommodate any misalignment.

7.4.1 Wear of Metals and Alloys

As indicated in Chapters 4 and 5, the clean metals and alloys in a solid contact exhibit high adhesion consequently high friction and wear. The wear rate of contacting metallic surfaces cleaned in a high vacuum can be very high. The slightest contamination mitigates contact or forms chemical films which reduce adhesion, resulting in the reduction of friction and wear (Buckley, 1981; Rigney, 1988; Bhushan, 1996). In the case of soft metals, such as In, Pb, and

Table 7.4.3 Wear coefficient of softer material for various metal-metal pairs at a normal load of 20 N and a sliding velocity of 1.8 m/s (*Source*: Archard, 1953). The stated value of hardness is that of the softer wearing material in each example.

Metal pair	Vickers hardness (kg/mm^2)	Wear coefficient, k ($\times 10^{-4}$)
Cadmium on cadmium	20	57
Zinc on zinc	38	530
Silver on silver	43	40
Copper on copper	95	110
Platinum on platinum	138	130
Mild steel on mild steel	158	150
Stainless steel on stainless steel	217	70
Cadmium on mild steel	20	0.3
Copper on mild steel	95	5
Platinum on mild steel	138	5
Mild steel on copper	95	1.7
Platinum on silver	43	0.3

Sn, the contact area is large, even at low loads, which results in high wear rates. Hexagonal metals such as Co and Mg as well as other non-hexagonal metals such as Mo and Cr exhibit low friction and wear; consequently Co, Mo, and Cr are common alloying elements in steels to reduce friction, wear and corrosion. Metallurgical compatibility determines the wear rates of a given metal pair. Lead-based white metals (babbitts), brass, bronze, and cast iron generally exhibit relatively low friction and wear in dry and lubricated conditions. In general, wear for alloys tends to be lower than that for pure components. Typical values of wear coefficients for various similar and dissimilar metals are presented in Table 7.4.3.

Steels form the most commonly used family of materials for structural and tribological applications. Based on chemical composition (percentage of alloying components and carbon) and processing, a variety of microstructures and physical properties of steel can be obtained. The wear resistance of different microstructures are summarized in Figure 7.4.2 (Moore, 1981; Zum Gahr, 1987; Glaeser, 1992).

In metal-to-metal wear tests, high stresses can result in catastrophic galling and eventual seizure even after a single cycle, so a committee of stainless steel producers of AISI devised a button and block galling test. In this test, a small button specimen and a large block specimen are machined and polished to provide parallel contacting surfaces. The specimens are dead-weight loaded in a Brinell hardness tester, and the button is rotated 360° against the block. Specimens are then examined for galling at 10x magnification, with new specimens being tested at progressively higher stress levels until galling just begins. This point is called the unlubricated threshold galling stress. Galling usually appears as a groove, or score mark, terminating in a mound of metal (Schumacher, 1977; Anonymous, 1978; Foroulis, 1984). Galling stress is a good measure of wear resistance of a given material pair.

Galling data show that identical metal couples usually do poorly in terms of galling compared with dissimilar metal couples. When stainless steels are coupled with each other, with the exception of some Nitronic steels, they exhibit worse galling resistance than all other steels by a factor of 2 or more.

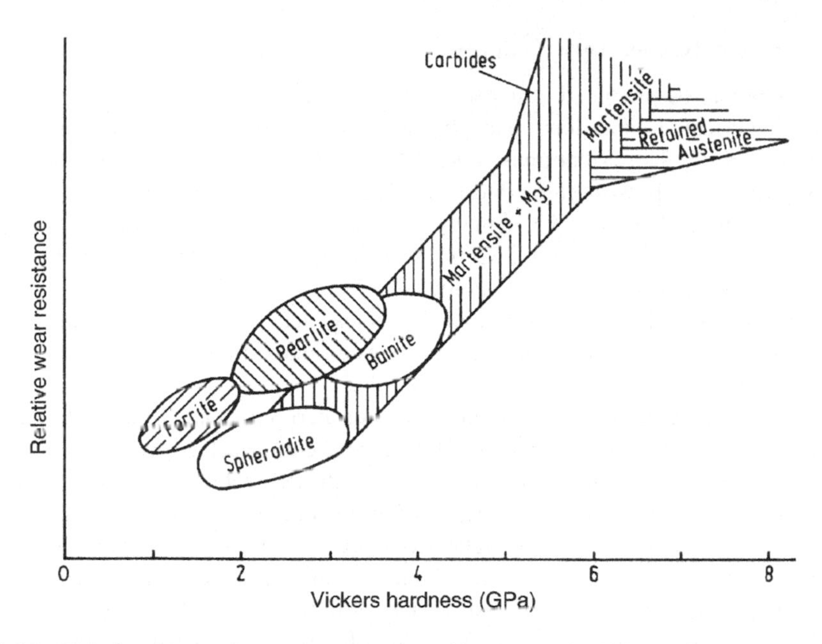

Figure 7.4.2 Relative wear resistance as a function of hardness of different microstructures of steels. Reproduced with permission from Zum Gahr, K.H. (1987), *Microstructure and Wear of Materials*, Elsevier, Amsterdam. Copyright 1987. Elsevier.

Cobalt-based alloys such as T-400 and Stellite 6B have, in general, good galling resistance. However, several nickel-based alloys exhibit a very low threshold galling stress when self-mated or coupled with other similar alloys (Bhansali, 1980). A nickel-based alloy such as Waukesha 88 can be modified specifically for galling resistance. Waukesha 88 exhibits extremely high galling resistance in combination with several stainless steels. It should also be noted that a high nickel content in steels has a detrimental effect on galling resistance. When compared with steels, with the exception of 316 stainless steel, the remaining steels exhibit moderate threshold galling stress when coupled with the cobalt-based alloy Stellite 6B. Type 316 steel probably exhibits a low galling stress because it has a higher nickel content than type 304 stainless steel.

Wear data (Anonymous, 1978; Bhansali, 1980; Foroulis, 1984) show that among the various steels tested, types 201, 301, and hardened 440C and the proprietary Nitronic austenitic grades provide good wear resistance when mated to themselves under unlubricated conditions. High-nickel alloys generally are rated as intermediate between the austenitic and martensitic stainless steels. Cobalt-based alloys also do well. Considerable improvement in wear resistance can be achieved when dissimilar metals are coupled, and this is especially true for steels coupled with silicon bronze and Stellite alloys. The wear data further suggest that improvement in wear resistance can be achieved by altering the surface characteristics, such as by surface treatment or by adding a coating.

Operating conditions – normal load, sliding velocity and environment – have a significant effect on the wear modes as well as wear rates. Their discussion follows.

7.4.1.1 Effect of Temperature (Oxidative Wear)

Interface temperatures produced at asperity contacts during sliding of metallic pairs under nominally unlubricated conditions result in thermal oxidation which produces oxide films several microns thick. The oxidation is generally a beneficial form of corrosion. A thick oxide film reduces the shear strength of the interface which suppresses the wear as a result of plastic deformation (Quinn, 1983a, 1983b). In many cases, tribological oxidation can reduce the wear rate of metallic pairs by as much as two orders of magnitude, as compared with that of the same pair under an inert atmosphere. Tribological oxidation can also occur under conditions of boundary lubrication when the oil film thickness is less than the combined surface roughness of the interface. The oxidation can prevent severe wear. In oxidative wear, debris is generated from the oxide film.

At low ambient temperatures, oxidation occurs at asperity contacts from frictional heating. At higher ambient temperatures, general oxidation of the entire surface occurs and affects wear. In the case of steels, the predominant oxide present in the debris depends on the sliding conditions. At low speeds and ambient temperatures, the predominant oxide is $\alpha - Fe_2O_3$, at intermediate conditions it is Fe_3O_4, and at high speeds and temperatures it is FeO (Quinn, 1983b).

Oxygen and other molecules are adsorbed on clean metals and ceramic surfaces, and form strong chemical bonds with them. The slow step inhibiting the continuation of this reaction is the diffusion of the reacting species through the film of reaction product. Oxidation of iron and many metals follows a parabolic law, with the oxide film thickness increasing with the square root of time,

$$h = Ct^{1/2} \tag{7.4.1}$$

where h is the oxide film thickness, t is the average growth time, and C is the parabolic rate constant at elevated temperatures.

Since diffusion is thermally activated, growth rate in oxide film thickness during sliding as a function of temperature, similar to thermal oxidation under static conditions, follows an Arrhenius type of relationship

$$K = A\exp(-Q/RT) \tag{7.4.2}$$

where K is the parabolic rate constant for the growth of the oxide film, A is the parabolic Arrhenius constant (kg^2/(m^4 s)) for the reaction, Q is the parabolic activation energy associated with oxide (kJ/mole), R is the universal gas constant and T is the absolute temperature of the surface. It has been reported that the Arrhenius constant for sliding is several orders of magnitude larger than that for static conditions, which means that oxidation under sliding conditions is much more rapid than that in the static oxidation condition. Increased oxidation during sliding may result from increased diffusion rates of ions through a growing oxide film which generally consists of high defect density due to mechanical perturbations.

7.4.1.2 Effect of Operating Conditions (Wear-Regime Maps)

The wear-regime maps elucidate the role of operating environment on wear mechanisms. No single wear mechanism operates over a wide range of conditions. There are several wear mechanisms which change in relative importance as the operating conditions are changed. The transitions in dominant wear mechanisms, consequently wear rates, occur as sliding loads and velocities are changed. In some cases, changes also occur as a function of sliding time (or distance). The dominant wear mechanisms are based on mechanical strength and interfacial adhesion. Increase in normal load results in an increase in mechanical damage due to high surface stresses. Increase in both normal load and sliding velocity results in a monotonic increase in interface temperature. High interface temperature results in the formation of chemical films, mostly the formation of oxide films in air. High temperatures result in a decrease in mechanical strength and in some cases in structural changes. At high load-velocity (PV) conditions, there may even be localized melting near the surface.

Various regimes of mechanical (plastically dominated) and chemical (oxidational) wear for a particular sliding material pair are observed on a single wear-regime map (or wear-mode map or wear-mechanism map) plotted on axes of normalized pressure and normalized sliding velocity on the macroscale (Lim and Ashby, 1987; Lim *et al.*, 1987) and nanoscale (Tambe and Bhushan, 2008). Here we present maps for macroscale and maps for nanoscale will be presented later in Chapter 10 on nanotribology. Normalized pressure is the nominal pressure divided by the surface hardness (p/H) and normalized velocity is the sliding velocity divided by the velocity of heat flow (given by the radius of the circular nominal contact area divided by the thermal diffusivity). As an example, a wear regime map for steel sliding on steel in air at room temperature in the pin-on-disk configuration is shown in Figure 7.4.3. The general form of the map would be similar for the sliding of most unlubricated metals in air. It can be seen that, in principle, the map can be divided into areas corresponding to different wear regimes, with boundaries of sliding velocities and contact pressure beyond which oxidative wear would be dominant, as compared to mechanical wear at low speeds. Prevailing wear mechanisms can give mild or severe wear. Mild wear gives a smooth surface and severe wear produces a surface that is rough and deeply torn and the wear rate is usually high. The wear rates may differ by as much as two or three orders of magnitude. The transition between mild and severe wear takes place over a wide range of sliding conditions. These are load-dependent, velocity-dependent or load- and velocity-dependent. In addition, some are sliding-distance dependent.

The mild wear occurs because direct metal–metal contacts are minimized mostly by the oxide layer produced as a result of frictional heating. Mild wear takes place under four distinct sets of conditions (Lim *et al.*, 1987). In the first set (i), at low contact pressures and sliding velocities, a thin (usually several nm thick) and ductile oxide film is formed which prevents direct metal–metal contact and is not ruptured at light loads. In the second set (ii), at higher velocities, a thicker and more brittle oxide film is continuously generated by high interface temperatures. Continuous oxidation replenishes the oxide film. In the third set (iii), at higher loads, a hard surface layer (martensite) formed on carbon–steel surfaces because of localized frictional heating followed by rapid quenching as the friction heat is dissipated. The higher interface temperatures also produce a thicker film of oxide, supported by the hardened substrate. In the fourth set (iv), at yet higher sliding velocities, the increased interface temperature produces thick films. Insulating oxide films reduce the heat flow from the surface to the underlying conducting substrate resulting in severe oxidation.

Severe wear occurs under conditions in which direct metal–metal contacts occur. Severe wear takes place under three distinct sets of conditions (Lim *et al.*, 1987). In the first set (i), at high contact pressures and low sliding velocities, contact pressures are high enough to rupture the thin oxide layer [as described in mild wear – (i)] which leads to direct metal–metal

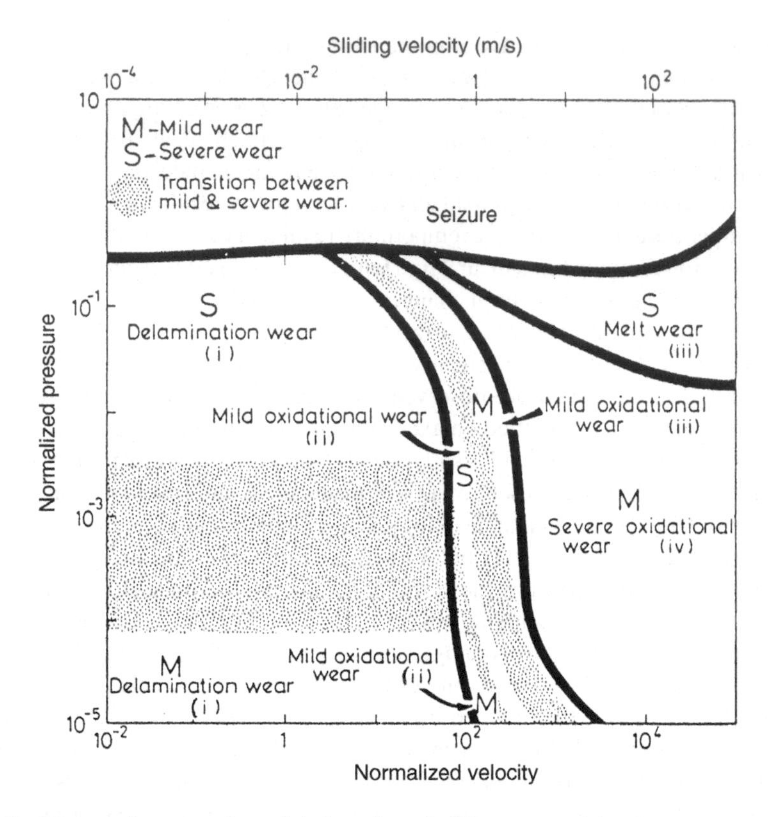

Figure 7.4.3 Wear-regime map for unlubricated steel sliding on steel in air at room temperature in the pin-on-disk configuration. Reproduced with permission from Lim, S.C., Ashby, M.F., and Brunton, J.H. (1987), "Wear-Rate Transitions and Their Relationship to Wear Mechanisms," *Acta Metall.* **35**, 1343–1348. Copyright 1987. Elsevier.

contact. In the second set (ii), at moderate contact pressures and sliding velocities, the load is high enough to penetrate the thicker but brittle oxide films generated [as described in mild wear – (ii)]. In the third set (iii), at high contact pressures and sliding velocities, the sliding conditions are so severe that local temperatures reach the melting point of the steel, resulting in a liquid film in contact which leads to severe wear.

All steels exhibit similar wear-regime maps as just discussed. Many other metals show similar behavior. These wear maps are useful to provide guidance with respect to the proper selection of materials and performance envelopes for metals.

7.4.2 Wear of Ceramics

Ceramics exhibit high mechanical strength, do not lose much mechanical strength or oxidize readily at elevated temperatures and are resistant to corrosive environments; therefore, ceramic couples are commonly used in extreme environmental applications, such as high loads, high speeds, high temperatures and corrosive environments. High mechanical properties result in very low real area of contact responsible for low friction and very low wear. Under clean

Table 7.4.4 Abrasive wear coefficient values of various ceramic pairs at room temperature (self-mated couples) in a three-body abrasive wear experiment at a normal load of 10 N, hydrocarbon-base lapping fluid and 600-grit SiC.

Material	k ($\times 10^{-3}$)
SiC	1.1
WC	3.3
B_4C	5.5
Si_3N_4	43

environments, the coefficients of friction and the wear rates of ceramic pairs do not reach the very high values observed in clean metals, especially in ultra-high vacuum or in the absence of oxygen. Ceramic materials show only limited plastic flow at room temperature and much less ductility than metals. The fracture toughness of ceramics is an important property in the wear of ceramics. Ceramic materials respond to conventional lubricants similar to metals.

A significant amount of wear data of ceramic pairs is found in the literature (e.g., Bhushan and Sibley, 1982; Anonymous, 1987; Jahanmir, 1994; Bhushan and Gupta, 1997). Wear data of various ceramic pairs at elevated temperatures are presented by Zeman and Coffin (1960). It is desirable to use self-mated ceramic pairs, unlike in metals. Table 7.4.4 presents the wear data of ceramic pairs against themselves in a three-body abrasive wear experiment conducted using a lapping machine (Bhushan and Gupta, 1997).

Operating conditions – normal load, sliding velocity and environment – have asignificant effect on the wear modes as well as the wear rates. A discussion of these follows.

7.4.2.1 Effect of the Operating Environment

Tribochemical interactions of ceramics with the liquid or gaseous environment control the wear and friction of ceramics. Depending on the chemical reaction between the ceramic and the environment, wear and friction can decrease or increase (Lancaster, 1990). It can change both the wear mechanisms and the wear rates. This chemical interaction can result in modification of the surface composition and decrease in the purely chemical form of wear by dissolution in the liquid environment (chemomechanical polishing or CMP), but it can induce chemical fracture, which increases wear rates. The CMP process has been described earlier.

Tribochemical Wear of Non-oxide Ceramics

The formation of oxide films in the case of non-oxide ceramics exposed to an oxidizing environment and the formation of hydrated layers in all ceramics exposed to humid environments are responsible for a change in friction as a function of the environment (Fischer, 1988). The formation of chemical films during sliding at interface temperatures is referred to as tribochemistry. Non-oxide ceramics such as silicon nitride, silicon carbide, titanium nitride and titanium carbide are all known to form oxide films during sliding in an oxidizing environment. Oxygen may be derived from oxygen in the air or from water vapor, e.g.,

$$Si_3N_4 + 3O_2 \rightarrow 3SiO_2 + 2N_2 \quad (7.4.3)$$

or

$$Si_3N_4 + 6H_2O \rightarrow 3SiO_2 + 4NH_3 \tag{7.4.4}$$

Oxide ceramics react with water, whether it is present as a liquid or as a vapor. In the case of non-oxide ceramics, oxidation (Equations 7.4.3 or 7.4.4) can be followed by hydration,

$$SiO_2 + 2H_2O \rightarrow Si(OH)_4 \tag{7.4.5}$$

During sliding at low temperatures (ambient and from frictional heating), the kinetics of chemical reactions does not allow a sufficient amount of oxide to form. The rate of oxidation is accelerated by the simultaneous action of friction. The hydrated layer [$Si(OH)_4$] exhibits low friction and wear and also provides the source of the wear debris. The removal of the hydrated material exposes the nascent surface and the hydrated layer is reformed from tribochemical reaction.

As an example, Figure 7.4.4 shows the effect of the absorbed water on friction and wear. The values of the coefficient of friction and the wear rates fall with the increasing availability of water for silicon nitride sliding against itself in dry nitrogen gas, air of two different humidity levels, and liquid water. The reactions outlined in Equations 7.4.3 to 7.4.5 lead to the formation and hydration of a silica film at the interface, which is soft with low shear strength and reduces the coefficient of friction and wear rate. A decrease in the wear rate of silicon nitride against itself as a function of increase in the relative humidity has also been reported by Fischer and Tomizawa (1985). A drop in the coefficient of friction and wear rate as a function of relative humidity has also been observed for silicon carbide, Figure 7.4.5 (Kapelski, 1989).

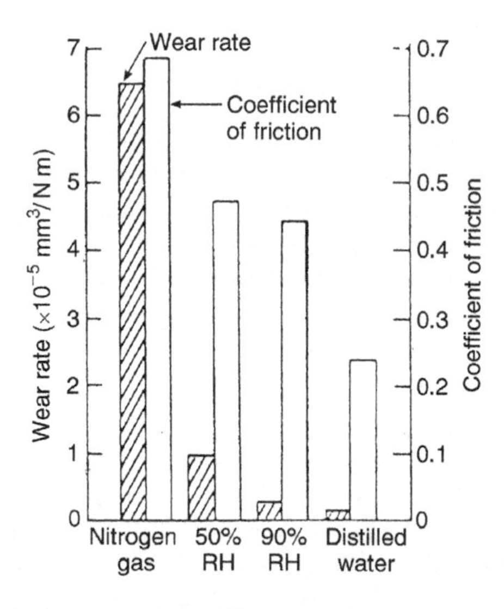

Figure 7.4.4 Effect of environment on the coefficient of friction and wear rate of hot-pressed silicon nitride on itself at a normal load of 10 N and sliding velocity of 150 mm/s, in a pin-on-disk configuration. Reproduced with permission from Ishigaki, H., Kawaguchi, I., Iwasa, M. and Toibana, Y. (1986), "Friction and Wear of Hot Pressed Silicon Nitride and Other Ceramics," *ASME J. Trib.* **108**, 514–521. Copyright 1986. ASME.

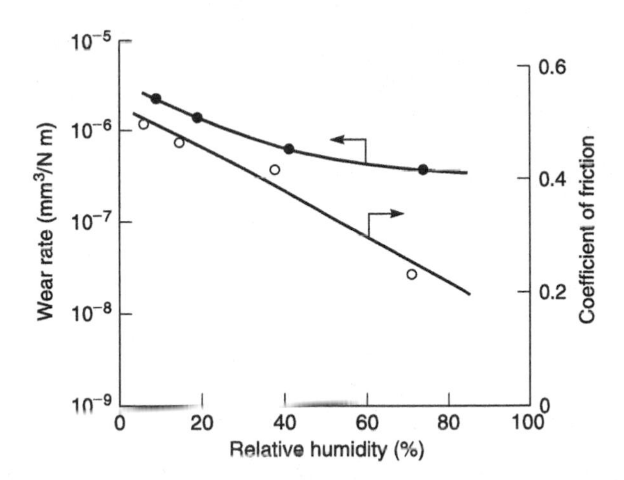

Figure 7.4.5 Coefficient of friction and wear rate of silicon carbide on itself as a function of relative humidity at a normal load of 10 N and sliding velocity of 0.1 m/s after a sliding distance of 1 km in a ball-on-disk configuration.

Figure 7.4.6 shows the effect of sliding velocity on the coefficient of friction and wear rate of hot-pressed silicon nitride on itself. Both the coefficient of friction and the wear rate are relatively constant at low velocities but increase considerably at transition velocities of about 200 mm/s. At low velocities, silicon nitride reacts with water vapor in the air and forms a hydrated film which is responsible for low friction and wear. At higher velocities, the interface

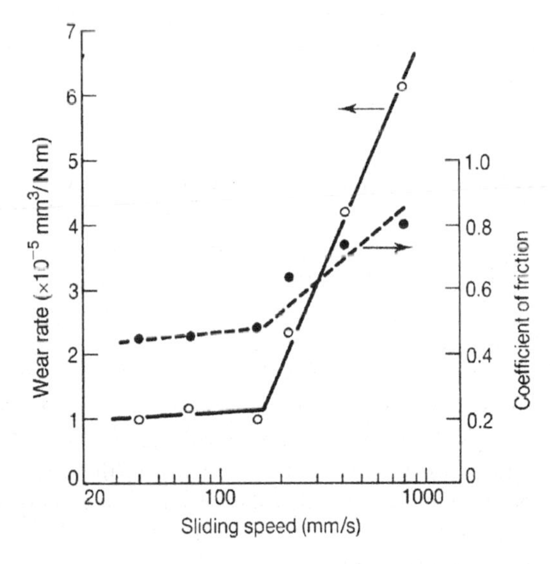

Figure 7.4.6 Coefficient of friction and wear rate as a function of sliding velocity of hot-pressed silicon nitride on itself at a normal load of 10 N and ambient air in a pin-on-disk configuration. Reproduced with permission from Ishigaki, H., Kawaguchi, I., Iwasa, M., and Toibana, Y. (1986), "Friction and Wear of Hot Pressed Silicon Nitride and Other Ceramics," *ASME J. Trib.* **108**, 514–521. Copyright 1986. ASME.

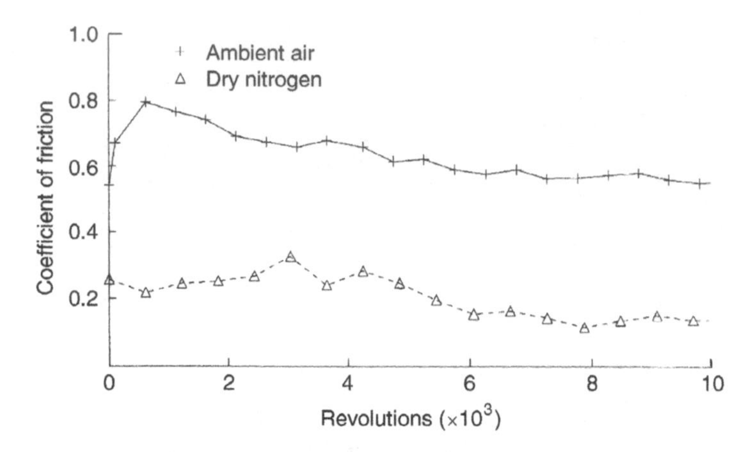

Figure 7.4.7 Coefficient of friction as a function of sliding revolutions for single-crystal silicon on itself in ambient air and dry nitrogen environment in a pin-on-disk configuration.

temperature increases, which reduces the amount of water vapor in the air. Reduction in the water vapor results in a reduced amount of tribochemical products, which is responsible for high friction and wear.

Friction tests conducted with bare single-crystal silicon and thermally oxidized single-crystal silicon sliding against themselves show that the coefficient of friction in dry nitrogen is about half that of in air, Figure 7.4.7 (Venkatesan and Bhushan, 1994). These data suggest that the interaction between the surfaces is strong in air (promoted by either oxygen or water vapor present in the ambient air), resulting in higher friction and consequently high wear in air compared to nitrogen.

Chemically-Induced Fracture (Static Fatigue) in Oxide Ceramics

In the case of many oxide ceramics such as alumina (Wallbridge *et al.*, 1983; Kapelski, 1989), and zirconia (Fischer *et al.*, 1988), friction and wear of these ceramics also show strong sensitivity to water, but in these cases the coefficient of friction and the wear rate increase with an increase in relative humidity, Figure 7.4.8 (Kapelski, 1989). This increase in wear rate occurs because of enhanced crack growth rate, which results from the attack of the bonds between the neighboring metal and oxide ions at a crack tip by water. This chemically induced fracture phenomenon is known as stress-corrosion cracking or static fatigue. Exposure to humidity may also increase surface plasticity as a result of change in the mobility of near-surface dislocations, with consequent wear. This chemomechanical effect, in which the mechanical properties of many materials change as a result of exposure to many liquids, is also known as the Joffe–Rehbinder effect (Rehbinder and Shchukin, 1972).

The effect of temperature on the coefficient of friction and wear is shown in Figure 7.4.9 (Dong *et al.*, 1991). At temperatures below 200°C and at temperatures above 800°C, the coefficient of friction and the wear volume are low. In the temperature range of 200–800°C, the coefficient of friction and the wear volume are very large.

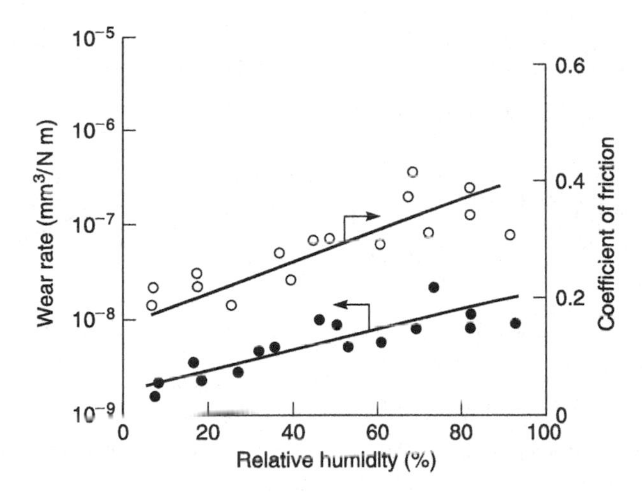

Figure 7.4.8 Coefficient of friction and wear rate of silicon carbide with alumina as a function of relative humidity at a normal load of 10 N and sliding velocity of 0.1 m/s after a sliding distance of 1 km in a ball-on-disk configuration.

7.4.2.2 Effect of Operating Conditions (Wear-Regime Maps)

As stated for metals, no single wear mechanism operates for ceramics either over a wide range of conditions. Various regimes of mechanical (plastic deformation or brittle fracture) and chemical (or tribochemical) wear for a particular sliding materials pair are observed on a single wear-regime map plotted on axes of Hertzian pressure and sliding velocity (Hsu and Shen, 1996). Wear mechanisms dominated by plastic flow or tribochemical reactions

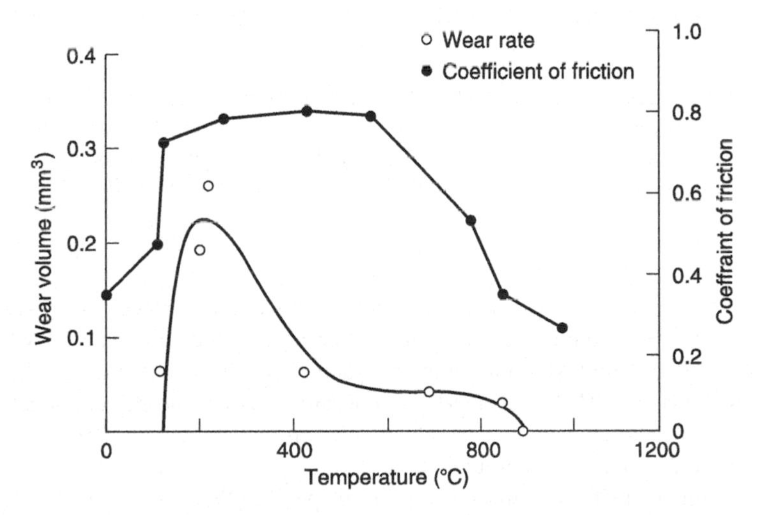

Figure 7.4.9 Coefficient of friction and wear volume as a function of temperature for α-alumina on itself at a normal load of 59 N and sliding velocity of 1.4 mm/s in a ball-on-a flat configuration.

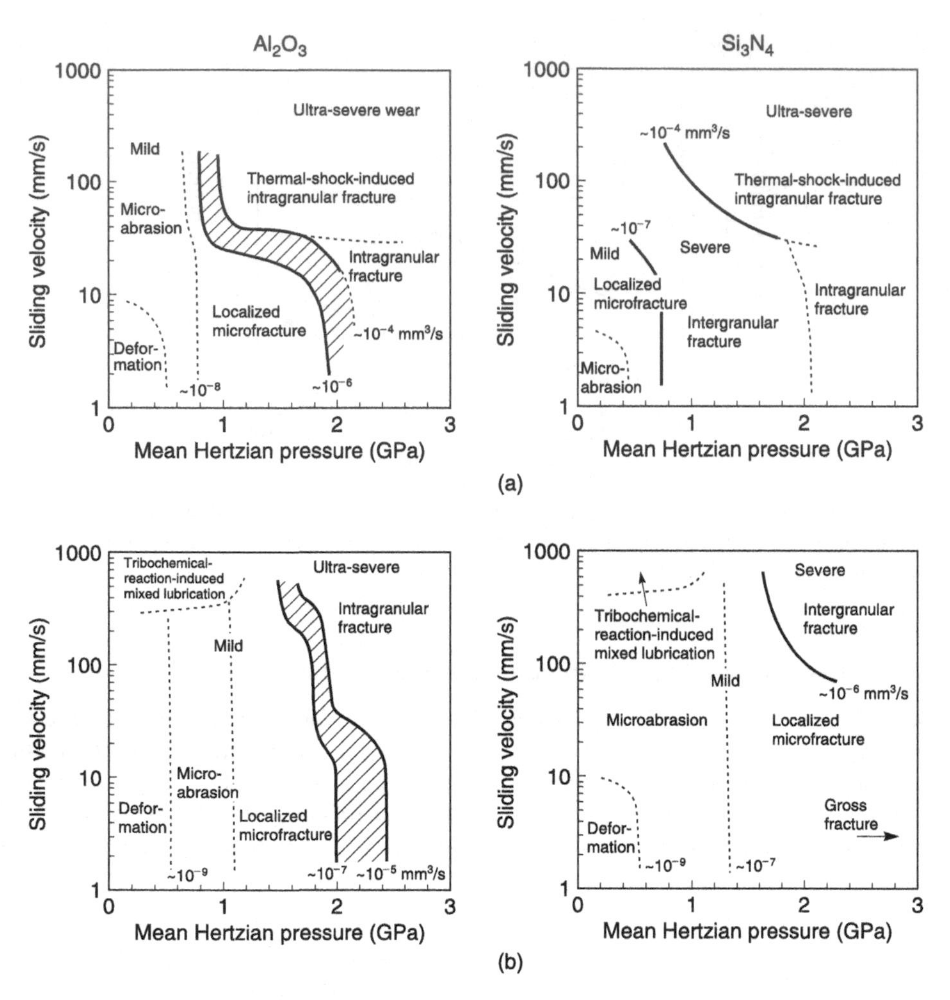

Figure 7.4.10 Wear-regime maps for Al_2O_3 and Si_3N_4 sliding on themselves in air at room temperature in a ball-on-three-flats geometry on a four-ball wear tester (a) under dry conditions and (b) under paraffin oil lubricated conditions. Wear volumes per unit time are also listed in the figures. Reproduced with permission from Hsu, S.M. and Shen, M.C. (1996), "Ceramic Wear Maps," *Wear* **200**, 154–175. Copyright 1996. Elsevier.

generally result in mild wear with low wear rates and smooth surfaces. The wear debris is generally finely divided and may be chemically different from the bulk material, whereas wear mechanisms dominated by brittle intergranular fracture result in severe wear with high wear rates and rough surfaces. The wear debris is generally angular and not chemically different from the substrate. Mild wear occurs at a combination of low pressures and velocities, whereas severe wear occurs at combinations of high pressures and velocities.

The wear-regime maps for ceramics are material specific. Wear maps for Al_2O_3 and Si_3N_4 under dry and paraffin-oil-lubricated conditions are shown in Figure 7.4.10. The tests were conducted by using a ball-on-three-flats geometry on a four-ball wear tester with identical materials. Wear volumes per unit time are also listed in the figures. Various wear mechanisms listed in the figures are self-explanatory. Note that the interaction of the lubricant with the ceramics extends the pressure-velocity boundary toward the higher values for a transition between mild to severe wear.

7.4.3 Wear of Polymers

Polymers include plastics and elastomers. Polymers generally exhibit low friction as compared to metal and ceramic couples but exhibit moderate wear. Most commonly used plastics in tribological applications include polytetrafluoroethylene (PTFE), acetal, high-density polyethylene (HDPE), polyamide (Nylon), poly (amide-imide), polyimide and polyphenylene sulfide (Bartenev and Lavrentev, 1981; Briscoe, 1981; Bhushan and Wilcock, 1982, Bhushan and Gupta 1997). Most commonly used elastomers include natural and synthetic rubber, butadiene – acrylonitrile (Buna-N or nitrile) rubber, styrene-butadiene rubber (SBR) and silicone rubber (Bhushan and Winn, 1981; Bhushan and Gupta, 1997). These polymers are a family of self-lubricating solids. The polymer composites, impregnated generally with fibers of carbon graphite or glass and powders of graphite, MoS_2, bronze and PTFE, are used for their desirable mechanical and tribological properties. Polymers are also used as additives to nonpolymeric solids and liquid lubricants.

The dominant wear mechanisms are adhesive, abrasive, and fatigue. If the mating surface is smooth, then the wear primarily occurs from adhesion between the mating surfaces. As stated in Chapter 5, wear of many polymers occurs first by the transfer of polymer to the harder mating surface followed by removal as wear particles (Steijn, 1967; Lancaster, 1973; Bhushan and Wilcock, 1982). During the initial run-in period, a steady-state condition is reached. If the steady-state condition is reached, the wear rate is generally small and stable. The transfer film thickness for PTFE composites sliding against smooth, steel surfaces is on the order of 0.5–2 μm, which is adherent and cannot be scraped off easily. For the cases of sliding of polymers against rough surfaces, the abrasive mechanism may be dominant. The fatigue mechanism is important in harder polymers such as many thermoset polymers sliding against smooth surfaces. Asperity deformation in polymers is primarily elastic and wear due to fatigue results from the formation of cracks associated predominantly with elastic deformation. Wear particles are produced by the propagation and intersection of cracks.

Polymers flow readily at modest pressures and temperatures. Therefore polymers and polymer composites are used at relatively low loads, speed and temperatures, lower than that in the case of metals and ceramics. Polymers generally have low thermal conductivities, therefore they result in high interface temperatures. The interface temperatures generated during sliding are a function of normal pressure x sliding velocity (PV), thus polymers and solid lubricants are classified based on a PV limit. Beyond the PV limit, polymers start to melt at the interface even at ambient temperature and wear rate increases rapidly. Methods generally used for establishing PV limit are described by Bhushan (1999).

Polymers have a high tolerance to abrasive particles (embeddability), resilience in distributing the load under misaligned conditions (thus preventing seizure), low cost and easy availability. Polymers are generally insensitive to corrosive environments unlike metals, but react with many fluids; they swell with degradation in mechanical properties.

Table 7.4.5 The PV limits, wear coefficients, and coefficients of friction of various unfilled and filled plastics sliding on steel, under dry conditions. (These are approximate values taken from various publications.)

Material	*PV* limits at (*V*) and 22°C, MPa·m/s (at m/s)	Maximum operating temperature (°C)	Wear coefficient, *k* (× 10^{-7} mm^3/Nm)	Coefficient of friction
PTFE (unfilled)	0.06 (0.5)	110–150	4000	0.05–0.1
PTFE (glass-fiber filled)	0.35 (0.05-5.0)	200	1.19	0.1–0.25
PTFE (graphite-fiber filled)	1.05 (5.0)	200	–	0.1
Acetal	0.14 (0.5)	85–105	9.5	0.2–0.3
Acetal (PTFE filled)	0.19 (0.5)	–	3.8	0.15–0.27
	0.09 (5.0)	–	–	–
UHMW polyethylene	0.10 (0.5)	105	–	0.15–0.3
UHMW polyethylene (glass-fiber filled)	0.19 (0.5)	105	–	0.15–0.3
Polyamide	0.14 (0.5)	110	38.0	0.2–0.4
Polyamide (graphite filled)	0.14 (0.5)	150	3.0	0.1–0.25
Polycarbonate	0.03 (0.05)	135	480	0.35
	0.01 (0.5)			
Polycarbonate (PTFE filled)	0.06 (0.5)	135	–	0.15
Polycarbonate (PTFE, glass fiber)	1.05 (0.5)	135	5.8	0.2
Polyphenylene sulfide	3.50 (0.5)	260–315	–	0.15–0.3
Polyphenylene sulfide (PTFE, carbon fibers)	3.50 (0.5)	260–315	–	0.1–0.3
Poly(amide-imide)	3.50 (0.5)	260	–	0.15–0.3
Poly(amide-imide)(PTFE, graphite)	1.75 (0.5)	260	–	0.08–0.3
Linear aromatic polyester (graphite filled)	1.75 (0.5)	260–315	–	0.2–0.4
Phenolic	0.17 (0.05)	260	–	0.9–1.1
Phenolic (PTFE filled)	1.38 (0.5)	–	–	0.1–0.45
Polyimide	3.50 (0.5)	315	30.0	0.15–0.3
Polyimide (graphite filled)	3.50 (0.5)	315	5.0	0.1–0.3
Epoxy (glass filled)	1.75 (0.5)	260	–	0.3–0.5

7.4.3.1 Plastics

The maximum operating temperature, wear coefficients, the coefficient of friction and the PV limits of various unfilled plastic and plastic composites are presented in Table 7.4.5. The PV limits of elastomers are generally lower than those of plastics. The PV limit of the polymers in the lubricated conditions (oils or water) can be up to an order of magnitude larger than that in dry conditions. A liquid medium removes frictional heat from the interface, thus allowing operation at high PV conditions. High-temperature polymers can be operated under lubricated conditions with a PV of 17.5 MPa × m/s (500,000 psi × fpm), comparable to the PV limit of carbon-graphites (manufactured carbon), commonly used in wear applications.

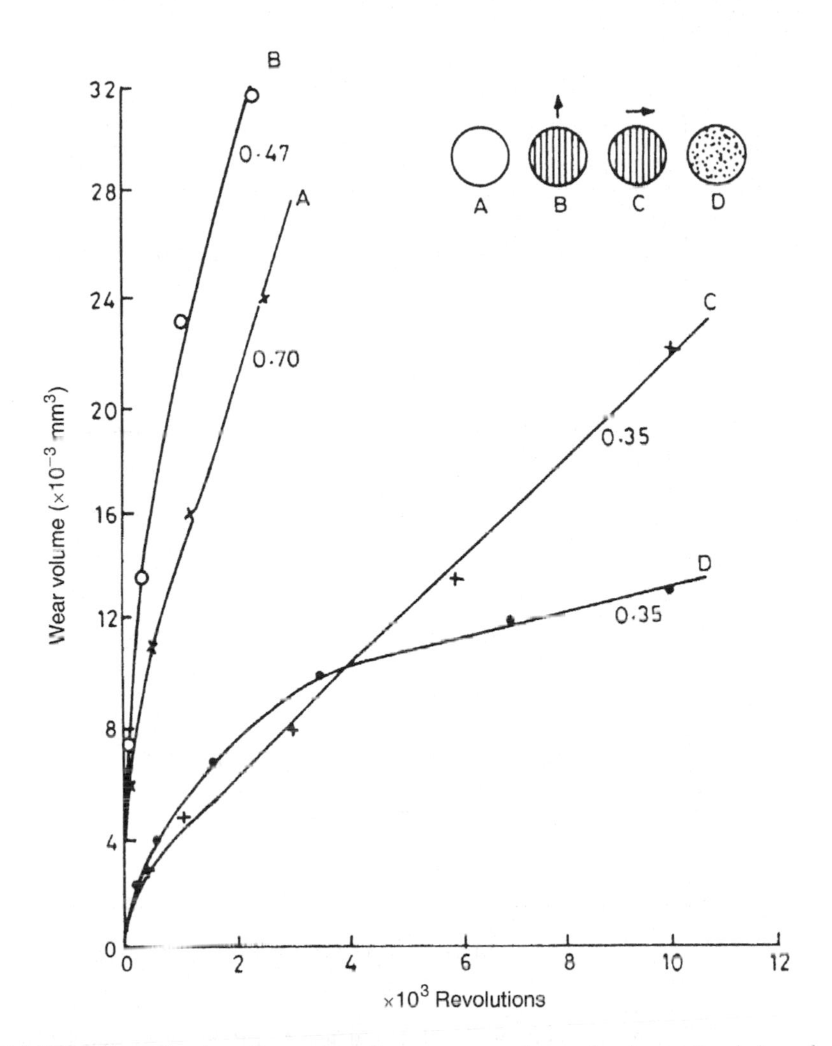

Figure 7.4.11 Effect of carbon fiber orientation on wear volume of polymer in reinforced polyester resin (25 wt. % fiber) slid against hardened tool steel. The coefficients of friction are given adjacent to each curve. Reproduced with permission from Lancaster, J.K. (1968), "The Effect of Carbon Fiber-Reinforcement on the Friction and Wear Behavior of Polymers," *Br. J. Appl. Phys.* **1**, 549–559. Copyright 1968. IOP Publishing.

In polymer composites, orientation of the fibers affects the wear rates. Figure 7.4.11 illustrates the effect of fiber orientation on carbon-fiber-reinforced polyester sliding against a relatively smooth hardened tool steel surface. Both the coefficient of friction and the wear rate of the polymers are lower when the fibers are oriented normal to the sliding surface. Certain fillers, such as glass and carbon, commonly used in polymer composites are harder than the mating metals, such as mild steel and cast iron, and thus may cause damage to the mating metal surface. Cumulative damage caused by the composite can be attributed to the abrasiveness of the particular filler used and is an important factor to consider when selecting a material for

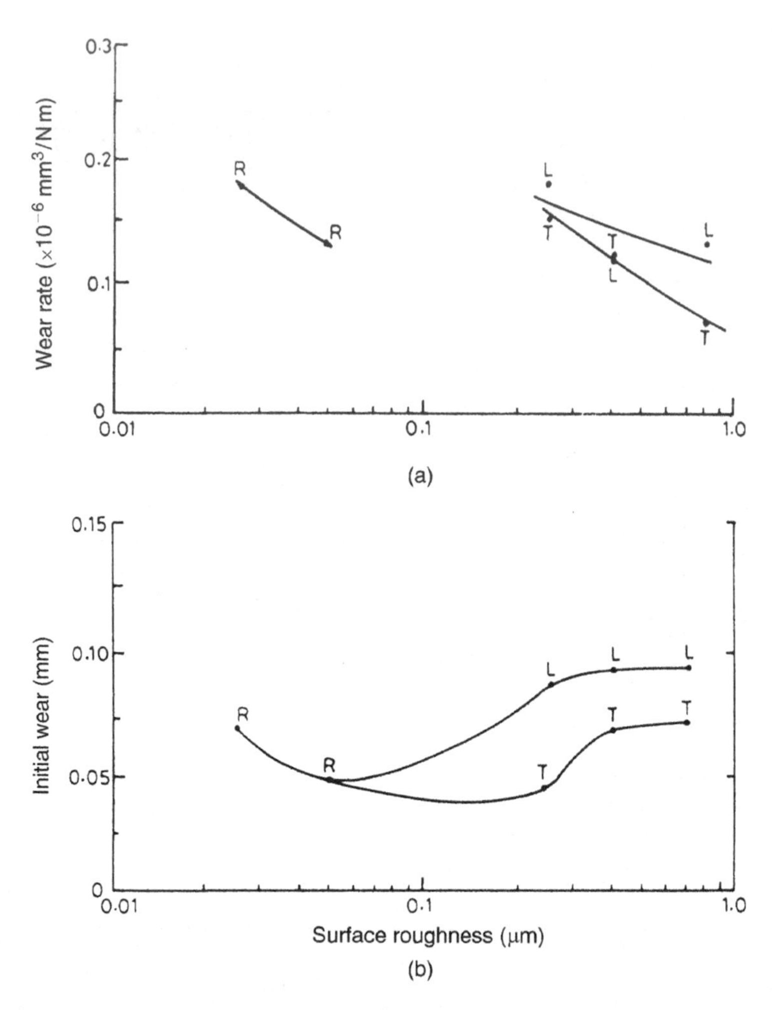

Figure 7.4.12 (a) Wear rate as a function of surface roughness of PTFE filled with 10% glass fiber plus 15 wt. % CdO-graphite-Ag slid against nitrided SAE 7140 steel in reciprocating mode in a nitrogen atmosphere, (b) initial wear versus roughness of same combination in nitrogen atmosphere (*R*, random; *L*, longitudinal; and *T*, transverse orientation of roughness of the mating metal with respect to the direction of sliding). Reproduced with permission from Bhushan, B. and Wilcock, D.F. (1982), "Wear Behavior of Polymeric Compositons in Dry Reciprocating Sliding," *Wear* **75**, 41–70. Copyright 1982. Elsevier.

use against a relatively soft metal such as aluminum. The hardness of the mating metal plays a role in determining abrasiveness of a particular filler.

The surface roughness of the mating metal and its orientation in relation to the direction of sliding have a significant influence on wear rate, Figure 7.4.12 (Bhushan and Wilcock, 1982). The rougher mating metal surface results in a thicker transfer film buildup, which may be responsible for the lower friction. Initial wear is high with high roughness because it takes more polymer material to pack the roughness grooves. Once the adherent transfer film is built up, the wear rate decreases and the subsequent wear rate of a rougher surface is lower than that of a smoother surface. Transverse grooves have more initial and steady-state wear than longitudinal grooves owing to a more abrading (cutting) action. Mating metals with random roughness show high initial wear and longer time for buildup of the transfer film. Therefore, a mating surface with a relatively high surface roughness (~0.8 μm) and roughness grooves on the metal surface in the direction of sliding are recommended.

7.4.3.2 Elastomers

Elastomers are generally used at PV limits lower than that for many plastics. Friction and wear properties of elastomers are modified by adding fillers such as carbon black, silica, graphite, MoS_2, and PTFE powders and sometimes glass fibers. Friction and wear of elastomer composites can be comparable to the plastic composites (Bhushan and Winn, 1981).

7.4.3.3 Effect of Operating Environment

Exposure to environment (gases and humidities) affects mechanical properties and friction and wear of polymers. In the data shown in Figure 7.4.13, the wear rate of PTFE composite sliding against cast iron at 70°C, exposed to different environments decreases with an increase of environmental humidity (Schubert, 1971). The exact trend depends on the particular environment. The wear rate in air and oxygen is up to 1000 times greater than in nitrogen.

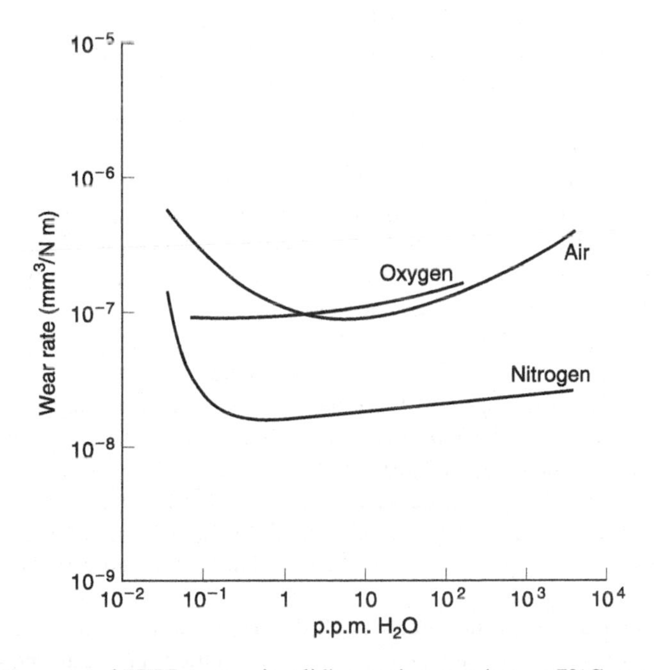

Figure 7.4.13 Wear rate of PTFE composite sliding against cast iron at 70°C as a function of water concentration in three different environments.

7.5 Closure

Wear is the surface damage or removal of material from one or both of two solid surfaces in a sliding, rolling or impact motion relative to one another. Wear damage precedes actual loss of material, and it may also occur independently. The definition of wear is generally based on loss of material from one or both of the mating surfaces. Strictly, the wear-like friction is not an inherent material property and it depends on the operating conditions and surface conditions. Wear rate does not necessarily relate to friction. Wear resistance of a material pair is generally classified based on a wear coefficient, a non-dimensional parameter or wear volume per unit load per unit sliding distance.

Wear occurs by mechanical and/or chemical means and is generally accelerated by frictional heating. Principal types of wear mechanism include: (1) adhesive; (2) abrasive; (3) fatigue; (4) impact by erosion and percussion; (5) chemical; and (6) electrical-arc-induced wear. Other, not distinct, mechanisms are fretting and fretting corrosion, a combination of adhesive, corrosive, and abrasive forms of wear. Wear by all mechanisms, except by fatigue mechanism, occurs by gradual removal of material. Of the aforementioned wear mechanisms, one or more may be operating in one particular machine. In many cases, wear is initiated by one mechanism and it may proceed by other wear mechanisms, thereby complicating failure analysis.

Adhesive wear occurs because of adhesion at asperity contacts at the interface. These contacts are sheared by sliding which may result in the detachment of a fragment from one surface to another surface. As the sliding continues, the transferred fragments may come off the surface on which they are transferred and be transferred back to the original surface, or else form loose particles. Some are fractured by a fatigue process during repeated loading and unloading action resulting in formation of loose particles. During sliding, surface asperities on or near undergo plastic deformation and/or fracture. The subsurface, up to several microns in thickness also undergoes plastic deformation and strain hardening with microhardness as much as factor of two higher than the bulk hardness. Based on Archard's equation, the volume of wear of contacts going through plastic deformation is proportional to the normal load and sliding distance and is inversely proportional to the hardness of the surface being worn away. Based on Bhushan's equation, the volume of wear of contacts going through primarily elastic deformations is proportional to the normal load and sliding distance and inversely proportional to the composite modulus of elasticity and roughness parameters ratio. Wear equations suggest that the wear coefficient is independent of normal load and sliding velocity, but this assumption holds only for a range of values of loads and velocities.

Abrasive wear occurs when the asperities of a rough, hard surface or hard particles slide on a softer surface, and damage the interface by plastic deformation or fracture in the case of ductile and brittle materials, respectively. In many cases, there are two general situations for abrasive wear. In the first case, the hard surface is the harder of two rubbing surfaces (two-body abrasion); and in the second case, the hard surface is a third body, generally a small particle of abrasive, caught between the two surfaces and sufficiently harder that it is able to abrade either one or both of the mating surfaces (three-body abrasion). In many cases, the wear mechanism at the start is adhesive, which generates wear particles that get trapped at the interface, resulting in a three-body abrasive wear. In most abrasive wear situations, scratching is observed with a series of grooves parallel to the direction of sliding. During sliding, like adhesive wear, asperities on or near the surface undergo plastic deformation and strain hardening with an increase in hardness. Abrasive wear rate is a function of surface roughness and, in contrast to adhesive wear mechanism, it increases with an increase in surface roughness. The wear equation for two-body abrasive wear is also valid for three-body abrasive wear. However, the wear rate will be lower by about an order of magnitude because many particles will tend to roll rather than slide. If the wear takes place with a fresh abrasive medium, wear continues at a steady rate, whereas, if a limited amount of abrasive medium is used as the sliding continues, the wear rate generally decreases as a function of time. A decrease in wear rate as a function of time is believed to occur primarily as a result of blunting of the abrasives. The relative hardness of the abrasive medium to the workpiece affects the wear rate. When the hardness ratio of the workpiece to the abrasive particles is less than unity, the wear coefficient remains approximately constant; however, if the ratio is equal to or greater than unity, the wear coefficient decreases rapidly with an increase in the hardness ratio.

Subsurface and surface fatigue are observed during repeated rolling (negligible friction) and sliding (coefficient of friction ≥ 0.3), respectively. The repeated loading and unloading cycles to which the materials are exposed may induce the formation of subsurface and surface cracks, which eventually, after a critical number of cycles will result in the breakup of the surface with the formation of large fragments, leaving large pits on the surface. Prior to this critical point, negligible wear takes place, which is in marked contrast to the wear caused by the adhesive or abrasive wear mechanism, where wear causes a gradual deterioration from the start of running. Therefore, the material removed by fatigue wear is not a useful parameter. Much more relevant is the useful life in terms of the number of revolutions or time before failure occurs. Another difference between adhesive and abrasive wear and fatigue wear is that fatigue wear does not require direct physical contact between two surfaces. Mating surfaces experience large stresses, transmitted through the lubricating film during the rolling motion such as in well-designed rolling element bearings. The failure time in fatigue wear is statistical in nature and is predicted based on Weibull analysis in terms of probability of survival.

Chemical-induced crack growth (most common in ceramics) is commonly referred to as static fatigue. In the presence of tensile stresses and water vapor at the crack tip in many ceramics, a chemically induced rupture of the crack-tip bonds occurs rapidly, which increases the crack velocity. Chemically enhanced deformation and fracture result in an increased wear of surface layers in static and dynamic (rolling and sliding) conditions.

Impact wear includes erosive and percussive wear. Erosion can occur by jets and streams of solid particles, liquid droplets, and implosion of bubbles formed in the fluid. Percussion occurs from repetitive solid body impacts. Repeated impacts result in progressive loss of solid material. Solid particle erosion is a form of abrasion that is generally treated rather differently because the contact stress arises from the kinetic energy of particles flowing in an air or liquid stream as it encounters a surface. The particle velocity and impact angle combined with the size of the abrasive give a measure of the kinetic energy of the impinging particles, that is, of the square of the velocity. As in the abrasive wear, erosive wear occurs by plastic deformation and/or fracture, dependent upon material being eroded away and operating parameters. In liquid impingement erosion, with small drops of liquid striking the surface of a solid at high speeds (as low as 300 m/s), very high pressures are experienced, exceeding the yield strength of most materials. Thus, plastic deformation or fracture can result from a single impact, and repeated impact leads to pitting and erosive wear. Cavitation erosion arises when a solid and fluid are in relative motion, and bubbles formed in the fluid become unstable and implode against the surface of the solid. Cavitation erosion is similar to surface fatigue wear. Percussion is a repetitive solid body impact. Percussion wear occurs by hybrid wear mechanisms which combine several of the following mechanisms: adhesive, abrasive, surface fatigue, fracture and tribochemical wear.

Chemical or corrosive wear occurs when sliding takes place in a corrosive environment. Corrosion can occur because of a chemical or electrochemical interaction of the interface with the environment. In air, the most corrosive medium is oxygen. Therefore, chemical wear in air is generally called oxidative wear. In the absence of sliding, the chemical products of the corrosion (e.g., oxides) would form a film typically less than a micrometer thick on the surfaces, which would tend to slow down or even arrest the corrosion, but the sliding action wears the chemical films away, so that the chemical attack can continue. Thus chemical wear requires both chemical reaction (corrosion) and rubbing. Frictional heating modifies the kinetics of chemical reactions of sliding bodies with each other, and with the gaseous or liquid environment, to the extent that reactions which normally occur at high temperatures occur at moderate or even ambient temperatures during sliding. The wear controlled by this reaction is referred to as tribochemical wear.

When a high potential is present over a thin air film in a sliding process, a dielectric breakdown results, leading to arcing. During arcing, a relatively high-power density occurs over a very short period of time. The heat-affected zone is usually very shallow (on the order of 50 μm) and the heating results in considerable melting and subsequent resolidification, corrosion, hardness changes, and other phase changes, and even in the direct ablation of material. Arcing causes large craters, and any sliding or oscillation after an arc either shears or fractures the lips, leading to three-body abrasion, corrosion, surface fatigue, and fretting.

Fretting occurs where low-amplitude oscillatory motion (a few tens of nanometers to a few tens of microns) takes place between contacting surfaces, which are nominally at rest. A rapid increase in wear rate occurs with slip amplitude over an amplitude range. Basically, fretting is a form of adhesive or abrasive wear, where the normal load causes adhesion between asperities and oscillatory movement causes ruptures, resulting in wear debris. Most commonly, fretting is combined with corrosion, in which case the wear mode is known as fretting corrosion.

Regarding the particles present in wear debris, these are generally classified based on their morphology: plate-shaped; ribbon-shaped' spherical; and irregular-shaped.

Finally, wear of a material is dependent on the mating material (or material pair), surface preparation and operating conditions. Clean metals and alloys exhibit high adhesion, and consequently high friction and wear. Any contamination mitigates contact, and chemically produced films which reduce adhesion result in reduction in friction and wear. In dry sliding, identical metals, particularly iron on iron, are metallurgically compatible and exhibit high friction and wear, so they must be avoided. Soft and ductile metals such as In, Pb, and Sn exhibit high friction and wear. Hexagonal metals such as Co and Mg as well as some non-hexagonal metals such as Mo and Cr exhibit low friction and wear. Lead-based white metals (babbitts), brass, bronze, and gray cast iron generally exhibit relatively low friction and wear, and are commonly used in dry and lubricated bearing and seal applications. For high-temperature applications, cobalt-based alloys are used which exhibit good galling resistance. (Galling resistance is a measure of the normal stress at which two materials loaded against each other gall or weld.) Nickel-based alloys are poor in unlubricated sliding because of generally catastrophic galling.

In dry sliding conditions, similar or dissimilar ceramic pairs are commonly used which exhibit moderate friction but maximum wear resistance. In ceramics, fracture toughness is an important mechanical property which affects friction. Ceramics react with the humidity from the environment. Non-oxides may form beneficial hydrides (from the tribochemical reaction) and result in low friction and wear. On the other hand, oxide ceramics, because of enhanced crack growth at high humidity (static fatigue) result in high friction and wear. In both metals and ceramics, no single wear mechanism operates over a wide range of operating conditions. Various regimes of mechanical and chemical wear for a sliding material pair are presented in a wear-regime map, plotted on axes of Hertzian pressure and sliding velocity.

Polymers, which include plastics and elastomers, generally exhibit very low friction and moderate wear. Among polymers, PTFE exhibits the lowest friction and low wear. Polymers flow at modest pressures and modest temperatures; therefore, polymer composites are commonly used. Polymers react with fluids in the environment and can swell and lose mechanical properties. Since polymers soften at moderate temperatures, they are classified based on the PV limit, which is a measure of the interface temperature rise.

Problems

7.1 A cylindrical bronze pin of 1 mm radius rests on a rotating steel disk at a mean radius of 25 mm. The normal load on the pin is 10 N. The rotational speed of the disk is 300 rpm and the test lasts for 10 hours. The mass losses of the pin and disk are 50 mg and 3 mg, respectively. Using the material data given below, calculate the wear coefficients and wear depths for the bronze pin and steel disk. (Hardness of bronze = 0.8 GPa, density of bronze = 8.5 Mg/m^3; hardness of steel = 2.5 GPa, density of steel = 7.8 Mg/m^3). Calculate the wear coefficients for a cylindrical steel pin on a bronze disk under the same test conditions.

7.2 A milling cutter was used to saw through a medium carbon steel bar (H = 3 GPa) of 10 mm diameter with a width of cut of 0.5 mm. It took 10 minutes to saw and the energy expended was 50 W (Nm/s). The coefficient of friction between the saw and the steel bar is 0.3. Calculate the wear coefficient of the steel bar during the cutting process.

7.3 A steel surface consisting of conical asperities with roughness angle of 10°, reciprocates on a soft lead surface (H = 75 MPa) under a load of 1 N with a reciprocating amplitude of 10 mm at 5 Hz. Given that the volume of lead material removed is 10^{-6} m^3 in 10 hours, calculate the abrasive wear coefficient of the lead material. Given that the roughness angle of the steel surface is 30°, calculate the wear coefficient of the lead material.

7.4 A nickel surface (hardness = 3 GPa) of a square block (10 mm × 10 mm) electroplated with ruthenium (hardness = 5 GPa) to a thickness of 5 μm is rubbed at a normal load of 5 N against an abrasive paper so that fresh abrasive paper always contacts the ruthenium. Estimate what distance of rubbing is needed before the abrasive first penetrates the ruthenium, thus exposing the nickel substrate. Assume that k_{abr} for ruthenium rubbing against the abrasive paper is 10^{-6}.

7.5 A cubic pin with a linear dimension of 1 mm and with hardness H of 0.2 GPa slides upon a surface at a constant velocity V of 0.1 m/s and apparent pressure p_a of 0.001 GPa. The wear coefficient k is = 4×10^{-6}. The failure occurs when the fraction of the volume of 0.1% is worn. Calculate sliding time until failure.

7.6 A body slides upon another body with a plastic contact. For a distance x of 1000 m, the volume v of 1 mm^3 is worn during sliding and the wear coefficient k is 10^{-6}. Calculate friction force, if adhesion shear strength τ_a is equal to 10^6 Pa.

7.7 In grinding of silicon carbide, the material removal rate by brittle fracture is 2 μm/h. Silicon carbide is processed to increase its fracture toughness by a factor of 1.5 with the same hardness and modulus of elasticity. Calculate the material removal rate for the processed silicon carbide.

7.8 Based on a tribological test of various materials, the following values of coefficient of friction and wear coefficient were reported for graphite-steel, steel-steel, steel-bronze, and alumina-alumina. Enter name of these material pairs in Table P.7.1.

Table P.7.1

Material pair	Coefficent of friction	Wear coefficient
?	0.2	10^{-6}
?	0.1	10^{-4}
?	0.3	10^{-9}
?	0.6	10^{-2}

7.9 Using standard AFBMA calculations, the life, L_{10}, of a roller bearing is 1000 h. Given that the material factor is 2.2, the processing factor is 3, and the lubrication factor is 4, calculate the expected bearing life.

7.10 The basic load rating of a roller bearing based upon the AFBMA calculations is 10 kN. Calculate the bearing catalog life for an applied radial load of 1 kN and shaft speed of 6000 rpm.

References

Aleinikov, F.K. (1957), "The Influence of Abrasive Powder Microhardness on the Values of the Coefficients of Volume Removal," *Soviet Physics: Technical Physics* **2**, 505–511.

Anonymous (1955), "Fretting and Fretting Corrosion," *Lubrication* **41**, 85–96.

Anonymous (1978), "Review of the Wear and Galling Characteristics of Stainless Steels," Committee of Stainless Steel Producers, AISI, Washington, DC.

Anonymous (1987), *Tribology of Ceramics*, Special Publications SP-23 and SP-24, STLE, Park Ridge, Illinois.

Archard, J.F. (1953), "Contact and Rubbing of Flat Surfaces," *J. Appl. Phys.* **24**, 981–988.

Archard, J.F. (1980), "Wear Theory and Mechanisms," in *Wear Control Handbook* (M.B. Peterson and W.O. Winer, eds), pp. 35–80, ASME, New York.

Archard, J.F. and Hirst, W. (1956), "The Wear of Metals Under Unlubricated Conditions," *Proc. R. Soc. Lond. A* **236**, 397–410.

Bamberger, E.N., Harris, T.A. Kacmarsky, W.M., Moyer, C.A., Parker, R.J., Sherlock, J.J., and Zaretsky, E.V. (1971), *Life Adjustment Factors for Ball and Roller Bearings: An Engineering Design Guide*, ASME, New York.

Bartenev, G.M. and Lavrentev, V.V. (1981), *Friction and Wear of Polymers*, Elsevier, Amsterdam.

Bayer, R.G. (1994), *Mechanical Wear Prediction and Prevention*, Marcel Dekker, New York.

Bellman, R. and Levy, A. (1981), "Erosion Mechanism in Ductile Metals," *Wear* **70**, 1–27.

Bhansali, K.J. (1980), "Wear Coefficients of Hard-Surfacing Materials," in *Wear Control Handbook* (M.B. Peterson and W.O. Winer, eds), pp. 373–383, ASME, New York.

Bhushan, B. (1985), "Assessment of Accelerated Head-Wear Test Methods and Wear Mechanisms," in *Tribology and Mechanics of Magnetic Storage Systems*, Vol. 2 (B. Bhushan and N.S. Eiss, eds), pp. 101–111, special publication SP-19, ASLE, Park Ridge, Illinois.

Bhushan, B. (1996), *Tribology and Mechanics of Magnetic Storage Devices*, Second edition, Springer-Verlag, New York.

Bhushan, B. (1999), *Principles and Applications of Tribology*, Wiley, New York.

Bhushan, B. (2001a), *Modern Tribology Handbook Vol. 1: Principles of Tribology*, CRC Press, Boca Raton, Florida.

Bhushan, B. (2001b), *Fundamentals of Tribology and Bridging the Gap Between the Macro- and Micro/Nanoscales*, NATO Science Series II-Vol. 10, Kluwer Academic Pub., Dordrecht, The Netherlands.

Bhushan, B. (2011), *Nanotribology and Nanomechanics I – Measurement Techniques and Nanomechanics, II – Nanotribology, Biomimetics, and Industrial Applications*, Third edition, Springer-Verlag, Heidelberg, Germany.

Bhushan, B. and Davis, R.E. (1983), "Surface Analysis Study of Electrical-Arc-Induced Wear," *Thin Solid Films* **108**, 135–156.

Bhushan, B. and Gupta, B.K. (1997), *Handbook of Tribology: Materials, Coatings and Surface Treatments*, McGraw-Hill, New York (1991); reprinted by Krieger, Malabar, Florida (1997).

Bhushan, B. and Hahn, F.W. (1995), "Stains on Magnetic Tape Heads," *Wear* **184**, 193–202.

Bhushan, B. and Khatavkar, D.V. (1996), "Role of Water Vapor on the Wear of Mn-Zn Ferrite Heads Sliding Against Magnetic Tapes," *Wear* **202**, 30–34.

Bhushan, B. and Sibley, L.B (1982), "Silicon Nitride Rolling Bearings for Extreme Operating Conditions," *ASLE Trans.* **25**, 417–428.

Bhushan, B. and Wilcock, D.F. (1982), "Wear Behavior of Polymeric Compositons in Dry Reciprocating Sliding," *Wear* **75**, 41–70.

Bhushan, B. and Winn, L.W. (1981), "Material Study for Advanced Stern-tube Lip Seals," *ASLE Trans.* **24**, 398–409.

Bhushan, B., Davis, R.E., and Gordon, M. (1985a), "Metallurgical Re-examination of Wear Modes I: Erosive, Electrical Arcing, and Fretting," *Thin Solid Films* **123**, 93–112.

Bhushan, B., Davis, R.E., and Kolar, H.R. (1985b), "Metallurgical Re-examination of Wear Modes II: Adhesive and Abrasive," *Thin Solid Films* **123**, 113–126.

Bhushan, B., Nelson, G.W., and Wacks, M.E. (1986), "Head-Wear Measurements by Autoradiography of the Worn Magnetic Tapes," *ASME J. Trib.* **108**, 241–255.

Bitter, J.G.A. (1963), "A Study of Erosion Phenomena," *Wear* **6**, Part I, 5–21; Part II, 169–190.

Blau, P.J. (1992), *ASM Handbook, Vol. 18: Friction, Lubrication, and Wear Technology*, Tenth edition, ASM International, Materials Park, Ohio.

Briscoe, B.J. (1981), "Wear of Polymers: An Essay of Fundamental Aspects", *Tribo. Int.* **24**, 231–243.

Buckley, D.H. (1981), *Surface Effects in Adhesion, Friction, Wear, and Lubrication*, Elsevier, Amsterdam.

Burwell, J.T. (1957/1958), "Survey of Possible Wear Mechanisms," *Wear* **1**, 119–141.

Chaikin, S.W. (1967), "On Friction Polymer," *Wear* **10**, 49–60.

Cook, N.H. and Bhushan, B. (1973), "Sliding Surface Interface Temperatures," *ASME J. Lub. Tech* **95**, 31–36.

Dong, X., Jahanmir, S., and Hsu, S.M. (1991), "Tribological Chracterization of α-Alumina at Elevated Temperature," *J. Am. Ceram. Soc.***74**, 1036–1044.

Engel, P.A. (1976), *Impact Wear of Materials*, Elsevier, Amsterdam.

Evans, A.G. and Marshall, D.B. (1981), "Wear Mechanisms in Ceramics," in *Fundamentals of Friction and Wear of Materials* (D.A. Rigney, ed), pp. 439–452, Amer. Soc. Metals, Metals Park, Ohio.

Eyre, T.S. (1976), "Wear Characteristics of Metals," *Tribol. Inter.* **9**, 203–212.

Finnie, I. (1960), "Erosion of Surfaces by Solid Particles," *Wear* **3**, 87–103.

Fischer, T.E. (1988), "Tribochemistry," *Ann. Rev. Mater. Sci.* **18**, 303–323.

Fischer, T.E. and Tomizawa, H. (1985), "Interaction of Tribochemistry and Microfracture in the Friction and Wear of Silicon Nitride," in *Wear of Materials* (K.C. Ludema, ed), pp. 22–32, ASME, New York.

Fischer, T.E., Anderson, M.P., Jahanmir, S., and Salher, R. (1988), "Friction and Wear of Tough and Brittle Fracture in Nitrogen, Air, Water, and Hexadecane Containing Stearic Acid," *Wear* **124**, 133–148.

Foroulis, Z.A. (1984), "Guidelines for the Selection of Hardfacing Alloys for Sliding Wear Resistant Applications," *Wear* **96**, 203–218.

Furey, M.J. (1973), "The Formation of Polymeric Films Directly on Rubbing Surfaces to Reduce Wear," *Wear* **26**, 369–392.

Glaeser, W.A. (1992), *Materials for Tribology*, Elsevier, Amsterdam.

Glaeser, W.A. (2001), "Wear Debris Classification," in *Modern Tribology Handbook Vol. 1: Principles of Tribology* (B. Bhushan, ed.), pp. 301–315, CRC Press, Boca Raton, Florida.

Goddard, J. and Wilman, M. (1962), "A Theory of Friction and Wear During the Abrasion of Metals," *Wear* **5**, 114–135.

Guile, A.E. and Juttner, B. (1980), "Basic Erosion Process of Oxidized and Clean Metal Cathodes by Electric Arcs," *IEEE Trans. Components, Hybrids, Manuf. Technol.* **PS-8**, 259–269.

Hansson, C.M. and Hansson, L.M. (1992), "Cavitation Erosion," in *ASM Handbook Vol. 18: Friction, Lubrication and Wear Technology*, pp. 214–220, ASM International, Metals Park, Ohio.

Harris, T.A. (1991), *Rolling Bearing Analysis*, Third edition, Wiley, New York.

Haymann, F.J. (1992), "Liquid Impact Erosion," in *ASM Handbook Vol. 18: Friction, Lubrication and Wear Technology*, pp. 221–232, ASM International, Metals Park, Ohio.

Heinicke, G. (1984), *Tribochemistry*, Carl Hanser Verlag, Munich.

Hokkirigawa, K. and Kato, K. (1988), "An Experimental and Theoretical Investigation of Ploughing, Cutting and Wedge Formation During Abrasive Wear," *Tribol. Inter.* **21**, 51–57.

Holm, R. (1946), *Electric Contacts*, H. Gerbers, Stockholm, Sweden.

Hsu, S.M. and Shen, M.C. (1996), "Ceramic Wear Maps," *Wear* **200**, 154–175.

Hurricks, P.L. (1970), "The Mechanism of Fretting – A Review," *Wear* **15**, 389–409.

Hutchings, I.M. (1992), *Tribology: Friction and Wear of Engineering Materials*, CRC Press, Boca Raton, Florida.

Hutchings, I.M. and Winter, R.E. (1974), "Particle Erosion of Ductile Metals: A Mechanism of Material Removal," *Wear* **27**, 121–128.

Hutchings, I.M., Winter, R.E., and Field, J.E (1976), "Solid Particle Erosion of Metals: The Removal of Surface Material by Spherical Projectiles," *Proc. R. Soc. Lond.* A **348**, 379–392.

Ioannides, E. and Harris, T.A. (1985), "A New Fatigue Life Model for Rolling Bearings," *ASME J. Trib.* **107**, 367–378.

Ishigaki, H., Kawaguchi, I., Iwasa, M., and Toibana, Y. (1986), "Friction and Wear of Hot Pressed Silicon Nitride and Other Ceramics," *ASME J. Trib.* **108**, 514–521.

Ives, L.K. and Ruff, A.W. (1979), in *Erosion: Prevention and Useful Applications* (W.F. Adler, ed), pp. 5–35, Special Tech. Pub. ASTM, Philadelphia.

Jahanmir, S. (ed) (1994), *Friction and Wear of Ceramics*, Marcel Dekker, New York.

Johnson, J.L. and Moberly, L.E. (1978), "High Current Brushes, Part I: Effect of Brush and Ring Materials," *IEEE Trans. Components, Hybrids, Manuf. Technol.* **CHMT-1**, 36–40.

Johnson, L.G. (1964), *The Statistical Treatment of Fatigue Experiments*, Elsevier, New York.

Kapelski, G. (1989), "Etudes des Proprietés Tribologiques de Céramiques Thermo-mécaniques en Fonction de la Temperature et pour Diffèrents Environnements," Thesis, University of Limoges.

Kayaba, T. and Kato, K. (1981), "Adhesive Transfer of the Slip-Tongue and the Wedge," *ASLE Trans.* **24**, 164–174.

Kehr, W.D., Meldrum, C.B., and Thornley, R.F.M. (1975), "The Influence of Grain Size on the Wear of Nickel-Zinc Ferrite by Flexible Media," *Wear* **31**, 109–117.

Kerridge, M. and Lancaster, J.K. (1956), "The Stages in a Process of Severe Metallic Wear," *Proc. R. Soc. Lond. A* **236**, 250–264.

Kragelski, I.V. (1965), *Friction and Wear*, Butterworths, London.

Kruschov, M.M. (1957), "Resistance of Metals to Wear by Abrasion, as Related to Hardness," *Proc. Conf. Lubrication and Wear*, pp. 655–659, Instn Mech. Engrs, London, UK.

Kruschov, M.M. (1974), "Principles of Abrasive Wear", *Wear* **28**, 69–88.

Kruschov, M.M. and Babichev, M.A. (1958), "Resistance to Abrasive Wear of Structurally Inhomogeneous Materials," *Friction and Wear in Machinery*, Vol. **12**, pp. 5–23, ASME, New York.

Lancaster, J.K. (1968), "The Effect of Carbon Fiber-Reinforcement on the Friction and Wear Behavior of Polymers," *Br. J. Appl. Phys.* **1**, 549–559.

Lancaster, J.K. (1973), "Dry Bearings: A Survey of Materials and Factors Affecting Their Performance," *Tribol. Inter.* **6**, 219–251.

Lancaster, J.K. (1990), "A Review of the Influence of Environmental Humidity and Water on Friction, Lubrication and Wear," *Tribol. Inter.* **23**, 371–389.

Larsen-Basse, J. (1975), "Influence of Atmospheric Humidity on Abrasive Wear-I.3-Body Abrasion," *Wear* **31**, 373–379.

Larsen-Basse, J. and Sokoloski, S.S. (1975), "Influence of Atmospheric Humidity on Abrasive Wear-II. 2-Body Abrasion," *Wear* **32**, 9–14.

Lauer, J.L. and Jones, W.R. (1986), "Friction Polymers," in *Tribology and Mechanics and Magnetic Storage Systems* (B. Bhushan and N.S. Eiss, eds), Vol. **3**, pp. 14–23, STLE, Park Ridge, Illinois.

Lawn, B.R. (1993), *Fracture of Brittle Solids*, Second edition, Cambridge University Press, Cambridge, UK.

Lawn, B.R. and Marshall, D.B. (1979), "Hardness, Toughness, and Brittleness: An Indentation Analysis," *J. Amer. Ceram. Soc.* **62**, 347–350.

Lim, S.C and Ashby, M.F. (1987), "Wear-Mechanism Maps," *Acta Metall.* **35**, 1–24.

Lim, S.C., Ashby, M.F., and Brunton, J.H. (1987), "Wear-Rate Transitions and Their Relationship to Wear Mechanisms," *Acta Metall.* **35**, 1343–1348.

Loomis, W.R. (ed) (1985), *New Directions in Lubrication, Materials, Wear, and Surface Interactions: Tribology in the 80s*, Noyes Publications, Park Ridge, New Jersey.

Lundberg, G. and Palmgren, A. (1947), "Dynamic Capacity of Rolling Bearings," *Acta Polytechnica*, Mech. Eng. Series 1, No. 3, 7, RSAEE.

Lundberg, G. and Palmgren, A. (1951), "Dynamic Capacity of Roller Bearings," *Acta Polytechnica*, Mech. Eng. Series 2, No. 4, 96, RSAEE.

Misra, A. and Finnie, I. (1981), "Some Observations on Two-Body Abrasive Wear," *Wear* **68**, 41–56.

Moore, M.A. (1981), "Abrasive Wear," in *Fundamentals of Friction and Wear of Materials* (D.A. Rigney, ed), pp. 73–118, Amer. Soc. Metals, Metals Park, Ohio.

Mulhearn, T.O. and Samuels, L.E. (1962), "In Abrasion of Metals: A Model of the Process," *Wear* **5**, 478–498.

Nelson, W. (1982), *Applied Life Data Analysis*, Wiley, New York.

Peterson, M.B. and Winer, W.O. (eds) (1980), *Wear Control Handbook*, ASME, New York.

Preece, C.M. (ed) (1979), *Treatise on Materials Science and Technology, Vol. 16: Erosion*, Academic Press, San Diego, California.

Quinn, T.F.J. (1983a), "Review of Oxidational Wear-Part I: The Origins of Oxidational Wear," *Tribol. Inter.* **16**, 257–271.

Quinn, T.F.J. (1983b), "Review of Oxidational Wear-Part II: Recent Developments and Future Trends in Oxidational Wear Research," *Tribol. Inter.* **16**, 305–315.

Rabinowicz, E. (1953), "A Quantitative Study of the Wear Process," *Proc. Phys. Soc. Lond. B* **66**, 929–936.

Rabinowicz, E. (1977), "Abrasive Wear Resistance as a Materials Test," *Lub. Eng.* **33**, 378–381.

Rabinowicz, E. (1980), "Wear Coefficients – Metals," *Wear Control Handbook* (M.B. Peterson and W.O. Winer, eds), pp. 475–506, ASME, New York.

Rabinowicz, E. (1983), "The Wear of Hard Surfaces by Soft Abrasives," *Proc. of Wear of Materials* (K.C. Ludema, ed), pp. 12–18, ASME, New York.

Rabinowicz, E. (1995), *Friction and Wear of Materials*, Second edition, Wiley, New York.

Rabinowicz, E. and Tabor, D. (1951), "Metallic Transfer Between Sliding Metals: An Autoradiographic Study," *Proc. R. Soc. Lond. A* **208**, 455–475.

Rabinowicz, E., Dunn, L.A., and Russell, P.G. (1961), "A Study of Abrasive Wear Under Three-Body Abrasion," *Wear* **4**, 345–355.

Rehbinder, P.A. and Shchukin, E.D. (1972), "Surface Phenomena in Solids During Deformation and Fracture Processes," *Prog. Surface Sci.* **3**, 97–188.

Richardson, R.C.D. (1968), "The Wear of Metals by Relatively Soft Abrasives," *Wear* **11**, 245–275.

Rigney, D.A. (ed) (1981), *Fundamentals of Friction and Wear of Materials*, Amer. Soc. Metals, Metals Park, Ohio.

Rigney, D.A. (1988), "Sliding Wear of Metals," *Ann. Rev. Mater., Sci.* **18**, 141–163.

Rigney, D.A. (1992), "The Role of Characterization in Understanding Debris Generation" in *Wear Particles* (D. Dowson, C.M. Taylor, T.H.C. Childs, M. Godet and G. Dalmaz, eds), pp. 405–412, Elsevier Science Publishers, Amsterdam.

Rigney, D.A. and Glaeser, W.A. (eds.) (1978), *Source Book on Wear Control Technology*, Amer. Soc. Metals, Metal Park, Ohio.

Ruff, A.W., Ives, L.K., and Glaeser, W.A. (1981), "Characterization of Worn Surfaces and Wear Debris," in *Fundamentals of Friction and Wear of Materials* (D.A. Rigney, ed), pp. 235–289, Amer. Soc. Metals, Metals Park, Ohio.

Samuels, L.E., Doyle, E.D., and Turley, D.M. (1981), "Sliding Wear Mechanisms," in *Fundamentals of Friction and Wear of Materials* (D.A. Rigney, ed), pp. 13–41, Amer. Soc. Metals, Metals Park, Ohio.

Schubert, R. (1971), "The Influence of a Gas Atmosphere and its Moisture on Sliding Wear in PTFE Compositions," *ASME J. Lub. Tech.* **93**, 216–223.

Schumacher, W.J. (1977), "Wear and Galling Can Knock Out Equipment," *Chem. Eng.*, Sept. 21, **88**, 155–160.

Scott, D. (1975), "Debris Examination – A Prognostic Approach to Failure Prevention," *Wear* **34**, 15–22.

Scott, D. (ed) (1979), *Wear Treatise on Materials Science and Technology*, Vol. **13**, Academic Press, San Diego, California.

Scott, D. and Westcott, V.C. (1977), "Predictive Maintenance by Ferrography," *Wear* **44**, 173–182.

Shipley, R.J. and Becker, W.T. (2002), *Metals Handbook, Vol. 11: Failure Analysis and Prevention*, ASM International, Metals Park, Ohio.

Smith, R.A. (1980), "Interfaces of Wear and Fatigue," in *Fundamentals of Tribology* (N.P. Suh and N. Saka, eds), pp. 605–616, MIT Press, Cambridge, Massachusetts.
Soderberg, S., Hogmark, S., and Swahn, H. (1983), "Mechanisms of Material Removal During Erosion of a Stainless Steel," *ASLE Trans.* **26**, 161–172.
Steijn, R.P. (1967), "Friction and Wear of Plastics," *Metals Eng. Q.* **7**, 371–383.
Suh, N.P. (1986), *Tribophysics*, Prentice-Hall, Englewood, New Jersey.
Suh, N.P. and Saka, N. (1980), *Fundamentals of Tribology*, MIT Press, Cambridge, Massachusetts.
Takadoum, J. (1993), "Tribological Behavior of Alumina Sliding on Several Kinds of Materials," *Wear* **170**, 285–290.
Tallian, T.E., Baile, G.H., Dalal, H., and Gustafsson, O.G. (1974), *Rolling Bearing Damage*, SKF Industries Inc., King of Prussia, Pennsylvania.
Tambe, N.S. and Bhushan, B. (2008), "Nanoscale Friction and Wear Maps," *Phil. Trans. R. Soc. A* **366**, 1405–1424.
Tsukamoto, Y., Yamaguchi, H., and Yanagisawa, M. (1988), "Mechanical Properties and Wear Characteristics of Various Thin Films for Rigid Magnetic Disks," *IEEE Trans. Magn.* **MAG-24**, 2644–2646.
Venkatesan, S. and Bhushan, B. (1994), "The Sliding Friction and Wear Behavior of Single-Crystal, Polycrystalline and Oxidized Silicon," *Wear* **171**, 25–32.
Wagner, C. and Traud, W. (1938), "Interpretation of Corrosion Phenomena by Superimposition of Electrochemical Partial Reaction and the Formation of Potentials of Mixed Electrodes," *Z. Elektrochem.* **44**, 391–402.
Wallbridge, N., Dowson, D., and Roberts, E.W. (1983), "The Wear Characteristics of Sliding Pairs of High Density Polycrystalline Aluminum Oxide Under Both Dry and Wet Conditions," in *Wear of Materials* (K.C. Ludema, ed), pp. 202–211, ASME New York.
Waterhouse, R.B. (1981), "Fretting Wear," in *Proc. Int. Conf. on Wear of Materials*, pp. 17–22, ASME, New York.
Waterhouse, R.B. (1992), "Fretting Wear," in *ASM Handbook, Vol. 18: Friction, Lubrication and Wear Technology*, pp. 242–256, ASM International, Metals Park, Ohio.
Weibull, W. (1951), "A Statistical Distribution Function of Wide Range of Applicability," *J. Appl. Mech.* **18**, 293–297.
Westwood, A.R.C. (1977), "Environment-Sensitive Fracture of Ionic and Ceramic Solids," *Proc. Int. Conf. on Mechanisms of Environment Sensitive Cracking of Materials* (A.R.C. Westwood et al., eds), pp. 283–297, Metals Soc., London.
Wiederhorn, S.M. (1967), "Influence of Water Vapor on Crack Propagation in Soda-Lime Glass," *J. Amer. Cer. Soc.* **50**, 407–414.
Wiederhorn, S.M. (1969), *Mechanical and Thermal Properties of Ceramics* (J.B. Wachtman, ed), p. 217, NBS Spec. Pub. 303, Gaithersburg, Maryland.
Xie, Y. and Bhushan, B. (1996a), "Effect of Particle Size, Polishing Pad and Contact Pressure in Free Abrasive Polishing," *Wear* **200**, 281–295.
Xie, Y. and Bhushan, B. (1996b), "Fundamental Wear Studies with Magnetic Particles and Head Cleaning Agents Used in Magnetic Tapes," *Wear* **202**, 3–16.
Yamamoto, T., Olsson, M., and Hogmark, S. (1994), "Three-Body Abrasive Wear of Ceramic Materials," *Wear* **174**, 21–31.
Zaretsky, E.V. (ed) (1992), *Life Factors for Rolling Bearings*, Special Publication SP-34, STLE, Park Ridge, Illinois.
Zeman, K.P and Coffin, L.F. (1960), "Friction and Wear of Refractory Compounds," *ASLE Trans.* **3**, 191–202.
Zum Gahr, K.H. (1987), *Microstructure and Wear of Materials*, Elsevier, Amsterdam.

Further Reading

Bartenev, G.M. and Lavrentev, V.V. (1981), *Friction and Wear of Polymers*, Elsevier, Amsterdam.
Bayer, R.G. (1994), *Mechanical Wear Prediction and Prevention*, Marcel Dekker, New York.
Bhushan, B. (1996), *Tribology and Mechanics of Magnetic Storage Devices*, Second edition, Springer-Verlag, New York.
Bhushan, B. (2001a), *Modern Tribology Handbook Vol. 1: Principles of Tribology*, CRC Press, Boca Raton, Florida.
Bhushan, B. (2001b), *Fundamentals of Tribology and Bridging the Gap Between the Macro- and Micro/Nanoscales*, NATO Science Series II-Vol. **10**, Kluwer, Dordrecht, The Netherlands.
Bhushan, B. (2011), *Nanotribology and Nanomechanics I – Measurement Techniques and Nanomechanics, II – Nanotribology, Biomimetics, and Industrial Applications*, Third edition, Springer-Verlag, Heidelberg, Germany.
Bhushan, B. and Gupta, B.K. (1997), *Handbook of Tribology: Materials, Coatings and Surface Treatments*, McGraw-Hill, New York (1991); reprinted by Krieger, Malabar, Florida (1997).

Blau, P.J. (1992), *ASM Handbook, Vol. 18: Friction, Lubrication, and Wear Technology*, Tenth edition, ASM International, Materials Park, Ohio.

Bruce, R.W. (2012), *Handbook of Lubrication and Tribology, Vol. II: Theory and Design*, Second edition, CRC Press, Boca Raton, Florida.

Buckley, D.H. (1981), *Surface Effects in Adhesion, Friction, Wear, and Lubrication*, Elsevier, Amsterdam.

Hutchings, I.M. (1992), *Tribology: Friction and Wear of Engineering Materials*, CRC Press, Boca Raton, Florida.

Kragelski, I.V. (1965), *Friction and Wear*, Butterworths, London.

Loomis, W.R. (ed) (1985), *New Directions in Lubrication, Materials, Wear, and Surface Interactions: Tribology in the 80s*, Noyes Publications, Park Ridge, New Jersey.

Peterson, M.B. and Winer, W.O. (eds) (1980), *Wear Control Handbook*, ASME, New York.

Rabinowicz, E. (1995), *Friction and Wear of Materials*, Second edition, Wiley, New York.

Rigney, D.A. (ed) (1981), *Fundamentals of Friction and Wear of Materials*, Amer. Soc. Metals, Metals Park, Ohio.

Rigney, D.A. and Glaeser, W.A. (eds) (1978), *Source Book on Wear Control Technology*, Amer. Soc. Metals, Metal Park, Ohio.

Scott, D. (ed) (1979), *Wear Treatise on Materials Science and Technology*, Vol. **13**, Academic Press, San Diego, California.

Shipley, R.J. and Becker, W.T. (2002), *Metals Handbook, Vol. 11: Failure Analysis and Prevention*, ASM International, Metals Park, Ohio.

Suh, N.P. (1986), *Tribophysics*, Prentice-Hall, Englewood, New Jersey.

Suh, N.P. and Saka, N. (1980), *Fundamentals of Tribology*, MIT Press, Cambridge, Massachusetts.

Zum Gahr, K.H. (1987), *Microstructure and Wear of Materials*, Elsevier, Amsterdam.

8

Fluid Film Lubrication

8.1 Introduction

Sliding between clean solid surfaces is generally characterized by a high coefficient of friction and severe wear due to the specific properties of the surfaces, such as low hardness, high surface energy, reactivity, and mutual solubility. Clean surfaces readily adsorb traces of foreign substances, such as organic compounds, from the environment. The newly formed surfaces generally have a much lower coefficient of friction and wear than the clean surface. The presence of a layer of foreign material at an interface cannot be guaranteed during a sliding process; therefore, lubricants are deliberately applied to produce low friction and wear. The term "lubrication" is applied to two different situations: solid lubrication and fluid (liquid or gaseous) film lubrication.

A solid lubricant is any material used as a powder or a thin, solid film on a surface to provide protection from damage during a relative movement by reducing friction and wear. Solid lubricants are used for applications in which any sliding contact occurs, for example, a bearing operating at high loads and low speeds and a hydrodynamically lubricated bearing requiring start/stop operations. The term solid lubricants embrace a wide range of materials that provide low friction and wear (Braithwaite, 1967; Clauss, 1972; Bhushan, 1987a, b; Bhushan and Gupta, 1997). Hard materials are also used for low wear and/or under extreme operating conditions. Friction and wear properties of solid lubricants have been presented in Chapters 5 and 7.

A thin film on the order of surface roughness of moving surfaces, results in relatively low friction and wear, as compared to solid–solid contact. A thick fluid film between two surfaces in relative motion prevents solid–solid contact and can provide very low friction (in the range of 0.001–0.003) and negligible wear. Fluid can be liquid or gaseous; even a thick film of air transposed between two moving surfaces is a method of good lubrication. In this chapter, we will describe various regimes of fluid film lubrication and present associated data and mathematical analyses and their application to bearing applications.

Introduction to Tribology, Second Edition. Bharat Bhushan.
© 2013 John Wiley & Sons, Ltd. Published 2013 by John Wiley & Sons, Ltd.

8.2 Regimes of Fluid Film Lubrication

A regime of lubrication, in which a thick film is maintained between two surfaces with little or no relative motion by an external pumping agency, is called hydrostatic lubrication.

A summary of the lubrication regimes observed in fluid lubrication without an external pumping agency (self-acting) can be found in the familiar Stribeck curve in Figure 8.2.1

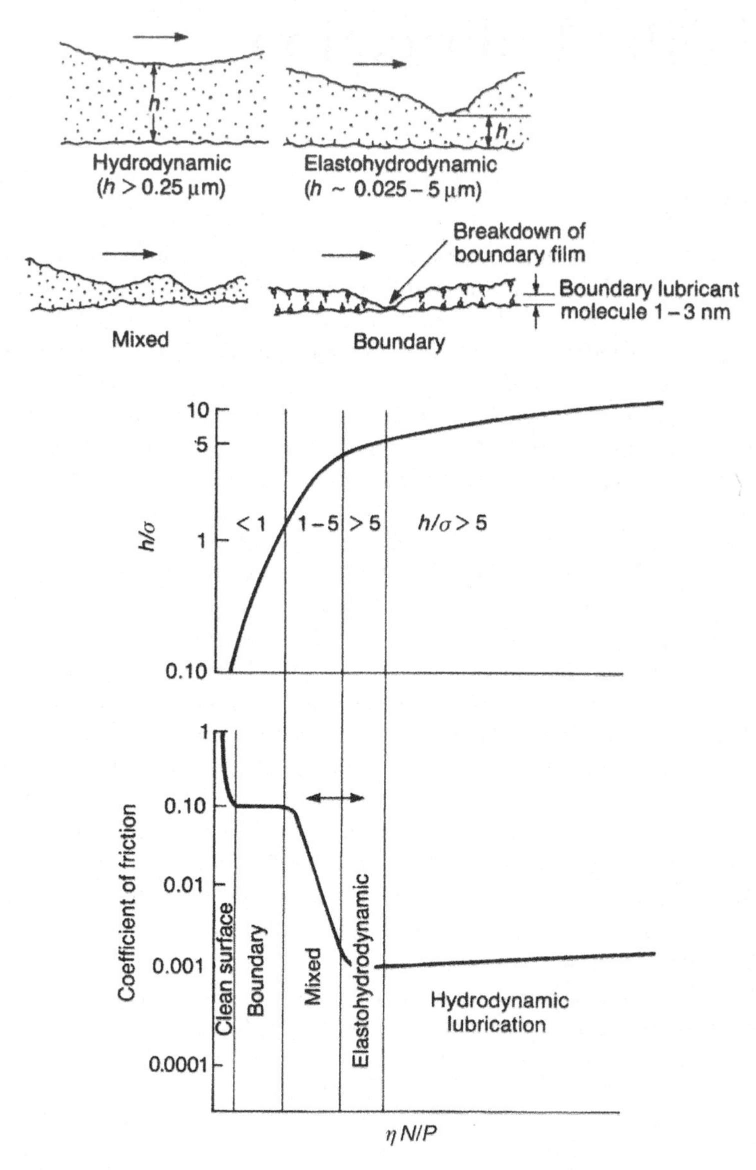

Figure 8.2.1 Lubricant film parameter (h/σ) and coefficient of friction as a function of ηN/P (Stribeck curve) showing different lubrication regimes observed in fluid lubrication without an external pumping agency.

(Stribeck, 1902). This plot for a hypothetical fluid-lubricated bearing system presents the coefficient of friction as a function of the product of absolute viscosity (η) and rotational speed in revolutions per unit second (N) divided by the load per unit projected bearing area (P). The curve has a minimum, which immediately suggests that more than one lubrication mechanism is involved. The regimes of lubrication are sometimes identified by a lubricant film parameter equal to h/σ – (mean film thickness)(composite standard deviation of surface heights of the two surfaces). Descriptions of the different regimes of lubrication follow (Bisson and Anderson, 1964; Wilcock, 1972; Booser, 1984; Fuller, 1984; Bhushan, 2001; Khonsari and Booser, 2001; Hamrock *et al.*, 2004; Stachowiak and Batchelor, 2005; Totten, 2006; Szeri, 2010; Bruce, 2012).

8.2.1 Hydrostatic Lubrication

Hydrostatic bearings support load on a thick film of fluid supplied from an external pressure source, a pump, which feeds pressurized fluid to the film. For this reason, these bearings are often called "externally pressurized." Hydrostatic bearings are designed for use with both incompressible and compressible fluids. Since hydrostatic bearings do not require relative motion of the bearing surfaces to build up the load-supporting pressures as necessary in hydrodynamic bearings, hydrostatic bearings are used in applications with little or no relative motion between the surfaces. Hydrostatic bearings may also be required in applications where, for one reason or another, touching or rubbing of the bearing surfaces cannot be permitted at startup and shutdown. In addition, hydrostatic bearings provide high stiffness. Hydrostatic bearings, however, have the disadvantage of requiring high-pressure pumps and equipment for fluid cleaning, which adds to space and cost.

In hydrostatic bearings, corrosive (chemical) wear of the bearing surfaces occurs as a result of interaction of the lubricant with the interface materials.

8.2.2 Hydrodynamic Lubrication

Hydrodynamic (HD) lubrication is sometimes called fluid-film or thick-film lubrication. As a bearing with convergent shape in the direction of motion starts to move in the longitudinal direction from rest, a thin layer of fluid is pulled through because of viscous entrainment and is then compressed between the bearing surfaces, creating a sufficient (hydrodynamic) pressure to support the load without any external pumping agency, Figure 8.2.1. This is the principle of hydrodynamic lubrication, a mechanism that is essential to the efficient functioning of the hydrodynamic journal and thrust bearings widely used in modern industry. A high load capacity can be achieved in the bearings that operate at high velocities in the presence of fluids of high viscosity. These bearings are also called self-acting bearings (Pinkus and Sternlicht 1961; Cameron, 1976; Gross *et al.*, 1980; Booser, 1984; Fuller, 1984; Frene *et al.*, 1997; Bhushan, 2001; Khonsari and Booser, 2001; Hamrock *et al.*, 2004; Szeri, 2010; Bruce, 2012).

Fluid film can also be generated solely by a reciprocating or oscillating motion in the normal direction towards each other (squeeze) which may be fixed or variable in magnitude (transient or steady state). This load-carrying phenomenon arises from the fact that a viscous fluid cannot be instantaneously squeezed out from the interface with two surfaces that are approaching each other. It takes a finite time for these surfaces to meet and during that interval, because of the fluid's resistance to extrusion, a pressure is built up and the load is actually supported by the fluid film. When the load is relieved or two surfaces move apart, the fluid is sucked in and the fluid film can often recover its thickness in time for the next application. The squeeze phenomenon controls the buildup of a water film under the tires of automobiles and airplanes on wet roadways or landing strips (commonly known as hydroplaning) which have virtually no relative slidng motion (Pinkus and Sternlicht, 1961; Gross *et al.*, 1980; Booser, 1984; Fuller, 1984; Frene *et al.*, 1997; Bhushan, 2001; Khonsari and Booser, 2001; Hamrock *et al.*, 2004; Szeri, 2010; Bruce, 2012). The squeeze-film effect is used to reduce friction at the interfaces (Tam and Bhushan, 1987).

HD lubrication is often referred to as the ideal lubricated contact condition because the lubricating films are normally many times thicker (typically 5–500 μm) than the height of the irregularities on the bearing surface, and solid contacts do not occur. The coefficient of friction in the HD regime can be as small as 0.001, Figure 8.2.1. The friction increases slightly with the sliding speed because of viscous drag. Physical contact occurs during start-stop operations at low surface speeds. The behavior of the contact is governed by the bulk physical properties of the lubricant, notably viscosity, and the frictional characteristics arise purely from the shearing of the viscous lubricant. The behavior of the contact is determined from the solution of the Reynolds equation. This will be discussed in detail later.

In HD lubrication, adhesive wear occurs during start-stop operations and corrosive (chemical) wear of the bearing surfaces can also occur as a result of interaction with the lubricant. One of the most effective ways to minimize corrosive wear is by the participation of the lubricant and bearing surface in the formation of a relatively complete and inert film on the bearing surface. In ferrous bearing systems, this can be accomplished with phosphate-containing additives or organo-metal salts. This mechanism produces a film that appears as a blue or brown stain.

8.2.3 *Elastohydrodynamic Lubrication*

Elastohydrodynamic (EHD) lubrication (EHL) is a subset of HD lubrication in which the elastic deformation of the contacting solids plays a significant role in the HD lubrication process. The film thickness in EHD lubrication is thinner (typically 0.5–5 μm) than that in conventional HD lubrication, Figure 8.2.1, and the load is still primarily supported by the EHD film. In isolated areas, asperities may actually touch. Therefore, in liquid lubricated systems, boundary lubricants that provide boundary films on the surfaces for protection against any solid–solid contact are used. Bearings with heavily loaded contacts fail primarily by a fatigue mode that may be significantly affected by the lubricant.

EHL is most readily induced in heavily loaded contacts (such as machine elements of low geometrical conformity), where loads act over relatively small contact areas (on the order of one-thousandth of the apparent area of a journal bearing), such as the point contacts of ball bearings and the line contacts of roller bearings and of gear teeth (Pinkus and Sternlicht, 1961; Dowson and Higginson, 1966; Cameron, 1976; Harris 1991; Bhushan, 2001; Khonsari and Booser, 2001; Hamrock *et al.*, 2004; Szeri, 2010; Bruce, 2012). EHD phenomena also occur in some low elastic modulus contacts of high geometrical conformity, such as lip seals, conventional journal and thrust bearings with soft liners, and head–tape interface in magnetic recording tape drives (Gross *et al.*, 1980; Bhushan, 1996).

In heavily loaded contacts, high pressures can lead to both changes in the viscosity of the lubricant and elastic deformation of the bodies in contact, with consequent changes in the geometry of the bodies bounding the lubricant film. Therefore, hydrodynamic solutions that are used to study journal and thrust bearings have to be modified. In EHL, one is faced with the simultaneous solutions of the Reynolds equation, the elastic deformation equation, and the equation relating viscosity and pressure. Thermal and shear rate effects also become important and need to be taken into account.

In EHL, adhesive wear occurs during start–stop operations and corrosive wear of the bearing surfaces can also occur as a result of interaction with the lubricant. In well-designed heavily loaded bearings, fatigue wear is most common.

8.2.4 Mixed Lubrication

The transition between the hydrodynamic/elastohydrodynamic and boundary lubrication regimes is a gray area known as a mixed lubrication in which two lubrication mechanisms may be functioning. There may be more frequent solid contacts, but at least a portion of the bearing surface remains supported by a partial hydrodynamic film, Figure 8.2.1. The solid contacts, between unprotected virgin metal surfaces, could lead to a cycle of adhesion, metal transfer, wear particle formation, and eventual seizure. However, in liquid lubricated bearings, physi- or chemisorbed or chemically reacted films (boundary lubrication) prevent adhesion during most asperity encounters. The mixed regime is also sometimes referred to as quasi-hydrodynamic, partial fluid, or thin-film (typically 0.025–2.5 μm) lubrication.

8.2.5 Boundary Lubrication

As the load increases, speed decreases or the fluid viscosity decreases in the Stribeck curve shown in Figure 8.2.1, and the coefficient of friction can increase sharply and approach high levels (about 0.1 or much higher). In this region, it is customary to speak of boundary lubrication. This condition can also occur in a starved contact. Boundary lubrication is that condition in which the solid surfaces are so close together that surface interaction between monomolecular or multimolecular films of lubricants (liquids or gases) and the solid asperities dominates the contact. (It does not apply to solid lubricants.) The concept is represented in Figure 8.2.1, which shows a microscopic cross section of films on two surfaces and areas of asperity contact (Bowden and Tabor, 1950; Ling *et al.*, 1969; Ku, 1970; Beerbower, 1972; Booser, 1984; Bhushan, 2001; Bruce, 2012). In the absence of boundary lubricants and gases (no oxide films), friction may become very high (> 1). All self-acting bearing interfaces during contact start–stops (CSS), before a fluid film as a result of HD or EHL is developed, operate in the boundary lubrication regime.

Failure in boundary lubrication occurs by adhesive and chemical (corrosive) wear. Boundary lubricants form an easily sheared film on the bearing surfaces, thereby minimizing adhesive wear and chemical wear. The important physical properties of the films are the melting point, shear strength, and hardness. Other properties are adhesion or tenacity, cohesion, and rates of formation. The bulk flow properties of the lubricant (such as viscosity) play little part in the friction and wear behavior.

8.3 Viscous Flow and Reynolds Equation

8.3.1 Viscosity and Newtonian Fluids

8.3.1.1 Definition of Viscosity

Sir Isaac Newton (1642–1727) was the first to propose that a force is necessary to shear a fluid film. This force resembles friction between two solid surfaces. The force is a measure of the internal friction of the fluid or its resistance to shear. For two surfaces separated by a fluid film of thickness h and having relative motion at a linear velocity, u_a, Figure 8.3.1, the force per unit swept area (F/A) or shear stress (τ) is proportional to the velocity gradient (du/dh) or shear strain rate $\dot{\gamma}$ (or simply shear rate) in the film,

$$\tau = \frac{F}{A} = \eta\dot{\gamma} = \eta\frac{du}{dh} \tag{8.3.1}$$

where η is known as absolute (dynamic) viscosity. If the velocity is a continuous function of the film thickness and there is no slip at the interface between the fluid film and the solid surfaces, $du/dh = u_a/h = \dot{\gamma}$, then Equation 8.3.1 reduces to

$$\tau = \eta\frac{u_a}{h} \tag{8.3.2}$$

The units of η are lb s/in^2 (Reyn) or dynes s/cm^2 (Poise or P). Conversion of unit are 1 cP = 1 mPa s and 1 Reyn = 68,750 P. Another measure of viscosity, kinematic viscosity (ν), equals η divided by density (ρ),

$$\nu = \frac{\eta}{\rho} \tag{8.3.3}$$

The unit for ν is mm^2/s (centiStoke or cSt). Viscosity is an important property of the lubricants in fluid film lubrication.

8.3.1.2 Types of Viscometers

Several types of viscometers are commonly used to measure absolute viscosity. Most commonly used instruments typically fall into three categories based on geometry: capillary,

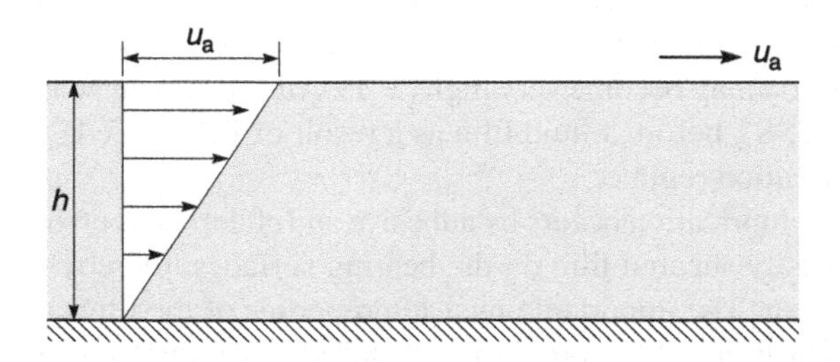

Figure 8.3.1 Schematic of two parallel plates in relative motion with a velocity of u_a, separated by a fluid film of thickness h with a linear velocity gradient.

rotational, and falling sphere viscometers (Van Wazer *et al.*, 1963; Walters, 1975; Fuller 1984). The oldest technique to measure viscosity is capillary viscometry which is based on measuring the rate at which fluid is forced though a fine-bore tube, and the viscosity of the fluid is determined from the measured volumetric flow rate for an applied pressure difference and tube dimensions. The most common type of rotational viscometer devised by Couette in 1890 is the coaxial-cylinder viscometer in which the viscosity is determined by shearing the fluid between two relatively rotating surfaces. In this technique, one of the cylinders moves concentrically with respect to the other with the space between the two members filled with test fluid. The viscosity measurements are made either by applying a fixed torque and measuring the speed of rotation, or by driving the rotating element at a constant speed and measuring the torque required. The cone and plate and parallel-plate rotational rheometers are the variations of the Couette technique. For viscosity measurements of non-Newtonian fluids, rotational types of viscometers involving shearing of the fluid are used. In a falling-sphere viscometer, the time taken for a ball to fall through a measured height of fluid in a glass tube is measured. The time required is a measure of absolute viscosity.

A capillary viscometer measures absolute viscosity when flow is caused by a constant pressure difference. If the flow is caused by means of a head of fluid, then the force produced to cause flow depends upon the density of liquid. And the viscosity measured is kinematic viscosity. The kinematic type of viscometer is cheaper and easier to operate and is commonly used. The most widely used instrument of the capillary type viscometer is the Saybolt Universal Viscometer. It measures the time required, in seconds, for 60 cm^3 of the sample to flow through the tube, known as SUS viscosity. SUS viscosity in seconds can be converted to kinematic viscosity in cSt by using empirical equations (Fuller, 1984). If the density of the fluid is known, absolute viscosity can be calculated from the kinematic viscosity.

SAE (Society of Automotive Engineers) and API (American Petroleum Institute) ratings are used to identify permissible ranges in viscosity, not a specific value of viscosity.

8.3.1.3 Effect of Temperature, Pressure, and Shear Rates on Viscosity

Viscosity of fluids changes as a function of temperature, pressure, and in many cases, shear strain rates. The viscosity of a liquid is primarily due to intermolecular forces. As the temperature is increased, the liquid expands, the molecules move farther apart and the intermolecular forces decrease which results in a decrease of viscosity, for example see Figure 8.3.2a. A simple expression for viscosity-temperature dependence of a liquid is given as

$$\eta = \eta_0 \exp\left[\beta\left(\frac{1}{T} - \frac{1}{T_0}\right)\right] \tag{8.3.4}$$

where η and η_0 are the viscosity at temperature T and reference temperature T_0, respectively, and both at ambient pressure, and β is the temperature-viscosity coefficient. An expression which fits the experimental data of liquids better is given by Roelands (1966). In the case of gases, the dominant contribution to their viscosity is the momentum transfer. As the gas temperature is raised, the velocity of molecules increases which results in an increase in momentum transfer and consequently an increase in viscosity. Thus the effect of temperature on viscosity of gases is opposite to that for liquids; for an example see Figure 8.3.2b.

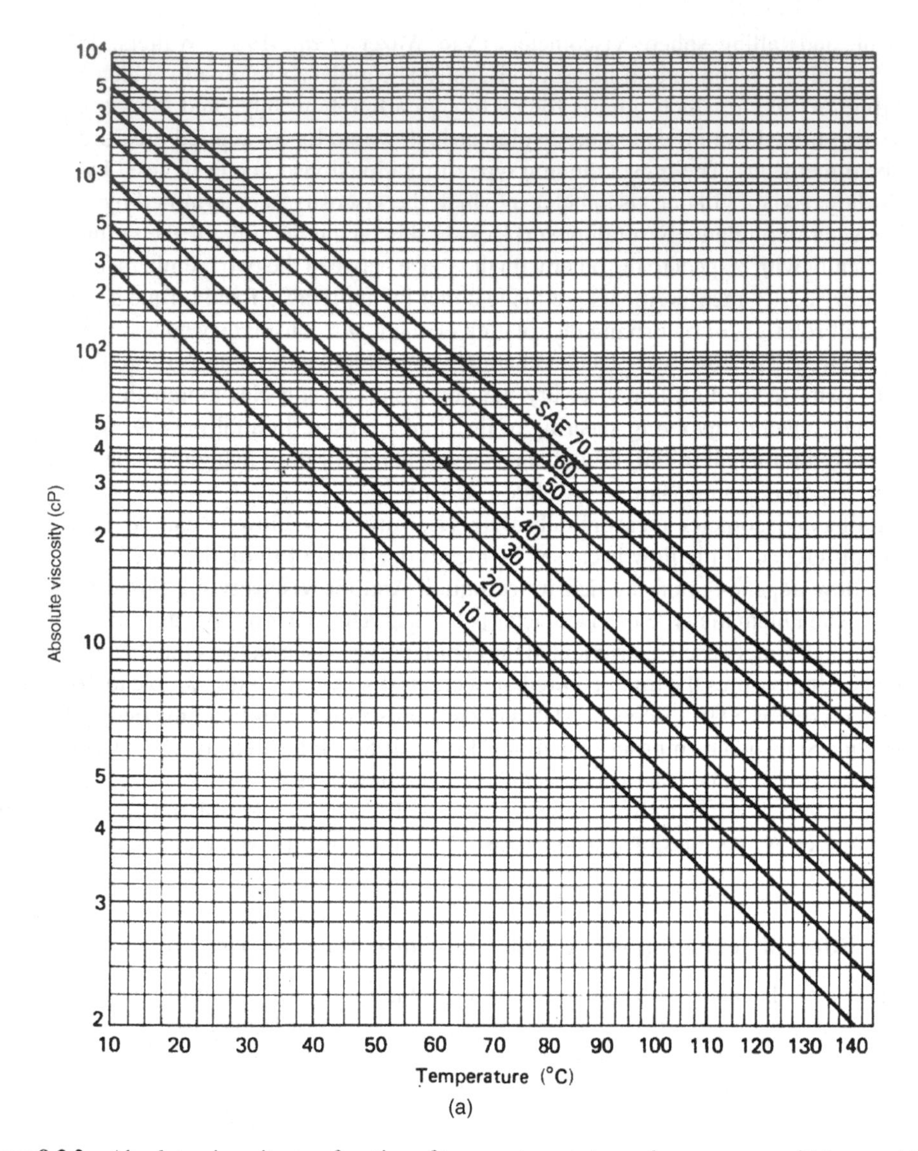

Figure 8.3.2 Absolute viscosity as a function of temperature at atmosphere pressure of (a) several SAE petroleum-based oils and (b) air. (*Continued*)

The relationship between viscosity and temperature for petroleum-based or mineral oils is identified based on an arbitrary system of comparison using the viscosity index (VI). This relates the change in viscosity of the sample lubricant at two temperatures, 38°C and 100°C, to two arbitrary oils. At the time of its introduction, the natural mineral oils which showed the least variation of viscosity with temperature came from Pennsylvania oil fields and were given a VI of 100 and the oils which suffered a greatest decrease of viscosity with temperature came from the Gulf of Mexico and were given a VI of 0. For calculations of viscosity index, viscosities of the two reference oils and the sample oil are assumed to be equal at 100°C and

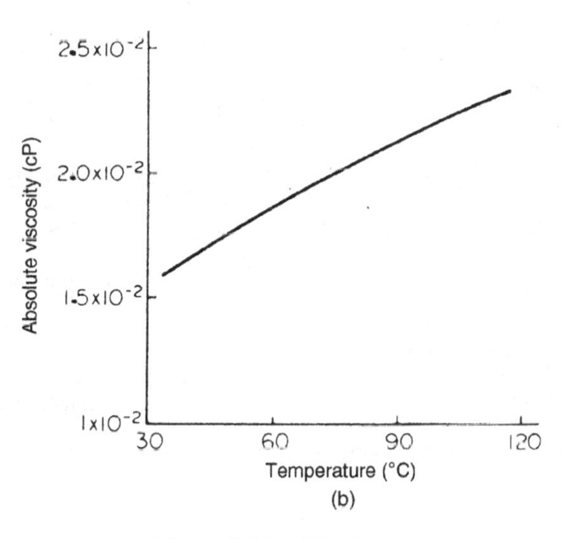

Figure 8.3.2 *(Continued)*

VI for the sample oil is calculated graphically based on the relative viscosity of the sample oil at a temperature of 38°C.

When the pressure of a liquid or gas is increased, the molecules are forced closer together. This increases the intermolecular forces and consequently viscosity. It is known that viscosity of petroleum-based oils increases rapidly with an increase in pressure. The viscosity may increase by several orders of magnitude. Some oils become plastic at pressures on the order of 200 MPa. In 1893, C. Barus proposed the following relationship for the isothermal viscosity—pressure dependence of liquids (Barus, 1893)

$$\eta = \eta_0 \exp(\alpha\, p) \tag{8.3.5a}$$

where η and η_0 are the viscosities at pressure p (above ambient) and normal atmosphere, respectively, α is viscosity-pressure coefficient in Pa^{-1} (m^2/N) and p is the normal pressure in Pa. α for petroleum-based oils at 38°C is on the order of $2 \times 10^{-8}\ \mathrm{Pa}^{-1}$. An expression for isothermal viscosity–pressure dependence has been proposed by Roelands (1966) which better fits the experimental data.

Equations 8.3.4 and 8.3.5a can be combined as follows:

$$\eta = \eta_0 \exp\left[\alpha\, p + \beta\left(\frac{1}{T} - \frac{1}{T_0}\right)\right] \tag{8.3.5b}$$

Study of viscous properties as a function of shear rate is referred to as fluid rheology. A fluid which follows Equation 8.3.2 is called a "Newtonian fluid," Figure 8.3.3. Fluids whose viscosities vary as a function of shear rate are known as non-Newtonian. Non-Newtonian behavior is, in general, a function of structural complexibility of the fluid. Liquids with loose molecular structure, such as water and highly dispersed suspensions of solids, may behave as Newtonian fluids. In the so-called pseudo-plastic fluids, thinning of the fluid occurs with an increase in the shear rate, known as shear-thinning. These fluids are usually composed of

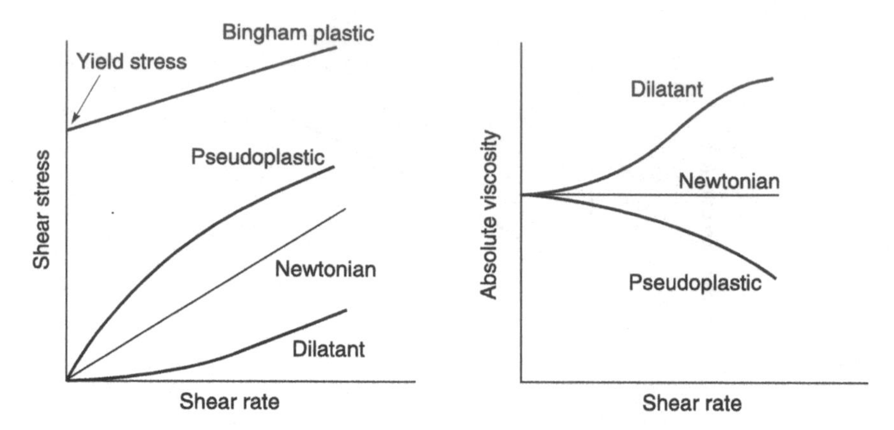

Figure 8.3.3 Schematic curves to show dependence of shear rate on shear stress and absolute viscosity for various fluids.

long molecules which are randomly oriented with no connecting structure. Application of a shear stress tends to align the molecules, giving a reduction in the apparent viscosity. In the so-called dilatant fluids, thickening of the fluid occurs with an increase in the shear stress, known as shear-thickening. Fluids which exhibit dilatancy are usually suspensions having a high solid content and their behavior can be related to the arrangement of particles. For some liquids, known as plastic fluids, or Bingham plastic fluid, some shear stress is required before flow begins, Figure 8.3.3. The Bingham fluids usually possess a three-dimensional structure, which can resist a certain value of shear stress known as yield stress. Many greases behave as Bingham fluids. Any material whose viscosity is dependent upon its previous shearing history is known as thixotropic material. All solid and liquid polymers are thixotropic to some extent.

Many liquids exhibit non-Newtonian behavior at high shear rates. Viscosity starts to drop above a certain strain rate and the fluid exhibits non-Newtonian behavior, known as "shear thinning." Thermal thinning as a result of viscous heating at high shear rates also results in a drop in viscosity. In some cases, at high shear rates, lubricant becomes plastic and can only support a constant stress known as the limiting shear strength, τ_L, at high shear rates, Figure 8.3.4. The limiting shear strength is a function of temperature and pressure; it increases at higher pressures and at lower temperatures. The value of shear rate at which the viscous-plastic transition occurs increases with a decrease in the pressure and an increase in the temperature. At high pressures, on the order of 0.1 to 1 GPa relevant for nonconforming contacts, such as in rolling element bearings, most liquid lubricants behave as a plastic solid at relatively low shear rates (on the order 10–100/s) and the data fits in the following rheological model (Bair and Winer, 1979)

$$\frac{\tau}{\tau_L} = 1 - \exp\left(-\frac{\eta_0 \dot{\gamma}}{\tau_L}\right) \tag{8.3.6a}$$

$$\tau_L = \tau_0 + \zeta\, p \tag{8.3.6b}$$

where τ_0 is the shear strength at normal atmosphere and ζ is the limiting-shear-strength proportionality constant, $\partial\tau_L/\partial p$. This model suggests that for $\eta_0\,\dot{\gamma}/\tau_L > 5$, the material becomes a plastic solid. The limiting shear strength is linearly dependent on pressure.

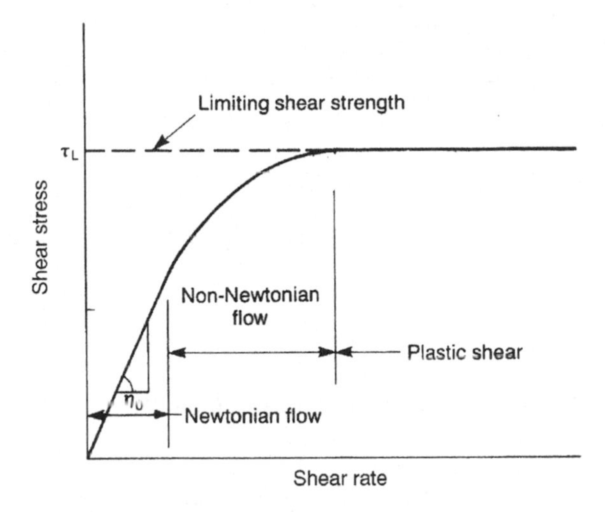

Figure 8.3.4 Schematic curve to show dependence of shear rate on shear stress for many liquid lubricants (non-Newtonian flow).

8.3.2 Fluid Flow

8.3.2.1 Turbulence and Laminar Flow

The analyses of fluid flow are mostly based on the existence of laminar viscous flow. Based on O. Reynolds' observations in 1886, laminar flow implies that the fluid flows as in a series of parallel or concentric surfaces or layers, with relative velocities but no mixing between the layers. Laminar flow occurs in bearings and machine elements at low relative velocities. At high velocities, turbulence occurs in the fluid film. The critical flow velocity at which turbulence is initiated is based on the dimensional Reynolds number, a ratio of interia to viscous forces, given as

$$\mathrm{Re} = \frac{\rho\, v\, d}{\eta} = \frac{vd}{\nu} \tag{8.3.7}$$

where v is the linear velocity and d is the diameter of a tube for flow through a tube or the film thickness for flow between the two surfaces. Generally, a Reynolds number of about 2000 is the critical value above which turbulence occurs.

8.3.2.2 Petroff's Equation

For a concentric (lightly loaded) journal bearing shown in Figure 8.3.5, the friction force in journal bearings for Newtonian flow in Equation 8.3.2 is given as

$$F = \eta_0 \frac{u_a}{h} A \tag{8.3.8a}$$

where A is the surface area of the bearing interface, h is the film thickness or bearing clearance, c, u_0 is the relative velocity, and η_0 is the viscosity at ambient pressure and constant temperature.

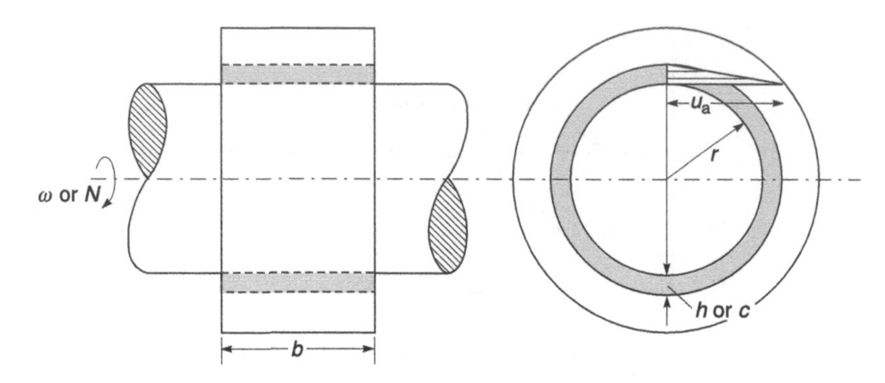

Figure 8.3.5 Schematic of a concentric journal bearing.

For a bearing of radius r, width b, and rotating at an angular velocity ω in radians per second

$$F = \frac{2\pi\,\eta_0\,r^2\,b\,\omega}{h} \tag{8.3.8b}$$

The coefficient of friction is given as

$$\mu = \frac{F}{W} = \frac{2\pi\eta_0 r^2 b\omega}{Wh} \tag{8.3.9}$$

where W is the normal applied load. Friction torque is given as

$$\begin{aligned} T &= Fr \\ &= \frac{2\pi\,\eta_0\,r^3\,b\,\omega}{h} \end{aligned} \tag{8.3.10}$$

Equation 8.3.10 was first proposed by Petroff (1883) and is known as Petroff's equation.

The power loss from viscous dissipation, H_v is the friction force times the velocity:

$$\begin{aligned} H_v &= F\,u_a \\ &= \mathrm{T}\omega \\ &= \frac{2\pi\,\eta_0\,r^3\,b\,\omega^2}{h} \quad (8.3.11a) \\ &= \frac{8\pi^3\,\eta_0\,r^3\,b\,N^2}{3600\ \mathrm{h}} \quad (8.3.11b) \end{aligned}$$

where N is the rotational velocity in rpm ($\omega = 2\pi\,N$). The power loss is expressed either in kilowatts (kN m/s) or horse power (550 ft lb/s).

The power loss results in a temperature rise of the fluid during viscous flow.

Example Problem 8.3.1

Consider two concentric cylinders filled with an SAE 30 oil at 38°C with an absolute viscosity of 100 cP (mPa s). The radius of the inner cylinder is 50 mm, radial clearance between cylinders is 0.5 mm, and their width is 100 mm. For the outer cylinder rotating at 300 rpm, calculate the friction torque and power loss in hp acting on the inner cylinder.

Solution

Given,

$$\eta_0 = 0.1 \text{ Pa s}$$

$$r = 50 \text{ mm}$$

$$b = 100 \text{ mm}$$

$$\omega = \frac{300}{60} \times 2\pi = 31.4 \text{ rad/s}$$

$$h = 0.5 \text{ mm}$$

$$T = \frac{2\pi\, \eta_0\, r^3\, b\, \omega}{h}$$

$$= \frac{2\pi \times 0.1 \times 0.05^3 \times 0.1 \times 31.4}{5 \times 10^{-4}} \text{ N m}$$

$$= 0.123 \text{ N m}$$

$$\text{Power loss} = T\omega$$

$$= 0.123 \times 31.4 \text{ W}$$

$$= 3.86\ W = 2.35 \text{ hp}$$

8.3.2.3 One-Dimensional Flow Between Parallel Plates

Consider flow through the clearance h between two parallel surfaces of width b and length ℓ along the x-axis, with the top surface moving with a velocity u_a and the bottom surface at rest, Figure 8.3.6a. If the width b is large compared to the length ℓ, side flow can be neglected and the fluid flow can be assumed as one-dimensional along the length axis. Assume that the fluid is Newtonian and the flow is laminar. Further, assume that inertia and body (gravity) forces can be neglected, viscosity of the fluid is constant, η_0, and the fluid film thickness is much smaller than other dimensions. For this case, the simplified Navier-Stokes equation (Bhushan, 2013) is given as

$$\eta_0 \frac{\partial^2 u}{\partial z^2} = \frac{dp}{dx} \tag{8.3.12}$$

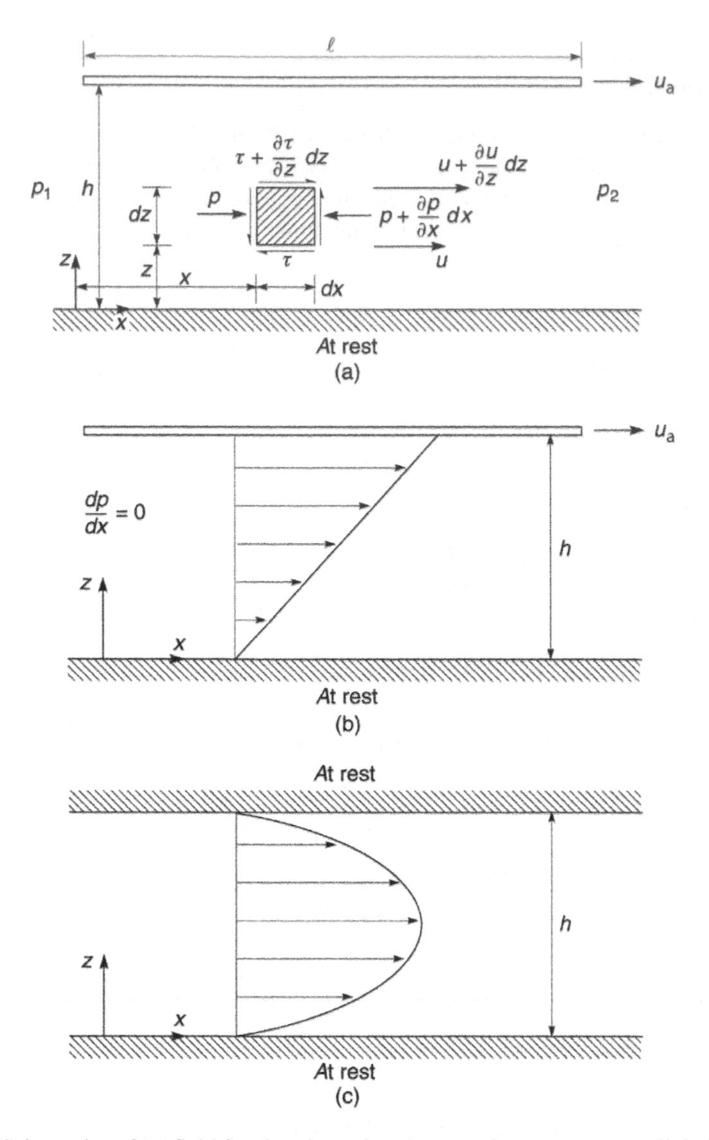

Figure 8.3.6 Schematics of (a) fluid flowing through a clearance between two parallel plates as a result of pressure difference acting on it and stresses acting on a fluid element, and (b) linear velocity profile with two plates in relative motion, and (c) parabolic velocity profile with both plates and at rest.

dp/dx is not a function of z. Equation 8.3.12 is integrated twice to get an expression for u. Using no-slip boundary conditions: at $z = 0$, $u = 0$ and at $z = h$, $u = u_a$, we get

$$u = \frac{1}{2\eta_0}\left(-\frac{dp}{dx}\right) z\,(h - z) + u_a \frac{z}{h} \tag{8.3.13}$$

It is clear that the pressure gradient must be negative if fluid flow proceeds to the right. The total velocity at any value of z is given by the sum of the pressure-induced (or Poiseuille) velocity and the shear (or Couette) velocity induced by the movement of the upper surface. The Poiseuille velocity has a parabolic distribution in z, Figure 8.3.6c and the Couette velocity has a linear distribution in z, Figure 8.3.6b.

For the case of $u_a = 0$, maximum velocity occurs at the center of clearance $z = h/2$,

$$u_{\max} = \frac{h^2}{8\eta_0}\left(-\frac{dp}{dx}\right) \tag{8.3.14a}$$

and the average velocity is two thirds of the maximum velocity,

$$u_{av} = \frac{2}{3}\,u_{\max} \tag{8.3.14b}$$

Now, we calculate the volumetric flow rate per unit width through the gap as

$$q = \int_0^h u\,dz = \frac{h^3}{12\eta_0}\left(-\frac{dp}{dx}\right) + \frac{u_a h}{2} \tag{8.3.15a}$$

The volumetric flow rate does not vary with x. Therefore, it can also be directly obtained by multiplying the average velocity by the cross-sectional area

$$q = u_{av}\,h \tag{8.3.15b}$$

Since, q does not vary with x, the pressure gradient is constant

$$\frac{dp}{dx} = \frac{p_2 - p_1}{\ell} \tag{8.3.16}$$

The expression for $\frac{dp}{dx}$ in Equation 8.3.16 can be used in Equations. 8.3.13, 8.3.14a and 8.3.15a for calculations of velocity and volumetric rate of flow, respectively.

The friction loss or power loss is given as

$$H_v = q\,b\,(p_1 - p_2) \tag{8.3.17}$$

Viscous resistance to flow during fluid being forced through a gap results in temperature rise. If all of the friction losses are dissipated as heat which is assumed to be carried away by the fluid, then the increase in the fluid temperature is

$$q\,(p_1 - p_2) = (q\rho)\,c_p\,\Delta t$$

where ρ is the mass density (in kg/m^3) c_p is the specific heat of the fluid (in J/g K) and Δt is the temperature rise (°C). Therefore,

$$\Delta t = \frac{p_1 - p_2}{\rho\,c_p} \tag{8.3.18}$$

Example Problem 8.3.2

Consider oil flow of absolute viscosity of 100 cP through a gap 200 mm wide, 2 m long, and with a pressure difference of 1 MPa. Calculate maximum and average velocities, volumetric flow rate in liters/s and the temperature rise of the oil. Mass density and specific heat of oil are 880 kg/m^3 and 1.88 J/g K, respectively.

Solution

Given

$$
\begin{aligned}
p_1 - p_2 &= 1 \text{ MPa} \\
h &= 200\ \mu\text{m} \\
\eta_0 &= 0.1 \text{ Pa s} \\
\ell &= 2 \text{ m} \\
b &= 200 \text{ mm} \\
u_{\max} &= \frac{(p_1 - p_2)\, h^2}{8\, \eta_0\, \ell} \\
&= \frac{10^6 \times 2^2 \times 10^{-8}}{8 \times 0.1 \times 2} \text{ m/s} \\
&= 25 \text{ mm/s} \\
u_{avg} &= \frac{2}{3} u_{\max} \\
&= 16.65 \text{ mm/s} \\
Q &= \text{qb} = u_{av}\, bh = 16.65 \times 200 \times 0.2 \text{ mm}^3/\text{s} \\
&= 16.65 \text{ mm/s} \\
&= 666 \text{ mm}^3/\text{s} \\
&= 6.66 \times 10^{-4} \text{ liters/s} \\
\Delta t &= \frac{p_1 - p_2}{\rho\, c_p} = \frac{10^6}{880 \times 10^3 \times 1.88} \,^\circ\text{C} \\
&= 0.60^\circ\text{C}
\end{aligned}
$$

8.3.2.4 Reynolds Equation

The differential equation governing the pressure distribution in fluid film lubrication was first derived by O. Reynolds in 1886, for incompressible fluid (Reynolds, 1886). This was an unnecessary restriction, and later the effects of compressibility were included. The Reynolds

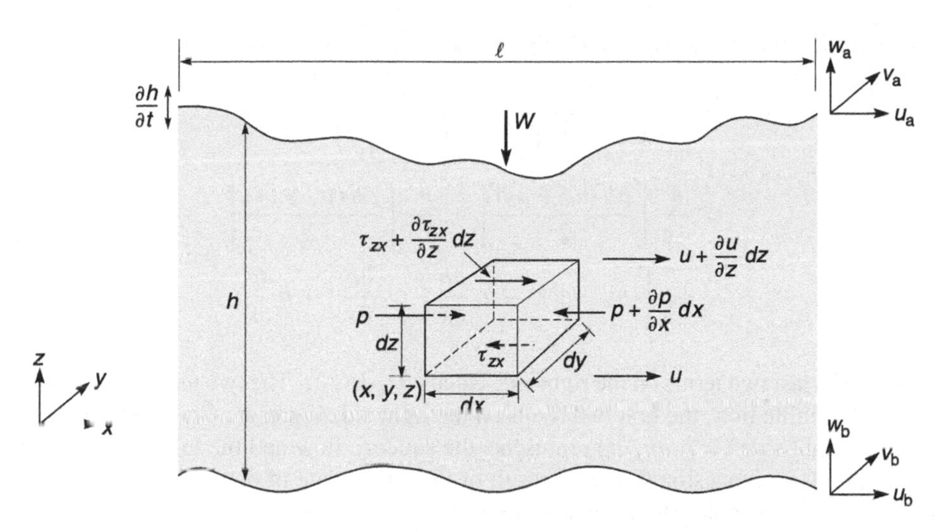

Figure 8.3.7 Schematic of fluid flowing between two surfaces and stresses acting on a fluid element and the velocities in the x-z plane. First of the two subscripts in the shear stress indicates the direction normal to the side of the element on which the component acts and the second subscripts indicates the axis to which the stress or strain component arrow is parallel.

equation forms the foundation of fluid film lubrication theory. This equation establishes a relation between the geometry of the surfaces, relative sliding velocity, the property of the fluid and the magnitude of the normal load the bearing can support. The Reynolds equation can be derived either from the Navier-Stokes equations of fluid motion and the continuity equation or from the laws of viscous flow and from the principles of mass conservation and the laws of viscous flow (Pinkus and Sternlicht, 1961; Cameron, 1976; Gross *et al.*, 1980; Frene *et al.*, 1997; Hamrock *et al.*, 2004; Khonsari and Booser, 2001; Bhushan, 2001, 2013; Szeri, 2010).

We analyze fluid flow between two surfaces with the upper surface moving at velocities u_a, v_a and w_a and the lower surface moving at velocities u_b, v_b and w_b along the x, y and z axes, respectively, Figure 8.3.7. We consider flow in a fluid element in the viscous fluid of length Δx at a distance x from the origin, of width Δy at a distance y from the origin and of thickness Δz at a height z from the origin. For simplications, a number of justifiable assumptions are made for the case of slow viscous motion in which pressure and viscous terms predominate. These assumptions are: (1) the surfaces are smooth, (2) the fluid is Newtonian and the flow is laminar, (3) inertia forces resulting from acceleration of the liquids ($\partial u/\partial t = 0$, $\partial v/\partial t = 0$, $\partial w/\partial t = 0$) and body forces are small compared with the surface (viscous shear) forces and may be neglected, (4) surface tension effects are negligible, (5) the fluid film thickness is much smaller than other bearing dimensions so that curvature of the fluid film can be ignored, (6) at any location, the pressure, density and viscosity are constant across the fluid film, i.e. $\partial p/\partial z = \partial \rho/\partial z = \partial \eta/\partial z = 0$, (7) nonslip boundary conditions are obeyed at the walls, i.e. at the bearing surfaces the velocity of the fluid is identical with the surface velocity, and (8) compared with the two velocity gradients $\frac{\partial u}{\partial z}$ *and* $\frac{\partial v}{\partial z}$, all other velocity gradients are negligible since u and v are usually much greater than w, and z is a much smaller dimension than x and y.

The generalized Reynold equation is derived as follows (Bhushan, 2013):

$$\frac{\partial}{\partial x}\left(\frac{\rho h^3}{12\eta}\frac{\partial p}{\partial x}\right)+\frac{\partial}{\partial y}\left(\frac{\rho h^3}{12\eta}\frac{\partial p}{\partial y}\right)$$
$$=\frac{\partial}{\partial x}\left[\frac{\rho h\left(u_a+u_b\right)}{2}\right]+\frac{\partial}{\partial y}\left[\frac{\rho h\left(v_a+v_b\right)}{2}\right]$$
$$+\rho\left[\left(w_a-w_b\right)-u_a\frac{\partial h}{\partial x}-v_a\frac{\partial h}{\partial y}\right]+h\frac{\partial \rho}{\partial t} \quad (8.3.19)$$

The sum of the last two terms on the right side equals $\partial\left(\rho h\right)/\partial t$. The two terms on the left side represent Poiseuille flow, the first two terms on the right side represent Couette flow, the third term on the right side ($=\rho\ \partial h/\partial t$) represents the squeeze flow and the last term in the right represent the local expansion flow as a result of local time rate of density. The squeeze flow term ($=\rho\,\partial h/\partial t$) includes the normal squeeze term, $\rho\left(w_a-w_b\right)$ and translational squeeze terms $-\rho u_a\,\partial h/\partial x-\rho v_a\partial h/\partial y$. The normal squeeze term results from the difference in the normal velocities and the translational squeeze term results from the translation of inclined surfaces.

The generalized Reynolds equation provides a relationship between the film thickness and the fluid pressure. Density and viscosity of the fluid are a function of pressure and temperature and their values at local conditions need to be used. There is no general closed form solution for this equation. Boundary conditions and other simplifications are required to solve the Reynolds equation by numerical methods. For relatively low interface pressures in hydrodynamic lubrication, the viscosity of fluids can be assumed to be constant.

We now look at *special cases*. First, consider the case of pure tangential motion under steady state conditions, where $\partial h/\partial t=0$ and $w_b=0$ or $w_a=u_a\ \partial h/\partial x+\ v_a\,\partial h/\partial y$ and there is no change in viscosity with time. The Reynolds equation for this case is given as

$$\frac{\partial}{\partial x}\left(\frac{\rho h^3}{\eta}\frac{\partial p}{\partial x}\right)+\frac{\partial}{\partial y}\left(\frac{\rho h^3}{\eta}\frac{\partial p}{\partial y}\right)=12\overline{u}\frac{\partial\left(\rho h\right)}{\partial x}+12\overline{v}\frac{\partial\left(\rho h\right)}{\partial y} \quad (8.3.20)$$

where $\overline{u}=\frac{u_a+u_b}{2}=$ constant and $\overline{v}=\frac{v_a+v_b}{2}=$ constant. $\overline{u}$ *and* $\overline{v}$ are known as entraining velocities. For rolling or sliding motion such that $\overline{v}$ is zero, the last term on the right hand side drops out.

For a gas-lubricated bearing with perfect gas,

$$p=\rho RT \quad (8.3.21)$$

where R is gas constant (= universal gas constant divided by molecular weight) and T is the absolute temperature. Therefore, ρ is replaced by p in the Reynolds equation. For unidirectional tangential (rolling or sliding) motion,

$$\frac{\partial}{\partial x}\left(\frac{ph^3}{\eta}\frac{\partial p}{\partial x}\right)+\frac{\partial}{\partial y}\left(\frac{ph^3}{\eta}\frac{\partial p}{\partial y}\right)=12\overline{u}\frac{\partial\left(ph\right)}{\partial x} \quad (8.3.22)$$

Liquids can be assumed to be incompressible, i.e. their density remains constant during flow. For an incompressible fluid and unidirectional tangential (rolling or sliding) motion, the Reynolds equation is given as

$$\frac{\partial}{\partial x}\left(\frac{h^3}{\eta}\frac{\partial p}{\partial x}\right)+\frac{\partial}{\partial y}\left(\frac{h^3}{\eta}\frac{\partial p}{\partial y}\right)=12\overline{u}\frac{\partial h}{\partial x} \tag{8.3.23}$$

For a compressible fluid with one-dimensional flow (in the x-direction) with unidirectional motion,

$$\frac{d}{dx}\left(\frac{\rho h^3}{\eta}\frac{dp}{dx}\right)=12\overline{u}\frac{d}{dx}(\rho h) \tag{8.3.24}$$

This equation can be integrated with respect to x to give

$$\frac{1}{\eta}\frac{dp}{dx}=\frac{12\overline{u}}{h^2}+\frac{C_1}{\rho h^3} \tag{8.3.25}$$

The integration constant C_1 can be calculated using the boundary condition that $dp/dx = 0$ at $x = x_m$, $h = h_m$, and $\rho = \rho_m$ (maximum pressure location). We get

$$C_1 = -12\overline{u}\,\rho_m\,h_m \tag{8.3.26}$$

Substituting the expression from Equation 8.3.26 into Equation 8.3.25, we get,

$$\frac{dp}{dx}=12\overline{u}\,\eta\frac{\rho h-\rho_m h_m}{\rho h^3} \tag{8.3.27}$$

No assumptions have been made regarding the density and viscosity. For a perfect gas ($\rho \propto p$)

$$\frac{dp}{dx}=12\overline{u}\,\eta\left(\frac{ph-p_m h_m}{p h^3}\right) \tag{8.3.28a}$$

For an incompressible fluid (constant ρ), Equation 8.3.27 reduces to

$$\frac{dp}{dx}=12\overline{u}\,\eta\left(\frac{h-h_m}{h^3}\right) \tag{8.3.28b}$$

The Reynolds equation in cylindrical polar coordinates for tangential motion is given as

$$\frac{\partial}{\partial r}\left(\frac{r\,\rho\,h^3}{\eta}\frac{\partial p}{\partial r}\right)+\frac{1}{r}\frac{\partial}{\partial\theta}\left(\frac{\rho h^3}{\eta}\frac{\partial p}{\partial\theta}\right)$$
$$=12\,\overline{v}_r\frac{\partial(\rho r h)}{\partial r}+12\,\overline{v}_\theta\frac{\partial(\rho h)}{\partial\theta} \tag{8.3.29}$$

where $\overline{v}_r = (v_{ra}+v_{rb})/2$, $\overline{v}_\theta = (v_{\theta a}+v_{\theta b})/2$.

8.4 Hydrostatic Lubrication

Hydrostatic, also called externally pressurized, lubricated bearings can operate at little or no relative tangential motion with a large film thickness. The bearing surfaces are separated by supplying a fluid (liquid or gaseous) under pressure at the interface using an external pressure source, providing a high bearing stiffness and damping. There is no physical contact during start-up and shut-down as in hydrodynamic lubrication. Hydrostatic bearings provide high load-carrying capacity at low speeds, and therefore are used in applications requiring operation at high loads and low speeds such as in large telescopes and radar tracking units. High stiffness and damping of these bearings also provide high positioning accuracy in high-speed, light-load applications, such as bearings in machine tools, high-speed dental drills, gyroscopes, and ultracentrifuges. However, the lubricating system in hydrostatic bearings is more complicated than that in a hydrodynamic bearing. Hydrostatic bearings require high-pressure pumps and equipment for fluid cleaning which adds to space and cost.

By supplying high-pressure fluid at a constant pressure or volume to a recess relief or pocket area at the bearing interface, the two surfaces can be separated and the frictional force reduced to a small viscous force, Figure 8.4.1. By proper proportioning of the recess area to the cross-sectional (land) area of the bearing surface, the appropriate bearing load capacity can be achieved (Wilcock, 1972; Gross *et al.*, 1980; Fuller, 1984; Bhushan, 2001; Hamrock *et al.*, 2004; Williams, 2005).

Figure 8.4.2a shows the essential features of a typical hydrostatic thrust bearing with a circular step pad, designed to carry thrust load (Williams, 2005). A pump is used to draw fluid from a reservoir to the bearing through a line filter. The fluid under pressure p_s supplied to the bearing before entering the central recess or pocket, passes through a compensating or restrictor element in which its pressure is dropped to some low value p_r. The fluid then passes out of the bearing through the narrow gap of thickness, h, between the bearing land and the opposing bearing surface, also known as the slider or runner. The depth of the recess is much larger than the gap. The purpose of the compensating element is to bring a pressurized fluid from the supply tank to the recess. The compensating element allows the pocket pressure p_r to be different from the supply pressure p_s; this difference between p_r and p_s depends on the applied load W. Three common types of compensating elements for hydrostatic bearings include capillary tube, the sharp-edge orifice and constant-flow-valve compensation.

We now analyze the bearing performance. The bearing has an outer radius $\mathrm{r_0}$ and the central recess of r_i with slider at rest, Figure 8.4.2a. The film thickness is the same in radial or angular positions and the pressure does not vary in the θ direction. We assume an incompressible fluid. In the land region, $r_i < r < r_0$, the simplified Reynolds equation in the polar coordinates is given as

$$\frac{\partial}{\partial r}\left(r\frac{\partial p}{\partial r}\right) = 0 \tag{8.4.1}$$

Integrating we get

$$\frac{\partial p}{\partial r} = C_1$$

$$p = C_1\,\ell n\, r + C_2 \tag{8.4.2}$$

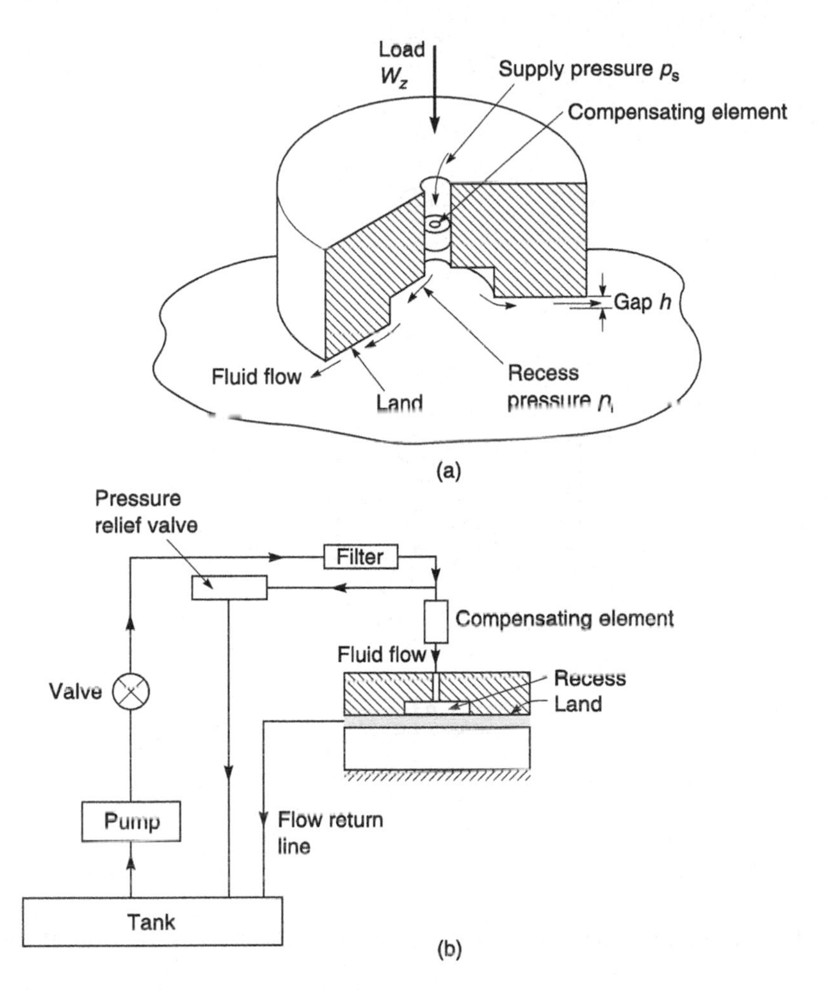

Figure 8.4.1 Schematics of (a) a hydrostatic thrust bearing with circular step pad, and (b) a fluid supply system. Reproduced with permission from Williams, J.A. (2005), *Engineering Tribology*, Second edition, Cambridge University Press, Cambridge. Copyright 2005. Cambridge University Press

We solve for constants C_1 and C_2 by using the boundary conditions that $p = p_r$ at $r = r_i$ and $p = 0$ at $r = r_0$. We get

$$\frac{p}{p_r} = \frac{\ell n\ (r_0/r)}{\ell n\ (r_0/r_i)} \tag{8.4.3}$$

and

$$\frac{dp}{dr} = -\frac{p_r}{r\ \ell n\ (r_0/r_i)} \tag{8.4.4}$$

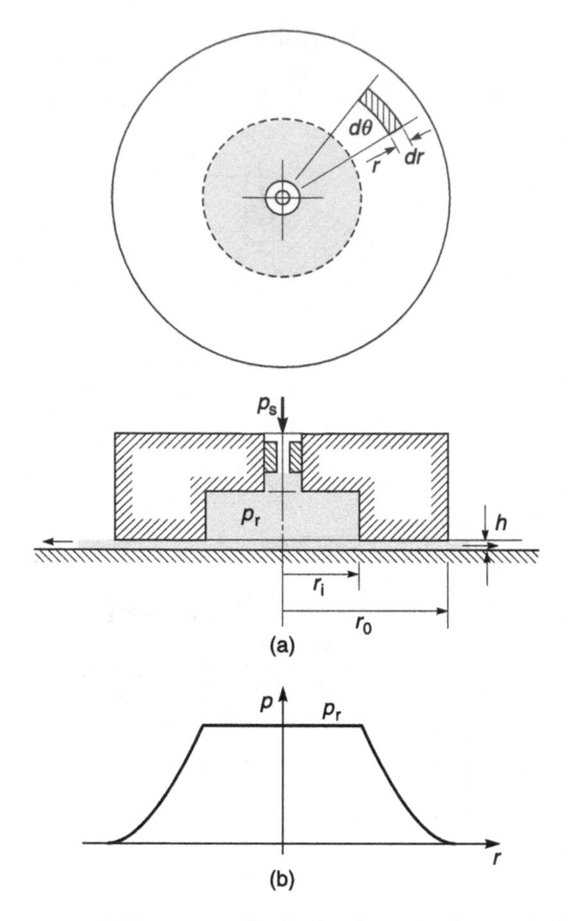

Figure 8.4.2 (a) Geometry and (b) pressure distribution for a circular step hydrostatic thrust bearing.

The radial volumetric flow rate per unit circumference in polar coordinates is given as

$$\begin{aligned} q &= \frac{h^3}{12\,\eta_0}\left(-\frac{dp}{dr}\right) \\ &= \frac{h^3\,p_r}{12\,\eta_0\,r\,\ell n\,(r_0/r_i)} \end{aligned} \tag{8.4.5a}$$

and the total volumetric flow rate is

$$Q = 2\pi\,r\,q \tag{8.4.5b}$$

Combining Equations. 8.4.3 and 8.4.5, we obtain an expression for p in terms of Q:

$$p = \frac{6\,\eta_0\,Q}{\pi\,h^3}\,\ell n\,(r_0/r) \tag{8.4.6}$$

The drop in fluid pressure across the land is shown in Figure 8.4.2b. It is generally assumed that the pressure of the fluid is uniform over the whole area of recess because the depth of the recess in a hydrostatic bearing is on the order of one hundred times greater than the mean film thickness of the fluid over its lands.

The normal load carried by the bearing, load-carrying capacity, is given as

$$W_z = \pi\, r_i^2\, p_r + \int_{r_i}^{r_0} p_r \frac{\ell n\,(r_0/r)}{\ell n\,(r_0/r_i)} 2\pi\, r\, dr = \frac{\pi\, p_r\,\left(r_0^2 - r_i^2\right)}{2\,\ell n\ (r_0/r_i)} \tag{8.4.7}$$

For a given bearing geometry, the load capacity linearly increases with an increase of fluid pressure. Note that the load capacity is not a function of the viscosity. Therefore, any fluid that does not damage the bearing materials can be used.

The load capacity in terms of Q can be obtained by combining an expression for Q from Equations. 8.4.5a and b,

$$W_z = \frac{3\,\eta_0\, Q}{h^3}\left(r_0^2 - r_i^2\right) \tag{8.4.8}$$

Fluid film bearings of the hydrostatic or hydrodynamic types have a stiffness characteristic and will act like a spring. In conjunction with the supported mass, they will have a natural frequency of vibration for the bearing. The frequency is of interest in the dynamic behavior of rotating machinery. To calculate the film stiffness of the bearing, we take the derivative of Equation 8.4.8 with respect to h, where h is considered a variable. For a bearing with a constant flow-valve compensation (constant feed rate of Q, and not a function of h), the film stiffness is given as

$$k_f \equiv \frac{dW_z}{dh} = -\frac{3}{h}\left[\frac{3\,\eta_0 Q\,\left(r_0^2 - r_i^2\right)}{h^3}\right] = -\frac{3W_z}{h} \tag{8.4.9}$$

The negative sign indicates that k_f decreases as h increases. The stiffness of the films in a hydrostatic bearing with capillary tube and orifice compensation is lower than for a bearing with a constant feed rate. Expressions for bearing stiffness for these bearings are presented by Fuller (1984). The oil film stiffness in a hydrostatic bearing can be extremely high, comparable to metal structures.

Next we calculate the frictional torque. Assume that the circumferential component of the fluid velocity varies linearly across the film and that viscous friction within the recess is negligible. From Equation 8.3.1, the shear force on a fluid element of area dA is written as

$$\begin{aligned} f &= \eta_0\, dA \frac{u}{h} \\ &= \eta_0\ (r\, d\theta\, dr)\frac{r\omega}{h} \\ &= \frac{\eta_0\, \omega r^2\, dr d\theta}{h} \end{aligned} \tag{8.4.10}$$

The friction torque is given by integrating over the entire land outside the recess area,

$$T = \frac{\eta_0 \, \omega}{h} \int_0^{2\pi} \int_{r_i}^{r_0} r^3 \, dr \, d\theta$$
$$= \frac{\pi \, \eta_0 \, \omega}{2\, h} \left(r_0^4 - r_i^4\right) \tag{8.4.11}$$

The total power loss consists of viscous dissipation, H_v, and pumping loss, H_p, which are given as

$$H_v = T \, \omega \tag{8.4.12a}$$

and

$$H_p = p_r \, Q \tag{8.4.12b}$$

Therefore, the total power loss

$$H_t = H_v + H_p$$
$$= \frac{\pi \, \eta_0 \, \omega^2}{2\, h} \left(r_0^4 - r_i^4\right) + \frac{\pi \, h^3 \, p_r^2}{6 \, \eta_0 \, \ell n \, (r_0/r_i)} \tag{8.4.12c}$$

Note that H_v is inversely proportional to h and proportional to the square of the sliding velocity, and H_p is proportional to h^3 and independent of velocity. Generally, the bearing velocities are low and only pumping power is significant.

It is generally assumed that total power loss is dissipated as heat. Further assuming that all of the heat appears in the fluid, then the temperature rise, Δt, is given as

$$H_t = Q \, \rho \, c_p \, \Delta t$$

or

$$\Delta t = \frac{H_t}{Q \, \rho \, c_p} \tag{8.4.13}$$

The load-carrying capacity, associated flow rate and pumping loss are often expressed in nondimensional terms by defining a normalized or nondimensional load $\overline{W}_z$, nondimensional flow rate $\overline{Q}$ and nondimensional pumping loss $\overline{H}_p$, known as bearing pad coefficients. These are given as,

$$\overline{W}_z = \frac{W_z}{A_p \, p_r} = \frac{1 - (r_i/r_0)^2}{2 \, \ell n \, (r_0/r_i)} \tag{8.4.14}$$

$$\overline{Q} = \frac{Q}{(W/A_p)(h^3/\eta)} = \frac{\pi}{3 \left[1 - (r_i/r_0)^2\right]} \tag{8.4.15}$$

$$\overline{H}_p = \frac{H_p}{(W/A_p)^2 (h^3/\eta)} = \frac{2\pi \, \ell n \, (r_0/r_i)}{3 \left[1 - (r_i/r_0)^2\right]^2} \tag{8.4.16}$$

where A_p is the total projected pad area $= \pi \, r_0^2$.

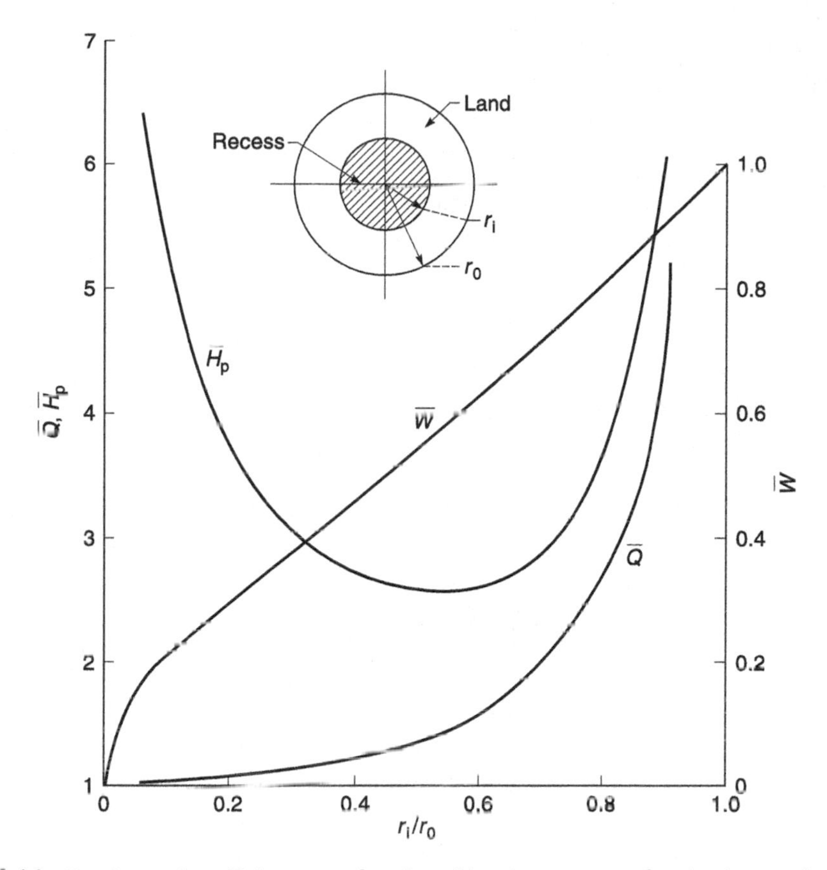

Figure 8.4.3 Bearing pad coefficients as a function of bearing geometry for circular step hydrostatic thrust bearing (*Source*: Rippel, 1963).

Figure 8.4.3 shows the three bearing pad coefficients for various ratios of recess radius to bearing radius. W^* is a measure of how efficiently the bearing uses the recess pressure to support the applied load. It varies from zero for relatively small recesses to unity for bearings with large recesses with respect to pad dimensions. Q^* varies from unity for relatively small recesses to a value approaching infinity for bearings with large recesses. $\overline{H}_p$ approaches infinity for extremely small recesses, decreases to a minimum as the recess size increases ($r_i/r_0 = 0.53$) then approaches to infinity again for large recesses.

Example Problem 8.4.1

A hydrostatic thrust bearing with a circular step pad has an outside diameter of 400 mm and recess diameter of 250 mm. (a) Calculate the recess pressure for a thrust load of 100,000 N, (b) calculate the volumetric flow rate of the oil which will be pumped to maintain the film thickness of 150 μm with an oil viscosity of 30 cP, (c) calculate the film stiffness for an applied load of 100,000 N and operating film thickness of 150 μm, and (d) calculate the pumping loss and the oil temperature rise. The mass density of the oil is 880 kg/m^3 and its specific heat is 1.88 J/g K.

Solution

(a) Given

$$r_0 = 200 \text{ mm}$$

$$r_i = 125 \text{ mm}$$

$$W_z = 100{,}000 \text{ N}$$

$$p_r = \frac{2W\ell n\,(r_0/r_i)}{\pi\left(r_0^2 - r_i^2\right)}$$

$$= \frac{2 \times 10^5 \times \ell n\,(200/125)}{\pi\left(0.2^2 - 0.125^2\right)} \text{ Pa}$$

$$= 1.23 \text{ MPa}$$

(b)

$$\eta_0 = 30 \text{ mPa s}$$

$$h = 150\ \mu\text{m}$$

$$Q = \frac{\pi\, h^3\, p_r}{6\,\eta_0\, \ell n\,(r_0/r_i)}$$

$$= \frac{\pi\left(1.5 \times 10^{-4}\right)^3 \times 1.23 \times 10^6}{6 \times 30 \times 10^{-3}\, \ell n\,(200/125)} \text{ m}^3/\text{s}$$

$$= 154.1 \times 10^3 \text{ mm}^3/\text{s}$$

(c)

$$k_f = -\frac{3W_z}{h}$$

$$= -\frac{3 \times 10^5}{150 \times 10^{-6}} \text{ N/m}$$

$$= -2 \times 10^9 \text{ N/m}$$

(d) Given

$$\rho = 880\,\text{kg/m}^3$$

$$c_p = 1.88 \text{ J/g K}$$

$$H_p = p_r\, Q$$

$$= 1.23 \times 10^6 \times 154.1 \times 10^{-6} \text{ N m/s}$$

$$= 189.5 \text{ W}$$

$$\Delta t = \frac{H_p}{Q\,\rho\, c_p}$$

$$= \frac{189.5}{154.1 \times 10^{-6} \times 880 \times 10^3 \times 1.88}\ ^\circ\text{C}$$

$$= 0.74\ ^\circ\text{C}$$

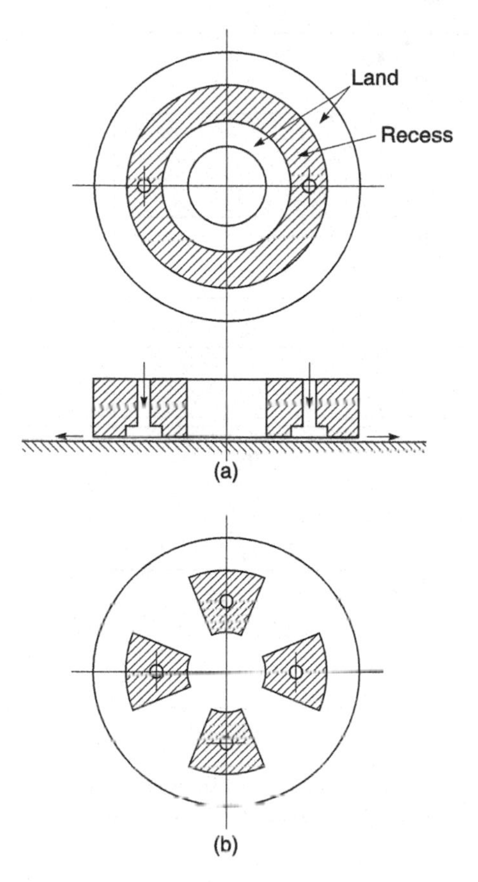

Figure 8.4.4 Schematics of a hydrostatic thrust bearing (a) with annular recess, and (b) four recess segments.

Hydrostatic bearings can have single or multiple recesses that are circular, or annular or rectangular in shape. Schematics of thrust bearings with annular recess, four recess segments and a rectangular recess are shown in Figures. 8.4.4a, b and 8.4.5. A schematic of a journal bearing with four rectangular recesses is shown in Figure 8.4.6.

In the case of a rectangular recess without the essential degree of symmetry of the circular pads, there are pressure gradients and so fluid flow in both the x and y directions in the bearing plane. For a bearing with constant film thickness along the x and y axes in the land region and for an incompressible fluid, the modified Reynolds equation is given as

$$\frac{\partial^2 p}{\partial x^2} + \frac{\partial^2 p}{\partial y^2} = 0 \qquad (8.4.17)$$

This is known as the Laplace equation in two dimensions. For the case of a bearing with the length much greater than the width of the lands, i.e. $\ell >> b$, most of the fluid which is supplied to the bearing by the pump leaves by flowing from the recess over the lands in the

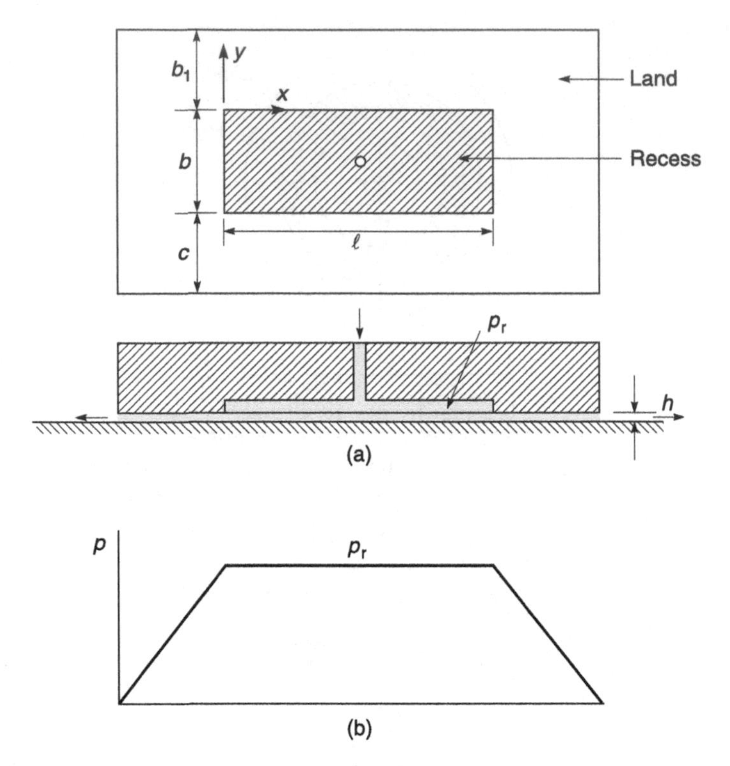

Figure 8.4.5 Schematics of (a) a hydrostatic thrust bearing with rectangular recess, and (b) pressure distribution within the recess along the horizontal axis, at the bearing interface.

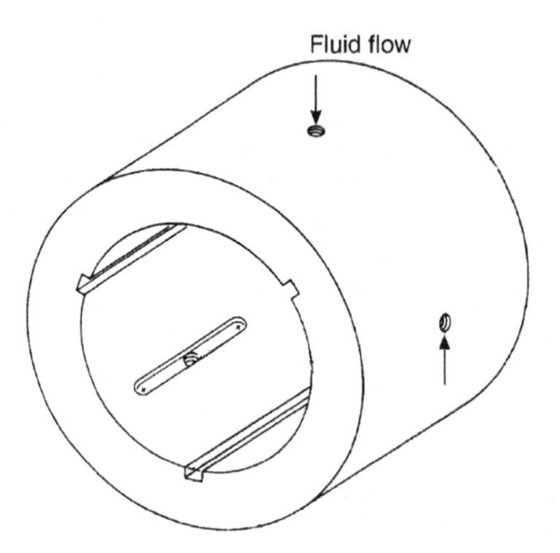

Figure 8.4.6 Schematic of a hydrostatic journal bearing with four rectangular recesses.

direction of the y axis. It is then evident that there is a negligible change of pressure over the lands in the direction of the x axis at the ends of the recess. Therefore, the Reynolds equation reduces to

$$\frac{d^2 p}{dy^2} = 0 \tag{8.4.18a}$$

The Laplace equation can be solved for the rectangular recess by analytical methods. For complex geometries this should be solved by numerical methods. Integrating Equation 8.4.18a we get

$$p = C_1 y + C_2 \tag{8.4.18b}$$

Integrating constants C_1 and C_2 are calculated by considering the boundary conditions that $p = p_r$ at $y = 0$ and $p = 0$ at $y = c$ (Figure 8.4.5). Therefore

$$p = p_r \left(1 - \frac{y}{c}\right) \tag{8.4.19a}$$

or

$$\frac{dp}{dy} = -\frac{p_r}{c} \tag{8.4.19b}$$

The pressure gradient is linear. From Equation 8.3.17, the volumetric flow rate along the y axis for uniform pressure along the x axis is $h^3 \ell\, p_r / 12\, \eta_0\, c$. Doubling this quantity must equal the total flow rate of flow of fluid into the bearing from the pump:

$$Q = \frac{h^3 \ell\, p_r}{6\, \eta_0\, c} \tag{8.4.20}$$

The load capacity of the bearing is given as

$$W_z = p_r\, b\ell + 2\ell \int_0^c p_r \left(1 - \frac{y}{c}\right) dy$$

$$= p_r\, \ell\, (b + c) \tag{8.4.21a}$$

$$= \frac{6\, \eta_0\, c}{h^3} \left(\frac{Q}{\ell}\right) [\ell\, (b + c)] \tag{8.4.21b}$$

The film stiffness is given as

$$k_f \equiv \frac{\partial W_z}{\partial h} = -\frac{18\, \eta_0\, c}{h^4} \left(\frac{Q}{\ell}\right) [\ell\, (b + c)] \tag{8.4.22}$$

8.5 Hydrodynamic Lubrication

Beauchamp Tower, employed by the British Railroad to study the friction in railroad journal bearings, was the first to observe the hydrodynamic effect in a partial sleeve lubricated with an oil bath, Figure 8.5.1 (Tower, 1883/84, 1885). He reported that oil lubrication produced a low coefficient of friction at relative sliding velocity. Tower later drilled a lubricator hole through the top. When the apparatus was set in motion, oil flowed out of this hole and a pressure gage connected to this hole indicated an oil film pressure as much as twice the average pressure of the oil based on the projected area.

Osborne Reynolds then considered this apparent phenomenon of Tower's experiments and suggested that film lubrication was a hydrodynamic action and depended on the viscosity of the lubricant (Reynolds, 1886). The lubricant adheres to both the stationary and moving surfaces of the bearing and is dragged into a wedge-shaped gap, converging in the direction of motion, where it develops a fluid pressure sufficient to carry the load. He developed a governing differential fluid flow equation for a wedge-shaped film, known as Reynolds equation, as presented earlier. This theory is the basis of hydrodynamic (HD) lubrication and elastohydrodynamic (EHD) lubrication (EHL).

Automobile engines, railroad locomotives, aircraft engines, domestic appliances, underwater vessels, machine tools, pumps, gearboxes, and computer peripheral devices are only a small number of machines which consist of machine components depending on hydrodynamic films for their operation. In a properly designed component, a hydrodynamic film is thick enough, compared to roughness of two sliding surfaces that there is no physical contact during steady operation. However, physical contact occurs during start and stop operations and mating materials need to be selected to prevent wear and minimize friction. In some cases hydrodynamic lubrication is undesirable such as air entrapment during winding of plastic webs at high speeds.

Hydrodynamic action occurs in bearings with a convergent clearance space through the length of the bearing. Loads carried by a rotating shaft in rotating machinery may have a component, in addition to the radial load, of an axial or thrust load in the direction of the shaft axis of rotation. The radial load is carried by a journal bearing and the thrust load is carried by a thrust bearing, Figure 8.5.2. The surfaces of a thrust bearing are perpendicular

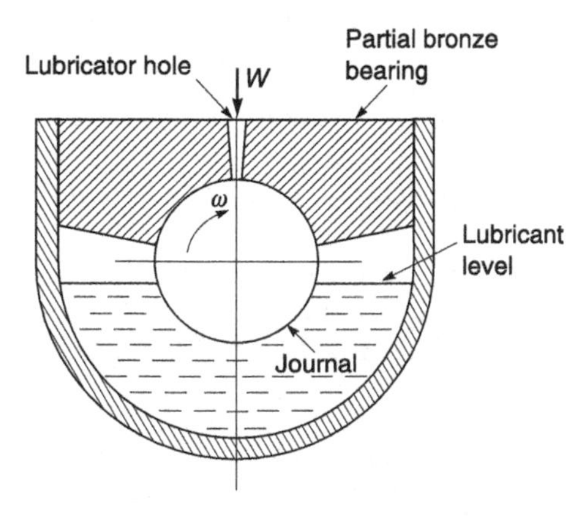

Figure 8.5.1 Schematic of the partial sleeve bearing having bath-type lubrication, used by Beauchamp Tower.

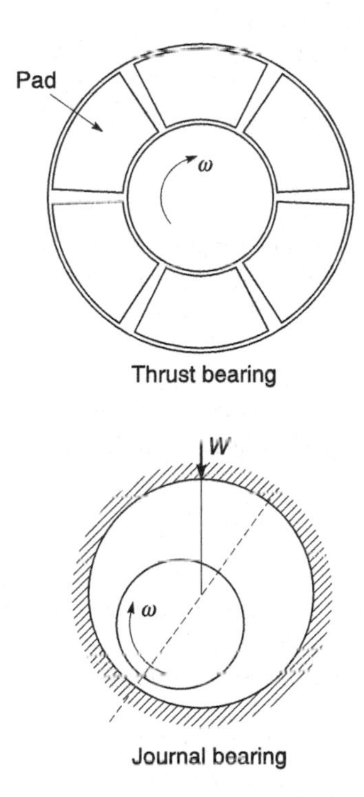

Figure 8.5.2 Schematics of typical thrust and journal bearing configurations.

to the axis of rotation, whereas that of a journal bearing are parallel to the axis of rotation. Thrust bearings consist of multiple pads. The pad geometry is selected such that it results in a convergent clearance. Eccentricity of the shaft with respect to the journal during rotation results in formation of convergent clearance. Nonparallel surfaces and long waves with small amplitudes in face seals and asperities on the lip seal and sheft surfaces develop hydrodynamic action. Hydrodynamic bearings developed in the case of face and lip seals are not very strong, with a small separation on the order of 200 nm.

The Reynolds equation is solved for a given bearing configuration to obtain bearing performance including pressure and film thickness distribution, film stiffness, fluid flow rate and viscous shear forces, viscous loss, and temperature rise. Numerical methods are used for the solution of real bearing configurations. Analytical solutions can only be obtained for very simple cases. For an infinitely wide bearing, side flow can be neglected. In addition to this simplification, it is assumed that the pressure and temperature effects on viscosity and density can be neglected. Liquids are essentially incompressible and their density can be assumed to be independent of pressure. For these simplifications, closed-form analytical solutions can be obtained. Analytical solutions for various thrust bearing configurations and a journal bearing are presented first followed by limited details on numerical solutions of finite-width bearings and gas (compressible fluid) bearings.

Bearings involving nonconforming contacts experience high contact stresses which result in local deformations, and high stresses also affect fluid viscosity. If the contact stresses are relatively low, the local deformations and changes in fluid viscosity can be neglected and the

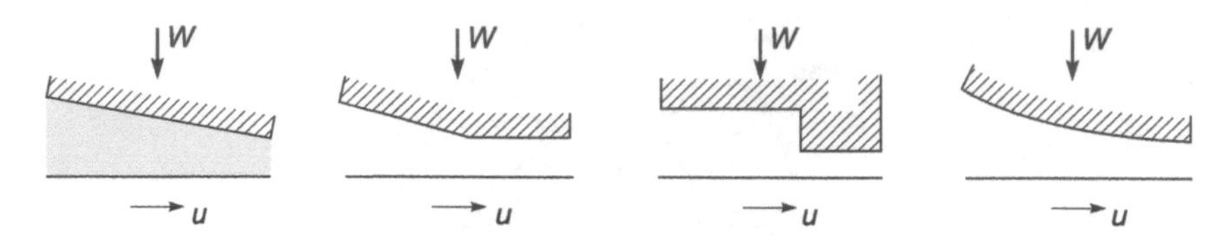

Figure 8.5.3 Schematics of various shapes for pads in thrust bearings. Reproduced with permission from Raimondi, A.A. and Boyd, J. (1955), "Applying Bearing Theory to the Analysis and Design of Pad-Type Bearings," *ASME Trans.* **77**, 287–309. Copyright 1955. ASME.

fluid flow in a bearing configuration can be analyzed using hydrodynamic lubrication. If the contact stresses are large, the elastic deformation of components and changes in fluid viscosity need to be taken into account. These bearings will be analyzed in the following section on elastohydrodynamic lubrication.

8.5.1 Thrust Bearings

Figure 8.5.3 illustrates some of the shapes for thrust pads which satisfy the conditions for successful hydrodynamic lubrication during sliding motion. These shapes occur in practice, either because they are manufactured, or are produced due to subsequent wear or deformation. One of the bearings shown in Figure 8.5.3 (third from left) is known as a Rayleigh step bearing, proposed by Lord Rayleigh in 1918 (Rayleigh, 1918). He concluded that for the same inlet and outlet film thickness, the Rayleigh step bearing produces the highest peak pressure.

Thrust bearings used to support thrust loads in rotating machinery consist of multiple pads, either fixed or pivoted, Figure 8.5.4. In this section, a single pad is analyzed with a straight line motion and the effect of curvature is neglected. The load capacity, film stiffness, volumetric flow rate, and power losses of a bearing would be equal to the value of a single pad times the number of pads.

8.5.1.1 Fixed-Inclined-Pad Thrust Bearing

A simple case of fixed-inclined-pad bearing is shown in Figure 8.5.5. It consists of two nonparallel plane surfaces separated by an incompressible (liquid) fluid film. The lower surface

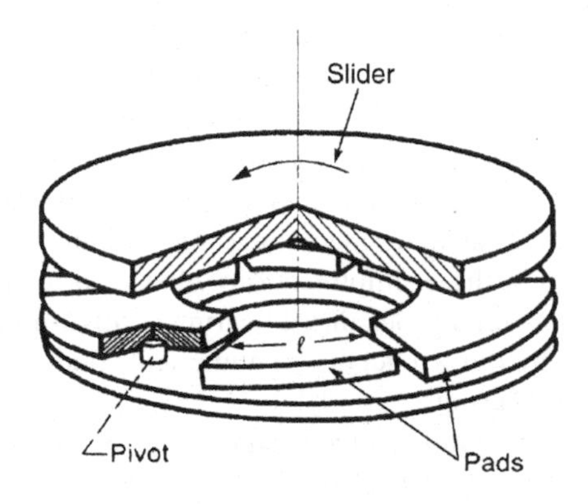

Figure 8.5.4 Schematic of a multiple-pivoted-pad thrust bearing.

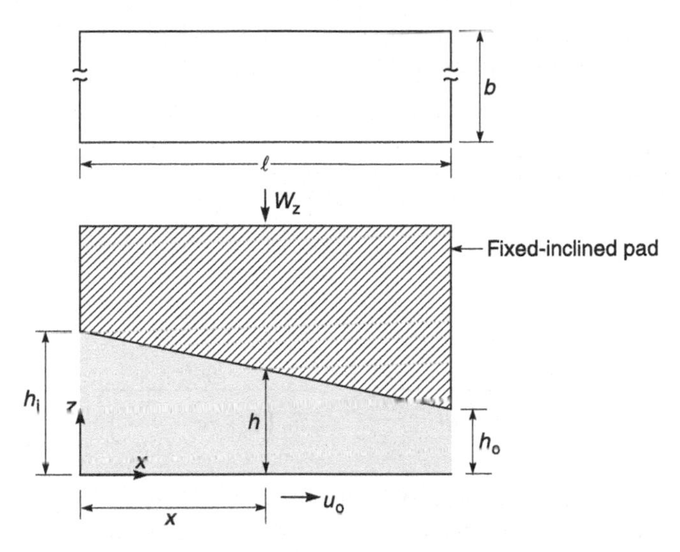

Figure 8.5.5 Schematic of a fixed-inclined-pad thrust bearing.

moves with a unidirectional velocity and the upper surface is stationary. The sliding direction is such that a convergent fluid film is formed between the surfaces to produce hydrodynamic pressure. It is assumed that the width of the bearing b, is much greater than its length $\ell(\frac{b}{\ell} > 4)$. Therefore, most of the flow through the gap between the two surfaces occurs in the direction of the the x axis. There will be significant flow in the direction of y axis, near the ends of the bearing, where $y = 0$ and b. There is no change in gap or film thickness in the direction of y axis but a uniform inclination is assumed to exist in the direction of the x axis. The gap is shown to vary uniformly from a maximum of h_i to a minimum of h_0 over the length, ℓ, of the bearing. The origin is taken to be at the bottom left end of the bearing. For the case of one surface sliding with a constant velocity u_0 ($\overline{u} = u_0/2$) over a stationary surface in one direction (x direction) with no normal motion and an incompressible fluid between them of viscosity η_0, the integrated form of the Reynolds equation from Equation 8.3.28b is given as

$$\frac{dp}{dx} = 6\,\eta_0\,u_0\left(\frac{h - h_m}{h^3}\right) \tag{8.5.1}$$

where h_m is the film thickness at the maximum pressure location ($dp/dx = 0$). Assume that viscosity remains constant to η_0 (isoviscous fluid). The film thickness h at any point may be expressed as

$$\frac{h}{h_0} = 1 + m\left(1 - \frac{x}{\ell}\right) \tag{8.5.2}$$

where

$$m = \frac{h_i}{h_0} - 1$$

Physically the slope m, of the tapered end, usually varies in the range 0.5–2, and is typically 1. Substituting the expression for h from Equation 8.5.2 into Equation 8.5.1,

$$\frac{dp}{dx} = 6\,\eta_0\,u_0 \left[\frac{1}{h_0^2\left(1+m-\dfrac{mx}{\ell}\right)^2} - \frac{h_m}{h_0^3\left(1+m-\dfrac{mx}{\ell}\right)^3} \right] \tag{8.5.3}$$

Integrating we get

$$p = \frac{6\,\eta_0\,u_0\ell}{m\,h_0^2} \left[\frac{1}{1+m-\dfrac{mx}{\ell}} - \frac{h_m}{2\,h_0\left(1+m-\dfrac{mx}{\ell}\right)^2} + C_1 \right] \tag{8.5.4}$$

The constant h_m and integration constant C_1 are unknown. These can be evaluated using the boundary conditions at $x = 0$, $p = 0$ and $x = \ell$, $p = 0$. These conditions give

$$h_m = 2h_0\left(\frac{1+m}{2+m}\right) \tag{8.5.5a}$$

$$C_1 = -\frac{1}{2+m} \tag{8.5.5b}$$

After substitution of these constants from Equation 8.5.5, we get an expression for p from Equation 8.5.4:

$$p = \frac{6\,\eta_0\,u_0\,\ell}{h_0^2} \left[\frac{m\dfrac{x}{\ell}\left(1-\dfrac{x}{\ell}\right)}{(2+m)\left(1+m-m\dfrac{x}{\ell}\right)^2} \right] \tag{8.5.6a}$$

$$= \frac{\eta_0\,u_0\,\ell}{h_0^2}\,\overline{p} \tag{8.5.6b}$$

where $\overline{p}$ is the dimensionless pressure, known as the pressure coefficient. The pressure profile is plotted as a function of x/ℓ for various values of m in Figure 8.5.6. Note that the pressure remains constant along the y axis. Note that in most thrust bearings the film is nondiverging and continuous and the problem of negative pressure does not arise, as will be seen later in the case of journal bearings. Further note that for a parallel-surface slider bearing ($m = 0$), $p = 0$. Hence, a parallel-surface slider bearing does not develop pressure due to the absence of converging channel.

Maximum pressure occurs at $dp/dx = 0$. From Equation 8.5.3, $h_m = h_0\,(1 + m - mx_m/\ell)$. By substituting this expression in Equation 8.5.5a, the maximum pressure is found to occur at a location x_m,

$$\frac{x_m}{\ell} = \frac{1+m}{2+m} \tag{8.5.7}$$

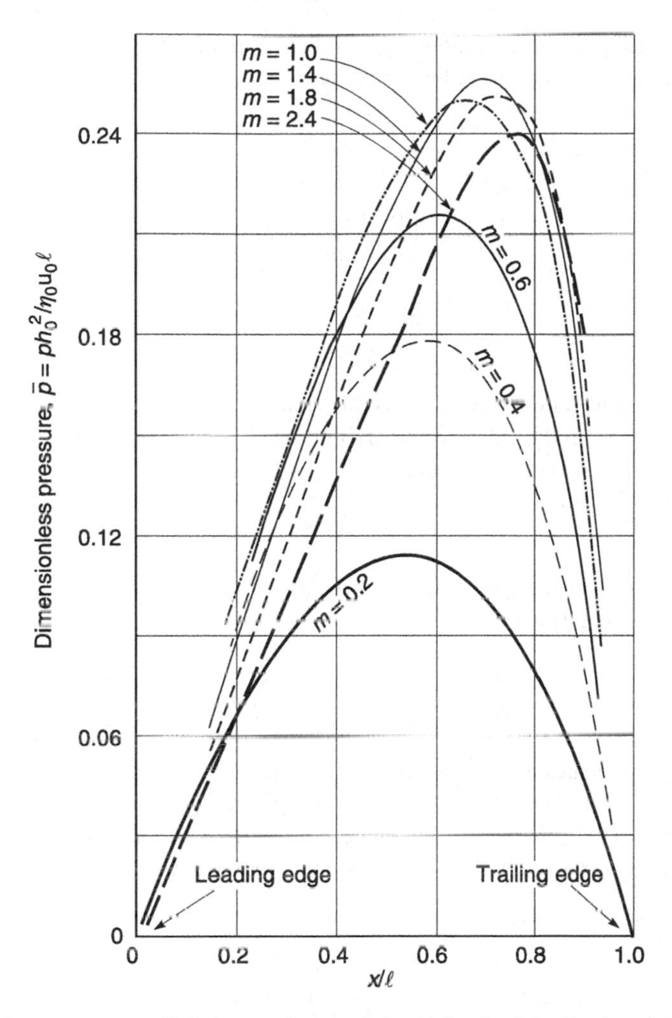

Figure 8.5.6 The pressure coefficient as a function of x/ℓ for fixed-inclined-pad thrust bearing for various values of slope m.

and after substitutions of x_m in Equation 8.5.6a

$$p_m = \frac{\eta_0 \, u_0 \, \ell}{h_0^2} \left[\frac{3m}{2\,(1+m)\,(2+m)} \right] \tag{8.5.8a}$$

$$= \frac{\eta_0 \, u_0 \, \ell}{h_0^2} \, \overline{p}_m \tag{8.5.8b}$$

where $\overline{p}_m$ is the dimensionless pressure. Note that maximum pressure always lies in the trailing half of the bearing pad.

The further integration of the pressure gives the normal load capacity per unit width, also given by the average pressure times bearing length,

$$w_z = p_{av}\,\ell = \int_0^{\ell} p\,dx$$

$$= \frac{\eta_0 u_0 \ell^2}{h_0^2}\left[\frac{6\,\ell n\,(1+m)}{m^2} - \frac{12}{m\,(2+m)}\right] \quad (8.5.9a)$$

$$= \frac{\eta_0 u_0 \ell^2}{h_0^2}\overline{W}_z \quad (8.5.9b)$$

where $\overline{W}_z$ is the dimensionless load capacity, also known as the load coefficient. The maximum load capacity depends on the values of m. By putting $dW/dm = 0$, we find that the value of m for which optimum load capacity occurs is

$$m = 1.1889 \quad (8.5.9c)$$

The load capacity is very insensitive to m for m larger than 1.1889.

The film stiffness is given as

$$k_f \equiv \frac{dW_z}{dh_0} = -\frac{2}{h_0}\left[\frac{\eta_0 u_0 \ell^2 b}{h_0^2}\overline{W}_z\right] = -\frac{2W_z}{h_0} \quad (8.5.9)$$

The negative sign indicates that k_f decreases as h_0 increases.

The volumetric flow rate per unit width through the bearing may be easily found at x_m $(dp/dx = 0)$ as there is no pressure-induced flow at this location. Therefore, from Equation 8.3.15b, the velocity- or shear-induced flow is given by the second term of this equation as

$$q = \frac{u_0 h_m}{2}$$

$$= u_0 h_0 \left(\frac{1+m}{2+m}\right) \quad (8.5.11a)$$

$$= u_0 h_0 \overline{Q} \quad (8.5.11b)$$

where $\overline{Q}$ is the dimensionless volumetric flow rate.

The shear force per unit width experienced by the lower sliding member, due to the shear stress distributed over the lower sliding surface is given as

$$f = \int_0^{\ell} \tau_{zx}\,|_{z=0}\,dx = \int_0^{\ell}\left(\eta_0 \frac{\partial u}{\partial z}\,|_{z=0}\right)dx = \int_0^{\ell}\left[-\frac{h}{2}\frac{dp}{dx} - \frac{\eta_0 u_0}{h}\right]dx$$

$$= \frac{\eta_0 u_0 \ell}{h_0}\left[\frac{4}{m}\,\ell n\,(1+m) - \frac{6}{2+m}\right] \quad (8.5.12a)$$

$$= \frac{\eta_0 u_0 \ell}{h_0}\overline{F} \quad (8.5.12b)$$

where $\overline{F}$ is the dimensionless friction force.

The dimensionless coefficiants $\overline{p}_m$, $\overline{W}_z$, $\overline{Q}$, and $\overline{F}$ as a function of m are presented in Figure 8.5.7. For $m = 1.1889$, the maximum pressure occurs.

The values of various coefficients are:

$$\overline{p}_m = 0.2555 \tag{8.5.13a}$$

$$\overline{W}_z = 0.1602 \tag{8.5.13b}$$

$$\overline{Q} = 0.6804 \tag{8.5.13c}$$

and

$$\overline{F} = 0.7542 \tag{8.5.13d}$$

The coefficient of friction can be expressed as

$$\mu = \frac{f}{w_z} = \frac{h_0}{6\ell}\left[\frac{\frac{4}{m}\,\ell n\,(1+m) - \frac{6}{2+m}}{\frac{1}{m^2}\,\ell n\,(1+m) - \frac{2}{m(2+m)}}\right] \tag{8.5.14}$$

The coefficient of friction is typically very low (0.002–0.01) for a self-acting hydrodynamic bearing.

The friction loss because of viscous stresses or power loss from Equation 8.5.12 is given as

$$H_v = F\,u_0 = \frac{\eta_0\,u_0^2\,b\ell}{h_0}\left[\frac{4}{m}\,\ell n\,(1+m) - \frac{6}{2+m}\right] \tag{8.5.15}$$

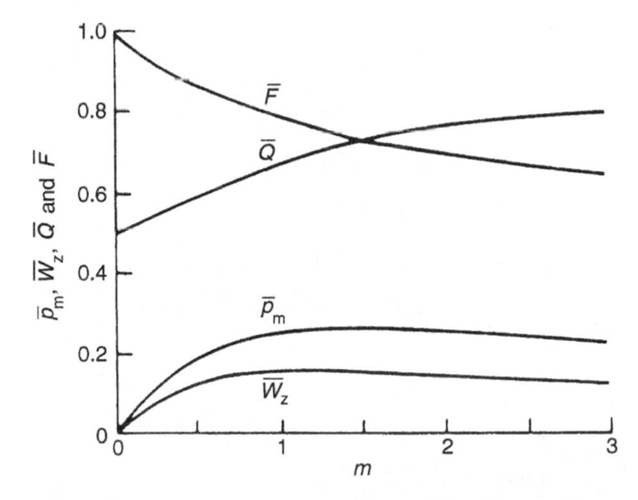

Figure 8.5.7 A plot of dimensionless coefficients $\overline{p}_m$, $\overline{W}_z$, $\overline{Q}$, and $\overline{F}$ as a function of m.

If all of the friction losses are assumed to be dissipated as heat which is assumed to be carried away by the fluid (convection), then the increase in the fluid temperature (known as adiabatic fluid temperature rise) is

$$\Delta t = \frac{H_v}{Q\,\rho\,c_p} \tag{8.5.16}$$

The temperature effects can be included in the analysis by an iterative process. We first calculate the temperature rise for the fluid viscosity at the inlet temperature, t_i. In the next step, we use the fluid viscosity obtained at the mean temperature,

$$t_m = t_i + \frac{\Delta t}{2} \tag{8.5.17}$$

We iterate until we get Δt comparable to the input values of the temperature rise.

We have presented analysis for a single pad. For multiple pads, the normal load capacity, film stiffness, volumetric flow rate, shear force, and power loss will be equal to the values of a single shoe multiplied by the number of shoes.

Example Problem 8.5.1

A fixed-inclined-pad thrust bearing of length 100 mm and width 500 mm, with a minimum film thickness of 50 μm, operates at a sliding velocity of 1 m/s with a mineral oil of absolute viscosity of 30 cP. Film thickness ratio is adjusted to produce the maximum load capacity. Calculate the maximum pressure and the location of the maximum pressure, normal load capacity, film stiffness, volumetric flow rate, the shear force experienced by the sliding surface, the coefficient of friction, the power loss and the average temperature rise of the fluid. The mass density and specific heat of oil are 880 kg/m^3 and 1.88 J/g K, respectively.

Solution

Given

$$\begin{aligned} \ell &= 100 \text{ mm} \\ b &= 500 \text{ mm} \\ h_0 &= 50\ \mu\text{m} \\ u_0 &= 1 \text{ m/s} \\ \eta_0 &= 0.03 \text{ Pa s} \end{aligned}$$

For maximum load capacity,

$$\begin{aligned} m &= 1.1889 \\ p_m &= \frac{\eta_0\, u_0\, \ell}{h_0^2}\left[\frac{3m}{2\,(1+m)\,(2+m)}\right] \end{aligned}$$

$$= \frac{0.03 \times 1 \times 0.1}{\left(5 \times 10^{-5}\right)^2} \left[\frac{3 \times 1.1889}{2 \times 2.1889 \times 3.1889}\right] \text{Pa}$$

$$= 0.307 \text{ MPa}$$

$$x_m = \frac{\ell (1+m)}{2+m}$$

$$= \frac{100 \times 2.1889}{3.1889} \text{ mm}$$

$$= 68.64 \text{ mm}$$

$$W_z = \frac{\eta_0 u_0 \ell^2 b}{h_0^2} \left[\frac{6 \, \ell n \, (1+m)}{m^2} - \frac{12}{m(2+m)}\right]$$

$$= \frac{0.03 \times 1 \times 0.1^2 \times 0.5}{\left(5 \times 10^{-5}\right)^2} \left[\frac{6 \, \ell n \, (2.1889)}{1.1889^2} - \frac{12}{1.1889 \times 3.1889}\right] \text{N}$$

$$= 9.62 \text{ kN}$$

$$k_f = -\frac{2W_z}{h_0} = -\frac{2 \times 9.62 \times 10^3}{5 \times 10^{-5}} \text{ N/m} = -385 \text{ N/}\mu\text{m}$$

$$Q = u_0 b h_0 \left(\frac{1+m}{2+m}\right)$$

$$= 1 \times 0.5 \times 5 \times 10^{-5} \frac{2.1889}{3.1889} \text{ m}^3\text{/s}$$

$$= 1.72 \times 10^{-5} \text{ m}^3\text{/s}$$

$$F = \frac{\eta_0 u_0 b \ell}{h_0} \left[\frac{4}{m} \ell n \, (1+m) - \frac{6}{2+m}\right]$$

$$= \frac{0.03 \times 1 \times 0.5 \times 0.1}{5 \times 10^{-5}} \left[\frac{4}{1.1889} \ell n \, (2.1889) - \frac{6}{3.1889}\right]$$

$$= 22.63 \text{ N}$$

$$\mu = \frac{F}{W_z} = \frac{22.63}{9.62 \times 10^3}$$

$$= 0.0024$$

$$H_v = F u_0$$

$$= 22.63 \times 1 \text{ Nm}$$

$$= 22.63 \text{ Nm}$$

$$\Delta t = \frac{H_v}{Q \rho c_p}$$

$$= \frac{22.63}{1.72 \times 10^{-5} \times 880 \times 10^3 \times 1.88} \,^\circ\text{C}$$

$$= 0.80^\circ\text{C}$$

Example Problem 8.5.2

A thrust bearing with a minimum film thickness of 50 μm is required for a rotating machinery to support a downward load of 45 kN. A fixed-inclined-pad thrust bearing with a pad design as in Example Problem 8.5.1 is selected for this application. Calculate the number of pads required to support the design load.

Solution

The normal load capacity of a single pad,

$$W_z = 9.62 \text{ kN}$$

Number of pads required to support the load of 45 kN,

$$= \frac{45}{9.62} = 4.68 \text{ or } 5$$

8.5.1.2 Pivot-Pad or Tilting-Pad Thrust Bearing

For bearings in most engineering applications, the minimum film thickness is on the order of 25 μm to 1 mm. Geometrical parameters such as tapered angle or the step required to transform a rigid surface into an efficient bearing with small film thicknesses are extremely small and are not easy to manufacture. Therefore, pivoted pad or tilted-shoe bearings are more commonly used, and take up their own taper angle with respect to the other surface. For a given ratio between the inlet and outlet film thicknesses, the center of pressure must coincide with the pivot position. Pivoted-pad or pivoted-shoe slider bearings represent a commonly used design.

The location of the center of pressure, x_c, indicates the position at which the resultant force acts. The expression is

$$w_z x_c = \int_0^\ell px\, dx \tag{8.5.18a}$$

By substituting the expression for p from Equation 8.5.6 into Equation 8.5.18 and integrating, and substituting an expression for w_z from Equation 8.5.9, we get

$$\frac{x_c}{\ell} = \frac{(1+m)\,(3+m)\ \ell n\ (1+m) - 3m - 2.5\,m^2}{m\,(2+m)\ \ell n\ (1+m) - 2m^2} \tag{8.5.18b}$$

Note that x_c/ℓ is a function of m only and increases with an increase in m. For example for $m = 1$, $x_c/\ell = 0.568$. The center of pressure is always more towards the trailing part of the pad ($x_c/\ell > 0.5$). The x_p locates the position of the pivot for a given m, Figure 8.5.8. Detailed analyses of pivoted-pad bearings are presented by Cameron (1976). For bidirectional operation of bearings, the pivot should be located at the center of the pad.

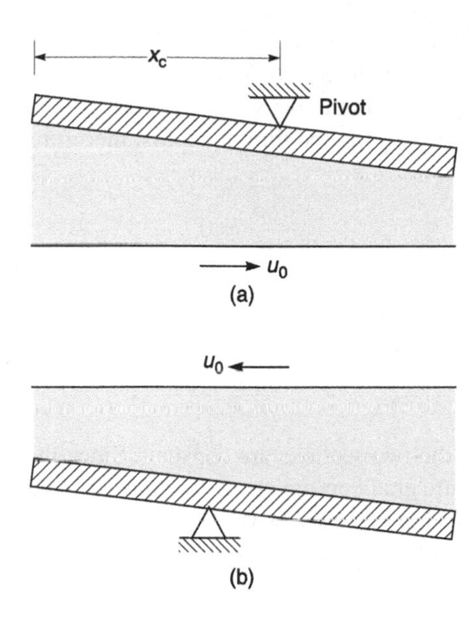

Figure 8.5.8 Schematics of thrust bearings with (a) upper member pivoted, and (b) lower member pivoted.

8.5.1.3 The Rayleigh Step Thrust Bearing

In 1918, Lord Rayleigh proposed a parallel step thrust bearing that has the greatest load capacity of all the slider shapes (Rayleigh, 1918), Figure 8.5.9. This geometry is not as popular as the pivoted pad because of difficulties in manufacturing the small step.

For simplified analysis we assume an infinitely wide bearing (one-dimensional fluid flow) with an incompressible fluid flowing through the bearing. The bottom surface moves in a

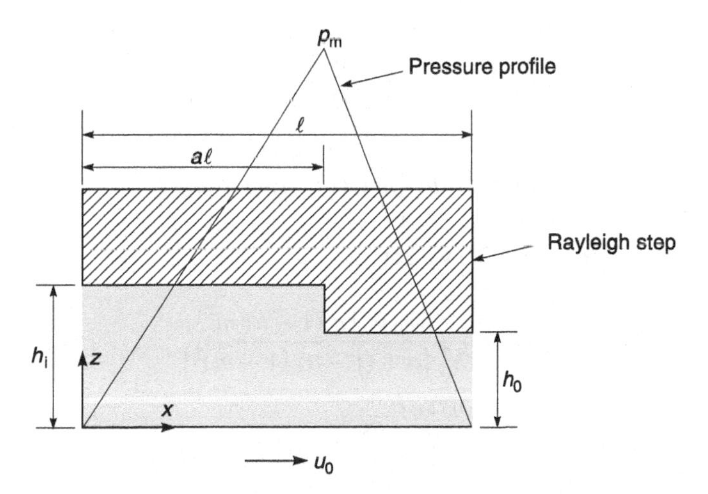

Figure 8.5.9 Schematic of a Rayleigh-step thrust bearing and associated pressure distribution.

direction with a linear velocity u_0 with no normal motion with respect to the top stationary surface. This bearing geometry is analyzed by considering two connected parallel-surface bearings. The Reynolds equation, Equation 8.3.24 is applied to each section of the bearing. For each section, there is no change in the film thickness; therefore:

$$\frac{d^2 p}{dx^2} = 0 \tag{8.5.19a}$$

Integrating we get,

$$p = C_1 x + C_2 \tag{8.5.19b}$$

Thus pressure gradients in the two sections are constant. Since the film thickness in the two sections are different, pressure gradients are also different with a maximum pressure occurring at the step. The boundary conditions are $p = 0$ at $x = 0$ and $x = \ell$, and $p = p_m$ at $x = a\ell$ for the regions where $h = h_0$ and for the region where $h = h_i$. Therefore

$$p_m = a\ell \left(\frac{dp}{dx}\right)_i = -(1-a)\,\ell \left(\frac{dp}{dx}\right)_0 \tag{8.5.20}$$

The flow rate per unit width at the step must be the same or $q_{x,\,0} = q_{x,\,i}$. From Equation 8.3.15a, we get

$$-\frac{h_0^3}{12\,\eta}\left(\frac{dp}{dx}\right)_0 + \frac{u_0\,h_0}{2} = -\frac{(1+m)^3\,h_0^3}{12\,\eta_0}\left(\frac{dp}{dx}\right)_i + \frac{(1+m)\,u_0\,h_0}{2} \tag{8.5.21}$$

where $m = \frac{h_i}{h_0} - 1$. By solving Equations 8.5.20 and 8.5.21 we get an expression for the two pressure gradients:

$$\left(\frac{dp}{dx}\right)_i = \frac{6\,\eta_0\,u_0\,(1-a)\,m}{h_0^2\left[a + (1-a)\,(1+m)^3\right]} \tag{8.5.22a}$$

$$\left(\frac{dp}{dx}\right)_0 = \frac{-6\,\eta_0\,a\,m}{h_0^2\left[a + (1-a)\,(1+m)^3\right]} \tag{8.5.22b}$$

Solving Equations 8.5.20 and 8.5.22, we get

$$p_m = \frac{6\,\eta_0\,u_0\,\ell a\,(1-a)\,m}{h_0^2\left[a + (1-a)\,(1+m)^3\right]} \tag{8.5.23a}$$

$$= \frac{\eta_0\,u_0\,\ell}{h_0^2}\,\overline{p}_m \tag{8.5.23b}$$

where $\overline{p}_m$ is the dimensionless pressure.

The normal load capacity is directly proportional to the triangular area formed by the pressure distribution. Therefore, normal load capacity per unit width is given as

$$w_z = \frac{p_m \,\ell}{2}$$

$$= \frac{\eta_0 \, u_0 \, \ell^2}{h_0^2} \overline{W}_z \tag{8.5.24a}$$

$$= \frac{\overline{p}_m}{2} \tag{8.5.24b}$$

where $\overline{W}_z$ is the dimensionless load capacity.

Volumetric flow rate per unit width is obtained, using Equations 8.5.21, 8.5.22, and 8.5.23, as

$$q_x = -\frac{(1+m)^3 \, p_m}{12 \, \eta_0 \, \ell a} + \frac{(1+m) \, u_0 \, h_0}{2} \tag{8.5.25}$$

$$= u_0 \, h_0 \, \overline{Q}$$

where $\overline{Q}$ is the dimensionless volumetric flow rate.

The bearing geometry to maximize p_m is obtained by putting derivatives of p_m with respect to a and m equal to zero. The maximum pressure is generated at $m = 0.866$ and $a = 0.7182$. For this optimum geometry,

$$p_m = 0.4104 \frac{\eta_0 \, u_0 \, \ell}{h_0^2} \tag{8.5.26a}$$

and

$$w_z = 0.2052 \frac{\eta_0 \, u_0 \, \ell^2}{h_0^2} \tag{8.5.26b}$$

Based on Equations 8.5.14 and 8.5.26, the load capacity of the step slider bearing is better than that of the fixed-inclined-plane bearing.

Step bearings with a variety of step designs have been extensively analyzed (Cameron, 1976). Bearings with shrouded steps, such as a semicircular step, have been analyzed and shown to retard side flow. In most cases, the step height is roughly equal to the minimum film thickness, which is difficult to fabricate, especially if it is less than 25 μm. Also, if the bearing touches the runner during use, a small amount of wear may reduce the step height or may remove it entirely. Etching and electroplating techniques are often used to produce the step. In spite of these disadvantages, step bearings are simple to fabricate as compared to tilted pad bearings, and these are commonly used in gas-bearing applications.

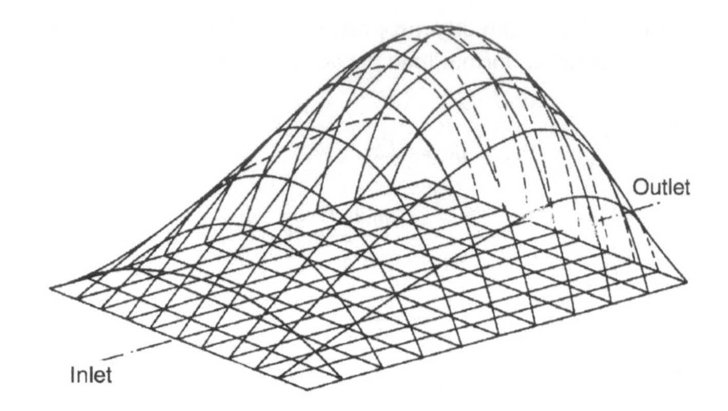

Figure 8.5.10 Pressure map in a thrust bearing of finite width with side flow.

8.5.1.4 Thrust Bearings of Finite Width

We now consider bearings of finite width. If b/ℓ is less than 4, the fluid flow in both x and y directions needs to be considered by solving a two-dimensional Reynolds equation. The effect of the flow of fluid in the direction orthogonal to the direction of sliding is to diminish the pressure in the fluid at the ends of the bearing, and pressure is not uniform across the width of the bearing, Figure 8.5.10. Thus the maximum load capacity of a bearing of finite width is less than that of a bearing of infinite width. For example, for a bearing of b/ℓ ratio equal to 2, the load capacity is about 30% less than that for a bearing with infinite width which operates under the same conditions.

For flow of an incompressible fluid with unidirectional motion under steady state conditions, a two-dimensional Reynolds equation given by Equation 8.3.23, is used. There is no general closed-form solution. The only film shape that can be solved analytically is that for a parallel-step slider bearing with constant film shapes within the inlet and outlet regions. For all other shapes, approximate solutions have been obtained by using electrical analogies, semi-analytical methods, numerical and graphical methods (Pinkus and Sternlicht, 1961; Cameron, 1976; Fuller, 1984; Frene *et al.*, 1997; Khonsari and Booser, 2001; Hamrock *et al.*, 2004; Szeri, 2010).

A. A. Raimondi and J. Boyd of Westinghouse Research Lab. (Raimondi and Boyd, 1955) solved the Reynolds equation for a fixed-inclined-pad slider bearing using a relaxation method in which derivatives are replaced by finite difference approximations and the functions are represented by a quadratic expression. They develop solutions as a function of bearing characteristic number, equivalent to the Sommerfeld number for journal bearings:

$$S = \left(\frac{\eta_0 u_0 b}{W}\right)\left(\frac{\ell}{h_i - h_0}\right)^2 \tag{8.5.27}$$

Another parameter in a finite-width bearing which affects bearing performance is length-to-width ratio (ℓ/b). An example of the bearing design curve showing minimum film thickness as a function of bearing characteristic number for various values of b/ℓ is shown in Figure 8.5.11. A square pad ($b/\ell = 1$) generally gives a good performance.

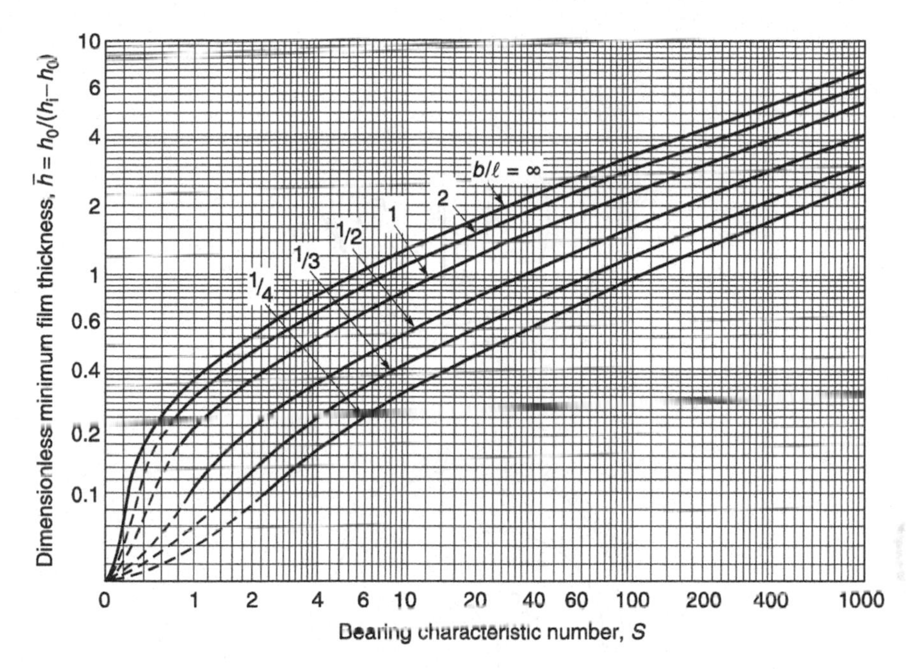

Figure 8.5.11 Dimensionless film thickness as a function of bearing characteristic number (S) for various values of b/ℓ of a fixed-inclined-pad thrust bearing. Reproduced with permission from Raimondi, A.A. and Boyd, J. (1955), "Applying Bearing Theory to the Analysis and Design of Pad-Type Bearings," *ASME Trans.* **77**, 287–309. Copyright 1955. ASME.

8.5.2 *Journal Bearings*

Journal bearings are commonly used machine components to carry radial loads both in dry and lubricated conditions. A loaded, rotating shaft (journal) is supported in a circular sleeve (bearing or bushing) with slightly larger diameter than that of a journal. The lubricant is supplied to the bearing through a hole or a groove. If the bearing extends around the full 360° of the journal, it is called a full journal bearing. If a wrap angle is less than 360°, it is called a partial journal bearing.

Figure 8.5.12 shows a schematic of a journal bearing. Letter o is the center of journal of a radius r whereas o' is the center of the bearing with c as the radial clearance, or simply clearance, which is the difference in radii of the journal and the bearing. The value of c/r is typically 10^{-4} to 10^{-3}. Based on Williams (2005), as the shaft, carrying a unidirectional load W_r, starts to rotate with an angular speed ω (or N in revolutions per unit time), the journal and the bearing surfaces are in contact at a point A, Figure 8.5.12a. At this point, the normal component of the contact force, F_R, is equal and opposite to the net normal load, W_r. The force F_R can be resolved in two components – friction force F_F and normal force F_N at the contact point. For steady sliding, F_F/F_N is equal to μ the coefficient of friction of the interface, which determines the location of contact point A. If the shaft rotates in the presence of a viscous fluid, the convergent channel formed by the clearance gap on the upstream side of the contact point A drags the fluid into the gap. If the journal speed is fast enough to develop hydrodynamic pressure larger than the applied load, two surfaces start to separate and the journal moves

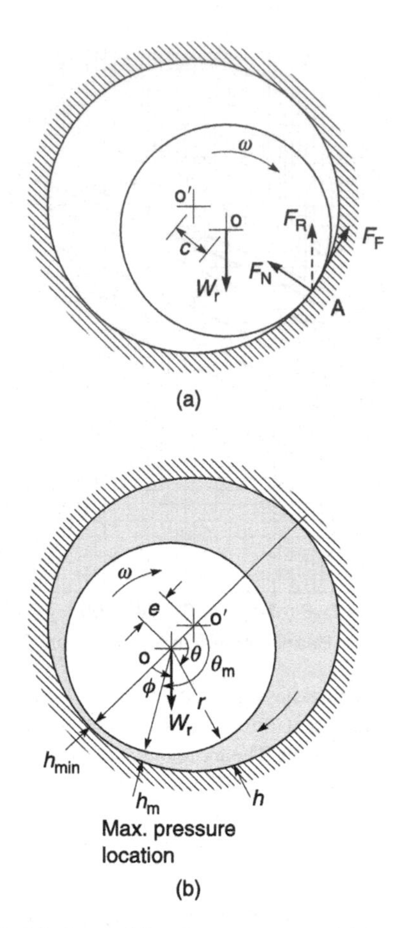

Figure 8.5.12 Schematics of plain journal bearing geometry with journal rotating at angular speed ω under (a) dry and (b) lubricated conditions.

around the bearing in the same sense as the rotation until it reaches an equilibrium with its center o to the left of the center of the bearing o' as shown in Figure 8.5.12b. The journal attains the position as a result of force equilibrium, including hydrodynamic pressure. The distance between the centers of the journal and bearing is known as the eccentricity, e. At very light or zero load, e is zero. As the load increases, the journal is forced downward and the limiting position is reached when $e = c$ and the journal touches the bearing.

The eccentricity ratio, ε, is defined as the ratio of eccentricity to clearance,

$$\varepsilon = \frac{e}{c} \tag{8.5.28}$$

Note that $0 \leq \varepsilon \leq 1$. The minimum film thickness h_{min} is given as

$$h_{min} = c - e = c\,(1 - \varepsilon) \tag{8.5.29}$$

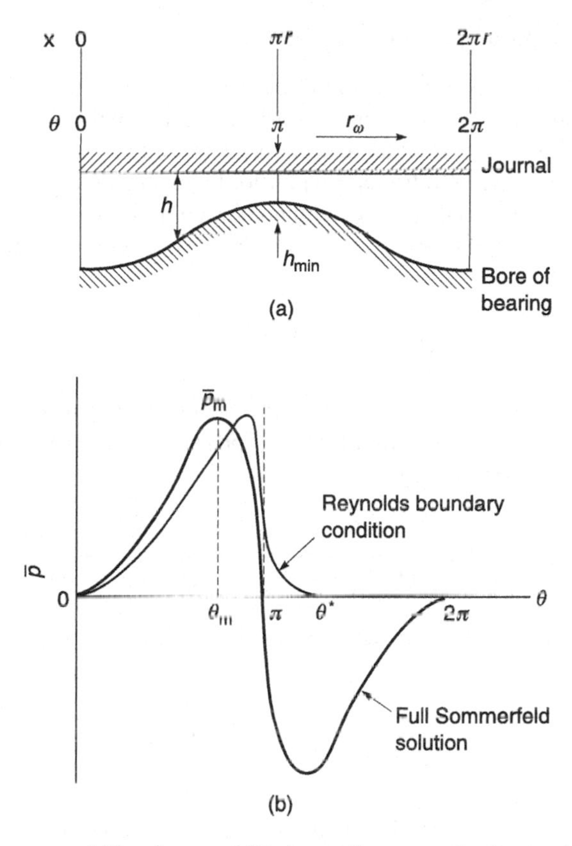

Figure 8.5.13 (a) Unwrapped film shape and (b) shape of pressure distribution in a journal bearing for full Sommerfeld solution and Reynolds boundary condition.

Since the film thickness is small compared to the shaft radius, the curvature of the film can be neglected. Therefore, the film can be unwrapped as shown in Figure 8.5.13a from around the shaft and it is a periodic stationary profile with wavelength $2\pi r$. The value of θ measures the angular position from the position of maximum film thickness. (The minimum film thickness occurs at $\theta = \pi$.) The distance x along the circumference is equal to $r\theta$. For most bearings, c/r is between 10^{-4} and 10^{-3}. For the case of $c/r << 1$, from geometry the film thickness at any point can be approximated as (see, e.g., Hamrock *et al.*, 2004),

$$h \sim c\ (1 + \varepsilon\ \cos\ \theta) \tag{8.5.30}$$

The angle between the line $O\ O'$, representing minimum film thickness location and the load axis ϕ, is known as the attitude angle.

Analytical solutions of infinitely wide and short width bearings can be obtained by assuming one-dimensional fluid flow. Numerical solutions are required for bearings with two-dimensional flow.

8.5.2.1 Infinitely-Wide Journal Bearing

If the width of the bearing is assumed to be much greater than its diameter d $(= 2r)$, then most of the flow through the gap between two surfaces occurs in the circumferential direction. The flow in the axial direction is small and the pressure in the axial direction is assumed to be constant. This assumption is valid for a width-to-diameter ratio, b/d, greater than 2 (or b/r greater than 4). For unidirectional motion with no normal motion and with constant viscosity, the integrated form of Reynolds equation, in cylindrical polar coordinates, is given as

$$\frac{dp}{d\theta} = 6\,\eta_0\, r^2 \omega \left(\frac{h - h_m}{h^3}\right) \tag{8.5.31}$$

where h_m is the film thickness when $dp/dx = 0$, corresponding to the maximum pressure location. Note that $dx = r\,d\theta$. Substituting the expression for h from Equation 8.5.30 into Equation 8.5.31 and integrating we get

$$p = 6\,\eta_0\,\omega \left(\frac{r}{c}\right)^2 \int \left[\frac{1}{(1+\varepsilon\,\cos\,\theta)^2} - \frac{h_m}{c\,(1+\varepsilon\,\cos\,\theta)^3}\right] d\theta + C_1 \tag{8.5.32}$$

The integral in Equation 8.5.32 is solved by using the Sommerfeld substitution (Sommerfeld, 1904),

$$1 + \varepsilon\,\cos\,\theta = \frac{1-\varepsilon^2}{1-\varepsilon\,\cos\,\gamma}$$

where γ is known as a Sommerfeld variable. In the full Sommerfeld solution, the periodic boundary condition with $p = p_0$ at $\theta = 0$ and 2π (pressure at the point of maximum film thickness), to solve for constants and pressure is given as

$$p - p_0 = 6\,\eta_0\,\omega \left(\frac{r}{c}\right)^2 \frac{6\varepsilon\,\sin\,\theta\,\,(2+\varepsilon\,\cos\,\theta)}{(2+\varepsilon^2)\,(1+\varepsilon\,\cos\,\theta)^2} \tag{8.5.33a}$$

$$= 6\,\eta_0\,\omega \left(\frac{r}{c}\right)^2 \overline{p} \tag{8.5.33b}$$

and

$$h_m = \frac{2c\,(1+\varepsilon^2)}{2+\varepsilon^2} \tag{8.5.34}$$

where $\overline{p}$ is the dimensionless pressure. The shape of the pressure distribution is shown in Figure 8.5.13b. The positive pressure is developed in the convergent film $(0 \le \theta \le \pi)$ and negative pressure in the divergent film $(\pi \le \theta \le 2\pi)$. The pressure distribution is skewed symmetrically.

Note that fluid pressure, where it is introduced into the clearance of the bearing, is equal to the supply pressure. This supply may or may not be equal to ambient pressure and it also need not coincide with $\theta = 0$ $(p = p_0)$.

From Equations 8.5.30 and 8.5.34, we get θ_m where dp/dx = 0 corresponds to the maximum pressure location.

$$\theta_m = \cos^{-1}\left(-\frac{3\varepsilon}{2+\varepsilon^2}\right) \tag{8.5.35}$$

The maximum pressure from Equations. 8.5.33 and 8.5.34 is given as

$$\overline{p}_m = \frac{3\,\varepsilon\,(4-5\,\varepsilon^2+\varepsilon^4)^{1/2}(4-\varepsilon^2)}{2\,(2+\varepsilon^2)\,(1-\varepsilon^2)^2} \tag{8.5.36}$$

Maximum pressure occurs in the second quadrant and the minimum pressure occurs in the third quadrant. If $\varepsilon \to 0$, $\theta_m \to \pm\pi/2$ and $\overline{p}_m = 0$, and if $\varepsilon \to 1$, $\theta_m = \pm\pi$ and $\overline{p}_m \to \infty$ (Hamrock *et al.*, 2004).

Based on practical experience, the negative pressures predicted in the divergent film cannot be supplied by liquids, and are rarely encountered. Thus, subambient pressures predicted by the analysis should be ignored. Experimental measurement indicates that the pressure in the fluid has the form illustrated by one of the curves (labeled Reynolds boundary condition) in Figure 8.5.13b. An approach which limits the analysis to the convergent film ($0 \le \theta \le \pi$) is known as the half Sommerfeld solution. The pressure p-p_0 can be assumed to be zero for $\pi < \theta \le 2\pi$. However, this assumption violates the continuity of mass flow at the outlet end of the pressure. A better boundary condition is a Reynolds cavitation boundary condition which is in good agreement with experience and states that the pressure curve terminates with zero gradient with unknown position in the divergent part of the film:

$$p = \frac{dp}{dx} = 0 \; at \; \theta = \theta^*, \;\; \pi < \theta^* \le 2\pi \tag{8.5.37}$$

Figure 8.5.13b shows a pressure profile using the Reynolds boundary condition.

An estimate of the load capacity per unit width can be obtained by integration of pressure over bearing area by adopting the half Sommerfeld solution by setting the film pressure equal to zero at values of θ between π and 2π. The forces per unit width acting on the journal in Figure 8.5.14, are given as

$$w_x = \int_0^{\pi} p\; r\; \sin\,\theta\; d\theta \tag{8.5.38a}$$

$$w_z = -\int_0^{\pi} p\; r\; \cos\,\theta\; d\theta \tag{8.5.38b}$$

By substituting for p from Equation 8.5.32, using the Sommerfeld substitution and integrating the resulting equation, we get

$$w_x = 6\,\eta_0\,\omega\,r\left(\frac{r}{c}\right)^2 \frac{\pi\,\varepsilon}{(2+\varepsilon^2)\,\left(1-\varepsilon^2\right)^{1/2}} \tag{8.5.39a}$$

$$= \eta_0\,\omega\,r\left(\frac{r}{c}\right)^2 \overline{W}_x \tag{8.5.39b}$$

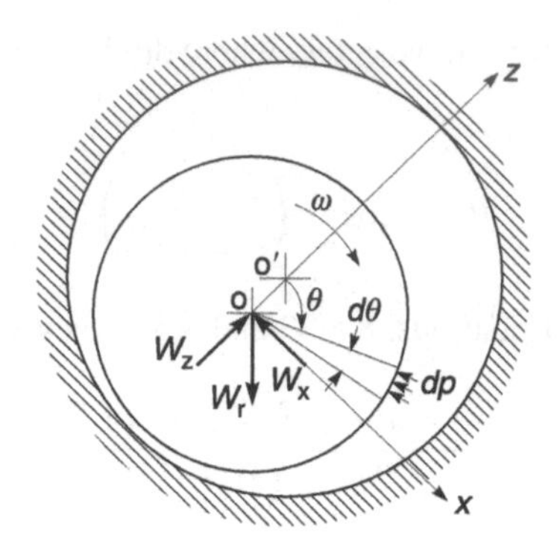

Figure 8.5.14 Coordinate system and force components acting on the journal in a journal bearing.

$$w_z = 6\,\eta_0\,\omega\,r\,(\frac{r}{c})^2 \frac{\varepsilon^2}{(2+\varepsilon^2)(1-\varepsilon^2)} \tag{8.5.39c}$$

$$= \eta_0\,\omega\,r\,\left(\frac{r}{c}\right)^2 \overline{W}_z \tag{8.5.39d}$$

The resultant load is given as

$$w_r = \left(w_x^2 + w_y^2\right)^{1/2}$$

$$= \eta_0\,\omega\,r\,\left(\frac{r}{c}\right)^2 \frac{6\,\varepsilon\left[\pi^2 - \varepsilon^2\,(\pi^2 - 4)\right]^{1/2}}{(2+\varepsilon^2)\,(1-\varepsilon^2)} \tag{8.5.40a}$$

$$= \eta_0\,\omega\,r\,\left(\frac{r}{c}\right)^2 \overline{W}_r \tag{8.5.40b}$$

where $\overline{W}_x$, $\overline{W}_y$ and $\overline{W}_r$ are the dimensionless loads in the x and y directions and resultant load, respectively. We can also write

$$S = \frac{\eta_0\,N}{P}\left(\frac{r}{c}\right)^2 = \frac{1}{\pi\,\overline{W}_r} = \frac{(2+\varepsilon^2)\,(1-\varepsilon^2)}{6\pi\,\varepsilon\,[\pi^2 - \varepsilon^2\,(\pi^2 - 4)]^{1/2}} \tag{8.5.40c}$$

where S is the Sommerfeld number, N is the angular speed in revolutions per second, and P is load per unit projected bearing area ($P = W_r/2\,r\,b$). Note that S is a function of ε only.

The attitude angle, ϕ, the angle between the minimum film thickness location and the resultant load axis, shown in Figure 8.5.12b, is given as

$$\phi = \tan^{-1}\left(\frac{w_x}{w_z}\right)$$

$$= \tan^{-1}\left[\frac{\pi}{2\varepsilon}\,(1-\varepsilon^2)^{1/2}\right] \tag{8.5.41}$$

Note that when $\varepsilon = 0$, $\phi = 90°$ and when $\varepsilon = 1$, $\phi = 0°$.

The shear force per unit width experienced by the journal and the bearing from Equation 8.3.16, are given as (Pinkus and Sternlicht, 1961),

$$f_j = \int_0^{2\pi} \left(\frac{h}{2r}\frac{dp}{d\theta} + \frac{\eta_0 \, r \, \omega}{h} \right) r \, d\theta \tag{8.5.42a}$$

and

$$f_b = \int_0^{2\pi} \left(-\frac{h}{2r}\frac{dp}{d\theta} + \frac{\eta_0 \, r \, \omega}{h} \right) r \, d\theta \tag{8.5.42b}$$

Viscous drag is provided by the entire bearing; therefore, by substituting for $dp/d\theta$ from Equation 8.5.31, using Sommerfeld substitution and integrating over the entire bearing, we get

$$f_j = -4\,\pi\,\eta_0\,\omega\,r \left(\frac{r}{c}\right) \frac{1+2\varepsilon^2}{\left(2+\varepsilon^2\right)\left(1-\varepsilon^2\right)^{1/2}} \tag{8.5.43a}$$

and

$$f_b = 4\,\pi\,\eta_0\,\omega\,r \left(\frac{r}{c}\right) \frac{\left(1-\varepsilon^2\right)^{1/2}}{2+\varepsilon^2} \tag{8.5.43b}$$

The friction torque is the shear force times the radius of the journal, r. The shear force on the journal, used for calculation of power loss H_v, is always greater than the shear force on the bearing, except for the concentric case. The difference in the journal and bearing torque is balanced by the external load, which exerts a moment through the eccentricity:

$$r f_j = r f_b + W_r \tag{8.5.44}$$

The coefficient of friction is given from Equations. 8.5.40 and 8.5.43 as,

$$\mu = \frac{f_j}{w_r} \tag{8.5.45a}$$

$$\mu\left(\frac{r}{c}\right) = \frac{4\,\pi\,\left(1+2\,\varepsilon^2\right)\left(1-\varepsilon^2\right)^{1/2}}{6\,\varepsilon\,\left[\pi^2-\varepsilon^2\left(\pi^2-4\right)\right]^{1/2}} \tag{8.5.45b}$$

Almost half the clearance of the full bearing is occupied by low pressure, which contributes very little to the load capacity but adds to the viscous drag. Therefore, the part of the bearing occupied by low pressure may be eliminated; this type of bearing is known as a partial-arc bearing, Figure 8.5.15. A partial bearing can be analyzed using Reynolds equation but the boundary conditions are different. The cyclic form is no longer present. The inlet boundary condition is the ambient pressure of the bearing and the outlet condition is also the ambient pressure. Dependent upon the length of arc, for very large divergent film, the Reynolds boundary condition may be required.

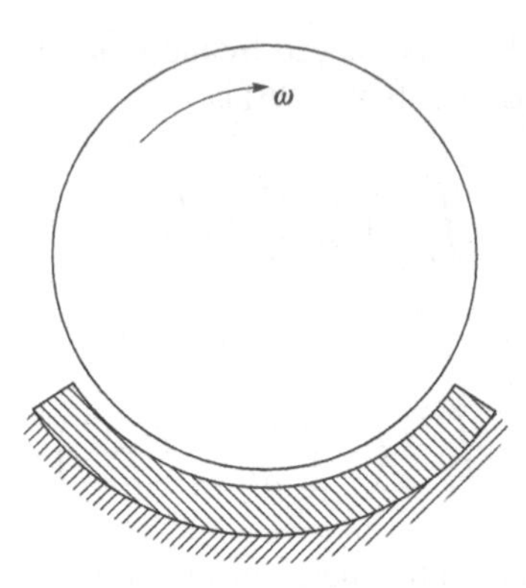

Figure 8.5.15 Schematic of a partial arc journal bearing.

Example Problem 8.5.3 45b

A journal bearing of width 1 m operates with a shaft of 200 mm diameter which rotates at 1200 rpm. The diametral clearance is 200 μm and absolute viscosity of the lubricating oil at am inlet temperature of 20°C is 40 cP. For an eccentricity ratio of 0.7, calculate the minimum film thickness, attitude angle, maximum film pressure, location of maximum film pressure, load capacity, and coefficient of friction.

Solution

$$\begin{aligned}
b &= 1 \text{ m} \\
d &= 2r = 200 \text{ mm},\ b/d = 5 \\
N &= 1200 \text{ rpm},\ \omega = 125.66 \text{ rad/s} \\
c &= 100\ \mu\text{m} \\
\eta_0 &= 0.04 \text{ Pa s} \\
\varepsilon &= 0.7 \\
h_{\min} &= c\,(1-\varepsilon) \\
&= 100(1-0.7) \text{ mm} \\
&= 30\ \mu\text{m} \\
\phi &= \tan^{-1}\left[\frac{\pi}{2\varepsilon}\left(1-\varepsilon^2\right)^{1/2}\right] \\
&= 58.03^\circ \\
p_m - p_0 &= 6\,\eta_0\,\omega\left(\frac{r}{c}\right)^2 \frac{3\,\varepsilon\left(4-5\,\varepsilon^2+\varepsilon^4\right)^{1/2}\left(4-\varepsilon^2\right)}{2\left(2+\varepsilon^2\right)\left(1+\varepsilon^2\right)^2} \\
&= 6\times 0.04\times 125.66\left(\frac{0.1}{1\times 10^{-4}}\right)^2 \\
&\quad \frac{3\times 0.7\left(4-5\times 0.7^2+0.7^4\right)^{1/2}\left(4-0.7^2\right)}{2\left(2+0.7^2\right)\left(1-0.7^2\right)^2}\ \text{Pa} \\
&= 30.72 \text{ MPa}
\end{aligned}$$

Location of film pressure,

$$\theta_m = \cos^{-1}\left(-\frac{3\,\varepsilon}{2+\varepsilon^2}\right)$$
$$= \cos^{-1}\left(-\frac{3\times 0.7}{2+0.7^2}\right)$$
$$= 147.5^\circ$$

Load capacity per unit width,

$$w_r = \eta_0\,\omega\,r\left(\frac{r}{c}\right)^2\frac{6\,\varepsilon\left[\pi^2-\varepsilon^2\left(\pi^2-4\right)\right]^{1/2}}{\left(2+\varepsilon^2\right)\left(1-\varepsilon^2\right)}$$
$$= 0.04\times 125.66\times 0.1\left(\frac{0.1}{1\times 10^{-4}}\right)^2$$
$$\frac{6\times 0.7\left[\pi^2-0.7^2\left(\pi^2-4\right)\right]^{1/2}}{\left(2+0.7^2\right)\left(1-0.7^2\right)}\ \mathrm{N/m}$$
$$= 4.4\times 10^6\ \mathrm{N/m}$$
$$\mu = \left(\frac{c}{r}\right)\frac{4\pi\left(1+2\,\varepsilon^2\right)\left(1-\varepsilon^2\right)^{1/2}}{6\,\varepsilon\left[\pi^2-\varepsilon^2\left(\pi^2-4\right)\right]^{1/2}}$$
$$= \left(\frac{1\times 10^{-4}}{0.1}\right)\frac{4\,\pi\left(1+2\times 0.7^2\right)\left(1-0.7^2\right)^{1/2}}{6\times 0.7\left[\pi^2-0.7^2\left(\pi^2-4\right)\right]^{1/2}}$$
$$= 1.6\times 10^{-3}$$

8.5.2.2 Short-Width Journal Bearing

If the diameter-to-width ratio (d/b) is greater than 2, the pressure-induced flow in the circumferential direction is small relative to that in the axial direction (Dubois and Ocvirk, 1953; Pinkus and Sternlicht, 1961). For this case, the Reynolds equation can be simplified as

$$h^3\frac{\partial^2 p}{\partial y^2} = 6\,\eta_0\,\omega\frac{dh}{d\theta} \tag{8.5.46}$$

As previously, subambient pressures are ignored (half Sommerfeld assumption). Integrating this equation and making use of the boundary conditions that $p = p_0$ at $y = \pm b/2$, we get (Pinkus and Sternlicht, 1961)

$$p = \frac{3\,\eta_0\,\omega\,\varepsilon}{c^2}\left(\frac{b^2}{4}-y^2\right)\frac{\sin\,\theta}{\left(1+\varepsilon\,\cos\,\theta\right)^3},\quad \text{for } 0\le\theta\le\pi \tag{8.5.47}$$

This equation shows that the axial variation of the pressure is parabolic.

The location of maximum pressure is obtained by solving $\partial p/\partial\theta = 0$. We get

$$\theta_m = \cos^{-1}\left[1 - \frac{\left(1+24\varepsilon^2\right)^{1/2}}{4\varepsilon}\right] \tag{8.5.48}$$

Maximum pressure occurs when $\theta = \theta_m$ *and* $y = 0$. From Equation 8.5.47, we get

$$p_m = \frac{3\,\eta_0\,\omega\,\varepsilon\,b^2\ \sin\theta_m}{4\,c^2\ (1+\varepsilon\ \cos\theta_m)^3} \tag{8.5.49}$$

The normal load components are obtained by integrating pressure over bearing area:

$$W_x = 2\int_0^{\pi}\int_0^{b/2} p\,r\ \sin\theta\,dy\,d\theta \tag{8.5.50a}$$

$$W_y = -2\int_0^{\pi}\int_0^{b/2} p\,r\ \cos\theta\,dy\,d\theta \tag{8.5.50b}$$

Substituting expression for p from Equation 8.5.42 and by using Sommerfeld substitution, we get

$$W_x = \eta_0\,\omega\,r\,b\left(\frac{r}{c}\right)\left(\frac{b}{d}\right)^2\frac{\pi\,\varepsilon}{\left(1-\varepsilon^2\right)^{3/2}} = \eta_0\,\omega\,r\,b\left(\frac{r}{c}\right)^2\overline{W}_x \tag{8.5.51a}$$

$$W_z = \eta_0\,\omega\,r\,b\left(\frac{r}{c}\right)\left(\frac{b}{d}\right)^2\frac{4\,\varepsilon^2}{\left(1-\varepsilon^2\right)^{1/2}} = \eta_0\,\omega\,r\,b\left(\frac{r}{c}\right)^2\overline{W}_z \tag{8.5.51b}$$

$$W_r = \eta_0\omega r\,b\left(\frac{r}{c}\right)\left(\frac{b}{d}\right)^2\frac{\varepsilon}{\left(1-\varepsilon^2\right)^2}\left[16\varepsilon^2+\pi^2\left(1-\varepsilon^2\right)\right]^{1/2} = \eta_0\omega r\,b\left(\frac{r}{c}\right)^2\overline{W}_r \tag{8.5.51c}$$

We can also write the dimensional resultant load in terms of Sommerfeld number,

$$S\left(\frac{b}{d}\right)^2 = \frac{1}{\pi\,\overline{W}_r}\left(\frac{b}{d}\right)^2 = \frac{\left(1-\varepsilon^2\right)^2}{\pi\,\varepsilon\left[16\,\varepsilon^2+\pi^2\left(1-\varepsilon^2\right)\right]^{1/2}} \tag{8.5.51d}$$

and

$$\phi = \tan^{-1}\left[\frac{\pi\left(1-\varepsilon^2\right)^{1/2}}{4\,\varepsilon}\right] \tag{8.5.52}$$

Since there is no pressure-induced shear, the shear force experienced by the journal (or the bearing) is simply given as

$$F_j = -F_b = \int_0^{2\pi}\frac{\eta_0\,\omega\,r\,br}{h}\,d\theta$$

$$= \eta_0\,\omega\,r\,b\left(\frac{r}{c}\right)\frac{2\,\pi}{\left(1-\varepsilon^2\right)^{1/2}} \tag{8.5.53}$$

Viscous drag is provided by the entire bearing.

The coefficient of friction is given as

$$\mu = \frac{F_j}{W}$$

$$\mu\left(\frac{r}{c}\right) = \frac{2\pi^2 S}{\left(1-\varepsilon^2\right)^{1/2}} \tag{8.5.54}$$

The fluid flow out the sides of the bearing is given from Equation 8.3.17 as

$$Q_y = -2r\int_0^{\pi}\left(\frac{h^3}{12\,\eta_0}\frac{\partial p}{\partial y}\right)_{y=b/2} d\theta \tag{8.5.55a}$$

Using Equation 8.5.47 in Equation 8.5.55a and integrating, we get

$$Q_y = \omega\, r\, b\, e = \frac{\omega\, r\, b\, c}{2\pi}\,\overline{Q}_y \tag{8.5.55b}$$

Note that as $e \to 0$, $Q_y \to 0$ (no side leakage) and as $e \to c$, $Q_y = \omega\, r\, b\, c$ (complete side leakage) (Hamrock *et al.*, 2004).

8.5.2.3 Journal Bearing with Various Slenderness Ratio

We now consider a journal bearing with various *b*/*d* values or slenderness ratios. A two-dimensional Reynolds equation is solved numerically to obtain solution of a bearing. Solutions are presented as a function of bearing characteristic number or Sommerfeld number described earlier. Sommerfeld (1904) found that

$$\mu\left(\frac{r}{c}\right) = \phi\,(S) \tag{8.5.56}$$

A. A. Raimondi and J. Boyd of the Westinghouse Research Laboratory solved the Reynolds equation for journal bearings of different *b*/*d* ratios (Raimondi and Boyd, 1958). They used a relaxation method in which derivatives are replaced by finite difference approximations and the functions are represented by quadratic expressions. They developed solutions as a function of bearing characteristic number or Sommerfeld number. Selected charts for performance of a journal bearing are presented in Figure 8.5.16. In Figure 8.5.16a, an optimum eccentricity ratio or minimum film thickness is indicated. The left boundary (dotted line) defines it the optimum eccentricity ratio for a minimum coefficient of friction and the right boundary defines for maximum load. The recommended eccentricity ratio is in between these two boundaries. The trend in pressure distribution in a journal bearing is shown in Figure 8.5.17. The figure also shows the definition of attitude angle, ϕ (the angle between the load axis and minimum film thickness), angle of maximum pressure, ϕ_m (the angle between the load axis and the

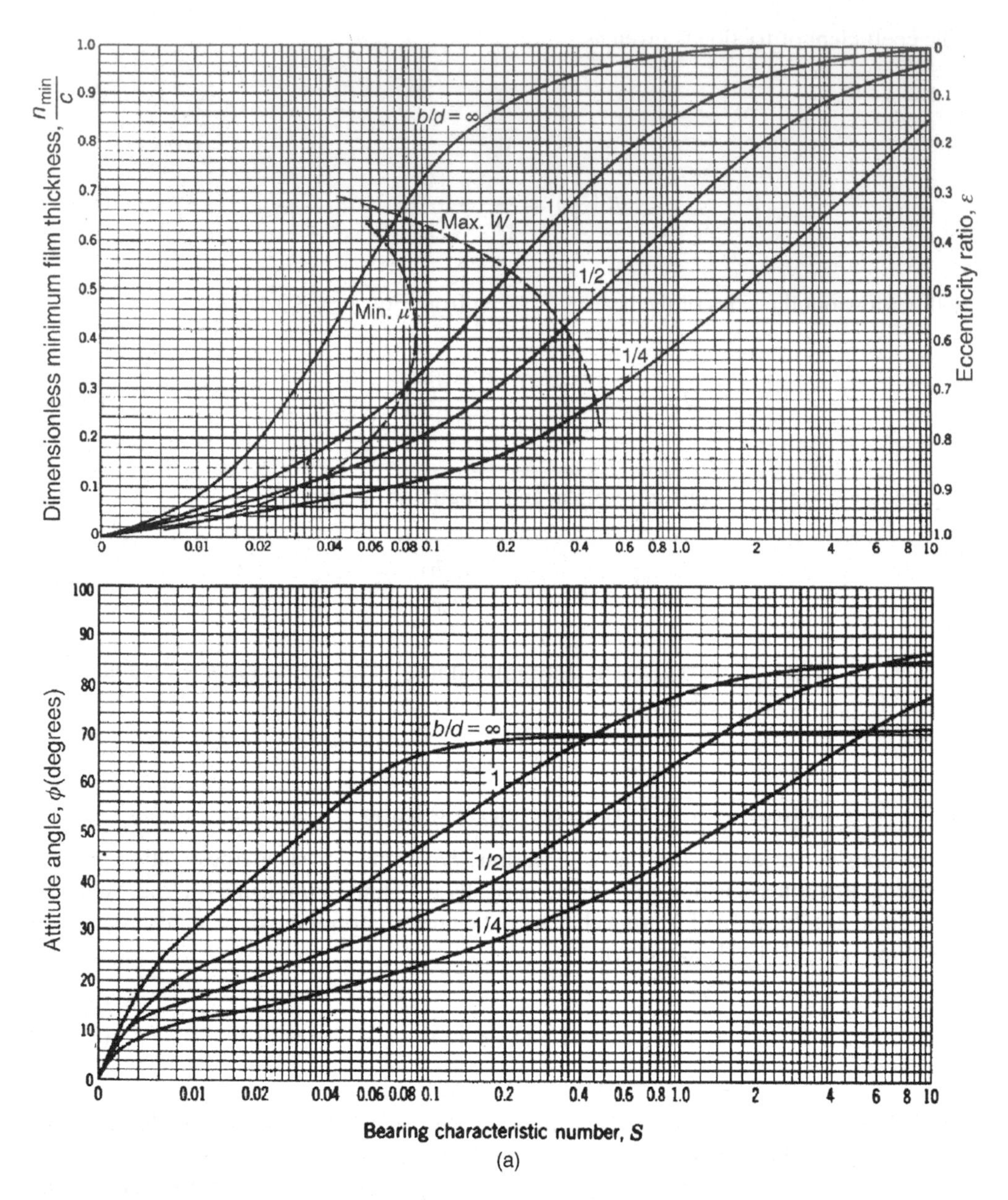

Figure 8.5.16 Effect of bearing characteristic number (S) on (a) dimensionless minimum film thickness (or eccentricity ratio) and attitude angle, (b) dimensionless maximum film pressure and terminating position of the fluid film and the position of maximum film pressure, and (c) dimensionless coefficient of friction variable, dimensionless volumetric flow rate and volumetric side flow ratio, for four width-to-diameter ratios in a liquid-lubricated journal bearing. Reproduced with permission from Raimondi, A.A. and Boyd, J. (1958), "A Solution for the Finite Journal Bearing and its Application to Analysis and Design – I, – II, and – III," *ASLE Trans.* **1**, 159–174; 175–193; 194–209. Copyright 1958. Taylor and Francis. (*Continued*)

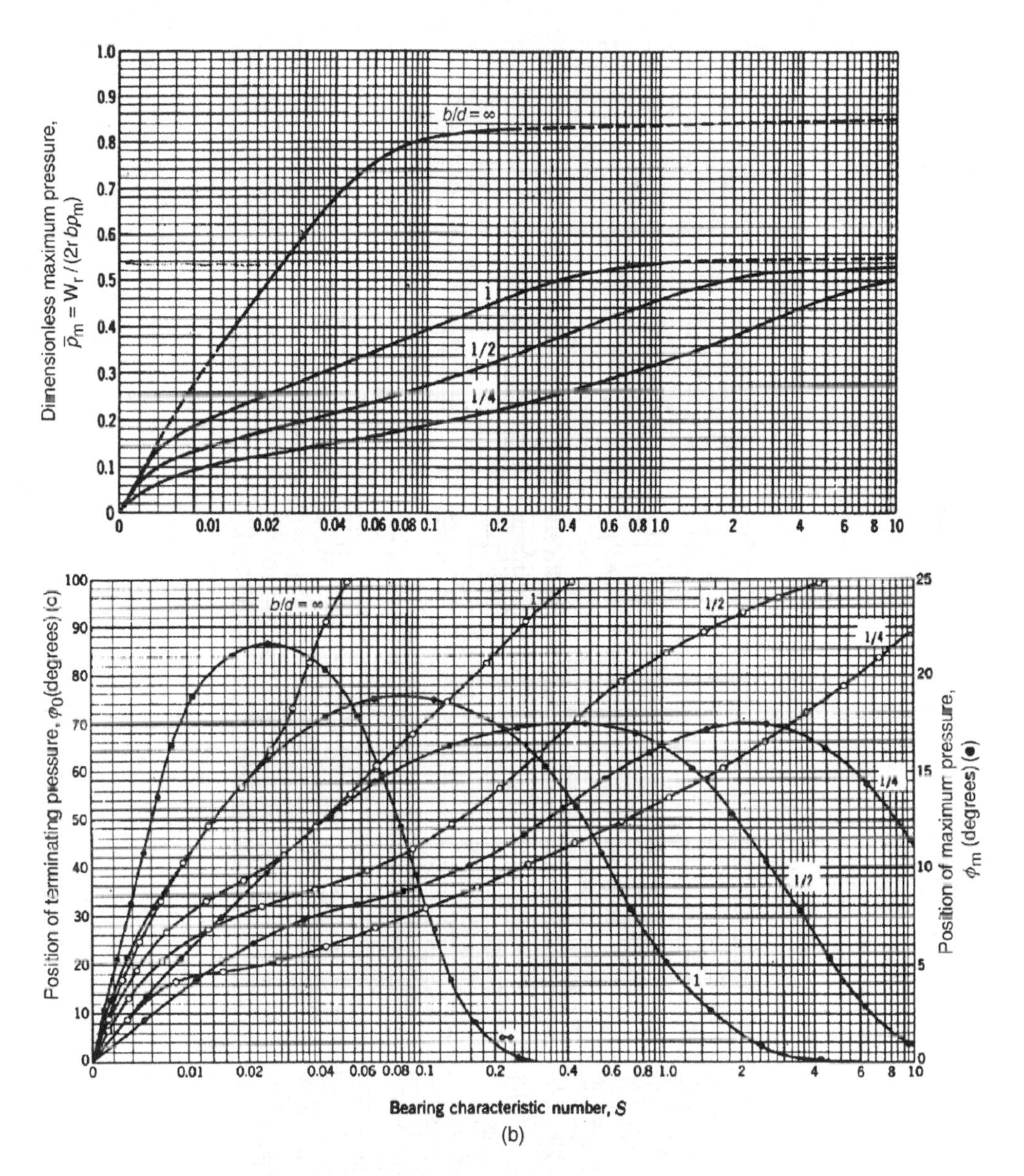

Figure 8.5.16 (*Continued*)

maximum pressure location), and location of terminating pressure, ϕ_0 (the angle between the load axis and terminating pressure location).

The film stiffness of a journal bearing is given as

$$k_f = \frac{dW_r}{dh_{\min}} \tag{8.5.57}$$

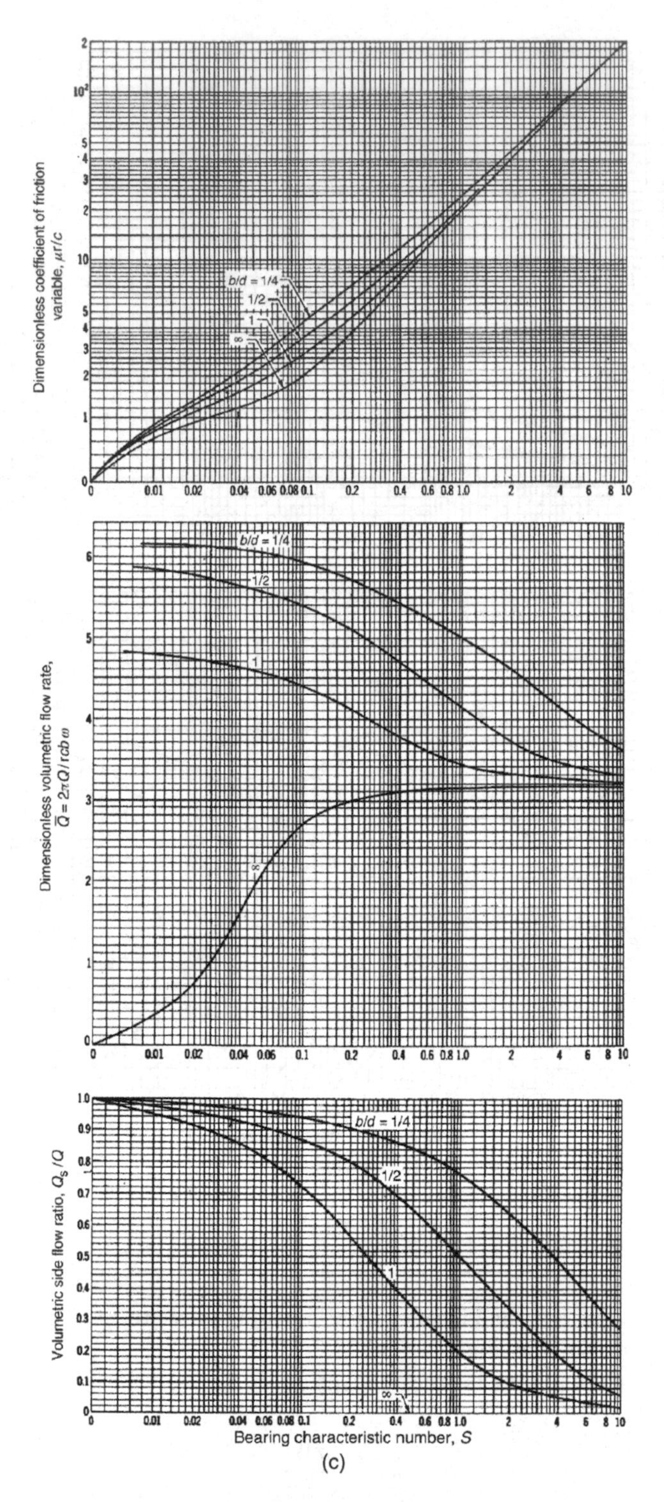

Figure 8.5.16 (*Continued*)

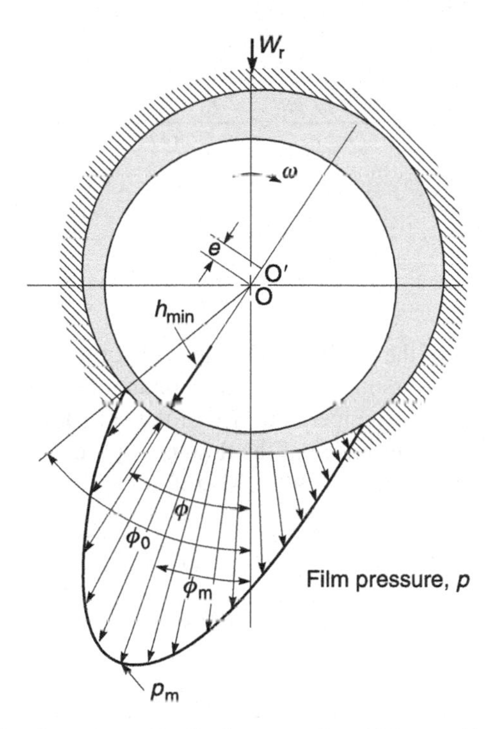

Figure 8.5.17 Schematic of pressure distribution and film thickness distribution and definition of various locations.

The stiffness of the bearing is nonlinear. It may be taken as linear for small displacements about an equilibrium position. In Figure 8.5.16a, a tangent can be drawn to the curve at the operating ε or $h_{\min}$ and its slope is evaluated to get stiffness in the appropriate units (N/m) at that point. The natural frequency of vibration in the vertical direction is

$$f_n = \frac{1}{2\pi}\left[\frac{k_f}{W_r/g}\right]^{1/2} Hz \tag{8.5.58}$$

This is known as the first critical speed (called synchronous whirl). The rotational speed of the shaft should be smaller than this frequency.

Work is done on the fluid due to viscous shear, which results in an increase in the temperature of fluid when it leaves the contact. The temperature rise can be calculated by assuming that all of the work on the fluid is dissipated as heat. We further assume that all the heat generated is carried away by the oil flow. Therefore,

$$\mu W_r\, r\, \omega = Q\, \rho\, c_p\, \Delta t$$

or

$$\Delta t = \frac{\mu W_r r \omega}{Q \rho c_p} = \frac{2P}{\rho c_p} \frac{(r/c)\mu}{Q/(\omega r b c)} \tag{8.5.59a}$$

where

$$P = \frac{W_r}{2rb}$$

We now consider that some of the oil flows out at the side of the bearing before the hydrodynamic film is terminated. If we assume that the temperature of the side flow is the mean of the inlet and outlet temperatures, the temperature rise of the side flow is $\Delta t/2$. This means that the heat generated raises the temperature of the flow $Q - Q_s$ an amount Δt, and the flow Q_s an amount $\Delta t/2$ (Shigley and Mitchell, 1993). Therefore,

$$\mu W_r r \omega = \rho c_p \left[(Q - Q_s) + \frac{Q_s}{2} \right] \Delta t$$

or

$$\begin{aligned} \Delta t &= \frac{2P}{\rho c_p} \frac{(r/c)\mu}{Q(1 - 0.5 Q_s/Q)/\omega r b c} \\ &= \frac{4\pi P}{\rho c_p} \frac{(r/c)\mu}{\overline{Q}(1 - 0.5 Q_s/Q)} \end{aligned} \tag{8.5.59b}$$

The temperature rise for a given load can be calculated using Equation 8.5.59 with data obtained from Figure 8.5.16. Viscosity in the analysis of bearing performance should be used as the mean of the inlet and outlet temperatures, t_m. If the value of t_m differs from the value initially assumed, then viscosity at the mean temperature should be used and bearing performance should be recalculated. This process should be iterated until predicted temperature rise is comparable to that assumed.

Note that with a decrease in bearing clearance, c, temperature rise increases as does minimum film thickness. However, if the clearance becomes too large, minimum film thickness begins to drop again. Therefore, bearing clearance should be optimized.

So far, we have considered constant loaded bearings. Many journal bearings are subjected to loads whose magnitude and direction vary with time, which results in change in size and position of the minimum film thickness in a cyclic manner. This results in variation of the center of the journal, which moves in an orbit about the geometric center of a rotating shaft, in the direction of the shaft rotation. This phenomenon is called "synchronous whirl." The intensity of this vibration is a function of the inertia of the rotor and the stiffness and damping capacity of the bearings. This is a more serious problem in gas-lubricated bearings.

Finally, we examine lubricant delivery methods. Fluid is pumped to the bearing by external means at a supply pressure to a supply groove. Various types of oil-supply grooves in journal bearings are used. Figure 8.5.18 shows two typical oil-supply grooves. The most common type is a single rectangular feed groove machined in the bearing liner. In another design, the

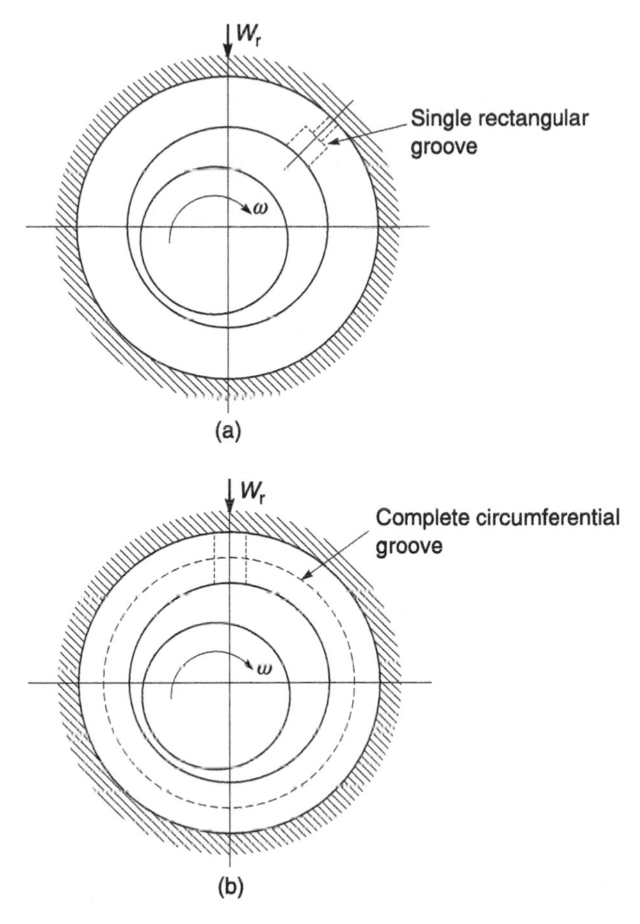

Figure 8.5.18 Schematics of typical oil-supply feed grooves: (a) single rectangular groove, and (b) complete circumferential groove.

groove could extend throughout the circumference with an inlet. The most common location of a single groove is about 45–90° to the load line, in the direction of shaft rotation.

Example Problem 8.5.4

A hydrodynamic journal bearing of width 200 mm operates with a shaft of 200 mm diameter which rotates at 1200 rpm. The diametral clearance is 200 μm and absolute viscosity of the lubricating oil at an inlet temperature of 20°C is 40 cP. For an eccentricity ratio of 0.7, calculate the minimum film thickness, attitude angle, load capacity, pressure P, maximum film pressure, location of maximum film pressure, volumetric flow rate, volumetric side flow rate, coefficient of friction, viscous power loss, and temperature rise. The mass density and specific heat of oil are 880 kg/m^3 and 1.88 J/g K, respectively.

Solution

Given

$$b = 200 \text{ mm}$$
$$d = 2r = 200 \text{ mm}, \; b/d = 1$$
$$N = 1200 \text{ rpm}, \; \omega = 125.66 \text{ rad/s}$$
$$c = 100 \; \mu\text{m}$$
$$\eta_0 = 0.04 \text{ Pa s}$$
$$\varepsilon = 0.7$$

From Figure 8.5.16, for b/d = 1,

$$S = \frac{1}{\pi \overline{W}_r} = 0.08$$
$$\phi = 45^\circ$$
$$\overline{p}_m \sim 0.38$$
$$\phi_m = 19^\circ$$
$$\overline{Q} \sim 4.95$$
$$\frac{Q_s}{Q} \sim 0.76$$
$$\frac{\mu r}{c} \sim 2.4$$

Therefore, minimum film thickness, $h_{\min} = c\,(1 - \varepsilon)$

$$= 100\;(1 - 0.7)\;\; \mu\text{m}$$
$$= 30 \; \mu\text{m}$$

Attitude angle, $\phi = 45^\circ$

Load capacity, $W_r = \eta_0\, \omega\, r\, b\, \left(\frac{r}{c}\right)^2 \overline{W}_r$

$$= 0.04 \times 125.66 \times 0.1 \left(\frac{0.1}{1 \times 10^{-4}}\right)^2 0.2 \frac{1}{\pi \times 0.08} \text{ N}$$
$$= 4 \times 10^5 \text{N}$$
$$P = \frac{W_r}{2\, r\, b}$$
$$= \frac{4 \times 10^5}{0.2 \times 0.2} \text{ Pa}$$
$$= 1 \text{ MPa}$$

Maximum film pressure, $p_m = \frac{P}{\bar{p}_m}$

$$\frac{1 \times 10^6}{0.38} \text{ Pa}$$
$$= 2.63 \text{ MPa}$$

Location of maximum film pressure, $\phi_m = 19^\circ$

Volumetric flow rate is $\frac{\overline{Q}rcb\omega}{2\pi}$

$$= \frac{4.95 \times 0.1 \times 1 \times 10^{-4} \times 0.2 \times 125.66}{2\,\pi} \text{ m}^3/\text{s}$$
$$= 1.98 \times 10^{-4} \text{ m}^3/\text{s}$$

Volumetric side flow $= 0.76 \times 1.98 \times 10^{-4} \text{ m}^3/\text{s}$

$$= 1.50 \text{ m}^3/\text{s}$$
$$\mu = \frac{2.4 \times 1 \times 10^{-4}}{0.1}$$
$$= 2.4 \times 10^{-3}$$

Viscous power loss is $\mu \; W_r \; r \; \omega$

$$= \frac{2.4 \times 10^{-4}}{0.1} 4 \times 10^5 \times 0.1 \times 125.66 \text{ W}$$
$$= 12.06 \text{ kW}$$

Temperature rise, Δt is $\frac{4\pi\,P}{\rho\,c_p}\frac{(r/c)\,\mu}{\overline{Q}(1-0.5Q_s/Q)}$

$$= \frac{4\,\pi \times 10^6}{880 \times 10^3 \times 1.88}\,\frac{2.4}{4.95\,(1-0.36)} \,^\circ\text{C}$$
$$= 5.75^\circ\text{C}$$

The mean temperature of the oil,

$$t_m = t_i + \frac{\Delta t}{2}$$
$$= 20 + \frac{5.75}{2}$$
$$= 22.88^\circ\text{C}$$

For more exact results, bearing performance parameters should be recalculated using viscosity of the oil at 22.88°C.

Example Problem 8.5.5

For the hydrodynamic journal bearing in Example Problem 8.5.4, is there any possibility of encountering critical speed during rotation.

Solution

The slope of the W_r as a function of $h_{\min}$ curve at $\varepsilon = 0.7$ and $\ell/d = 1$ from Figure 8.5.16a is calculated as follows. S at $\varepsilon = 0.6$ and 0.8 are given as 0.12 and 0.045, respectively.

At $\varepsilon = 0.6$,
$$h_{\min} = 40\ \mu\text{m}$$
$$W_r = 2.66 \times 10^5\ \text{N}$$

At $\varepsilon = 0.8$,
$$h_{\min} = 20\ \mu\text{m}$$
$$W_r = 7.11 \times 10^5\ \text{N}$$

Therefore,
$$k_f = \frac{dW_r}{dh_{\min}} = \frac{(2.66 - 7.11) \times 10^5}{(40 - 20) \times 10^{-6}}\ \text{N/m} = -2.225 \times 10^4\ \text{N}/\mu\text{m}$$

Natural frequency of vibration is
$$f_n = \frac{1}{2\pi}\left[\frac{k_f}{W_r/g}\right]^{1/2} = \frac{1}{2\pi}\left[\frac{2.225 \times 10^{10}}{4 \times 10^5/9.81}\right]^{1/2}\ \text{Hz} = 117.6\ \text{Hz}$$

With a shaft speed of 1200 rpm (20 Hz), there is no possibility of encountering critical speed during rotation.

8.5.3 Squeeze Film Bearings

Fluid film can also be generated by an oscillating motion in the normal direction towards each other (squeeze), Figure 8.5.19. This load-carrying phenomenon arises from the fact that a viscous fluid cannot be instantaneously squeezed out from the interface with two surfaces that are approaching each other, and this action provides a cushioning effect in bearings. When the load is relieved or two surfaces move apart, the fluid is sucked in and the fluid film can often recover its thickness in time for the next application (Pinkus and Sternlicht, 1961; Fuller, 1984; Khonsari and Booser, 2001; Hamrock *et al.*, 2004; Szeri, 2010). This squeeze film effect is efficient in oscillations with high frequencies in the kHz to MHz range at submillimeter amplitudes (Tam and Bhushan, 1987).

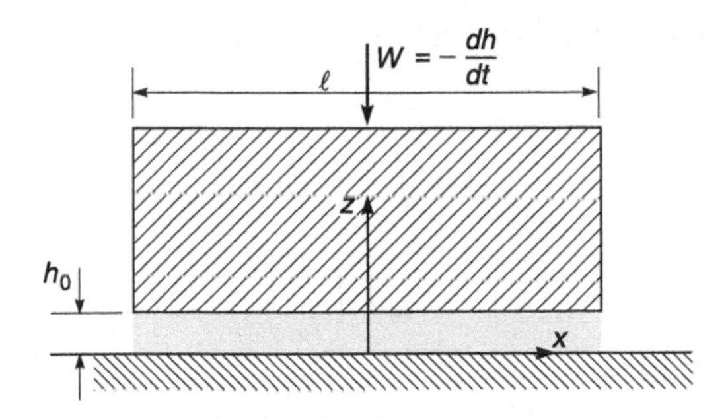

Figure 8.5.19 Schematic of parallel-surface squeeze film bearing.

For only normal motion, the generalized Reynolds equation 8.3.31 reduces to

$$\frac{\partial}{\partial x}\left(\frac{\rho\, h^3}{12\,\eta}\frac{\partial p}{\partial x}\right)+\frac{\partial}{\partial y}\left(\frac{\rho\, h^3}{12\,\eta}\frac{\partial p}{\partial y}\right)=\frac{\partial\,(\rho h)}{\partial t} \tag{8.5.60}$$

We consider a simple case of an infinitely wide parallel-surface bearing (side leakage neglected) moving with a vertical velocity w with a constant film thickness h_0 as shown in Figure 8.5.19. If the density and viscosity are assumed to be constant, then the Reynolds equation reduces to

$$\frac{\partial^2 p}{\partial x^2}=-\frac{12\,\eta_0\, w}{h_0^3} \tag{8.5.61a}$$

and

$$w=-\frac{dh}{dt} \tag{8.5.61b}$$

Integrating Equation 8.5.61 and using boundary conditions of $p = 0$ at the bearing ends, $\pm\,\ell/2$, we get

$$\frac{dp}{dx}=-\frac{12\,\eta_0\, w\, x}{h_0^3} \tag{8.5.62a}$$

$$p=\frac{3\,\eta_0\, w}{2h_0^3}\left(\ell^2-4\,x^2\right) \tag{8.5.62b}$$

and

$$p_m=\frac{3\,\eta_0\, w\, \ell^2}{2\, h_0^3} \tag{8.5.62c}$$

The pressure distribution is parabolic and symmetrical about the bearing center.

This bearing has no tangential load capacity, $w_x = 0$. The normal load capacity per unit width is given as

$$w_z = \int_{-\ell/2}^{\ell/2} p\,dx$$
$$= \eta_0 w \left(\frac{\ell}{h_0}\right)^3 \quad (8.5.63)$$

Note that load capacity of squeeze film bearing is proportional to $(\ell/h_0)^3$ whereas for sliding bearing it is proportional to $(\ell/h_0)^2$.

From Equation 8.3.16, the shear stresses acting on the solid surfaces are given as

$$(\tau_{zx})_{z=0} = -(\tau_{zx})_{z=h} = \left(\eta_0 \frac{\partial u}{\partial z}\right)_{z=0} = \frac{6\,\eta_0\,wx}{h_0^2} \quad (8.5.64a)$$

The shear forces at the solid surfaces are given as

$$f\,|_{z=0} = f\,|_{z=h} = \int_{\ell/2}^{-\ell/2} (\tau_{zx})_{z=0}\,dx = 0 \quad (8.5.64b)$$

The volume flow is given from Equations 8.3.17a and 8.5.62a as

$$q_x = -\frac{h_0^3}{12\,\eta_0}\frac{dp}{dx} = wx \quad (8.5.65)$$

Note that the volumetric flow rate is zero at the bearing center and is maximum at the bearing edges.

For a time-independent load, w_z, Equation 8.5.63 can be integrated to calculate the time taken for the film to change in its thickness from one value to another (Hamrock *et al.*, 2004). Using the expression from Equation 8.5.61 in 8.5.63, we get

$$\frac{w_z}{\eta_0\,\ell^3}\int_{t_1}^{t_2} dt = \int_{h_{0,1}}^{h_{0,2}} \frac{dh_0}{h_0^3}$$

or

$$t_2 - t_1 = \Delta t = \frac{\eta_0\,\ell^3}{2\,w_z}\left[\frac{1}{h_{0,2}^2} - \frac{1}{h_{0,1}^2}\right] \quad (8.5.66)$$

Note that the time taken to squeeze the entire film out to a zero film is infinite. Thus, the film will never be squeezed out. Under dynamically loaded conditions in hydrodynamic sliding bearings, bearings are subjected to loads which constantly change. As a result of these fluctuations, as the bearing surfaces approach each other, the squeeze action keeps them apart and many failures are so avoided.

Example Problem 8.5.6

A normal load of 10 kN is applied to a parallel-plate squeeze film bearing with plates 10 mm long and 1 m wide and a film thickness of 10 μm. The bearing is lubricated with an oil film of viscosity of 40 cP. Calculate (a) the time required to reduce the film thickness to 1 μm, and (b) the film thickness after 1 second.

Solution

Given

$$W_z = 10 \text{ kN}$$
$$\ell = 10 \text{ mm}$$
$$b = 1 \text{ m}$$
$$h_{0,1} = 10 \ \mu\text{m}$$
$$\eta_0 = 0.04 \ \text{Pa s}$$

(a) $h_{0,2} = 1 \ \mu\text{m}$

$$\Delta t = \frac{\eta_0 \ell^3 b}{2 W_z}\left[\frac{1}{h_{0,2}^2} - \frac{1}{h_{0,1}^2}\right]$$
$$= \frac{0.04 \times 0.01^3 \times 1}{2 \times 10^4}\left[\frac{1}{(10^{-6})^2} - \frac{1}{(10^{-5})^2}\right]$$
$$= 1.98 \text{ s}$$

(b) $\Delta t = 1 \ s$

$$h_{0,2} = \frac{h_{0,1}}{\left[1 + 2W_z \Delta t \, h_{0,1}^2 / (\eta_0 \ell^3 b)\right]^{1/2}}$$
$$= \frac{10 \ \mu\text{m}}{\left[1 + 2 \times 10^4 \times 1 \times (10^{-5})^2 / (0.04 \times 0.01^3 \times 1)\right]^{1/2}}$$
$$= 1.4 \ \mu\text{m}$$

8.5.4 Gas-Lubricated Bearings

The first gas journal bearing was demonstrated by Kingsbury (1897). Gas-lubricated bearings are used in many industrial applications in which the hydrodynamic film of gaseous fluid is produced by hydrodynamic action. The gas is generally air. This avoids the need for a liquid lubrication system, simplifies the bearing design, and reduces maintenance. Gas bearings are used in gyroscopes where precision and constant torque are required, machine tool spindles, turbomachinery, dental drills, food and textile machinery and tape and disk drives as part of magnetic storage devices. Gas bearings are also called aerodynamic or self-acting gas bearings.

So far the special case of liquid-lubricated bearings has been considered because the density of liquids can be assumed to be constant, which simplifies the Reynolds equation. In gas-lubricated bearings, the gas is compressible and the change in density as a function of pressure cannot be neglected in the solution of Reynolds equation. The viscosity of air is about 0.0185 cP $(1.85 \times 10^{-5}$ Pa s$)$ at ambient temperature, which is much lower than liquid lubricants – on the order of 1/1000 of that of liquid lubricants. Therefore the film thickness, pressures, and load capacities of gas bearings, which are proportional to the fluid viscosity, are much lower than with a liquid. Pressures in self-acting gas bearings are typically 0.1 MPa, whereas these are on the order of 100 MPa in liquid bearings. The frictional force is reduced in roughly the same proportion; therefore the value of friction force in gas bearings is very low. However, the coefficient of friction is comparable to that of liquid bearings. Since the energy dissipated by friction losses is low, the temperature rise is low in the gas bearings as compared to liquid bearings.

A generalized Reynolds equation for a gas-lubricated interface under unidirectional rolling or sliding and with vertical motion is given from Equations. 8.3.19 and 8.3.21 as

$$\frac{\partial}{\partial x}\left(\frac{p\,h^3}{\eta}\frac{\partial p}{\partial x}\right)+\frac{\partial}{\partial y}\left(\frac{p\,h^3}{\eta}\frac{\partial p}{\partial y}\right)=12\,\overline{u}\frac{\partial\,(ph)}{\partial x}+12\frac{\partial\,(ph)}{\partial t} \tag{8.5.67}$$

The viscosity of gases varies little with pressure so it can be assumed to be constant ($\eta = \eta_0$ at atmospheric pressure), and is assumed to be a function of temperature only. This equation for a thrust bearing is written in nondimensional form as

$$\frac{\partial}{\partial X}\left(PH^3\frac{\partial P}{\partial X}\right)+\lambda^2\frac{\partial}{\partial Y}\left(PH^3\frac{\partial P}{\partial Y}\right)$$
$$=\Lambda\frac{\partial\,(PH)}{\partial X}+S\frac{\partial\,(PH)}{\partial T} \tag{8.5.68a}$$

The bearing number,

$$\Lambda=\frac{12\,\eta_0\,\overline{u}\,\ell}{p_a\ h_{\min}^2} \tag{8.5.68b}$$

and the squeeze number,

$$S=\frac{12\,\eta_0\,\omega\,\ell^2}{p_a\ h_{\min}^2} \tag{8.5.68c}$$

where $X = x/\ell$, $Y = y/b$, $\lambda = \ell/b$, $T = \omega t$, $P = p/p_a$, and $H = h/h_{\min}$. The ℓ and b are the bearing length (in the direction of motion) and width, respectively; ω is the frequency of vertical motion, p_a is the ambient pressure, and $h_{\min}$ is the minimum film thickness. The bearing number is also called the compressibility number. When Λ approaches zero, the operation of the bearing approaches that of the incompressible case. As Λ gets larger, as with lower ambient pressures or higher speed, the compressibility effects become very significant and must be included.

The Reynolds equation is used in analysis of gas lubrication. Gas compressibility makes the left side of Reynolds equation nonlinear in the variable P. The equation is solved by numerical methods (Gross *et al.*, 1980). The equation can be solved analytically for one-dimensional fluid flow under steady state conditions.

8.5.4.1 Slip Flow

The Reynolds equation is based on the continuum theory of fluid mechanics. If the mean free path of the molecules is small compared to the film thickness, continuum flow occurs. However, if the mean free path of the molecules (λ) becomes comparable to the film thickness (h), the fluid does not behave entirely as a continuous fluid but rather exhibits some characteristics of its molecular chaos. The layer of fluid immediately adjacent to the solid surface has a finite relative slip velocity, producing an apparent diminution in the viscosity of the fluid (rarefaction). The ratio of the mean free path of the molecules to the film thickness is measure of the degree of rarefaction. Slip flow can be an issue in gas lubrication with ultra-thin films (Gross *et al.*, 1980; Bhushan, 1996).

The Knudsen number based on local flow parameters is defined as

$$M_\ell = \lambda / h, \tag{8.5.69a}$$

where λ is the local mean free path. (For example, λ for air is 0.064 μm.) The criteria for the boundaries of the regimes between continuum flow, slip flow, and free-molecular flow with respect to the M_ℓ values for gaseous fluid can be approximately defined as follows (Bhushan 1996):

$$\text{Continuum flow: } M_\ell < 0.01$$

$$\text{Slip flow: } 0.01 < M_\ell < 3$$

$$\text{Transition flow: } 3 < M_\ell, \quad M_\ell/(\text{Re})^{1/2} < 10$$

$$\text{Free-molecular flow: } 10 < M_\ell/(\text{Re})^{1/2}$$

where *Re* is the Reynolds number based on the film thickness. We note that the rarefaction effects are not dependent solely on M_ℓ because these effects are weakened in lubricating films supporting heavy loads.

In the slip-flow regime, as a first approximation, the flow may still be treated by conventional continuum theories but with modified boundary conditions. Instead of velocities vanishing at the boundaries, the concept of slip velocities is introduced. In his original derivation of the slip-flow boundary condition, Burgdorfer (1959) suggested that the equation might only be valid to the limit where the minimum spacing of the bearing equals the mean free path of the gas. It was only intended as a first-order correction to the velocity field at the boundary. A more recent theoretical analysis by Gans (1985) attempts to derive the Reynold's slip-flow equation from the kinetic theory of gases, where the average molecular motion is considered along with mass flow. The results suggest that the equations derived previously by Burgdorfer

(1959) are, in fact, valid for even closer spacing than previously anticipated. These have been experimentally verified.

From Burgdorfer (1959), the local viscosity (η) is

$$Z\eta = \frac{\eta_0}{1 + (6a\lambda/h)} \tag{8.5.70}$$

where a is the surface correction coefficient for λ. We further note that the molecular mean free path of a gas is inversely proportional to the density ρ. For a perfect gas under isothermal conditions,

$$p/p_a = \rho/\rho_a = \lambda_a/\lambda \tag{8.5.71}$$

where λ_a is the molecular mean free path at ambient conditions, and p_a is the ambient pressure. We now define the Knudsen number at ambient conditions:

$$M = \lambda_a/h_{\min} \tag{8.5.69b}$$

where h_m is a reference film thickness (usually a minimum mean). Therefore, the expression for effective viscosity becomes

$$\eta = \frac{\eta_0}{1 + \left(6aMp_a h_{\min}/ph\right)} \tag{8.5.72a}$$

$$= \frac{\eta_0}{1 + \dfrac{6aM}{PH}} \tag{8.5.72b}$$

This expression for η should be used in the Reynolds equation for perfect gas-lubricating films.

8.5.4.2 Surface Roughness Effects

An important aspect of lubrication with very thin films, more common in gas-film lubrication, in the presence of roughness is the possible breakdown of the Reynolds equation. The Reynolds equation cannot be employed if the roughness slope is too large and/or the wavelength is too short compared to the film thickness, because one of the important assumption behind the Reynolds equation is the near parallelism of the surfaces. The validity of the Reynolds equation is based on a parameter, h/σ, where h is the fluid film thickness and σ is the composite standard derivation of surface heights. The roughness effects can be neglected if $h/\sigma > 6$. Patir and Cheng (1978) developed a method known as the average flow model. They proposed an ensemble-averaged Reynolds equation for incompressible lubrication in which two-dimensional roughness effects are built into a number of special film thickness averages or pressure and shear flow factors. In these formulations, the actual flow between rough surfaces is equated to an averaged flow between nominally smooth surfaces, while parameters describing the roughness are included in the Reynolds equation through the flow factors. The problem is thus reduced to a formulation that is much easier to solve numerically. Tonder (1985) has shown analytically that the flow factor approach is also valid for compressible lubrication. The modified Reynolds equation for compressible lubrication for a constant viscosity is given as (Bhushan and Tonder, 1989a, 1989b)

$$\frac{\partial}{\partial X}\left(\phi_X \overline{P}H^3\frac{\partial \overline{P}}{\partial X}\right)+\lambda^2\frac{\partial}{\partial Y}\left(\phi_Y \overline{P}H^3\frac{\partial \overline{P}}{\partial Y}\right)$$
$$=\Lambda\left[\frac{\partial\left(\phi_u \overline{P}H\right)}{\partial X}+\frac{\partial\left(\overline{P}H\right)}{\partial T'}\right]+S\frac{\partial\left(\overline{P}H\right)}{\partial T} \qquad (8.5.73)$$

The expression involving T' allows for the squeeze film effects due to moving roughness; $T' = \overline{u}t/\ell$.

The ϕ_X, ϕ_Y, and ϕ_u are nondimensional film thickness weighing functions pertinent to pressure flow in the X, Y, and shear flow directions, respectively. If both surfaces have the same roughness, $\phi_u = 1$.

Patir and Cheng (1978) obtained the flow factors by numerical flow simulation on numerically generated surfaces. They defined the directional properties of a surface by a parameter γ^P called the Peklenik number defined as the ratio of correlation lengths in the X and Y directions at which the value of the autocorrelation function is 0.5 of the value at origin,

$$\gamma P = \frac{\beta^*(0.5X)}{\beta^*(0.5Y)} \qquad (8.5.74a)$$

γ^P can be visualized as the length-to-width ratio of a representative asperity. Purely transverse, isotropic and longitudinal roughness patterns correspond to $\gamma = 0$, 1, and ∞, respectively. Note that the parameter γ cannot give a full account of two-dimensional roughness distribution when roughnesses are not oriented along the X and Y axes. For example, γ for a surface with unidirectional striation aligned at 45° to the X or Y axes is equal to unity. This result based on this definition of $\gamma^P = 1$ will not be distinguished from the isotropic case. We define an isotropic surface with γ^B (Bhushan number) (Bhushan and Tonder, 1989a),

$$\gamma^B = \frac{\beta^*_{\max}}{\beta^*_{\min}} \qquad (8.5.74b)$$

For an isotropic surface, $\gamma^B = 1$. The effect of different roughness structures (given by the parameter γ^P) on the air flow is shown schematically in Figure 8.5.20. Simple expressions for various ϕ at any γ are given by Tripp (1983). Figure 8.5.21 shows the dimensionless load capacity in the z direction for a rough surface, $\overline{W}_{rough} = W_{rough}/p_a\,\ell b$, as a function of h/σ for an oil-lubricated thrust bearing (incompressible lubrication) of finite width. Note that roles of transversely and longitudinally oriented roughnesses are switched for narrow and wide bearings, whereas there is no effect for isotropic roughness. For example, in the case of a very wide bearing, a transverse roughness orientation leads to a thicker mean film and higher load capacity than the longitudinal roughness orientation.

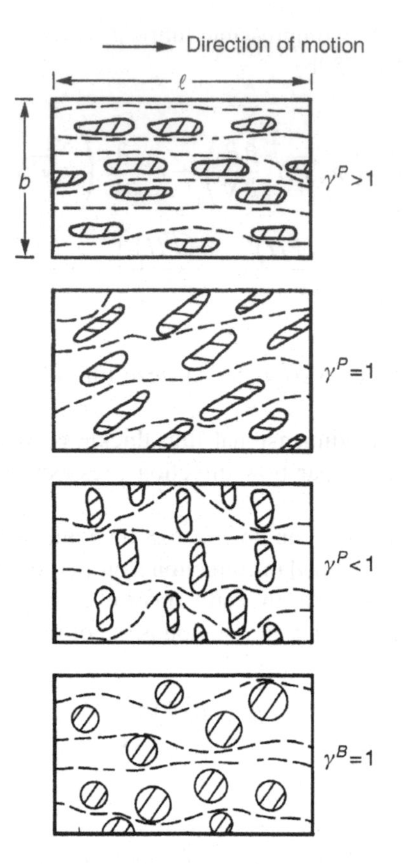

Figure 8.5.20 Schematic of fluid flow for a bearing with various roughness orientations and motion in the horizontal direction.

8.5.4.3 Fixed-Inclined-Pad Thrust Bearings

We consider the inclined-pad slider bearing shown in Figure 8.5.22. From Equation 8.3.28a, for one-dimensional fluid flow (in the x direction) with unidirectional motion

$$\frac{dp}{dx} = 6\,\eta_0\,u_0\left[\frac{h-(p_m/p)\,h_m}{h^3}\right] \tag{8.5.75}$$

or

$$\frac{dp}{dx} = \Lambda\left[\frac{H-(P_m/P)\,H_m}{H^3}\right] \tag{8.5.76a}$$

where $P_m = p_m/p_a$ *and* $H_m = h_m/h_0$. p_m/p can be replaced with p_m/p, if desired. For an inclined pad bearing, from Equation 8.5.2

$$H = 1 + m\,(1 - X) \tag{8.5.76b}$$

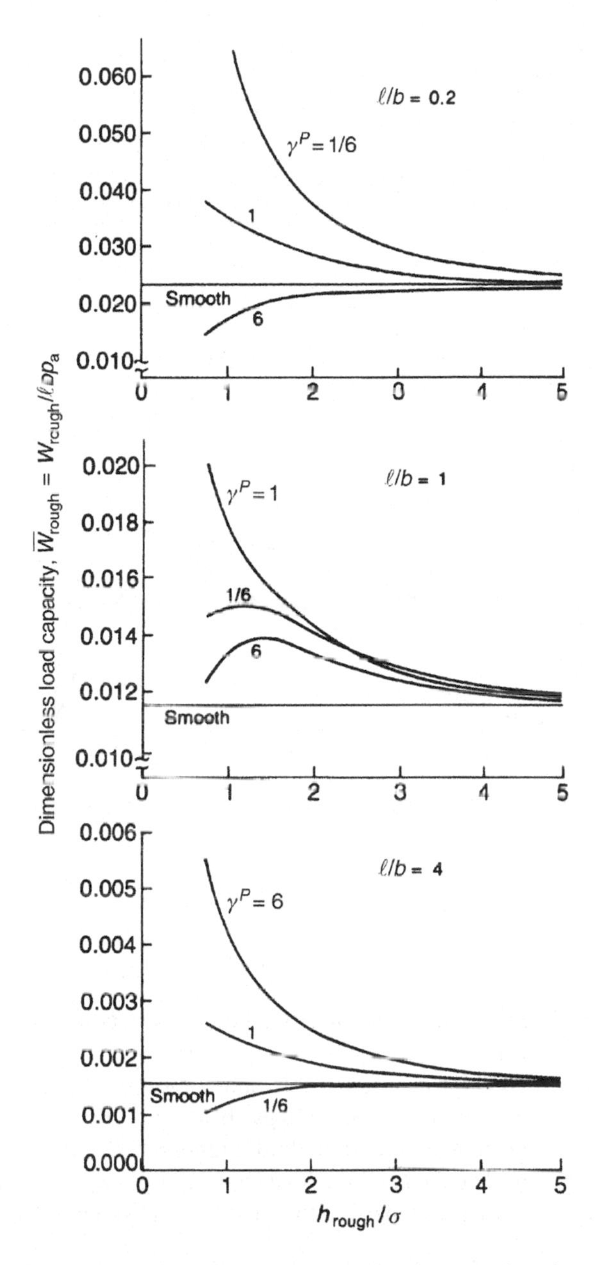

Figure 8.5.21 Dimensionless load capacity in the z direction for a rough surface moving over another rough surface as a function of h_{rough}/σ at different roughness orientations (γ^P) for a finite thrust bearing with different ℓ/b ratios. Reproduced with permission from Patir, N. and Cheng, H.S. (1979), "Application of Average Flow Model to Lubrication Between Rough Sliding Surfaces," *ASME J. Lub. Tech.* **101**, 220–230. Copyright 1979. ASME.

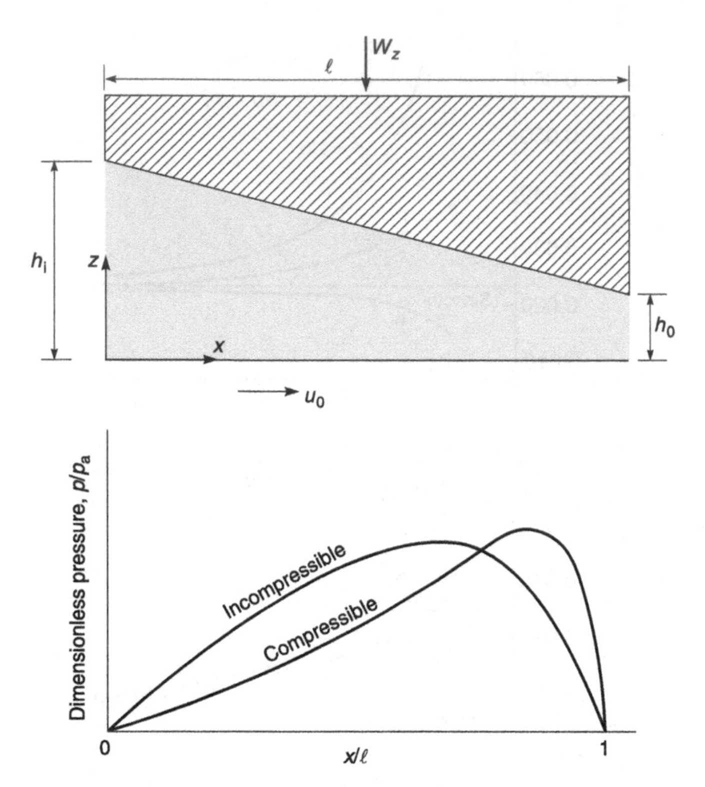

Figure 8.5.22 Schematic of a fixed-inclined-pad thrust bearing and pressure distribution trends for incompressible and compressible fluids.

where

$$m = (h_i/h_0) - 1$$

This equation is solved by integrating and inserting boundary condition that $P = 1$ $(p = p_a)$ at inlet and outlet ($X = 0$ and 1).

Pressure distribution trends for incompressible and compressible fluids are shown in Figure 8.5.22. Note that the maximum pressure for a gas bearing occurs more towards the trailing edge as compared to that in an incompressible fluid. Figure 8.5.23 shows the effect of bearing number on the pressure profile (p/p_a) and the load capacity ($W_z/p_a\,\ell b$) of a gas bearing.

In magnetic storage disk drives, a taper-flat two-or-three-rail slider is used (Bhushan, 1996).

8.5.4.4 Journal Bearings

As stated earlier, for a journal bearing, it is convenient to write Reynolds equation in polar coordinates. The equation, for unidirectional rolling or sliding under steady-state conditions, nondimensional form, is written as

$$\frac{\partial}{\partial\theta}\left(PH^3\frac{\partial P}{\partial\theta}\right) + \frac{\partial}{\partial\phi}\left(PH^3\frac{\partial P}{\partial\phi}\right) = \Lambda\frac{\partial (PH)}{\partial\theta} \tag{8.5.77a}$$

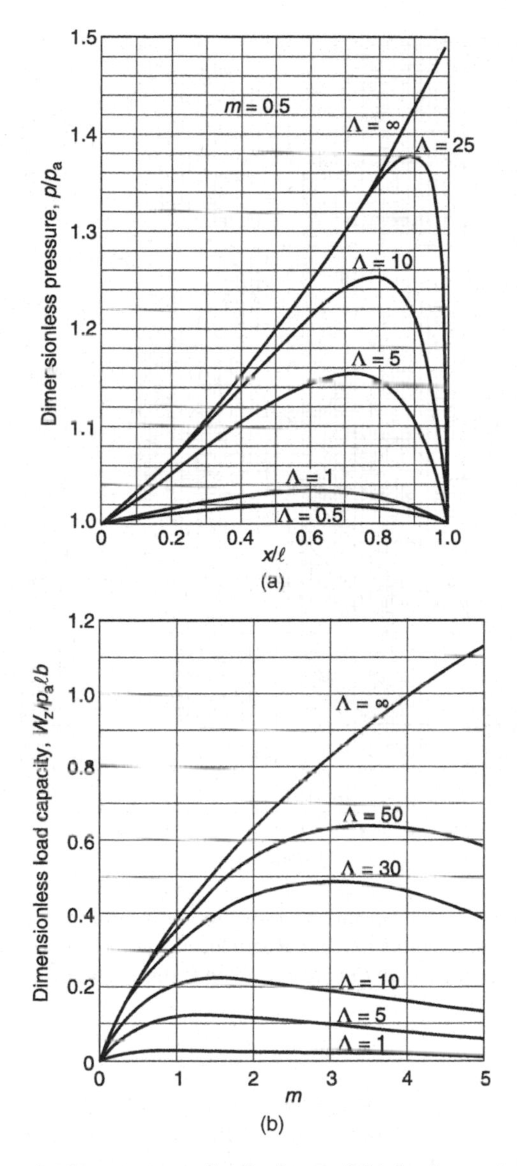

Figure 8.5.23 (a) Dimensionless pressure distribution and (b) load capacity of a fixed-inclined-pad thrust gas bearing as a function of *m*. Reproduced with permission from Gross, W.A., Matsch, L.A., Castelli, V., Eshel, A., Vohr, J.H., and Wildmann, M. (1980), *Fluid Film Lubrication*, Wiley, New York. Copyright 1980. Wiley.

where $x = r\theta,\ \ y = r\phi,\ \ \overline{u} = r\omega/2, H = h/c, P = p/p_a$, and

$$\Lambda = \frac{6\,\eta_0\,\omega\,r^2}{p_a\,c^2} \tag{8.5.77b}$$

A bearing with a small bearing number represents incompressible conditions.

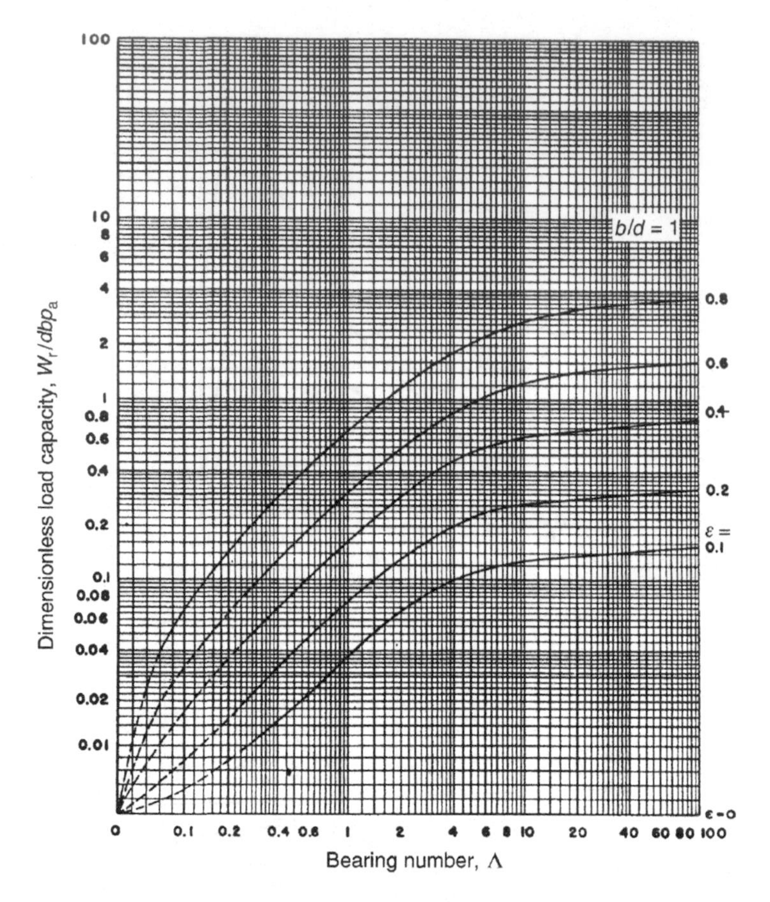

Figure 8.5.24 Dimensionless load capacity as a function of bearing number (Λ) for different eccentricity ratios for a gas lubricated journal bearing. Reproduced with permission from Fuller, D.D. (1984), *Theory and Practice of Lubrication for Engineers*, Second edition, Wiley, New York. Copyright 1984. Wiley.

Many analytical and semi-analytical solutions are available for very high and very low bearing numbers (Gross *et al.*, 1980; Fuller, 1984; Hamrock *et al.*, 2004). Similarly the solution for compressible fixed-inclined-pad slider bearings, Raimondi (1961) used relaxation methods to solve for the bearing performance. As an example, load capacity as a function of bearing number for different eccentricities for a full journal bearing with a width/diameter ratio of 1 is presented in Figure 8.5.24.

Example Problem 8.5.7

Two full, air-lubricated journal bearings support a rotor, weighing 2 N, which rotates at 24,000 rpm. The width, diameter and radial clearance of the bearing are 10 mm, 10 mm, and 5 μm, respectively. The bearing operates at an ambient pressure of 101 kPa and the absolute viscosity of air is 1.84×10^{-5} Pa s. Calculate the minimum film thickness during the bearing operation.

Solution

Given

$$W(\text{per bearing}) = 1\ \text{N}$$
$$N = 24000\ \text{rpm or } 2513\ \text{rad/s}$$
$$b = d = 2r = 10\ \text{mm}$$
$$c = 5\ \mu\text{m}$$
$$p_a = 101\ \text{kPa}$$
$$\eta_0 = 1.84 \times 10^{-5}\ \text{Pa s}$$

The bearing number, $\Lambda = \frac{6\,\eta_0\,\omega\,r^2}{p_a\,c^2}$

$$= \frac{6 \times 1.84 \times 10^{-5} \times 2513 \times (5 \times 10^{-3})^2}{101 \times 10^3 \times (5 \times 10^{-6})^2}$$
$$= 2.75$$

and the nondimensional load is $\frac{W_r}{d\,b\,p_a}$

$$= \frac{1}{0.01 \times 0.01 \times 101 \times 10^3} = 0.099$$

For a bearing with $\Lambda = 2.75$, $b/d = 1$, and $W = 0.0495$, we get ε from Figure 8.5.24 as

$$\varepsilon \sim 0.11$$

and

$$h_{\min} = c\,(1 - \varepsilon)$$
$$= 5\,(1 - 0.11)\ \mu\text{m}$$
$$= 4.45\ \mu\text{m}$$

8.5.4.5 Other Gas Bearing Types

In addition to fixed-inclined-pad, tilted-pad and Rayleigh-step thrust bearings, and full 360° and partial arc journal bearings, commonly used with both liquid and gas lubrication, many other types of gas bearings are used in industrial applications.

Variations of step bearings, which are in widespread use, are inward-pumping, spiral-grooved thrust bearings (Fuller, 1984). In a grooved thrust plate, operating on the principle

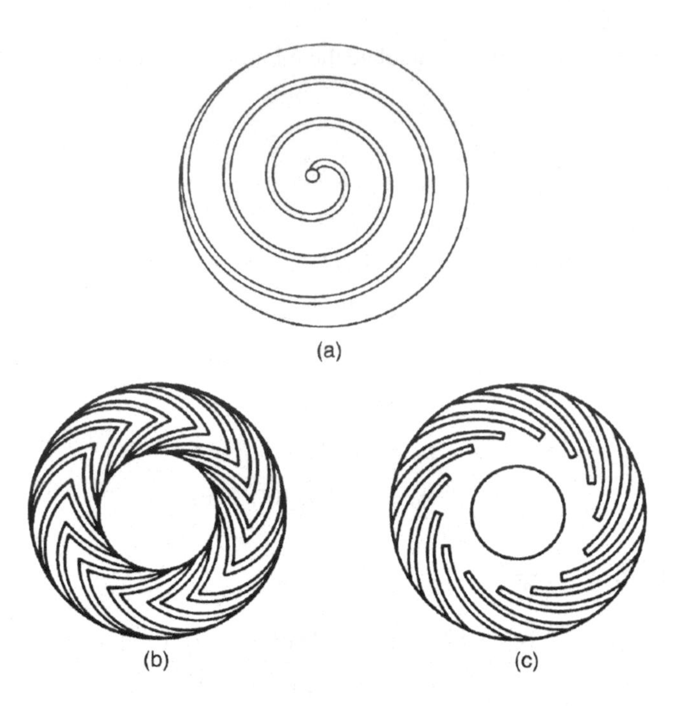

Figure 8.5.25 Schematics of grooved thrust bearings, (a) grooved with either logarithmic or archimedian spiral, (b) herringbone grooved, and (c) spiral grooved. Reproduced with permission from Fuller, D.D. (1984), *Theory and Practice of Lubrication for Engineers*, Second edition, Wiley, New York. Copyright 1984. Wiley.

of the Rayleigh step bearing, lubricant is dragged into a slot or groove by a moving runner, Figure 8.5.25a. The exit end of the slot has a restrictor or dam so that exit flow is retarded. The rotation of the runner continues to pump fluid into the entrance of the slot due to viscous drag, and as a consequence the pressure in the slot builds up. This enables the bearing to carry load with a fluid film separating the surfaces. An optimization of the single groove thrust plate is shown in Figure 8.5.25b. In a so-called herringbone configuration, the fluid is pumped into the grooves from both the outside edge and the inside edge. In a spiral-grooved configuration, it can be made as an inward pumping spiral, as shown in Figure 8.5.25c, or as an outward pumping spiral. For analyses of these bearings, see for example, Gross *et al.* (1980) and Hamrock *et al.* (2004).

Synchronous whirl of shafts, as mentioned earlier, is a more serious issue in gas-lubricated journal bearings because of lower stiffness as compared to lubricated bearings. For increased stability, non-plain tilted-pad journal bearings are used, Figure 8.5.26a. The pads are mounted on pivots so that pads can pivot axially as well as circumferentially to the shaft surface. Although these bearings are more expensive to manufacture, these can be used at high speeds without synchronous whirl as compared to the full 360° journal bearings. A journal bearing akin to the spiral-grooved thrust bearing is the herringbone grooved bearing shown in Figure 8.5.26b. The gas is pumped through the grooves from the bearing ends to the center.

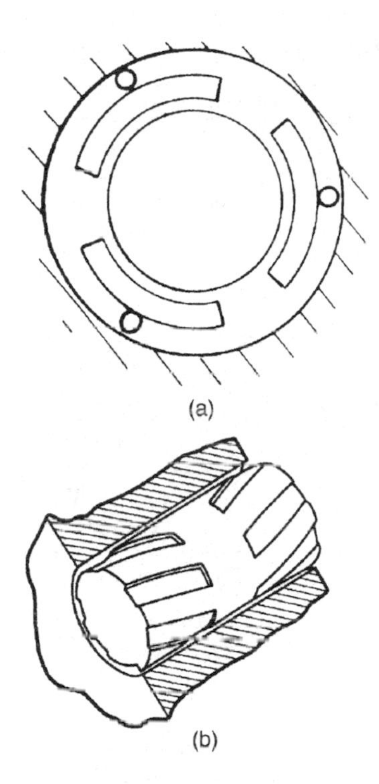

Figure 8.5.26 Schematics of (a) tilted-pad, and (b) spiral-grooved (herringbone) journal bearings. Reproduced with permission from Hamrock, B.J., Schmid, S.R., and Jacobson, B.O. (2004), *Fundamentals of Fluid Film Lubrication*, Second edition, Marcel Dekker, New York. Copyright 2004. Taylor and Francis.

Another form of journal and thrust bearings for high-speed applications is the compliant or foil bearing (Walowit and Anno, 1975; Gross *et al.*, 1980). In this bearing, a flexible metal foil comprise the bearing surface, Figure 8.5.27.

We consider the geometry of an infinitely wide, perfectly flexible foil wrapped around a cylindrical journal of radius r under tension per unit width T and moving at a linear velocity u_0 as shown in Figure 8.5.28a. Blok and van Rossum (1953) assumed that the foil is rigid in the inlet region and neglected the large pressure gradients in the exit region. However, three regions are important in a foil bearing (Gross *et al.*, 1980), as shown in Figure 8.5.28a. In the inlet region, the pressure increases from ambient to the pressure in the constant film thickness region. This requires a positive pressure gradient with a decrease in film thickness from infinity to constant film thickness, h_0. The central region is a region of constant pressure and a constant film thickness h_0. In the exit region, the pressure decreases from the pressure in the constant gap region to ambient while the film thickness increases from h_0 to infinity. This requires a negative pressure gradient. From the integrated Reynolds equation, a negative pressure gradient can exist only if $h < h_0$, which is incompatible with an increasing film thickness. Therefore, the increase in film thickness is preceded by a region where $h < h_0$ in which pressure decreases to below ambient, followed by a region of increasing film thickness and increasing pressure (Gross *et al.*, 1980).

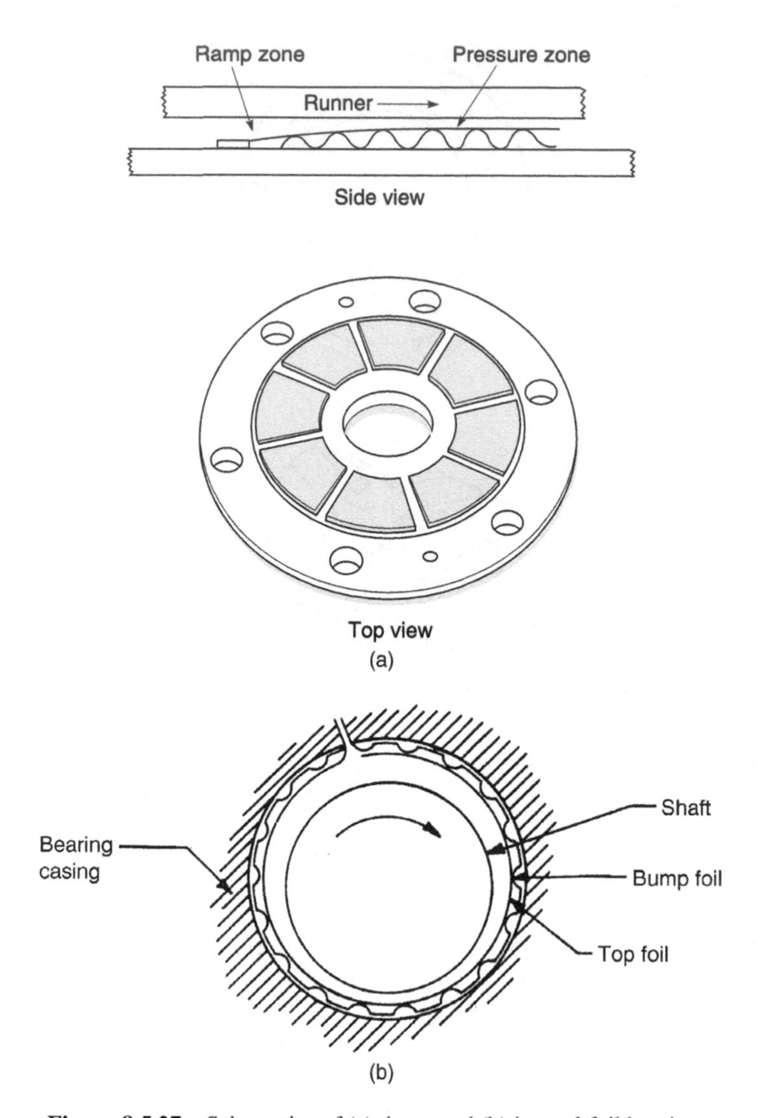

Figure 8.5.27 Schematics of (a) thrust and (b) journal foil bearings.

If the film thickness is small compared with the radius of the cylinder r, the curvature of the foil in the inlet region can be expressed as

$$\kappa = \frac{1}{r} - \frac{d^2h}{dx^2} \tag{8.5.78}$$

For a perfectly flexible foil, pressure distribution is given as

$$p(x) = T\kappa = \frac{T}{r}\left(1 - r\frac{d^2h}{dx^2}\right) \tag{8.5.79}$$

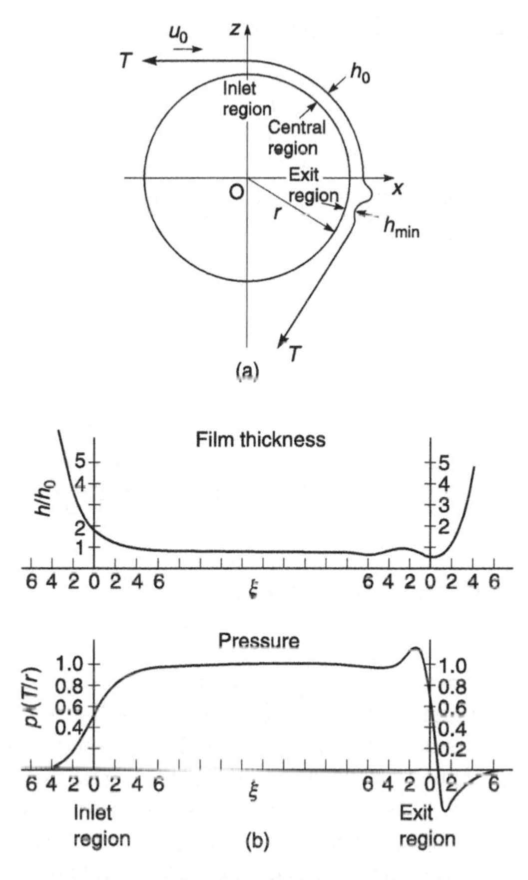

Figure 8.5.28 (a) Cross-sectional view of an infinitely-wide, self-acting foil bearing with three regions, and (b) dimensionless film thickness and pressure profiles in the bearing. Reproduced with permission from Gross, W.A., Matsch, L.A., Castelli, V., Eshel, A., Vohr, J.H., and Wildmann, M. (1980), *Fluid Film Lubrication*, Wiley, New York. Copyright 1980. Wiley.

We combine Equation 8.5.79 with the integrated Reynolds equation 8.3.28b, and we get

$$\frac{d^3h}{dx^3} = \frac{6\,\eta_0\,u_0}{T}\left(\frac{h-h_0}{h^3}\right) \tag{8.5.80a}$$

Let

$$H = \frac{h}{h_0}$$

and

$$\xi = \frac{x}{h_0}\left(\frac{6\,\eta_0\,u_0}{T}\right)^{1/3}$$

Using these parameters, Equation 8.5.80a reduces to

$$\frac{d^3\,H}{d\,\xi^3} = \left(\frac{1-H}{H^3}\right) \tag{8.5.80b}$$

The equation is nonlinear in the third degree. The solution results from a simple linearization. We assume that the gap has little variation from a constant gap, $H \sim 1 \quad (\varepsilon << 1)$. Thus

$$\frac{\partial^3\,\varepsilon}{\partial\,\xi^3} + \varepsilon = 0 \tag{8.5.80c}$$

which has the solution

$$\varepsilon = A\ \exp\ (-\xi) + B\ \exp\ (\xi/2)\ \cos\ \left(\frac{\sqrt{3}}{2}\xi\right) \\ + C\ \exp\ (\xi/2)\ \sin\ \left(\frac{\sqrt{3}}{2}\xi\right) \tag{8.5.81}$$

where A, B, and C are constants. This solution contains both positive and negative exponents. If ε is to be bounded in the region under consideration, then the constants A, B, and C have to be very small. The result is that sinusoidal terms containing the positive exponents can be present only near the exit region of the bearing, while the simple exponential will be present only near the entrance region. This leaves the central region of the foil bearing a region of constant film thickness. Constant pressure also means constant curvature. Thus, the pressure in the constant film thickness region is T/r. The film thickness and pressure profiles are shown in Figure 8.5.28b. The film thickness in the central region is (Gross *et al.*, 1980)

$$h_0 = 0.643\ r \left(\frac{6\,\eta_0\,u_0}{T}\right)^{2/3} \tag{8.5.82a}$$

A foil bearing number, Λ is defined as

$$\Lambda = \left(\frac{h_0}{r}\right) \left(\frac{6\,h_0\,u_0}{T}\right)^{2/3} \tag{8.5.82b}$$

Then in the central film thickness region,

$$\Lambda = 0.643 \tag{8.5.82c}$$

The minimum film thickness is given as

$$h_{\min} \sim 0.72\ h_0 \tag{8.5.83}$$

The effect of foil stiffness has been investigated theoretically by Eshel and Elrod (1967) and experimentally by Licht (1968). Similar trends in theoretical and experimental results have been reported for the head–tape interface in magnetic storage tape drives (Bhushan, 1996).

8.6 Elastohydrodynamic Lubrication

In the elastohydrodynamic (EHD) lubrication (EHL) regime, the elastic deformation of the bounding solids is large and affects the hydrodynamic lubrication process. EHL is important in nonconforming, heavily-loaded contacts such as point contacts of ball bearings (Figure 8.6.1a), line contacts of roller bearings, point and line contacts of gear teeth, Figure 8.6.1b, and compliant bearings and seals at moderate loads. (See the last chapter for examples.) The EHL phenomena also occur in some low elastic modulus contacts of high geometrical conformity, such as lip seals, conventional journal and thrust bearings with soft liners, and head–tape interfaces in magnetic recording tape drives.

In heavily loaded contacts, high pressures can lead both to changes in the viscosity of the lubricant and elastic deformation of the bodies in contact, with consequent changes in the geometry of the bodies bounding the lubricant film. For example, Hertzian contact stresses

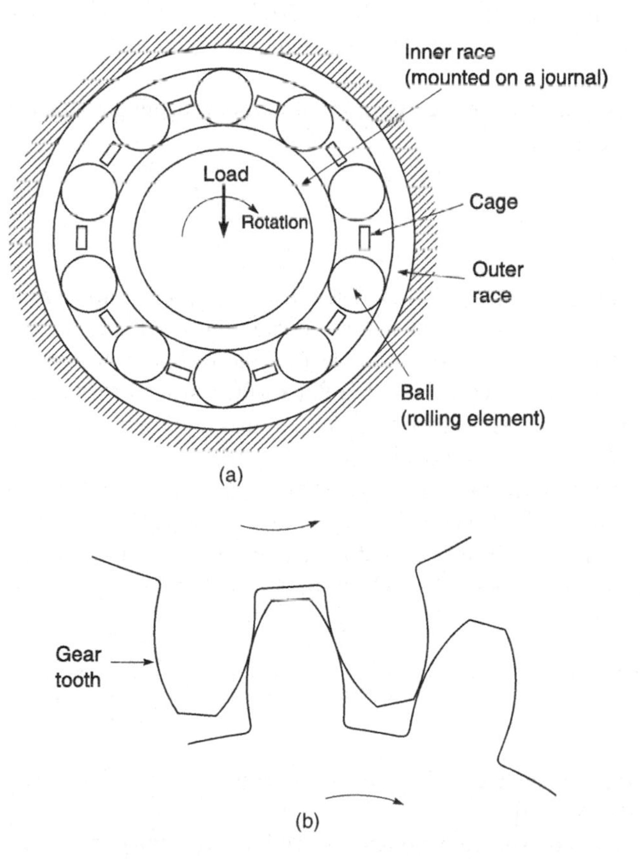

Figure 8.6.1 Schematics of (a) a radial ball bearing, and (b) contact of two spur gears.

in rolling element bearings and gears range from 0.5 to 3 GPa. The viscosity of most liquid lubricants at these pressures can be very high, as much as 10^8 times, so that a change to a solid phase occurs in the conjunction. The elastic deflection of steel surfaces can be several orders of magnitude larger than the minimum film thickness by a factor as much as 10^3. In EHL, we are faced with simultaneous solution of the Reynolds equation, the elastic deformation equations, and the equation relating viscosity and pressure. Shear-rate and thermal effects also become important and need to be taken into account. Significant developments in the analyses of EHL contacts have taken place in the second half of the twentieth century, which is more recent compared to HD lubrication analyses.

In a pioneering work, Grubin (1949) developed an analytical approach to incorporate both the elastic deformation of the solids and the viscosity–pressure properties of the lubricants in order to solve the EHL problem, and he obtained the film shape and pressure distribution in the cylindrical (line) contact. He reported that film thickness using EHL analysis is one to two orders of magnitude larger than that obtained using HD lubrication analysis (rigid body and viscosity independent of pressure). In 1959, Dowson and Higginson (1959, 1966) developed an iterative procedure to solve a variety of contact problems and derived an empirical formula for minimum film thickness. They found that load has little effect but speed has a significant effect on the film thickness. Later, Cheng (1970) developed a Grubin type of inlet analysis applicable to elliptical Hertzian contact areas. In the late 1970s, B. J. Hamrock and D. Dowson developed numerical methods applied to EHL of rolling element bearings (Hamrock and Dowson, 1981; Hamrock *et al.*, 2004).

In this section, we discuss simple examples of non-conforming contacts relevant for rolling element bearings and gears.

8.6.1 Forms of Contacts

The most commonly encountered forms of contacts, commonly known as footprints, are point and line contacts. When a sphere comes into contact with a flat surface, it initially forms a point contact with a circular shape and the size of the footprint grows as a function of load. When a cylinder comes into contact with a flat surface, it forms a line contact and it grows into a rectangular footprint as the load is increased. Incidentally a point contact between a ball and raceway develops into an elliptical footprint.

Two nonconforming surfaces with simple shapes can be analyzed by considering an equivalent nonconforming surface in contact with a plane surface, Figure 8.6.2. For example, two cylinders or two spheres with radii r_a and r_b can be represented by an equivalent cylinder or sphere, respectively, in contact with a plane surface with the radius of the equivalent surface as

$$R = \frac{r_a \, r_b}{r_a + r_b} \tag{8.6.1}$$

In the case of elliptical contacts, the equivalent radius of curvature is calculated in the two principal directions, r_x and r_y.

If the two cylinders move with velocities u_a and u_b, then the entraining velocity of interest in lubrication, also called the rolling velocity, is $\overline{u} = (u_a + u_b)/2$. For $u_a \neq u_b$, there is a

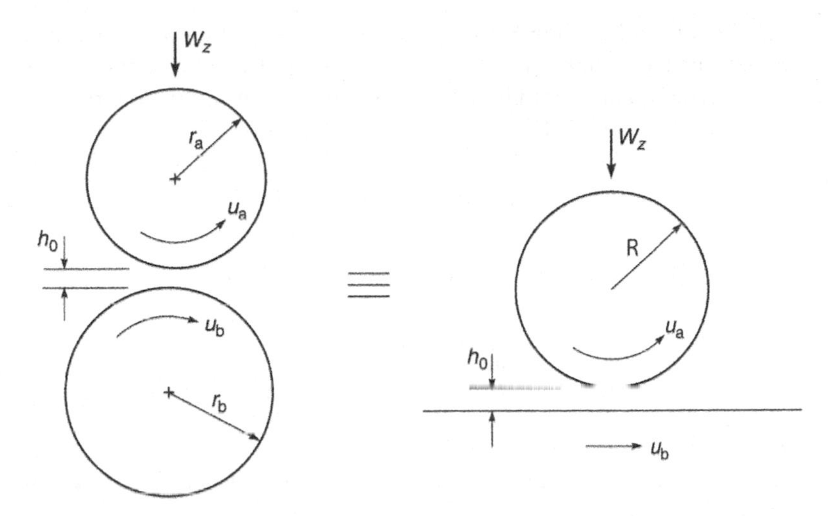

Figure 8.6.2 Schematics of two cylinders of radii r_a and r_b and with a film thickness of h_0 between them and equivalent cylinder of radius r against a plane surface.

relative sliding velocity between the two surfaces equal to $|u_a - u_b|$. The relative amounts of sliding and rolling are expressed by a slide-to-roll ratio, *S*, also known as slip ratio:

$$S = \frac{sliding\ velocity}{2 \times rolling\ velocity} = \frac{u_a - u_b}{u_a + u_b} \tag{8.6.2}$$

In pure rolling condition, $u_a = u_b,\ S = 0.\ If\ u_a = -u_b$, then rolling velocity is zero and hydrodynamic pressure is not developed.

8.6.2 Line Contact

8.6.2.1 Rigid Cylinder Contact

Isoviscous

Consider an infinitely wide, rigid cylinder moving over a plane surface in one direction (the x direction) under steady-state conditions in the presence of an incompressible fluid, Figure 8.6.3 (undeformed cylinder). A relevant integrated Reynolds equation is given by

$$\frac{dp}{dx} = 12\,\eta\bar{u}\left(\frac{h - h_m}{h^3}\right) \tag{8.6.3}$$

Based on Martin (1916), for the case of a rigid cylinder approaching a plane surface with a minimum film thickness h_0, the film thickness can be expressed as a parabolic function,

$$h \sim h_0 + \frac{x^2}{2R}\ for\ x\ << R \tag{8.6.4}$$

The expression for h from Equation 8.6.4 is substituted in Equation 8.6.3 and the Reynolds equation is solved with the boundary conditions of the inlet and outlet pressures being ambient; $p = 0$ at $x = \pm\infty$. The equation must be solved numerically. An expression for the maximum pressure is given by (Dowson and Higginson, 1966; Williams, 2005)

$$p_{\max} \sim 2.15\ \eta_0 \bar{u} \left(R/h_0^3\right)^{1/2} \tag{8.6.5a}$$

and the load capacity per unit width can be found by integration by

$$w_z \sim 4.9 \frac{\eta_0 \bar{u}\ R}{h_0} \tag{8.6.5b}$$

Variable Viscosity

Viscosity, η, in the Reynolds equation 8.6.3, is a function of pressure. We use a simple Barus equation,

$$\eta = \eta_0\ \exp\ (\alpha p) \tag{8.6.6}$$

where α is the viscosity–pressure coefficient and η_0 is the absolute viscosity at ambient pressure. We define a variable known as reduced pressure:

$$p = -\frac{1}{\alpha}\ \ell n\ (1 - \alpha p_r) \tag{8.6.7}$$

Note as $p \to 0,\ p_r \to 0$ and as $p \to \infty,\ p_r \to 1/\alpha$. Substituting Equations 8.6.6 and 8.6.7 into Equation 8.6.3, we get

$$\frac{dp_r}{dx} = 12\eta_0 \bar{u} \left(\frac{h - h_m}{h^3}\right) \tag{8.6.8}$$

Equation 8.6.8 has the same form as the Reynolds equation for a constant viscosity fluid. The advantage of solving this equation is that reduced pressure is the only variable rather than two pressure and viscosity variables. From Equation 8.6.5a,

$$(p_r)_{\max} \sim 2.15\eta_0 \bar{u} \left(R/h_0^3\right)^{1/2} \tag{8.6.9a}$$

For a large p, the p_r approaches $1/\alpha$. Therefore, from Equation 8.6.9a, h_0 corresponding to this maximum pressure is given as

$$h_0 = 1.66\,(\alpha \eta_0 \bar{u})^{2/3}\, R^{1/3} \tag{8.6.9b}$$

The load capacity per unit width for this case can be evaluated by integrating the area under the pressure curve.

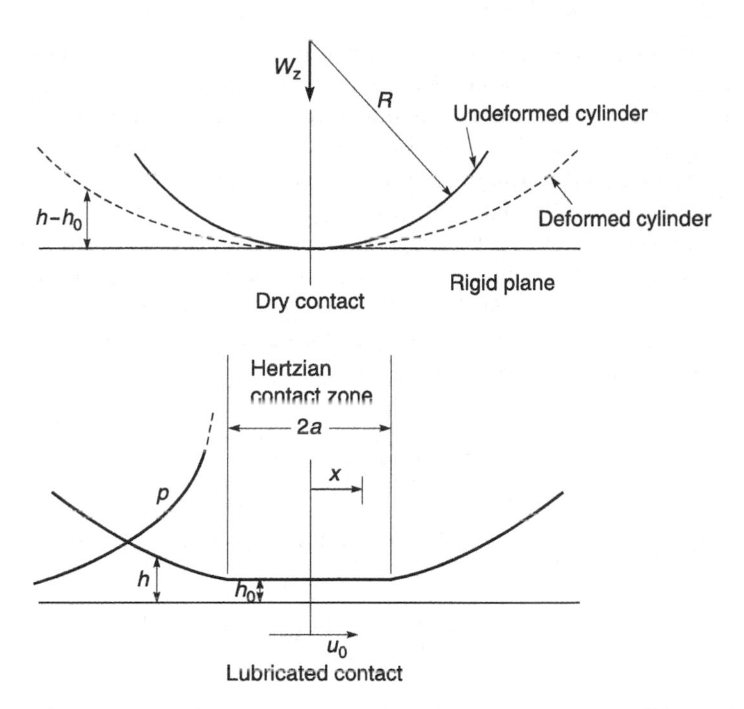

Figure 8.6.3 Schematic of a cylinder in contact with a rigid plane in dry conditions (rigid cylinder – undeformed profile; elastic cylinder-deformed profile) and lubricated conditions in Grubin's model.

8.6.2.2 Elastic Cylinder Contact (Rectangular Contact) and Variable Viscosity

In a nonconformal contact, elastic bodies will deform even at light loads. These small deformations will affect the generation of hydrodynamic pressure. The EHL problem is to solve the Reynolds equation and the elasticity equations.

We first present the analytical approach developed by Grubin (1949) who took into account only the entry region to calculate film thickness. Consider an infinitely wide elastic cylinder against a plane as shown in Figure 8.6.3 (deformed cylinder). The cylinder will be flattened against the plane over the Hertzian contact zone and the contour of the cylinder outside the zone will also change. For a small film thickness, h_0, if the local elastic flattening in the contact is large compared with h_0, the pressure distribution in the contact must be near Hertzian (elliptical distribution). Then, the geometry of the fluid film in a lubricated contact must be close to the form calculated for dry contact. The pressure distribution for any value of h_0 can be calculated by using the shape of the deformed cylinder $(h - h_0)$ outside the contact zone. Grubin observed that the pressure builds up to very high values in the inlet, and remains high through the Hertzian region. Pressure over most of the contact length is so high that the viscosity is orders of magnitude bigger than its atmospheric value, so if dp/dx is to have realistic values, $h - h_m$ must be very small, in fact approximately zero. Therefore, the film thickness must be constant over most of the high pressure zone.

The first task is to calculate stresses and displacements in solids in contact. These are assumed to be uniform along the contact length, except near the ends, and the solids are in a condition of plane strain. In addition, it is necessary to determine the stresses and displacements for a semi-infinite flat solid and to add the displacements to the curved surface of the roller. Tangential displacement of the surface has little effect when the two surfaces are separated by a film. The Boussinesq function is used for the stresses due to a normal line load on the surface of a semi-infinite solid. The stresses and displacements under this load are integrated to give corresponding quantities under a distributed pressure.

The integrated form of the Reynolds equation, in terms of reduced pressure for an incompressible fluid, is solved. The boundary conditions are: at the inlet $p = 0$ at large distance from the high pressure zone and at the outlet $p = \frac{\partial p}{\partial x} = 0$. Thermal effects are neglected. Assuming that around the contact region, the undeformed cylindrical profile can be modeled as a parabola, the magnitude of the gap h is given by

$$h = h_0 + \frac{x^2}{2r} + w \qquad (8.6.10a)$$

where w is the combined deformation of the two solids (Dowson and Higginson, 1966),

$$w = -\frac{1}{\pi E^*} \int_{-a}^{a} p(s)\ \ell n\,(x - s)^2\ ds$$

and

$$\frac{1}{E^*} = \frac{1 - \nu_1^2}{E_1} + \frac{1 - \nu_2^2}{E_2} \qquad (8.6.10b)$$

where E^*, E_1 and E_2 are effective modulus, and moduli of bodies 1 and 2, respectively, and ν_1 and ν_2 are Poisson's ratios of bodies 1 and 2, respectively. $p(s)$ is the normal load per unit width over a strip of length $2a$ (along the x axis). The reduced form of Reynolds equation 8.6.8 together with deformation equation 8.6.10 must be solved simultaneously to yield expressions for the film shape h and the pressure with position throughout the bearing along the x axis.

Grubin integrated the Reynolds equation numerically for a range of values of h_0 and found an expression which gave a good fit:

$$\frac{h_0}{R} \sim 2.08 \left(\frac{\eta_0 \bar{u}\,\alpha}{R}\right)^{8/11} \left(\frac{E^*\,R}{w_z}\right)^{1/11} \qquad (8.6.11a)$$

where w_z is the normal load per unit width. This equation can be rewritten in terms of dimensionless parameters:

$$\mathrm{U} = \text{dimensionless speed parameter } \frac{\eta_0 \bar{u}}{E^*\,R}$$

$$\mathrm{G} = \text{dimensionless materials parameter} = \alpha\,E^*$$

$$\mathrm{W} = \text{dimensionless load parameter } = \frac{w_z}{E^*\,R}$$

and

$$\frac{h_0}{R} = 2.08\,(UG)^{8/11}\ W^{-1/11} \tag{8.6.11b}$$

From Equations. 8.6.5b and 8.6.11b, the ratio of film thicknesses for the two cases is given by

$$\frac{h_0\ \text{(elastic, viscosity function of pressure)}}{h_0\ \text{(rigid, constant viscosity)}} = 0.424\frac{\alpha^{8/11}\ E^{*1/11}\,(w_z)^{10/11}}{R^{7/11}\ (\eta_0\overline{u})^{3/11}} \tag{8.6.12}$$

Example Problem 8.6.1

Two cylindrical gears with 50 mm radius, made of steel and separated by an incompressible film of mineral oil, roll together at surface velocities of 10 m/s each under a normal load per unit width of 1×10^6 N/m. The effective modulus of elasticity of the gears is 456 GPa and absolute viscosity and α for the mineral oil are 50 mPa s and 0.022 MPa^{-1}, respectively. Calculate the film thickness for the case of rigid teeth and constant viscosity and film thickness for elastic teeth lubricated with variable viscosity using Grubin's analysis (based on Dowson and Higginson, 1966).

Solution

Given

$$\begin{aligned}
R &= 50/2\ \text{mm} = 25\ \text{mm} \\
\overline{u} &= \frac{u_a + u_b}{2} = 10\ \text{m/s} \\
w_z &= 1 \times 10^6\ \text{N/m} \\
E^* &= 456\ \text{GPa} \\
\eta_0 &= 50\ \ \text{mPa s} \\
\alpha &= 2.2 \times 10^{-8}\ \ \text{Pa}^{-1}
\end{aligned}$$

For the case of rigid teeth and constant viscosity,

$$\begin{aligned}
h_0 &\sim \frac{4.9\,\eta_0\,\overline{u}\,R}{w_z} \\
&= \frac{4.9 \times 50 \times 10^{-3} \times 10 \times 25 \times 10^{-3}}{1 \times 10^6}\text{m} \\
&= 0.061\ \mu\text{m}
\end{aligned}$$

For the case of elastic teeth with variable viscosity,

$$h_0 = 2.08\left(\frac{\eta_0 \bar{u} \alpha}{R}\right)^{8/11}\left(\frac{E^* R}{w_z}\right)^{1/11} R$$

$$= 2.08\left(\frac{50 \times 10^{-3} \times 10 \times 2.2 \times 10^{-8}}{25 \times 10^{-3}}\right)^{8/11}$$

$$\left(\frac{456 \times 10^{9} \times 25 \times 10^{-3}}{1 \times 10^{6}}\right)^{1/11} \times 25 \times 10^{-3}\ \text{m}$$

$$= 29.07\ \mu\text{m}$$

Note that the ratio of film thicknesses for elastic gear teeth with variable viscosity to that for rigid gear teeth with constant viscosity is

$$= \frac{29.07}{0.061} = 476.6$$

This analysis suggests that isoviscous and rigid body assumptions underestimate the predicted film thickness.

One of the assumptions by Grubin is that the surfaces have the deformed shape of an unlubricated contact given by Hertz analysis, but are moved apart by a displacement h_0. However, the fluid film boundaries will diverge rapidly beyond the thin film zone. Consequently, a large pressure gradient must exist near the end of the zone to reduce the pressure to the ambient value. The steep pressure gradients in the outlet region would increase the flow rate. In order to maintain flow continuity, a reduction in film thickness and abrupt rise in pressure, known as a pressure spike, occurs near the outlet. Thus, a significant local reduction in the film thickness with corresponding pressure spike will occur near the outlet of the thin film zone. The reduction in film thickness is on the order of about 25% which means the minimum film thickness is about 75% of h_0 calculated by Grubin's method. Later analyses included the exit constriction (Dowson and Higginson, 1959, 1966).

Typical dimensionless pressure and film thickness profiles for an EHL contact are presented in Figure 8.6.4 (Hamrock *et al.*, 2004). For comparisons, the results of the isoviscous case are also presented. The viscous film thickness is more than three times the isoviscous film thickness in the contact region. Also for the isoviscous case, the reduction in film thickness near the outlet of the contact region is much smaller. Based on numerical results of various EHL contacts, Hamrock *et al.* (2004) presented the following expressions for minimum and center film thicknesses:

$$\frac{h_{min}}{R} = 1.714\ U^{0.694}\ G^{0.568}\ W^{-0.128} \tag{8.6.13a}$$

$$\frac{h_c}{R} = 2.922\ U^{0.692}\ G^{0.470}\ W^{-0.166} \tag{8.6.13b}$$

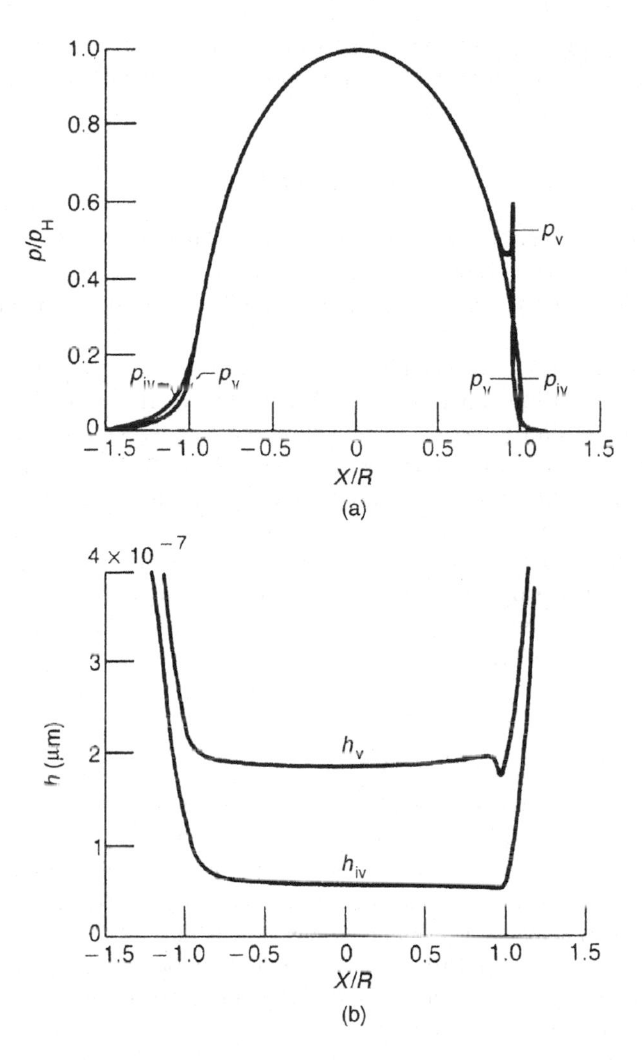

Figure 8.6.4 (a) Dimensionless pressure (p/p_H, where p_H is the maximum Hertz pressure), and (b) film thickness profiles for isoviscous (iv) and viscous (v) solutions at $U = 1.0 \times 10^{-11}$, $\overline{W} = 1.3 \times 10^{-4}$ and $G = 5007$. Reproduced with permission from Hamrock, B.J., Schmid, S.R., and Jacobson, B.O. (2004), *Fundamentals of Fluid Film Lubrication*, Second edition, Marcel Dekker, New York. Copyright 2004. Taylor and Francis.

Note that the minimum film thickness is only slightly dependent on the normal load and the effective modulus of elasticity. It is primarily dependent on the velocity, viscosity, viscosity–pressure coefficient and radius. For good lubrication, the film thickness should be much greater than the roughnesses of the bearing surfaces.

The influence of compressibility of the lubricant should be included because of the high pressures involved. The variation in density with pressure for lubricants is roughly linear at low pressure and the rate of increase of density falls away at high pressure. For mineral oils, density as a function of pressure is given as (Dowson and Higginson, 1966),

$$\frac{\rho}{\rho_0} = 1 + \frac{0.6p}{1 + 1.7p} \tag{8.6.14}$$

where p is the pressure in GPa. This gives a maximum density increase of 33%. Dowson and Higginson reported that the general form of the film shape is not altered and the minimum film thickness is not significantly changed. The pressure spike moves downstream and is reduced in height.

Minimum film thickness in rectangular contacts was first measured by Crook (1958) and later by Orcutt (1965) using capacitance technique and experimental results confirmed the order of magnitude of film thickness predicted by the analyses. Orcutt (1965) and Kannel (1965–66) measured the pressure distribution in the EHL contact using a surface deposited manganin transducer on the disk surfaces used in the experiments. They reported that the spike is difficult to find experimentally because it is very narrow.

Finally, EHL situations can arise both in materials of high elastic modulus, known as hard EHL (such as in roller bearings and gears) and materials of low elastic modulus, known as soft EHL (such as in elastomeric bearings and seals and head–tape interface in magnetic storage tape drives, see Bhushan, 1996).

8.6.3 *Point Contact*

When two spheres (or an equivalent sphere and plane) come into contact, a point contact with a circular shape occurs. A sphere in contact with a raceway develops an elliptical contact. High pressures are generated in the contact zone, resulting in a significant increase in lubricant viscosity and significant elastic deformation of solid surfaces. As in line contacts, the deformed surfaces in lubricated contacts are similar to Hertzian contact with an interposed lubricant film. A minimum film thickness will occur in the outlet region. A two-dimensional Reynolds equation is solved in conjunction with elastic deformation equations and the viscosity-pressure characteristics, to predict the film thickness.

Based on calculations of many cases, the minimum film thickness for materials of high elastic modulus in EHL contacts (hard EHL) is given in terms of dimensionless parameters (Hamrock *et al.*, 2004),

$$U = \frac{\eta_0 \overline{u}}{E^* R_x}$$

$$G = \alpha E^*$$

$$W = \frac{w_z}{E^* R_x^2}$$

and

$$\frac{h_{\min}}{R_x} = 3.63\ U^{0.68}\ G^{0.49}\ W^{-0.073}\ [1 - \exp(-0.68k)] \tag{8.6.15a}$$

where the ellipticity parameter $k = R_y/R_x$, R_y and R_x are effective radii in the y and x directions, respectively, and x is the sliding direction. The center film thickness is given as (Hamrock *et al.*, 2004),

$$\frac{h_c}{R_x} = 2.69\ U^{0.67}\ G^{0.53}\ W^{-0.067}\ [1 - 0.61\ \exp\ (-0.73k)] \tag{8.6.15b}$$

Representative contour plots of dimensionless pressure and film thickness for an ellipticity parameter, k, are shown in Figure 8.6.5. A good agreement has been found between predicted and measured values of film thicknesses. Detailed film thickness measurements have been made between steel balls and sapphire plates ($k = 1$) using the optical interference technique (Archard and Kirk, 1963; Cameron and Gohar, 1966; Foord *et al.*, 1969–70).

The film thicknesses for EHL situations with materials of low elastic modulus (soft EHL) are given as (Hamrock *et al.*, 2004)

$$\frac{h_{\min}}{R_x} = 7.43\ U^{0.65}\ W^{-0.21}[1 - 0.85\ \exp(-0.31k)] \tag{8.6.16a}$$

and

$$\frac{h_c}{R_x} = 7.32\ U^{0.64}\ W^{-0.22}\ \left[1 - 0.72\ \exp\ (-0.28k)\right] \tag{8.6.16b}$$

8.6.4 Thermal Correction

Analyses presented so far have been developed for isothermal conditions and are applicable to pure rolling. However, in applications with some sliding, a temperature rise occurs as a result of shear heating. Interface temperature rise has significant effect on the fluid viscosity, which affects hydrodynamic action. The temperature rise in EHL contacts is up to 100°C. Generally, the lubricant temperature in the conjunction inlet is calculated, which is then used to obtain the absolute viscosity η_0. Based on numerical calculations from a large number of cases, the expression for the thermal correction factor is presented in the literature which gives the percent of film thickness reduction due to inlet heating (Hamrock *et al.*, 2004). The temperature profiles for a steel ball sliding against a sapphire disk have been successfully measured using infrared microscopy methods (Ausherman *et al.*, 1976).

8.6.5 Lubricant Rheology

As stated earlier, lubricant viscosity is a strong function of pressure and temperature. In the case of most lubricating oils, an exponential increase in viscosity occurs with pressure and an exponential decrease with temperature (Equations. 8.3.4 and 8.3.5). For EHL analyses, η_0, α and β of the lubricant are required. The most direct method to determine these parameters is to measure viscosity at different temperatures and pressures and to calculate parameters. However, viscosity measurements at high pressures are difficult. A commonly used method of obtaining these parameters is from analysis of traction data obtained using a so-called rolling disk-type apparatus. In this apparatus, traction behavior of lubricants is measured by using two

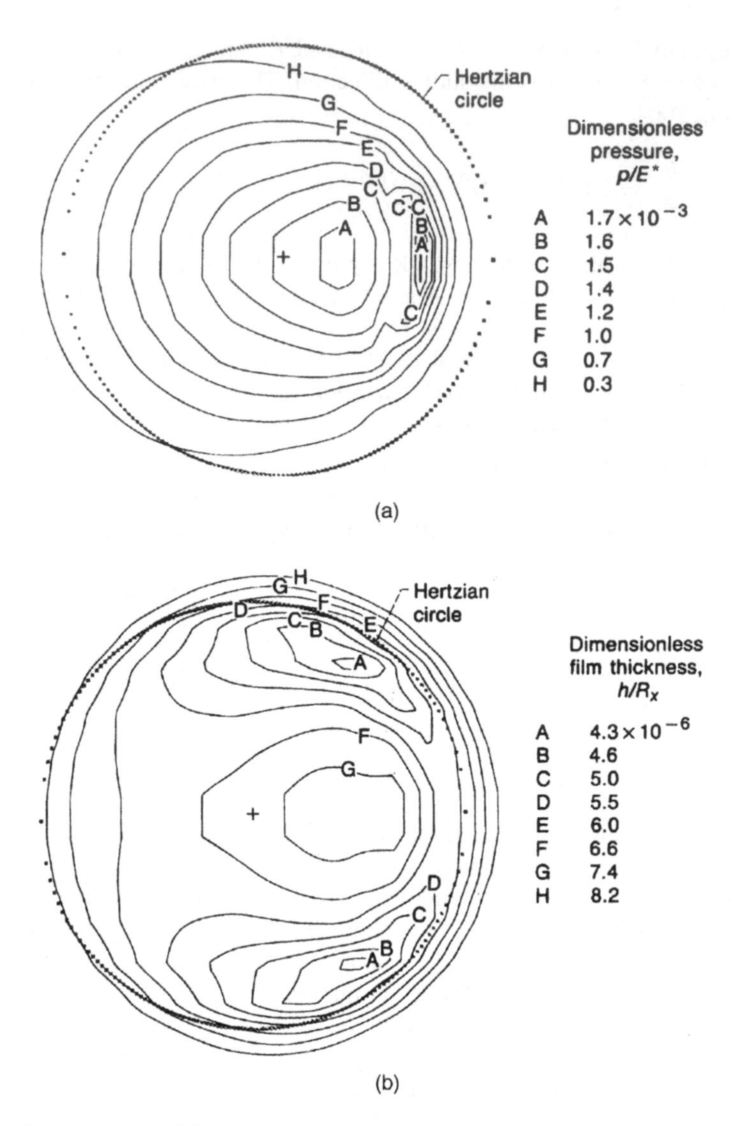

Figure 8.6.5 Contour plots of dimensionless pressure (p/E^*) and dimensionless film thickness (h/R_x) with ellipticity parameter $k = 1.25$ and dimensionless speed, load and materials parameters held at $U = 0.168 \times 10^{-11}$, $W = 1.11 \times 10^{-7}$, and $G = 4522$. Reproduced with permission from Hamrock, B.J., Schmid, S.R., and Jacobson, B.O. (2004), *Fundamentals of Fluid Film Lubrication*, Second edition, Marcel Dekker, New York. Copyright 2004. Taylor and Francis.

or four crowned disks flooded with a lubricant. Using this apparatus, the traction coefficient (shear or traction force divided by the normal load) of the lubricant is measured as a function of slip at different contact pressures, temperatures, and rolling and sliding velocities (Gupta, 1984; Harris, 1991; Zaretsky, 1992).

A typical traction-slip behavior for an oil is shown in Figure 8.6.6. The traction coefficient (coefficient of friction) initially increases with increasing slip ratio, peaks to a maximum value

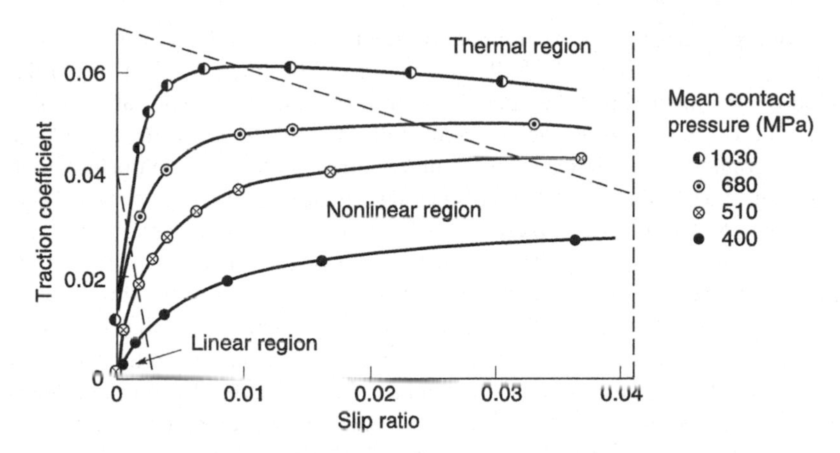

Figure 8.6.6 Typical traction-slip curves for an oil at different mean contact pressures measured on a rolling disk machine in line contact. Reproduced with permission from Harris, T.A. (1991), *Rolling Bearing Analysis*, Third edition, Wiley, New York. Copyright 1991. Wiley.

and then starts to drop with a further increase in the slip ratio and levels off at a high slip ratio. The decrease in traction coefficient occurs because of shear heating in the lubricant at high slip ratios. We further note that the traction coefficient-slip ratio has a linear relationship at very low values of slip ratios, less than 0.003 for most lubricants above which it increases, nonlinearly until it peaks. Therefore, a lubricant behaves as a Newtonian fluid at low slip ratios. The traction coefficient at a given slip ratio increases with an increase in contact pressure. The traction coefficient decreases with an increase in rolling velocity because of an increase in the film thickness.

8.7 Closure

Lubricants are deliberately applied to provide low friction and wear. Lubricants can be solid or fluid (liquid or gaseous). In this chapter, we have focused on fluid film lubrication. A thick fluid film between two surfaces in relative motion prevents solid-solid contact and can provide very low friction and wear. There are various regimes of lubrication.

A regime of lubrication in which a thick fluid film is maintained, between two surfaces with little or no tangential motion, by an external pumping agency, is called hydrostatic lubrication. A summary of the lubrication regimes observed in fluid lubrication without an external pumping agency (self-acting) can be found in the familiar Stribeck curve. This plot for a hypothetical fluid-lubricated bearing system presents the coefficient of friction as a function of the product of absolute viscosity and rotational speed divided by the load per unit projected bearing area (η N/P). Various regimes include hydrodynamic (HD) lubrication, elastohydrodynamic lubrication (EHL), mixed and boundary lubrication regimes which occur at decreasing values of ηN/P. In the case of HD lubrication, as a bearing with convergent shape in the direction of motion starts to move in the longitudinal direction from rest, a thin layer of fluid is pulled through because of viscous entrainment and is then compressed between the bearing surfaces, creating a sufficient hydrodynamic pressure to support the load without an external pumping agency. A HD lubrication regime is relevant for conforming solids or nonconforming solids at low loads. EHL is a subset of HD lubrication in which the elastic deformation of the contacting solids plays a significant role in the HD lubrication process. An EHL regime is relevant for lubricated contact of nonconforming, heavily-loaded contacts (hard EHL) or contacts with low elastic modulus and high geometrical conformity (soft EHL).

The fluid film in HD lubrication/EHL is thick and there is no physical contact between the two surfaces except during start-stop operation at low surface speeds. Because of local elastic deformation, the film thickness in EHL is generally lower than in classical HD lubrication. A high load capacity can be achieved in bearings that operate at high velocities in the presence of fluids of high viscosity. On the other hand, in boundary lubrication, the solid surfaces are so close together that surface interaction between monomolecular or multimolecular films of lubricants and solid asperities dominate the contact. All self-acting bearing interfaces during start-stop operations operate in the boundary lubrication regime, before a fluid film is developed as a result of HD lubrication or EHL. Incidentally, fluid film can also be generated simply by oscillating motion in the normal direction, known as the squeeze effect.

The coefficient of friction in the hydrostatic and hydrodynamic/elastohydrodynamic lubrication regimes is in the range of 0.001 to 0.003, whereas in the boundary lubrication regime, it is on the order of 0.1. In the HD lubrication/EHL, adhesive wear occurs during the start-stop operation and corrosive (chemical) wear of the bearing surfaces can also occur from interaction with the lubricant. In EHL, fatigue wear is the common mode in well-designed heavily loaded bearings. In the hydrostatic lubrication regime, corrosive wear is common.

The major advantage of hydrostatic bearings over self-acting bearings is that they can be used for applications at little or no tangential motion. There is no physical contact during the start-stop operation. The bearing stiffness is very high; however, these require high-pressure pumps and equipment for fluid cleaning which adds to space and cost.

Liquids and gases are used as lubricant media. Viscosity is an important property which determines the load-carrying capacity of a self-acting bearing. Viscosity is a strong function of temperature, pressure, and shear rate. The viscosity of air is about five orders of magnitude lower than that of the liquid lubricants. Hence the load-carrying capacity and stiffness of air bearings are much less than those of liquid bearings.

In hydrostatic, hydrodynamic, and elastohydrodynamic lubrication, the Reynolds equation is used to obtain a relation between the geometry of the surfaces, relative sliding velocity, the property (viscosity and density) of the fluid and the magnitude of the normal load. In the EHL regime, the Reynolds equation, the elastic deformation equation and the equation relating viscosity and pressure are simultaneously solved.

In hydrostatic thrust or journal bearings, high pressure fluid at a constant pressure or volume to a recess, relief, or pocket area is supplied at the bearing interface to maintain a fluid film. In hydrodynamic or elastohydrodynamic thrust or journal bearings, a bearing with convergent shape in the direction of motion is required. The hydrodynamic thrust bearings consist of multiple pads with various shapes including fixed-inclined-pad, tilted pad, and Rayleigh step. The Rayleigh step bearing has the greatest load capacity of all the slider shapes with the same inlet and outlet film thicknesses. In the case of journal bearings, eccentricity of the journal with respect to the bearing produces the convergent shape necessary for production of hydrodynamic pressure. Thrust bearings, except for some tilted-pad bearings, are unidirectional, whereas journal bearings are bidirectional. For bearings with finite width ($b/\ell < 4$ for thrust bearings, where ℓ is the length in the direction of motion and b is the width, and $b/d < 2$ for journal bearings, where d is the being journal diameter), side flow occurs resulting in reduced load capacity. Liquids can be assumed to be incompressible, whereas compressibility should be considered in the case of gaseous films. Analytical solutions are available for hydrostatic and hydrodynamic bearings with infinite width or short width, and with incompressible fluids. Other bearings are analyzed using numerical methods.

In machine components with nonconforming contacts, such as in contact of gear teeth and rolling element bearings, lubrication occurs by EHL. Contacts can be either line or point contacts. Numerical methods are used for solution of the EHL problem. It is found that a significant local reduction in the film thickness with a corresponding pressure spike occurs near the outlet of the thin film zone. In machine components with low elastic modulus contacts of high conformity, such as lip seals, conventional bearings with soft liners and head–tape interface, elastic deformation needs to be taken into account.

The Reynolds equation is based on the continuum theory of fluid mechanics. If the mean free path of the molecules is small compared to the film thickness, continuum flow occurs. However, if the mean free path of the molecules becomes comparable to the film thickness, the fluid does not behave entirely as continuous fluid and the layer of fluid immediately adjacent to the solid surface has a finite slip velocity, producing an apparent diminution in the viscosity of the fluid (rarefaction). The ratio of the mean free path of the molecules to the film thickness is a measure of the degree of rarefaction. Slip flow occurs if this ratio is greater than 0.01. In the case of slip flow, effective viscosity is obtained by using a Knudsen number, a ratio of molecular mean free path at ambient conditions to the reference film thickness. This viscosity is then used in the solution of the Reynolds equation.

The Reynolds equation is based on the assumption of two surfaces being parallel to each other. If the ratio of film thickness to the composite σ roughness of the two surfaces (h/σ) is less than 6, the assumption of parallelism is violated. A flow factor approach is generally used to take into account the effect of surface roughness in the Reynolds equation. Flow factors are obtained by numerical flow simulations in which the actual flow between rough surfaces is equated to an averaged flow between nominally smooth surfaces. Flow factors are a function of h/σ and directional properties of the surfaces.

Temperature rise occurs during relative sliding. This rise will affect viscosity and this effect should be taken into account. Shear rate effects also become important at high relative velocities.

Problems

8.1 A concentric journal bearing is driven by an electrical motor which delivers a net power of 1 kW. The bearing is lubricated with an SAE 10 oil at 40.5°C (absolute viscosity = 31.5 cP). The bearing is 25 mm in width, 25 mm in diameter, 0.2 mm in radial clearance. What is the maximum rotational speed in rpm at which the bearing can be operated?

8.2 A slot connects two oil reservoirs filled with an SAE 30 oil at 37°C (absolute viscosity = 105 cP) with pressures of 1000 kPa and 300 kPa. For a slot with a width of 200 mm, and a length of 300 mm, what thickness would support a volumetric flow rate of 1 liter/min?

8.3 A circular hydrostatic pad thrust bearing of a turbine generator is designed for a thrust load of 10 kN. The outside diameter is 100 mm and the diameter of the recess is 40 mm. A pump with constant feed rate of 10 mm^3/s is available. (a) Select the absolute viscosity of a mineral oil such that the film thickness does not drop below 100 μm. (b) Calculate the oil pressure in the recess. (c) Calculate the film stiffness. (d) Assuming that the generator is running at 750 rpm, calculate the frictional torque absorbed by the bearing. (e) Calculate the power loss. (f) Calculate the power loss due to viscous friction and also due to pumping, and calculate the temperature rise of the oil with a mass density of 880 kg/m^3 and specific heat of 1.88 J/gK.

8.4 A circular hydrostatic pad thrust bearing is used to support a centrifuge weighing 50 N and rotating at 100,000 rpm. The air recess pressure is 50 kPa, the ratio of recess to outside diameter is 0.6, and the film thickness is 50 μm. (a) Calculate the bearing dimensions and the air flow rate. (b) Calculate the frictional torque. Assume that the air is incompressible and its absolute viscosity is 18.2×10^{-6} Pa s at normal pressure and at 20°C.

8.5 A square hydrostatic pad bearing with a long rectangular pocket of width 0.8 m and length of 8 m and a land width of 0.1 m, is used to support a load of 100 kN. Calculate the pressure required and the flow capacity of the pump to maintain a film thickness of 0.1 mm with SAE 30 oil at 37°C (absolute viscosity = 105 cP).

8.6 A fixed inclined-plane hydrodynamic thrust bearing of length 100 mm and width 500 mm operates at a sliding velocity of 1 m/s and a normal load of 10 kN. Select a mineral oil such that the minimum film thickness is at least 50 μm for a bearing operating at maximum load capacity. Calculate the volumetric flow rate. For an application requiring bearing stiffness of 16 N/μm, how many pads are required in the bearing?

8.7 A fixed inclined-plane hydrodynamic thrust bearing of length 100 mm, width 500 mm, operates at a sliding velocity of 1 m/s and a normal load of 10 kN with a mineral oil of absolute viscosity of 10 cP. Calculate the minimum film thickness for m = 2. What is the taper?

8.8 A fixed-incline-pad thrust bearing is designed to carry a total normal load of 50 kN with the following specifications: $r_o = 100$ mm, $r_i = 50$ mm, N = 3600 rpm, $\ell/b = 1$, b = 40 mm, h_i-h_o = 25 μm and SAE 20 oil is used with an inlet temperature of 55°C ($\eta = 29$ cP). Determine the minimum film thickness assuming that the change in oil temperature is negligible.

8.9 A hydrodynamic journal bearing of width 40 mm operates with a shaft of 40 mm diameter which rotates at 1800 rpm and carries a load of 2220 N. The diametral clearance is 80 μm and the absolute viscosity of the lubricant is 28 cP. Calculate the minimum film thickness, attitude angle, volumetric flow rate, volumetric side flow rate, maximum film pressure and location of maximum film pressure.

8.10 A shaft of total mass of 10 kN rotates at 1200 rpm and is supported by two identical hydrodynamic journal bearings of width 100 mm, diameter 100 mm and diametral clearance 150 μm. The absolute viscosity of the lubricant is 40 cP. Calculate the minimum film thickness and maximum film pressure.

8.11 A sleeve bearing of 38 mm diameter, a clearance ratio of 1000, and *b/d* of unity carries a radial load of 2.5 kN. The journal speed is 20 rev/s. The bearing is supplied with SAE 40 lubricant at an inlet temperature of 35°C. Mass density and specific heat of the oil are 880 kg/m^3 and 1.88 J/gK, respectively. Calculate the average oil temperature, minimum film thickness, and the maximum oil-film pressure.

8.12 In a steel ball bearing, balls and inner race of effective radii 25 mm and 10 mm along and transverse to the bearing axis of rotation, respectively, are lubricated with an incompressible film of mineral oil. The bearing components roll together at surface velocities of 10 m/s each under a normal load of 1 kN. The effective modulus of elasticity of the bearing components is 456 GPa and absolute viscosity and α for the mineral oil are 50 mPa s and 0.022/MPa, respectively. Calculate the minimum film thickness.

References

Archard, J.F. and Kirk, M.T. (1963), "Influence of Elastic Modulus on the Lubrication of Point Contacts," Lubrication and Wear Convention, Paper 15, pp. 181–189, Institution of Mechanical Engineers, London.

Ausherman, V.K., Nagaraj, H.S., Sanborn, D.M., and Winer, W.O. (1976), "Infrared Temperature Mapping in Elastohydrodynamic Lubrication," *ASME J. Lub. Tech.* **98**, 236–243.

Bair, S. and Winer, W.O. (1979), "Shear Strength Measurements of Lubricants at High Pressure," *ASME J. Lub. Tech.* **101**, 251–257.

Barus, C. (1893), "Isothermals, Isopiestics, and Isometrics Relative to Viscosity," *Am. J. Sci.* **45**, 87–96.

Beerbower, A. (1972), *Boundary Lubrication*, AD-747 336, Office of the Chief of Research and Development, Department of the Army, Washington, DC.

Bhushan, B. (1987a), "Overview of Coating Materials, Surface Treatments, and Screening Techniques for Tribological Applications – Part I: Coating Materials and Surface Treatments," In *Testing of Metallic and Inorganic Coatings* (W.B. Harding and G.A. DiBari, eds), pp. 289–309, STP947, ASTM, Philadelphia, Pennsylvania.

Bhushan, B. (1987b), "Overview of Coating Materials, Surface Treatments, and Screening Techniques for Tribological Applications – Part II: Screening Techniques," In *Testing of Metallic and Inorganic Coatings* (W.B. Harding and G.A. DiBari, eds), pp. 310–319, STP947, ASTM, Philadelphia, Pennsylvania.

Bhushan, B. (1996), *Tribology and Mechanics of Magnetic Storage Devices*, Second edition, Springer-Verlag, New York.

Bhushan, B. (2001), *Modern Tribology Handbook, Vol. 1: Principles of Tribology*, CRC Press, Boca Raton, Florida.

Bhushan, B. (2013), *Principles and Applications of Tribology*, Second edition, Wiley, New York.

Bhushan, B. and Gupta, B.K. (1997), *Handbook of Tribology: Materials, Coatings and Surface Treatments*, McGraw-Hill, New York (1991), Reprinted with corrections, Krieger, Malabar, Florida (1997).

Bhushan, B. and Tonder, K. (1989a), "Roughness-Induced Shear- and Squeeze-Film Effects in Magnetic Recording Part I: Analysis," *ASME J. Trib.* **111**, 220–227.

Bhushan, B. and Tonder, K. (1989b), "Roughness-Induced Shear- and Squeeze-Film Effects in Magnetic Recording Part II: Analysis," *ASME J. Trib.* **111**, 228–237.

Bisson, E.E. and Anderson, W.J. (1964), *Advanced Bearing Technology*, SP-38, NASA, Washington, DC.

Blok, H. and van Rossum, J.J. (1953), "The Foil Bearing – A New Departure in Hydrodynamic Lubrication", *Lub. Eng.* **9**, 316–320.

Booser, E.R. (1984), *CRC Handbook of Lubrication, Vol. 2 Theory and Design*, CRC Press, Boca Raton, Florida.

Bowden, F.P. and Tabor, D. (1950), *Friction and Lubrication Solids*, Part 1, Clarendon Press, Oxford, UK.

Braithwaite, E.R. (1967), *Lubrication and Lubricants*, Elsevier, Amsterdam.

Bruce, R.W. (2012), *Handbook of Lubrication and Tribology, Vol. II: Theory and Design*, Second edition, CRC Press, Boca Raton, Florida.

Burgdorfer, A. (1959), "The Influence of the Molecular Mean Free Path on the Performance of Hydrodynamic Gas Lubricated Bearings," *ASME J. Basic Eng.* **81**, 94–100.

Cameron, A. (1976), *Basic Lubrication Theory*, Second edition, Wiley, New York.

Cameron, A. and Gohar, R. (1966), "Theoretical and Experimental Studies of the Oil Film in Lubricated Point Contacts," *Proc. R. Soc. Lond. A* **291**, 520–536.

Cheng, H.S. (1970), "A Numerical Solution to the Elastohydrodynamic Film Thickness in an Elliptical Contact," *ASME J. Lub. Tech.* **92**, 155–162.

Clauss, F.J. (1972), *Solid Lubricants and Self-Lubricating Solids*, Academic Press, New York.

Crook, A.W. (1958), "The Lubrication of Rollers," *Phil. Trans. R. Soc. Lond. A* **250**, 387–409.

Dowson, D. and Higginson, G.R. (1959), "A Numerical Solution to the Elastohydrodynamic Problem," *J. Mech. Eng. Sci.* **1**, 6–15.

Dowson, D. and Higginson, G.R. (1966), *Elastohydrodynamic Lubrication*, Pergamon, Oxford.

DuBois, G.B. and Ocvirk, F.W. (1953), "Analytical Derivation and Experimental Evaluation of Short-Bearing Approximation for Full Journal Bearings," NACA Report 1157.

Eshel, A. and Elrod, H.G. (1967), "Stiffness Effects on the Infinitely Wide Foil Bearing," *ASME J. Lub. Tech.* **89**, 92–97.

Foord, C.A., Wedevan, L.D., Westlake, F.J., and Cameron, A. (1969–1970), "Optical Elastohydrodynamics," *Proc. Instn Mech. Engrs* **184**, Part I.

Frene, J., Nicolas, D., Degueurce, B., Berthe, D., and Godet, M. (1997), *Hydrodynamic Lubrication – Bearings and Thrust Bearings*, Elsevier, Amsterdam.

Fuller, D.D. (1984), *Theory and Practice of Lubrication for Engineers*, Second edition, Wiley, New York.

Gans, R.F. (1985), "Lubrication Theory at Arbitrary Knudsen Number," *ASME J. Trib.* **107**, 431–433.

Gross, W.A., Matsch, L.A., Castelli, V., Eshel, A., Vohr, J.H., and Wildmann, M. (1980), *Fluid Film Lubrication*, Wiley, New York.

Grubin, A.N. (1949), "Fundamentals of the Hydrodynamic Theory of Lubrication of Heavily Loaded Cylindrical Surfaces, *Investigation of the Contact Machine Components* (Kh. F. Ketova, ed). Translation of Russian Book No. 30 Central Scientific Institute for Technology and Mechanical Engineering, Moscow, chap. 2. (Available From Dept. of Scientific and Industrial Research, Great Britain, Trans. CTS-235 and Special Libraries Association, Trans. R-3554).

Gupta, P.K. (1984), *Advanced Dynamics of Rolling Elements*, Springer-Verlag, New York.

Hamrock, B.J. and Dowson, D. (1981), *Ball Bearing Lubrication – The Elastohydro-dynamics of Elliptical Contacts*, Wiley, New York.

Hamrock, B.J., Schmid, S.R., and Jacobson, B.O. (2004), *Fundamentals of Fluid Film Lubrication*, Second edition, Marcel Dekker, New York.

Harris, T.A. (1991), *Rolling Bearing Analysis*, Third edition, Wiley, New York.

Kannel, J.W. (1965–66), "Measurement of Pressure in Rolling Contact," *Proc. Instn Mech. Engrs* **180**, Pt. 3B, 135.

Khonsari, M.M. and Booser, E.R (2001), *Applied Tribology – Bearing Design and Lubrication*, Wiley, New York.

Kingsbury, A. (1897), "Experiments with an Air-Lubricated Journal," *J. Am. Soc. Nav. Engrs* **9**, 267–292.

Ku, P.M. (1970), *Interdisciplinary Approach to Friction and Wear*, pp. 335–379, SP-181, NASA, Washington, DC.

Licht, L. (1968), "An Experimental Study of Elasto-Hydrodynamic Lubrication of Foil Bearings," *ASME J. Lub. Tech.* **90**, 199–220.

Ling, F.F., Klaus, E.E., and Fein, R.S. (1969), *Boundary Lubrication – An Appraisal of World Literature*, ASME, New York.

Martin, H.M. (1916), "Lubrication of Gear Teeth," *Engineering, London* **102**, 199.

Orcutt, F.K. (1965), "Experimental Study of Elastohydrodynamic Lubrication," *ASLE Trans.* **8**, 321–326.

Patir, N. and Cheng, H.S. (1978), "An Average Flow Model for Determining Effects of Three-Dimensional Roughness on Partial Hydrodynamic Lubrication," *ASME J. Lub. Tech.* **100**, 12–17.

Patir, N. and Cheng, H.S. (1979), "Application of Average Flow Model to Lubrication Between Rough Sliding Surfaces," *ASME J. Lub. Tech.* **101**, 220–230.

Petroff, N.P. (1883), "Friction in Machines and the Effect of Lubricant," *Inzh. Zh. St. Petersburg*, **1**, 71–140; **2**, 227–279; **3**, 377–463; **4**, 535–564.

Pinkus, O. and Sternlicht, B. (1961), *Theory of Hydrodynamic Lubrication*, McGraw-Hill, New York.

Raimondi, A.A. (1961), "A Numerical Solution for the Gas-Lubricated, Full Journal Bearing of Finite Length," *ASLE Trans.* **4**, 131–155.

Raimondi, A.A. and Boyd, J. (1955), "Applying Bearing Theory to the Analysis and Design of Pad-Type Bearings," *ASME Trans.* **77**, 287–309.

Raimondi, A.A. and Boyd, J. (1958), "A Solution for the Finite Journal Bearing and its Application to Analysis and Design – I, - II, and – III," *ASLE Trans.* **1**, 159–174; 175–193; 194–209.

Rayleigh, L. (1918), "Notes on the Theory of Lubrication," *Philos. Mag.* **35**, 1–12.

Reynolds, O. (1886), "On the Theory of Lubrication and its Application to Mr. Beauchamp Tower's Experiments, Including an Experimental Determination of the Viscosity of Olive Oil," *Philos. Trans. R. Soc. Lond.* **177**, 157–234.

Rippel, H.C. (1963) *Cast Bronze Hydrostatic Bearing Design Manual*, Second edition, Cast Bronze Bearing Institute Inc., Evanston, Illinois.

Roelands, C.J.A. (1966), *Correlation Aspects of the Viscosity–Temperature-Pressure Relationship of Lubricating Oils*, Druk, V.R.B., Groningen, Netherlands.

Shigley, J.E. and Mitchell, L.D. (1993), *Mechanical Engineering Design*, Fourth edition, McGraw-Hill, New York.

Sommerfeld, A. (1904), "Zur Hydrodynamischen Theorie der Schmiermittelreibung," *Z. Angew. Math. Phys.* **50**, 97–155.

Stachowiak, G.W. and Batchelor, A.W. (2005), *Engineering Tribology*, Third edition, Elesevier Butterworth-Heinemann, Dordrecht, Netherlands.

Stribeck, R. (1902), "Characteristics of Plain and Roller Bearings," *Zeit. Ver. Deut. Ing.* **46**, 1341–1348, 1432–1438, 1463–1470.

Szeri, A.Z. (2010), *Fluid Film Lubrication – Theory and Design*, Second edition, Cambridge University Press, Cambridge, UK.

Tam, A.C. and Bhushan, B. (1987), "Reduction of Friction Between a Tape and a Smooth Surface by Acoustic Excitation," *J. Appl. Phys.* **61**, 1646–1648.

Tonder, K. (1985), "Theory of Effects of Striated Roughness on Gas Lubrication," *Proc. JSLE International Trib. Conf.*, Tokyo, Japan, pp. 761–766.

Tripp, J.H. (1983), "Surface Roughness Effects in Hydrodynamic Lubrication: The Flow Factor Method," *ASME J. Lub. Tech.* **105**, 458–465.

Totten, G.E. (2006), *Handbook of Lubrication and Tribology: Vol. 1 – Applications and Maintenance*, Second edition, CRC Press, Boca Raton, Florida.

Tower, B. (1883/84), "First Report on Friction Experiments," *Proc. Instn Mech. Engrs*, **1883**, 632–659; **1884**, 29–35.

Tower, B. (1885), "Second Report on Friction Experiments," *Proc. Instn Mech. Engrs*, **1885**, 58–70.

Van Wazer, J.R., Lyons, J.W., Kim, K.Y., and Colwell, R.E. (1963), *Viscosity and Flow Measurement*, Wiley, New York.

Walowit, J.A. and Anno, J.N. (1975), *Modern Developments in Lubrication Mechanics*, Applied Science, London, UK.

Walters, K. (1975), *Rheometry*, Wiley, New York.

Wilcock, D.F. (1972), *Design of Gas Bearings*, Vols. **1 and 2**, Mechanical Technology Inc., Latham, New York.

Williams, J.A. (2005), *Engineering Tribology*, Second edition, Cambridge University Press, Cambridge, UK.

Zaretsky, E.V. (ed.) (1992), *Life Factors for Rolling Bearings*, STLE, Park Ridge, Illinois.

Further Reading

Bassani, R. and Piccigallo, B. (1992), *Hydrostatic Lubrication*, Elesvier, Amsterdam.

Bhushan, B. (2001a), *Modern Tribology Handbook, Vol. 1: Principles of Tribology*, CRC Press, Boca Raton, Florida.

Bhushan, B. (2001b), *Fundamentals of Tribology and Bridging the Gap Between the Macro- and Micro/Nanoscales*, NATO Science Series II – Vol. **10**, Kluwer, Dordrecht, Netherlands.

Bhushan, B. (2011), *Nanotribology and Nanomechanics I*, Third edition, Springer-Verlag, Heidelburg, Germany.

Bhushan, B. (2013), *Principles and Applications of Tribology*, Second edition, Wiley, New York.

Bisson, E.E. and Anderson, W.J. (1964), *Advanced Bearing Technology*, SP-38, NASA, Washington, DC.

Booser, E.R. (1984), *CRC Handbook of Lubrication, Vol. 2 Theory and Design*, CRC Press, Boca Raton, Florida.

Bruce, R.W. (2012), *Handbook of Lubrication and Tribology, Vol. II: Theory and Design*, Second edition, CRC Press, Boca Raton, Florida.

Cameron, A. (1976), *Basic Lubrication Theory*, Second edition, Wiley, New York.

Frene, J., Nicolas, D., Degueurce, B., Berthe, D., and Godet, M. (1997), *Hydrodynamic Lubrication – Bearings and Thrust Bearings*, Elsevier, Amsterdam, Holland.

Fuller, D.D. (1984), *Theory and Practice of Lubrication for Engineers*, Second edition, Wiley, New York.

Gohar, R. and Rahnejat, H. (2012), *Fundamentals of Tribology*, Second edition, Imperial College Press, London, UK.

Gross, W.A., Matsch, L.A., Castelli, V., Eshel, A., Vohr, J.H., and Wildmann, M. (1980), *Fluid Film Lubrication*, Wiley, New York.

Hamrock, B.J., Schmid, S.R., and Jacobson, B.O. (2004), *Fundamentals of Fluid Film Lubrication*, Second edition, Marcel Dekker, New York.

Khonsari, M.M. and Booser, E.R (2001), *Applied Tribology – Bearing Design and Lubrication*, Wiley, New York.

Pinkus, O. and Sternlicht, B. (1961), *Theory of Hydrodynamic Lubrication*, McGraw-Hill, New York.

Stachowiak, G.W. and Batchelor, A.W. (2005), *Engineering Tribology*, Third edition, Elesevier Butterworth-Heinemann, Dordrecht, Netherlands.

Szeri, A.Z. (2010), *Fluid Film Lubrication – Theory and Design*, Second edition, Cambridge University Press, Cambridge, UK.
Totten, G.E. (2006), *Handbook of Lubrication and Tribology: Vol. 1 – Applications and Maintenance*, Second edition, CRC Press, Boca Raton, Florida.
Wilcock, D.F. (1972), *Design of Gas Bearings*, Vols. 1 and 2, Mechanical Technology Inc., Latham, New York.
Williams, J.A. (2005), *Engineering Tribology*, Second edition, Cambridge University Press, Cambridge, UK.

9

Boundary Lubrication and Lubricants

9.1 Introduction

Lubricants are commonly used for reducing friction at, and wear of, interfaces. In some applications, the solid surfaces are so close together that some asperities come in contact and others are mitigated by a thin film of lubricant, Figure 9.1.1. Under these conditions, the lubricant viscosity is relatively unimportant and the physical and chemical interactions of the lubricant with the solid bodies controls friction and wear. Even a monolayer of adsorbed molecules may provide some protection against wear. Lubrication in some situations can be achieved by the use of multimolecular lubricant films. Monolayer lubrication is referred to as boundary lubrication and multimolecular lubrication is referred to as mixed lubrication. Boundary lubrication usually occurs under high-load and low-speed conditions in machine components such as bearings, gears, cam and tappet interfaces, and piston ring and liner interfaces. Boundary lubrication forms a last line of defense. In many cases, it is the regime which controls the component life (Godfrey, 1968; Ling *et al.*, 1969; Ku, 1970; Beerbower, 1972; Iliuc, 1980; Booser, 1984; Loomis, 1985; Anon., 1997).

Various lubricants and greases are used for lubrication of machine components operating in various lubrication regimes (Gunderson and Hart, 1962; Bisson and Anderson, 1964; Braithwaite, 1967; Gunther, 1971; Evans *et al.*, 1972; McConnell, 1972; Boner, 1976; Booser, 1984; Loomis, 1985; Bhushan and Zhao, 1999). Additives are commonly used to provide the desirable properties and interaction with the interface. In this chapter, we will describe mechanisms of boundary lubrication and an overview of various types of lubricants and their properties.

9.2 Boundary Lubrication

For the case of two contacting bodies coated with a continuous solid monolayer of lubricant with a load too small to cause plastic deformation, the interface is in equilibrium under load for some time since the films prevent contact between the substrates. The films do not allow

Introduction to Tribology, Second Edition. Bharat Bhushan.
© 2013 John Wiley & Sons, Ltd. Published 2013 by John Wiley & Sons, Ltd.

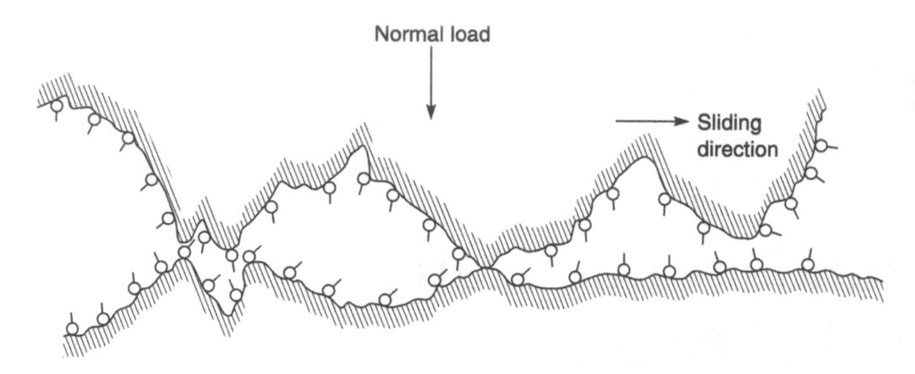

Figure 9.1.1 Schematic of two surfaces separated by a boundary layer of lubricant.

any state of lower free surface energy than the initial state. The films will thus lubricate over a considerable sliding distance, if the bodies are subjected to low-speed sliding, although they will eventually be worn away. Alternatively, if the temperature is raised slightly above the melting point of the films, the admolecules will acquire some mobility and no state involving more than one complete monolayer can be stable. Activation energy for migrating away from the loaded region is provided thermally—and mechanically also, if a low speed sliding is imposed. The two layers initially present thus penetrate each other and adsorbed molecules tend to move away from the loaded interface. An equilibrium between a partial monolayer at the loaded interface and a surrounding vapor, if it exists, is generally metastable, since the state for which the bodies are in direct contact usually has lower energy than any lubricated one.

The boundary films are formed by physical adsorption, chemical adsorption, and chemical reaction (Chapter 2); for typical examples, see Figure 9.2.1. The physisorbed film can be of either monomolecular (typically < 3 nm) or polymolecular thickness. The chemisorbed films are monomolecular, but stoichiometric films formed by chemical reaction can have a large film thickness. In general, the stability and durability of surface films decrease in the following order: chemical reaction films, chemisorbed films, and physisorbed films.

A good boundary lubricant should have a high degree of interaction between its molecules and the sliding surface. As a general rule, liquids are good lubricants when they are polar and thus able to grip solid surfaces (or be adsorbed) (Bhushan and Zhao, 1999). Polar lubricants contain reactive functional groups with low ionization potential or groups having high polarizability. The boundary lubrication properties of lubricants are also dependent upon the molecular conformation and lubricant spreading. Examples of nonpolar and polar molecules are shown in Figure 9.2.2. In the case of Z-Dol, a hydrogen atom, covalently bonded with oxygen atom in the O-H bond, exposes a bare proton on the end of the bond. This proton can be easily attracted to the negative charge of other molecules because the proton is not shielded by electrons, and this is responsible for the polarity of the O-H ends. Likewise, the lone pairs of electrons in the oxygen and fluorine atoms in both molecules are unshielded, and can be attracted to positive charges of other molecules, and thus exhibit electronegativity which is responsible for some polarity. The CF_3 end in Z-15 is symmetric and its polarity is low.

In addition to the polarity of liquids, the shape of their molecules governs the effectiveness, which determines whether they can form a dense, thick layer on the solid surface. Ring molecules or branch chain molecules tend to be poorer than straight chain molecules because

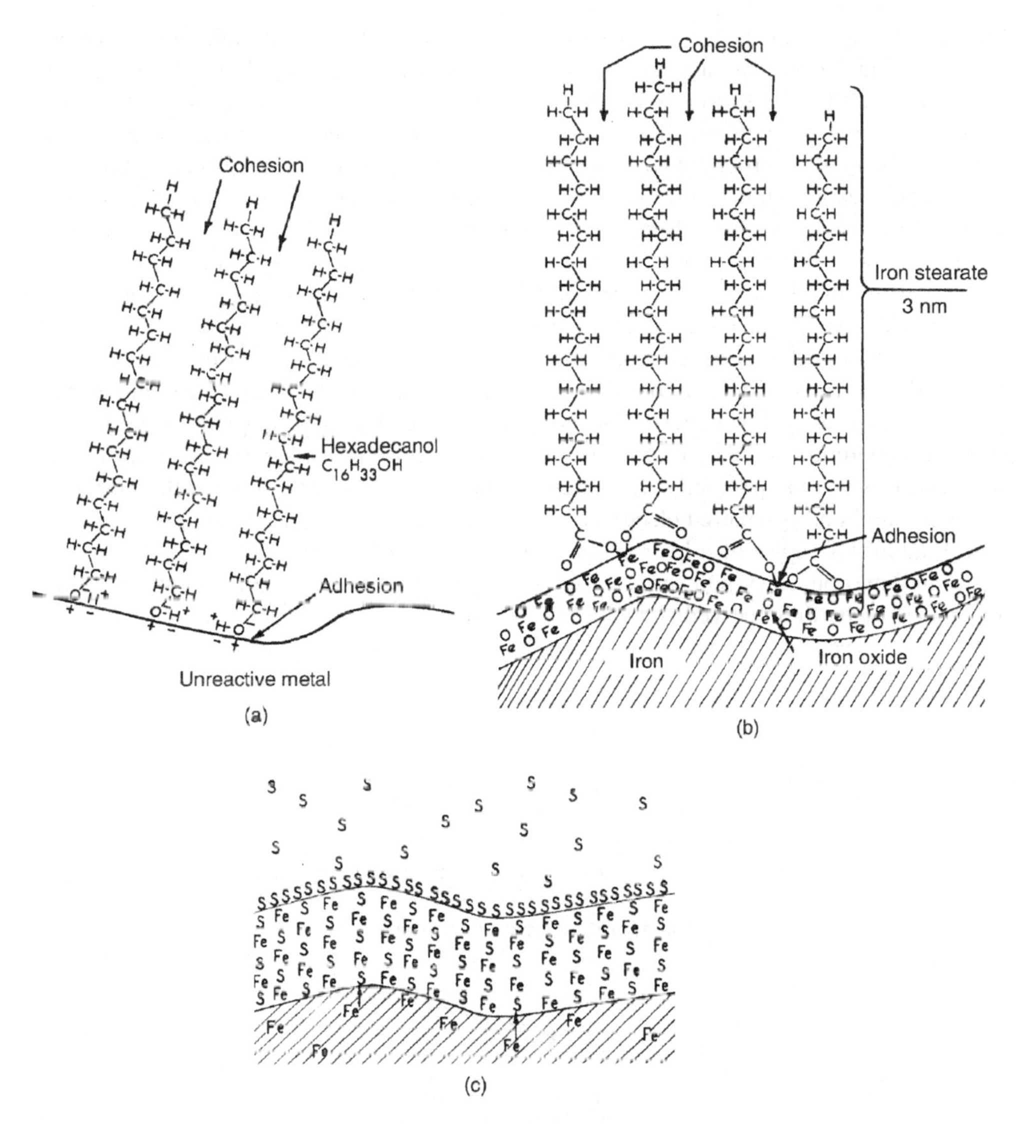

Figure 9.2.1 (a) Schematic diagram representing the physisorption with preferred orientation of three polar molecules of hexadecanol to a metal surface; (b) schematic diagram representing the chemisorption of stearic acid on an iron surface to form a monolayer of iron stearate, a soap; (c) schematic representation of an inorganic film formed by chemical reaction of sulfur with iron to form iron sulfide (*Source*: Ku, 1970).

Nonpolar molecule

Fomblin Z-15 $CF_3-O-(CF_2-CF_2-O)_m-(CF_2-O)_n-CF_3$ $(m/n \sim 2/3)$

Polar molecule

Fomblin Z-DOL $HO-CH_2-CF_2-O-(CF_2-CF_2-O)_m-(CF_2-O)_n-CF_2-CH_2-OH$

Figure 9.2.2 Structures of non-polar and polar (-OH) organic lubricant molecules.

there is no way in which they can achieve a high packing density. Straight chain molecules with one polar end, such as alcohols and soaps of fatty acids, are highly desirable, because they enable a thick film to be formed with the polar end tightly held on the surface and the rest of the molecule normal to the surface. If the sliding surface has to operate under humid conditions, the lubricant should be hydrophobic (i.e., it should not absorb water or be displaced by the water).

The most readily observed cause of breakdown of thin film lubrication is the melting of a solid film and degradation of liquid films, but some degree of lubrication may persist to a higher temperature. Sliding speed and load influence the performance of multilayers.

We now discuss the properties of the solid surface that are desirable for good lubrication. A solid should have a high surface energy, so that there will be a strong tendency for molecules to adsorb on the surface. Consequently, metals tend to be the easiest surfaces to be lubricated. The solid surfaces should have a high wetting (or low contact angle) so that the liquid lubricant wets the solid easily. For better lubrication, the surface should be reactive to the lubricant under test conditions so that durable, chemically reacted films can form. Another property of solid surfaces is hydrophobicity. The surfaces should be highly functional with polar groups and dangling bonds (unpaired electrons) so that they can react with lubricant molecules and adsorb them. Examples of a hydrophilic silicon oxide surface and its reactivity to ambient water and amorphous carbon surface with polar groups and dangling bonds which promote adsorption of perfluoropolyether molecules are shown in Figure 9.2.3 (Bhushan and Zhao, 1999). Additives to the lubricants can also enhance the formation of chemically reacted films.

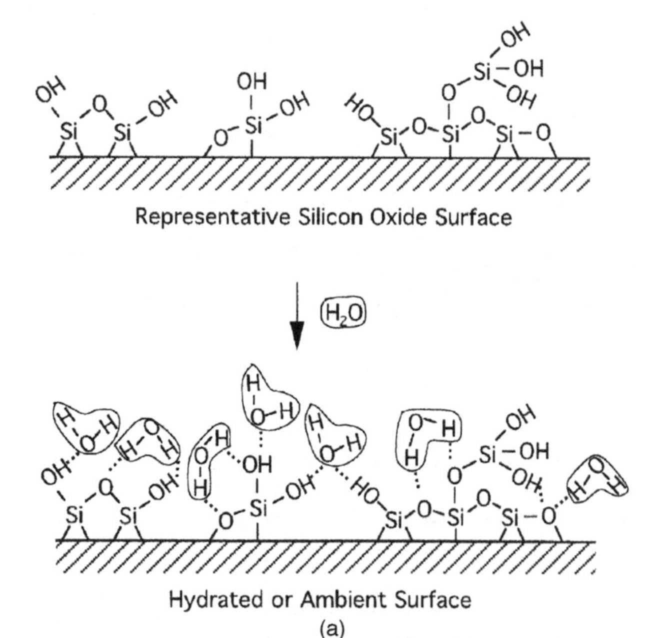

Figure 9.2.3 Schematic illustrations of (a) a hydrophilic silicon oxide surface before and after adsorption of water molecules; hydrogen bonding occurs between the solid surface and water molecules, (b) a hydrogenerated diamondlike carbon surface with adsorbed polar perfluoropolyether (Z-Dol) lubricant molecules; the symbol • represents dangling bonds. (*Continued*)

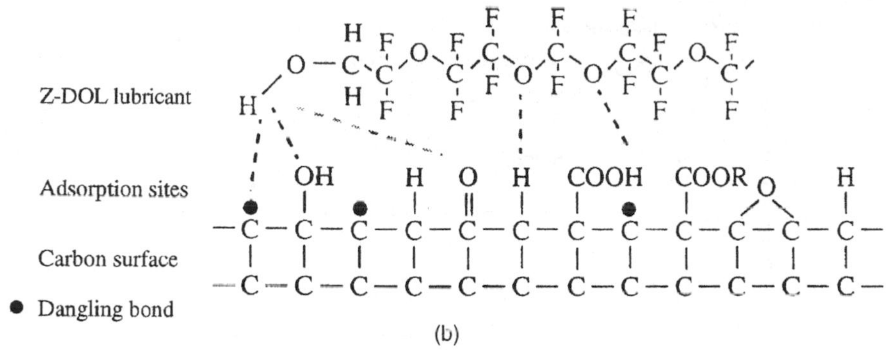

Z-DOL lubricant

Adsorption sites

Carbon surface

● Dangling bond

(b)

Figure 9.2.3 *(Continued)*

We now present data showing the effect of environment and types of lubricants and their interaction with solid surfaces on boundary lubrication behavior.

9.2.1 *Effect of Adsorbed Gases*

Boundary films occur on almost all surfaces because they reduce the surface energy and are thus thermodynamically favored. Normally, air covers any surface with an oxidized film plus adsorbed moisture and organic material. Inadvertent lubrication by air is the most common boundary lubrication. Figure 9.2.4a shows the reduction in the coefficient of friction that is obtained by the adsorption and/or chemical reaction of oxygen on clean iron surfaces outgassed in a vacuum (roughly 10^{-6} torr or mm Hg). The coefficient of friction is markedly reduced by admission of oxygen gas, though the oxygen pressure is very low (roughly 10^{-4} torr). As oxygen pressure is allowed to increase, the friction is reduced still more. Finally, if the surfaces are allowed to stand for some period of time, the adsorbed oxygen film becomes more complete and the friction drops still further. We note that the seizure of clean metals is prevented by even a trace of oxygen, as obtained at 10^{-4} torr (Bowden, 1951).

Figure 9.2.4b shows the effect of the addition of hydrogen sulfide on the coefficient of friction of outgassed (clean) iron surfaces; the friction reduced abruptly and appreciably. It is necessary to heat the surface to over 790°C before the decomposition of the film takes place and friction rises. It is probable that hydrogen sulfide reacts with the clean iron surfaces to form an iron sulfide (FeS) film.

9.2.2 *Effect of Monolayers and Multilayers*

It is possible to show by use of monomolecular layers and multimolecular layers that a very thin film of lubricant at the surface can be effective in reducing friction. In studying the effects of monolayers or multilayers, it is convenient to use the well-known Langmuir–Blodgett (L-B) technique. This technique involves floating an insoluble monolayer on the surface of water and then transferring it from the surface of the water to the surface of the solid (by successive dippings) to which the monolayer or multilayer is to be applied. This technique is convenient for deposition of films of known and controllable thicknesses.

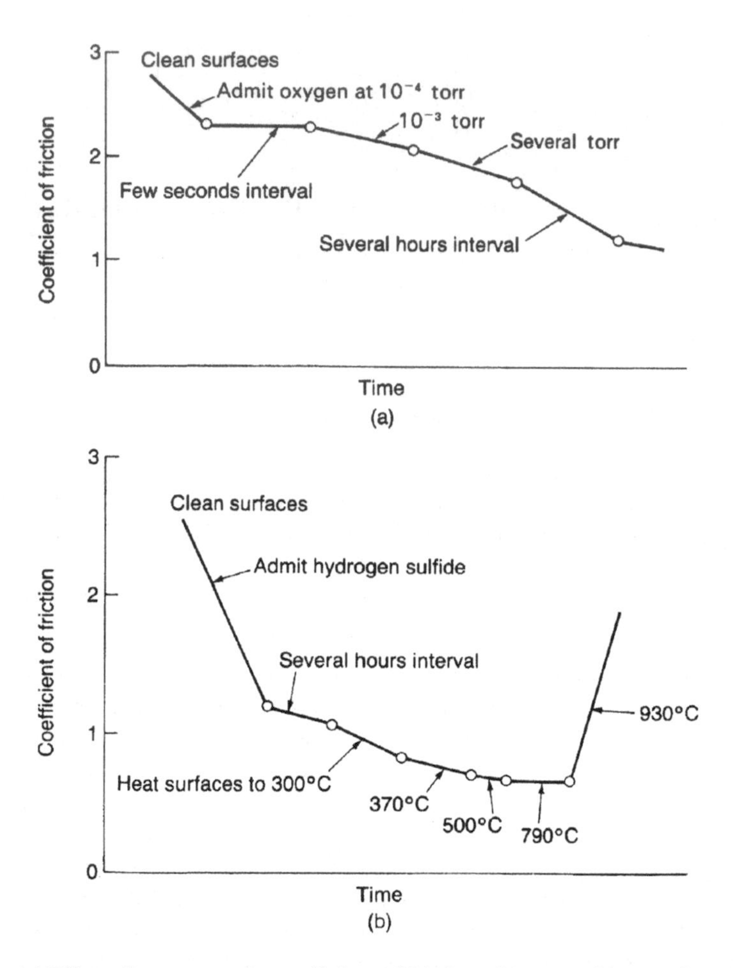

Figure 9.2.4 (a) Effect of oxygen on the coefficient of friction of outgassed iron surfaces and (b) effect of hydrogen sulfide on the coefficient of friction of outgassed iron surfaces. Reproduced with permission from Bowden, F.P. (1951), "The Influence of Surface Films on the Friction, Adhesion and Surface Damage of Solid," *The Fundamental Aspects of Lubrication,* Annals of NY Academy of Sciences, 53, Art 4, June 27, 753–994. Copyright 1951. New York Academy of Sciences).

Bowden and Tabor (1950) deposited the films of a long-chain fatty acid (stearic acid) on a stainless steel surface. The lubricated surface was slid against an unlubricated surface at 10 mm/s, and the coefficient of friction was recorded from the beginning of sliding. Data shown in Figure 9.2.5 for a monolayer and with multilayers of 3, 9, or 53 films show that the greater the number of films, the longer it takes to wear off or displace this protective film and, consequently, the longer the time in which the film is an effective boundary lubricant. The films deposited by the L-B technique are not entirely equivalent to the type of protective film developed from lubricants in practice, with respect to either molecular packing or composition. The stearic acid films, however, were close packed and regularly oriented with the polar group in the water surface.

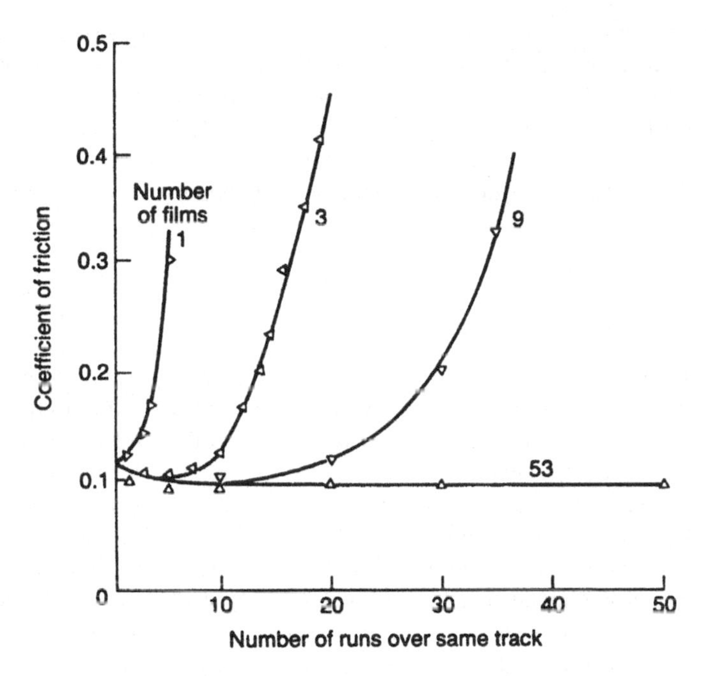

Figure 9.2.5 Wear behavior of a number of stearic acid films deposited on stainless steel sliding against unlubricated stainless steel surface. Reproduced with permission from Bowden, F.P. and Tabor, D. (1950), *Friction and Lubrication of Solids*, Part I, Clarendon Press, Oxford, UK. Fig. 72 p. 188. Copyright 1950 Oxford University Press.

Israelachvili *et al.* (1988) have elegantly measured the frictional force or shear stress (frictional force divided by the apparent area of contact) required to sustain sliding (shearing) of two molecularly smooth mica surfaces, with various molecular layer(s) of a liquid film in between. The two liquids used were octamethylcyclotetrasiloxane (OMCTS) and cyclohexane, and have mean molecular diameters of 0.85 and 0.5 nm, respectively. Measurements were made at sliding velocities ranging from 0.25 to 2 μm/s after steady-state sliding was attained. They found that the shear stress depended on the number of boundary liquid layers, Table 9.2.1;

Table 9.2.1 Shear stress as a function of number of boundary layers trapped between two mica surfaces for octamethylcyclotetrasiloxane (OMCTS)[a] and cyclohexane[b].

	Shear Stress (MPa)	
Number of layers	OMCTS	Cyclohexane
1	8.0 ± 0.5	$2.3 \pm 0.6 \times 10$
2	6.0 ± 1.0	1.0 ± 0.2
3	3.0 ± 1.0	$4.3 \pm 1.5 \times 10^{-1}$
4	Not measured	$2.0 \pm 1.0 \times 10^{-2}$

[a]Molecular diameter ~ 0.85 nm.
[b]Molecular diameter ~ 0.5 nm.

in cyclohexane, for example, shear stress fell by about an order of magnitude per additional layer. In other words, the friction is quantized depending on the number of molecular layers separating the surfaces. By extrapolation, one may infer that when 7–10 layers are present, the shear stress of the liquid film would have fallen to the value expected for bulk continuum Newtonian flow. It is noteworthy that this is about the same number of layers as when the forces across a thin film and the whole concept of viscosity begin to be described by continuum theories (Israelachvili, 1992).

9.2.3 Effect of Chemical Films

The addition of a small trace of a fatty acid (polar lubricant) to a nonpolar mineral oil or to a pure hydrocarbon can bring about a considerable reduction in the friction and wear of chemically reactive surfaces. Typical results taken from Bowden and Tabor (1950) are given in Table 9.2.2. In these experiments, friction was measured using identical materials sliding against each other and lubricated with a (nonpolar) paraffin oil or with 1% (polar) lauric acid added to the paraffinic oil. They found that with unreactive metals (such as Pt, Ag, Ni, and Cr) and glass the fatty acid is no more effective than a paraffin oil. In contrast, the results for lubrication of reactive metals (such as Zn, Cd, Cu, Mg, and Fe) show that very effective lubrication can be obtained with a 1% solution of lauric acid in paraffin oil. Bowden and Tabor (1950) have further shown that a 0.01% solution of lauric acid in paraffin oil reduces the coefficient of friction of chemically reactive cadmium surfaces from 0.45 to 0.10. Even

Table 9.2.2 Efficiency of lubrication by paraffin oils or 1% lauric acid in paraffin oil compared with reactivity of metal to lauric acid (*Source*: Bowden and Tabor, 1950).

	Coefficient of friction at 20°C				
Metal	Clean	Paraffin oil	1% Lauric acid in paraffin oil	Transition temperature (°C)	%Acid reactive[a]
Unreactive					
Platinum	1.2	0.28	0.25	20	None
Silver	1.4	0.8	0.7	20	None
Nickel	0.7	0.3	0.28	20	None
Chromium	0.4	0.3	0.3	20	None
Glass	0.9	0.4	0.4	20	None
Reactive					
Zinc	0.6	0.2	0.04	94	10
Cadmium	0.5	0.45	0.05	103	9.3
Copper	1.4	0.3	0.08	97	4.6
Magnesium	0.6	0.5	0.08	80	Trace
Iron	1.0	0.3	0.2		Trace

[a]Estimated amount of acid involved in the reaction assuming formation of a normal salt.

Table 9.2.3 Lubrication of various metal surfaces by layers of stearic acid and metal stearates deposited by the Langmuir–Blodgett technique (*Source*: Bowden and Tabor, 1950).

Metal	Number of layers for effective lubrication	
	Stearic acid	Metal soap (Cu orAg stearates)
Unreactive		
Platinum	>10	7–9
Silver	7	3
Nickel	3	3
Reactive		
Copper	3	3
Stainless Steel	3	1

smaller concentrations (0.001%) of lauric acid can reduce friction slowly with time (after a few hours). In the case of the less reactive metals, such as iron, which can not be lubricated by a 1% solution of fatty acid, they are well lubricated by a more concentrated solution. These results indicate the strong effects of films formed by chemical reaction on friction and wear.

Friction and wear measurements on monolayers of various substances show that whereas a single monolayer of a given polar compound on a reactive surface provides low friction and wear, on unreactive surfaces it may be completely ineffective, and as many as 10 or more layers may be needed. Bowden and Tabor (1950) deposited multilayers of stearic acid and metal soaps (believed to be responsible for good lubrication behavior when stearic acid is used as a lubricant on a metal surface) on unreactive and reactive metal surfaces and made the friction measurements. Table 9.2.3 shows the number of layers required for effective lubrication. We find that a larger number of films are needed for unreactive surfaces (see also Ling *et al.*, 1969).

Paraffins, alcohols, ketones, and amides become ineffective lubricants at the bulk melting point of the lubricant. When the melting occurs, adhesion between the molecules in the boundary film is diminished and breakdown of the film takes place. The increased metallic contact through the lubricant film leads to increased friction and wear. With saturated fatty acids on reactive metals, however, the breakdown does not occur at their melting points but at considerably higher temperatures. This is shown in Figure 9.2.6 for a series of fatty acids on steel surfaces, and it is seen that breakdown (transition temperature) occurs at 50 to 70°C above the melting point. The transition temperature corresponds approximately to the melting point of the metallic soaps formed by chemical reaction (Bowden and Tabor, 1950). The actual value of the breakdown temperature depends on the nature of the metals and on the load and speed of sliding. The esters of saturated fatty acids also behave like acids except that the difference between the transition and melting temperatures (T_t-T_m) decreases with increasing ester group length, approaching zero at 26 carbon atoms.

Thus, lubrication is affected not by the fatty acid itself but by the metallic soap formed as a result of the chemical reaction between the metal and the fatty acid. Also, T_t-T_m is a measure of the strength of adsorption that is due to a dipole–metal interaction.

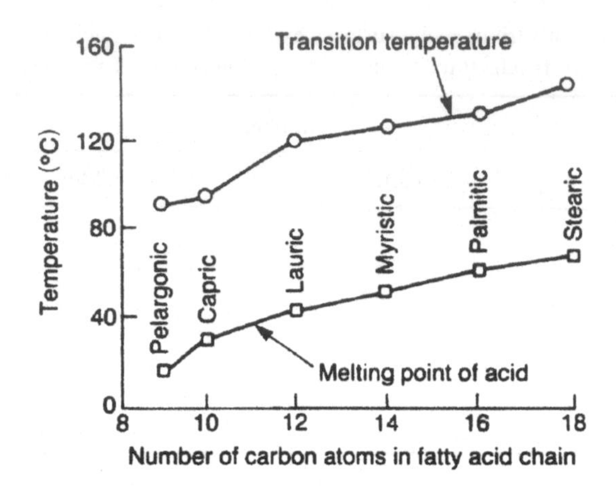

Figure 9.2.6 Breakdown or transition temperature of fatty acids on steel surfaces and their melting points as a function of chain length. Reproduced with permission from Bowden, F.P. and Tabor, D. (1950), *Friction and Lubrication of Solids*, Part I, Clarendon Press, Oxford, UK. Fig. 72 p. 188. Copyright 1950 Oxford University Press.

9.2.4 Effect of Chain Length (or Molecular Weight)

The effect of chain length of the carbon atoms of paraffins (nonpolar), alcohols (nonpolar), and fatty acids (polar) on the coefficient of friction was studied by Bowden and Tabor (1950) and Zisman (1959). Figure 9.2.7 shows the coefficient of friction of a stainless steel surface sliding against a glass surface lubricated with a monolayer of fatty acid. It is seen that there is a steady decrease in friction with increasing chain length. At a sufficiently long chain

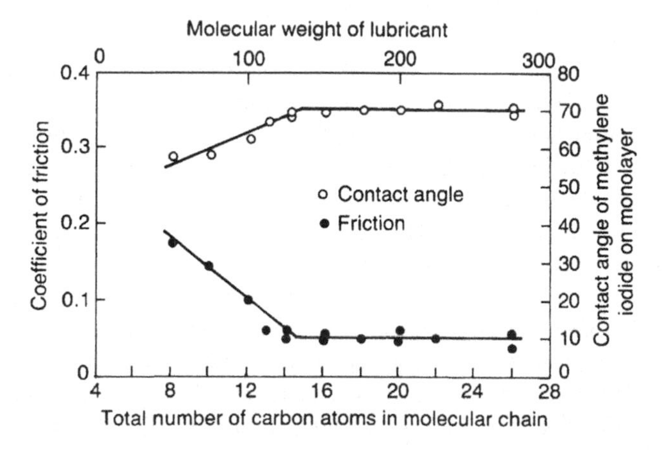

Figure 9.2.7 Effect of chain length (or molecular weight) on coefficient of friction (of stainless steel sliding on glass lubricated with a monolayer of fatty acid) and contact angle (of methyl iodide on condensed monolayers of fatty acids on glass). Reproduced with permission from Zisman, W.A. (1959), "Durability and Wettability Properties of Monomolecular Films on Solids," in *Friction and Wear* (R. Davies, ed), pp. 110–148, Elsevier, Amsterdam. Copyright 1959. Elsevier.

length, the coefficient of friction reaches a lower limit of above 0.07. Similar trends were found for other lubricant films (paraffins and alcohols). Zisman (1959) also found that the contact angle of methylene iodide on a monolayer rises to a high constant value for polar, long-chain molecules of paraffins, alcohols, and fatty acids, indicating an increase in packing to an optimum condition with an increase in the number of carbon atoms in the chain. Owens (1964) has confirmed that in polymer lubrication a complete surface coverage occurs with fatty acids having long chain lengths.

Zisman (1959) and Owens (1964) have shown that the durability of the lubricant film also increases with an increase in the film chain length. These results suggest that monolayers having a chain length below 12 carbon atoms behave as liquids (poor durability), those with chain lengths of 12–15 carbon atoms behave like a plastic solid (medium durability), whereas those with chain lengths above 15 carbon atoms behave like a crystalline solid (high durability).

9.3 Liquid Lubricants

9.3.1 Principal Classes of Lubricants

Liquid lubricants (oils) include natural organics consisting of animal fat, vegetable oils, mineral (or petroleum) fractions, synthetic organics, and mixtures of two or more of these materials. Various additives are used to improve the specific properties (Gunderson and Hart, 1962; Bisson and Anderson, 1964; Braithwaite, 1967; Gunther, 1971; Evans *et al.*, 1972; McConnell, 1972; Booser, 1984; Loomis, 1985). A partial list of the lubricant types is shown in Table 9.3.1.

Common industrial lubricants include natural and synthetic organics. However, these lubricants exhibit low electrical conductivity, making them undesirable in some nanotechnology applications. Ionic liquids (ILs) have been explored as lubricants for various device applications due to their excellent electrical conductivity as well as good thermal conductivity, where the latter allows frictional heating dissipation. Since they do not emit volatile organic compounds, they are regarded as "green" lubricants. For further details, see Palacio and Bhushan (2010) and Bhushan (2013).

9.3.1.1 Natural Oils

Animal fats (naturally occurring esters, long-chain organic acids combined with a tertiary alcohol), shark oil, whale oil, and vegetable oils (such as castor and rape seed oils) were for many centuries the only commonly used oils. They are usually good boundary lubricants, but they are much less oxidatively and thermally stable than mineral oils and tend to break down to give sticky deposits. These can be used up to a maximum temperature of 120°C.

Mineral (petroleum) oils are excellent boundary lubricants and by far the most widely used. These can be used up to a maximum temperature of 130°C and super-refined oils can be used up to 200°C. The chemical compounds making up mineral oils are mainly hydrocarbons, which contain only carbon and hydrogen. The majority in any oil consists of paraffins as shown in Figure 9.3.1a, b, in which the carbon atoms are straight or branched chains. The second most common type consists of naphthenes, in which some of the carbon atoms form rings, as shown in Figure 9.3.1c. Finally, there is usually a small proportion, perhaps 2% of aromatics, in which carbon rings are again present, but the proportion of hydrogen is reduced, as shown

Table 9.3.1 Types of liquid lubricants (oils).

Natural organics	Synthetic organics
Animal fat	Synthetic hydrocarbons (polybutene)
Shark oil	Chlorinated hydrocarbons
Whale oil	Chlorofluorocarbons
Vegetable oils	Esters
Mineral (petroleum) oils	Organic acid
Paraffinic	Fatty acid
Naphthenic	Dibasic acid (di)
Aromatic	Neopentyl polyol
	Polyglycol ethers
	Fluoro
	Phosphate
	Silicate
	Disiloxane
	Silicones
	Dimethyl
	Phenyl methyl
	Chlorophenyl methyl
	Alkyl methyl
	Fluoro
	Silanes
	Polyphenyl ethers
	Perfluoropolyethers

in Figure 9.3.1d. The number of carbon atoms in a ring and the alternate single and double bonds give special properties to aromatic compounds. If the amount of carbon present in a paraffin chain is much higher than the amount in naphthene rings, the oil is called paraffinic oil. If the proportion in the naphthene ring is only a little less than the proportion in paraffin chains, the oil is called naphthenic. Although the amounts of aromatics present are very small, they play an important role in boundary lubrication.

9.3.1.2 Synthetic Oils

The demand for lubricants of improved performance has been created principally by developments in aviation, initially by requirements of higher speeds and performance of gas turbine engines. This demand has led to the development of synthetic lubricants that can withstand environments with extremes of temperatures and pressures, high vacuum, and high humidity. These are less of a fire hazard. The synthetic lubricants can be used up to a maximum temperature of 370°C and up to 430°C for short periods. However, these are more expensive than mineral oils.

Figure 9.3.2 shows the chemical structure of the principal classes of synthetic lubricants. The typical structural formulas are intended merely to illustrate the typical chemical structure employed, not the extent of the variations in structural symmetry, and the usually large number of possible alternative substituent groups.

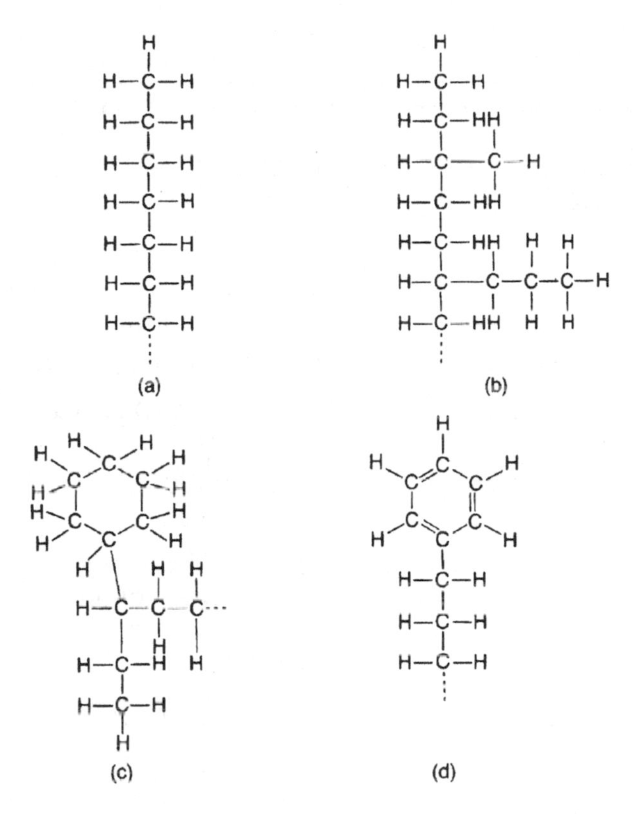

Figure 9.3.1 Main types of mineral (petroleum) oils: (a) straight paraffin; (b) branched paraffin; (c) naphthene; (d) aromatic.

Synthetic Hydrocarbon

The stability of mineral oils depends on the structure of the hydrocarbon chain (Figure 9.3.1). Improvements in stability can be made by the replacement of weakly bonded fractions of hydrocarbon material by branched hydrocarbon chain material and by the use of various inhibitors of oxidative degradation as additives. Synthetic polymeric hydrocarbons tend to produce smaller oxidation products than the original molecules.

Synthetic hydrocarbons prepared by polymerizing specific olefin monomers can be prepared to optimize their viscosity-temperature, low-temperature, and volatility properties. Polybutene (low molecular weight) and alpha olefin oligomers of decene-1 are examples of these lubricants. However, the bond energy of the C-C linkage (85 kcal/mol) remains a fundamental limitation.

Chlorofluorocarbons

Some improvements in stability are made by using a mixed substitution of chlorine and fluorine for the hydrogen in a hydrocarbon to protect the C-C bond as demonstrated in chlorinated and fluorinated compounds. In these compounds, hydrogen in hydrocarbon compounds is replaced completely or in part by chlorine or fluorine. The chlorinated hydrocarbons are more common since chlorination is more easily achieved than fluorination. Recently, much emphasis has

Class	Typical structural formula
Synthetic hydrocarbon (Polybutene)	$(-CH_2-CH_2-CH_2-CH_2-)_n$
Chlorofluorocarbon	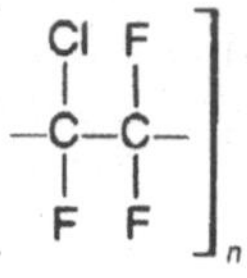 $\left[-\overset{Cl}{\underset{F}{C}}-\overset{F}{\underset{F}{C}}-\right]_n$
Diester	$C_8H_{17}-O-CO-C_8H_{16}-CO-O-C_8H_{17}$
Neopentyl polyol ester	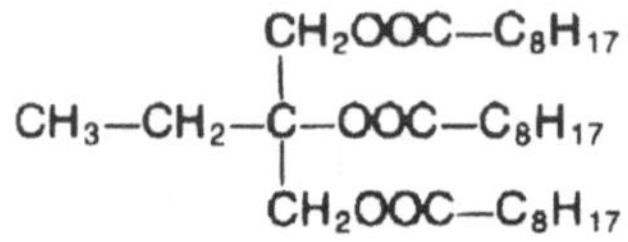 $CH_3-CH_2-\overset{CH_2OOC-C_8H_{17}}{\underset{CH_2OOC-C_8H_{17}}{C}}-OOC-C_8H_{17}$
Fatty acid ester	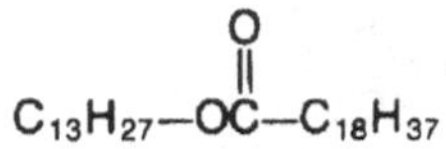 $C_{13}H_{27}-O\overset{O}{\overset{\Vert}{C}}-C_{18}H_{37}$
Polyglycol ether	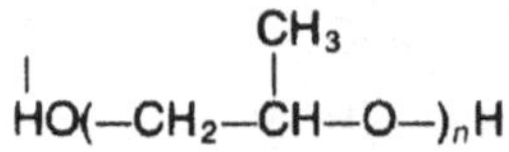 $HO(-CH_2-\overset{CH_3}{CH}-O-)_nH$
Fluoroester	$F(CF_2)_4CH_2OOC(CF_2)_4F$
Phosphate ester	$(CH_3-C_6H_4-O)_3P=O$
Silicate ester	$Si(O-C_8H_{17})_4$
Disiloxane	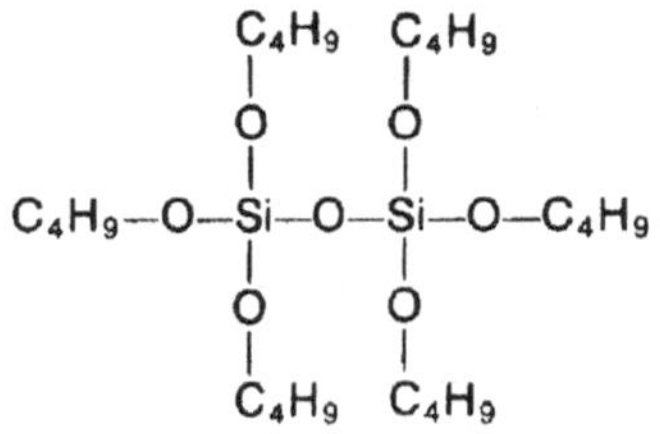 $C_4H_9-O-\overset{O-C_4H_9}{\underset{O-C_4H_9}{Si}}-O-\overset{O-C_4H_9}{\underset{O-C_4H_9}{Si}}-O-C_4H_9$
Silicone	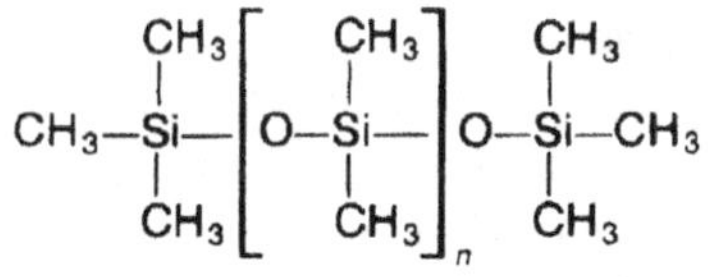 $CH_3-\overset{CH_3}{\underset{CH_3}{Si}}-\left[O-\overset{CH_3}{\underset{CH_3}{Si}}-\right]_n O-\overset{CH_3}{\underset{CH_3}{Si}}-CH_3$
Silane	$(C_{12}H_{25})Si(C_6H_{13})_3$
Polyphenyl ether	 $C_6H_5-O-C_6H_4-O-C_6H_5$
Perfluoroalkyl polyether	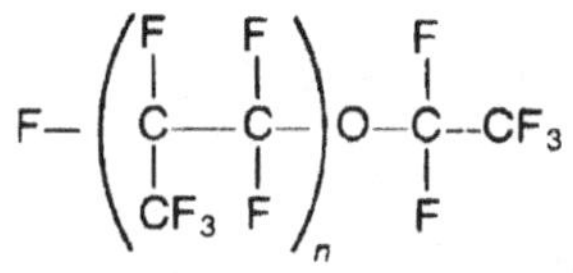 $F-\left(\overset{F}{\underset{CF_3}{C}}-\overset{F}{\underset{F}{C}}-\right)_n O-\overset{F}{\underset{F}{C}}-CF_3$

Figure 9.3.2 Typical chemical structure of principal classes of synthetic lubricants. Reproduced with permission from Braithwaite, E.R. (1967), *Lubrication and Lubricants*, Elsevier, Amsterdam. Copyright 1967. Elsevier.

been placed on fluorinated hydrocarbons and chlorofluorocarbons. These are chemically inert and have outstanding oxidation and thermal stability but they have high volatility, high pour point, and poor viscosity-temperature properties. These are excellent lubricants. Since they are difficult and expensive to make, they find limited applications.

Esters

By far the most important extension of the use of synthetic lubricants has been to employ compounds containing an ester linkage [the product of reacting an alcohol ($R\text{-}CH_2OH$) with an organic acid ($R\text{-}COOH$)]. The ester linkage may be regarded as

$$\begin{array}{c} \quad\quad\quad O \quad\quad\;\; O \\ \quad\quad\quad \| \quad\quad\;\; \| \\ R' - O - C - R - C - O - R', \end{array}$$

in which the organic groups R' come from the alcohol used and R from the acid used. The ester linkages are stable to heat, more so than the $C\text{-}C$ linkage of hydrocarbons in mineral oils because of the higher bond energy of ester linkages. The main advantages of esters as lubricants include excellent viscosity-temperature and volatility properties and a good additive response for oxidation and lubrication behavior. The weakest link in ester-based lubricants is the acid-catalyzed degradation reactions that can proceed by oxidative, thermal, or hydrolytic mechanisms. Phenothiazine and aromatic amines as oxidation inhibitors, copper salts and tricresyl phosphate as surface coaters (metal deactivators), and benzothiazole and epoxy compounds as hydrolytic inhibitors are commonly used for organic acid ester lubricants.

The combination of different acids and alcohols makes possible an enormous variety of esters, such as organic acid esters (fatty acid esters, dibasic acid esters, neopentyl polyol esters, and polyalkylene glycols), fluoro esters, phosphate esters, silicate esters, and disiloxanes.

Fatty acids esters are made from a reaction between saturated fatty acids and monoesters of alcohols. Solid fatty acids whose melting points range from 50° to 80°C are also used as lubricants. The fatty acid esters have moderately low volatility and low oxidation and thermal resistance. They have good lubrication properties with metals and metal oxides, which are reactive to fatty acids. They have not become very popular because they cannot be made into big molecules. They are used as internal lubricants in most magnetic tapes and floppy disks (Bhushan, 1996).

The dibasic esters (diesters) represent one of the most widely researched and commonly used synthetic lubricants for aircraft engines. These esters are a reaction product of dibasic and monobasic acids with primary, secondary, or tertiary alcohols. Castor oil and fats are naturally occurring diesters. The diesters have slightly better volatility characteristics than fatty acid esters because diesters usually have a higher molecular weight. The neopentyl polyol esters are a group of hindered esters formed from monoacids and polyfunctional alcohols. Their oxidative and thermal stability and volatility are somewhat superior to that of dibasic acid and fatty acid esters. Neopentyl glycol, trimethylol propane (TMP), and pentaerythritol are examples of neopentyl polyol containing two, three, and four alcohols, respectively, and these provide progressively lower volatility. The boiling point of pentaerythritol ester is 490°C compared to 384°C for a fatty acid ester. These lubricants are also used as a gas turbine engine lubricant and as a hydraulic fluid in supersonic transport. Polyglycol ethers or polyalkylene glycols and derivatives are either derived from the reaction of an alcohol and a propylene oxide that results in a water-insoluble series or the reaction of an alcohol and a mixture of propylene and ethylene oxide that produces a water-soluble series. Both types are used as lubricants and have fair lubricating properties. They have very good thermal properties and are used as hydraulic fluids. Fluoro esters are derived from organic carboxylic acids and fluoroalcohols. They have good oxidative characteristics and low flammability but have poor viscosity temperature characteristics. They are used as lubricants and hydraulic fluids.

Another important class of synthetics is the phosphate esters derived from orthophosphoric acid and various alcohols. The phosphates are generally described as organic-inorganic esters since the acid involved is an inorganic acid. Aromatic-type phosphate esters have a better thermal stability than diesters, but have high surface tension, have excellent fire-resistant properties, and are used as special hydraulic fluids. Phosphate esters, such as tricresyl phosphate (TCP), have excellent lubricity and have been widely used as antiwear additives for petroleum and some other synthetic lubricants. Other organic-inorganic esters are silicate esters (or orthosilicate esters), which are the reaction product of silicic acid and an alkyl or aryl alcohol. The presence of the silicon-oxygen-carbon bonds distinguishes the silicate esters from silanes (which have direct silicon-carbon bonds) and the silicones (which also have silicon-carbon bonds). Another group of compounds closely related to silicate esters is known as the disiloxanes. The desirable feature of silicate esters is high thermal stability, low viscosity, relatively low volatility, and fair lubricity, but poor hydrolytic stability. Their principal use is for low-temperature ordinance lubrication and corrosion prevention.

Silicones

Silicones (polysiloxanes) are a diversified class of synthetics. Pure members of the silicone lubricant family exist, but the most common ones used as lubricants are probably the dimethyl silicones and the phenyl methyl silicones. All silicone polymers have extreme chemical inertness, thermal stability, low volatility, and low surface tension. Because of the foregoing characteristics, the silicones are ideal hydrodynamic lubricants, but are poor boundary lubricants because of their chemical characteristics. Another class of silicone fluids, in which hydrocarbon groups containing substituent fluorine atoms replace a portion of the methyl groups of dimethyl silicone fluid, displays much improved lubricating properties. Their largest use is in various types of grease formulations for space applications. The silicone fluids are more expensive than the mineral oils and most other synthetic lubricants. Therefore, their application is largely confined to extreme operating temperatures where most other lubricants are unsuitable.

Silanes

A silane differs from a silicone in that it is not a polymer and lacks the familiar silicon-oxygen linkage that provides the central backbone of the silicone polymer fluid. Silanes have a high temperature stability comparable to polyphenyl ether.

Polyphenyl Ethers

The polyphenyl ethers (PPEs) consist of three or more benzene rings linked together in a linear chain through oxygen atoms. Being aromatic, they are very resistant to oxidation to 290°C and thermal degradation to 430°C; they have very low volatility (lower than silicones), but poor viscosity properties. They are fair boundary lubricants, better than silicone oils. The commercial product 5P4E is the most widely used fluid in this class of materials. They are used in aircraft hydraulic pumps and with the addition of tricresyl phosphate can be used up to 480°C.

Perfluoropolyethers

The perfluoropolyether (PFPE) family consists of perfluoroalkyl polyethers, perfluoroisopropyl polyethers, perfluoroethylene oxide, etc. PFPEs have been one of the most promising classes of materials for high-temperature applications. The perfluoroalkyl polyethers can be prepared in a number of different molecular weights with different viscosity ranges. They have greatest oxidative stability to 320°C, have thermal stability to 370°C, have very low surface tension, and they are chemically inert. Polar groups can be easily attached to them. Compared to PPEs, although PPEs have greatest thermal stability, PFPEs have greatest oxidative stability (Table 9.3.2), have slightly lower volatility because they can be made in bigger molecules, and have better boundary lubrication characteristics (Jones and Snyder, 1980; Bhushan, 1996). Some PFPEs (e.g., Fomblin-Z type) have much lower volatility than others (e.g., Fomblin-Y type). The major applications of PFPEs are as oils in greases for high-temperature, high vacuum, and chemical-resistant applications, hydraulic fluids, and gas turbine engine oils, and they are used in magnetic rigid disks (Bhushan, 1996).

9.3.2 Physical and Chemical Properties of Lubricants

Liquid lubricants provide a substantial range of physical and chemical properties. The physical properties are attributable primarily to the structure of the lubricant base stock. Chemical properties of the finished or formulated lubricants result primarily from the additives used with the base stock. Selected properties of interest are: viscosity, surface tension, thermal properties, volatility, oxidative stability, thermal stability, hydrolytic stability, gas solubility, and inflammability (Bhushan, 2013).

Typical properties of some of the classes of synthetic lubricants and petroleum lubricants are given in Table 9.3.2. The characteristics and typical applications of several individual classes of synthetic lubricants are summarized in Table 9.3.3 (based on Bisson and Anderson, 1964). The broad descriptions used to characterize the functional properties of these materials are generalizations and should be used with caution. Properties that are indicated as deficient may or may not be improved by using additives.

A liquid with low surface tension and a low contact angle would spread easily on the solid surface and provide good lubrication. The surface tension of several base oils is shown in Table 9.3.4. The surface tension for the finished lubricants is sensitive to the additives. For example, less than 0.1% of a methyl silicone in mineral oil will reduce the surface tension to essentially that of the silicone.

9.3.3 Additives

In boundary lubrication, one of the most important properties of a lubricant is its chemical function or polarity, which governs the ability of the lubricant molecules to be physisorbed, chemisorbed, or chemically reacted with the surfaces. Modified surfaces minimize the damage that can occur in intermittent asperity contacts. Additives between 0.1 and 0.5% are added to boundary lubricants (used in boundary lubrication conditions) to produce the protective films. Lubricants are classified as follows, based on the behavior of the additives: nonreactive, low-friction (lubricity), anti-wear, and extreme pressure (EP).

Nonreactive lubricants include nonadditive mineral oils and esters. Low friction or lubricity is defined as the ability of a lubricant to reduce friction below that of the base oil. The additives are adsorbed on, or react with, the metal surface or its oxide to form monolayers of low shear

Table 9.3.2 Typical properties of commonly used classes of synthetic lubricants (oils).

Lubricants	Thermal stability (°C)	Kinematic viscosity (cSt) at °C					Specific gravity at 20°C	Thermal conductivity (cal/h m °C)	Specific heat at 38°C (cal/g °C)	Flash point (°C)	Pour point (°C)	Oxidative stability (°C)	Vapor pressure at 20°C (torr)
		−20	0	40	100	200							
Mineral oils	135	170	75	19	5.5		0.86	115	0.39	105	−57		10^{-6} to 10^{-2}
Diesters	210	193	75	13	3.3	1.1	0.90	132	0.46	230	−60		10^{-6}
Neopentyl polyol esters	230	16	16	15	4.5		0.96			250	−62		10^{-7}
Phosphate esters	240	85	38	11	4		1.09	109	0.42	180	−57		10^{-7}
Silicate esters	250	115	47	12	4	1.3	0.89			185	−65		10^{-7}
Disiloxanes	230	200	100	33	11	3.8	0.93			200	−70		
Silicones													5×10^{-8}
Phenyl methyl	280	850	250	74	25	22	1.03	124	0.34	260	−70	240	
Fluoro	260	20,000		190	30	24	1.20			290	−50	220	
Polyphenyl ethers													
4P-3E	430		2500	70	6.3	1.4	1.18	133	0.43	240	−7	290	10^{-8}
5P-4E	430			363	13.1	2.1				290	+4	290	
Perfluoropolyethers													
Fomblin YR	370		8000	515	35		1.92	82	0.24	none	−30	320	10^{-9}
Fomblin Z-25	370	1000	440	150	41		1.87		0.20	none	−67	320	3×10^{-12}

Table 9.3.3 General properties of some classes of synthetic lubricants (oils)[a].

Lubricant	Viscosity temperature characteristics	Volatility	Oxidation resistance	Thermal stability	Resistance to hydrolysis	Flammability characteristics	Lubrication characteristics	Solvent effect on paints, rubber, etc.	Solubility in petroleum and other synthetics	Improvement additive compatibility	Price range per liter ($)	Typical applications
Chloro fluorocarbons	Poor	Some lower, some higher	Excellent	Poor to excellent	Poor to excellent	Excellent	Good to excellent	Widely variable	Generally poor	Widely variable	10–100	Nonflammable, extreme oxidation-resistant lubricants for plant processes or devices handling reactive materials
Dibastic-acid ethers	Good to excellent	Generally lower	Fair to good	Fair to good	Fair to good	Poor to fair	Fair to good	Pronounced effect	Good to excellent	Generally good	2–5	Instrument oils, low-volatility grease bases, special hydraulic fluids, gas turbine lubricants
Nonpentyl polyol ethers	Good to excellent	Lower	Good	Good	Fair to good	Poor to fair	Fair to good	Pronounced effect	Good to excellent	Generally good	2–5	Instrument oils, low-volatility grease bases, special hydraulic fluids, gas turbine lubricants
Polyglycol ethers	Good to excellent	Generally lower	Poor to fair	Fair to good	Fair to good	Fair to good	Good	Pronounced effect on paint, small effect on rubber	Fair to good	Generally good	1–3	Special hydraulic fluids, forming and drawing lubricants, low-temperature grease base, vacuum-pump lubricants, components of other synthetic lubricant formulations
Phosphate esters	Excellent	Generally lower	Good	Fair to good	Poor to good	Good to excellent	Good to excellent	Pronounced effect	Good to excellent	Generally good	2–6	Fire-resistant hydraulic fluids, low-volatility, high-lubricity grease base, lubrication additives in other synthetics, special low-temperature lubricants
Silicate esters	Excellent	Lower	Poor to fair	Good	Poor to fair	Poor to fair	Fair to good	Some effect	Fair to good	Fair to good	4–10	Heat-transfer fluids, high-temperature hydraulic fluids, low-volatility, low-viscosity grease bases, components for low-viscosity hydraulic fluids
Silicones	Excellent	Much lower	Good to excellent	Good to excellent	Excellent	Poor to good	Poor to fair	Generally small effect	Poor	Poor	20–40	High-temperature bearings, condensation pump lubricant, low-volatility grease base for lightly loaded bearings, damping fluids, devices requiring minimum viscosity change with temperature
Silanes	Fair to good	Much lower	Fair to good	Good to excellent	Good	Poor to fair	Fair to good	Generally small effect	Fair	Fair		Base stocks for high-temperature greases, hydraulic fluids, and engine lubricants; require extensive formulation
Polyphenyl ethers	Poor to fair, generally high pour points	Much lower	Excellent	Excellent	Excellent	Poor to fair	Fair to good	Generally small effect	Fair		30–60	High-temperature fluid for reactor coolant, hydraulic applications at very high temperatures
Perfluoro poly-ethers	Good to excellent	Lowest	Excellent	Excellent	Excellent	Poor to fair	Good	Generally small effect	Fair		30–60	Base stocks for greases for high-temperature, vacuum and chemical resistance applications; hydraulic fluids for very high temperatures, magnetic rigid disks

[a]Properties are for typical class members, comparing synthetics with well-refined petroleum products in equivalent service. For comparative purposes, the petroleum product would rate as fair to good. Exceptions are flammability and volatility where petroleum products would rate poor and lubrication characteristics and hydrolysis where petroleum products would be rated excellent.

Table 9.3.4 Surface tension of several base oils.

Liquid	Surface tension (dynes/cm) (=mN/m)
Water	72
Mineral oils	30–35
Esters	30–35
Methyl silicone	20–22
Perfluoropolyethers	19–21

strength material. A friction polymer (condensation polymerized oxidation product) of organic material appears to be the chemically formed, easily sheared film on the bearing surface. Most common additives are long chain (greater than 12 carbon atoms), alcohol, amines, and fatty acids. As an example, oleic acid reacts with iron oxide to form a film of the iron oleate soap which exhibits low friction.

Anti-wear additives function by reacting with the bearing surface to form a relatively thick, tenacious coating on the bearing surface that is not easily removed by shear or cavitational forces which control wear. They form organic, metallo-organic or metal salt films on the surface. Anti-wear additives, zinc dialkyl dithiophosphate (ZDDP), organic phosphates such as tricresyl phosphate (TCP), and ethyl stearate are the typical examples for esters and other synthetic lubricants. The most common additive ZDDP decomposes to deposit metallo-organic species, zinc sulfide or zinc phosphate or reacts with the steel surface to form iron sulfide or iron phosphate. In heavily loaded bearing and gear contacts (EHL), associated loads and speeds are high, resulting in scuffing (damage caused by solid-phase welding between sliding surfaces). In such cases, more powerful antiwear, extreme pressure additives are needed. EP lubricants are chemically corrosive additives that have a strong affinity for the bearing surfaces and form thick films of high melting point metal salts on the surface which prevent metal to metal contact. With frictional heating, they are readily removed, but another additive molecule is rapidly readsorbed. Common EP additives are: organo sulfur compounds such as sulfurized olefins and ZDDP and phosphorous compounds such as TCP. These additives are more chemically active than anti-wear additives for esters and other synthetic lubricants. No anti-wear additives are necessary with the phosphate esters. Silicate esters and silicones, being more inert, are relatively poor lubricants. Anti-wear additives with these fluids are of lesser use. The oxygen dissolved in oil forms metal oxide films, such as Fe_3O_4 on steel, which exhibit anti-wear and limited EP properties.

The introduction of these additives into the fluid reduces the thermal stability. Thus, the controlled use of expendable additives has become the acceptable practice to provide surface activity for boundary lubrication as well as the additives used to improve or modify the bulk properties of fluids. The definitions of different additives apply to typical metal-bearing surfaces; similar behavior could be expected for oxide or ceramic materials but does not apply to plastic lubrication.

9.4 Greases

Greases are used where circulating liquid lubricant cannot be contained because of space and cost and where cooling by the oil is not required or the application of a liquid lubricant is not feasible. A grease is a semisolid lubricant produced by the dispersion of a thickening agent in a liquid lubricant that may contain other ingredients that impart special properties (Braithwaite, 1967; Boner, 1976; Booser, 1984). The majority of greases are composed of petroleum and synthetic oils thickened with metal soaps and other agents such as clay, silica, carbon black, and polytetrafluoroethylene (PTFE). Most lubricant greases used in industry have petroleum oils as their liquid base. However, almost every lubricating liquid can serve with a suitable thickener or gelling agent to make a grease. All synthetic lubricants are eligible, but in practice, the cost of such materials restricts their use to applications having special requirements.

The petroleum oil-based greases can be used at temperatures up to 175°C depending on the thickener, e.g., barium metal soap gives the maximum usable temperature of 175°C. The synthetic oil-based greases can be used at higher temperatures. The best materials in these greases are silicones thickened with ammeline and silica, and they can be used at temperatures up to 300°C. However, there is some evidence that diester-based greases are superior in performance to silicon-based greases.

9.5 Closure

Boundary lubrication is accomplished by mono- or multimolecular films. The films are so thin that their behavior is controlled by interaction with the substrates. Molecular structure of the lubricants and functionality of the substrate affect the type and degree of bonding of the lubricant to the substrate.

Liquid lubricants include mineral (or petroleum) and synthetic organics. Various additives are used to improve specific properties. Mineral oils are excellent boundary lubricants and by far the most used lubricants. Synthetic lubricants can be used at greater extremes of environment including temperature, humidity and vapor pressure. Mineral oils are typically used up to a maximum temperature of about 130°C and some synthetic oils up to about 370°C. However, synthetic lubricants are more expensive than mineral oils. There are several properties of lubricants which are important for lubrication; their relative importance depends upon the industrial application. Additives are commonly used to modify friction and wear of lubricants and greases. These are classified as friction modifier, anti-wear and extreme pressure.

Greases are used where circulating liquid lubricant can not be contained because of space and cost and where cooling by the oil is not required or the application of a liquid lubricant is not feasible.

References

Anonymous (1997), "Limits of Lubrication," Special issue of *Tribology Letters* **3**, No. 1.

Beerbower, A. (1972), *Boundary Lubrication*, Report No. AD-747336, US Dept of Commerce, Office of the Chief of Research and Development, Department of the Army, Washington, DC.

Bhushan, B. (1996), *Tribology and Mechanics of Magnetic Storage Devices*, Second edition, Springer-Verlag, New York.

Bhushan, B. (2013), *Principles and Applications of Tribology*, Second edition, Wiley, New York.

Bhushan, B. and Zhao, Z. (1999), "Macro- and Microscale Tribological Studies of Molecularly-Thick Boundary Layers of Perfluoropolyether Lubricants for Magnetic Thin-Film Rigid Disks," *J. Info. Storage Proc. Syst.* **1**, 1–21.

Bisson, E.E. and Anderson, W.J. (1964), *Advanced Bearing Technology*, SP-38, NASA Washington, DC.
Boner, C.J. (1976), *Modern Lubricating Greases*, Scientific Publications, Broseley, UK.
Booser, E.R. (1984), *CRC Handbook of Lubrication*, Vol. **2** Theory and Design, CRC Press, Boca Raton, Florida.
Bowden, F.P. (1951), "The Influence of Surface Films on the Friction, Adhesion and Surface Damage of Solid," *The Fundamental Aspects of Lubrication, Annals of NY Academy of Sciences*, **53**, Art 4, June 27, 753–994.
Bowden, F.P. and Tabor, D. (1950), *Friction and Lubrication of Solids*, Part I, Clarendon Press, Oxford, UK.
Braithwaite, E.R. (1967), *Lubrication and Lubricants*, Elsevier, Amsterdam.
Evans, G.G., Galvin, V.M., Robertson, W.S., and Walker, W.F. (1972), *Lubrication in Practice*, Macmillan, Basingstoke, UK.
Godfrey, D. (1968), "Boundary Lubrication," in *Interdisciplinary Approach to Friction and Wear* (P.M. Ku, ed.), SP-181, pp. 335–384, NASA, Washington, DC.
Gunderson, R.C. and Hart, A.W. (1962), *Synthetic Lubricants*, Reinhold, New York.
Gunther, R.C. (1971), *Lubrication*, Bailey Brothers and Swinfen Ltd, Folkestone, UK.
Iliuc, I. (1980), *Tribology of Thin Films*, Elsevier, New York.
Israelachvili, J.N. (1992), *Intermolecular and Surface Forces*, Second edition, Academic Press, San Diego, California.
Israelachvili, J.N., McGuiggan, P.M., and Homola, A.M. (1988), "Dynamic Properties of Molecularly Thin Liquid Films," *Science* **240**, 189–191.
Jones, W.R. and Snyder, C.E. (1980), "Boundary Lubrication, Thermal and Oxidative Stability of a Flourinated Polyether and a Perfluoropolyether Triazine," *ASLE Trans*. **23**, 253–261.
Ku, P.M. (1970), *Interdisciplinary Approach to the Lubrication of Concentrated Contacts*, SP-237, NASA, Washington, DC.
Ling, F.F., Klaus, E.E., and Fein, R.S. (1969), *Boundary Lubrication – An Appraisal of World Literature*, ASME, New York.
Loomis, W.R. (1985), *New Directions in Lubrication, Materials, Wear, and Surface Interactions – Tribology in the 80's*, Noyes Publications, Park Ridge, New Jersey.
McConnell, B.D. (1972), *Assessment of Lubricant Technology*, ASME, New York.
Owens, D.K. (1964), "Friction of Polymers I. Lubrication," *J. Appl. Poly. Sci.* **8**, 1465–1475.
Palacio, M. and Bhushan, B. (2010), "A Review of Ionic Liquids for Green Molecular Lubrication in Nanotechnology," *Tribol. Lett*. **40**, 247–268.
Zisman, W.A. (1959), "Durability and Wettability Properties of Monomolecular Films on Solids," in *Friction and Wear* (R. Davies, ed), pp. 110–148, Elsevier, Amsterdam.

Further Reading

Anonymous (1997), "Limits of Lubrication," Special issue of *Tribology Letters* **3**, No. 1.
Beerbower, A. (1972), *Boundary Lubrication*, Report No. AD-747336, US Dept of Commerce, Office of the Chief of Research and Development, Department of the Army, Washington, DC.
Bhushan, B. (2013), *Principles and Applications of Tribology*, Second edition, Wiley, New York.
Bhushan, B. and Zhao, Z. (1999), "Macro- and Microscale Tribological Studies of Molecularly-Thick Boundary Layers of Perfluoropolyether Lubricants for Magnetic Thin-Film Rigid Disks," *J. Info. Storage Proc. Syst.* **1**, 1–21.
Boner, C.J. (1976), *Modern Lubricating Greases*, Scientific Publications, Broseley, Shropshire, UK.
Booser, E.R. (1984), *CRC Handbook of Lubrication*, Vol. **2** Theory and Design, CRC Press, Boca Raton, Florida.
Braithwaite, E.R. (1967), *Lubrication and Lubricants*, Elsevier, Amsterdam.
Evans, G.G., Galvin, V.M., Robertson, W.S., and Walker, W.F. (1972), *Lubrication in Practice*, Macmillan, Basingstoke, UK.
Godfrey, D. (1968), "Boundary Lubrication," In *Interdisciplinary Approach to Friction and Wear* (P.M. Ku, ed.), SP-181, pp. 335–384, NASA, Washington, DC.
Gohar, R. and Rahnejat, H. (2008), *Fundamentals of Tribology*, Imperial College Press, London, UK.
Gunderson, R.C. and Hart, A.W. (1962), *Synthetic Lubricants*, Reinhold, New York.
Gunther, R.C. (1971), *Lubrication*, Bailey Brothers and Swinfen Ltd., Folkestone, UK.
Iliuc, I. (1980), *Tribology of Thin Films*, Elsevier, New York.
Khonsari, M.M. and Booser, E.R (2001), *Applied Tribology – Bearing Design and Lubrication*, Wiley, New York.

Ku, P.M. (1970), *Interdisciplinary Approach to the Lubrication of Concentrated Contacts*, SP-237, NASA, Washington, DC.

Ling, F.F., Klaus, E.E., and Fein, R.S. (1969), *Boundary Lubrication – An Appraisal of World Literature*, ASME, New York.

Loomis, W.R. (1985), *New Directions in Lubrication, Materials, Wear, and Surface Interactions – Tribology in the 80's*, Noyes Publications, Park Ridge, New Jersey.

Palacio, M. and Bhushan, B. (2010), "A Review of Ionic Liquids for Green Molecular Lubrication in Nanotechnology," *Tribol. Lett.* **40**, 247–268.

Stachowiak, G. and Batchelor, A. (2005), *Engineering Tribology*, Third edition, Elsevier Butterworth-Heinemann, Burlington, Maine.

10

Nanotribology

10.1 Introduction

The mechanisms and dynamics of the interactions of two contacting solids during relative motion, ranging from atomic-scale to microscale, need to be understood in order to develop a fundamental understanding of adhesion, friction, wear, indentation, and lubrication processes. For most solid–solid interfaces of technological relevance, contact occurs at multiple asperities. Consequently the importance of investigating single asperity contacts in studies of the fundamental micro/nanomechanical and micro/nanotribological properties of surfaces and interfaces has long been recognized. The recent emergence and proliferation of proximal probes, in particular, scanning probe microscopies (the scanning tunneling microscope and the atomic force microscope), the surface force apparatus, and of computational techniques for simulating tip-surface interactions and interfacial properties, have allowed systematic investigations of interfacial problems with high resolution as well as ways and means for modifying and manipulating nanoscale structures. These advances have led to the appearance of the new field of nanotribology, which pertains to experimental and theoretical investigations of interfacial processes on scales ranging from the atomic-scale and molecular-scale to the microscale, occurring during adhesion, friction, scratching, wear, indentation, and thin-film lubrication at sliding surfaces (Singer and Pollock, 1992; Bhushan *et al.*, 1995a; Guntherodt *et al.*, 1995; Persson and Tosatti, 1996; Bhushan, 1997, 1999a, b, 2001a, b, c, 2005, 2008, 2011, 2012). Proximal probes have also been used for mechanical and electrical characterization, *in situ* characterization of local deformation, and other nanomechanics studies (Bhushan, 1999c, 2001c, 2011, 2012).

Nanotribological and nanomechanics studies are needed to develop a fundamental understanding of interfacial phenomena on a small scale and to study interfacial phenomena in nanostructures used in magnetic storage devices, micro/electromechanical systems (MEMS/NEMS), and other applications (Bhushan *et al.*, 1995a; Bhushan, 1996, 1997, 1998, 1999a, b, 2001a, b, c, 2003, 2005, 2008, 2011, 2012). Friction and wear of lightly loaded micro/nanocomponents are highly dependent on the surface interactions (few atomic layers). These structures are generally coated with molecularly thin films. Nanotribological and nanomechanics studies are also valuable in the fundamental understanding of interfacial phenomena in macrostructures, and provide a bridge between science and engineering.

Introduction to Tribology, Second Edition. Bharat Bhushan.
© 2013 John Wiley & Sons, Ltd. Published 2013 by John Wiley & Sons, Ltd.

Table 10.1.1 Comparison of typical operating parameters in SFA, STM, and AFM/FFM used for micro/nanotribological studies.

Operating parameter	SFA	STM[a]	AFM/FFM
Radius of mating surface/tip	~10 mm[b]	5–100 nm	5–100 nm
Radius of contact area	10–40 μm	N/A	0.05–0.5 nm
Normal load	10–100 mN	N/A	<0.1 nN-500 nN
Sliding velocity	0.001–100 μm/s	0.02–200 μm/s (scan size ~1 nm × 1 nm to 125 μm × 125 μm; scan rate < 1–122 Hz)	0.02–200 μm/s (scan size ~1 nm × 1 nm to 125 μm × 125 μm; scan rate < 1–122 Hz)
Sample limitations	Typically atomically smooth, optically transparent mica; opaque ceramic, smooth surfaces can also be used	Electrically conducting samples	None of the above

[a]Can be used for atomic-scale imaging
[b]Since stresses scale inverse of tip radius, SFA can provide very low stress measurement capabilities

The surface force apparatus (SFA), the scanning tunneling microscopes (STM), and atomic force and friction force microscopes (AFM and FFM) are widely used in nanotribological and nanomechanics studies. Typical operating parameters are compared in Table 10.1.1. The SFA was developed in 1968 and is commonly employed to study both static and dynamic properties of molecularly thin films sandwiched between two molecularly smooth surfaces. The STM, developed in 1981, allows the imaging of electrically conducting surfaces with atomic resolution, and has been used for the imaging of clean surfaces as well as of lubricant molecules. The introduction of the AFM in 1985 provided a method for measuring ultra-small forces between a probe tip and an engineering (electrically conducting or insulating) surface, and has been used for morphological and surface roughness measurements of surfaces on the nanoscale, as well as for adhesion measurements. Subsequent modifications of the AFM led to the development of the FFM, designed for atomic-scale and microscale studies of friction. This instrument measures forces in the scanning direction. The AFM is also used for various investigations including scratching, wear, indentation, detection of transfer of material, boundary lubrication, and fabrication and machining (Bhushan *et al.*, 1995a; Bhushan, 1999a, 2011, 2012). Meanwhile, significant progress in understanding the fundamental nature of bonding and interactions in materials, combined with advances in computer-based modeling and simulation methods, has allowed theoretical studies of complex interfacial phenomena with high resolution in space and time. Such simulations provide insights into atomic-scale energetics, structure, dynamics, thermodynamics, transport, and rheological aspects of tribological processes.

The nature of interactions between two surfaces brought close together, and those between two surfaces in contact as they are separated, have been studied experimentally with the surface force apparatus. This has led to a basic understanding of the normal forces between surfaces

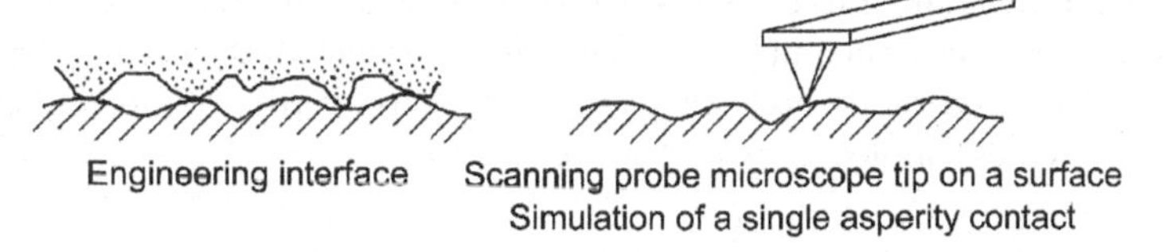

Figure 10.1.1 Schematics of an engineering interface and scanning probe microscope tip in contact with an engineering interface.

and the way in which these are modified by the presence of a thin liquid or a polymer film. The frictional properties of such systems have been studied by moving the surfaces laterally, and such experiments have provided insights into the molecular-scale operation of lubricants such as thin liquid or polymer films. Complementary to these studies are those in which the AFM tip is used to simulate a single asperity contact with a solid or lubricated surface, Figure 10.1.1. These experiments have demonstrated that the relationship between friction and surface roughness is not always simple or obvious. AFM studies have also revealed much about the nanoscale nature of intimate contact during wear, indentation, and lubrication.

In this chapter, we present a review of significant experimental and theoretical aspects of nanotribology.

10.2 SFA Studies

The SFA was originally developed to measure static normal (van der Waals) forces between two molecularly smooth mica surfaces (Tabor and Winterton, 1969; Israelachvili and Tabor, 1972) and normal forces between two surfaces immersed in a liquid (Israelachvili, 1989) as a function of separation in air or vacuum. Later these were developed to study dynamic shear (sliding) response of the molecularly thin liquid films sandwiched between two molecularly smooth macroscopic surfaces (Chan and Horn, 1985; Israelachvili *et al.*, 1988; Van Alsten and Granick, 1988; Israelachvili, 1989; Granick, 1991; Klein *et al.*, 1991; Georges *et al.*, 1994; Bhushan *et al.*, 1995a; Luengo *et al.*, 1997; Bhushan, 1999a, 2001a, b, 2011). Sliding experiments at constant velocities and varying sliding velocities or oscillating frequencies have been performed. In the latter experiments, viscous dissipation and elasticity of confined liquids are measured by using periodic sinusoidal oscillations over a range of amplitudes and frequencies. Very weak forces are determined from the deflection of a spring and ultra-small surface separation levels are measured using some optical, electrical, capacitive or strain gage technique. Submicroscope surface geometries and surface separations at the 0.1 nm level are mostly measured using optical interference techniques; therefore, optically transparent surfaces (typically cleaved mica sheets with cross-cylinder geometry) are required. Because the mica surfaces are molecularly smooth, the real area of contact is well defined and measurable, and asperity deformation does not complicate the analysis. During sliding experiments, the area of parallel surfaces is very large compared to the thickness of the sheared film and this provides an ideal condition for studying shear behavior, because it permits one to study molecularly thin liquid films whose thickness is well defined to the resolution of 0.1 nm.

Molecularly thin liquid films cease to behave as a structural continuum with properties different from that of the bulk material. Using SFA, the structure and dynamics of fluid molecules in confined geometries can be understood which are of interest, for example, during flow through narrow pores or in thin lubricating films between two shearing surfaces. The SFA is commonly employed to study both static and dynamic properties of molecularly thin films sandwiched between two molecularly smooth surfaces.

10.2.1 Description of an SFA

A schematic of an SFA with a sliding attachment is shown in Figure 10.2.1a (Luengo *et al.*, 1997). (For a historical overview and description of various SFA designs, see Bhushan, 1999a.) The apparatus consists of a small airtight stainless steel chamber in which two molecularly smooth curved mica surfaces are brought into contact. The two mica surfaces can be translated toward or away from each other and separation and normal forces are measured. To study the shear response of confined liquid films, both normal and friction forces are measured during simultaneous lateral sliding and normal motion of the surfaces. In addition, the surface separation is measured and local surface geometry can be visualized at all times during dynamic interactions.

The technique utilizes two molecularly smooth mica sheets, each about 2 μm thick, coated with a semireflecting 50–60 nm layer of pure silver glued to rigid cylindrical silica disks of radius 10 mm (silver side down) mounted facing each other with their axes mutually at right angles (crossed cylinder position), which is geometrically equivalent to a sphere contacting a flat surface. The radius of contact area typically ranges from 10 to 40 μm at normal loads from 10 to 100 mN. Under the action of adhesive forces or applied load, mica flattens to produce a contact zone in which surfaces are locally parallel and planar. They flatten elastically so that the contact zone is circular for the duration of the static or sliding interactions. The surface separation is measured by using optical interference fringes of equal chromatic order (FECO). From the positions and shapes of the colored FECO fringes the area of molecular contact and the surface separation (including the quantity of material deposited or adsorbed on the surfaces) can be measured within 0.1 nm.

The lower surface is supported at the end of a double-cantilever spring (S1, Figure 10.2.1a) used to measure the normal forces between the surfaces. Lateral movement of the lower surface is accomplished with two or four parallel piezoelectric bimorph strips (S2, Figure 10.2.1a.). Lateral motion of the lower surface (A, Figure 10.2.1a) is produced by applying a DC or AC voltage to the bimorphs for constant velocity (typically ranging from 0.001 to 100 μm/s) and oscillating shear (on the order of 1 Hz) experiments, Figure 10.2.1b. The AC voltage difference applied by a signal generator (driver) across one of the bimorphs bends it in an oscillating fashion at frequencies ranging from 10^{-6} Hz to 200 kHz, while the friction force resists that motion. A constant velocity or a constant shear rate can be obtained by applying a triangular voltage signal. A travel distance up to 1 mm p-p can be obtained by increasing the active bimorph length L using clamps C1 and C2, Figure 10.2.1a. Vertical motion of the whole slider assembly and lower surface is produced by a three-stage mechanical translation mechanism composed of micromotors and springs located in the upper chamber.

The steel plate supporting the upper silica disk is mounted to two vertical double-cantilever springs used to measure the friction forces between the surfaces during sliding. One of the vertical springs acts as a friction force detector by having four semiconductor strain gages attached to it, forming the four arms of a Wheatstone bridge, and signal output is fed to a chart recorder or a computer. The forces are measured with a sensitivity of 10 nN. If the upper

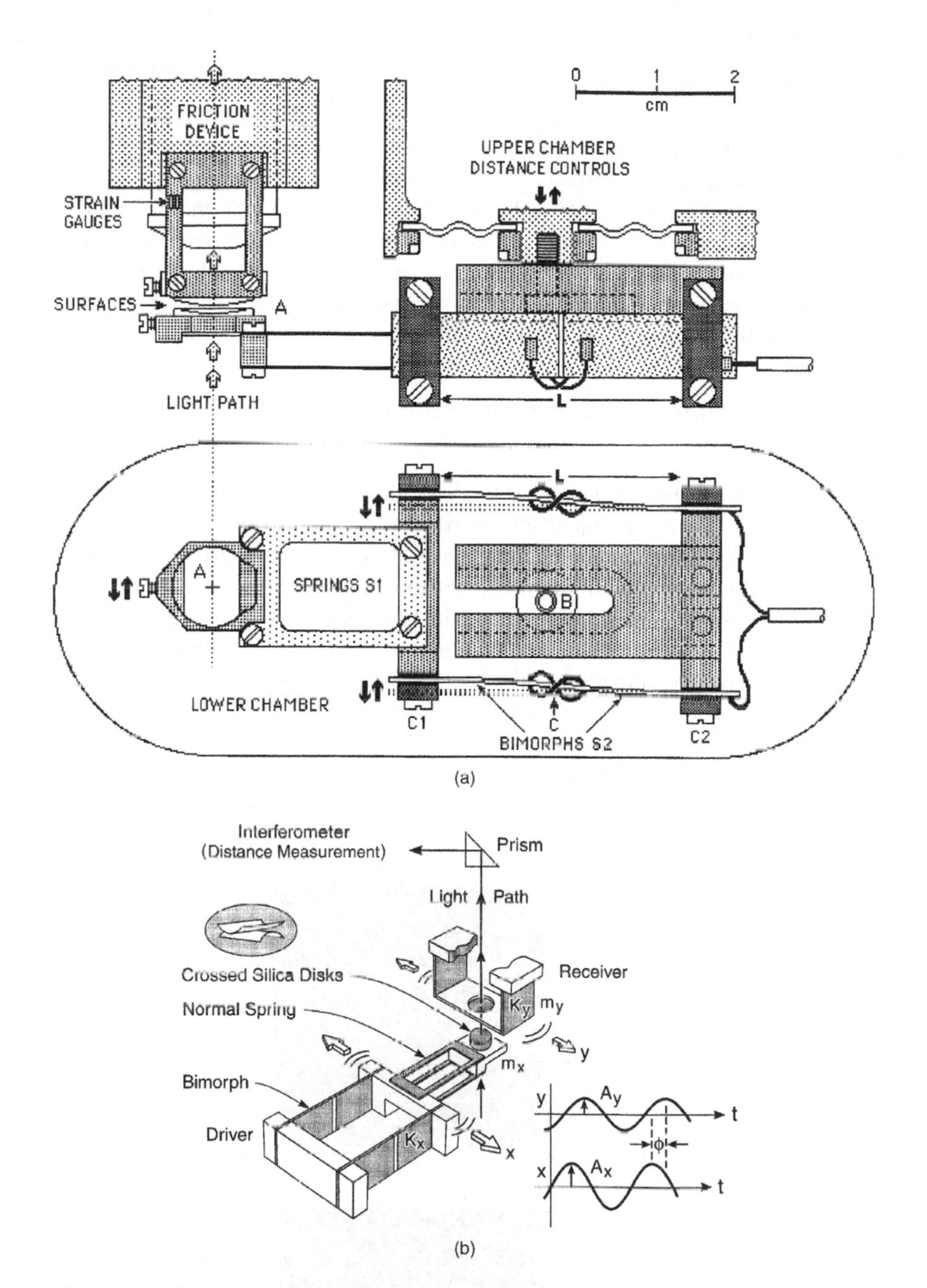

Figure 10.2.1 (a) Schematics of a surface force apparatus that employs the cross-cylinder geometry, with sliding attachment. The bimorph slider is a displacement transducer allowing for steady, sinusoidal, or triangular motion of the lower surface (up to 1 mm) and frequencies (from 10^{-6} Hz to 200 kHz). The friction device measures the friction forces produced on the upper surface via four semiconductor strain gages. (b) Schematic of shear elements of the dynamic SFA. The lower surface rests on a normal spring that is attached to two piezoelectric bimorph strips of stiffness. The vibration amplitudes A_x and A_y and phase delay ϕ characterize the rheological properties of the liquid. The surface shape and gap or film thickness is continuously monitored by an optical interferometric technique. Adapted with permission from Luengo, G., Schmitt, F.J., Hill, R. and Israelachvili, J. (1997), "Thin Film Rheology and Tribology of Confined Polymer Melts: Contrasts with Bulk Properties," *Macromolecules* **30**, 2482–2494. Copyright 1997 American Chemical Society.

mica surface experiences a transverse frictional or viscous shearing force, this will cause the vertical springs to deflect, and the deflection is measured by the strain gages.

In the SFA developed by Tonck *et al.* (1988) and Georges *et al.* (1994), the static and dynamic properties in normal and shear sliding of liquid films introduced between a macroscopic ceramic spherical body and a plane can be studied. Three piezoelectric elements controlled by a three-capacitance sensor permit accurate motion control and force measurement along three orthogonal axes. This design does not require optically transparent surfaces.

10.2.2 Static (Equilibrium), Dynamic and Shear Properties of Molecularly Thin Liquid Films

When a liquid is confined between two surfaces or within any narrow space whose dimensions are less than five to ten molecular diameters (~0.4 nm), thin films become ordered into layers (out-of-plane ordering), and within each layer they can also have lateral order (in-plane ordering). In the confinement of molecules between two structured solid surfaces, there is generally little opposition to any lateral or vertical displacement of the two surface lattices relative to each other. This means that the two lattices can shift to accommodate the trapped molecules in the most crystallographically commensurate or "epitaxial" way, which would favor an ordered, solid-like state. Such films may be thought of as behaving more like a liquid crystal or a solid than a liquid in thick films, Figure 10.2.2.

Confinement and load can produce a greater variety of interfacial (non-bulk-like) structures – amorphous, solid or liquid-crystal-like – each of which gives rise to different static and dynamic

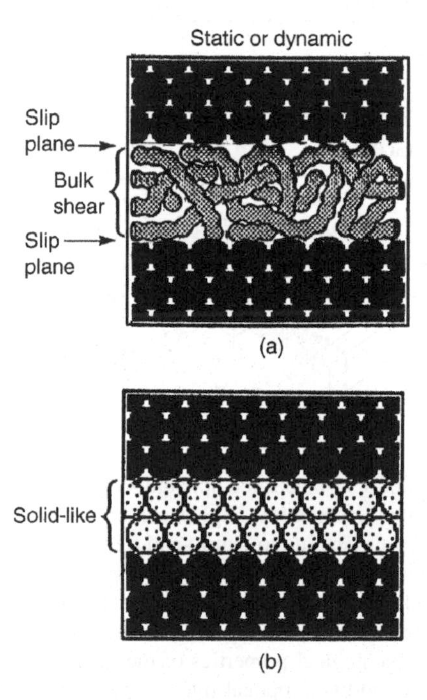

Figure 10.2.2 Schematic illustration of molecular arrangement of (a) a thick film (>10 molecular diameters), and (b) a molecularly thin film under normal stress.

properties, for example, density and viscosity. Both the static (equilibrium) and dynamic properties of the liquid can no longer be described even qualitatively in terms of the bulk properties, and molecular relaxation times can be longer by as much as an order of magnitude than in the bulk or more (Israelachvili *et al.*, 1988, 1990; Van Alsten and Granick, 1988, 1990a, b; Granick, 1991; Hu and Granick, 1992; Bhushan *et al.*, 1995a). For the data on viscosity as a function of the number of molecular layers, see Chapter 9. The shear stress (or viscosity) falls by an order of magnitude per additional layer. In other words, the friction is quantized depending on the number of layers separating the surface. It is reported that when seven to ten layers are present, the shear stress would fall to the value expected for bulk continuum Newtonian flow. Therefore, a liquid film with thickness in excess of ten molecular diameters (layers) can be described by their bulk properties.

A substantial increase in the viscosity of confined multimolecular films has important consequences. Consider a droplet of liquid between a ball and a table and let the ball fall, Figure 10.2.3. The liquid squirts out, initially rapidly, then slower and slower as the liquid thickness becomes less than the radius of the ball. Experiments show that the film eventually stabilizes at a finite thickness of a few molecular diameters. The liquid film supports the weight of the ball to an ultimate thickness of the liquid film, dependent upon the weight of the ball. An extraordinarily large pressure is needed to squeeze out the final few layers of liquid between two solid surfaces, and it is this film which can be crucial to minimize failure under high loads. Thus, for the thickness of a liquid film to be comparable to molecular dimensions, classical intuition based on continuum properties no longer applies (Granick, 1991).

10.2.2.1 Transition from Liquid-like to Solid-like

The dynamic properties of a liquid film undergoing shear are very complex. With a decrease in film thickness and an increase in normal stress, the film develops a yield stress, indicating a transition to a solid-like structure that must be broken down in order for sliding to occur. Figure 10.2.4 shows the shear stress as a function of time for films of different thicknesses

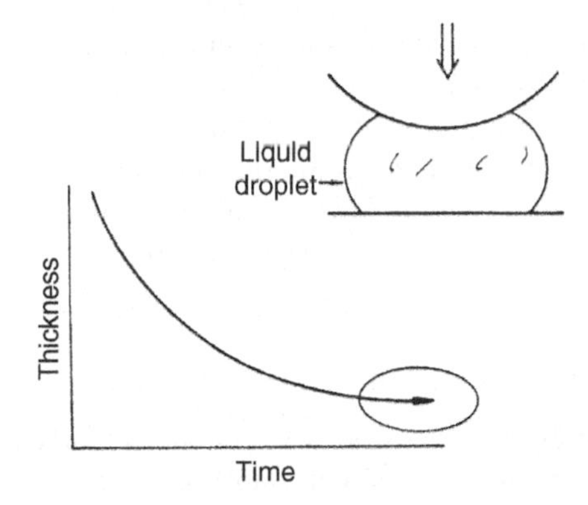

Figure 10.2.3 Schematic of a liquid droplet placed between a ball and a flat surface showing that a multimolecular layer of liquid can support a normal load. In the graph, liquid thickness is plotted schematically against time after the ball has begun to fall, it shows that the film thickness remains finite (a few molecular dimensions) even at equilibrium. Reproduced with permission from Granick, S. (1991), "Motions and Relaxations of Confined Liquids," *Science* **253**, 1374–1379. Copyright 1991 AAAS.

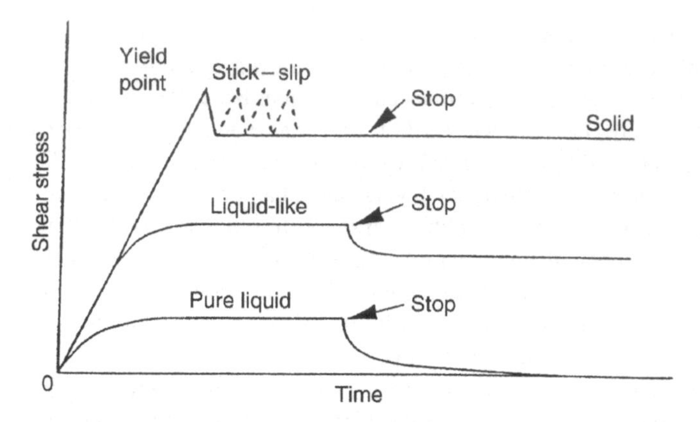

Figure 10.2.4 Shear stress as a function of sliding time for liquid films of different thicknesses. Reproduced with permission from Gee, M.L., McGuiggan, P.M., Israelachvili, J.N., and Homola, A.M. (1990), "Liquid to Solidlike Transitions of Molecularly Thin Films Under Shear," *J. Chem. Phys.* 93, 1895–1906. Copyright 1990 American Institute of Physics.

(Gee *et al.*, 1990). Film with a thickness less than five molecular diameters (uppermost curve) exhibits a yield point before it begins to flow. Such films can therefore sustain a finite shear stress, in addition to a finite normal stress. This behavior is indicative of solid-like films. The value of the yield stress depends on the number of layers comprising the film and the normal stress. Beyond the yield point, solid-like films show either a smooth yield-elongation zone, or a single spike having upper and lower yield points, or more typically a series of successive spikes known as the stick-slip pattern. For a film between five and ten molecular diameters in thickness, the film (middle curve) exhibits a liquid-like behavior. The main difference between the liquid-like and solid-like film is that in the former the stress relaxes somewhat, as a result of changes in molecular ordering or entanglement, when the velocity is set to zero. If the film thickness is greater than about ten molecular diameters, the film (bottom curve) is liquid and exhibits bulk properties, i.e. during steady sliding the stress slowly builds up to a constant value with no apparent yield point and then falls off to zero after sliding stops.

The static friction force increases with the length of time for which the surfaces have been in contact, at rest with respect to each other. Shorter sticking time and higher relative sliding velocities produce lower static friction force, because the lubricant has less time to fully solidify between the surfaces (Ruths and Israelachvili, 2011).

10.2.2.2 Smooth Sliding and Stick-Slip

Based on the data just presented, two surfaces with a molecularly thin liquid film in steady-state sliding still prefer to remain in one of their stable potential energy minima, i.e., a sheared film of liquid can retain its basic layered structure. Thus, even during motion the film may not become totally liquid-like. Depending on whether the film is more liquid-like or solid-like, the motion will be smooth or of the stick-slip type (repetitive transitions between solid-like and liquid-like). During sliding, transitions can occur between n layers and $(n - 1)$ or $(n + 1)$ layers, and the details of the motion depend critically on the externally applied load, the

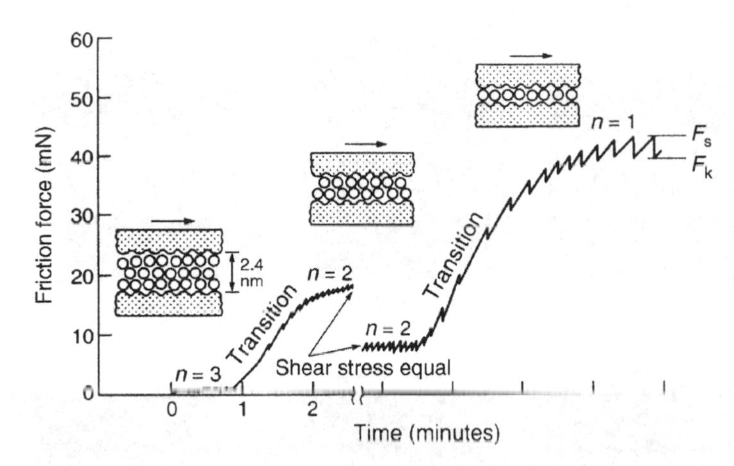

Figure 10.2.5 Measured change in the friction force during interlayer transitions of the silicone liquid octamethylcyclotetrasiloxane (OMCTS, an inert liquid whose quasi spherical molecules have a diameter of 0.8 nm), where n is number of molecular layers. In this system, the shear stress (friction force per unit area) is found to be constant as long as the number of layers n remained constant. Reproduced with permission from Gee, M.L., McGuiggan, P.M., Israelachvili, J.N., and Homola, A.M. (1990). "Liquid to Solidlike Transitions of Molecularly Thin Films Under Shear," *J. Chem. Phys.* **93**, 1895–1906. Copyright 1990 American Institute of Physics.

temperature, the sliding velocity, the twist angle between the two surface lattices and the sliding direction relative to the lattices (Ruths and Israelachvili, 2011). Figure 10.2.5 shows typical results for the friction traces measured as a function of time (after commencement of sliding) between two molecularly smooth mica surfaces separated by three molecular layers of the octamethylcyclotetrasiloxane (OMCTS) liquid, and how the friction increases to higher values in a quantized way when the number of molecular layers falls from $n = 3$ to $n = 2$ and then to $n = 1$ (Gee *et al.*, 1990). Gee *et al.* (1990) reported that the shear stresses are only weakly dependent on the sliding velocity. However, for sliding velocities above the critical value, the stick-slip disappears and the sliding proceeds smoothly at the kinetic value.

With the added insights provided by recent computer simulations of such systems to be presented later (Thompson and Robbins, 1990a; Robbins and Thompson, 1991), a number of distinct molecular processes have been identified during smooth and stick-slip sliding. These are shown schematically in Figure 10.2.6 for the case of spherical liquid molecules between two solid crystalline surfaces. Various regimes are identified in Figure 10.2.6 (Gee *et al.*, 1990; Israelachvili *et al.*, 1990; Ruths and Israelachvili, 2011). With surfaces at rest (Figure 10.2.6a) even with no externally applied load, film-surface epitaxial interactions can induce the liquid molecules in the film to solidify. Thus at rest the surfaces are stuck to each other through the film. When a progressively increasing lateral shear stress is applied, the film, being solid, responds elastically with a small lateral displacement and a small increase or dilatency in film thickness (less than a lattice spacing or molecular dimension, σ). In this so-called sticking regime (Figure 10.2.6b), the film retains its frozen, solid-like state – all the strains are elastic and reversible, and the surfaces remain effectively stuck to each other. However, slow creep may occur over long time periods. When the applied shear stress or

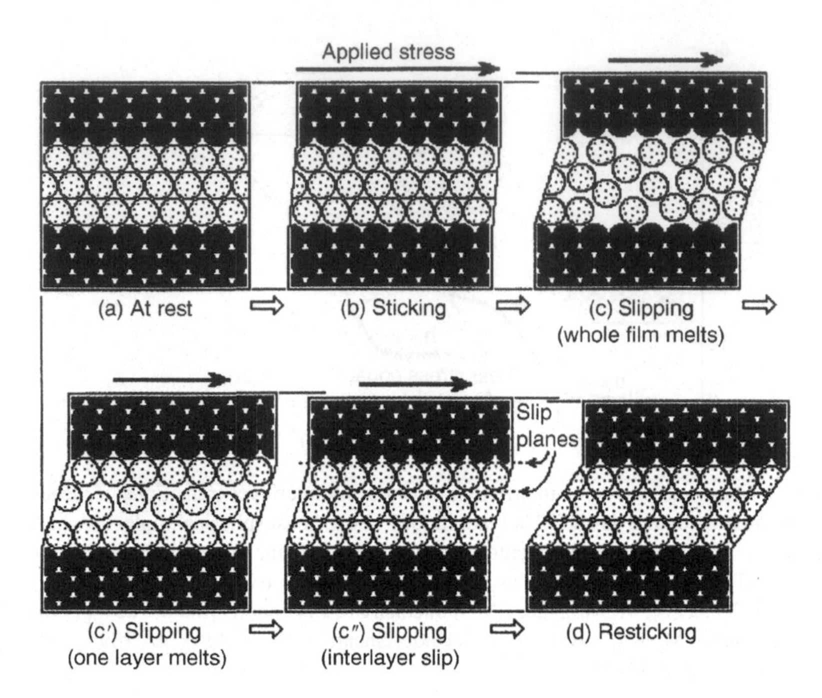

Figure 10.2.6 Schematic illustration of molecular rearrangements occurring in a molecularly thin film of spherical or simple chain molecules between two solid surfaces during shear. Reproduced with permission from Ruths, M. and Israelachvili, J.N. (2011), "Surface Forces and Nanorheology of Molecularly Thin Films." In *Nanotechtnology and Nanomechanics II*, Third edition (B. Bhushan, ed.), pp. 107–202, Springer-Verlag, Heidelberg, Germany. Copyright 2011 Springer.

force has reached a certain critical value F_s, the static friction force, the film suddenly melts (known as "shear melting") or rearranges to allow for wall-slip or film-slip to occur, at which point the two surfaces begin to slip rapidly past each other (Figure 10.2.6c). If the applied stress is kept at a high value the upper surface will continue to slide indefinitely, and this regime is called the slipping or sliding regime. Note that, depending on the system, a number of different molecular configurations within the film are possible during slipping and sliding, shown here as stages (c) – total disorder as whole film melts, (c') – partial disorder, and (c'') – order persists even during sliding with slip occurring at a single slip-plane either within the film or at the walls. The molecular configuration depends on the shapes of the molecules (e.g., whether spherical or linear or branched), sliding velocity, and other experimental conditions. In many practical cases, the rapid slip of the upper surface relieves some of the applied force, which eventually falls below another critical value F_k, the kinetic friction force, at which point the film resolidifies and resticks, and the whole stick-slip cycle is repeated (Figure 10.2.6d). On the other hand, if the slip rate is smaller than the rate at which the external stress is applied, the surfaces will continue to slide smoothly in the kinetic state and there will be no more stick-slip.

In addition to film thickness and experimental conditions, molecular shape and liquid structure have an effect on values of static and kinetic friction forces as well as propensity to stick-slip. Figure 10.2.7 shows the friction data for a linear perfluoropolyether (PFPE) (Fomblin

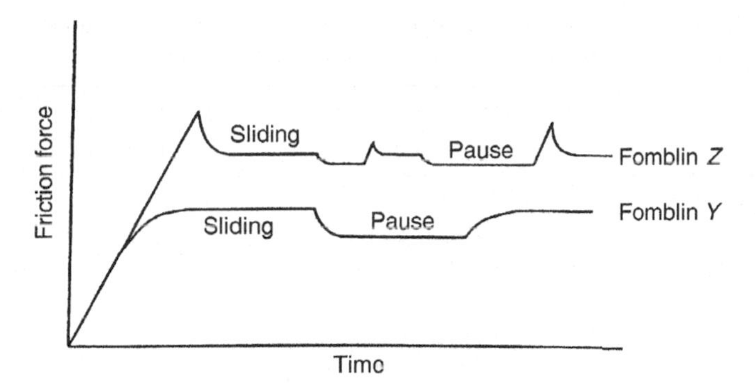

Figure 10.2.7 Friction force as a function of sliding time for Fomblin Z and Y lubricant films, measured at 1 μm/s. Reproduced with permission from Homola, A.M., Nguyen, H.V., and Hadziioannou, G. (1991), "Influence of Monomer Architecture on the Shear Properties of Molecular Thin Polymer Melts," *J. Chem. Phys.* **94**, 2346–2351. Copyright 1991 American Institute of Physics.

Z) and a branched PFPE (Fomblin Y) lubricants (Homola *et al.*, 1991). Figure 10.2.7 shows changes in friction during different phases of sliding. The linear lubricant exhibits a larger value of static friction followed by a lower value of kinetic friction, as compared to the branched lubricant. A pause in sliding is followed by another friction peak, with a difference between the static and kinetic friction being a function of the rest time. The friction of the branched lubricant is more typical of a Newtonian fluid with the static friction equal to the kinetic friction. We examine the reasons for different behavior of the two lubricants. As the sliding is initiated, the surfaces remain pinned until a stress equal to the yield stress is applied when melting followed by slip occurs. Branched molecules are less ordered during static contact than linear molecules, responsible for low static friction. As the sliding starts, the branched molecules are liquid or liquid-like during static contact and readily behave as liquid resulting in low kinetic friction force and with less propensity for stick-slip as compared to linear molecules. Based on Homola *et al.* (1991), during sliding, the random polymer orientation is disturbed and the chains deform and stretch in the direction of shear with an associated increase in the potential energy of the molecules (a source of the elastic force). The new molecular arrangement does not correspond to the minimum energy configuration, and the elastic energy is balanced by the shear force. During a pause in sliding, the elastic force tries to bring the molecules back to the state of lower energy and less ordering. The longer the pause duration, the more the system becomes disordered, and the higher the friction when the sliding is reinitiated.

Table 10.2.1 shows the trends observed with some organic and polymeric liquids between smooth mica surfaces. Also listed are the bulk viscosities of the liquids. From the data of Table 10.2.1, it appears that there is a direct correlation between the shapes of molecules and their coefficient of friction or effectiveness as lubricants (at least at low shear rates). Small spherical or chain molecules have high friction with stick-slip because they can pack into ordered solid-like layers. In contrast, longer chained and irregularly shaped molecules remain in an entangled, disordered, fluid-like state even in very thin films and these give low friction and smoother sliding. It is probably for this reason that irregularly shaped branched chain molecules are usually better lubricants. Examples of such liquids are octadecane, PDMS, PBD and perfluoropolyethers. It is interesting to note that the coefficient of friction generally

Table 10.2.1 Effect of molecular shape on friction properties for molecularly-thin liquid films between two shearing mica surfaces at 20°C (*Source*: Ruths and Israelachvili, 2011).

Liquid (dry)	Type of friction	Coefficient of friction	Bulk absolute viscosity (cP)
Spherical Molecules			
Cyclohexane ($\sigma = 0.5$ nm)[a]	Quantized stick-slip	>1	0.6
OMCTS[b] ($\sigma = 0.9$ nm)	Quantized stick-slip	>1	2.3
Chain Molecules			
Octane	Quantized stick-slip	1.5	0.5
Tetradecane	Stick-slip ←→ smooth	1.0	2.3
Octadecane (branched)	Stick-slip ←→ smooth	0.3	5.5
PDMS[b] ($M = 3700$, melt)[c]	Smooth	0.4	50
PBD[b] ($M = 3500$, branched)	Smooth	0.03	800
Water			
Water (KCl solution)	Smooth	0.01–0.03	1.0

[a] σ = Molecular dimension (diameter)
[b] OMCTS: Octamethylcyclotetrasiloxane, PDMS: Polydimethylsiloxane, PBD: Polybutadiene
[c] M = Molecular weight

decreases as the bulk viscosity of the liquids increases. This unexpected trend occurs because the factors that are conducive to low friction are generally conducive to high viscosity. Thus, molecules with side groups such as branched alkanes and polymer melts usually have higher bulk viscosities than their linear homologues for obvious reasons. However, in thin films, the linear molecules have higher shear stresses because of their ability to become ordered. The only exception to the above correlation is water, which has been found to exhibit both low viscosity and low friction. In addition, the presence of water can drastically lower the friction and eliminate the stick-slip of hydrocarbon liquids when the sliding surfaces are hydrophilic.

The effective viscosity for the liquids in the thin film form of Table 10.2.1 can be calculated. The values are 100–10^6 times that of the bulk viscosities. Thin film viscosity is of interest.

10.2.2.3 Phase Transitions Model

Molecular dynamic simulations have confirmed that the molecularly thin films undergo first-order phase transitions between solid- and liquid-like states during sliding and have suggested that this is responsible for the observed stick-slip behavior of simple isotropic liquids confined between solid surfaces (Thompson and Robbins, 1990a; Robbins and Thompson, 1991). The stick-slip occurs because of the abrupt change in the flow properties of a film at a transition (Israelachvili *et al.*, 1990; Thompson *et al.*, 1992).

The melting process during a slip takes a finite time but appears to be much faster than the freezing process during the stick regime. The velocity dependence of the stick-slip is thus dominated by the freezing time. Based on their computer simulation, Thompson and Robbins (1990a) presented an explanation for the phenomenon of decreasing coefficient of friction with an increase in the sliding velocity. They suggested that it is not the coefficient of friction that changes with sliding velocity, but the time various parts of the system spend in the sticking and sliding states. Therefore, at any instant during sliding, the friction at any local region is always F_s or F_k, corresponding to the static or kinetic values, and the measured frictional force is the sum of all these discrete values averaged over the whole contact area. As velocity increases, each local region spends more time in the sliding regime (F_k) and less in the sticking regime (F_s), thus, the overall coefficient of friction falls. This interpretation contradicts the traditional view that stick-slip occurs if friction decreases with velocity. It is the stick-slip phenomenon that results in a decrease of friction with an increase of sliding velocity.

10.2.2.4 Dynamic Phase Diagram Representation

Next, we examine the effect of experimental conditions on shear-induced ordering transitions. Both friction and adhesion hysteresis vary nonlinearly with temperature, often peaking at some particular temperature, T_0. The temperature-dependence of these forces can therefore be represented on a dynamic phase diagram such as shown in Figure 10.2.8. Experiments have shown that T_0, and the whole bell-shaped curve, are shifted along the temperature axis (as well as in the vertical direction) in a systematic way when the load, sliding velocity, and other experimental conditions are varied. These shifts also appear to be highly correlated with one another, for example, an increase in temperature producing effects that are similar to decreasing the sliding speed or load (Yoshizawa and Israelachvili, 1993; Bhushan *et al.*, 1995a).

Such effects are also commonly observed in other energy-dissipating phenomena such as polymer viscoelasticity, and it is likely that a similar physical mechanism is at the heart of all such phenomena. A possible molecular process underlying the energy dissipation of chain molecules during boundary layer sliding is illustrated in Figure 10.2.9, which shows the three

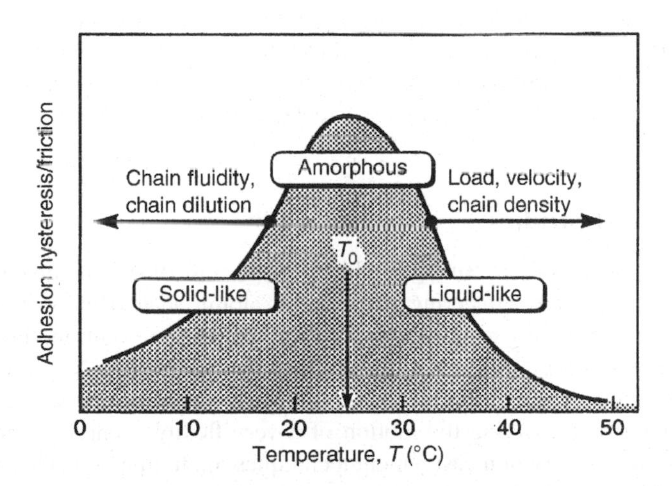

Figure 10.2.8 Schematic of friction phase diagram representing the trends observed in the boundary friction of a variety of different surfactant monolayers. The characteristic bell-shaped curve also correlates with the monolayers' adhesion energy hysteresis. Reproduced with permission from Yoshizawa, H. and Israelachvili, J.N. (1993), "Fundamental Mechanisms of Interfacial Friction II: Stick-Slip Friction of Spherical and Chain Molecules," *J. Phys. Chem.* **97**, 11300–11313. Copyright 1993 American Chemical Society.

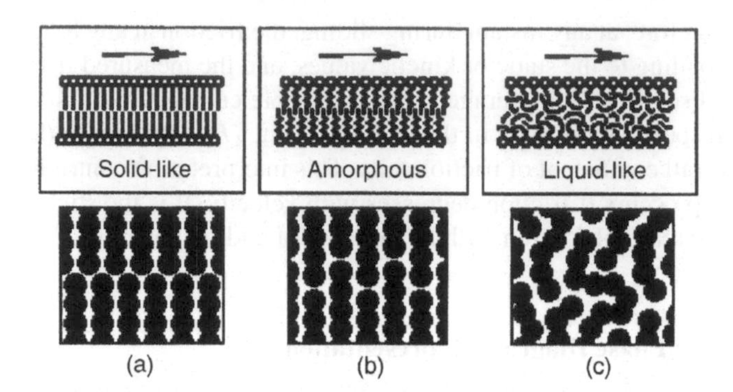

Figure 10.2.9 Different dynamic phase states of boundary monolayers during adhesive contact and/or frictional sliding. Reproduced with permission from Yoshizawa, H. and Israelachvili, J.N. (1993), "Fundamental Mechanisms of Interfacial Friction II: Stick-Slip Friction of Spherical and Chain Molecules," *J. Phys. Chem.* *97*, 11300–11313. Copyright 1993 American Chemical Society.

main dynamic phase states of boundary monolayers. Increasing the temperature generally shifts a system from the left to the right, (a)–(c). Changing the load, sliding velocity and other experimental conditions can also change the dynamic phase state of surface layers as shown in Figure 10.2.8. There is little interpenetration of chains across the solid-like crystalline interface (Figure 10.2.9a) and little energy is transferred (dissipated) between the two surfaces during adhesion or sliding. The adhesion hysteresis and friction forces are relatively low. The liquid-like chains of the two monolayers are significantly interdigitated across the interface but, unless the sliding velocity is very high, the system is always close to equilibrium. The friction force (similar to the viscous drag) and the adhesion hysteresis are again low. However, in the amorphous chains (Figure 10.2.9b), significant interdigitations occur across the interface with time both at rest (giving rise to adhesion energy hysteresis) and during sliding (giving rise to a large friction force) (Yoshizawa and Israelachvili, 1993).

10.3 AFM/FFM Studies

An AFM was developed by Gerd Binnig and his colleagues in 1985. It is capable of investigating surfaces of scientific and engineering interest on an atomic scale (Binnig *et al.*, 1986, 1987). The AFM relies on a scanning technique to produce very high-resolution, three-dimensional images of sample surfaces. It measures ultrasmall forces (less than 1 nN) present between the AFM tip surface mounted on a flexible cantilever beam and a sample surface. These small forces are obtained by measuring the motion of a very flexible cantilever beam having an ultrasmall mass, by a variety of measurement techniques, including optical deflection, optical interference, capacitance, and tunneling current. The deflection can be measured to within 0.02 nm, so for a typical cantilever spring constant of 10 N/m, a force as low as 0.2 nN can be detected. To put these numbers in perspective, individual atoms and a human hair are typically a fraction of a nanometer and about 75 μm in diameter, respectively, and a drop of water and an eyelash have a mass of about 10 μN and 100 nN, respectively. In the operation of high-resolution AFM, the sample is generally scanned rather than the tip because any cantilever movement would add vibrations. AFMs are available for the measurement of large samples, where the tip is scanned and the sample is stationary. To obtain an atomic resolution with the AFM, the spring constant of the cantilever should be weaker than the equivalent spring between atoms. A cantilever beam with a spring constant of about 1 N/m or lower is desirable. For high lateral resolution, tips should be as sharp as possible. Tips with a radius ranging from 5 to 50 nm are commonly available. Interfacial forces, adhesion, and surface roughness, including atomic-scale imaging, are routinely measured using the AFM.

A modification to the AFM providing a sensor to measure the lateral force led to the development of the friction force microscope (FFM) or the lateral force microscope (LFM), designed for atomic-scale and microscale studies of friction (Mate *et al.*, 1987; Bhushan and Ruan, 1994; Ruan and Bhushan, 1994a, b, c; Bhushan *et al.*, 1994, 1995a; Bhushan and Kulkarni, 1996; Bhushan, 1997, 1999a, b, 2001a, b, 2011, Bhushan and Sundararajan, 1998; Scherer *et al.*, 1998, 1999; Reinstaedtler *et al.*, 2003, 2005a, b; Bhushan and Kasai, 2004; Tambe and Bhushan, 2005a) and lubrication (Bhushan *et al.*, 1995b, 2005, 2006, 2007; Koinkar and Bhushan, 1996a, b; Bhushan and Liu, 2001; Liu *et al.*, 2001; Liu and Bhushan, 2002, 2003a; Kasai *et al.*, 2005; Lee *et al.*, 2005; Tambe and Bhushan, 2005b; Tao and Bhushan, 2005a, b; Palacio and Bhushan, 2007a, b; Bhushan, 2011). This instrument measures lateral or friction forces (in the plane of sample surface and in the scanning direction). By using a standard or a sharp diamond tip mounted on a stiff cantilever beam, AFM is used in investigations of scratching and wear (Bhushan *et al.*, 1994, 1995a; Bhushan and Koinkar, 1994a; Koinkar and Bhushan, 1996c, 1997a; Bhushan, 1999a, b, c, 2001c, 2005, 2008, 2011; Sundararajan and Bhushan, 2001), indentation (Ruan and Bhushan, 1993; Bhushan *et al.*, 1994, 1995a, 1996; Bhushan and Koinkar, 1994b; Bhushan, 1999c, 2001c; Bhushan and Li, 2003), and fabrication/machining (Bhushan *et al.*, 1994, 1995a; Bhushan, 1995, 1999a, 2011). An oscillating cantilever is used for localized surface elasticity and viscoelastic mapping, referred to as dynamic AFM (Maivald *et al.*, 1991; Anczykowski *et al.*, 1996; DeVecchio and Bhushan, 1997; Scherer *et al.*, 1997; Amelio *et al.*, 2001; Scott and Bhushan, 2003; Bhushan and Qi, 2003; Kasai *et al.*, 2004; Chen and Bhushan, 2005; Reinstaedtler *et al.*, 2005b; Bhushan, 2011). *In-situ* surface characterization of local deformation of materials and thin coatings has been carried out by imaging the sample surfaces using an AFM, during tensile deformation using a tensile stage (Bobji and Bhushan, 2001a, b; Tambe and Bhushan, 2004b; Bhushan, 2011).

10.3.1 Description of AFM/FFM and Various Measurement Techniques

Two commercial AFM/FFMs commonly used for measurements of nanotribological and nanomechanical properties ranging from micro- to atomic scales are shown in Figure 10.3.1 (Bhushan, 2010, 2011).

10.3.1.1 Surface Roughness and Friction Force Measurements

Surface height imaging down to atomic resolution of electrically-conducting surfaces is carried out using an STM. An AFM is also used for surface height imaging and roughness characterization down to nanoscale. Commercial AFM/FFM are routinely used for simultaneous measurements of surface roughness and friction force (Bhushan, 1999a, 2011). These

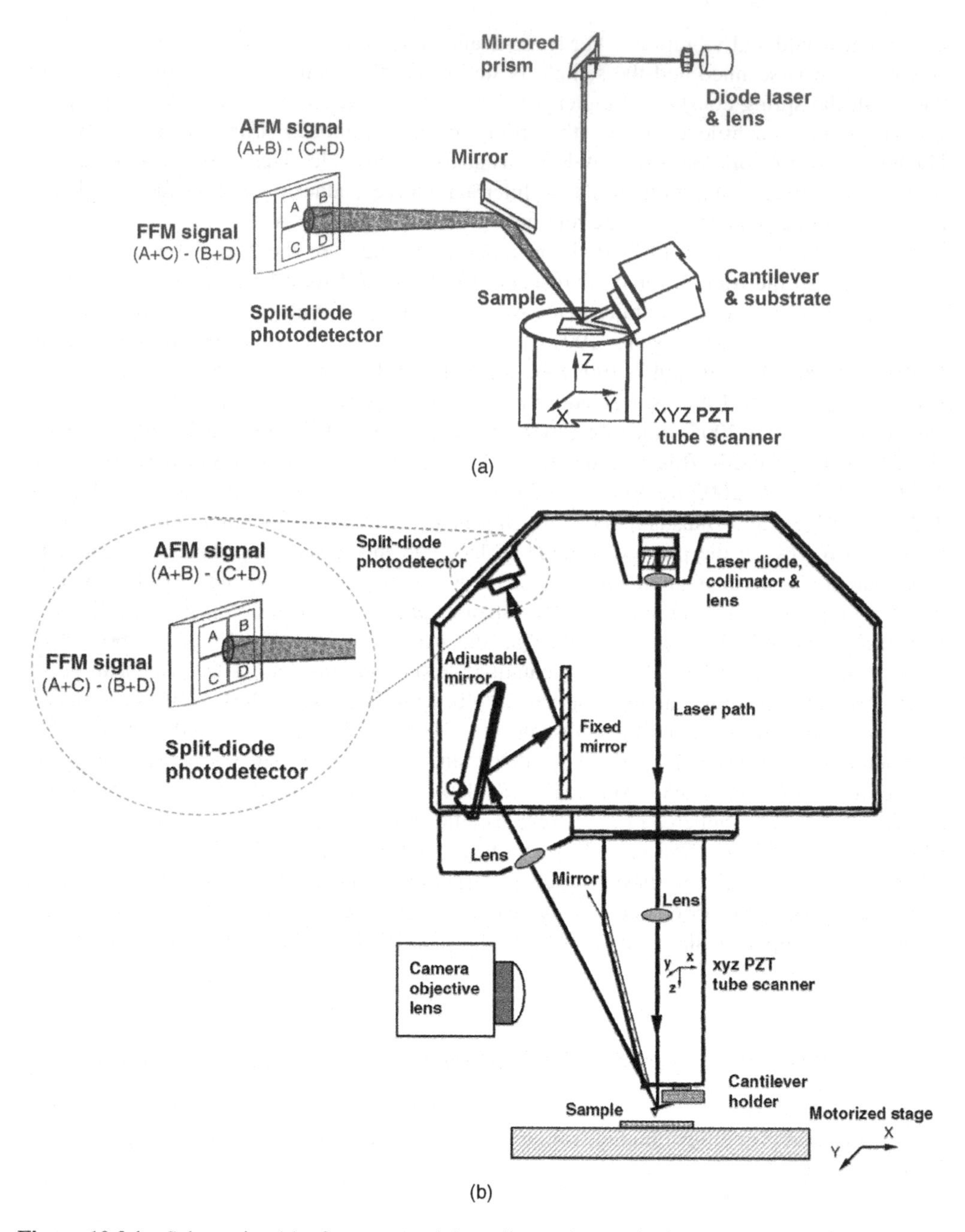

Figure 10.3.1 Schematics (a) of a commercial small sample atomic force microscope/friction force microscope (AFM/FFM), and (b) of a large sample AFM/FFM.

instruments are available for the measurement of small samples and large samples. In a small sample AFM shown in Figure 10.3.1(a), the sample, generally no larger than 10 mm × 10 mm, is mounted on a piezoelectric crystal in the form of a cylindrical tube (referred to as a PZT tube scanner) which consists of separate electrodes to scan the sample precisely in the X-Y plane in a raster pattern and to move the sample in the vertical (Z) direction. A sharp tip at the free end of a flexible cantilever is brought into contact with the sample. The normal and frictional forces being applied at the tip-sample interface are measured using a laser beam deflection technique. A laser beam from a diode laser is directed by a prism onto the back of a cantilever near its free end, tilted downward at about 10° with respect to the horizontal plane. The reflected beam from the vertex of the cantilever is directed through a mirror onto a quad photodetector (a split photodetector with four quadrants). The differential signal from the top and bottom photodiodes provides the AFM signal which is a sensitive measure of the cantilever vertical deflection. Topographic features of the sample cause the tip to deflect in the vertical direction as the sample is scanned under the tip. This tip deflection will change the direction of the reflected laser beam, changing the intensity difference between the top and bottom sets of photodetectors (the AFM signal). In the AFM operating mode called the height mode, for topographic imaging or for any other operation in which the applied normal force is to be kept constant, a feedback circuit is used to modulate the voltage applied to the PZT scanner to adjust the height of the PZT, so that the cantilever vertical deflection (given by the intensity difference between the top and bottom detector) will remain constant during scanning. The PZT height variation is thus a direct measure of the surface roughness of the sample.

In a large sample AFM, both force sensors using optical deflection method and scanning unit are mounted on the microscope head, Figure 10.3.1(b). Because of vibrations caused by the cantilever movement, lateral resolution of this design can be somewhat poorer than the design in Figure 10.3.1(a) in which the sample is scanned instead of cantilever beam. The advantage of the large sample AFM is that large samples can be measured readily.

Most AFMs can be used for surface roughness measurements in the so-called tapping mode (intermittent contact mode), also referred to as dynamic (atomic) force microscopy. In the tapping mode, during scanning over the surface, the cantilever/tip assembly with a normal stiffness of 20–100 N/m (DI tapping mode etched Si probe or TESP) is sinusoidally vibrated at its resonance frequency (350–400 kHz) by a piezo mounted above it, and the oscillating tip slightly taps the surface. The piezo is adjusted using the feedback control in the Z direction to maintain a constant (20–100 nm) oscillating amplitude (setpoint) and constant average normal force, Figure 10.3.2 (Bhushan, 1999a, 2011). The feedback signal to the Z-direction sample piezo (to keep the setpoint constant) is a measure of surface roughness. The cantilever/tip assembly is vibrated at some amplitude, here referred to as the free amplitude, before the tip engages the sample. The tip engages the sample at some setpoint, which may be thought of as the amplitude of the cantilever as influenced by contact with the sample. The setpoint is defined as a ratio of the vibration amplitude after engagement to the vibration amplitude in free air before engagement. A lower setpoint gives a reduced amplitude and closer mean tip-to-sample distance. The amplitude should be kept large enough so that the tip does not get stuck to the sample because of adhesive attractions. Also the oscillating amplitude applies a less average (normal) load as compared to the contact mode and reduces the sample damage. The tapping mode is used in topography measurements to minimize the effects of friction and other lateral forces and to measure the topography of soft surfaces.

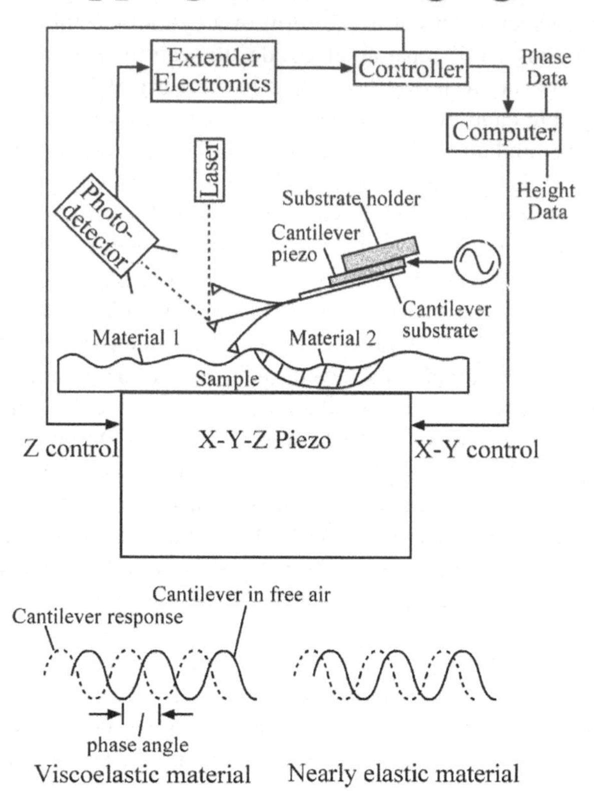

Figure 10.3.2 Schematic of tapping mode used to obtain height and phase data and definitions of free amplitude and setpoint. During scanning, the cantilever is vibrated at its resonance frequency and the sample X-Y-Z piezo is adjusted by feedback control in the Z-direction to maintain a constant setpoint. The computer records height (which is a measure of surface roughness) and phase angle (which is a function of the viscoelastic properties of the sample) data.

To measure the friction force at the tip surface during sliding, the left-hand and right-hand sets of quadrants of the photodetector are used. In the so-called friction mode, the sample is scanned back and forth in a direction orthogonal to the long axis of the cantilever beam. A friction force between the sample and the tip will produce a twisting of the cantilever. As a result, the laser beam will be reflected out of the plane defined by the incident beam and the beam reflected vertically from an untwisted cantilever. This produces an intensity difference of the laser beam received in the left-hand and right-hand sets of quadrants of the photodetector. The intensity difference between the two sets of detectors (FFM signal) is directly related to the degree of twisting and hence to the magnitude of the friction force. One problem associated with this method is that any misalignment between the laser beam and the photodetector axis would introduce error in the measurement. However, by following the procedures developed by Ruan and Bhushan (1994a), in which the average FFM signal for the sample scanned in two opposite directions is subtracted from the friction profiles of each of the two scans, the misalignment effect is eliminated. This method provides three-dimensional maps of friction force. By following the friction force calibration procedures developed by Ruan and Bhushan (1994a), voltages corresponding to friction forces can be converted to force units (Palacio and Bhushan, 2010). The coefficient of friction is obtained from the slope of friction force data measured as a function of normal loads typically ranging from 10 to 150 nN. This approach eliminates any contributions due to the adhesive forces (Bhushan *et al.*, 1994). To calculate the coefficient of friction based on a single point measurement, the friction force should be divided by the sum of the applied normal load and the intrinsic adhesive force. Furthermore it should be pointed out that for a single asperity contact; the coefficient of friction is not independent of load (see discussion later).

Surface roughness measurements in the contact mode are typically made using a sharp, microfabricated square-pyramidal Si_3N_4 tip with a radius of 30–50 nm on a triangular cantilever beam (Figure 10.3.3(a)) with normal stiffness on the order of 0.06–0.58 N/m with a normal natural frequency of 13–40 kHz (DI silicon nitride probe or NP) at a normal load of about 10 nN, and friction measurements are carried out in the load range of 1–100 nN. Surface roughness measurements in the tapping mode utilize a stiff cantilever with high resonance frequency; typically a square-pyramidal etched single-crystal silicon tip, with a tip radius of 5–10 nm, integrated with a stiff rectangular silicon cantilever beam (Figure 10.3.3(a)) with a normal stiffness on the order of 17–60 N/m and a normal resonance frequency of 250–400 kHz (DI TESP), is used. Multiwalled carbon nanotube tips having a small diameter (few nm) and a length of about 1 μm (high aspect ratio) attached on the single-crystal silicon, square-pyramidal tips are used for high resolution imaging of surfaces and of deep trenches in the tapping mode (noncontact mode) (Bhushan *et al.*, 2004a). The MWNT tips are hydrophobic. To study the effect of radius of a single asperity (tip) on adhesion and friction, microspheres of silica with radii ranging from about 4 to 15 μm are attached at the end of cantilever beams. Optical micrographs of a commercial Si_3N_4 tip and a modified tip showing 14.5 μm radius SiO_2 sphere mounted over the sharp tip at the end of the triangular Si_3N_4 cantilever beam are shown in Figure 10.3.3(b).

The tip is scanned in such a way that its trajectory on the sample forms a triangular pattern, Figure 10.3.4. Scanning speeds in the fast and slow scan directions depend on the scan area and scan frequency. Scan sizes ranging from less than 1 nm × 1 nm to 125 μm × 125 μm and scan rates from less than 0.5 to 122 Hz typically can be used. Higher scan rates are used for smaller scan lengths. For example, scan rates in the fast and slow scan directions for an area of 10 μm × 10 μm scanned at 0.5 Hz are 10 μm/s and 20 nm/s, respectively.

10.3.1.2 Adhesion Measurements

Adhesive force measurements are performed in the so-called force calibration mode. In this mode, force-distance curves are obtained, for an example see Figure 10.3.5. The horizontal axis gives the distance that the piezo (and hence the sample) travels, and the vertical axis gives the tip deflection. As the piezo extends, it approaches the tip, which is at this point in free air and hence shows no deflection. This is indicated by the flat portion of the curve. As the tip approaches the sample within a few nanometers (point A), an attractive force exists between the atoms of the tip surface and the atoms of the sample surface. The tip is pulled towards the sample and contact occurs at point B on the graph. From this point on, the tip is in contact with the surface and as the piezo further extends, the tip gets further deflected. This is represented

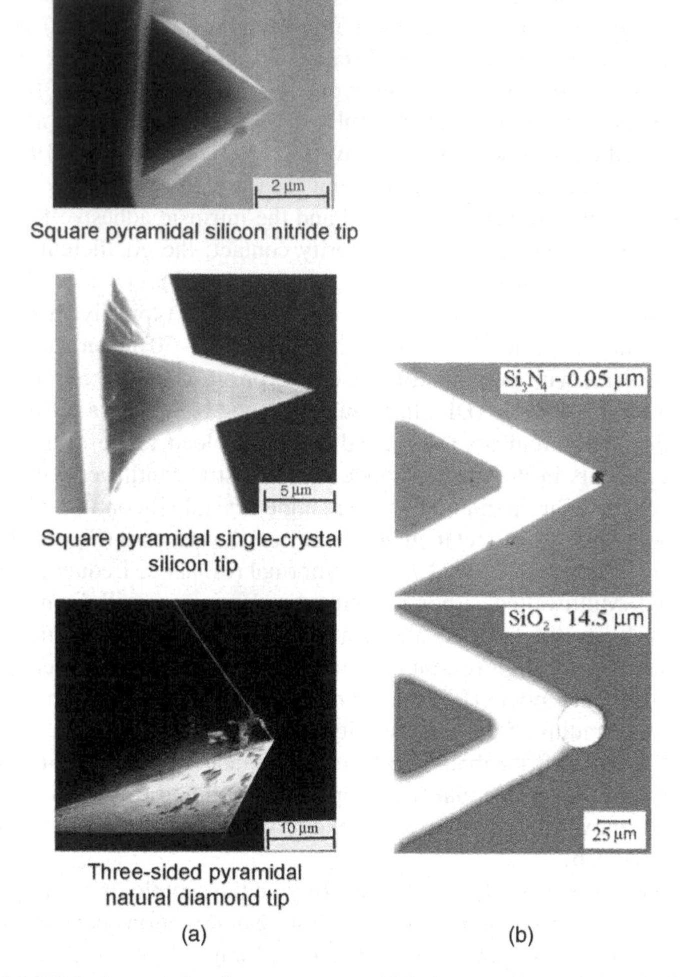

Figure 10.3.3 (a) SEM micrographs of a square-pyramidal plasma-enhanced chemical vapor deposition (PECVD) Si_3N_4 tip with a triangular cantilever beam, a square-pyramidal etched single-crystal silicon tip with a rectangular silicon cantilever beam, and a three-sided pyramidal natural diamond tip with a square stainless steel cantilever beam, and (b) optical micrographs of a commercial Si_3N_4 tip and a modified tip with a 14.5 μm radius SiO_2 sphere mounted over the sharp tip at the end of the triangular Si_3N_4 cantilever beams.

by the sloped portion of the curve. As the piezo retracts, the tip goes beyond the zero deflection (flat) line because of attractive forces (van der Waals forces and long-range meniscus forces), into the adhesive regime. At point C in the graph, the tip snaps free of the adhesive forces and is again in free air. The horizontal distance between points B and C along the retrace line gives the distance moved by the tip in the adhesive regime. This distance multiplied by the stiffness of the cantilever gives the adhesive force. Incidentally, the horizontal shift between the loading and unloading curves results from the hysteresis in the PZT tube (Bhushan, 1999a, 2011).

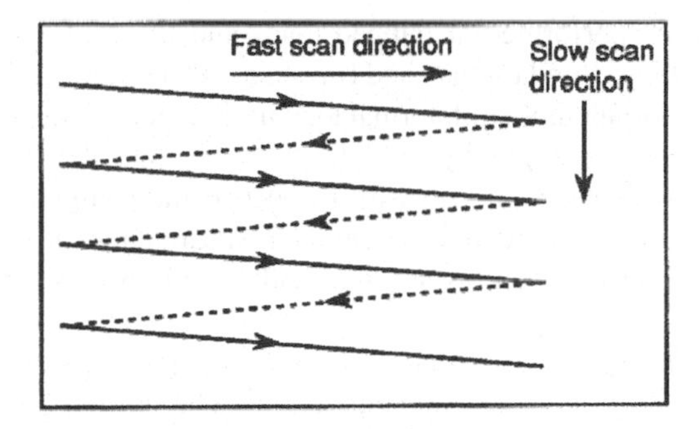

Figure 10.3.4 Schematic of triangular pattern trajectory of the tip as the sample (or the tip) is scanned in two dimensions. During scanning, data are recorded only during scans along the solid scan lines.

10.3.1.3 Scratching, Wear and Fabrication/Machining

For microscale scratching, microscale wear, nanofabrication/nanomachining, and nanoindentation hardness measurements, an extremely hard tip is required. A three-sided pyramidal single-crystal natural diamond tip with an apex angle of 80° and a radius of about 100 nm mounted on a stainless steel cantilever beam with normal stiffness of about 25 N/m is used at relatively higher loads (1–150 μN), Figure 10.3.3(a). For scratching and wear studies, the sample is generally scanned in a direction orthogonal to the long axis of the cantilever beam (typically at a rate of 0.5 Hz) so that friction can be measured during scratching and wear. The tip is mounted on the cantilever such that one of its edges is orthogonal to the long axis of the beam; therefore, wear during scanning along the beam axis is higher (about 2x to 3x) than that during scanning orthogonal to the beam axis. For wear studies, an area on the order of 2 μm × 2 μm is scanned at various normal loads (ranging from 1 to 100 μN) for a selected number of cycles (Bhushan *et al.*, 1994; Bhushan, 1999a, 2011).

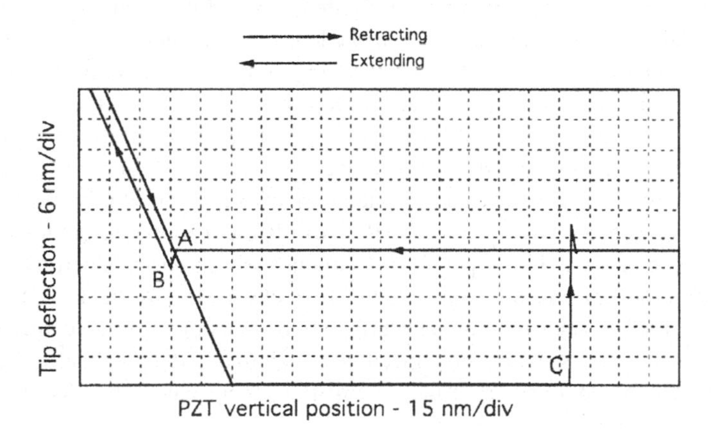

Figure 10.3.5 Typical force-distance curve for a contact between Si_3N_4 tip and single-crystal silicon surface in measurements made in the ambient environment. Snap-in occurs at point A; contact between the tip and silicon occurs at point B; tip breaks free of adhesive forces at point C as the sample moves away from the tip.

Scratching can also be performed at ramped loads and the coefficient of friction can be measured during scratching (Sundararajan and Bhushan, 2001). A linear increase in the normal load approximated by a large number of normal load increments of small magnitude is applied using a software interface (lithography module in Nanoscope III) that allows the user to generate controlled movement of the tip with respect to the sample. The friction signal is tapped out of the AFM and is recorded on a computer. A scratch length on the order of 25 μm and a velocity on the order of 0.5 μm/s are used and the number of loading steps is usually taken to be 50.

Nanofabrication/nanomachining is conducted by scratching the sample surface with a diamond tip at specified locations and scratching angles. The normal load used for scratching (writing) is on the order of 1–100 μN with a writing speed on the order of 0.1–200 μm/s (Bhushan *et al.*, 1994, 1995a; Bhushan, 1995, 1999a, 2011).

10.3.1.4 Nanoindentation Measurements

For nanoindentation hardness measurements the scan size is set to zero, and then a normal load is applied to make the indents using the diamond tip. During this procedure, the tip is continuously pressed against the sample surface for about two seconds at various indentation loads. The sample surface is scanned before and after the scratching, wear, or indentation to obtain the initial and the final surface topography, at a low normal load of about 0.3 μN using the same diamond tip. An area larger than the indentation region is scanned to observe the indentation marks. Nanohardness is calculated by dividing the indentation load by the projected residual area of the indents (Bhushan and Koinkar, 1994b).

Direct imaging of the indent allows one to quantify piling up of ductile material around the indenter. However, it becomes difficult to identify the boundary of the indentation mark with great accuracy. This makes the direct measurement of contact area somewhat inaccurate. A technique with the dual capability of depth-sensing as well as *in-situ* imaging, which is most appropriate in nanomechanical property studies, is used for accurate measurement of hardness with shallow depths (Bhushan *et al.*, 1996; Bhushan, 1999a, c, 2011). This nano/picoindentation system is used to make load-displacement measurement and subsequently carry out *in-situ* imaging of the indent, if required. The indentation system, shown in Figure 10.3.6, consists of a three-plate transducer with electrostatic actuation hardware used for the direct application of a normal load and a capacitive sensor used for the measurement of vertical displacement. The AFM head is replaced with this transducer assembly while the specimen is mounted on the PZT scanner, which remains stationary during indentation experiments. The transducer consists of a three (Be-Cu) plate capacitive structure, and the tip is mounted on the center plate. The upper and lower plates serve as drive electrodes, and the load is applied by applying appropriate voltage to the drive electrodes. Vertical displacement of the tip (indentation depth) is measured by measuring the displacement of the center plate relative to the two outer electrodes using the capacitance technique. Indent area and consequently hardness value can be obtained from the load-displacement data. The Young's modulus of elasticity is obtained from the slope of the unloading curve.

10.3.1.5 Boundary Lubrication Measurements

To study nanoscale boundary lubrication properties, adhesive forces are measured in the force calibration mode, as previously described. The adhesive forces are also calculated from the

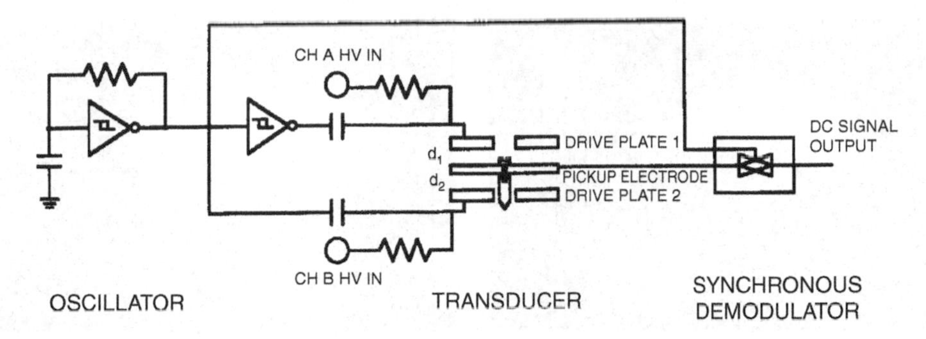

Figure 10.3.6 Schematic of a nano/picoindentation system with three-plate transducer with electrostatic actuation hardware and capacitance sensor. Reproduced with permission from Bhushan, B., Kulkarni, A.V., Bonin, W. and Wyrobek, J.T. (1996), "Nano/Picoindentation Measurement Using a Capacitance Transducer System in Atomic Force Microscopy," Philos. Mag. 74, 1117-1128. Copyright 1996 Taylor and Francis.

horizontal intercept of friction versus normal load curves at a zero value of friction force. For friction measurements, the samples are typically scanned using a Si_3N_4 tip over an area of 2×2 μm at the normal load ranging from 5 to 130 nN. The samples are generally scanned with a scan rate of 0.5 Hz resulting in a scanning speed of 2 μm/s. Velocity effects on friction are studied by changing the scan frequency from 0.1 to 60 Hz, while the scan size is maintained at 2×2 μm, which allows velocity to vary from 0.4 to 240 μm/s. To study the durability properties, the friction force and coefficient of friction are monitored during scanning at a normal load of 70 nN and a scanning speed of 0.8 μm/s, for a desired number of cycles (Koinkar and Bhushan, 1996a, b; Liu and Bhushan, 2003a).

10.3.2 Surface Imaging, Friction, and Adhesion

10.3.2.1 Atomic-Scale Imaging and Friction

Surface height imaging down to atomic resolution of electrically conducing surfaces can be carried out using an STM. An AFM can also be used for surface height imaging and roughness characterization down to the nanoscale. Figure 10.3.7 shows a sequence of STM images at various scan sizes of solvent deposited C_{60} film on 200-nm thick gold-coated freshly cleaved mica (Bhushan *et al.*, 1993). The film consists of clusters of C_{60} molecules of 8 nm in diameter. The C_{60} molecules within a cluster appear to pack into a hexagonal array with a spacing of about 1 nm; however, they do not follow any long range order. The measured cage diameter of the C_{60} molecule is about 0.7 nm, very close to the projected diameter of 0.71 nm.

In an AFM measurement during surface imaging, the tip comes into intimate contact with the sample surface and leads to surface deformation with finite tip-sample contact area (typically a few atoms). The finite size of the contact area prevents the imaging of individual point defects, and only the periodicity of the atomic lattice can be imaged. Figure 10.3.8 shows the topography image of a freshly-cleaved surface of highly oriented pyrolytic graphite (HOPG) (Ruan and Bhushan, 1994b). The periodicity of the graphite is clearly observed.

To study the friction mechanisms on an atomic scale, a freshly cleaved HOPG was studied by Mate *et al.* (1987) and Ruan and Bhushan (1994b). Figure 10.3.9(a) shows the atomic-scale friction force map (raw data) and Figure 10.3.8 shows the friction force maps (after

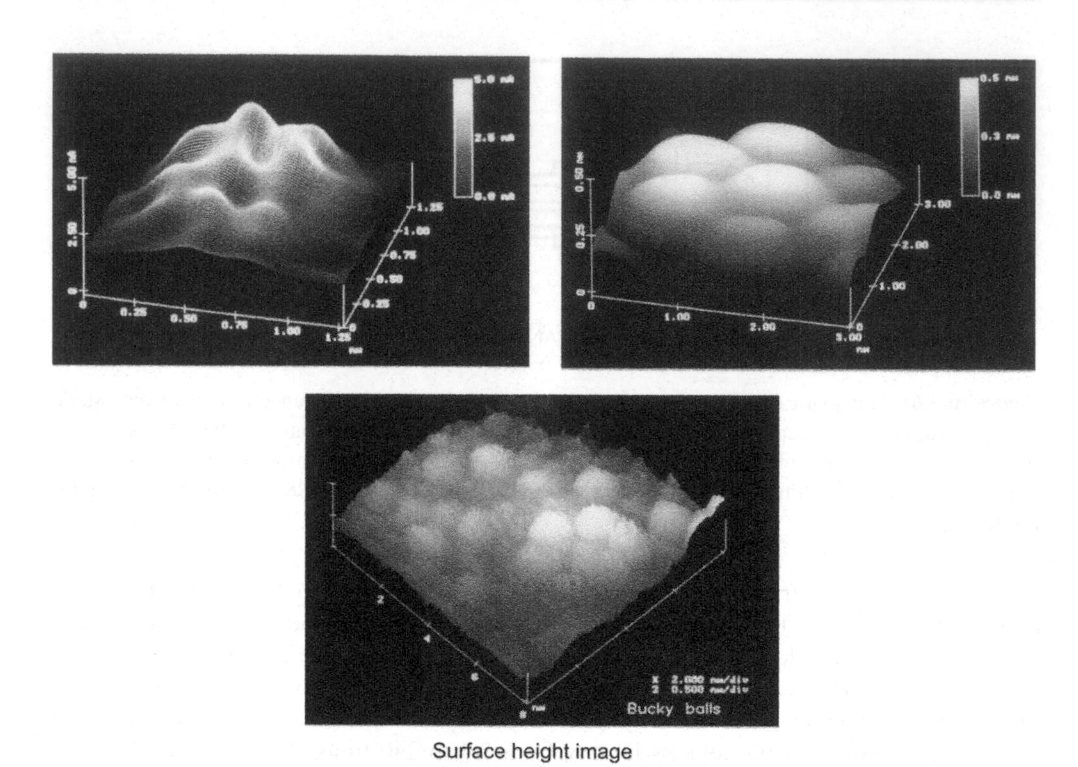

Figure 10.3.7 STM images of solvent deposited C_{60} film on a gold-coated freshly-cleaved mica at various scan sizes. Reproduced with permission from Bhushan, B., Mokashi, P.S., and Ma, T. (2003), "A New Technique to Measure Poisson's Ratio of Ultrathin Polymeric Films Using Atomic Force Microscopy," *Rev. Sci. Instrum.* **74**, 1043–1047. Copyright 2003 American Institute of Physics.

2D spectrum filtering with high frequency noise truncated) (Ruan and Bhushan, 1994b). Figure 10.3.9(a) also shows a line plot of friction force profiles along some crystallographic direction. The actual shape of the friction profile depends upon the spatial location of axis of tip motion. Note that a portion of atomic-scale lateral force is conservative. Mate *et al.* (1987) and Ruan and Bhushan (1994b) reported that the average friction force linearly increased with normal load and was reversible with load. Friction profiles were similar while sliding the tip in either direction.

During scanning, the tip moves discontinuously over the sample surface and jumps with discrete steps from one potential minimum (well) to the next. This leads to a saw-tooth-like pattern for the lateral motion (force) with a periodicity of the lattice constant. This motion is called the stick-slip movement of the tip (Mate *et al.*, 1987; Ruan and Bhushan, 1994b; Bhushan, 1999a, 2011). The observed friction force includes two components – conservative and periodic, and nonconservative and constant. If the relative motion of the sample and tip were simply that of two rigid collections of atoms, the effective force would be a conservative force oscillating about zero. Slow reversible elastic deformation would also contribute to conservative force. The origin of the nonconservative direction-dependent force component would be phonon generation, viscous dissipation, or plastic deformation.

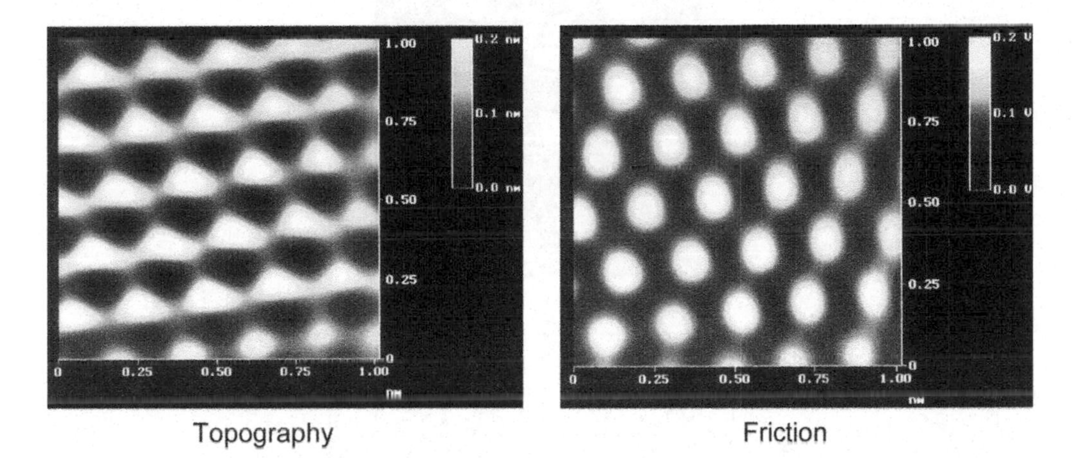

Figure 10.3.8 (a) Gray-scale plots of surface topography and friction force maps (2D spectrum filtered), measured simultaneously, of a 1 nm × 1 nm area of freshly cleaved HOPG, showing the atomic-scale variation of topography and friction. Reproduced with permission from Ruan, J. and Bhushan, B. (1994b), "Atomic-scale and Microscale Friction of Graphite and Diamond Using Friction Force Microscopy," *J. Appl. Phys.* 76, 5022–5035. Copyright 1994, American Institute of Physics.

Stick-slip on the atomic scale, discussed above, is the result of the energy barrier required to be overcome to jump over the atomic corrugations on the sample surface. It corresponds to the energy required for the jump of the tip from a stable equilibrium position on the surface to a neighboring position. The perfect atomic regularity of the surface guarantees the periodicity of the lateral force signal, independent of the actual atomic structure of the tip apex. A few atoms (based on the magnitude of the friction force, less than 10) on a tip sliding over an array of atoms on the sample are expected to undergo stick-slip. For simplicity, Figure 10.3.9(b) shows a simplified model for one atom on a tip with a one-dimensional spring mass system. As the sample surface slides against the AFM tip, the tip remains "stuck" initially until it can overcome the energy (potential) barrier, which is illustrated by a sinusoidal interaction potential as experienced by the tip. After some motion, there is enough energy stored in the spring which leads to "slip" into the neighboring stable equilibrium position. During the slip and before attaining stable equilibrium, stored energy is converted into vibrational energy of the surface atoms in the range of 10^{13} Hz (phonon generation) and decays within the range of 10^{-11} s into heat. (A wave of atoms vibrating in concert are termed a phonon.) The stick-slip phenomenon, resulting from irreversible atomic jumps, can be theoretically modeled with classical mechanical models (Tomlinson, 1929; Tomanek *et al.*, 1991). The Tomanek–Zhong–Thomas model (Tomanek *et al.*, 1991) is the starting point for determining friction force during atomic-scale stick-slip. The AFM model describes the total potential as the sum of the potential acting on the tip due to interaction with the sample and the elastic energy stored in the cantilever. Thermally activated stick-slip behavior can explain the velocity effects on friction, to be presented later.

Finally, based on Figure 10.3.8, the atomic-scale friction force of HOPG exhibited the same periodicity as that of the corresponding topography, but the peaks in friction and those in the topography are displaced relative to each other (superimposed images are not reported here).

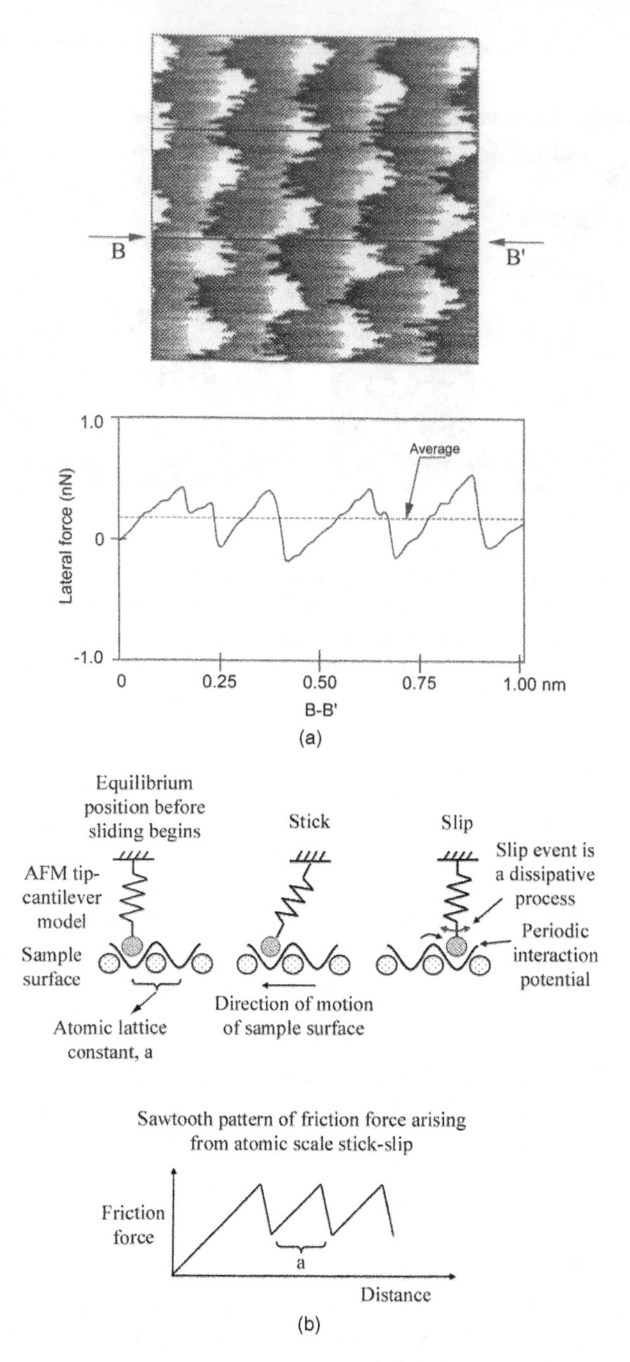

Figure 10.3.9 (a) Gray scale plot of friction force map (raw data) of a 1×1 nm^2 area of freshly cleaved HOPG, showing atomic-scale variation of friction force. High points are shows by lighter color. Also shown is line plot of friction force profile along the line indicated by arrows. The normal load was 25 nN and the cantilever normal stiffness was 0.4 N/m. Reproduced with permission from Ruan, J. and Bhushan, B. (1994b), "Atomic-scale and Microscale Friction of Graphite and Diamond Using Friction Force Microscopy," *J. Appl. Phys.* **76**, 5022–5035. Copyright 1994. American Institute of Physics; and (b) Schematic of a model for a tip atom sliding on an atomically flat periodic surface. The schematic shows the tip jumping from one potential minimum to another, resulting in stick-slip behavior.

A Fourier expansion of the interatomic potential was used by Ruan and Bhushan (1994b) to calculate the conservative interatomic forces between the atoms of the FFM tip and those of the graphite surface. Maxima in the interatomic forces in the normal and lateral directions do not occur at the same location, which explains the observed shift between the peaks in the lateral force and those in the corresponding topography.

10.3.2.2 Microscale Friction

Microscale friction is defined as the friction measured with a scan size equal to or larger than 1 μm × 1 μm. Local variations in the microscale friction of cleaved graphite and multi-phase materials have been observed (Meyer *et al.*, 1992; Ruan and Bhushan, 1994c; Koinkar and Bhushan, 1996c). These variations may occur due to the variations in the local phase of the surfaces. These measurements suggest that the FFM can be used for structural mapping of the surfaces. FFM measurements can also be used to map chemical variations, as reported by the use of the FFM with a modified probe tip to map the spatial arrangement of chemical functional groups in mixed organic monolayer films by Frisbie *et al.* (1994). In this study, sample regions that had stronger interactions with the functionalized probe tip exhibited larger friction.

Local variations in the microscale friction of nominally rough, homogeneous-material surfaces can be significant, and are seen to depend on the local surface slope rather than the surface height distribution. This dependence was first reported by Bhushan and Ruan (1994), Bhushan *et al.* (1994), and Bhushan (1995) and later discussed in more detail by Koinkar and Bhushan (1997b) and Sundararajan and Bhushan (2000). In order to elegantly show any correlation between local values of friction and surface roughness, surface roughness, surface slope, and friction force maps of a gold-coated ruler with somewhat rectangular grids and a silicon grid with square pits were obtained (Figure 10.3.10) (Sundararajan and Bhushan, 2000). There is a strong correlation between the surface slopes and friction forces. In Figure 10.3.10(b), the friction force is high locally at the edge of the pits with a positive slope and is low at the edges with a negative slope.

We now examine the mechanism of microscale friction, which may explain the resemblance between the slope of surface roughness maps and the corresponding friction force maps (Bhushan and Ruan, 1994; Bhushan *et al.*, 1994; Ruan and Bhushan, 1994b, c; Koinkar and Bhushan, 1997b; Bhushan, 1999a, 2011; Sundararajan and Bhushan, 2000). There are three dominant mechanisms of friction; adhesive, ratchet, and plowing (Bhushan, 1999a, 2011). As a first order, we may assume these to be additive. The adhesive mechanism cannot explain the local variation in friction. Next we consider the ratchet mechanism. We consider a small tip sliding over an asperity making an angle θ with the horizontal plane (Figure 10.3.11). The normal force W (normal to the general surface) applied by the tip to the sample surface is constant. The friction force F on the sample would be a constant for a smooth surface if the friction mechanism does not change. For a rough surface shown in Figure 10.3.11, if the adhesive mechanism does not change during sliding, the local value of the coefficient of friction remains constant

$$\mu_0 = S/N \tag{10.3.1}$$

where S is the local friction force and N is the local normal force. However, the friction and normal forces are measured with respect to global horizontal and normal axes, respectively.

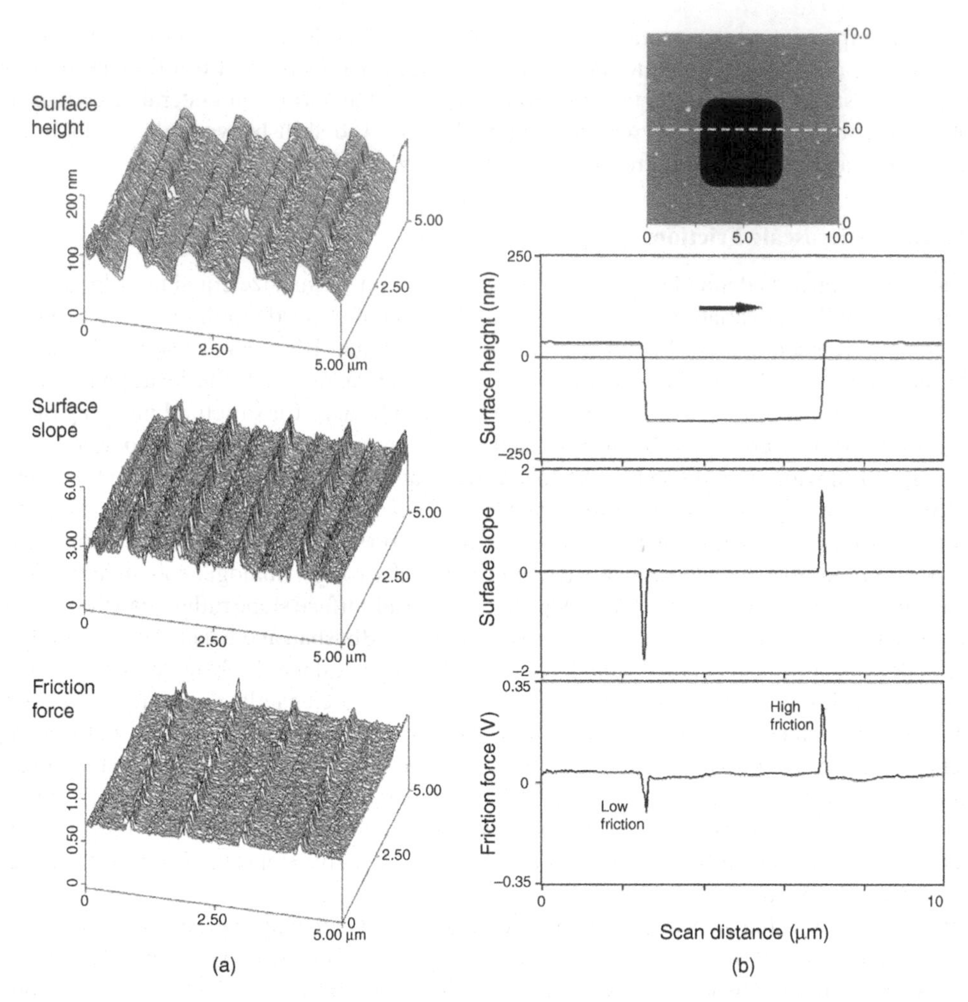

Figure 10.3.10 Surface roughness map, surface slope map taken in the sample sliding direction (the horizontal axis), and friction force map for (a) a gold-coated ruler (with somewhat rectangular grids with a pitch of 1 µm and a ruling step height of about 70 µm) at a normal load of 25 nN and (b) a silicon grid (with 5 µm square pits of depth 180 nm and a pitch of 10 µm). Reproduced with permission from Sundararajan, S. and Bhushan, B. (2000), "Topography-Induced Contributions to Friction Forces Measured Using an Atomic Force/Friction Force Microscope," *J. Appl. Phys.* **88**, 4825–4831. Copyright 2000 American Institute of Physics.

The measured local coefficient of friction μ_1 in the ascending part is

$$\mu_1 = F/W = (\mu_0 + \tan\theta)/(1 - \mu_0 \tan\theta) \sim \mu_0 + \tan\theta, \text{ for small } \mu_0 \tan\theta \qquad (10.3.2)$$

indicating that in the ascending part of the asperity one may simply add the friction force and the asperity slope to one another. Similarly, on the right-hand side (descending part) of the asperity

$$\mu_2 = (\mu_0 - \tan\theta)/(1 + \mu_0 \tan\theta) \sim \mu_0 - \tan\theta, \text{ for small } \mu_0 \tan\theta \qquad (10.3.3)$$

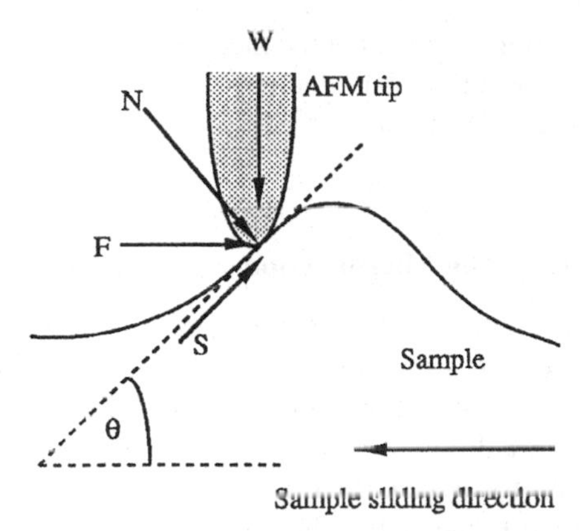

Figure 10.3.11 Schematic illustration showing the effect of an asperity (making an angle θ with the horizontal plane) on the surface in contact with the tip on local friction in the presence of adhesive friction mechanism. W and F are the normal and friction forces, respectively, and S and N are the force components along and perpendicular to the local surface of the sample at the contact point, respectively.

For a symmetrical asperity, the average coefficient of friction experienced by the FFM tip traveling across the whole asperity is

$$\begin{aligned}\mu_{ave} &= (\mu_1 + \mu_2)/2 \\ &= \mu_0(1+\tan^2\theta)/(1-\mu_0^2\tan^2\theta) \sim \mu_0(1+\tan^2\theta), \text{ for small } \mu_0\tan\theta \end{aligned} \tag{10.3.4}$$

Finally we consider the plowing component of friction with tip sliding in either direction, which is (Bhushan, 1999a, 2011)

$$\mu_p \sim \tan\theta \tag{10.3.5}$$

Because in FFM measurements we notice little damage of the sample surface, the contribution by plowing is expected to be small, and the ratchet mechanism is believed to be the dominant mechanism for the local variations in the friction force map. With the tip sliding over the leading (ascending) edge of an asperity, the surface slope is positive; it is negative during sliding over the trailing (descending) edge of an asperity. Thus, measured friction is high at the leading edge of asperities and low at the trailing edge. In addition to the slope effect, the collision of the tip when encountering an asperity with a positive slope produces additional torsion of the cantilever beam leading to higher measured friction force. When encountering an asperity with the same negative slope, however, there is no collision effect and hence no effect on torsion. This effect also contributes to the difference in friction forces when the tip scans up and down on the same topography feature. The ratchet mechanism and the collision effects thus semi-quantitatively explain the correlation between the slopes of the roughness maps and friction force maps observed in Figure 10.3.10. We note that in the ratchet mechanism, the FFM tip is assumed to be small compared to the size of asperities. This is valid since the typical radius of curvature of the tips is about 10–50 nm. The radii of curvature of the asperities of the samples measured here (the asperities that produce most of the friction variation) are found to typically be about 100–200 nm, which is larger than that of the FFM tip (Bhushan and

Blackman, 1991). It is important to note that the measured local values of friction and normal forces are measured with respect to global (and not local) horizontal and vertical axes, which are believed to be relevant in applications.

10.3.2.3 Directionality Effect on Microfriction

During friction measurements, the friction force data from both the forward (trace) and backward (retrace) scans are useful in understanding the origins of the observed friction forces. Magnitudes of material-induced effects are independent of the scanning direction whereas topography-induced effects are different between forward and backward scanning directions. Since the sign of the friction force changes as the scanning direction is reversed (because of the reversal of torque applied to the end of the tip), addition of the friction force data of the forward and backward scan eliminates the material-induced effects while topography-induced effects still remain. Subtraction of the data between forward and backward scans does not eliminate either effect, Figure 10.3.12 (Sundararajan and Bhushan, 2000).

Owing to the reversal of the sign of the retrace (R) friction force with respect to the trace (T) data, the friction force variations due to topography are in the same direction (peaks in trace correspond to peaks in retrace). However, the magnitudes of the peaks in trace and retrace at a given location are different. An increase in the friction force experienced by the tip when scanning up a sharp change in topography is more than the decrease in the friction force experienced when scanning down the same topography change, partly because of the collision effects discussed earlier. Asperities on engineering surfaces are asymmetrical, which also affect the magnitude of friction force in the two directions. Asymmetry in tip shape may also have an effect on the directionality effect of friction. We will note later that the magnitude of surface slopes are virtually identical, therefore, the tip shape asymmetry should not have much effect.

Figure 10.3.13 shows the surface height and friction force data for a silicon grid in the trace and retrace directions. Subtraction of two friction data yields a residual peak because of the differences in the magnitudes of friction forces in the two directions. This effect is observed at all locations of significant changes in topography.

In order to facilitate comparison of the directionality effect on friction, it is important to take into account the sign change of the surface slope and friction force in the trace and retrace directions. Figure 10.3.14 shows surface height, surface slope, and friction force data for a silicon grid in the trace and retrace directions. The correlations between surface slope and friction forces are clear. The third column in the figures shows retrace slope and friction data with an inverted sign (-retrace). Now we can compare trace data with -retrace data. It is clear that the friction experienced by the tip is dependent upon the scanning direction because of surface topography. In addition to the effect of topographical changes discussed earlier, during surface-finishing processes, material can be transferred preferentially onto one side of the asperities, which also causes asymmetry and direction dependence. Reduction in local variations and in the directionality of friction properties requires careful optimization of surface roughness distributions and of surface-finishing processes.

The directionality as a result of surface asperities effect will be also manifested in macroscopic friction data, i.e., the coefficient of friction may be different in one sliding direction than that in the other direction. The asymmetrical shape of the asperities accentuates this

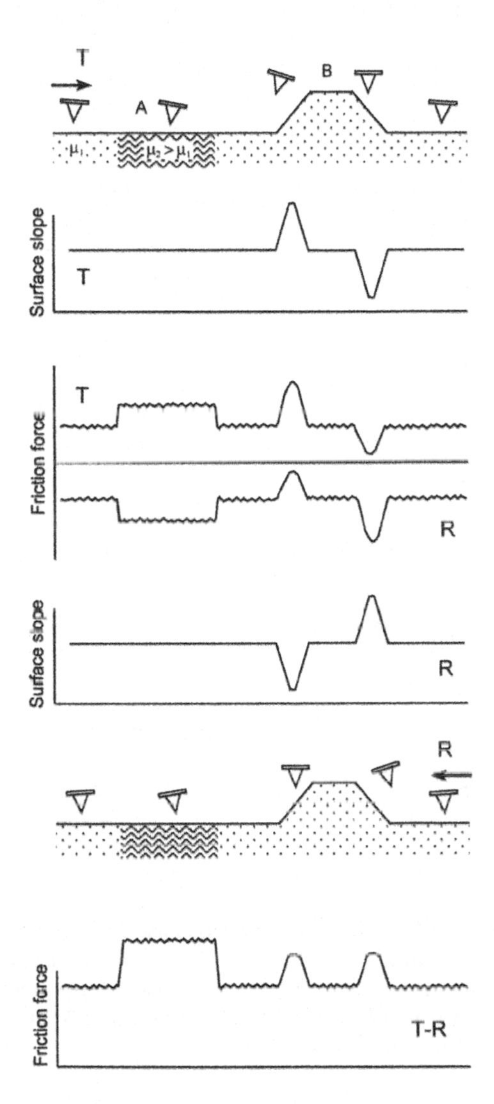

Figure 10.3.12 Schematic of friction forces expected when a tip traverses a sample that is composed of different materials and sharp changes in topography. A schematic of surface slope is also shown.

effect. The frictional directionality can also exist in materials with particles having a preferred orientation. The directionality effect in friction on a macroscale is observed in some magnetic tapes. In a macroscale test, a 12.7-mm wide polymeric magnetic tape was wrapped over an aluminum drum and slid in a reciprocating motion with a normal load of 0.5 N and a sliding speed of about 60 mm/s (Bhushan, 1995). The coefficient of friction as a function of sliding distance in either direction is shown in Figure 10.3.15. We note that the coefficient of friction on a macroscale for this tape is different in different directions. Directionality in friction is sometimes observed on the macroscale; on the microscale this is the norm (Bhushan, 1996, 1999a, 2011). On the macroscale, the effect of surface asperities normally is averaged out over a large number of contacting asperities.

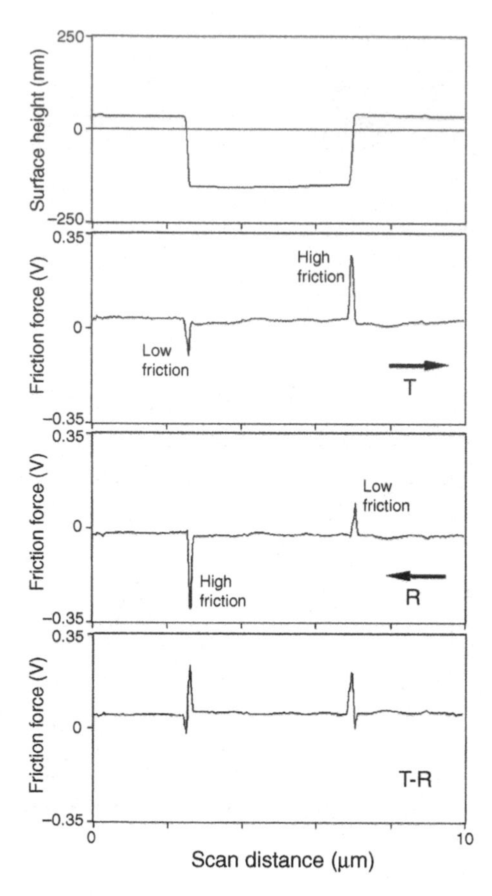

Figure 10.3.13 Two dimensional profiles of surface height and friction forces across a silicon grid pit. Friction force data in trace and retrace directions, and substrated force data are presented. Reproduced with permission from Sundararajan, S. and Bhushan, B. (2000), "Topography-Induced Contributions to Friction Forces Measured Using an Atomic Force/Friction Force Microscope," *J. Appl. Phys.* **88**, 4825–4831. Copyright 2000 American Institute of Physics.

10.3.2.4 Surface Roughness–Independent Microscale Friction

As just reported, the friction contrast in conventional friction measurements is based on interactions dependent upon interfacial material properties superimposed by roughness-induced lateral forces, and the cantilever twist is dependent on the sliding direction because of the local surface slope. Hence it is difficult to separate the friction-induced from the roughness-induced cantilever twist in the image. To obtain the roughness-independent friction, lateral or torsional modulation techniques are used in which the tip is oscillated in-plane with a small amplitude at a constant normal load, and change in shape and magnitude of cantilever resonance is used as a measure of friction force (Yamanaka and Tomita, 1995; Scherer *et al.*, 1997, 1998, 1999; Reinstaedtler *et al.*, 2003, 2005a, b; Bhushan and Kasai, 2004). These techniques also allow measurements over a very small area (few nm to few μm).

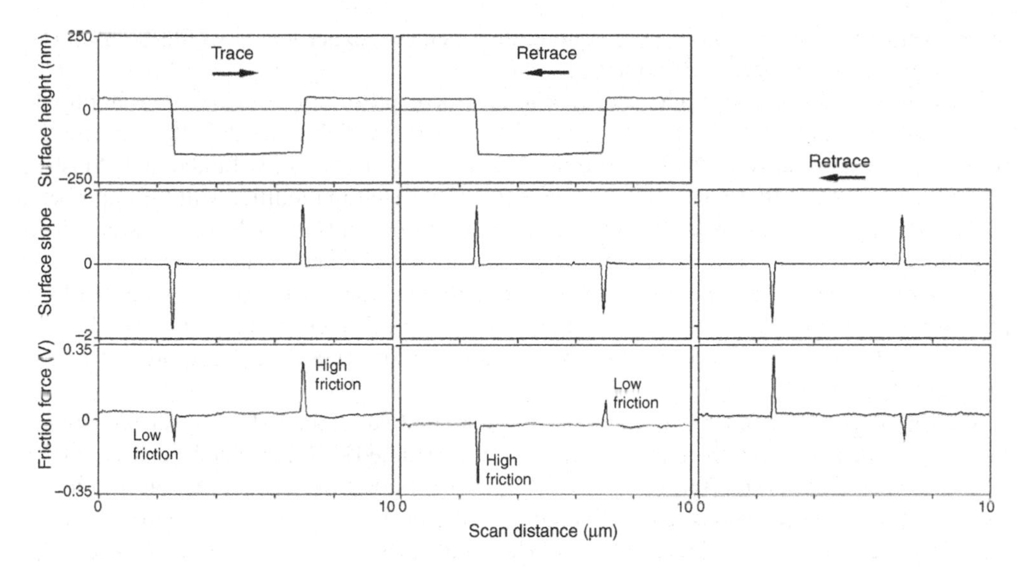

Figure 10.3.14 Two dimensional profiles of surface heights, surface slopes and friction forces for scans across the silicon grid pit. Arrows indicate the tip sliding direction. Reproduced with permission from Sundararajan, S. and Bhushan, B. (2000), "Topography-Induced Contributions to Friction Forces Measured Using an Atomic Force/Friction Force Microscope," *J. Appl. Phys.* **88**, 4825–4831. Copyright 2000 American Institute of Physics.

Bhushan and Kasai (2004) performed friction measurements on a silicon ruler and demonstrated that friction data in torsional resonance (TR) mode is essentially independent of surface-roughness and sliding direction.

10.3.2.5 Velocity Dependence on Micro/Nanoscale Friction

AFM/FFM experiments can be generally conducted at relative velocities as high as about 100–250 μm/s. To simulate applications, it is of interest to conduct friction experiments at

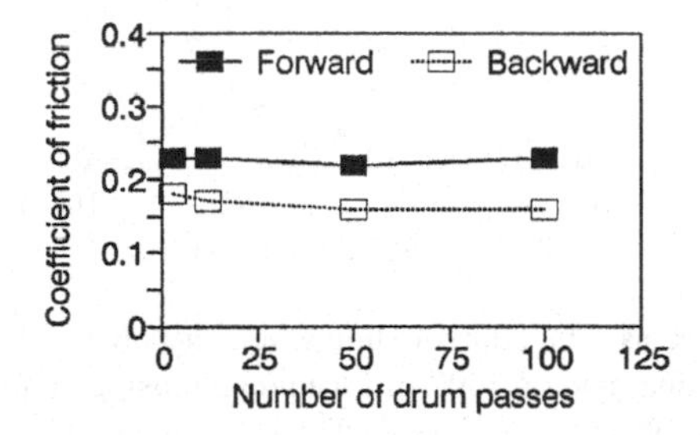

Figure 10.3.15 Coefficient of macroscale friction as a function of drum passes for a polymeric magnetic tape sliding over an aluminum drum in a reciprocating mode in both directions. Normal load = 0.5 N over 12.7-mm wide tape, sliding speed = 60 mm/s. Reproduced with permission from Bhushan, B. (1995), "Micro/Nanotribology and its Applications to Magnetic Storage Devices and MEMS," *Tribol. Int.* **28**, 85–95. Copyright 1995 Elsevier.

higher velocities (up to 1 m/s). Furthermore, high velocity experiments would be useful to study velocity dependence on friction and wear.

An approach to achieve high velocities is to utilize piezo stages with large amplitude (~10–100 μm) and relatively low resonance frequency (few kHz) and directly measure the friction force on microscale using the FFM signal. In the research by Tambe and Bhushan (2005a) and Tao and Bhushan (2006), a commercial AFM setup modified with this approach yielded sliding velocities up to 200 mm/s. During the experiments, the AFM cantilever is held stationary by maintaining a scan size of zero. The mounted sample is scanned below the AFM tip by moving stages, and the normal and torsional deflections of the tip are recorded by a photodiode detector. The raw deflection signals from the optical detection system are directly routed to a high speed data acquisition A/D board. Raw friction data is acquired at a high sampling rate up to 80 kilosamples/s.

Velocity dependence on friction for Si(100), diamondlike carbon (DLC), self-assembled monolayer, and perfluoropolyether lubricant films were studied by Tambe and Bhushan (2004a, 2005a, b, c, e) and Tao and Bhushan (2006, 2007). The friction force as a function of velocity for Si (100) and DLC (deposited by filtered cathodic arc) is shown in Figure 10.3.16 on a logarithm velocity scale (middle column). The solid lines in Figure 10.3.16 represent the results on a scan length of 1000 μm with a velocity ranging from 1000 μm/s to 2×10^5 μm/s using the ultrahigh velocity stage. The dotted lines represent results on a 25 μm scan length with velocity ranging from 5 μm/s to 500 μm /s using the high velocity stage. To clearly

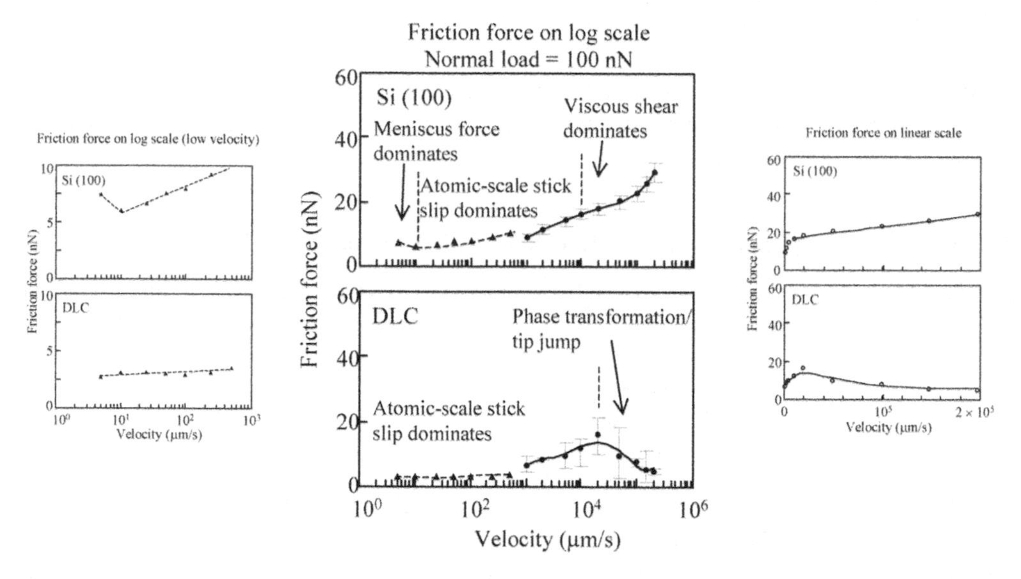

Figure 10.3.16 Friction force as a function of sliding velocity obtained on 25 μm scan length using a high velocity stage (dotted line) and on 1000 μm scan length using an ultrahigh velocity stage (solid line). In the left and middle graphs, velocity is plotted on log scale. Left column shows at lower range of the velocity – between 1 and 500 μm/s. Right column shows the data at higher range of velocity on the linear scale. Reproduced with permission from Tao, Z. and Bhushan, B. (2007), "Velocity Dependence and Rest Time Effect in Nanoscale Friction of Ultrathin Films at High Sliding Velocities," *J. Vac. Sci. Technol. A* **25**, 1267–1274. Copyright 2007. American Vacuum Society.

show the friction force dependence to velocity in the lower range, the test results with velocity varying from 5 μm/s to 500 μm/s on 25 μm are shown on a magnified scale in the left column of Figure 10.3.16.

On the Si (100) sample, the friction force decreased with velocity at low velocities ($v <$ 10 μm/s) and then increased linearly with log(v) on a 25 μm scan length. On the 1000 μm scan length, the friction force increased linearly with log(v) when velocity was lower than 2×10^4 μm/s. When velocities were higher than 2×10^4 μm/s, the friction force increased linearly with velocity. For DLC, the friction force increased linearly with log(v) from 5 to 500 μm/s on 25 μm scan length. On 1000 μm scan length, the friction force increases with velocity until about 2×10^4 μm/s where the friction force reaches a maximum, then the friction force decreased with velocity.

For different samples, the change in the friction force with velocity involves different mechanisms due to the sample surface conditions. The silicon surface is hydrophilic, and DLC surface is nearly hydrophobic. Under the ambient condition, a thin layer of water film is condensed on a hydrophilic sample surface. On a hydrophobic surface with high contact angle, the water film would be difficult to form on the sample surface, and the effect of the water film on the adhesive force and friction force could be neglected.

On the silicon surface, when the velocity is lower than 10 μm/s, the friction force decreased with velocity. This can be explained as follows. The water meniscus bridges develop as a function of time around the tip until reaching the equilibrium condition and are the dominant contributor to the friction force (Bhushan, 1999a, 2011). The motion of the tip results in continuous breaking and reforming of the meniscus bridges. As the tip sliding velocity exceeds a critical velocity (10 μm/s), there is not sufficient time for the menisci to reform, and the meniscus force would no longer play a dominant role. Between 10 and 2×10^4 μm/s, the friction increases linearly with log (v) on both 25 μm and 1000 μm scan lengths. This logarithmic dependence can be explained by the atomic-scale stick slip (Tambe and Bhushan, 2005b; Tao and Bhushan, 2007). At a velocity larger than 2×10^4 μm/s, the friction increases linearly with the velocity, and this trend can be explained by viscous shear; see the friction force plotted as a function of velocity on a linear scale on a magnified scale in the right column of Figure 10.3.16.

For the DLC film, since the surface is nearly hydrophobic, a uniform water film would not form on the surface. When sliding at a velocity lower than 1000 μm/s, the friction force increased linearly with log(v), which could also be explained by atomic-scale stick slip. At velocities higher than 1000 μm/s, the friction force increased with velocity until the local maximum at the velocity of 2×10^4 μm/s, then decreased with velocity. The decreasing trend in friction at higher velocities could be due to tip jump during sliding. The tip jump results in the reduction of lateral force during sliding. Variation of friction force with distance, indicative of the tip jump, was observed from the lateral force signal (not shown). When damping is low and velocity is high, the tip could jump several periodical cycles or several peaks (Fusco and Fasolino, 2005). At a given low damping coefficient, the slip results in a low transient lateral force, as discussed by Fusco and Fasolino (2005). Thus the average lateral force (friction force) over the scan length is low. The tip jump could also cause high velocity impact of asperities on DLC surface, resulting in the phase transformation of DLC from sp^3 to sp^2, as explained by Tambe and Bhushan (2005b). The layer of sp^2 phase can act as lubricant and reduce the interfacial friction.

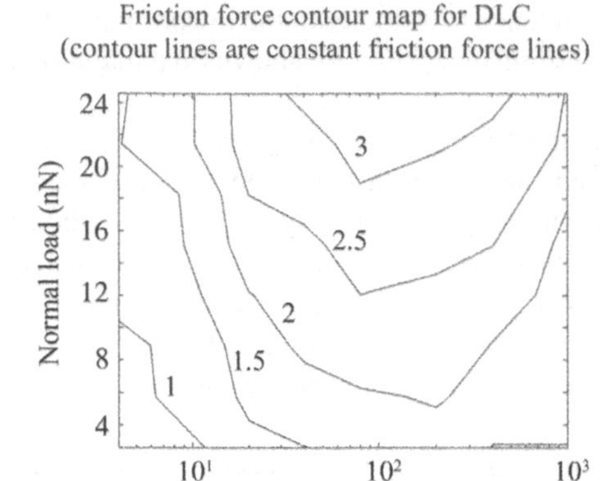

Figure 10.3.17 Contour map showing friction force dependence on normal load and sliding velocity for DLC. Reproduced with permission from Tambe, N.S. and Bhushan, B. (2005f), "Nanoscale Friction Mapping," *Appl. Phys. Lett.* **86**, 193102-1 to -3. Copyright 2005 American Institute of Physics.

10.3.2.6 Nanoscale Friction and Wear Mapping

Contrary to the classical friction laws postulated by Amontons and Coulomb centuries ago, nanoscale friction force is found to be strongly dependent on the normal load and sliding velocity. Many materials, coatings, and lubricants that have wide applications show reversals in friction behavior corresponding to transitions between different friction mechanisms (Tambe and Bhushan, 2004a, 2005a, b, c, 2008). Most of the analytical models developed to explain nanoscale friction behavior have remained limited in their focus and have left investigators at a loss when trying to explain friction behavior scaling multiple regimes. Nanoscale friction maps provide fundamental insights into friction behavior. They help identify and classify the dominant friction mechanisms, as well as determine the critical operating parameters that influence transitions between different mechanisms (Tambe and Bhushan, 2005b, c). Figure 10.3.17 shows a nanoscale friction map for DLC with the friction mapped as a function of the normal load and the sliding velocity (Tambe and Bhushan, 2005f). The contours represent constant friction force lines. The friction force is seen to increase with normal load as well as velocity. The increase in friction force with velocity is the result of atomic scale stick-slip. This is a result of thermal activation of the irreversible jumps of the AFM tip that arise from overcoming the energy barrier between the adjacent atomic positions, as described earlier. The concentric contour lines corresponding to the constant friction force predict a peak point, a point where the friction force reaches maxima and beyond which point any further increase in the normal load or the sliding velocity results in a decrease in friction force. This characteristic behavior for DLC is the result of phase transformation of DLC into a graphite-like phase by sp^3 to sp^2 phase transition, as described earlier. During the AFM experiments, the Si_3N_4 tip gives rise to contact pressures in the range of 1.8–4.4 GPa for DLC for normal loads of 10–150 nN (Tambe and Bhushan, 2005d). A combination of the high contact pressures that are encountered on the nanoscale and the high frictional energy dissipation arising from the asperity impacts at the tip–sample interface due to the high sliding velocities accelerates a phase transition process whereby a low shear strength graphite-like layer is formed at the sliding interface.

AFM image showing wear map as a function of normal load and sliding velocity

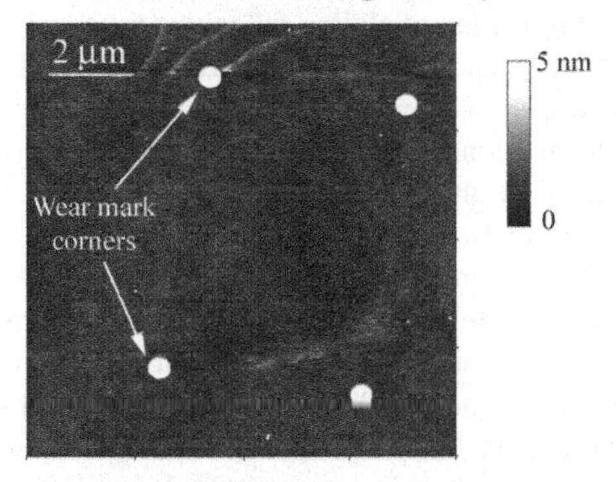

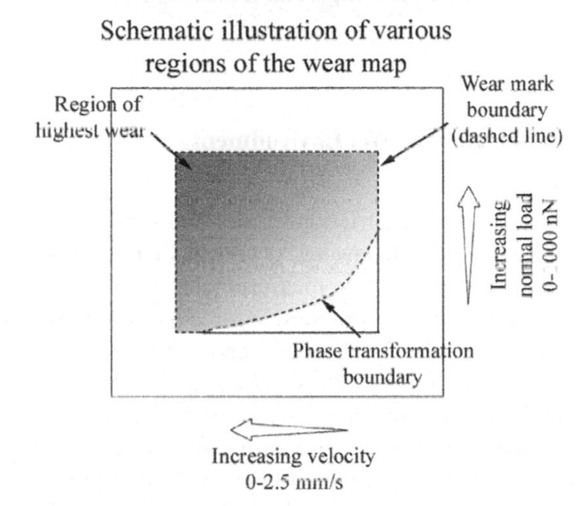

Figure 10.3.18 Nanowear map (AFM image and schematic) illustrating the effect of sliding velocity and normal load on the wear of DLC resulting from phase transformation. Curved area shows debris lining and is indicative of the minimum frictional energy needed for phase transformation. For clarity, the wear mark corners are indicated by white dots in the AFM image and the various zones of interest over the entire wear mark are schematically illustrated. Reproduced with permission from Tambe, N.S. and Bhushan, B. (2005g), "Nanowear Mapping: A Novel Atomic Force Microscopy Based Approach for Studying Nanoscale Wear at High Sliding Velocities," *Tribol. Lett.* **20**, 83–90. Copyright 2005 Springer.

Similar to friction mapping, one way of exploring the broader wear patterns is to construct wear mechanism maps that summarize data and models for wear, thereby showing mechanisms for any given set of conditions to be identified (Lim and Ashby, 1987; Lim *et al.*, 1987; Tambe and Bhushan 2005g, 2008). Wear of sliding surfaces can occur through one or more wear mechanisms, including adhesive, abrasive, fatigue, impact, corrosive, and fretting. Tambe and Bhushan (2005d, g) performed AFM experiments to develop nanoscale wear maps. Figure 10.3.18 shows a nanowear map generated for a DLC sample by simultaneously varying the normal load and the sliding velocity over the entire scan area. The wear map was generated for a normal load range of 0–1000 nN and sliding velocity range of 0–2.5 mm/s. Wear debris, believed to be resulting from phase transformation of DLC by sp^3 to sp^2 phase transition, was seen to form only for a high value of sliding velocities times normal loads, i.e., only beyond a certain threshold of friction energy dissipation (Tambe and Bhushan, 2005d, g). Hence the wear region exhibits a transition line indicating that for low velocities and low normal loads there is no phase transformation. For clarity, the wear mark corners are indicated by white dots in the AFM image (top) and the two zones of interest over the entire wear mark are schematically illustrated in Figure 10.3.18 (top).

Nanoscale friction and wear mapping are novel techniques for investigating friction and wear behavior on the nanoscale over a range of operating parameters. By simultaneously varying the sliding velocity and normal load over a large range of values, nanoscale friction and wear behavior can be mapped, and the transitions between different wear mechanisms can be investigated. These maps help identify and demarcate critical operating parameters for different wear mechanisms and are very important tools in the process of design and selection of materials/coatings.

10.3.2.7 Adhesion and Friction in Wet Environment

Experimental Observations

Relative humidity affects adhesion and friction for dry and lubricated surfaces (Bhushan and Sundararajan, 1998; Bhushan and Dandavate, 2000; Bhushan, 2003). Figure 10.3.19 shows the variation of single point adhesive force measurements as a function of a tip radius on a Si(100) sample for several humidities. The adhesive force data are also plotted as a function of relative humidity for several tip radii. The general trend at humidities up to the ambient is that a 50-nm radius Si_3N_4 tip exhibits a lower adhesive force as compared to the other microtips of larger radii; however, in the latter case, values are similar. Thus, for the microtips there is no appreciable variation in adhesive force with tip radius at a given humidity up to the ambient. The adhesive force increases as relative humidity increases for all tips.

The sources of adhesive force between a tip and a sample surface are van der Waals attraction and meniscus formation (Bhushan, 2003). The relative magnitudes of the forces from the two sources are dependent upon various factors including the distance between the tip and the sample surface, their surface roughness, their hydrophobicity, and the relative humidity (Stifter *et al.*, 2000). For most rough surfaces, the meniscus contribution dominates at moderate to high humidities which arise from capillary condensation of water vapor from the environment. If enough liquid is present to form a meniscus bridge, the meniscus force should increase with an increase in the tip radius (proportional to the tip radius for a spherical tip). In addition, an increase in the tip radius results in an increased contact area leading to higher values of van der Waals forces. However, if nanoasperities on the tip and the sample are considered then the number of contacting and near- contacting asperities forming meniscus bridges increases with an increase of humidity leading to an increase in meniscus forces. These explain the trends observed in Figure 10.3.19. From the data, the tip radius has little effect on the adhesive forces at low humidities but increases with tip radius at high humidity. Adhesive force also increases with an increase in humidity for all tips. This observation suggests that thickness of the liquid film at low humidities is insufficient to form continuous meniscus bridges to affect adhesive forces in the case of all tips.

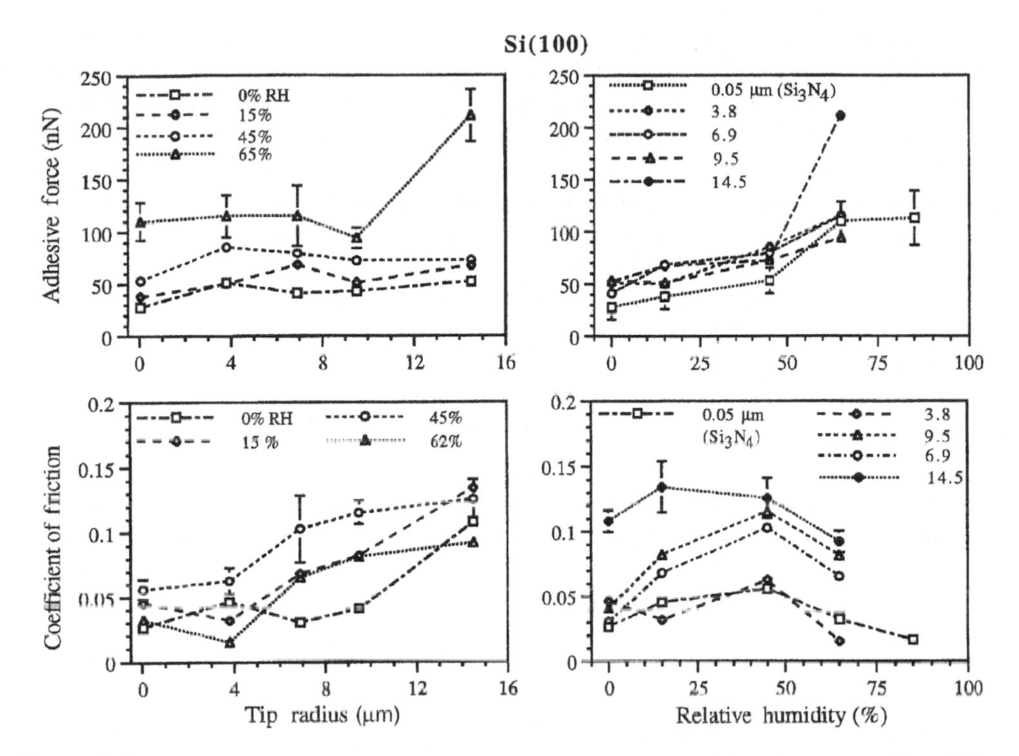

Figure 10.3.19 Adhesive force and coefficient of friction as a function of tip radius at several humidities and as a function of relative humidity at several tip radii on Si(100). Reproduced with permission from Bhushan, B. and Sundararajan, S. (1998), "Micro/nanoscale Friction and Wear Mechanisms of Thin Films Using Atomic Force and Friction Force Microscopy," *Acta Mater.* **46**, 3793–3804. Copyright 1998. Elsevier.

Figure 10.3.19 also shows the variation in the coefficient of friction as a function of the tip radius at a given humidity, and as a function of the relative humidity for a given tip radius for Si(100). It can be observed that for 0% RH, the coefficient of friction is about the same for the tip radii except for the largest tip, which shows a higher value. At all other humidities, the trend consistently shows that the coefficient of friction increases with the tip radius. An increase in friction with tip radius at low to moderate humidities arises from the increased contact area (due to higher van der Waals forces) and higher values of the shear forces required for a larger contact area. At high humidities, similar to adhesive force data, an increase with tip radius occurs because of both contact area and meniscus effects. Although the AFM/FFM measurements are able to measure the combined effect of the contribution of van der Waals and meniscus forces towards the friction force or adhesive force, it is difficult to measure their individual contributions separately. It can be seen that for all tips, the coefficient of friction increases with humidity to about ambient, beyond which it starts to decrease. The initial increase in the coefficient of friction with humidity arises from the fact that the thickness of the water film increases with an increase in the humidity, which results in a larger number of nanoasperities forming meniscus bridges and leads to higher friction (a larger shear force).

The same trend is expected with the microtips beyond 65% RH. This is attributed to the fact that at higher humidities, the adsorbed water film on the surface acts as a lubricant between the two surfaces. Thus, the interface is changed at higher humidities, resulting in lower shear strength and hence lower friction force and coefficient of friction.

Adhesion and Friction Force Expressions for a Single Asperity Contact

We now obtain the expressions for the adhesive force and coefficient of friction for a single asperity contact with a meniscus formed at the interface. For a spherical asperity of radius R in contact with a flat and smooth surface with the composite modulus of elasticity E^* and in the presence of liquid with a concave meniscus, the attractive meniscus force (adhesive force), designated as F_m or W_{ad}, is given as (Chapter 4)

$$W_{ad} = 2\pi R\gamma(\cos\theta_1 + \cos\theta_2) \tag{10.3.6}$$

where γ is the surface tension of the liquid, and θ_1 and θ_2 are the contact angles of the liquid with surfaces 1 and 2, respectively. For an elastic contact for both extrinsic (W) and intrinsic (W_{ad}) normal load, the friction force is given as,

$$F_e = \pi\tau\left[\frac{3(W + W_{ad})R}{4E^*}\right]^{2/3} \tag{10.3.7}$$

where W is the external load, and τ is the average shear strength of the contacts. (The surface energy effects are not considered here.) Note that adhesive force increases linearly with an increase in the tip radius, and the friction force increases with an increase in the tip radius as $R^{2/3}$ and with normal load as $(W + W_{ad})^{2/3}$. The experimental data in support of $W^{2/3}$ dependence on the friction force can be found in various references (see e. g., Schwarz *et al.*, 1997). The coefficient of friction μ_e is obtained from Equation 10.3.7 as

$$\mu_e = \frac{F_e}{(W + W_{ad})} = \pi\tau\left[\frac{3R}{4E^*}\right]^{2/3}\frac{1}{(W + W_{ad})^{1/3}} \tag{10.3.8}$$

In the plastic contact regime, the coefficient of friction μ_p is obtained as

$$\mu_p = \frac{F_p}{(W + W_{ad})} = \frac{\tau}{H_s} \tag{10.3.9}$$

where H_s is the hardness of the softer material. Note that in the plastic contact regime, the coefficient of friction is independent of external load, adhesive contributions and surface geometry.

For comparisons, for multiple asperity contacts in the elastic contact regime the total adhesive force W_{ad} is the summation of adhesive forces at n individual contacts,

$$W_{ad} = \sum_{i=1}^{n}(W_{ad})_i \tag{10.3.10}$$

and

$$\mu_e \approx \frac{3.2\tau}{E^* \left(\sigma_p/R_p\right)^{1/2} + (W_{ad}/W)}$$

where σ_p and R_p are the standard deviation of summit heights and average summit radius, respectively. Note that the coefficient of friction depends upon the surface roughness. In the plastic contact regime, the expression for μ_p in Equation 10.3.9 does not change.

The source of the adhesive force in a wet contact in the AFM experiments being performed in an ambient environment includes mainly attractive meniscus force due to capillary condensation of water vapor from the environment. The meniscus force for a single contact increases with an increase in the tip radius. A sharp AFM tip in contact with a smooth surface at low loads (on the order of a few nN) for most materials can be simulated as a single-asperity contact. At higher loads, for rough and soft surfaces, multiple contacts would occur. Furthermore, at low loads (nN range) for most materials, the local deformation would be primarily elastic. Assuming that the shear strength of contacts does not change, the adhesive force for smooth and hard surfaces at low normal load (on the order of few nN) (for a single asperity contact in the elastic contact regime) would increase with an increase in tip radius, and the coefficient of friction would decrease with an increase in total normal load as $(W + W_{ad})^{-1/3}$ and would increase with an increase of tip radius as $R^{2/3}$. In this case, Amontons' law of friction which states that the coefficient of friction is independent of the normal load and is independent of the apparent area of contact, does not hold. For a single-asperity plastic contact and multiple-asperity plastic contacts, neither the normal load nor tip radius comes into play in the calculation of the coefficient of friction. In the case of multiple-asperity contacts, the number of contacts increases with an increase of normal load, therefore the adhesive force increases with an increase in load.

In the data presented earlier in this section, the effect of tip radius and humidity on the adhesive forces and the coefficient of friction are investigated for experiments with Si(100) surface at loads in the range of 10–100 nN. The multiple asperity elastic-contact regime is relevant for this study involving large tip radii. An increase in humidity generally results in an increase in the number of meniscus bridges, which would increase the adhesive force. As was suggested earlier, that increase in humidity also may decrease the shear strength of contacts. A combination of an increase in adhesive force and a decrease in shear strength would affect the coefficient of friction. An increase in th tip radius would increase the meniscus force (adhesive force). A substantial increase in the tip radius may also increase the interatomic forces. These effects influence the coefficient of friction with an increase in the tip radius.

10.3.2.8 Scale Dependence in Friction

Table 10.3.1 presents the adhesive force and the coefficient of friction data obtained on the nanoscale and microscale (Ruan and Bhushan, 1994a; Liu and Bhushan, 2003b; Bhushan *et al.*, 2004b; Tambe and Bhushan, 2004a). Adhesive force and coefficient of friction values on the nanoscale are about half to one order of magnitude lower than that on the microscale. Scale dependence is clearly observed in this data.

Table 10.3.1 Micro- and nanoscale values of adhesive force and coefficient of friction in micro- and nanoscale measurements (*Source*: Bhushan *et al.*, 2004b).

	Adhesive force		Coefficient of friction	
Sample	Microscale[a] (μN)	Nanoscale[b] (nN)	Microscale[a]	Nanoscale[b]
Si(100)	685	52	0.47	0.06
DLC	325	44	0.19	0.03
Z-DOL	315	35	0.23	0.04
HDT	180	14	0.15	0.006

[a]Versus 500-μm radius Si(100) ball
[b]Versus 50-nm radius Si_3N_4 tip

There are several factors responsible for the differences in the coefficients of friction at the micro- and nanoscale. Among them are the contributions from wear and contaminant particles, transition from elasticity to plasticity, and meniscus effect (Bhushan and Nosonovsky, 2003, 2004a, b; Nosonovsky and Bhushan, 2005). The contribution of wear and contaminant particles is more significant at the macro/microscale because of the larger number of trapped particles, referred to as third body contribution. It can be argued that for the nanoscale AFM experiments the asperity contacts are predominantly elastic (with average real pressure being less than the hardness of the softer material), and adhesion is the main contribution to the friction, whereas for the microscale experiments, the asperity contacts are predominantly plastic, and deformation is an important factor. It will be shown later that hardness has scale effect; it increases with decreasing scale and is responsible for less deformation on a smaller scale. The meniscus effect results in an increase of friction with increasing tip radius (Figure 10.3.19). Therefore, third body contribution, scale-dependent hardness, and other properties transition from elastic contacts in nanoscale contacts to plastic deformation in microscale contacts, and meniscus contribution plays an important role.

To demonstrate the load dependence of friction at the nano/microscale, the coefficient of friction as a function of the normal load is presented in Figure 10.3.20. The coefficient of friction was measured by Bhushan and Kulkarni (1996) for Si_3N_4 tip versus Si, SiO_2, and natural diamond using an AFM. They reported that for low loads, the coefficient of friction is independent of load and increases with increasing load after a certain load. It is noted that the critical value of loads for Si and SiO_2 correspond to stresses equal to their hardness values, which suggests that transition to plasticity plays a role in this effect. The friction values at higher loads for Si and SiO_2 approach that of macroscale values.

10.3.3 Wear, Scratching, Local Deformation, and Fabrication/Machining

10.3.3.1 Nanoscale Wear

Bhushan and Ruan (1994) conducted nanoscale wear tests on polymeric magnetic tapes using conventional silicon nitride tips at two different loads of 10 and 100 nN (Figure 10.3.21). For a low normal load of 10 nN, measurements were made twice. There was no discernible difference between consecutive measurements for this load. However, as the load was increased from

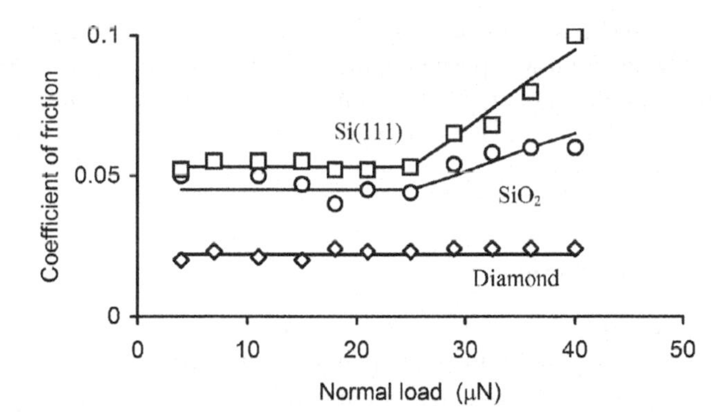

Figure 10.3.20 Coefficient of friction as a function of normal load and for Si(111), SiO_2 coating and natural diamond. Inflections in the curves for silicon and SiO_2 correspond to the contact stresses equal to the hardnesses of these materials. Reproduced with permission from Bhushan, B. and Kulkarni, A.V. (1996), "Effect of Normal Load on Microscale Friction Measurements," *Thin Solid Films* **278**, 49–56; 293, 333. Copyright 1996. Elsevier.

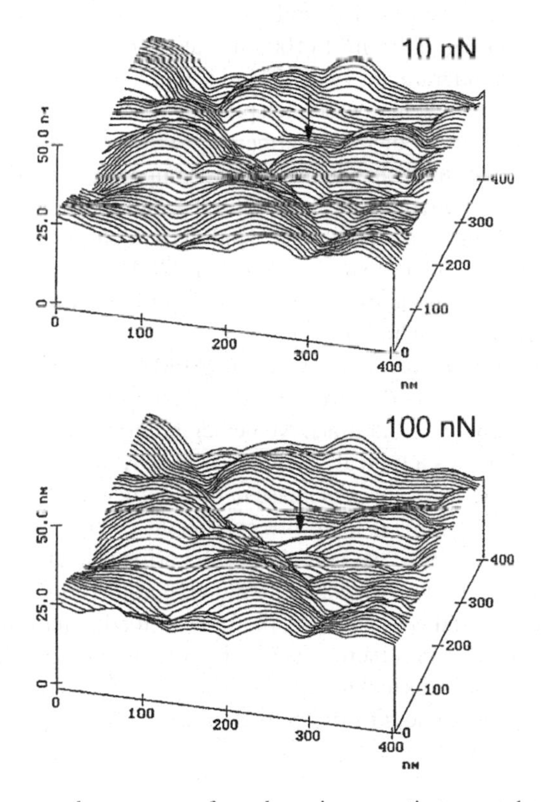

Figure 10.3.21 Surface roughness maps of a polymeric magnetic tape at the applied normal load of 10 nN and 100 nN. Location of the change in surface topography as a result of nanowear is indicated by arrows. Reproduced with permission from Bhushan, B. and Ruan, J. (1994), "Atomic-scale Friction Measurements Using Friction Force Microscopy: Part II – Application to Magnetic Media," *ASME J. Trib.* **116**, 389–396.

10 nN to 100 nN, topographical changes were observed during subsequent scanning at a normal load of 10 nN; material was pushed in the sliding direction of the AFM tip relative to the sample. The material movement is believed to occur as a result of plastic deformation of the tape surface. Thus, deformation and movement of the soft materials on a nanoscale can be observed.

10.3.3.2 Microscale Scratching

The AFM can be used to investigate how surface materials can be moved or removed on micro- to nanoscales, for example, in scratching and wear (Bhushan, 1999a, 2011) (where these things are undesirable) and nanofabrication/nanomachining (where they are desirable). Figure 10.3.22(a) shows microscratches made on Si(111) at various loads and a scanning velocity of 2 μm/s after 10 cycles (Bhushan *et al.*, 1994). As expected, the scratch depth increases linearly with load. Such microscratching measurements can be used to study failure mechanisms on the microscale and to evaluate the mechanical integrity (scratch resistance) of ultra-thin films at low loads.

To study the effect of scanning velocity, unidirectional scratches 5 μm in length were generated at scanning velocities ranging from 1 to 100 μm/s at various normal loads ranging from 40 to 140 μN (Bhushan and Sundararajan, 1998). There is no effect of scanning velocity obtained at a given normal load. For representative scratch profiles at 80 μN, see Figure 10.3.22(b). This may be because of a small effect of frictional heating with the change in scanning velocity used here. Furthermore, for a small change in interface temperature, there is a large underlying volume to dissipate the heat generated during scratching.

Scratching can be performed under ramped loading to determine the scratch resistance of materials and coatings (Sundararajan and Bhushan, 2001). The coefficient of friction is measured during scratching, and the load at which the coefficient of friction increases rapidly is known as the "critical load," which is a measure of scratch resistance. In addition, post-scratch imaging can be performed *in-situ* with the AFM in tapping mode to study failure mechanisms. Figure 10.3.23 shows data from a scratch test on Si(100) with a scratch length of 25 μm and a scratching velocity of 0.5 μm/s. At the beginning of the scratch, the coefficient of friction is 0.04, which indicates a typical value for silicon. At about 35 μN (indicated by the arrow in Figure 10.3.23), there is a sharp increase in the coefficient of friction, which indicates the critical load. Beyond the critical load, the coefficient of friction continues to increase steadily. In the post-scratch image, we note that at the critical load, a clear groove starts to form. This implies that Si(100) was damaged by plowing at the critical load, associated with the plastic flow of the material. At and after the critical load, small and uniform debris is observed, and the amount of debris increases with increasing normal load. Sundararajan and Bhushan (2001) have also used this technique to measure the scratch resistance of diamondlike carbon coatings ranging in thickness from 3.5 to 20 nm.

10.3.3.3 Microscale Wear

By scanning the sample in two dimensions with the AFM, wear scars are generated on the surface. Figure 10.3.24 shows the effect of normal load on wear depth on Si(100). We note that wear depth is very small below 20 μN of normal load (Koinkar and Bhushan, 1997c; Zhao

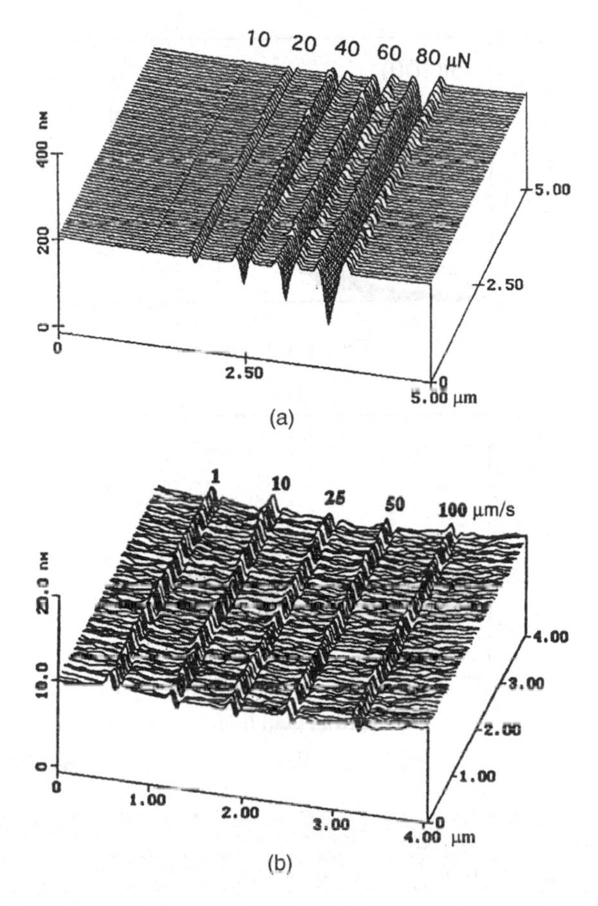

Figure 10.3.22 Surface plots of (a) Si(111) scratched for ten cycles at various loads and a scanning velocity of 2 μm/s (*Source*: Reproduced with permission from Bhushan, B., Koinkar V.N., and Ruan, J. (1994), "Microtribology of Magnetic Media," *Proc. Inst. Mech. Eng., Part J: J. Eng. Tribol.* **208**, 17–29. Copyright 1994 Sage Publications). Note that x and y axes are in μm and z axis is in nm, and (b) Si(100) scratched in one unidirectional scan cycle at a normal force of 80 μN and different scanning velocities (*Source*: Reproduced with permission from Bhushan, B. and Sundararajan, S. (1998), "Micro/nanoscale Friction and Wear Mechanisms of Thin Films Using Atomic Force and Friction Force Microscopy," *Acta Mater.* **46**, 3793–3804. Copyright 1998. Elsevier).

and Bhushan, 1998). A normal load of 20 μN corresponds to contact stresses comparable to the hardness of silicon. Primarily, elastic deformation at loads below 20 μN is responsible for low wear (Bhushan and Kulkarni, 1996).

A typical wear mark of the size 2 μm × 2 μm generated at a normal load of 40 μN for one scan cycle and imaged using AFM with scan size of 4 μm × 4 μm at 300 nN load is shown in Figure 10.3.25(a). The inverted map of wear marks shown in Figure 10.3.25(b) indicates the uniform material removal at the bottom of the wear mark (Koinkar and Bhushan, 1997c). An AFM image of the wear mark shows debris at the edges, probably swiped during AFM

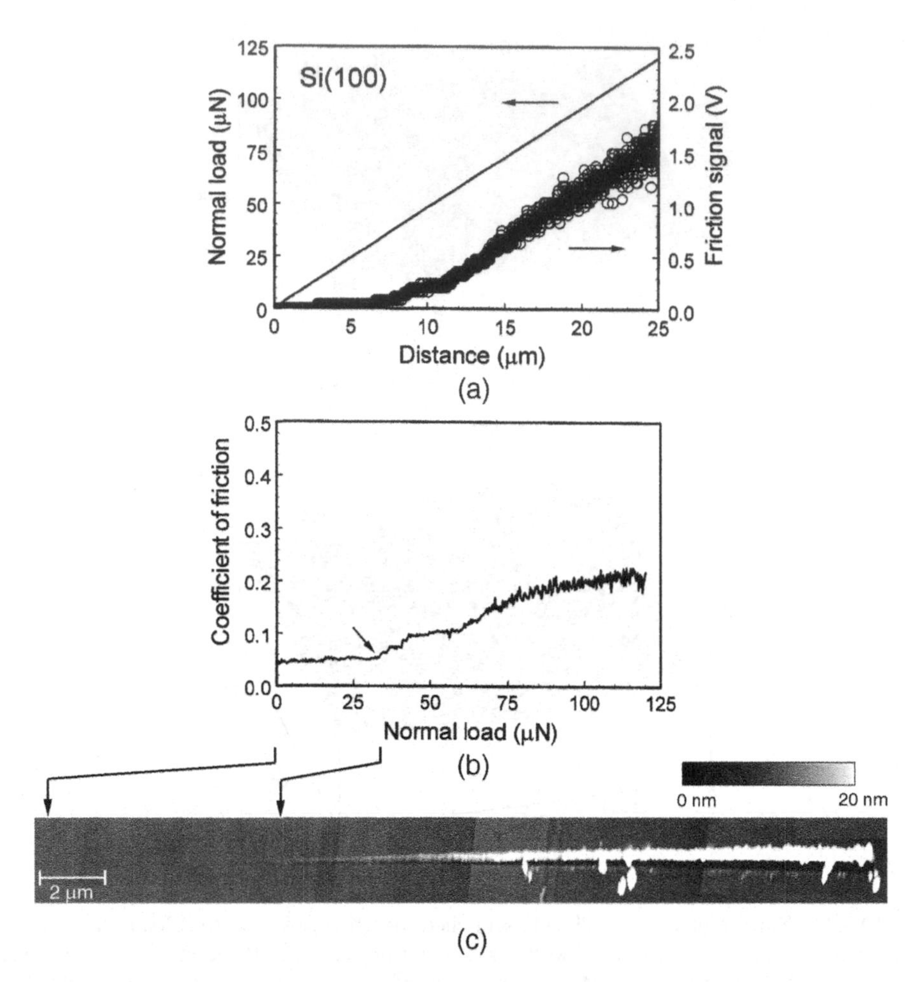

Figure 10.3.23 (a) Applied normal load and friction signal measured during the microscratch experiment on Si(100) as a function of scratch distance, (b) friction data plotted in the form of coefficient of friction as a function of normal load, and (c) AFM surface height image of scratch obtained in tapping mode. Reproduced with permission from Sundararajan, S. and Bhushan, B. (2001), "Development of a Continuous Microscratch Technique in an Atomic Force Microscope and its Application to Study Scratch Resistance of Ultra-Thin Hard Amorphous Carbon Coatings," *J. Mater. Res.* **16**, 75–84. Copyright 2001 Cambridge University Press.

scanning. This indicates that the debris is loose (not sticky) and can be removed during the AFM scanning.

Next, the mechanism of material removal on the microscale in AFM wear experiments is examined (Koinkar and Bhushan, 1997c; Bhushan and Sundararajan, 1998; Zhao and Bhushan, 1998). Figure 10.3.26(a) shows a secondary electron image of the wear mark and associated wear particles. The specimen used for the scanning electron microscope (SEM) was not scanned with the AFM after initial wear, in order to retain wear debris in the wear region. Wear debris is clearly observed. In the SEM micrographs, the wear debris appears to

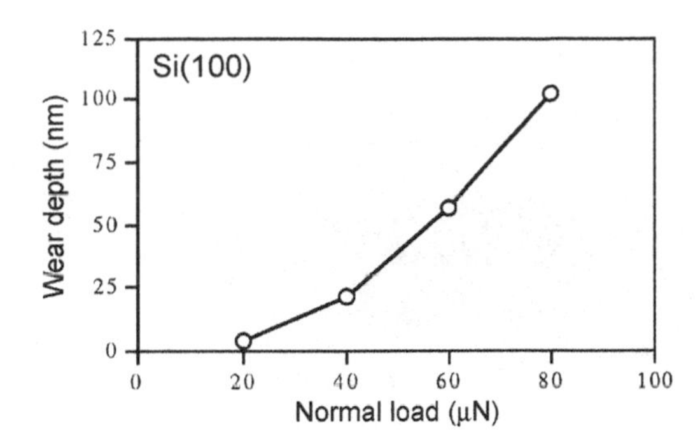

Figure 10.3.24 Wear depth as a function of normal load for Si(100) after one cycle. Reproduced with permission from Zhao, X. and Bhushan, B. (1998), "Material Removal Mechanism of Single-Crystal Silicon on Nanoscale and at Ultralow Loads," *Wear* **223**, 66–78. Copyright 1998. Elsevier.

be agglomerated because of the high surface energy of the fine particles. Particles appear to be a mixture of rounded and so called cutting type (feather-like or ribbon-like material). Zhao and Bhushan (1998) reported an increase in the number and size of cutting type particles with the normal load. The presence of cutting type particles indicates that the material is removed primarily by plastic deformation.

To better understand the material removal mechanisms, Zhao and Bhushan (1998) used transmission electron microscopy (TEM). The wear debris in Figure 10.3.26(b) shows that much of the debris is ribbon like, indicating that material is removed by a cutting process via plastic deformation, which is consistent with the SEM observations. The diffraction pattern from inside the wear mark was reported to be similar to that of virgin silicon, indicating no evidence of any phase transformation (amorphization) during wear. Diffraction patterns of the wear debris indicated the existence of amorphous material in the wear debris, confirmed as silicon oxide products from chemical analysis. It is known that plastic deformation occurs by generation and propagation of dislocations.

To understand wear mechanisms, evolution of wear can be studied using AFM. Figure 10.3.27 shows the evolution of wear marks of a DLC-coated disk sample. The data illustrate how the microwear profile for a load of 20 μN develops as a function of the number of scanning cycles (Bhushan *et al.*, 1994). Wear is not uniform, but is initiated at the nanoscratches. Surface defects (with high surface energy) present at the nanoscratches act as initiation sites for wear. Coating deposition also may not be uniform on and near nanoscratches which may lead to coating delamination. Thus, scratch-free surfaces will be relatively resistant to wear.

Wear precursors (precursors to measurable wear) can be studied by making surface potential measurements (DeVecchio and Bhushan, 1998; Bhushan and Goldade, 2000a, b). The contact potential difference, or simply the surface potential between two surfaces, depends on a variety of parameters such as electronic work function, adsorption, and oxide layers. The surface potential map of an interface gives a measure of changes in the work function which is sensitive to both physical and chemical conditions of the surfaces including structural and chemical changes. Before material is actually removed in a wear process, the surface experiences stresses that result in surface and subsurface changes of structure and/or chemistry. These can cause

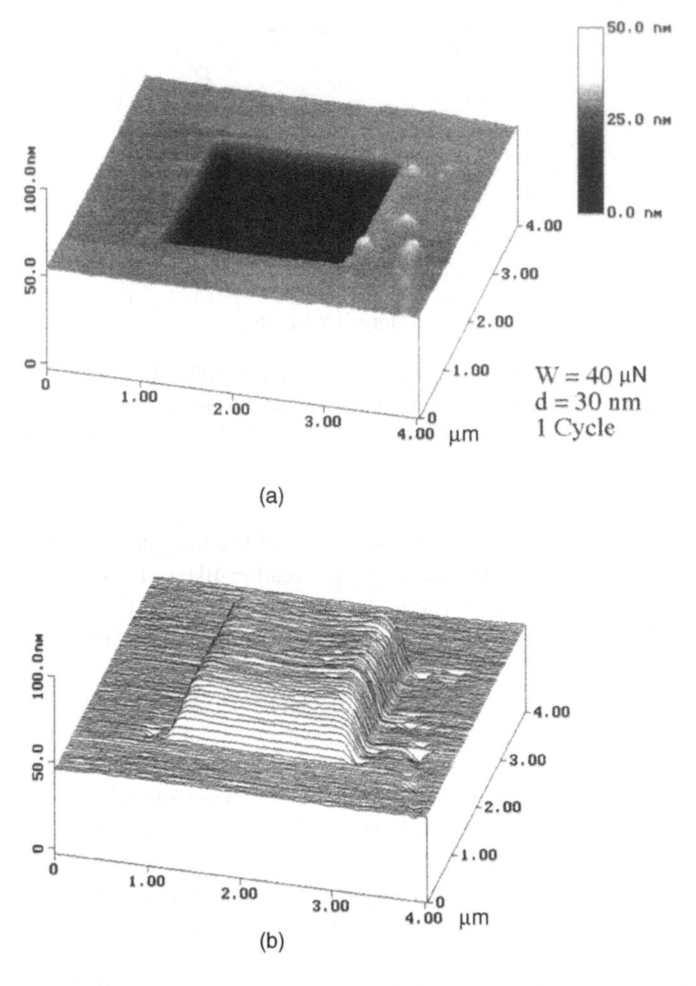

Figure 10.3.25 (a) Typical gray scale and (b) inverted AFM images of wear mark created using a diamond tip at a normal load of 40 µN and one scan cycle on Si(100) surface. Reproduced with permission from Koinkar, V.N. and Bhushan, B. (1997c), "Scanning and Transmission Electron Microscopies of Single-Crystal Silicon Microworn/Machined Using Atomic Force Microscopy," *J. Mater. Res.* **12**, 3219–3224. Copyright 1997 Cambridge University Press.

changes in the measured potential of a surface. An AFM tip allows the mapping of the surface potential with nanoscale resolution. Surface height and change in surface potential maps of a polished single-crystal aluminum (100) sample, abraded using a diamond tip at loads of 1 µN and 9 µN, are shown in Figure 10.3.28(a). It is evident that both abraded regions show a large potential contrast (~0.17 V), with respect to the non-abraded area. The black region in the lower right-hand part of the topography scan shows a step that was created during the polishing phase. There is no potential contrast between the high region and the low region of the sample, indicating that the technique is independent of surface height. Figure 10.3.28(b) shows a close up scan of the upper (low load) wear region in Figure 10.3.28(a). Notice that while there is

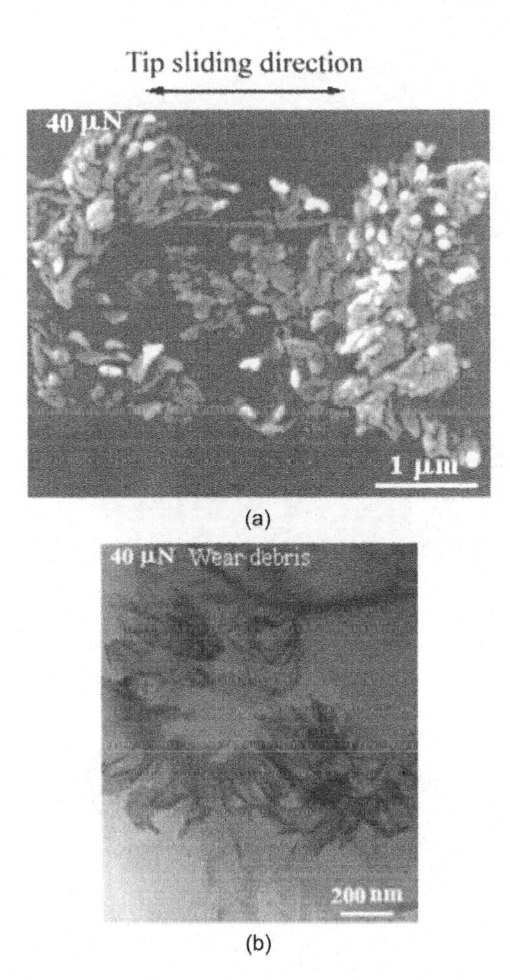

Figure 10.3.26 (a) Secondary electron micrograph and (b) bright field TEM micrograph of wear mark and debris for Si(100) produced at a normal load of 40 μN and one scan cycle. Reproduced with permission from Zhao, X. and Bhushan, B. (1998), "Material Removal Mechanism of Single-Crystal Silicon on Nanoscale and at Ultralow Loads," *Wear* **223**, 66–78. Copyright 1998. Elsevier.

no detectable change in the surface topography, there is nonetheless, a large change in the potential of the surface in the worn region. Indeed, the wear mark of Figure 10.3.28(b) might not be visible at all in the topography map were it not for the noted absence of wear debris generated nearby and then swept off during the low load scan. Thus, even in the case of zero wear (no measurable deformation of the surface using AFM), there can be a significant change in the surface potential inside the wear mark which is useful for the study of wear precursors. It is believed that the removal of the thin contaminant layer including the natural oxide layer gives rise to the initial change in surface potential. The structural changes, which precede the generation of wear debris and/or measurable wear scars, occur under ultra-low loads in the top few nanometers of the sample, and are primarily responsible for the subsequent changes in surface potential.

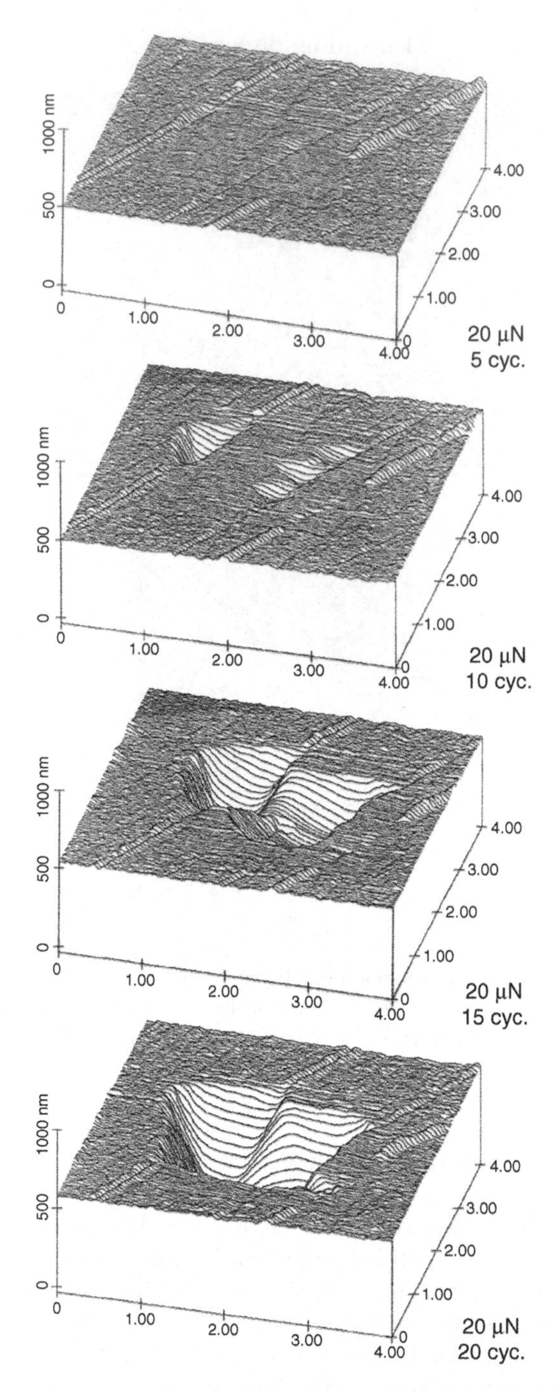

Figure 10.3.27 Surface plots of diamond-like carbon-coated thin-film disk showing the worn region; the normal load and number of test cycles are indicated. Reproduced with permission from Bhushan, B., Koinkar V.N., and Ruan, J. (1994), "Microtribology of Magnetic Media," *Proc. Inst. Mech. Eng., Part J: J. Eng. Tribol.* **2008**, 17–29. Copyright 1994 Sage Publications.

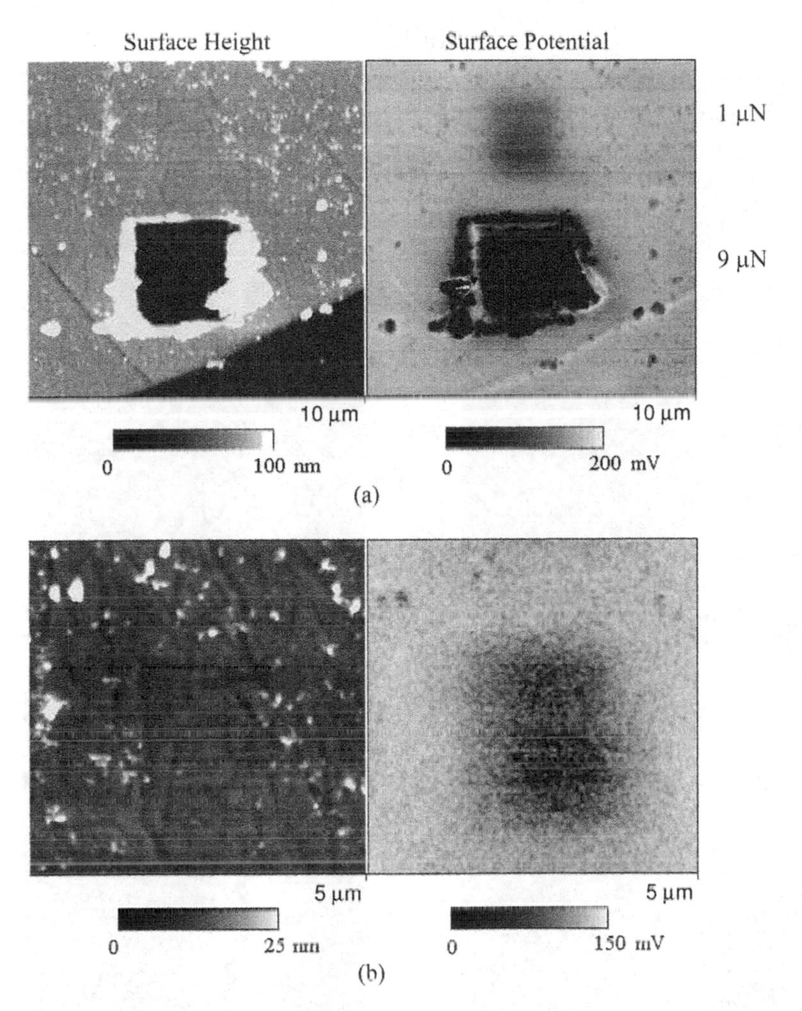

Figure 10.3.28 (a) Surface height and change in surface potential maps of wear regions generated at 1 μN (top) and 9 μN (bottom) on a single crystal aluminum sample showing bright contrast in the surface potential map on the worn regions. (b) Close up of upper (low load) wear region. Reproduced with permission from DeVecchio, D. and Bhushan, B. (1997), "Localized Surface Elasticity Measurements Using an Atomic Force Microscope," *Rev. Sci. Instrum.* **68**, 4498–4505. Copyright 1997 American Institute of Physics.

10.3.3.4 *In Situ* Characterization of Local Deformation

In situ surface characterization of local deformation of materials and thin films is carried out using a tensile stage inside an AFM. Failure mechanisms of coated polymeric thin films under tensile load were studied by Bobji and Bhushan (2001a, b). The specimens were strained at a rate of $4 \times 10^{-3}\%$ per second, and AFM images were captured at different strains up to about 10% to monitor generation and propagation of cracks and deformation bands.

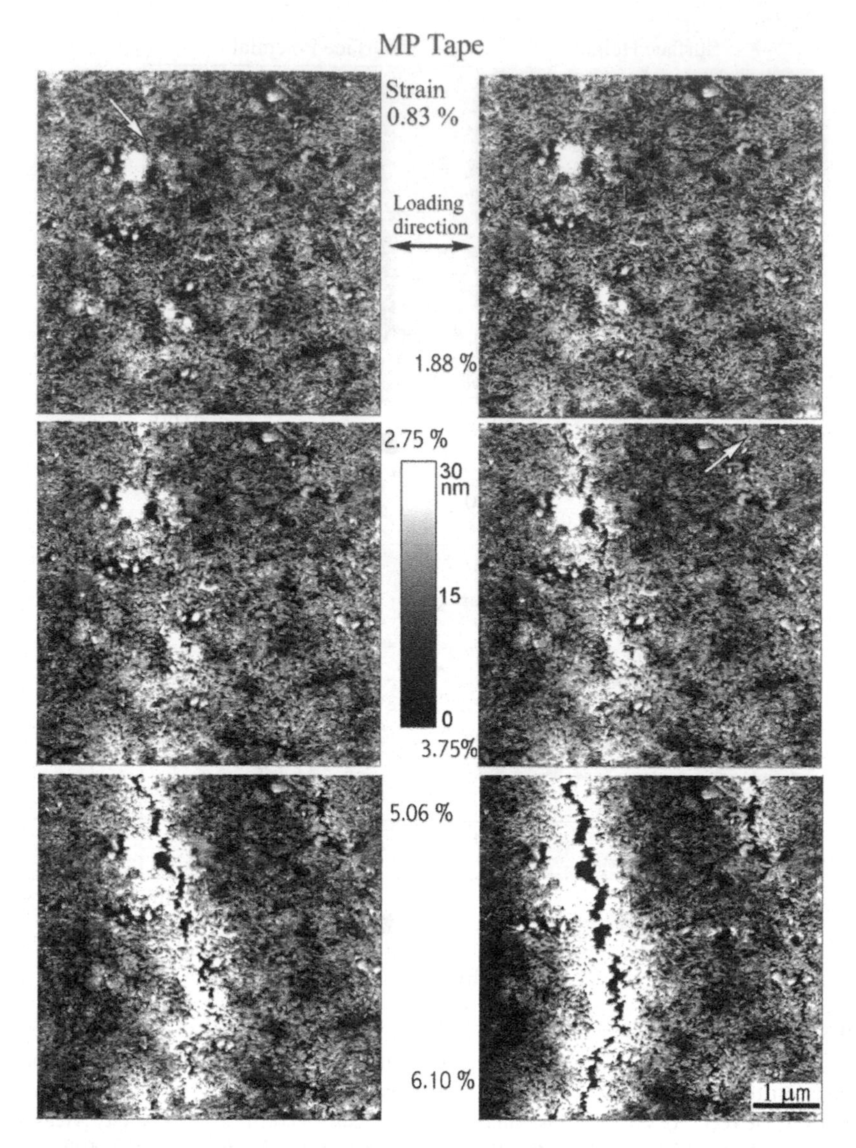

Figure 10.3.29 Topographical images of the MP magnetic tape at different strains. Reproduced with permission from Bobji, M. S. and Bhushan, B. (2001a), "Atomic Force Microscopic Study of the micro-Cracking of Magnetic Thin Films Under Tension," *Scripta Mater.* **44**, 37–42. Copyright 2001. Elsevier.

Bobji and Bhushan (2001a, b) studied three magnetic tapes of thickness ranging from 7 to 8.5 μm. One of these was with acicular-shaped metal particle (MP) coating and the other two with metal-evaporated (ME) coating and with and without a thin diamondlike carbon (DLC) overcoat both on a polymeric substrate and all with particulate back coating (Bhushan, 1996). They reported that cracking of the coatings started at about 1% strain for all tapes much before the substrate starts to yield at about 2% strain. As an example, Figure 10.3.29 shows the topographical images of the MP tape at different strains. At 0.83% strain, a crack can be seen, originating at the marked point. As the tape is further stretched along the direction, as shown in Figure 10.3.29, the crack propagates along the shorter boundary of the ellipsoidal particle. However, the general direction of the crack propagation remains perpendicular to the direction of the stretching. The length, width, and depth of the cracks increase with strain, and at the same time newer cracks keep nucleating and propagating with reduced crack spacing. At 3.75% strain, another crack can be seen nucleating. This crack continues to grow parallel to the first one. When the tape is unloaded after stretching up to a strain of about 2%, that is, within the elastic limit of the substrate, the cracks close perfectly, and it is impossible to determine the difference from the unstrained tape.

In-situ surface characterization of unstretched and stretched films has been used to measure Poisson's ratio of polymeric thin films by Bhushan *et al.* (2003). Uniaxial tension is applied by the tensile stage. Surface height profiles obtained from the AFM images of unstretched and stretched samples are used to monitor the changes in displacements of the polymer films in the longitudinal and lateral directions simultaneously.

10.3.3.5 Nanofabrication/Nanomachining

An AFM can be used for nanofabrication/nanomachining by extending the microscale scratching operation (Bhushan, 1995, 1999a, b, 2011; Bhushan *et al.*, 1994, 1995a). Figure 10.3.30 shows two examples of nanofabrication. The patterns were created on a single-crystal silicon (100) wafer by scratching the sample surface with a diamond tip at specified locations and scratching angles. Each line is inscribed manually at a normal load of 15 μN and a writing speed of 0.5 μm/s. The separation between lines is about 50 nm, and the variation in line width is due to the tip asymmetry. Nanofabrication parameters – normal load, scanning speed, and tip geometry – can be controlled precisely to control the depth and length of the devices.

Nanofabrication using mechanical scratching has several advantages over other techniques. Better control over the applied normal load, scan size, and scanning speed can be used for the nanofabrication of devices. Using the technique, nanofabrication can be performed on any engineering surface. Chemical etching or reactions is not required, and this dry nanofabrication process can be employed where the use of chemicals and electric field is prohibited. One disadvantage of this technique is the formation of debris during scratching. At light loads, debris formation is not a problem compared to high-load scratching. However, debris can be removed easily from the scan area at light loads during scanning.

10.3.4 Indentation

Mechanical properties on the relevant scales are needed for analysis of friction and wear mechanisms. Mechanical properties, such as hardness and Young's modulus of elasticity, can be determined on the micro- to picoscales using the AFM (Bhushan and Ruan, 1994; Bhushan *et al.*, 1994; Bhushan and Koinkar, 1994a, b) and a depth-sensing indentation system used in conjunction with an AFM (Bhushan *et al.*, 1996; Kulkarni and Bhushan, 1996a, b, 1997).

10.3.4.1 Picoindentation

Indentability on the scale of subnanometers of soft samples can be studied in the force calibration mode (Figure 10.3.5) by monitoring the slope of cantilever deflection as a function of the sample traveling distance after the tip is engaged and the sample is pushed against the

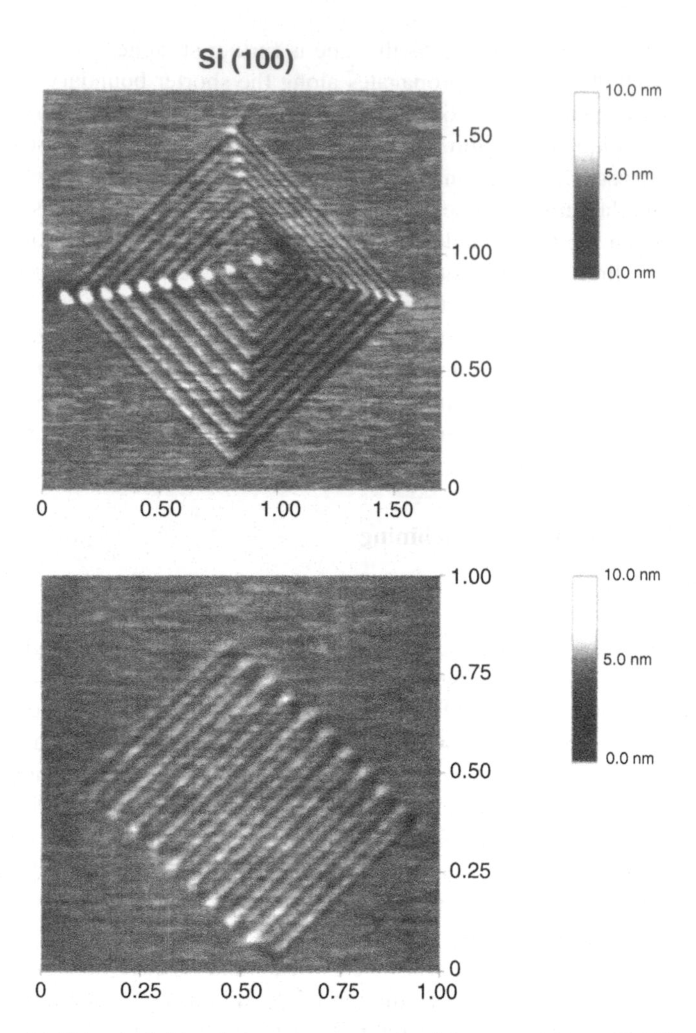

Figure 10.3.30 (a) Trim and (b) spiral patterns generated by scratching a Si(100) surface using a diamond tip at a normal load of 15 μN and writing speed of 0.5 μm/s. Reproduced from Bhushan, B. (1999a), *Handbook of Micro/Nanotribology*, Second edition, CRC Press, Boca Raton, Florida. Copyright 1999 CRC Press and from Bhushan, B. (1999a), Handbook of Micro/Nanotribology, Second edition, CRC Press, Boca Raton, Florida. Copyright 1999 CRC Press.

tip. For a rigid sample, the cantilever deflection equals the sample traveling distance, but the former quantity is smaller if the tip indents the sample.

10.3.4.2 Nanoscale Indentation

The indentation hardness of surface films with an indentation depth of as small as about 1 nm can be measured using an AFM (Bhushan and Koinkar, 1994b; Bhushan *et al.*, 1995a, 1996). To make accurate measurements of hardness at shallow depths, a depth-sensing

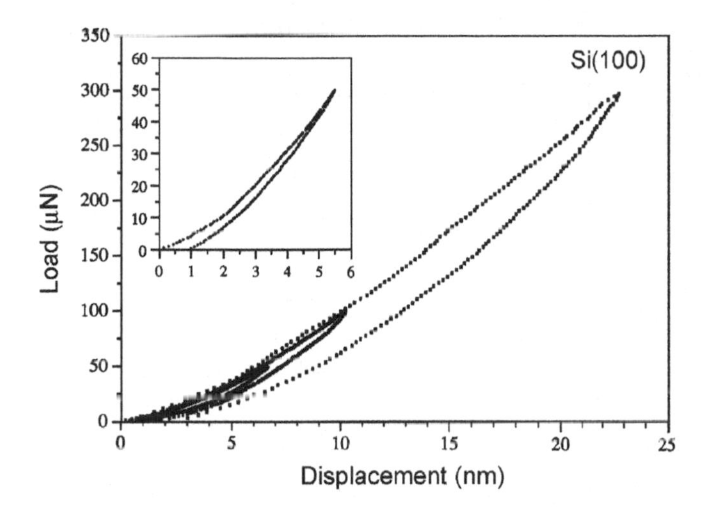

Figure 10.3.31 Load-displacement curves at various peak loads for Si(100); inset shows the magnified curve for peak load of 50 μN. Reproduced with permission from Bhushan, B., Kulkarni, A.V., Bonin, W., and Wyrobek, J.T. (1996), "Nano/Picoindentation Measurement Using a Capacitance Transducer System in Atomic Force Microscopy," *Philos. Mag.* **74**, 1117–1128. Copyright 1996 Taylor and Francis.

nano/picoindentation system (Figure 10.3.6) is used (Bhushan *et al.*, 1996). Figure 10.3.31 shows the load-displacement curves at different peak loads for Si(100). Loading/unloading curves often exhibit sharp discontinuities, particularly at high loads. Discontinuities, also referred to as pop-ins, occurring during the initial loading part of the curve mark a sharp transition from pure elastic loading to a plastic deformation of the specimen surface, thus corresponding to an initial yield point. The sharp discontinuities in unloading part of the curves are believed to be due to the formation of lateral cracks which form at the base of the median crack, which results in the surface of the specimen being thrust upward. Load-displacement data at residual depths as low as about 1 nm can be obtained. The indentation hardness of surface films has been measured for various materials at a range of loads including Si(100) up to a peak load of 500 μN and Al(100) up to a peak load of 2000 μN by Bhushan *et al.* (1996) and Kulkarni and Bhushan (1996a, b, 1997). The hardnesses of single-crystal silicon and single-crystal aluminum at shallow depths on the order of few nm (on a nanoscale) are found to be higher than at depths on the order of few hundred nm (on a microscale), Figure 10.3.32. Microhardness has also been reported to be higher than that on the millimeter scale by several investigators. The data reported to date show that hardness exhibits scale (size) effect.

During loading, the generation and propagation of dislocations are responsible for plastic deformation. A strain gradient plasticity theory has been developed for micro/nanoscale deformations, and is based on randomly created statistically stored and geometrically necessary dislocations (Fleck, *et al.*, 1994; Nix and Gao, 1998). Large strain gradients inherent in small indentations lead to the accumulation of geometrically necessary dislocations, located in a certain sub-surface volume, for strain compatibility reasons. The large strain gradients in small indentations require these dislocations to account for the large slope at the indented surface. These dislocations become obstacles to other dislocations that cause enhanced hardening.

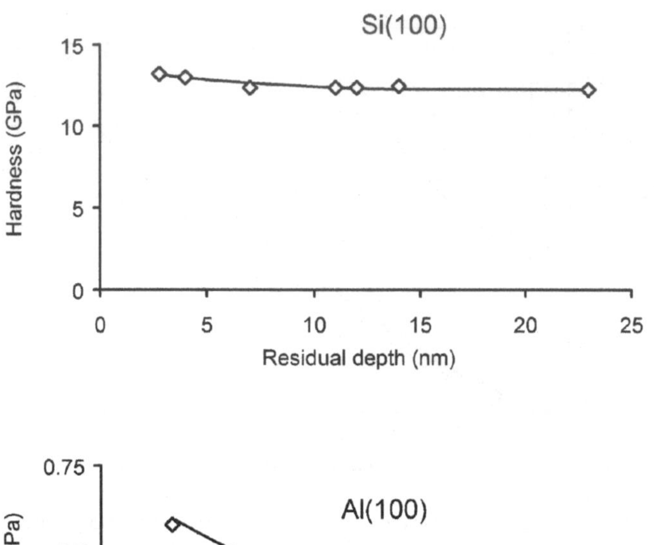

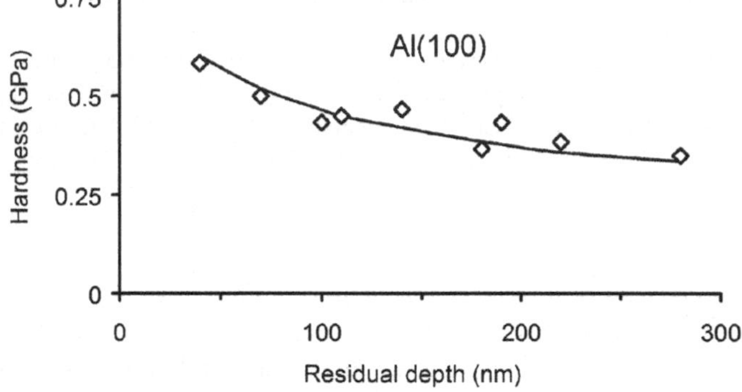

Figure 10.3.32 Indentation hardness as a function of residual indentation depth for Si(100) (Reproduced with permission from Bhushan, B., Kulkarni, A.V., Bonin, W., and Wyrobek, J.T. (1996), "Nano/Picoindentation Measurement Using a Capacitance Transducer System in Atomic Force Microscopy," *Philos. Mag.* **74**, 1117–1128. Copyright 1996 Taylor and Francis) and Al(100) (Reproduced with permission from Kulkarni, A.V. and Bhushan, B. (1996a), "Nanoscale Mechanical Property Measurements Using Modified Atomic Force Microscopy," *Thin Solid Films* **290-291**, 206–210. Copyright 1996. Elsevier).

These are a function of strain gradient, whereas statistically stored dislocations are a function of strain. Based on this theory, scale–dependent hardness is given as

$$H = H_0\sqrt{1 + \ell_d/a} \tag{10.3.11}$$

where H_0 is the hardness in the absence of strain gradient or macrohardness, ℓ_d is the material-specific characteristic length parameter, and a is the contact radius. In addition to the role of strain gradient plasticity theory, an increase in hardness with a decrease in indentation depth can possibly be rationalized on the basis that as the volume of deformed material decreases, there is a lower probability of encountering material defects.

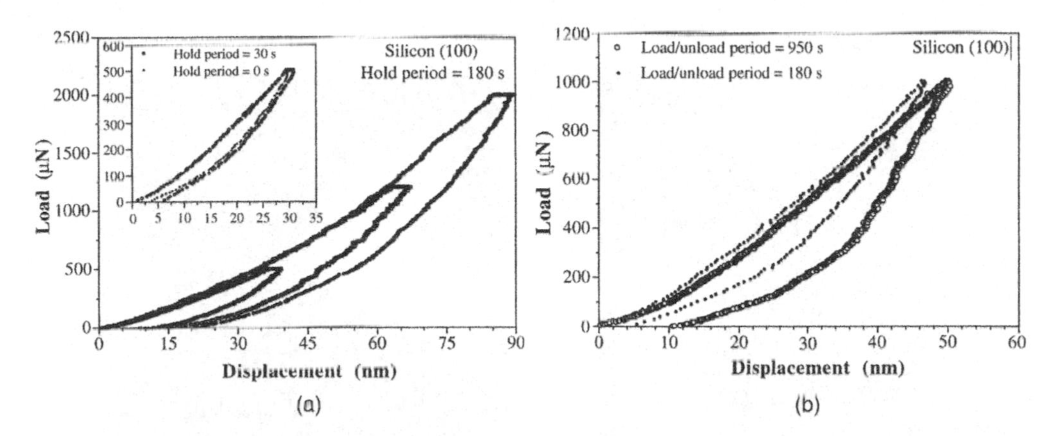

Figure 10.3.33 (a) Creep behavior and (b) strain-rate sensitivity of Si(100). Reproduced with permission from Bhushan, B., Kulkarni, A.V., Bonin, W. and Wyrobek, J.T. (1996), "Nano/Picoindentation Measurement Using a Capacitance Transducer System in Atomic Force Microscopy," *Philos. Mag. 74*, 1117–1128. Copyright 1996 Taylor and Francis.

Bhushan and Koinkar (1994a) used AFM measurements to show that ion implantation of silicon surfaces increases their hardness and thus their wear resistance. Formation of surface alloy films with improved mechanical properties by ion implantation is of growing technological importance as a means of improving the mechanical properties of materials. Hardness of 20 nm thick DLC films have been measured by Kulkarni and Bhushan (1997).

The creep and strain-rate effects (viscoelastic effects) of ceramics can be studied using a depth-sensing indentation system. Bhushan *et al.* (1996) and Kulkarni and Bhushan (1996, b, 1997) have reported that ceramics (single-crystal silicon and diamond-like carbon) exhibit significant plasticity and creep on a nanoscale. Figure 10.3.33(a) shows the load displacement curves for single-crystal silicon at various peak loads held at 180 s. To demonstrate the creep effects, the load-displacement curves for a 500 μN peak load held at 0 and 30 s are also shown as an inset. Note that significant creep occurs at room temperature. Nanoindenter experiments conducted by Li *et al.* (1991) exhibited significant creep only at high temperatures (greater than or equal to 0.25 times the melting point of silicon). The mechanism of dislocation glide plasticity is believed to dominate the indentation creep process on the macroscale. To study the strain-rate sensitivity of silicon, data at two different (constant) rates of loading are presented in Figure 10.3.33(b). Note that a change in the loading rate by a factor of about five results in a significant change in the load-displacement data. The viscoelastic effects observed here for silicon at ambient temperature could arise from the size effects mentioned earlier. Most likely, creep and strain rate experiments are being conducted on the hydrated films present on the silicon surface in an ambient environment, and these films are expected to be viscoelastic.

10.3.4.3 Localized Surface Elasticity and Viscoelasticity Mapping

The Young's modulus of elasticity can be calculated from the slope of the indentation curve during unloading. However, these measurements provide a single-point measurement. By

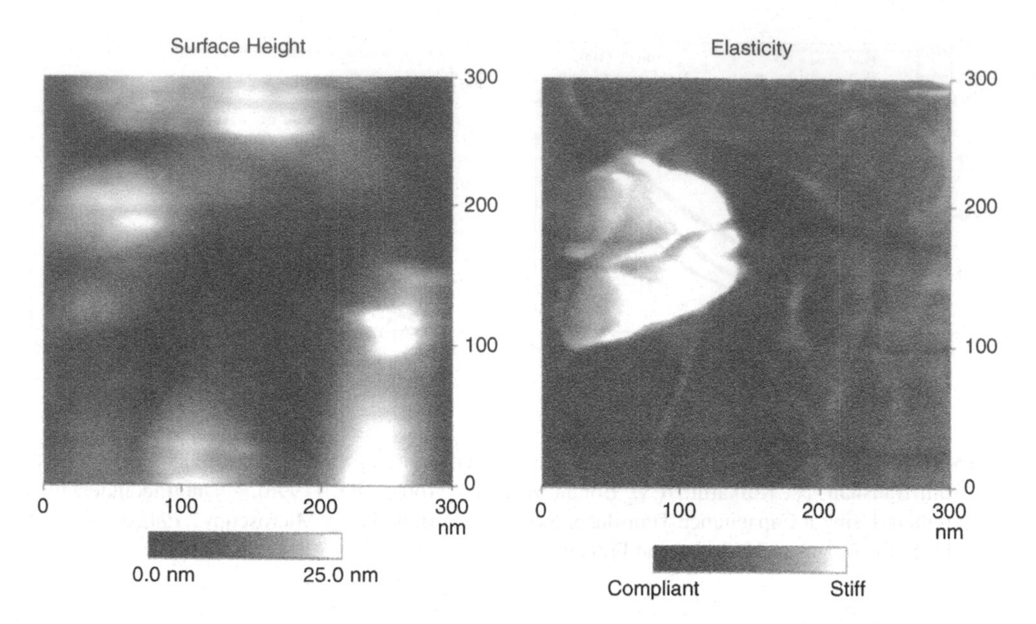

Figure 10.3.34 Surface height and elasticity maps on a polymeric magnetic tape ($\sigma = 6.7$ nm and P-V = 32 nm; σ and P-V refer to standard deviation of surface heights and peak-to-valley distance, respectively). The gray scale on the elasticity map is arbitrary. Reproduced with permission from DeVecchio, D. and Bhushan, B. (1997), "Localized Surface Elasticity Measurements Using an Atomic Force Microscope," *Rev. Sci. Instrum.* **68**, 4498–4505. Copyright 1997 American Institute of Physics.

using the force modulation technique, it is possible to obtain localized elasticity maps of soft and compliant materials of near surface regions with nanoscale lateral resolution. This technique has been successfully used for polymeric magnetic tapes, which consist of magnetic and nonmagnetic ceramic particles in a polymeric matrix. Elasticity maps of a tape can be used to identify the relative distribution of the hard magnetic and nonmagnetic ceramic particles on the tape surface, which has an effect on friction and stiction at the head–tape interface (Bhushan, 1996). Figure 10.3.34 shows the surface height and elasticity maps on a polymeric magnetic tape (DeVecchio and Bhushan, 1997). The elasticity image reveals sharp variations in the surface elasticity due to the composite nature of the film. As can be clearly seen, regions of high elasticity do not always correspond to high or low topography. Based on a Hertzian elastic-contact analysis, the static indentation depth of these samples during the force modulation scan is estimated to be about 1 nm. We conclude that the contrast seen is influenced most strongly by material properties in the top few nanometers, independent of the composite structure beneath the surface layer.

By using phase contrast microscopy in the tapping mode or torsional resonance mode, it is possible to obtain phase contrast maps or the contrast in viscoelastic properties of near surface regions with nanoscale lateral resolution. This technique has been successfully used for polymeric films and magnetic tapes which consist of ceramic particles in a polymeric matrix (Scott and Bhushan, 2003; Bhushan and Qi, 2003; Kasai *et al.*, 2004; Chen and Bhushan, 2005; Bhushan, 2011).

10.3.5 Boundary Lubrication

10.3.5.1 Perfluoropolyether Lubricants

The classical approach to lubrication uses freely supported multimolecular layers of liquid lubricants (Bowden and Tabor, 1950; Bhushan, 1996, 1999a, 2001a, 2011). The liquid lubricants are sometimes chemically bonded to improve their wear resistance (Bhushan, 1996). Partially chemically-bonded, molecularly-thick perfluoropolyether (PFPE) films are used to lubricate magnetic storage media because of their thermal stability and extremely low vapor pressure (Bhushan, 1996). Chemically-bonded lubricants are considered potential candidate lubricants for MEMS/NEMS. Molecularly-thick PFPEs are well suited because of the following properties: low surface tension and low contact angle which allows easy spreading on surfaces and provides hydrophobic properties; chemical and thermal stability which minimizes degradation under use; low vapor pressure which provides low out-gassing; high adhesion to substrate via organic functional bonds; and good lubricity which reduces contact surface wear.

For boundary lubrication studies, friction, adhesion, and durability experiments have been performed on virgin Si (100) surfaces and silicon surfaces lubricated with various PFPE lubricants (Koinkar and Bhushan, 1996a, b; Liu and Bhushan, 2003a; Tao and Bhushan, 2005a; Bhushan *et al.*, 2007; Palacio and Bhushan, 2007a, b). More recently, there has been interest in selected ionic liquids for lubrication (Bhushan *et al.*, 2008, Palacio and Bhushan, 2008, 2009). They possess efficient heat transfer properties. They are also electrically conducting, which is of interest in various MEMS/NEMS applications. Results of the following two PFPE lubricants will be presented here: Z-15 (with $-CF_3$ nonpolar end groups), CF_3-O-$(CF_2$-CF_2-$O)_m$-$(CF_2$-$O)_n$-CF_3 (m/n~2/3) and Z-DOL (with -OH polar end groups), HO-CH_2-CF_2-O-$(CF_2$-CF_2-$O)_m$-$(CF_2$-$O)_n$-CF_2-CH_2-OH (m/n~2/3). Z-DOL film was thermally bonded at 150°C for 30 minutes, and the unbonded fraction was removed by a solvent (fully bonded) (Bhushan, 1996). The thicknesses of Z-15 and Z-DOL films were 2.8 nm and 2.3 nm, respectively. The lubricant chain diameters of these molecules is about 0.6 nm, and molecularly thick films generally lie flat on surfaces with high coverage.

The adhesive forces of Si(100), Z-15, and Z-DOL (fully bonded) measured by force calibration plot and friction force versus normal load plot are summarized in Figure 10.3.35 (Liu and Bhushan, 2003a). The data obtained by these two methods are in good agreement. Figure 10.3.35 shows that the presence of mobile Z-15 lubricant film increases the adhesive force as compared to that of Si(100) by meniscus formation. Whereas the presence of the solid-like phase of the Z-DOL (fully bonded) film reduces the adhesive force as compared that of Si(100) because of the absence of mobile liquid. The schematic (bottom) in Figure 10.3.35 shows the relative size and sources of the meniscus. It is well known that the native oxide layer (SiO_2) on the top of Si(100) wafer exhibits hydrophilic properties, and some water molecules can be adsorbed on this surface. The condensed water will form a meniscus as the tip approaches the sample surface. The larger adhesive force in Z-15 is not only caused by the Z-15 meniscus alone, the non-polarized Z-15 liquid does not have good wettability and strong bonding with Si(100). Consequently, in the ambient environment, the condensed water molecules from the environment will permeate through the liquid Z-15 lubricant film and compete with the lubricant molecules present on the substrate. The interaction of the liquid lubricant with the substrate is weakened, and a boundary layer of the liquid lubricant forms puddles (Koinkar and Bhushan, 1996a, b). This dewetting allows water molecules to

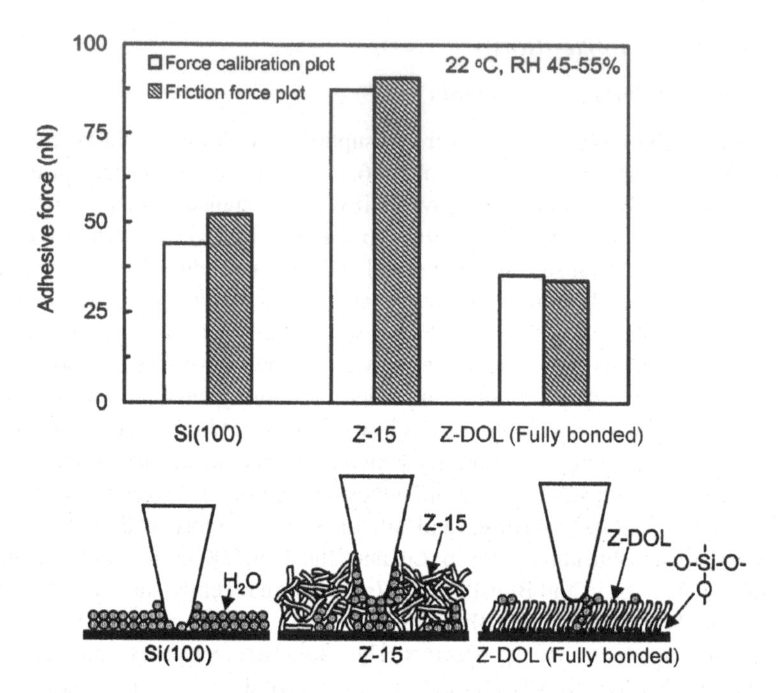

Figure 10.3.35 Summary of the adhesive forces of Si(100) and Z-15 and Z-DOL (fully bonded) films measured by force calibration plots and friction force versus normal load plots in ambient air. The schematic (bottom) showing the effect of meniscus, formed between AFM tip and the surface sample, on the adhesive and friction forces. Reproduced with permission from Liu, H. and Bhushan, B. (2003a), "Nanotribological Characterization of Molecularly-Thick Lubricant Films for Applications to MEMS/NEMS by AFM," *Ultramicroscopy* **97**, 321–340. Copyright 2003. Elsevier.

be adsorbed on the Si(100) surface along with Z-15 molecules, and both of them can form meniscus while the tip approaches the surface. Thus the dewetting of liquid Z-15 film results in a higher adhesive force and poorer lubrication performance. In addition, the Z-15 film is soft compared to the solid Si(100) surface, and penetration of the tip in the film occurs while pushing the tip down. This results in the large area of the tip being wetted by the liquid to form the meniscus at the tip–liquid (mixture of Z-15 and water) interface. It should also be noted that Z-15 has a higher viscosity compared to water, therefore the Z-15 film provides a higher resistance to motion and coefficient of friction. In the case of Z-DOL (fully bonded) film, both of the active groups of Z-DOL molecules are mostly bonded on Si(100) substrate, thus the Z-DOL (fully bonded) film has low free surface energy and cannot be displaced readily by the water molecules or readily adsorb the water molecules. Thus, the use of Z-DOL (fully bonded) can reduce the adhesive force.

To study the velocity effect on friction and adhesion, the variation of friction force, adhesive force, and the coefficient of friction of Si(100), Z-15 and Z-DOL(fully bonded) as a function of velocity are summarized in Figure 10.3.36 (Liu and Bhushan, 2003a). It indicates that for a silicon wafer, the friction force decreases logarithmically with increasing velocity. For Z-15, the friction force decreases with increasing velocity up to 10 μm/s, after which it remains

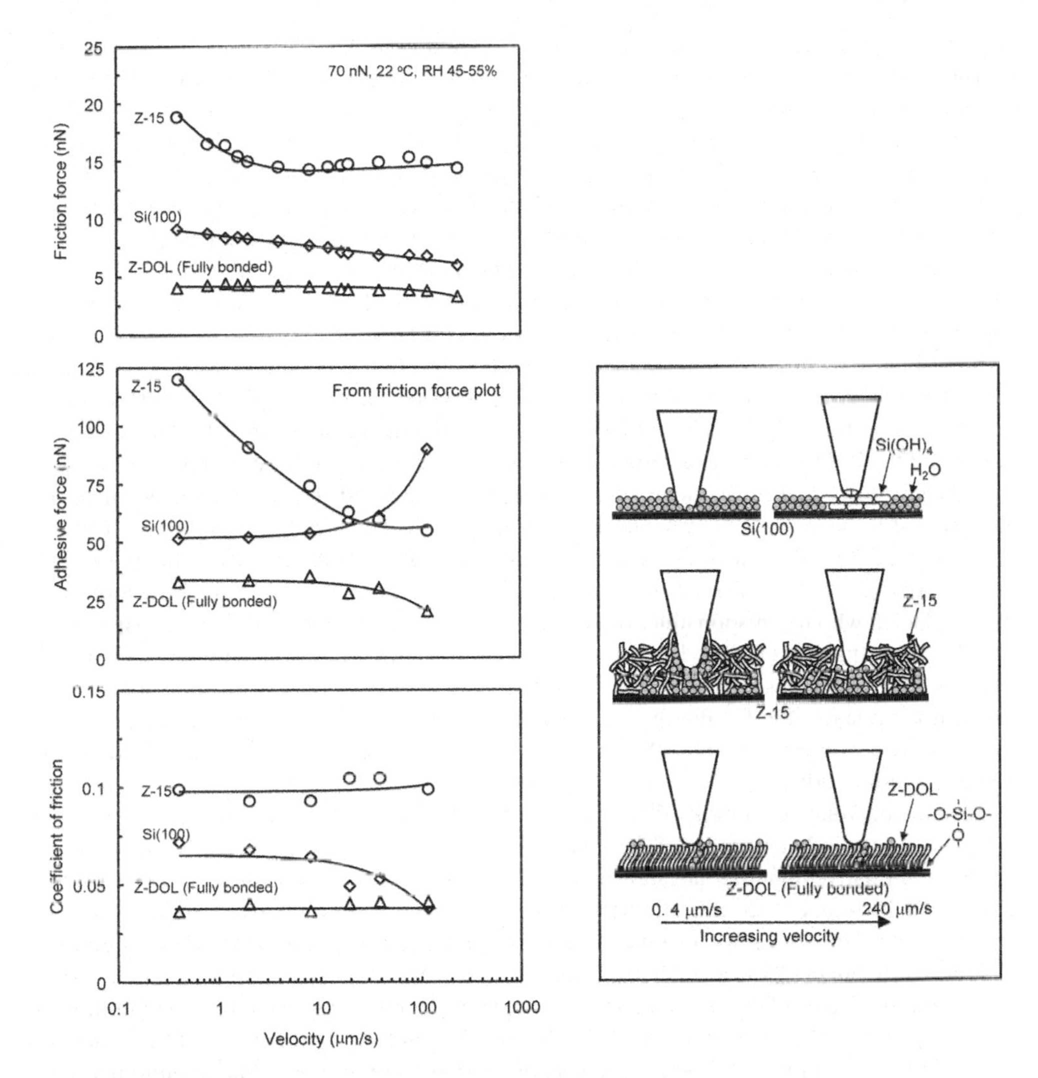

Figure 10.3.36 The influence of velocity on the friction force, adhesive force and coefficient of friction of Si(100) and Z-15 and Z-DOL (fully bonded) films at 70 nN, in ambient air. The schematic (right) shows the change of surface composition (by tribochemical reaction) and formation of meniscus while increasing the velocity. Reproduced with permission from Liu, H. and Bhushan, B. (2003a), "Nanotribological Characterization of Molecularly-Thick Lubricant Films for Applications to MEMS/NEMS by AFM," *Ultramicroscopy* **97**, 321–340. Copyright 2003. Elsevier.

almost constant. The velocity has a very small effect on the friction force of Z-DOL (fully bonded); it reduced slightly only at very high velocity. Figure 10.3.36 also indicates that the adhesive force of Si(100) is increased when the velocity is higher than 10 μm/s. The adhesive force of Z-15 is reduced dramatically with a velocity increase up to 20 μm/s, after which it is reduced slightly, and the adhesive force of Z-DOL (fully bonded) is also decreased at high velocity. In the testing range of velocity, only the coefficient of friction of Si(100) decreases with velocity, but the coefficients of friction of Z-15 and Z-DOL (fully bonded) almost remain constant. This implies that the friction mechanisms of Z-15 and Z-DOL (fully bonded) do not change with the variation of velocity.

The mechanisms of the effect of velocity on adhesion and friction are explained based on schematics shown in Figure 10.3.36 (right) (Liu and Bhushan, 2003a). For Si(100), a tribochemical reaction plays a major role. Although, at high velocity, the meniscus is broken and does not have enough time to rebuild, the contact stresses and high velocity lead to tribochemical reactions of the Si(100) wafer (which has native oxide (SiO_2)) and the Si_3N_4 tip with water molecules and they form $Si(OH)_4$. The $Si(OH)_4$ is removed and continuously replenished during sliding. The $Si(OH)_4$ layer between the tip and the Si(100) surface is known to be of low shear strength and causes a decrease in friction force and coefficient of friction (Bhushan, 1999c). The chemical bonds of Si-OH between the tip and Si(100) surface induce large adhesive force. For Z-15 film, at high velocity the meniscus formed by the condensed water and the Z-15 molecules is broken and does not have enough time to rebuild, therefore, the adhesive force and consequently the friction force is reduced. The friction mechanisms for Z-15 film still shears the same viscous liquid even at high velocity range, thus the coefficient of friction of the Z-15 does not change with velocity. For Z-DOL (fully bonded) film, the surface can adsorb a few water molecules in an ambient condition, and at high velocity these molecules are displaced, which is responsible for the slight decrease in friction force and adhesive force. Koinkar and Bhushan (1996a, 1996b) have suggested that in the case of samples with mobile films, such as condensed water and Z-15 films, alignment of the liquid molecules (shear thinning) is responsible for the drop in friction force with an increase in scanning velocity. This could be another reason for the decrease in friction force for Si(100) and Z-15 film with velocity in this study.

To study the relative humidity effect on friction and adhesion, the variation of friction force, adhesive force, and coefficient of friction of Si(100), Z-15, and Z-DOL (fully bonded) as a function of relative humidity are shown in Figure 10.3.37 (Liu and Bhushan, 2003a). It shows that for Si(100) and Z-15 film, the friction force increases with a relative humidity increase up to 45%, and then it shows a slight decrease with a further increase in the relative humidity. Z-DOL (fully bonded) has a smaller friction force than Si(100) and Z-15 in the whole testing range, and its friction force shows a relative apparent increase when the relative humidity is higher than 45%. For Si(100), Z-15 and Z-DOL (fully bonded), their adhesive forces increase with relative humidity, and their coefficients of friction increase with relative humidity up to 45%, after which they decrease with any further increase in the relative humidity. It is also observed that the humidity effect on Si(100) really depends on the history of the Si(100) sample. As the surface of Si(100) wafer readily adsorbs water in air, without any pre-treatment the Si(100) used in our study almost reaches its saturate stage of adsorbed water, and is responsible for less effect during increasing relative humidity. However, once the Si(100) wafer was thermally treated by baking at 150°C for 1 hour, a greater effect was observed.

The schematic (right) in Figure 10.3.37 shows that for Si(100), because of its high free surface energy, it can adsorb more water molecules during an increase in relative humidity (Liu and Bhushan, 2003a). As discussed earlier, for Z-15 film in the humid environment, the condensed water from the humid environment competes with the lubricant film present on the sample surface, and the interaction of the liquid lubricant film with the silicon substrate is weakened and a boundary layer of the liquid lubricant forms puddles. This dewetting allows the water molecules to be adsorbed on the Si(100) substrate mixed with the Z-15 molecules

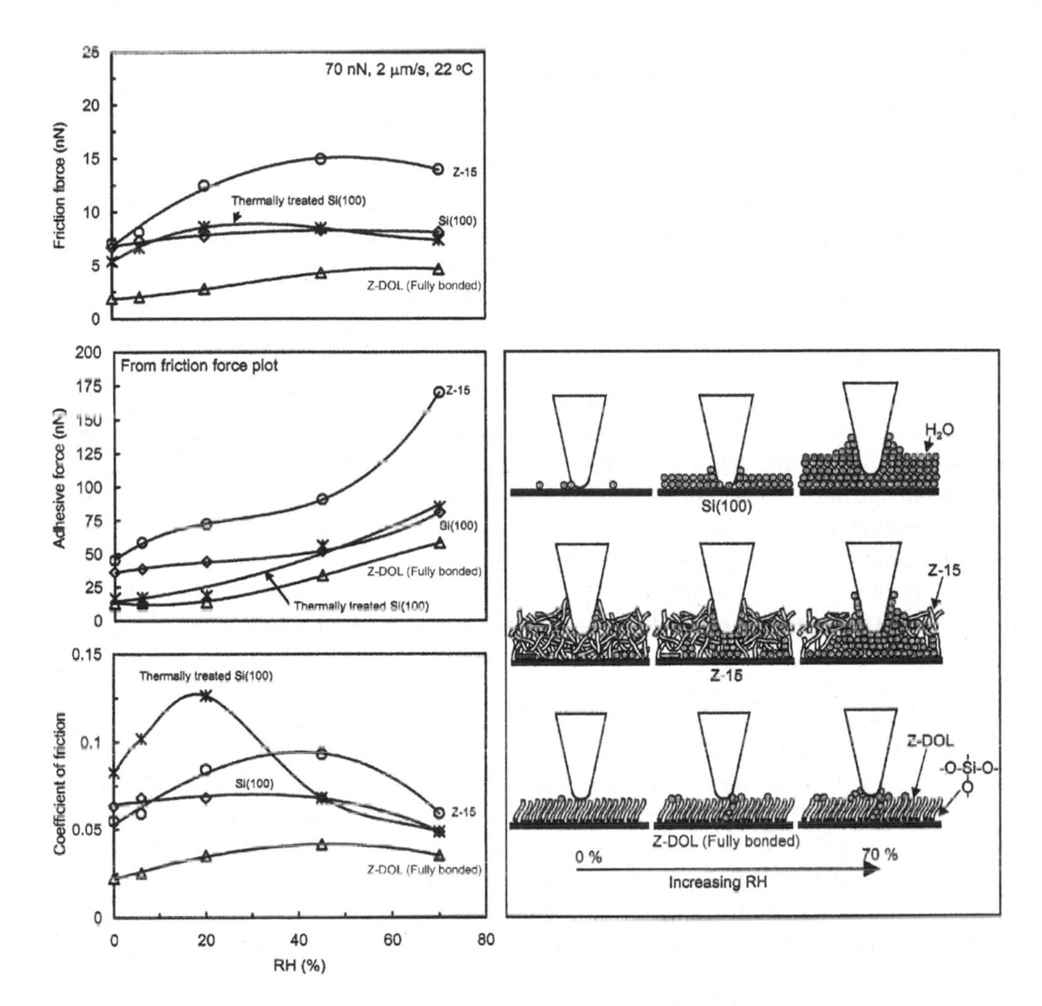

Figure 10.3.37 The influence of relative humidity on the friction force, adhesive force, and coefficient of friction of Si(100) and Z-15 and Z-DOL (fully bonded) films at 70 nN, 2 μm/s, and in 22°C air. Schematic (right) shows the change of meniscus while increasing the relative humidity. In this figure, the thermally treated Si(100) represents the Si(100) wafer that was baked at 150°C for 1 hour in an oven (in order to remove the adsorbed water) just before it was placed in the 0% RH chamber. Reproduced with permission from Liu, H. and Bhushan, B. (2003a), "Nanotribological Characterization of Molecularly-Thick Lubricant Films for Applications to MEMS/NEMS by AFM," *Ultramicroscopy* **97**, 321–340. Copyright 2003. Elsevier.

(Koinkar and Bhushan, 1996a, b). Obviously, more water molecules can be adsorbed on the Z-15 surface while increasing the relative humidity. The more adsorbed water molecules in the case of Si(100), along with lubricant molecules in Z-15 film case, form a bigger water meniscus which leads to an increase of friction force, adhesive force, and coefficient of friction of Si(100) and Z-15 with humidity, but at very high humidity of 70%, large quantities of adsorbed water can form a continuous water layer that separates the tip and sample surface and acts as a kind of lubricant, which causes a decrease in the friction force and the coefficient of friction. For

Z-DOL (fully bonded) film, because of their hydrophobic surface properties, water molecules can be adsorbed at humidity higher than 45%, and causes an increase in the adhesive force and friction force.

To study the temperature effect on friction and adhesion, the variation of friction force, adhesive force, and coefficient of friction of Si(100), Z-15 and Z-DOL (fully bonded) as a function of temperature are summarized in Figure 10.3.38 (Liu and Bhushan, 2003a). It shows that the increasing temperature causes a decrease of friction force, adhesive force, and coefficient of friction of Si(100), Z-15 and Z-DOL (fully bonded). The schematic (right) in Figure 10.3.38 indicates that at high temperature, desorption of water leads to the decrease of the friction force, the adhesive forces and the coefficient of friction for all of the samples. For Z-15 film, the reduction of viscosity at high temperature also contributes to the decrease of friction force and the coefficient of friction. In the case of Z-DOL (fully bonded) film, molecules are easier oriented at high temperature, which may be partly responsible for the low friction force and coefficient of friction.

In summary, the influence of velocity, relative humidity, and temperature on the friction force of mobile Z-15 film is presented in Figure 10.3.39 (Liu and Bhushan, 2003a). The changing trends are also addressed in this figure.

To study the durability of lubricant films at the nanoscale, the friction of Si(100), Z-15, and Z-DOL (fully bonded) as a function of the number of scanning cycles are shown in Figure 10.3.40 (Liu and Bhushan, 2003a). As observed earlier, the friction force of Z-15 is higher than that of Si(100) with the lowest values for Z-DOL(fully bonded). During cycling, the friction force and the coefficient of friction of Si(100) show a slight decrease during the initial few cycles then remain constant. This is related to the removal of the native oxide. In the case of Z-15 film, the friction force and coefficient of friction show an increase during the initial few cycles and then approach higher stable values. This is believed to be caused by the attachment of the Z-15 molecules to the tip. After several scans, the molecular interaction reaches equilibrium, and after that the friction force and coefficient of friction remain constant. In the case of Z-DOL (fully bonded) film, the friction force and coefficient of friction start out low and remain low during the entire test for 100 cycles. It suggests that Z-DOL (fully bonded) molecules do not get attached or displaced as readily as Z-15.

10.3.5.2 Self-Assembled Monolayers

For the lubrication of MEMS/NEMS, another effective approach involves the deposition of organized and dense molecular layers of long-chain molecules. Two common methods to produce monolayers and thin films are the Langmuir-Blodgett (L-B) deposition and self-assembled monolayers (SAMs) by chemical grafting of molecules. LB films are physically bonded to the substrate by weak van der Waals attraction, while SAMs are chemically bonded via covalent bonds to the substrate. Because of the choice of chain length and terminal linking group that SAMs offer, they hold great promise for boundary lubrication of MEMS/NEMS. A number of studies have been conducted to study the tribological properties of various SAMs deposited on Si, Al, and Cu substrates (Bhushan *et al.*, 1995b, 2005, 2006, 2007; Bhushan and Liu, 2001; Liu *et al.*, 2001; Liu and Bhushan, 2002; Kasai *et al.*, 2005; Lee *et al.*, 2005; Tambe and Bhushan, 2005h; Tao and Bhushan, 2005b; Hoque *et al.*, 2006a, b, 2007a, b, 2008, 2009; DeRose *et al.*, 2008).

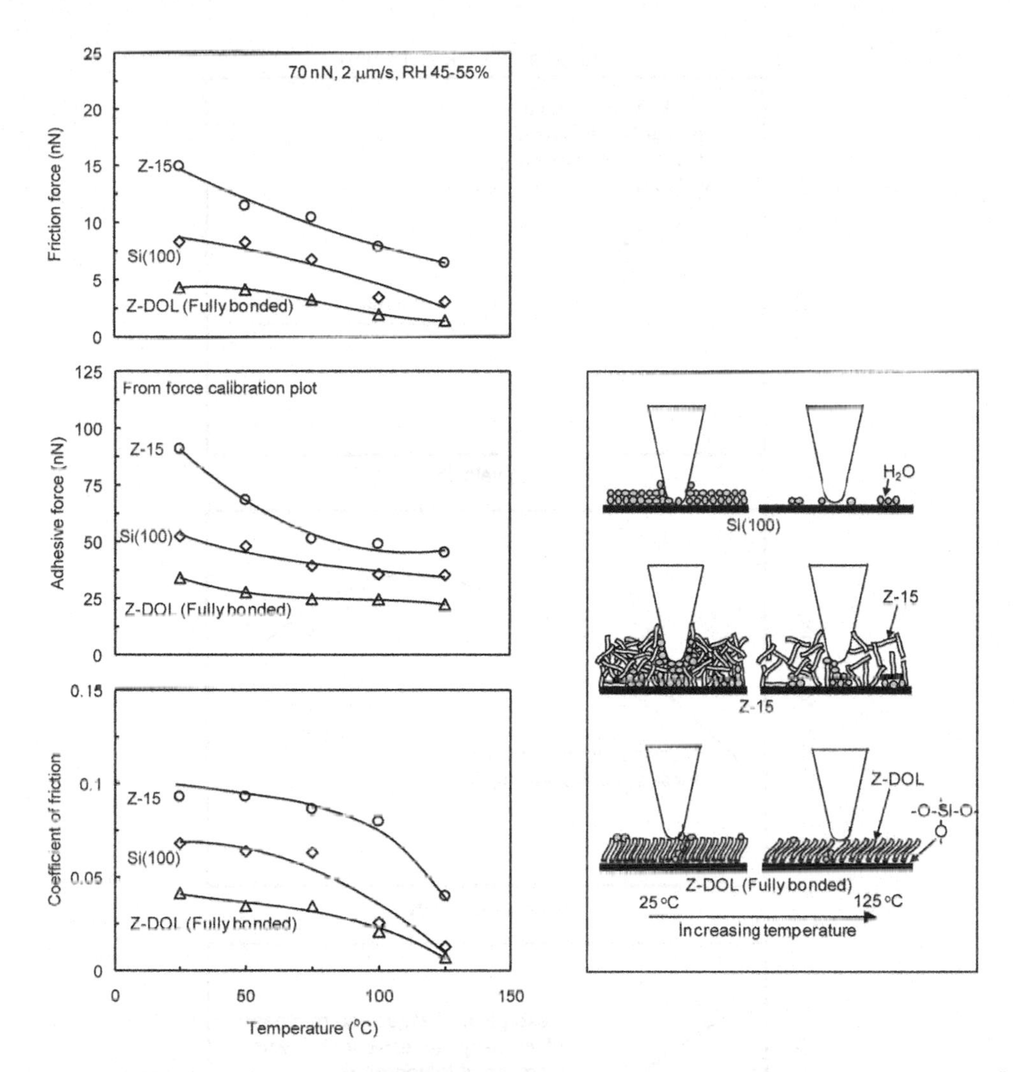

Figure 10.3.38 The influence of temperature on the friction force, adhesive force, and coefficient of friction of Si(100) and Z-15 and Z-DOL (fully bonded) films at 70 nN, at 2 μm/s, and in RH 40-50% air. The schematic (right) shows that at high temperature, desorption of water decreases the adhesive forces. And the reduced viscosity of Z-15 leads to the decrease of coefficient of friction. High temperature facilitates orientation of molecules in Z-DOL (fully bonded) film which results in lower coefficient of friction. Reproduced with permission from Liu, H. and Bhushan, B. (2003a), "Nanotribological Characterization of Molecularly-Thick Lubricant Films for Applications to MEMS/NEMS by AFM," *Ultramicroscopy* **97**, 321–340. Copyright 2003. Elsevier.

Bhushan and Liu (2001) studied the effect of film compliance on adhesion and friction. They used hexadecane thiol (HDT), 1,1,biphenyl-4-thiol (BPT), and crosslinked BPT (BPTC) solvent deposited on Au(111) substrate, Figure 10.3.41(a). The average values and standard deviation of the adhesive force and coefficient of friction are presented in Figure 10.3.41(b). Based on the data, the adhesive force and coefficient of friction of SAMs are less than

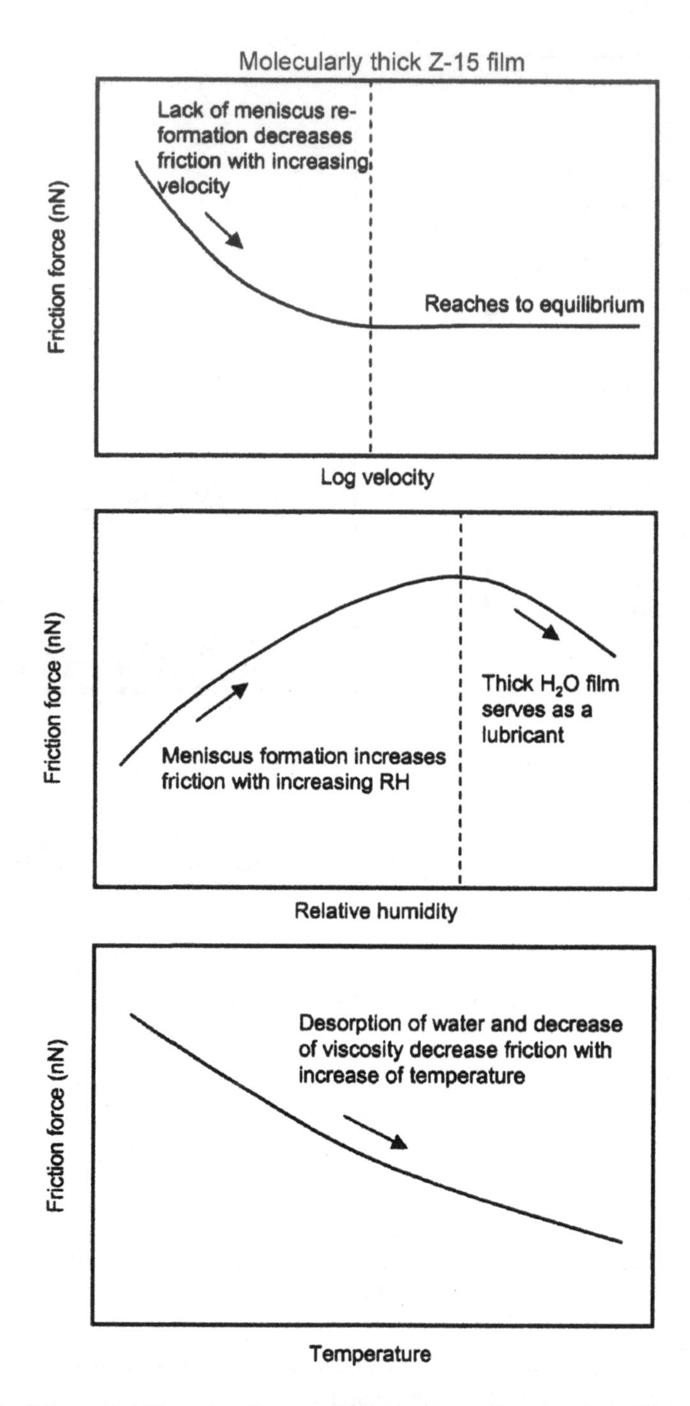

Figure 10.3.39 Schematic shows the change of friction force of molecularly thick Z-15 films with log velocity, relative humidity, and temperature. The changing trends are also addressed in this figure. Reproduced with permission from Liu, H. and Bhushan, B. (2003a), "Nanotribological Characterization of Molecularly-Thick Lubricant Films for Applications to MEMS/NEMS by AFM," *Ultramicroscopy* **97**, 321–340. Copyright 2003. Elsevier.

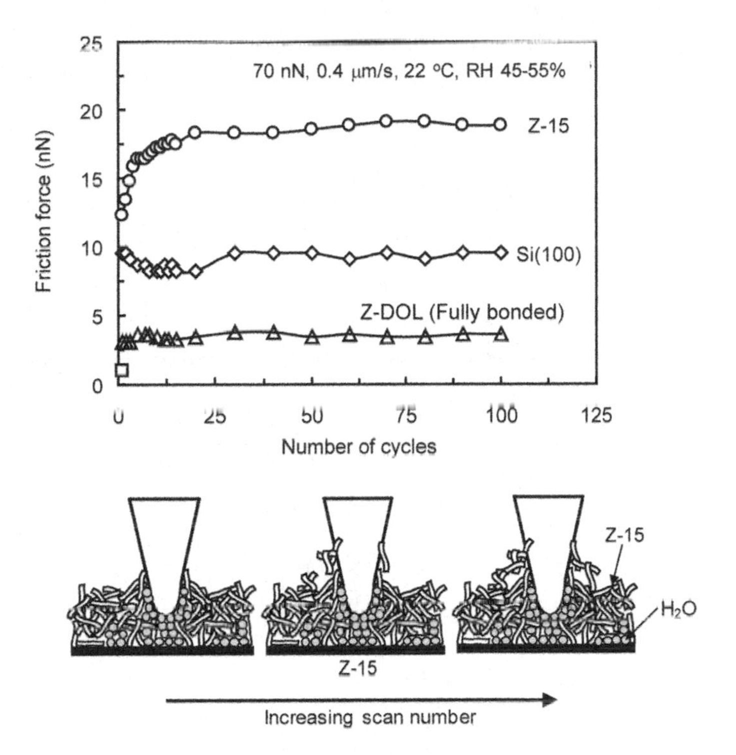

Figure 10.3.40 Friction force versus number of sliding cycles for Si(100) and Z-15 and Z-DOL (fully bonded) films at 70 nN, 0.8 μm/s, and in ambient air. Schematic (bottom) shows that some liquid Z-15 molecules can be attached onto the tip. The molecular interaction between the attached molecules onto the tip with the Z-15 molecules in the film results in an increase of the friction force with multi scanning. Reproduced with permission from Liu, H. and Bhushan, B. (2003a), "Nanotribological Characterization of Molecularly-Thick Lubricant Films for Applications to MEMS/NEMS by AFM," *Ultramicroscopy* **97**, 321–340. Copyright 2003. Elsevier.

the corresponding substrates. Among various films, HDT exhibits the lowest values. Based on stiffness measurements of various SAMs, HDT was most compliant, followed by BPT and BPTC. Based on friction and stiffness measurements, SAMs with high-compliance long carbon chains exhibit low friction; chain compliance is desirable for low friction. The friction mechanism of SAMs is explained by a so-called "molecular spring" model, Figure 10.3.42. According to this model, the chemically adsorbed self-assembled molecules on a substrate are like assembled molecular springs anchored to the substrate. An asperity sliding on the surface of SAMs is like a tip sliding on the top of "molecular springs or brush." The molecular spring assembly has compliant features and can experience orientation and compression under load. The orientation of the molecular springs or brush under normal load reduces the shearing force at the interface, which in turn reduces the friction force. The orientation is determined by the spring constant of a single molecule as well as by the interaction between the neighboring molecules, which can be reflected by the packing density or the packing energy. It should be noted that the orientation can lead to conformational defects along the molecular chains, which lead to energy dissipation.

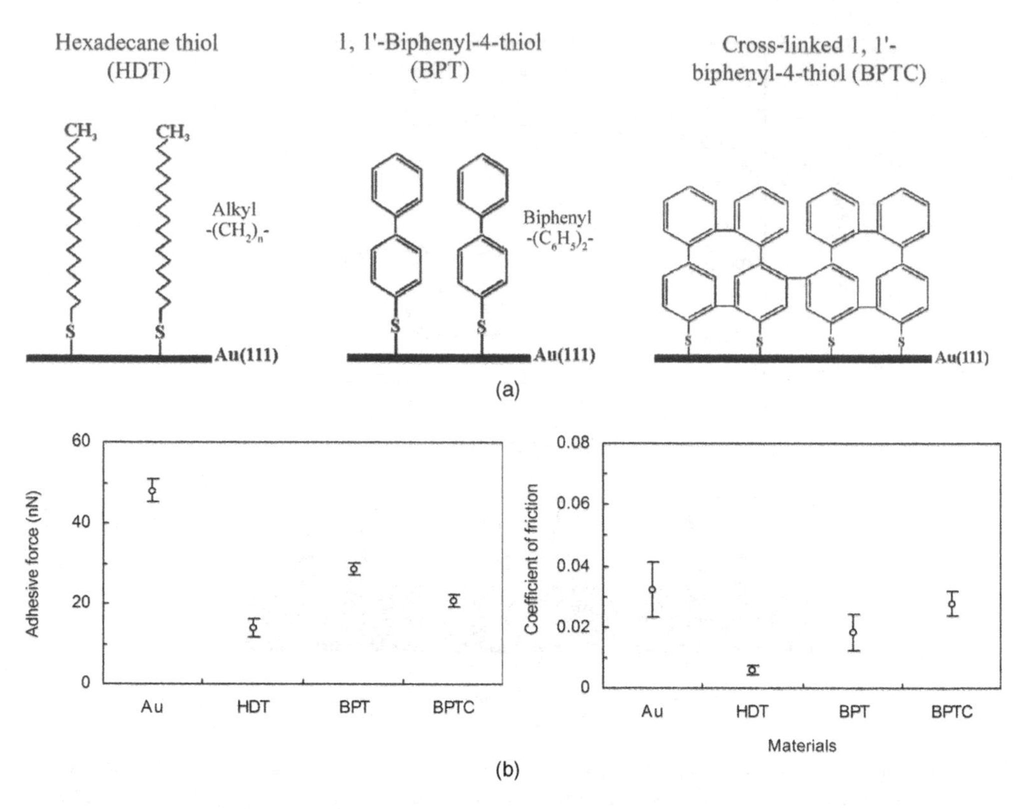

Figure 10.3.41 (a) Schematics of structures of hexadecane thiol and biphenyl thiol SAMs on Au(111) substrates, and (b) adhesive force and coefficient of friction of Au(111) substrate and various SAMs. Reproduced with permission from Bhushan B. and Liu, H. (2001), "Nanotribological Properties and Mechanisms of Alkylthiol and Biphenyl Thiol Self-Assembled Monolayers Studied by AFM," *Phys. Rev. B* **63**, 245412-1–245412-11. Copyright 2001 American Physical Society.

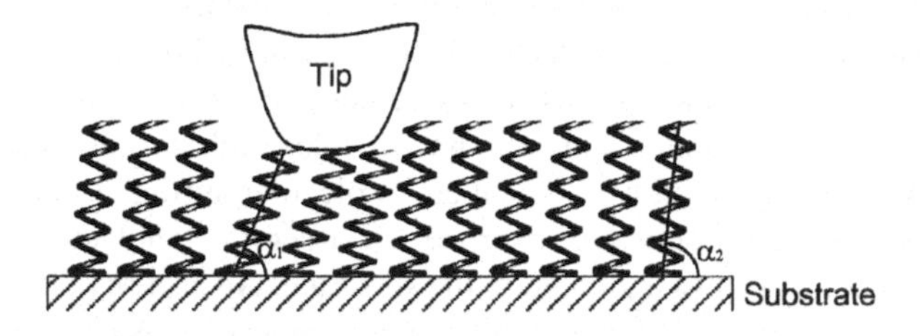

Figure 10.3.42 Molecular spring model of SAMs. In this figure, $\alpha_1 < \alpha_2$, which is caused by the further orientation under the normal load applied by an asperity tip. Reproduced with permission from Bhushan B. and Liu, H. (2001), "Nanotribological Properties and Mechanisms of Alkylthiol and Biphenyl Thiol Self-Assembled Monolayers Studied by AFM," *Phys. Rev. B* **63**, 245412-1–245412-11. Copyright 2001 American Physical Society.

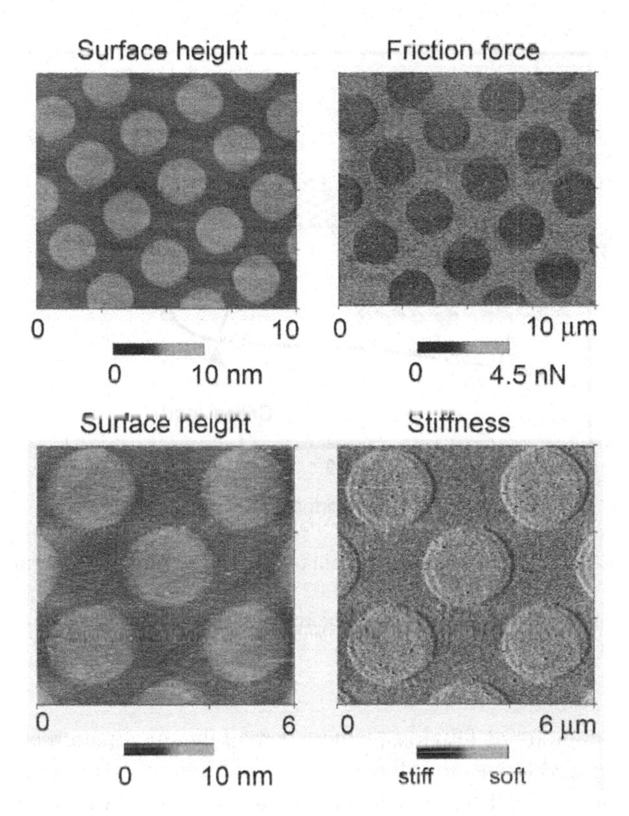

Figure 10.3.43 (a) AFM Grayscale surface height and stiffness images, and (b) AFM grayscale surface height and friction force images of micropatterned BDCS. Reproduced with permission from Liu, H. and Bhushan, B. (2002), "Investigation of Nanotribological Properties of Self-Assembled Monolayers with Alkyl and Biphenyl Spacer Chains," *Ultramicroscopy* **91**, 185–202. Copyright 2002. Elsevier.

An elegant way to demonstrate the influence of molecular stiffness on friction is to investigate SAMs with different structures on the same wafer. For this purpose, a micropatterned SAM was prepared. First the biphenyldimethylchlorosilane (BDCS) was deposited on silicon by a typical self-assembly method (Liu and Bhushan, 2002). Then the film was partially crosslinked using a mask technique by low-energy electron irradiation. Finally the micropatterned BDCS films were realized, which had the as-deposited and cross-linked coating regions on the same wafer. The local stiffness properties of this micropatterned sample were investigated by force modulation AFM technique (DeVecchio and Bhushan, 1997). The variation in the deflection amplitude provides a measure of the relative local stiffness of the surface. Surface height, stiffness, and friction images of the micropatterned biphenyldimethylchlorosilane (BDCS) specimen are obtained and presented in Figure 10.3.43 (Liu and Bhushan, 2002). The circular areas correspond to the as-deposited film, and the remaining area to the cross-linked film. Figure 10.3.43(a) indicates that cross-linking caused by the low energy electron irradiation leads to about 0.5 nm decrease of the surface height of BDCS films. The corresponding stiffness images indicate that the cross-linked area has a higher stiffness than the as-deposited area. Figure 10.3.43(b) indicates that the as-deposited area (higher surface height

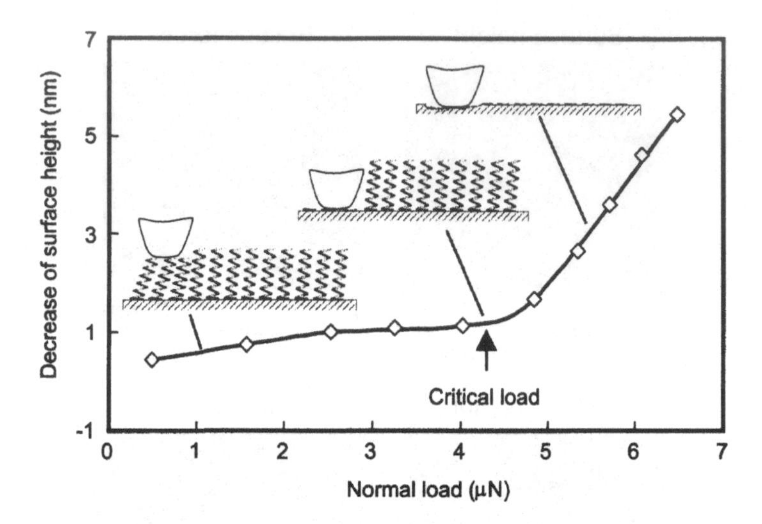

Figure 10.3.44 Illustration of the wear mechanism of SAMs with increasing normal load. Reproduced with permission from Liu, H. and Bhushan, B. (2002), "Investigation of Nanotribological Properties of Self-Assembled Monolayers with Alkyl and Biphenyl Spacer Chains," *Ultramicroscopy* **91**, 185–202. Copyright 2002. Elsevier.

area) has a lower friction force. Obviously, these data of the micropatterned sample prove that the local stiffness of SAMs has an influence on their friction performance. Higher stiffness leads to larger friction force. These results provide a strong proof of the suggested molecular spring model.

The SAMs with high-compliance long carbon chains also exhibit the best wear resistance (Bhushan and Liu, 2001; Liu and Bhushan, 2002). In wear experiments, the wear depth as a function of normal load curves shows a critical normal load, at which the film wears rapidly. A representative curve is shown in Figure 10.3.44. Below the critical normal load, SAMs undergo orientation; at the critical load SAMs wear away from the substrate due to the relatively weak interface bond strengths, while above the critical normal load severe wear takes place on the substrate.

10.3.5.3 Liquid Film Thickness Measurements

Liquid film thickness mapping of ultra-thin films (on the order of couple of 2 nm) can be obtained using friction force microscopy (Koinkar and Bhushan, 1996a) and adhesive force mapping (Bhushan and Dandavate, 2000). Figure 10.3.45 shows the gray-scale plots of the surface topography and friction force obtained simultaneously for unbonded Demnum S-100 type PFPE lubricant film on silicon. Demnum-type PFPE lubricant (Demnum, Daikin, Japan) chains have $-CF_2$-CH_2-OH, a reactive end group on one end, whereas Z-DOL chains have the hydroxyl groups on both ends, as described earlier. The friction force plot shows well-distinguished low and high friction regions roughly corresponding to high and low regions in surface topography (thick and thin lubricant regions). A uniformly lubricated sample does not

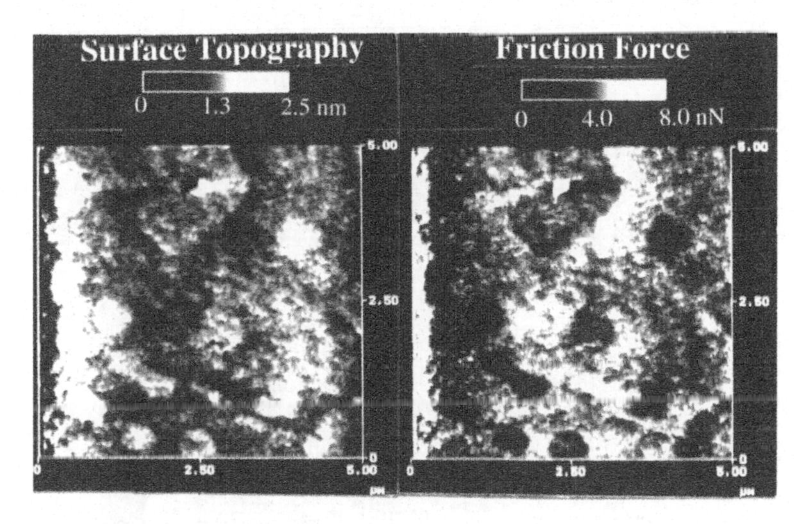

Figure 10.3.45 Gray scale plots of the surface topography and friction force obtained simultaneously for unbonded Demnum type perfluoropolyether lubricant film on silicon. Reproduced with permission from Koinkar, V.N. and Bhushan, B. (1996a), "Micro/nanoscale Studies of Boundary Layers of Liquid Lubricants for Magnetic Disks," J. Appl. Phys. 79, 8071-8075. Copyright 1996, American Institute of Physics.

show such a variation in the friction. Friction force imaging can thus be used to measure the lubricant uniformity on the sample surface, which cannot be identified by surface topography alone. Figure 10.3.46 shows the gray-scale plots of the adhesive force distribution for silicon samples coated uniformly and nonuniformly with Z-DOL type PFPE lubricant. It can be clearly seen that there exists a region which has adhesive force distinctly different from the other regions for the nonuniformly coated sample. This implies that the liquid film thickness is nonuniform, giving rise to a difference in the meniscus forces.

Quantitative measurements of liquid film thickness of thin lubricant films (on the order of few nm) with nanometer lateral resolution can be made with the AFM (Bhushan and Blackman, 1991; Bhushan, 1999a, 2011; Chen and Bhushan, 2005). The liquid film thickness is obtained by measuring the force on the tip as it approaches, contacts, and pushes through the liquid film and ultimately contacts the substrate. The distance between the sharp snap-in (owing to the formation of a liquid meniscus and van der Waals forces between the film and the tip) at the liquid surface and the hard repulsion at the substrate surface is a measure of the liquid film thickness. Figure 10.3.47 shows a plot of forces between tip and virgin and treated hair with hair conditioner. The hair sample was first brought into contact with the tip and then pulled away at a velocity of 400 nm/s. The zero tip–sample separation is defined to be the position where the force on the tip is zero, and the tip is not in contact with the sample. As the tip approaches the sample, a negative force exists which indicates an attractive force. The treated hair surface shows a much longer range of interaction with the tip compared to the very short range of interaction between virgin hair surfaces and the tip. Typically, the tip suddenly snaps into contact with the conditioner layer at a finite separation H (about 30 nm), which is proportional to conditioner thickness h. As the tip contacts the substrate, the tip travels

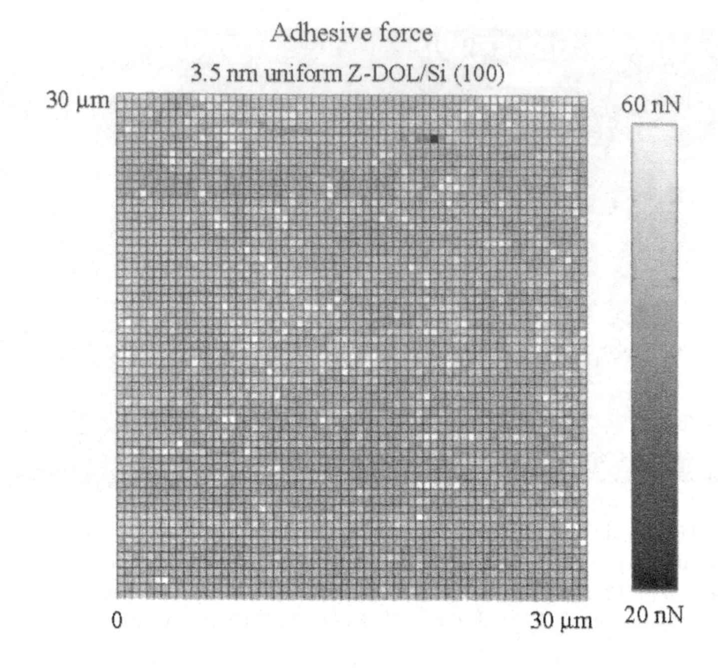

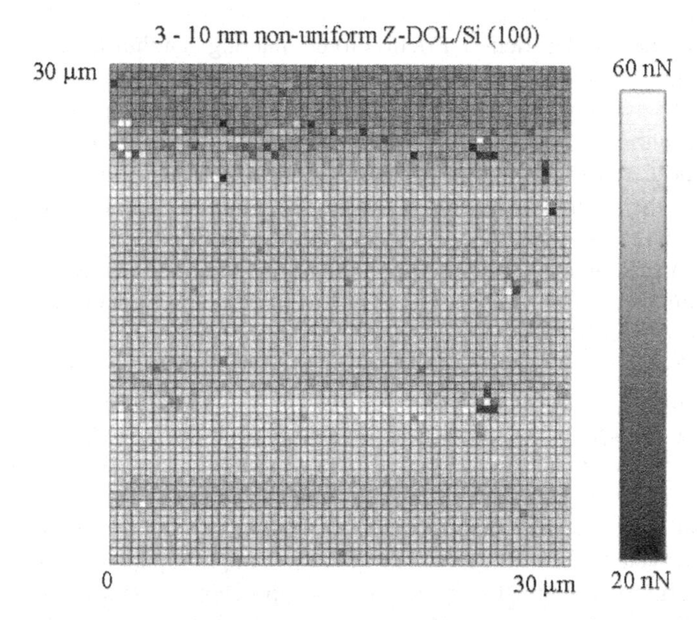

Figure 10.3.46 Gray-scale plots of the adhesive force distribution of a uniformly-coated, 3.5-nm thick unbonded Z-DOL film on silicon and 3- to 10-nm thick unbonded Z-DOL film on silicon that was deliberately coated nonuniformly by vibrating the sample during the coating process. Reproduced with permission from Bhushan, B. and Dandavate, C. (2000), "Thin-film Friction and Adhesion Studies Using Atomic Force Microscopy" *J. Appl. Phys.* **87**, 1201–1210. Copyright 2000 American institute of Physics.

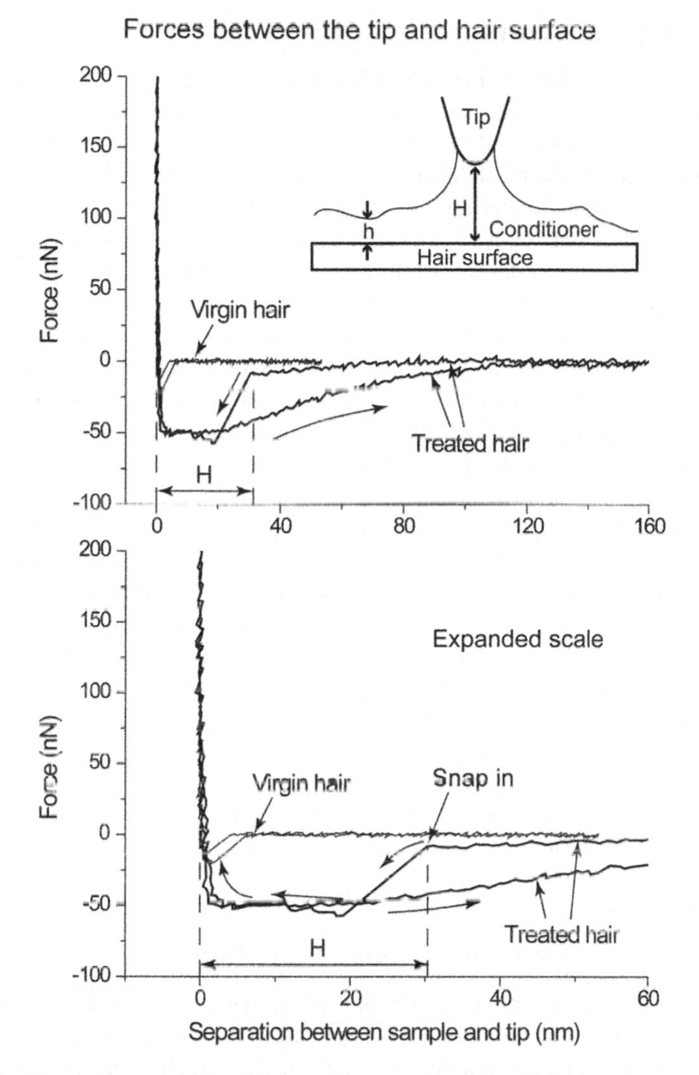

Figure 10.3.47 Forces between tip and hair surface as a function of tip sample separation for virgin hair and the conditioner treated hair. A schematic of measurement for localized conditioner thickness is shown in the inset at the top. The expanded scale view of force curve at small separation is shown at the bottom. Reproduced with permission from Chen, N. and Bhushan, B. (2005), "Morphological, Nanomechanical and Cellular Structural Characterization of Human Hair and Conditioner Distribution Using Torsional Resonance Mode in an AFM," *J. Micros.* **220**, 96–112. Copyright 2005 Wiley.

with the sample. When the sample is withdrawn, the forces on the tip slowly decrease to zero once the meniscus of liquid is drawn out from the hair surface. It should be noted that the distance H between the sharp snap-in at the liquid surface and the hard wall contact with the substrate is not the real conditioner thickness h. Due to the interaction of the liquid with the tip at some spacing distance, H tends to be thicker than the actual film thickness, but can still provide an estimate of the actual film thickness and upper limit of thickness.

10.4 Atomic-Scale Computer Simulations

Most theoretical approaches to contact problems on the macroscale are based on the continuum elasticity (e.g., see Johnson, 1985). However, continuum mechanics is not fully applicable as the scale of the material bodies and the characteristic dimensions of the contact between them are reduced. Furthermore, the mechanical properties of materials exhibit a strong dependence on the size of the sample; materials are stronger at the smaller scales. Concurrent with the development and use of innovative experimental techniques using SFA and AFM/FFM, theoretical methods have been developed for atomic-scale studies. Theoretical studies include analytical methods and large-scale molecular dynamics (MD) computer simulations (Hoover, 1986; Heermann, 1986; Haile, 1992). For many years, analytical methods have been used to study atomistic mechanisms of friction (Tomlinson, 1929; McClelland and Glosli, 1992; Ruan and Bhushan, 1994b; Sokoloff, 1996; Persson and Tosatti, 1996). The limitations of the analytical approaches are that simplifying assumptions must be made and, for example, unanticipated defect structures must be neglected. Advances in the theoretical understanding of the nature of interatomic interactions in materials and computer-based modeling of complex systems have led to MD computer simulations to explore atomic-scale interactions. In MD simulations, for a given set of initial conditions and a way of describing interatomic forces, classical equations of motions are integrated. The spatial and temporal motion of atoms and molecules as a function of time with high spatial and temporal resolution is obtained from the simulations through analyses of relative positions, velocities, and forces; by visual inspection of the trajectories through animated movies; or through a combination of both. Simulations often reveal unanticipated events which are analyzed (Landman *et al.*, 1990; Singer and Pollock, 1992; Bhushan *et al.*, 1995a; Guntherodt *et al.*, 1995; Persson and Tosatti, 1996; Bhushan, 1997, 1999a, 2001a, b, 2011; Sinnott *et al.*, 2011).

In this section, we present an overview of MD simulation modeling and selected results dealing with various friction, wear, and indentation studies.

10.4.1 Interatomic Forces and Equations of Motion

In MD computer simulations, first, a set of initial conditions (relative positions and velocities of particles from Boltzmann distribution) are described and the interatomic forces are calculated using classical potential energy functions from electronic structure calculations. The motion of atoms and molecules in a phase space of systems consisting of thousands of atoms is simulated with high spatial and temporal resolution by the numerical solution of a set of coupled differential equations based on the particles' classical equations of motion. Consider Newton's equation of motion

$$\vec{F} = m\frac{d\vec{v}}{dt} \tag{10.4.1}$$

where $\vec{F}$ is the force on a particle, m is its mass, $\vec{v}$ is the particle velocity, and t is time. For n particles, a set of $3n$ second-order differential equations governs the dynamics. These can be solved with finite difference interaction methods with finite time steps on the order of $1/25$ of a vibrational frequency (typically a tenth to a few femtoseconds). Integration is generally carried out for a total time of a few picoseconds to a few nanoseconds (Bhushan *et al.*, 1995a; Sinnott *et al.*, 2011).

Interatomic forces ($\vec{F}$) are calculated from the spatial derivatives of the classical potential energy functions. These functions are typically based on quantum-mechanical processes. There are two approaches to the potential. In the first approach, the potential energy of the atoms is represented as a function of their relative atomic positions. Two common forms of the pair potentials are the Morse potential and the Lennard-Jones (LJ) potential. Interaction of fluid molecules of a thin film of mass m with the two solids is generally modeled with a Lennard-Jones (LJ) potential.

$$V(x) = 4\varepsilon\left[\left(\frac{\sigma}{x}\right)^{12} - \left(\frac{\sigma}{x}\right)^{6}\right] \tag{10.4.2}$$

where ε, σ and $\tau[= (m\sigma^2/\varepsilon)^{1/2}]$ are characteristic energy, length and time scales for the interaction, and the distance x is the center-to-center distance between the interacting molecules. The form presented in Equation 10.4.2 is called the 6-12 potential.

In the second approach for the potential, the calculations of interatomic forces explicitly include electrons, and are more commonly used in computer simulations of solid–solid interactions. As an example for metals, the embedded-atom method (EAM) is commonly used. In this method, the cohesive energy of the material is viewed as the energy to embed an atom into the local electron density provided by the other atoms of the system. This background density is determined for each atom as the superposition of electron densities from the other atoms, evaluated at the location of the atoms in question. Thus the cohesive energy is represented in the EAM by a many-body embedding functional, supplemented by parametrized short-range pair interactions due to inter-core repulsion. The parameters of the pair potentials are determined via fitting to a number of bulk equilibrium properties of the metals and their alloys, such as lattice constants, cohesive energy, elastic constants, and vacancy formation energy (Foiles *et al.*, 1986; Sutton, 1993).

During relative motion (sliding or indentation), work performed on the system raises its energy and causes an increase in the temperature. In the simulation, the system temperature is controlled in the canonical ensemble by using a number of thermostats. In any thermostat, the atom velocities are altered in the process of controlling the temperature (Hoover, 1986).

10.4.2 Interfacial Solid Junctions

MD simulation studies have been conducted to study adhesion, friction, wear and indentation processes. Landman *et al.* (1990) and Landman and Luedtke (1991) used MD to simulate the indentation of a metallic substrate with a metallic tip. In the simulations (at a constant temperature of 300 K), a clean Ni tapered and faceted tip was used to indent a clean Au (001) substrate, and the situation was reversed, using an Ni surface and an Au tip. After equilibrium of the tip and substrate to 300 K, the tip was brought into contact with the surface by moving the tip 25 nm closer to the surface every 1525 fs (a tip velocity of about 16 m/s), Figure 10.4.1. Simulations revealed the onset of instability as the tip approaches the sample at a distance of about 0.4 nm. At this point there occurs a jump to contact (with gold atoms being displaced by about 0.2 nm in about 1 ps), with adhesive bonding between the two

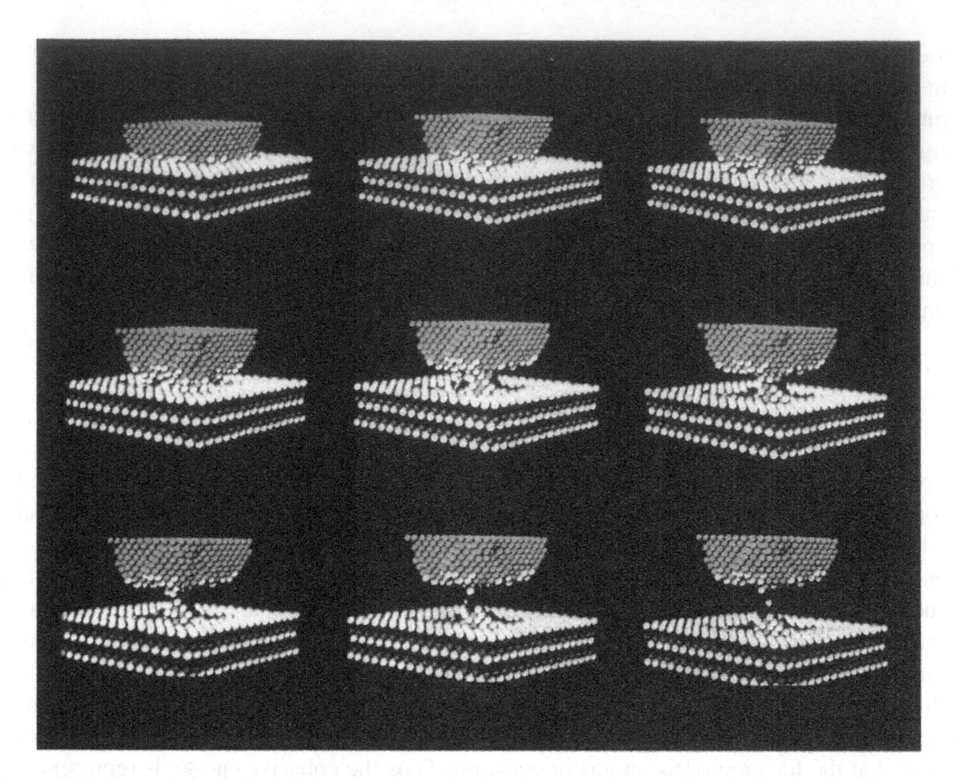

Figure 10.4.1 Sequence of atomic configuration starting from a Ni tip indented in an Au(001) substrate (top left) and during the process of retraction of the tip (from left to right) accompanied by formation of a connective solid gold junction (bottom right). Reproduced with permission from Bhushan, B., Israelachvili, J.N., and Landman, U. (1995a), "Nanotribology: Friction, Wear and Lubrication at the Atomic Scale," *Nature* **374**, 607–616. Copyright 1995. Nature Publishing Group.

materials driven and accompanied by atomic-scale wetting of nickel by gold atoms. The latter is the result of differences in their surface energies, just as it is for the case of surface wetting by a liquid film. Retraction of the tip from the surface after contact causes significant inelastic deformation of the sample, involving ductile extension, formation of a connective neck of atomic dimensions, and eventual rupture. The final result on separation is a gold-coated nickel tip and a damaged gold surface (Bhushan *et al.*, 1995a). This behavior has been seen experimentally in AFM studies of solid junctions between the tip and a gold surface (Landman *et al.*, 1990).

Belak and Stowers (1992) simulated orthogonal cutting of Cu(111) substrate using a rigid diamond cutting tool. Cutting was performed by continuously moving the tool closer to the plane of the surface while the surface was moved in a direction perpendicular to the surface normal at 100 m/s. This process formed a chip in front of the tool (Figure 10.4.2). This chip was crystalline but possessed a different orientation than the Cu substrate. Regions of disorder were formed in front of the tool tip and on the substrate surface in front of the chip. Additionally, dislocations originating from the tool tip contact point were formed on the substrate.

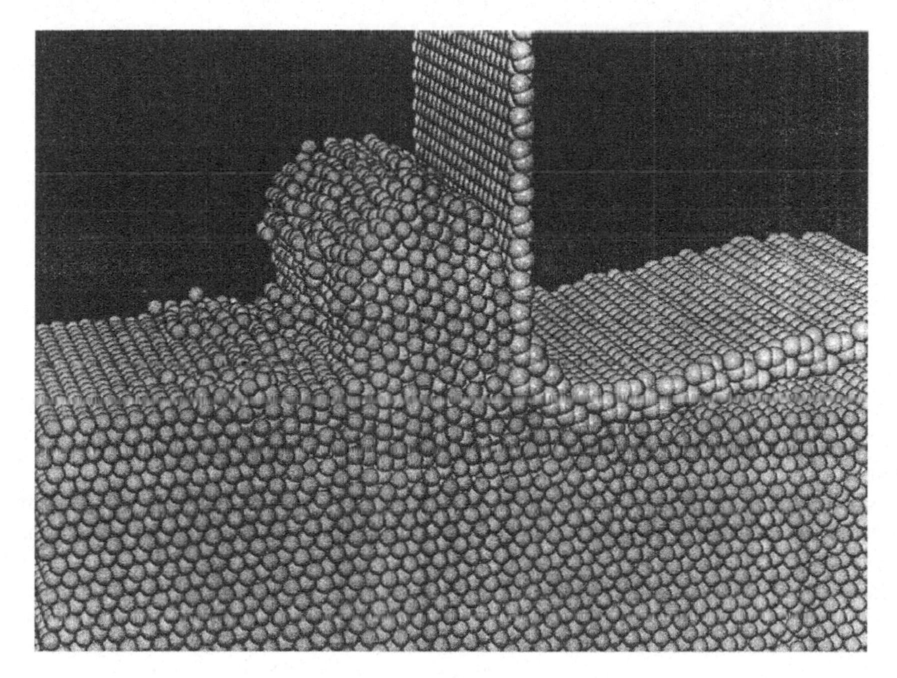

Figure 10.4.2 Atomic configuration of the orthogonal cutting of a Cu(111) substrate with a rigid diamond cutting tool. Reproduced with permission from Belak, J. and Stowers, I.F. (1992), "The Indentation and Scraping of a Metal Surface: A Molecular Dynamic Study," In *Fundamentals of Friction: Macroscopic and Microscopic Processes* (I.L. Singer and H.M. Pollock, eds.), Vol. E220, pp. 511–520, Kluwer, Dordrecht, The Netherlands. Copyright 1992 Springer.

10.4.3 Interfacial Liquid Junctions and Confined Films

MD simulations of interfacial liquid junctions and confined films have been carried out to understand the physical properties and response of model lubricants and their molecular characteristics, such as chain length and molecular structure (for example, straight versus branched chains) (Thompson and Robbins, 1990a, b; Robbins and Thompson, 1991; Ribarsky and Landman, 1992; Thompson *et al.*, 1992; Landman *et al.*, 1993; Wang *et al.*, 1993; Bhushan *et al.*, 1995a; Bhushan, 2011). The confinement of n-octane ranging in thickness from 1 to 2.4 nm, between parallel, crystalline solid walls (rigid Langmuir-Blodgett layers), was studied by Wang *et al.* (1993). They reported that for a small thickness of liquid film of 1 nm, the film formed a layered structure with molecules lying parallel to the solid walls. At larger thicknesses there was always an ordering structure on each wall surface and more poorly defined layers in the center of the gap. These results confirm the observation made from the SFA experimental data presented earlier.

The behavior of the viscosity of the film of different thicknesses as a function of shear rate was studied by Thompson *et al.* (1992). The predicted viscosity of films of different thicknesses as a function of shear rate is shown in Figure 10.4.3. The response of films that were 6-8 molecular diameters thick was approximately the same as the bulk viscosity. When the film thickness was reduced, the viscosity of the film increased dramatically, particularly

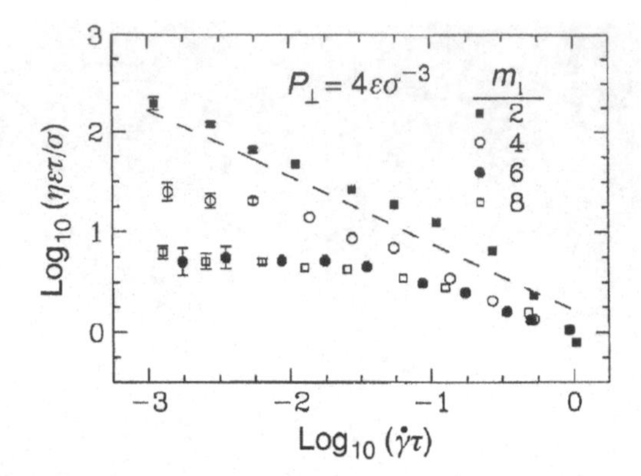

Figure 10.4.3 Averaged normalized absolute viscosity of liquid film as a function of shear rate for different number of fluid layers (m_l). The variable $P_{\perp}$ represents the pressure on the walls and ε, σ, and τ $\left(= (m\sigma^2/\varepsilon)^{1/2}\right)$ are characteristic energy, length and time scales for the interaction; m is the mass of fluid molecules. The dashed line has a slope of −2/3. Reproduced with permission from Thompson, P.A., Grest, G.S., and Robbins, M.O. (1992), "Phase Transitions and Universal Dynamics in Confined Films," *Phys. Rev. Lett.* **68**, 3448–3451. Copyright 1992 American Physical Society.

at low shear rates. These predictions are consistent with SFA experimental observations that molecularly thick liquid films confined between two solid surfaces do not behave as thick liquids; these often behave more like solids in terms of structure and flow.

Thompson and Robbins (1990a) and Robbins and Thompson (1991) studied the origins of stick-slip motion when LJ liquid film is sheared between two solid walls [fcc solids with (1 1 1) surfaces and shear direction (100)] observed in SFA experiments (reported earlier). In their simulation of LJ liquid between two solid walls (Figure 10.4.4a), the upper wall was coupled through a spring to a stage that advanced at constant velocity. Initially, the force on the spring is zero and the upper wall is at rest. As the stage moves forward, the spring stretches and the force increases. When the film is in the liquid state, the upper wall accelerates until the force applied by the spring balances the viscous dissipation. If the film is crystalline in nature, stick-slip behavior is observed, Figure 10.4.4.b. The film initially responds elastically, the wall remains stationary, and the force increases linearly. When the force exerted by the spring exceeds the yield stress of the film, the film melts and the top wall begins to slide. The wall accelerates to catch up with the stage, decreasing the force. This sawtooth-like behavior of the force is indicative of stick-slip behavior. The stick-slip behavior is dependent on the velocity of the stage, with the degree of stick-slip increasing as the velocity decreases. Analysis of the two-dimensional structure factor during the course of the simulation confirms that the film undergoes solid–liquid phase transitions as it proceeds from the static to the sliding states.

10.5 Closure

Innovative experimental techniques including SFA and AFM/FFM, and MD computer simulations are used to study the interaction of materials ranging from atomic scales to microscales. SFA experiments are used to study adhesion and friction of a molecularly thick liquid film

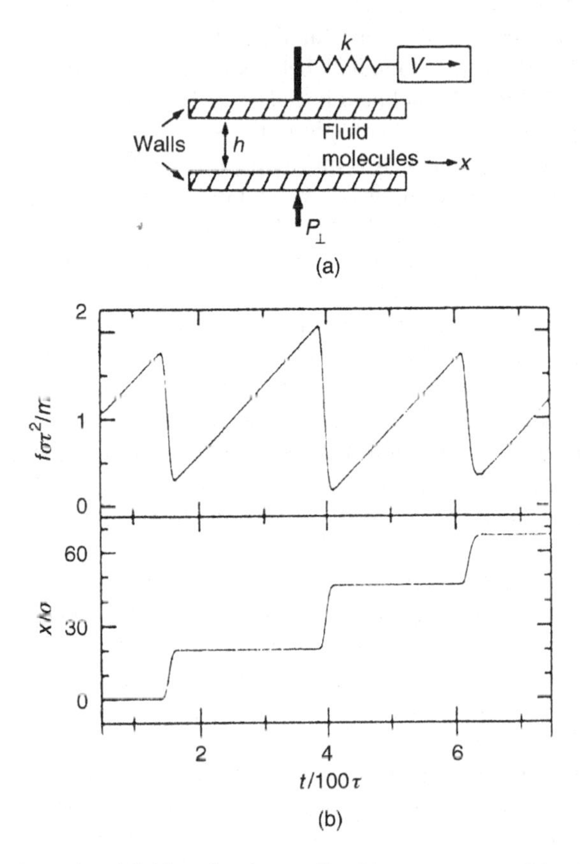

Figure 10.4.4 (a) Schematic of fluid molecules confined between two solid walls held together by a constant load $P_{\perp}$. The upper wall is attached to a stage which moves at constant velocity V by a spring. (b) Normalized shear force per unit area (f) and wall displacement (x) as a function of normalized sliding time (t) during stick-slip motion for a given wall velocity. Reproduced with permission from Thompson, P.A. and Robbins, M.O. (1990a), "Origin of Stick-Slip Motion in Boundary Lubrication," *Science* **250**, 792–794. Copyright 1990 AAAS.

confined between two smooth surfaces. At most solid–solid interfaces of technological relevance, contact occurs at multiple asperities. A sharp AFM/FFM tip sliding on a surface simulates just one such contact. However, asperities come in all shapes and sizes. The effect of radius of a single asperity (tip) on the friction performance can be studied using tips of different radii. AFM/FFM are used to study various tribological phenomena. In large-scale MD simulations, the trajectories in the phase space of systems consisting of thousands of atoms (and subject to appropriate boundary conditions) are calculated from the particles' Newtonian equation of motion. Analyses of particle trajectories help in studying tribological interactions and the determination of mechanical properties of the system on nanometer distance scales and femtosecond time scales.

SFA experiments and molecular dynamic computer simulations have found that when a liquid is confined between two surfaces or within any narrow space whose dimensions are less that 5–10 molecular diameters (~0.4 nm), thin liquid films behave as a crystalline solid (they solidify or freeze). Differences from the bulk behavior are more prominent for an increasingly thinner film. For liquid film with thickness in excess of 10 molecular diameters (layers), both static and dynamic properties are describable in terms of bulk properties. Once the number of molecular layers in the gap falls below this number, local density and viscosity, molecular orientation and structuring of molecules show significant deviations from bulk continuum properties. During sliding, a molecularly thin film will exhibit a yield point before it begins to flow. Such films can therefore sustain a finite shear stress, in addition to a finite normal stress. The value of yield stress depends on the number of layers comprising the film and normal stress. A shorter sticking time and higher relative sliding velocities produce lower static friction force, because the lubricant has less time to fully solidify between the surfaces. After sliding has initiated, depending on whether the film is solid-like or liquid-like, the motion will be smooth or of the stick-slip type. Successive first-order transitions between solid-like and liquid-like states are responsible for the observed stick-slip behavior of simple isotropic liquids between two solid surfaces, during which, instead of two solids in contact moving smoothly, the solids may alternatively stick and then slip past one another. With this interpretation, stick-slip is seen to arise because of the abrupt change in the flow properties of a film, a transition rather than the gradual or continuous change. In cases exhibiting stick-slip behavior, it is the successive freezing and melting transitions of the shearing film that give rise to the negative friction-velocity dependence, which is in contrast to the traditional picture of stick-slip motion ascribed to a negative slope in the friction-velocity function.

SFA experiments with molecularly thin films of complex fluids and polymers have also shown why branched-chain molecules are better lubricants than straight-chain molecules, even though the former have much higher bulk viscosities; the symmetrically shaped straight-chain molecules are prone to ordering and freezing, which dramatically increases their resistance to shear, whereas the irregularly shaped, branched molecules remain in the liquid state even under high loads.

AFM/FFM are used to study various tribological phenomena, which include surface roughness, adhesion, friction, scratching, wear, indentation, detection of material transfer, and boundary lubrication. Measurement of the atomic-scale friction of a freshly-cleaved highly-oriented pyrolytic graphite exhibits the same periodicity as that of the corresponding topography. However, the peaks in friction and those in the corresponding topography are displaced relative to each other. Variations in atomic-scale friction and the observed displacement can be explained by the variation in interatomic forces in the normal and lateral directions. The relevant friction mechanism is the atomic-scale stick-slip. Local variations in microscale friction occur and are found to correspond to the local slopes, suggesting that a ratchet mechanism and collision effects are responsible for this variation. Directionality in the friction is observed on both micro- and macroscales which results from the surface roughness and surface preparation. Anisotropy in surface roughness accentuates this effect. The friction contrast in conventional frictional measurements is based on interactions dependent upon interfacial material properties superimposed by roughness-induced lateral forces. To obtain roughness-independent friction, lateral or torsional modulation techniques can be used. These techniques also allow measurements over a small region. AFM/FFM experiments are generally conducted at relative velocities up to about 200 μm/s. High velocity experiments can be performed by either mounting a sample on a shear wave transducer driven at very high frequencies or mounting a sample on a high velocity piezo stage. By using these techniques, friction and wear experiments can be performed at a range of sliding velocities as well as normal loads, and the data have been used to develop nanoscale friction and wear maps. Relevant friction mechanisms are different for different ranges of sliding velocities and normal loads.

The adhesion and friction in wet environment depend on the tip radius, surface roughness, and relative humidity. Superhydrophobic surfaces can be designed by roughness optimization.

Nanoscale friction is generally found to be smaller than microscale friction. There are several factors responsible for the differences which include wear and contaminant particles, the transition from elasticity to plasticity, scale-dependent roughness and mechanical properties, and meniscus effects. Nanoscale friction values increase with an increase in the normal load above a certain critical load (pressure), approaching the macroscale friction. The critical contact pressure corresponds to the hardness of the softer of the two contacting materials.

Wear rate on the microscale for single-crystal silicon is negligible below 20 μN and is much higher and remains approximately constant at higher loads. Elastic deformation at low loads is responsible for negligible wear. Most of the wear debris is loose. SEM and TEM studies of the wear region suggest that the material on the microscale is removed by plastic deformation with a small contribution from elastic fracture; this observation corroborates with the scratch data. Evolution of wear has also been studied using AFM. Wear is found to be initiated at nanoscratches. For a sliding interface requiring near-zero friction and wear, contact stresses should be below the hardness of the softer material to minimize plastic deformation and surfaces should be free of nanoscratches. Further, wear precursors can be detected at early stages of wear by using surface potential measurements. It is found that even in the case of zero wear (no measurable deformation of the surface using AFM), there can be a significant change in the surface potential inside the wear mark which is useful for study of wear precursors. Detection of material transfer on a nanoscale is possible with AFM.

In situ surface characterization of the local deformation of materials and thin coatings can be carried out using a tensile stage inside an AFM. An AFM can also be used for nanofabrication/nanomachining.

A modified AFM can be used to obtain load-displacement curves and for the measurement of nanoindentation hardness and Young's modulus of elasticity, with a depth of indentation as low as 1 nm. Hardness of ceramics on nanoscales is found to be higher than that on the microscale. Ceramics exhibit significant plasticity and creep on a nanoscale. By using the force modulation technique, localized surface elasticity maps of composite materials with penetration depth as low as 1 nm, can be obtained. By using phase contrast microscopy in tapping or torsional mode, it is possible to obtain phase contrast maps or the contrast in viscoelastic properties of near surface regions. Scratching and indentation on nanoscales are powerful ways to screen for adhesion and resistance to deformation of ultrathin films.

Boundary lubrication studies and measurement of lubricant-film thickness with a lateral resolution on the nanoscale can be conducted using AFM. Chemically-bonded lubricant films and self-assembled monolayers are superior in friction and wear resistance. For chemically bonded lubricant films, the adsorption of water, the formation of meniscus and its change during sliding, and surface properties play an important role on the adhesion, friction, and durability of these films. Sliding velocity, relative humidity, and temperature affect adhesion and friction. For SAMs, their friction mechanism is explained by a so-called "molecular spring" model. The films with high-compliance long carbon chains exhibit low friction and wear. Also perfluoroalkylsilane SAMs on Si appear to be more hydrophobic with lower adhesion than alkylsilane SAMs on Si.

Investigations of adhesion, friction, wear, scratching, and indentation on the nanoscale using the AFM can provide insights into the failure mechanisms of materials. Coefficients of friction, wear rates and mechanical properties such as hardness have been found to be different on the nanoscale than on the macroscale; generally, the coefficients of friction and wear rates on micro- and nanoscales are smaller, whereas the hardness is greater. Therefore, nanotribological studies may help define the regimes for ultra-low friction and near zero wear. These studies also provide insight into the atomic origins of adhesion, friction, wear, and lubrication mechanisms.

The MD simulation results correlate well with the experimental observations at the nanoscale. Simulations consistent with experimental observation often reveal the physical behavior of materials at interfaces to be very different from that in the bulk. Atomic-scale simulation can be of great value for interpreting the results of micro- and nanotribological experiments, particularly those involving the surface-force apparatus and scanning-probe microscopies. Moreover, understanding the behavior of materials of very small dimensions is relevant to the development of nano- and microfabricated devices and to the atomic-scale manipulation of materials.

References

Amelio, S., Goldade, A.V., Rabe, U., Scherer, V., Bhushan, B., and Arnold, W. (2001), "Measurements of Elastic Properties of Ultra-thin Diamond-like Carbon Coatings Using Atomic Force Acoustic Microscopy," *Thin Solid Films* **392**, 75–84.

Anczykowski, B., Kruger, D., Babcock, K.L., and Fuchs, H. (1996), "Basic Properties of Dynamic Force Microscopy with the Scanning Force Microscope in Experiment and Simulation," *Ultramicroscopy* **66**, 251–259.

Belak, J. and Stowers, I.F. (1992), "The Indentation and Scraping of a Metal Surface: A Molecular Dynamic Study," In *Fundamentals of Friction: Macroscopic and Microscopic Processes* (I.L. Singer and H.M. Pollock, eds.), Vol. E220, pp. 511–520, Kluwer, Dordrecht, The Netherlands.

Bhushan, B. (1995), "Micro/Nanotribology and its Applications to Magnetic Storage Devices and MEMS," *Tribol. Int.* **28**, 85–95.

Bhushan, B. (1996), *Tribology and Mechanics of Magnetic Storage Devices*, Second edition, Springer-Verlag, New York.

Bhushan, B. (1997), *Micro/Nanotribology and its Applications*, Vol. E330, Kluwer, Dordrecht, The Netherlands.

Bhushan, B. (1998), *Tribology Issues and Opportunities in MEMS*, Kluwer, Dordrecht, Netherlands.

Bhushan, B. (1999a), *Handbook of Micro/Nanotribology*, Second edition, CRC Press, Boca Raton, Florida.

Bhushan, B. (1999b), "Nanoscale Tribophysics and Tribomechanics," *Wear* **225–229**, 465–492.

Bhushan, B. (1999c), "Wear and Mechanical Characterisation on Micro- to Picoscales Using AFM," *Int. Mat. Rev.* **44**, 105–117.

Bhushan, B. (2001a), *Modern Tribology Handbook, Vol. 1: Principles of Tribology*, CRC Press, Boca Raton, Florida.

Bhushan, B. (2001b), *Fundamentals of Tribology and Bridging the Gap Between the Macro- and Micro/Nanoscales*, NATO Science Series II – Vol. 10, Kluwer, Dordrecht, Netherlands.

Bhushan, B. (2001c), "Nano- to Microscale Wear and Mechanical Characterization Studies Using Scanning Probe Microscopy," *Wear* **251**, 1105–1123.

Bhushan, B. (2003), "Adhesion and Stiction: Mechanisms, Measurement Techniques, and Methods for Reduction," *J. Vac. Sci. Technol. B* **21**, 2262–2296.

Bhushan, B. (2005), "Nanotribology and Nanomechanics," *Wear* **259**, 1507–1531.

Bhushan, B. (2008), "Nanotribology, Nanomechanics and Nanomaterials Characterization," *Phil. Trans. R. Soc. A* **366**, 1351–1381.

Bhushan, B. (2010), *Springer Handbook of Nanotechnology*, Third edition, Springer-Verlag, Heidelberg, Germany.

Bhushan, B. (2011), *Nanotribology and Nanomechanics I & II*, Third edition, Springer-Verlag, Heidelberg, Germany.

Bhushan, B. (2012), *Encyclopedia of Nanotechnology*, Springer-Verlag, Heidelberg, Germany.

Bhushan, B. and Blackman, G.S. (1991), "Atomic Force Microscopy of Magnetic Rigid Disks and Sliders and its Applications to Tribology," *ASME J. Tribol.* **113**, 452–458.

Bhushan, B. and Dandavate, C. (2000), "Thin-film Friction and Adhesion Studies Using Atomic Force Microscopy" *J. Appl. Phys.* **87**, 1201–1210.

Bhushan, B. and Goldade, A.V. (2000a), "Measurements and Analysis of Surface Potential Change During Wear of Single Crystal Silicon (100) at Ultralow Loads Using Kelvin Probe Microscopy," *Appl. Surf. Sci.* **157**, 373–381.

Bhushan, B. and Goldade, A.V. (2000b), "Kelvin Probe Microscopy Measurements of Surface Potential Change Under Wear at Low Loads," *Wear* **244**, 104–117.

Bhushan, B. and Kasai, T. (2004), "A Surface Topography-Independent Friction Measurement Technique Using Torsional Resonance Mode in an AFM," *Nanotechnology* **15**, 923–935.

Bhushan, B. and Koinkar, V.N. (1994a), "Tribological Studies of Silicon for Magnetic Recording Applications," *J. Appl. Phys.* **75**, 5741–5746.

Bhushan, B. and Koinkar, V.N. (1994b), "Nanoindentation Hardness Measurements Using Atomic Force Microscopy," *Appl. Phys. Lett.* **64**, 1653–1655.

Bhushan, B. and Kulkarni, A.V. (1996), "Effect of Normal Load on Microscale Friction Measurements," *Thin Solid Films* **278**, 49–56; **293**, 333.

Bhushan, B. and Li, X. (2003), "Nanomechanical Characterisation of Solid Surfaces and Thin Films," *Intern. Mat. Rev.* **48** 125–164.

Bhushan, B. and Liu, H. (2001), "Nanotribological Properties and Mechanisms of Alkylthiol and Biphenyl Thiol Self-Assembled Monolayers Studied by AFM," *Phys. Rev. B* **63**, 245412-1–245412-10.

Bhushan, B. and Nosonovsky, M. (2003), "Scale Effects in Friction Using Strain Gradient Plasticity and Dislocation-Assisted Sliding (Microslip)," *Acta Mater.* **51**, 4331–4345.

Bhushan, B. and Nosonovsky, M. (2004a), "Comprehensive Model for Scale Effects in Friction Due to Adhesion and Two- and Three-Body Deformation (Plowing)," *Acta Mater.* **52**, 2461–2474.

Bhushan, B. and Nosonovsky, M. (2004b), "Scale Effects in Dry and Wet Friction, Wear, and Interface Temperature," *Nanotechnology* **15**, 749–761.

Bhushan, B. and Qi, J. (2003), "Phase Contrast Imaging of Nanocomposites and Molecularly-Thick Lubricant Films in Magnetic Media," *Nanotechnology* **14**, 886–895.

Bhushan, B. and Ruan, J. (1994), "Atomic-scale Friction Measurements Using Friction Force Microscopy: Part II – Application to Magnetic Media," *ASME J. Trib.* **116**, 389–396.

Bhushan, B. and Sundararajan, S. (1998), "Micro/nanoscale Friction and Wear Mechanisms of Thin Films Using Atomic Force and Friction Force Microscopy," *Acta Mater.* **46**, 3793–3804.

Bhushan, B., Ruan, J., and Gupta, B.K. (1993), "A Scanning Tunnelling Microscopy Study of Fullerene Films," *J. Phys. D: Appl. Phys.* **26**, 1319–1322.

Bhushan, B., Koinkar, V.N., and Ruan, J. (1994), "Microtribology of Magnetic Media," *Proc. Inst. Mech. Eng., Part J: J. Eng. Tribol.* **208**, 17–29.

Bhushan, B., Israelachvili, J.N., and Landman, U. (1995a), "Nanotribology: Friction, Wear and Lubrication at the Atomic Scale," *Nature* **374**, 607–616.

Bhushan, B., Kulkarni, A.V., Koinkar, V.N., Boehm, M., Odoni, L., Martelet, C., and Belin, M. (1995b), "Microtribological Characterization of Self-Assembled and Langmuir-Blodgett Monolayers by Atomic and Friction Force Microscopy," *Langmuir* **11**, 3189–3198.

Bhushan, B., Kulkarni, A.V., Bonin, W., and Wyrobek, J.T. (1996), "Nano/Picoindentation Measurement Using a Capacitance Transducer System in Atomic Force Microscopy," *Philos. Mag.* **74**, 1117–1128.

Bhushan, B., Mokashi, P.S., and Ma, T. (2003), "A New Technique to Measure Poisson's Ratio of Ultrathin Polymeric Films Using Atomic Force Microscopy," *Rev. Sci. Instrum.* **74**, 1043–1047.

Bhushan, B., Kasai, T., Nguyen, C.V., and Meyyappan, M. (2004a), "Multiwalled Carbon Nanotube AFM Probes for Surface Characterization of Micro/Nanostructures," *Microsys. Technol.* **10**, 633–639.

Bhushan, B., Liu, H., and Hsu, S.M. (2004b), "Adhesion and Friction Studies of Silicon and Hydrophobic and Low Friction Films and Investigation of Scale Effects," *ASME J. Tribol.* **126**, 583–590.

Bhushan, B., Kasai, T., Kulik, G., Barbieri, L., and Hoffmann, P. (2005), "AFM Study of Perfluorosilane and Alkylsilane Self-Assembled Monolayers for Anti-Stiction in MEMS/NEMS," *Ultramicroscopy* **105**, 176–188.

Bhushan, B., Hansford, D., and Lee, K.K. (2006), "Surface Modification of Silicon and Polydimethylsiloxane Surfaces with Vapor-Phase-Deposited Ultrathin Fluorosilane Films for Biomedical Nanodevices," *J. Vac. Sci. Technol. A* **24**, 1197–1202.

Bhushan, B., Cichomski, M., Tao, Z., Tran, N.T., Ethen, T., Merton, C., and Jewett, R.E. (2007), "Nanotribological Characterization and Lubricant Degradation Studies of Metal-Film Magnetic Tapes Using Novel Lubricants," *ASME J. Tribol.* **129**, 621–627.

Bhushan, B., Palacio, M., and Kinzig, B. (2008) "AFM-Based Nanotribological and Electrical Characterization of Ultrathin Wear-Resistant Ionic Liquid Films," *J. Colloid Interf. Sci.* **317**, 275–287.

Binnig, G., Quate, C.F., and Gerber, Ch. (1986), "Atomic Force Microscopy," *Phys. Rev. Lett.* **56**, 930–933.

Binnig, G., Gerber, Ch., Stoll, E., Albrecht, T.R., and Quate, C.F. (1987), "Atomic Resolution with Atomic Force Microscope," *Europhys. Lett.* **3**, 1281–1286.

Bobji, M.S. and Bhushan, B. (2001a), "Atomic Force Microscopic Study of the Micro-Cracking of Magnetic Thin Films Under Tension," *Scripta Mater.* **44**, 37–42.

Bobji, M.S. and Bhushan, B. (2001b), "In-Situ Microscopic Surface Characterization Studies of Polymeric Thin Films During Tensile Deformation Using Atomic Force Microscopy," *J. Mater. Res.* **16**, 844–855.

Bowden, F.P. and Tabor, D. (1950), *The Friction and Lubrication of Solids*, Part 1, Clarendon Press, Oxford, UK.

Chan, D.Y.C. and Horn, R.G. (1985), "The Drainage of Thin Liquid Films Between Solid Surfaces," *J. Chem. Phys.* **83**, 5311–5324.

Chen, N. and Bhushan, B. (2005), "Morphological, Nanomechanical and Cellular Structural Characterization of Human Hair and Conditioner Distribution Using Torsional Resonance Mode in an AFM," *J. Micros.* **220**, 96–112.

DeRose, J.A., Hoque, E., Bhushan, B., and Mathieu, H.J. (2008), "Characterization of Perfluorodecanote Self-Assembled Monolayers on Aluminum and Comparison of Stability with Phosphonate and Siloxy Self-Assembled Monolayers," *Surface Science* **602**, 1360–1367.

DeVecchio, D. and Bhushan, B. (1997), "Localized Surface Elasticity Measurements Using an Atomic Force Microscope," *Rev. Sci. Instrum.* **68**, 4498–4505.

DeVecchio, D. and Bhushan, B. (1998), "Use of a Nanoscale Kelvin Probe for Detecting Wear Precursors," *Rev. Sci. Instrum.* **69**, 3618–3624.

Fleck, N.A., Muller, G.M., Ashby, M.F., and Hutchinson, J.W. (1994), "Strain Gradient Plasticity: Theory and Experiment," *Acta Metall. Mater.* **42**, 475–487.

Foiles, S.M., Baskes, M.I., and Daw, M.S. (1986), "Embedded-Atom-Method Functions for the FCC Metals Cu, Ag, Au, Ni, Pd, Pt and Their Alloys," *Phys. Rev. B* **33**, 7983–7991.

Frisbie, C.D., Rozsnyai, L.F., Noy, A., Wrighton, M.S., and Lieber, C.M. (1994), "Functional Group Imaging by Chemical Force Microscopy," *Science* **265**, 2071–2074.

Fusco, C. and Fasolino, A. (2005), "Velocity Dependence of Atomic-Scale Friction: A Comparative Study of the One- and Two-Dimensional Tomlinson Model," *Phys. Rev. B* **71**, 045413.

Gee, M.L., McGuiggan, P.M., Israelachvili, J.N., and Homola, A.M. (1990), "Liquid to Solidlike Transitions of Molecularly Thin Films Under Shear," *J. Chem. Phys.* **93**, 1895–1906.

Georges, J.M., Tonck, A., and Mazuyer, D. (1994), "Interfacial Friction of Wetted Monolayers," *Wear* **175**, 59–62.

Granick, S. (1991), "Motions and Relaxations of Confined Liquids," *Science* **253**, 1374–1379.

Guntherodt, H.J., Anselmetti, D., and Meyer, E. (1995), *Forces in Scanning Probe Methods*, Vol. E286, Kluwer, Dordrecht, The Netherlands.

Haile, J.M. (1992), *Molecular Dynamics Simulation: Elementary Methods*, Wiley, New York.

Heermann, D.W. (1986), *Computer Simulation Methods in Theoretical Physics*, Springer-Verlag, Berlin.

Homola, A.M., Nguyen, H.V., and Hadziioannou, G. (1991), "Influence of Monomer Architecture on the Shear Properties of Molecular Thin Polymer Melts," *J. Chem. Phys.* **94**, 2346–2351.

Hoover, W.G. (1986), *Molecular Dynamics*, Springer-Verlag, Berlin.

Hoque, E., DeRose, J.A., Hoffmann, P., Mathieu, H.J., Bhushan, B., and Cichomski, M. (2006a), "Phosphonate Self-Assembled Monolayers on Aluminum Surfaces," *J. Chem. Phys.* **124**, 174710.

Hoque, E., DeRose, J.A., Kulik, G., Hoffmann, P., Mathieu, H.J., and Bhushan, B. (2006b), "Alkylphosphonate Modified Aluminum Oxide Surfaces," *J. Phys. Chem. B* **110**, 10855–10861.

Hoque, E., DeRose, J.A., Hoffmann, P., Bhushan, B., and Mathieu, H.J. (2007a), "Alkylperfluorosilane Self-Assembled Monolayers on Aluminum: A Comparison with Alkylphosphonate Self-Assembled Monolayers," *J. Phys. Chem. C* **111**, 3956–3962.

Hoque, E., DeRose, J.A., Hoffmann, P., Bhushan, B., and Mathieu, H.J. (2007b), "Chemical Stability of Nonwetting, Low Adhesion Self-Assembled Monolayer Films Formed by Perfluoroalkylsilazation of Copper," *J. Chem. Phys.* **126**, 114706.

Hoque, E., DeRose, J.A., Bhushan, B., and Mathieu, H.J. (2008), "Self-Assembled Monolayers on Aluminum and Copper Oxide Surfaces: Surface and Interface Characteristics, Nanotribological Properties, and Chemical Stability," *Applied Scanning Probe Methods Vol. IX – Characterization* (B. Bhushan, H. Fuchs and M. Tomitori, eds.), Springer-Verlag, Heidelberg, Germany, pp. 235–281.

Hoque, E., DeRose, J.A., Bhushan, B., and Hipps, K.W. (2009), "Low Adhesion, Non-Wetting Phosphonate Self-Assembled Monolayer Films Formed on Copper Oxide Surfaces," *Ultramicroscopy* **109**, 1015–1022.

Hu, H.W. and Granick, S. (1992), "Viscoelastic Dynamics of Confined Polymer Melts," *Science* **258**, 1339–1342.

Israelachvili, J.N. (1989), "Techniques for Direct Measurements of Forces between Surfaces in Liquid at the Atomic Scale," *Chemtracts Anal. Phys. Chem.* **1**, 1–12.

Israelachvili, J.N. and Tabor, D. (1972), "The Measurement of van der Waals Dispersion Forces in the Range of 1.5 to 130 nm," *Proc. R. Soc. Lond. A* **331**, 19–38.

Israelachvili, J.N., McGuiggan, P.M., and Homola, A.M. (1988), "Dynamic Properties of Molecularly Thin Liquid Films," *Science* **240**, 189–191.

Israelachvili, J.N., McGuiggan, P.M., Gee, M., Homola, A., Robbins, M., and Thompson, P. (1990), "Liquid Dynamics of Molecularly Thin Films," *J. Phys.: Condens. Matter* **2**, SA89–SA98.

Johnson, K.L. (1985), *Contact Mechanics*, Cambridge University Press, Cambridge, UK.

Kasai, T., Bhushan, B., Huang, L., and Su, C. (2004), "Topography and Phase Imaging Using the Torsional Resonance Mode," *Nanotechnology* **15**, 731–742.

Kasai, T., Bhushan, B., Kulik, G., Barbieri, L., and Hoffmann, P. (2005), "Nanotribological Study of Perfluorosilane SAMs for Anti-Stiction and Low Wear," *J. Vac. Sci. Technol. B* **23**, 995–1003.

Klein, J., Perahia, D., and Warburg, S. (1991), "Forces Between Polymer-Bearing Surfaces Undergoing Shear," *Nature* **352**, 143–145.

Koinkar, V.N. and Bhushan, B. (1996a), "Micro/nanoscale Studies of Boundary Layers of Liquid Lubricants for Magnetic Disks," *J. Appl. Phys.* **79**, 8071–8075.

Koinkar, V.N. and Bhushan, B. (1996b), "Microtribological Studies of Unlubricated and Lubricated Surfaces Using Atomic Force/Friction Force Microscopy," *J. Vac. Sci. Technol. A* **14**, 2378–2391.

Koinkar, V.N. and Bhushan, B. (1996c), "Microtribological Studies of Al_2O_3-TiC, Polycrystalline and Single-Crystal Mn-Zn Ferrite and SiC Head Slider Materials," *Wear* **202**, 110–122.

Koinkar, V.N. and Bhushan, B. (1997a), "Microtribological Properties of Hard Amorphous Carbon Protective Coatings for Thin Film Magnetic Disks and Heads," *Proc. Inst. Mech. Eng. Part J: J. Eng. Tribol.* **211**, 365–372.

Koinkar, V.N. and Bhushan, B. (1997b), "Effect of Scan Size and Surface Roughness on Microscale Friction Measurements," *J. Appl. Phys.* **81**, 2472–2479.

Koinkar, V.N. and Bhushan, B. (1997c), "Scanning and Transmission Electron Microscopies of Single-crystal Silicon Microworn/machined Using Atomic Force Microscopy," *J. Mater. Res.* **12**, 3219–3224.

Kulkarni, A.V. and Bhushan, B. (1996a), "Nanoscale Mechanical Property Measurements Using Modified Atomic Force Microscopy," *Thin Solid Films* **290–291**, 206–210.

Kulkarni, A.V. and Bhushan, B. (1996b), "Nano/Picoindentation Measurements on Single-crystal Aluminum Using Modified Atomic Force Microscopy," *Materials Letters* **29**, 221–227.

Kulkarni, A.V. and Bhushan, B. (1997), "Nanoindentation Measurement of Amorphous Carbon Coatings," *J. Mater. Res.* **12**, 2707–2714.

Landman, U. and Luedtke, W.D. (1991), "Nanomechanics and Dynamics of Tip-Substrate Interactions," *J. Vac. Sci. Technol. B* **9**, 414–423.

Landman, U., Luedtke, W.D., Burnham, N.A., and Colton, R.J. (1990), "Atomistic Mechanisms and Dynamics of Adhesion, Nanoindentation and Fracture," *Science* **248**, 454–461.

Landman, U., Luedtke, W.D., Ouyang, J., and Xia, T.K. (1993), "Nanotribology and the Stability of Nanostructures," *Jpn J. Appl. Phys.* **32**, 1444–1462.

Lee, K.K., Bhushan, B., and Hansford, D. (2005), "Nanotribological Characterization of Perfluoropolymer Thin Films for Biomedical Micro/Nanoelectromechanical Systems Applications," *J. Vac. Sci. Technol. A* **23**, 804–810.

Li, W.B., Henshall, J.L., Hooper, R.M., and Easterling, K.E. (1991), "The Mechanism of Indentation Creep," *Acta Metall. Mater.* **39**, 3099–3110.

Lim, S.C. and Ashby, M.F. (1987), "Wear Mechanism Maps," *Acta Metall.* **35**, 1–24.

Lim, S.C., Ashby, M.F., and Brunton, J.H. (1987), "Wear-Rate Transitions and Their Relationship to Wear Mechanisms," *Acta Metall.* **35**, 1343–1348.

Liu, H. and Bhushan, B. (2002), "Investigation of Nanotribological Properties of Self-Assembled Monolayers with Alkyl and Biphenyl Spacer Chains," *Ultramicroscopy* **91**, 185–202.

Liu, H. and Bhushan, B. (2003a), "Nanotribological Characterization of Molecularly-Thick Lubricant Films for Applications to MEMS/NEMS by AFM," *Ultramicroscopy* **97**, 321–340.

Liu, H. and Bhushan, B. (2003b), "Adhesion and Friction Studies of Microelectromechanical Systems/ Nanoelectromechanical Systems Materials Using a Novel Microtriboapparatus," *J. Vac. Sci. Technol. A* **21** 1528–1538.

Liu, H., Bhushan, B., Eck, W., and Staedtler, V. (2001), "Investigation of the Adhesion, Friction, and Wear Properties of Biphenyl Thiol Self-Assembled Monolayers by Atomic Force Microscopy," *J. Vac. Sci. Technol. A* **19**, 1234–1240.

Luengo, G., Schmitt, F.J., Hill, R., and Israelachvili, J. (1997), "Thin Film Rheology and Tribology of Confined Polymer Melts: Contrasts with Bulk Properties," *Macromolecules* **30**, 2482–2494.

Maivald, P., Butt, H.J., Gould, S.A.C., Prater, C.B., Drake, B., Gurley, J.A., Elings, V.B., and Hansma, P.K. (1991), "Using Force Modulation to Image Surface Elasticities with the Atomic Force Microscope," *Nanotechnology* **2**, 103–106.

Mate, C.M., McClelland, G.M., Erlandsson, R., and Chiang, S. (1987), "Atomic-scale Friction of a Tungsten Tip on a Graphite Surface," *Phys. Rev. Lett.* **59**, 1942–1945.

McClelland, G.M. and Glosli, J.N. (1992), "Friction at the Atomic Scale," In *Fundamentals of Frictions: Macroscopic and Microscopic Processes* (I.L. Singer and H.M. Pollock, eds.) Vol. E220, pp. 405–422, Kluwer, Dordrecht, The Netherlands.

Meyer, E., Overney, R., Luthi, R., Brodbeck, D., Howald, L., Frommer, J., Guntherodt, H.J., Wolter, O., Fujihira, M., Takano, T., and Gotoh, Y. (1992), "Friction Force Microscopy of Mixed Langmuir-Blodgett Films," *Thin Solid Films* **220**, 132–137.

Nix, W.D. and Gao, H. (1998) "Indentation Size Effects in Crystalline Materials: A Law for Strain Gradient Plasticity," *J. Mech. Phys. Solids* **46**, 411–425.

Nosonovsky, M. and Bhushan, B. (2005), "Scale Effects in Dry Friction During Multiple-Asperity Contact," *ASME J. Tribol.* **127**, 37–46.

Palacio, M. and Bhushan, B. (2007a), "Surface Potential and Resistance Measurements for Detecting Wear of Chemically-Bonded and Unbonded Molecularly-Thick Perfluoropolyether Lubricant Films Using Atomic Force Microscopy," *J. Colloid Interf. Sci.* **315**, 261–269.

Palacio, M. and Bhushan, B. (2007b), "Wear Detection of Candidate MEMS/NEMS Lubricant Films Using Atomic Force Microscopy-Based Surface Potential Measurements," *Scripta Mater.* **57**, 821–824.

Palacio, M. and Bhushan, B. (2008), "Ultrathin Wear-Resistant Ionic Liquid Films for Novel MEMS/NEMS Applications," *Adv. Mater.* **20**, 1194–1198.

Palacio, M. and Bhushan, B. (2009), "Molecularly Thick Dicationic Liquid Films for Nanolubrication," *J. Vac. Sci. Technol. A* **27**, 986–995.

Palacio, M. and Bhushan, B. (2010), "Normal and Lateral Force Calibration Techniques for AFM Cantilevers," *Critical Rev. Solid State Mater. Sci.* **35**, 73–104.

Persson, B.N.J. and Tosatti, E. (1996), *Physics of Sliding Friction*, Vol. E311, Kluwer, Dordrecht, Netherlands.

Reinstaedtler, M., Rabe, U., Scherer, V., Hartmann, U., Goldade, A., Bhushan, B., and Arnold, W. (2003), "On the Nanoscale Measurement of Friction Using Atomic-Force Microscope Cantilever Torsional Resonances," *Appl. Phys. Lett.* **82**, 2604–2606.

Reinstaedtler, M., Rabe, U., Goldade, A., Bhushan, B., and Arnold, W. (2005a), "Investigating Ultra-Thin Lubricant Layers Using Resonant Friction Force Microscopy," *Tribol. Inter.* **38**, 533–541.

Reinstaedtler, M., Kasai, T., Rabe, U., Bhushan, B., and Arnold, W. (2005b), "Imaging and Measurement of Elasticity and Friction Using the TR Mode," *J. Phys. D: Appl. Phys.* **38**, R269–R282.

Ribarsky, M.W. and Landman, U. (1992), "Structure and Dynamics of n-alkanes Confined by Solid Surfaces I. Stationary Crystalline Boundaries," *J. Chem. Phys.* **97**, 1937–1949.

Robbins, M.O. and Thompson, P.A. (1991), "Critical Velocity of Stick-Slip Motion," *Science* **253**, 916.

Ruan, J. and Bhushan, B. (1993), "Nanoindentation Studies of Fullerene Films Using Atomic Force Microscopy," *J. Mater. Res.* **8**, 3019–3022.

Ruan, J. and Bhushan, B. (1994a), "Atomic-scale Friction Measurements Using Friction Force Microscopy: Part I – General Principles and New Measurement Techniques," *ASME J. Tribol.* **116**, 378–388.

Ruan, J. and Bhushan, B. (1994b), "Atomic-scale and Microscale Friction of Graphite and Diamond Using Friction Force Microscopy," *J. Appl. Phys.* **76**, 5022–5035.

Ruan, J. and Bhushan, B. (1994c), "Frictional Behavior of Highly Oriented Pyrolytic Graphite," *J. Appl. Phys.* **76**, 8117–8120.

Ruths, M. and Israelachvili, J.N. (2011), "Surface Forces and Nanorheology of Molecularly Thin Films." In *Nanotechnology and Nanomechanics II*, Third edition (B. Bhushan, ed.), pp. 107–202, Springer-Verlag, Heidelberg, Germany.

Scherer, V., Bhushan, B., Rabe, U., and Arnold, W. (1997), "Local Elasticity and Lubrication Measurements Using Atomic Force and Friction Force Microscopy at Ultrasonic Frequencies," *IEEE Trans. Magn.* **33**, 4077–4079.

Scherer, V., Arnold, W., and Bhushan, B. (1998), "Active Friction Control Using Ultrasonic Vibration," in *Tribology Issues and Opportunities in MEMS* (B. Bhushan, ed.), pp. 463–469, Kluwer, Dordrecht, The Netherlands.

Scherer, V., Arnold, W., and Bhushan, B. (1999), "Lateral Force Microscopy Using Acoustic Friction Force Microscopy," *Surf. Interface Anal.* **27**, 578–587.

Schwarz, U.D., Zwoerner, O., Koester, P., and Wiesendanger, R. (1997), "Friction Force Spectroscopy in the Low-load Regime with Well-defined Tips," in *Micro/Nanotribology and Its Applications* (B. Bhushan, ed.), pp. 233–238, Kluwer, Dordrecht, The Netherlands.

Scott, W.W. and Bhushan, B. (2003), "Use of Phase Imaging in Atomic Force Microscopy for Measurement of Viscoelastic Contrast in Polymer Nanocomposites and Molecularly-Thick Lubricant Films," *Ultramicroscopy*, **97**, 151–169.

Singer, I.L. and Pollock, H.M. (1992), *Fundamentals of Friction: Macroscopic and Microscopic Processes*, Vol. E220, Kluwer, Dordrecht, The Netherlands.

Sinnott, S.B., Heo, S.J., Brenner, D.W., Harrison, J.A., and Irving, D.L. (2011), "Computer Simulations of Nanoscale Indentation." In *Nanotechnology and Nanomechanics I*, Third edition (B. Bhushan, ed.), pp. 439–525, Springer-Verlag, Heidelberg, Germany.

Sokoloff, J.B. (1996), "Theory of Electron and Phonon Contributions to Sliding Friction." In *Physics of Sliding Friction* (B.N.J. Persson and E. Tosatti, eds.), Vol. E331, pp. 217–229, Kluwer, Dordrecht, The Netherlands.

Stifter, T., Marti, O., and Bhushan, B. (2000), "Theoretical Investigation of the Distance Dependence of Capillary and van der Waals Forces in Scanning Probe Microscopy," *Phys. Rev. B* **62**, 13667–13673.

Sundararajan, S. and Bhushan, B. (2000), "Topography-Induced Contributions to Friction Forces Measured Using an Atomic Force/Friction Force Microscope," *J. Appl. Phys.* **88**, 4825–4831.

Sundararajan, S. and Bhushan, B. (2001), "Development of a Continuous Microscratch Technique in an Atomic Force Microscope and its Application to Study Scratch Resistance of Ultra-Thin Hard Amorphous Carbon Coatings," *J. Mater. Res.* **16**, 75–84.

Sutton, A.P. (1993), *Electronic Structure of Materials*, Clarendon Press, Oxford, UK.

Tabor, D. and Winterton, R.H.S. (1969), "The Direct Measurement of Normal and Retarded van der Waals Forces," *Proc. R. Soc. Lond. A* **312**, 435–450.

Tambe, N.S. and Bhushan, B. (2004a), "Scale Dependence of Micro/Nano-friction and Adhesion of MEMS/NEMS Materials, Coatings and Lubricants," *Nanotechnology* **15**, 1561–1570.

Tambe, N.S. and Bhushan, B. (2004b), "In Situ Study of Nano-cracking of Multilayered Magnetic Tapes Under Monotonic and Fatigue Loading Using an AFM," *Ultramicroscopy* **100**, 359–373.

Tambe, N.S. and Bhushan, B. (2005a), "A New Atomic Force Microscopy Based Technique for Studying Nanoscale Friction at High Sliding Velocities," *J. Phys. D: Appl. Phys.* **38**, 764–773.

Tambe, N.S. and Bhushan, B. (2005b), "Friction Model for the Velocity Dependence of Nanoscale Friction," *Nanotechnology* **16**, 2309–2324.

Tambe, N.S. and Bhushan, B. (2005c), "Durability Studies of Micro/Nanoelectromechanical System Materials, Coatings, and Lubricants at High Sliding Velocities (up to 10 mm/s) Using a Modified Atomic Force Microscope," *J. Vac. Sci. Technol. A* **23**, 830–835.

Tambe, N.S. and Bhushan, B. (2005d), "Nanoscale Friction-Induced Phase Transformation of Diamond-like Carbon," *Scripta Materiala* **52**, 751–755.

Tambe, N.S. and Bhushan, B. (2005e), "Identifying Materials with Low Friction and Adhesion for Nanotechnology Applications," *Appl. Phys. Lett* **86**, 061906; *Nature Mater. Nanozone*, Feb. 17, 2005.

Tambe, N.S. and Bhushan, B. (2005f), "Nanoscale Friction Mapping," *Appl. Phys. Lett.* **86**, 193102-1–193102-3.

Tambe, N.S. and Bhushan, B. (2005g), "Nanowear Mapping: A Novel Atomic Force Microscopy Based Approach for Studying Nanoscale Wear at High Sliding Velocities," *Tribol. Lett.* **20**, 83–90.

Tambe, N.S. and Bhushan, B. (2005h), "Nanotribological Characterization of Self Assembled Monolayers Deposited on Silicon and Aluminum Substrates," *Nanotechnology* **16**, 1549–1558.

Tambe, N.S. and Bhushan, B. (2008), "Nanoscale Friction and Wear Maps," *Philos. Trans. R. Soc. A* **366**, 1405–1424.

Tao, Z. and Bhushan, B. (2005a), "Bonding, Degradation, and Environmental Effects on Novel Perfluoropolyether Lubricants," *Wear* **259**, 1352–1361.

Tao, Z. and Bhushan, B. (2005b), "Degradation Mechanisms and Environmental Effects on Perfluoropolyether, Self Assembled Monolayers, and Diamondlike Carbon Films," *Langmuir* **21** 2391–2399.

Tao, Z. and Bhushan, B. (2006), "A New Technique for Studying Nanoscale Friction at Sliding Velocities up to 200 mm/s Using Atomic Force Microscope," *Rev. Sci. Instrum.* **71**, 103705.

Tao, Z. and Bhushan, B. (2007), "Velocity Dependence and Rest Time Effect in Nanoscale Friction of Ultrathin Films at High Sliding Velocities," *J. Vac. Sci. Technol. A* **25**, 1267–1274.

Thompson, P.A. and Robbins, M.O. (1990a), "Origin of Stick-Slip Motion in Boundary Lubrication," *Science* **250**, 792–794.

Thompson, P.A. and Robbins, M.O. (1990b), "Shear Flow Near Solids: Epitaxial Order and Flow Boundary Conditions," *Phys. Rev. A* **41**, 6830–6837.

Thompson, P.A., Grest, G.S., and Robbins, M.O. (1992), "Phase Transitions and Universal Dynamics in Confined Films," *Phys. Rev. Lett.* **68**, 3448–3451.

Tomanek, D., Zhong, W., and Thomas, H. (1991), "Calculation of an Atomically Modulated Friction Force in Atomic Force Microscopy," *Europhys. Lett.* **15**, 887–892.

Tomlinson, G.A. (1929), "A Molecular Theory of Friction," *Philos. Mag.* **7**, 905–939.

Tonck, A., Georges, J.M., and Loubet, J.L. (1988), "Measurement of Intermolecular Forces and the Rheology of Dodecane between Alumina Surfaces," *J. Colloid Interf. Sci.* **126**, 1540–1563.

Van Alsten, J. and Granick, S. (1988), "Molecular Tribology of Ultrathin Liquid Films," *Phys. Rev. Lett.* **61**, 2570–2573.

Van Alsten, J. and Granick, S. (1990a), "The Origin of Static Friction in Ultrathin Liquid Films," *Langmuir* **6**, 876–880.

Van Alsten, J. and Granick, S. (1990b), "Shear Rheology in a Confined Geometry-Polysiloxane Melts," *Macromolecules* **23**, 4856–4862.

Wang, Y., Hill, K., and Harris, J.G. (1993), "Thin Films of n-Octane Confined Between Parallel Solid Surfaces, Structures and Adhesive Forces vs Film Thickness from Molecular Dynamics Simulations," *J. Phys. Chem.* **97**, 9013–9021.

Yamanaka, K. and Tomita, E (1995), "Lateral Force Modulation Atomic Force Microscope for Selective Imaging of Friction Forces," *Jpn. J. Appl. Phys.* **34**, 2879–2882.

Yoshizawa, H. and Israelachvili, J.N. (1993), "Fundamental Mechanisms of Interfacial Friction II: Stick-Slip Friction of Spherical and Chain Molecules," *J. Phys. Chem.* **97**, 11300–11313.

Zhao, X. and Bhushan, B. (1998), "Material Removal Mechanism of Single-Crystal Silicon on Nanoscale and at Ultralow Loads," *Wear* **223**, 66–78.

Further Reading

Bhushan, B. (1996), *Tribology and Mechanics of Magnetic Storage Devices*, Second edition, Springer-Verlag, New York.

Bhushan, B. (1997), *Micro/Nanotribology and its Applications*, Vol. E330, Kluwer, Dordrecht, The Netherlands.

Bhushan, B. (1998), *Tribology Issues and Opportunities in MEMS*, Kluwer, Dordrecht, Netherlands.

Bhushan, B. (1999), *Handbook of Micro/Nanotribology*, Second edition, CRC Press, Boca Raton, FL.

Bhushan, B. (2001a), *Modern Tribology Handbook, Vol. 1: Principles of Tribology*, CRC Press, Boca Raton, Florida.

Bhushan, B. (2001b), *Fundamentals of Tribology and Bridging the Gap Between the Macro- and Micro/Nanoscales*, NATO Science Series II – Vol. 10, Kluwer, Dordrecht, Netherlands.

Bhushan, B. (2008), "Nanotribology, Nanomechanics and Nanomaterials Characterization," *Phil. Trans. R. Soc. A* **366**, 1351–1381.

Bhushan, B. (2010), *Springer Handbook of Nanotechnology*, Third edition, Springer-Verlag, Heidelberg, Germany.

Bhushan, B. (2011), *Nanotribology and Nanomechanics I & II*, Third edition, Springer-Verlag, Heidelberg, Germany.

Bhushan, B. (2012), *Encyclopedia of Nanotechnology*, Springer-Verlag, Heidelberg, Germany.

Bhushan, B., Israelachvili, J.N., and Landman, U. (1995), "Nanotribology: Friction, Wear and Lubrication at the Atomic Scale," *Nature* **374**, 607–616.

Guntherodt, H.J., Anselmetti, D., and Meyer, E. (1995), "Forces in Scanning Probe Methods," Vol. E286, Kluwer, Dordrecht, The Netherlands.

Helman, J.S., Baltensperger, W., and Holyst, J.A. (1994), "Simple-Model for Dry Friction," *Phys. Rev. B* **49**, 3831–3838.

Persson, B.N.J. and Tosatti, E. (1996), *Physics of Sliding Friction*, Vol. E311, Kluwer, Dordrecht, The Netherlands.

Singer, I.L. and Pollock, H.M. (1992), *Fundamentals of Friction: Macroscopic and Microscopic Processes*, Vol. E220, Kluwer, Dordrecht, The Netherlands.

Zworner, O., Holscher, H., Schwarz, U.D., and Wiesendanger, R. (1998), "The Velocity Dependence of Frictional Forces in Point-Contact Friction," *Appl Phys. A.: Mater. Sci. Process.* **66**, S263–S267.

11

Friction and Wear Screening Test Methods

11.1 Introduction

Screening tests have to be conducted during validation of the design of a machine component and/or during the development and selection of materials, coatings, and surface treatments for a particular application. These screening tests include accelerated friction and wear tests (including corrosion tests) and functional tests (Bhushan and Gupta, 1997). Simulated accelerated friction and wear tests are conducted to rank the candidate designs of a machine component or candidate materials. Accelerated tests are inexpensive and fast. After the designs and/or materials have been ranked by accelerated friction and wear tests, the most promising candidates (typically from 1 to 3) are tested in the actual machine under actual operating conditions (functional tests). In order to reduce test duration in functional tests, the tests can be conducted for times shorter than the end-of-life. By collecting friction and wear data at intermediate intervals, end-of-life can be predicted.

Accelerated friction and wear tests should accurately simulate the operating conditions to which the component will be subjected. If these tests are properly simulated, an acceleration factor between the simulated test and the functional test can be empirically determined so that the subsequent functional tests can be minimized, saving considerable test time. Standardization, repeatability, short testing time, and simple measuring and ranking techniques are desirable in these accelerated tests.

This chapter presents a review of accelerated friction and wear-test methods. It presents the design methodology and typical test geometries for friction and wear tests.

11.2 Design Methodology

The design methodology of a friction and wear test consists of four basic elements: simulation; acceleration; specimen preparation; and friction and wear measurements. Simulation is the most critical, but no other elements should be overlooked.

Introduction to Tribology, Second Edition. Bharat Bhushan.
© 2013 John Wiley & Sons, Ltd. Published 2013 by John Wiley & Sons, Ltd.

11.2.1 Simulation

Proper simulation ensures that the wear a mechanism experienced in the test is identical to that of the actual system. Given the complexity of wear processes and the incomplete understanding of wear mechanisms, test development is subject to trial and error and is dependent on the capabilities of the developer. The starting point in simulation is the collection of available data on the actual system and test system. A successful simulation requires similarity between the functions of the actual system and those of the test system, i.e., similarity of inputs and outputs and of the functional input–output relations.

To obtain this similarity, selection of the test geometry is a critical factor in simulating wear conditions. Generally, in laboratory testing for sliding contacts, three types of contact are employed: point contact (such as ball-on-disk), line contact (such as cylinder-on-disk), and conforming contact (such as flat-on-flat). Selection of the geometry depends on the geometry of the function to be simulated. Each of these contact geometries has its advantages and disadvantages. Point-contact geometry eliminates alignment problems and allows wear to be studied from the initial stages of the test. However, the stress level changes as the mating surfaces wear out. Conforming-contact tests generally allow the mating parts to wear in to establish a uniform and stable contact geometry before taking data. As a result, it is difficult to identify wear-in phenomena, because there is no elaborate regular monitoring of wear behavior.

Other factors besides contact type that significantly influence the success of a simulation include type of motion, load, speed, lubrication condition, and operating environment (contamination, temperature, and humidity). The type of motion that exists in the actual system is one of four basic types: sliding; rolling; spin; and impact. These motions can be simulated by performing wear tests under unidirectional, reciprocating, and oscillating (reciprocating with a high frequency and low amplitude) motions and combinations thereof. Load conditions are simulated by applying static or dynamic load by dead weight, spring, hydraulic means, or electromagnetic means. Lubrication (or lack thereof), temperature, and humidity also considerably influence the friction and wear characteristics of certain materials. The ambient temperature and contact temperature determine the thermal state of the system and should be precisely simulated and controlled during the test. Measurements at the surface of a specimen can be made with thermocouples, thermistors, or infrared (IR) methods. Thermocouples are the cheapest and simplest to use, and IR methods are the most accurate but complicated. Humidity affects chemical reactions which occur at a moving interface.

11.2.2 Acceleration

Accelerated tests are extremely inexpensive and fast. However, if the acceleration is not done properly, the wear mechanism to be simulated may change. Accelerated wear is normally caused by increasing load, speed, or temperature, by decreasing the amount of lubricant in lubricated interfaces, and by continuous operation.

11.2.3 Specimen Preparation

Specimen preparation plays a key role in obtaining repeatable/reproducible results. However, specimen preparation may vary depending on the type of material tested. For metals, surface roughness, geometry of the specimen, microstructure, homogeneity, hardness, and the presence of surface layers must be controlled carefully for both the mating materials. Similar controls are necessary for the wear-causing medium. For instance, in an abrasive wear test, purity, particle size, particle shape, and moisture content of the abrasive must be controlled.

11.2.4 Friction and Wear Measurements

The coefficient of friction is calculated from the ratio of friction force to applied normal force. The stationary member of the material pair is mounted on a flexible member, and the frictional force (force required to restrain the stationary member) is measured using the strain gages (known as strain-gage transducers) or displacement gages (based on capacitance or optical methods) (Doeblin, 1990). Under certain conditions, piezoelectric force transducers (mostly for dynamic measurements) are also used for friction-force measurements (Bhushan, 1980).

Examples of two accelerated pin/ball on flat/disk tests are shown in Figure 11.2.1. In these tests, a flat or a disk can reciprocate or rotate. Figure 11.2.1a shows the reciprocating stage which can be replaced with a rotating stage for unidirectional sliding motion. In both examples, strain-gage beams are used to measure friction force during a sliding test. In Figure 11.2.1a, normal load is applied by dead weight loading and a strain-gage ring is used to measure the friction force. In Figure 11.2.1b, normal load is applied by a microactuator and both normal and friction forces are measured by using a structure with two crossed I beams. For high sensitivity, semiconductor strain-gages with a gage factor of 115 or larger, as compared to 2.1 for resistive gages, can be used.

In the case of fibers, belts, and tapes wrapped around a cylinder, the coefficient of friction (μ) is measured by using the belt equation, $\mu = (1/\theta)\ell n\,(T/T_0)$, where θ is the wrap angle and T_0 and T are the inlet and exit tensions ($T > T_0$) (Bhushan, 1996). The coefficient of static friction of particles against a flat surface can be measured by utilization of the centrifugal force experienced by a rotating body (e.g., see Dunkin and Kim, 1996). A particle is placed on a disk which is rotated at increasing speeds until the particle flies off due to centrifugal action. At an angular speed at which the particle starts to fly off, the centrifugal force just exceeds the friction force. For a particle flying off at an angular speed (ω) and radial location of the particle (r), $\mu_s = \omega^2 r/g$. The angular speed can be obtained by videotaping the particle during the test.

Commonly used wear measurement techniques are weight loss, volume loss or wear scar width or depth, or other geometric measures and indirect measurements such as time required to wear through a coating or load required to cause severe wear or a change in surface finish. Scanning electron microscopy (SEM), scanning tunneling microscopy (STM), and atomic force microscopy (AFM) of worn surfaces are commonly used to measure microscopic wear. Other less commonly used techniques include radioactive decay. The resolutions of several techniques are presented in Table 11.2.1 (Bhushan, 1996). For applications requiring low particulate contamination, particle counts are measured by using a particle counter (Bhushan, 1996).

Weight-loss measurements are suitable for large amounts of wear. However, weight-loss measurements have two major limitations. First, wear is related primarily to the volume of material removed or displaced. Thus such methods may furnish different results if materials to be compared differ in density. Second, this measurement does not account for wear by

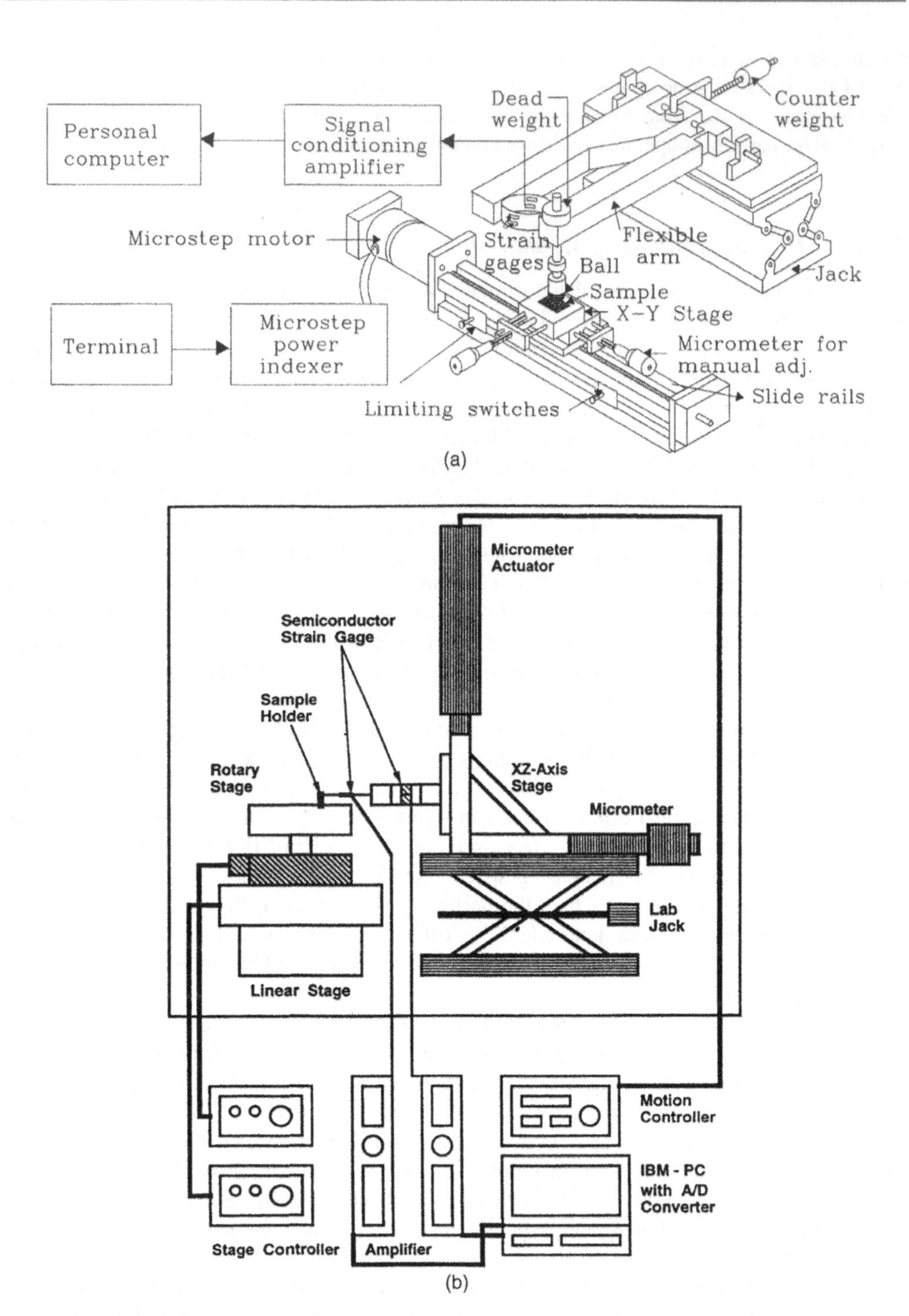

Figure 11.2.1 Schematics of accelerated friction and wear tests (a) with a strain-gage ring and dead weight normal loading, and (b) with a crossed I-beam strain gage transducer and microactuator loading.

Table 11.2.1 Resolutions of several wear-measurement techniques.

Measurement technique	Resolution
Weight loss	10–100 μg
Radioactive decay	~1 pg
Stylus and optical profilers	0.5–10 nm
Microhardness indentation	25–50 nm
Nanoscratch technique	1–10 nm
Scanning electron microscope	0.1 nm
Scanning tunneling microscope/atomic force microscope	0.05–0.1 nm

material displacement, that is, a specimen may gain weight by transfer. Thus weight-loss measurements are valid only when density remains constant and transfer does not occur during the wear process. This technique is not sensitive enough for tests being conducted at low load and/or short time or in the case of thin wear-resistant coatings, where wear is very small.

A stylus or noncontact optical profiler and Vickers or Knoop microhardness indentation techniques are easy to use and are commonly used to measure depth of wear with a resolution of up to a fraction of a nanometer. An example of a wear-track profile obtained with a stylus profiler is shown in Figure 11.2.2. Three-dimensional worn-surface profiles also can be obtained with fully automated profilers. In the microhardness indentation technique, Vickers or Knoop indentations are made on the wear surface. By measuring the width of the indentations before and after the wear test under a microscope, the wear depth can be calculated (Bhushan and Martin, 1988). In the nanoscratch technique, a nanoscratch is made with a conical tip using a nanoindenter at low loads (Bhushan and Lowry, 1995). Measurement of the depth of the nanoscratch before and after the wear test, using an AFM, gives the wear depth.

For measurements of microscopic wear, SEM, STM, and AFM of worn surfaces are commonly used (Bhushan, 1996, 1999). Radioactive decay (also called autoradiography) is very sensitive, but it requires facilities to irradiate one of the members and to measure the changes in radiation (Rabinowicz and Tabor, 1951; Bhushan *et al.*, 1986). If particulate contamination is an issue such as in gas-lubricated bearings, a particle count is also made during the test. Particle counters measure the number of particles per unit volume and their size distribution. Laser particle counters measure the particles in the range of 0.1–7.5 μm by the principle of light scattering (Bhushan, 1996). A small volume of air (a few cc/s) containing particles to be sampled is brought into contact with the particle-detecting optical system that measures the scattered light.

11.3 Typical Test Geometries

11.3.1 Sliding Friction and Wear Tests

Many accelerated test apparatuses are commercially available that allow control of such factors as sample geometry, applied load, sliding velocity, ambient temperature, and humidity. Benzing *et al.* (1976), Bayer (1976, 1979, 1982), Clauss (1972), Nicoll (1983), Bhushan

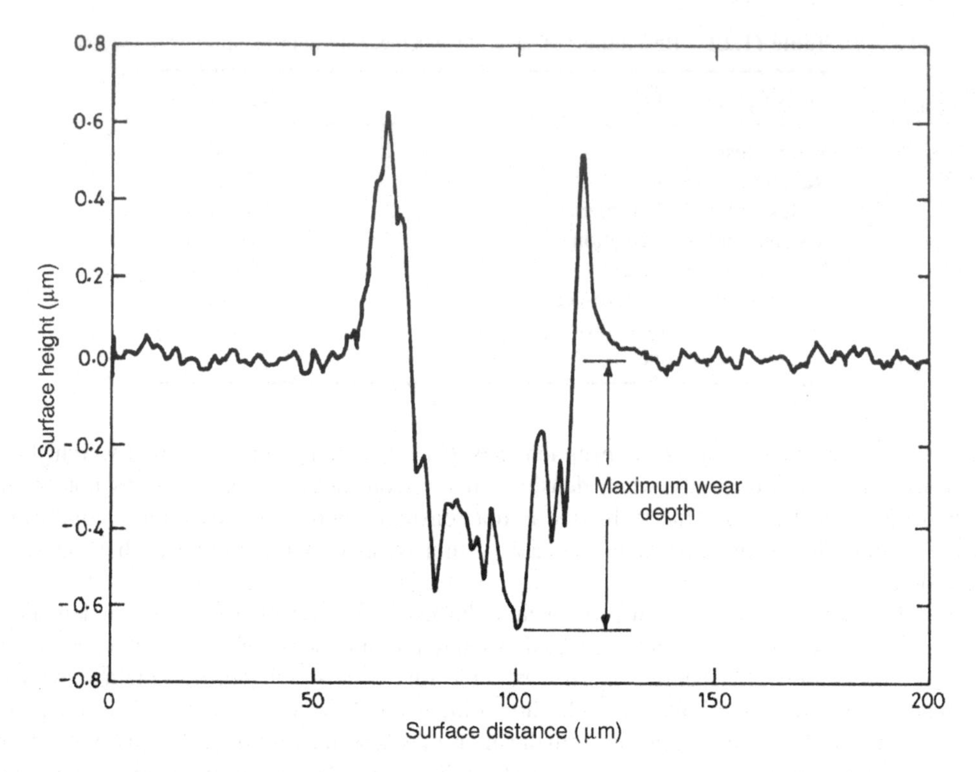

Figure 11.2.2 Example of a wear-track profile obtained by a stylus profiler.

(1987), Yust and Bayer (1988), Bhushan and Gupta (1997), and Bhushan (2001) have reviewed the various friction and wear test apparatuses that have been used in various tribological applications. The most commonly used interface geometries for screening component designs and materials are shown in Figure 11.3.1 and are compared in Table 11.3.1 Many of the test configurations are one-of-a-kind machines; others are available as commercial units from such companies as Falex-Le Valley Corporation, Downers Grove, Illinois, Cameron Plint Tribology, Berkshire, UK, Swansea Tribology Center, Swansea, UK, Optimol Instruments GmbH, Munich, Germany, and CSEM, Neuchatel, Switzerland. However, testing is not limited to such equipment; tests often are performed with replicas and facsimiles of actual devices. Brief descriptions of the typical test geometries, illustrated in Figure 11.3.1 and compared in Table 11.3.1, are presented in the following subsections. Static or dynamic loading can be applied in any of the test geometries.

11.3.1.1 Pin-on-Disk (Face Loaded)

In the pin-on-disk test apparatus, the pin is held stationary and the disk rotates, Figure 11.3.1a. The pin can be a nonrotating ball, a hemispherically tipped rider, a flat-ended cylinder, or even a rectangular parallelepiped. This test apparatus is probably the most commonly used during the development of materials for tribological applications.

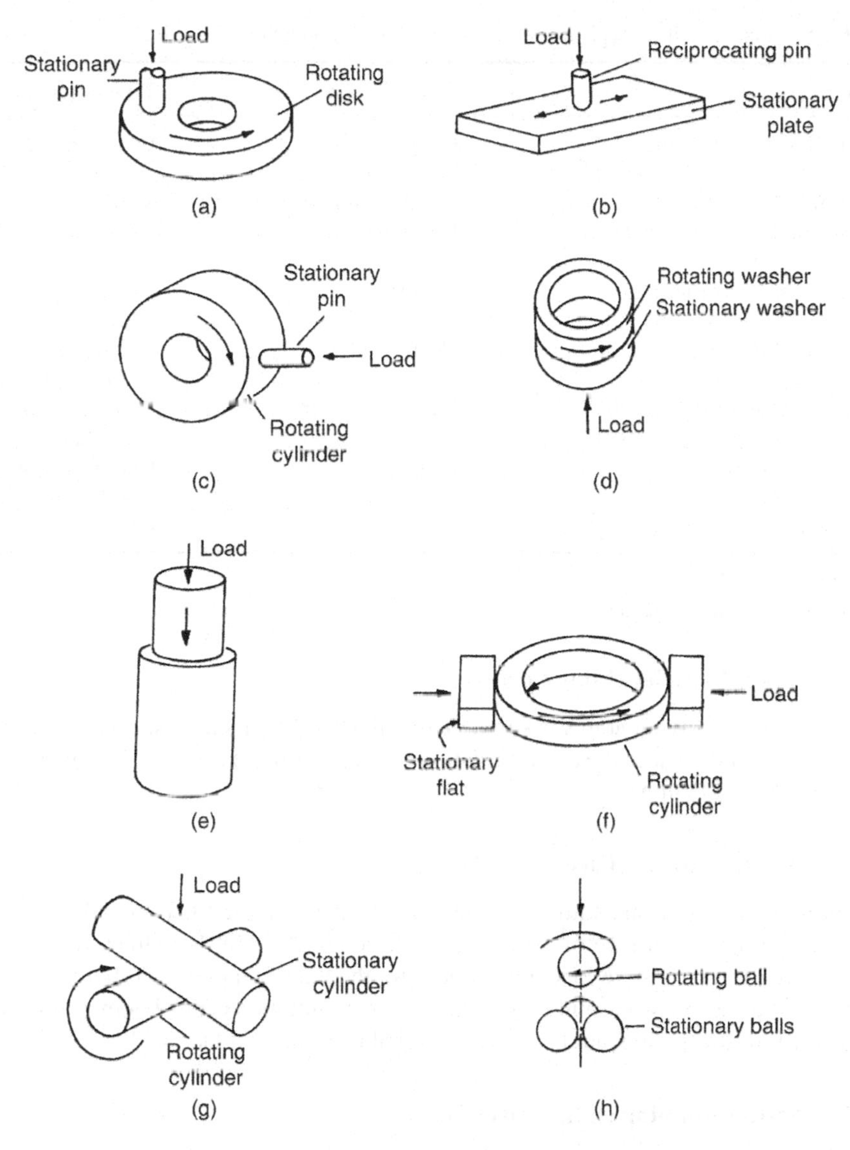

Figure 11.3.1 Schematic illustrations of typical interface geometries used for sliding friction and wear tests: (a) pin-on-disk, (b) pin-on-flat, (c) pin-on-cylinder, (d) thrust washers, (e) pin-into-bushing, (f) rectangular flats on rotating cylinder, (g) crossed cylinders, and (h) four-ball.

11.3.1.2 Pin-on-Flat (Reciprocating)

In the pin-on-flat test apparatus, a flat moves relative to a stationary pin in reciprocating motion, such as in a Bowden and Leben apparatus, Figure 11.3.1b. In some cases, the flat is stationary and the pin reciprocates. The pin can be a ball, a hemispherically tipped pin, or a flat-ended cylinder. By using a small oscillation amplitude at high frequency, fretting wear experiments can be conducted.

Table 11.3.1 Some details of typical test geometries for friction and wear testing.

Geometry[a]	Type of contact	Type of motion
1. Pin-on-disk (face loaded)	Point/conformal	Unidirectional sliding, oscillating
2. Pin-on-flat (reciprocating)	Point/conformal	Reciprocating sliding
3. Pin-on-cylinder (edge loaded)	Point/conformal	Unidirectional sliding, oscillating
4. Thrust washers (face loaded)	Conformal	Unidirectional sliding, oscillating
5. Pin-into-bushing	Conformal	Unidirectional sliding, oscillating
6. Flat-on-cylinder (edge loaded)	Line	Unidirectional sliding, oscillating
7. Crossed cylinders	Elliptical	Unidirectional sliding, oscillating
8. Four balls	Point	Unidirectional sliding

[a]see Figure 11.3.1
Type of Loading: Static, dynamic.

11.3.1.3 Pin-on-Cylinder (Edge Loaded)

The pin-on-cylinder test apparatus is similar to the pin-on-disk apparatus, except that loading of the pin is perpendicular to the axis of rotation or oscillation, Figure 11.3.1c. The pin can be flat or hemispherically tipped.

11.3.1.4 Thrust Washers (Face Loaded)

In the thrust-washer test apparatus, the flat surface of a washer (disk or cylinder) rotates or oscillates on the flat surface of a stationary washer, such as in the Alpha model LFW-3, Figure 11.3.1d. The testers are face loaded because the load is applied parallel to the axis of rotation. The washers may be solid or annular. This configuration is most common for testing materials for low-stress applications, such as journal bearings and face seals.

11.3.1.5 Pin-into-Bushing (Edge Loaded)

In the pin-into-bushing test apparatus, the axial force necessary to press an oversized pin into a bushing is measured, such as in the Alpha model LFW-4, Figure 11.3.1e. The normal (axial) force acts in the radial direction and tends to expand the bushing; this radial force can be calculated from the material properties, the interference, and the change in the bushing's outer diameter. Dividing the axial force by the radial force gives the coefficient of friction.

11.3.1.6 Rectangular Flats on a Rotating Cylinder (Edge Loaded)

In the rectangular-flats-on-a-rotating-cylinder test apparatus, two rectangular flats are loaded perpendicular to the axis of rotation or oscillation of the disk, Figure 11.3.1f. This apparatus includes some of the most widely used configurations, such as the Hohman A-6 tester. In the

Alpha model LFW-1 or the Timken tester, only one flat is pressed against the cylinder. The major difference between Alpha and Timken testers is in the loading system. In the Falex tester, a rotating pin is sandwiched between two V-shaped (instead of flat) blocks so that there are four lines of contact with the pin. In the Almon-Wieland tester, a rotating pin is sandwiched between two conforming bearing shells.

11.3.1.7 Crossed Cylinders

The crossed-cylinders test apparatus consists of a hollow (water-cooled) or solid cylinder as the stationary wear member and a solid cylinder as the rotating or oscillating wear member that operates at 90° to the stationary member, such as in the Reichert wear tester, Figure 11.3.1g.

11.3.1.8 Four Ball

The four-ball test apparatus, also called the Shell four-ball tester, consists of four balls in the configuration of an equilateral tetrahedron, Figure 11.3.1h. The upper ball rotates and rubs against the lower three balls, which are held in a fixed position.

11.3.2 Abrasion Tests

Abrasion tests include two-body and three-body tests. In a two-body abrasion test, one of the moving members is abrasive. In a three-body abrasion test, abrasive particles are introduced at the interface. Abrasion tests can be conducted using any of the so-called conventional test geometries just described, with one of the surfaces being made of abrasive material or in the presence of abrasive particles. A few commonly used specialized tests are described here.

11.3.2.1 Taber Abrasion Test

The Taber tester (manufactured by Teledyne Taber, North Tonawanda, NY) is widely used for determining the abrasion resistance of various materials and coatings. Test specimens (typically 100 mm square or 110 mm in diameter) are placed on the abrader turntable and are subjected to the rubbing action of a pair of rotating abrasive wheels (resilient calibrade, nonresilient calibrade, wool felt, plain rubber, and tungsten carbide) at known weights (250, 500, or 1000 g), Figure 11.3.2a. Wear action results when a pair of abrasive wheels is rotated in opposite directions by a turntable on which the specimen material is mounted. The abrading wheels travel on the material about a horizontal axis displaced tangentially from the axis of the test material, which results in a sliding action. Results are evaluated by four different methods: end point or general breakdown of the material, comparison of weight loss between materials of the same specific gravity, volume loss in materials of different specific gravities, and measuring the depth of wear.

The Taber abrasion test provides a technique for conducting comparative wear performance evaluations with an intralaboratory precision of $\pm 15\%$. These tests are commonly used by industry, government agencies, and research institutions for product development, testing, and evaluation.

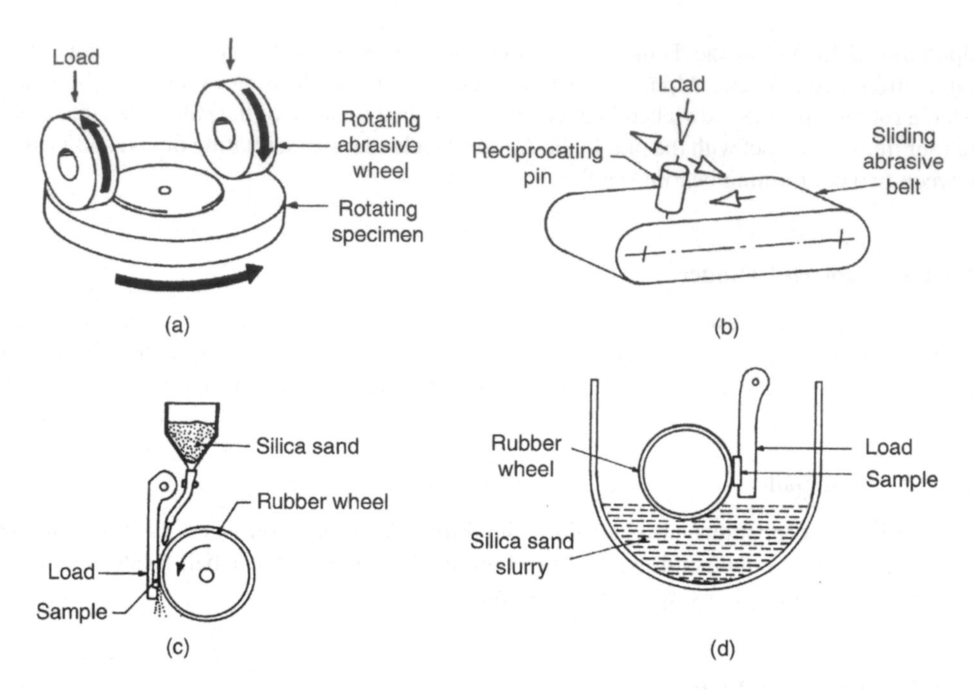

Figure 11.3.2 Schematic illustrations of abrasion test apparatuses: (a) two abradent wheels weighted on test specimen driven in opposite directions in the Taber abrasion test apparatus, (b) abrasive belt test apparatus, (c) dry-sand abrasion test apparatus, (d) wet-sand abrasion test apparatus.

11.3.2.2 Abrasive Belt Test

A flat-ended block or cylindrical specimen is abraded by sliding against an abrasive belt, Figure 11.3.2b. The belt runs horizontally, while the specimen runs transversely across the belt. The specimen also can be rotated during this abrasion test (Benzing *et al.*, 1976).

11.3.2.3 Dry-Sand Abrasion Test

In the dry-sand abrasion or dry-sand rubber-wheel abrasion test apparatus (ASTM G65), the specimen is loaded against the rotating rubber wheel, Figure 11.3.2c. The load is applied along the horizontal diametral line of the wheel. The abrasive is gravity-fed into the vee formed at the contact between the sample block and the wheel. The abrasive is typically 50–70 mesh (200–300 μm) dry American Foundry Society (AFS) test sand. Specimen weight loss is used as a measure of abrasive wear (Bayer, 1982).

11.3.2.4 Wet-Sand Abrasion Test

In the wet-sand abrasion test apparatus, also known as the SAE wet-sand rubber-wheel test apparatus, the specimen is pressed against the rubber wheel, Figure 11.3.2d. It consists of a neoprene rubber rim on a steel hub that rotates through a silica-sand slurry. The wheel has stirring paddles on each side to agitate the slurry as it rotates. Sand is carried by the rubber wheel to the interface between it and the test specimen. The slurry consists of 940 g of deionized water and 1500 g of AFS 50–70 mesh silica test sand. Specimen weight loss is used as a measure of abrasive wear (Bayer, 1982).

11.3.2.5 Mar-Resistance Abrasion Test

The mar-resistance abrasion test apparatus, also called the falling silicon carbide test apparatus (ASTM D673), simulates abrasive wear resulting from the impingement or impact of coarse, hard silicon carbide particles. The test involves allowing a weighed amount of no. 80 silicon carbide grit to fall through a glass tube and strike the surface of a test specimen at a 45° angle. The abrasion resistance is determined by measuring the percent change in haze of the abraded test specimen by ASTM D1003 (Bayer, 1982).

11.3.3 Rolling-Contact Fatigue Tests

A number of rolling-contact fatigue (RCF) tests are used for testing materials and lubricants for rolling-contact applications such as antifriction bearings and gears.

11.3.3.1 Disk-on-Disk

The disk-on-disk test apparatus uses two disks or a ball-on-disk rotating against each other on their outer surfaces (edge loaded), Figure 11.3.3a. The disk samples may be crowned or flat. Usually, the samples rotate at different sliding speeds to produce some relative sliding (slip) at the interface (Benzing *et al.*, 1976).

11.3.3.2 Rotating Four Ball

The rotating four-ball test apparatus consists of four balls in the configuration of an equilateral tetrahedron, Figure 11.3.3b. The rotating upper ball is dead weight-loaded against the three support balls (positioned 120° apart), which orbit the upper ball in rotating contact. In some tests, five balls instead of four balls are used. Also, in some studies, the lower balls are clamped (Bhushan and Sibley, 1982).

11.3.3.3 Rolling-Element-on-Flat

The rolling-element-on-flat test apparatus consists of three balls or rollers equispaced by a retainer that are loaded between a stationary flat washer and a rotating grooved washer, Figure 11.3.3c. The rotating washer produces ball motion and serves to transmit load to the balls and the flat washer (Bhushan and Sibley, 1982).

11.3.4 Solid-Particle Erosion Test

Erosion testing is generally conducted at room temperature using an air-blast test apparatus, shown in Figure 11.3.4. The tester is operated by feeding the eroding particles from a vibrating

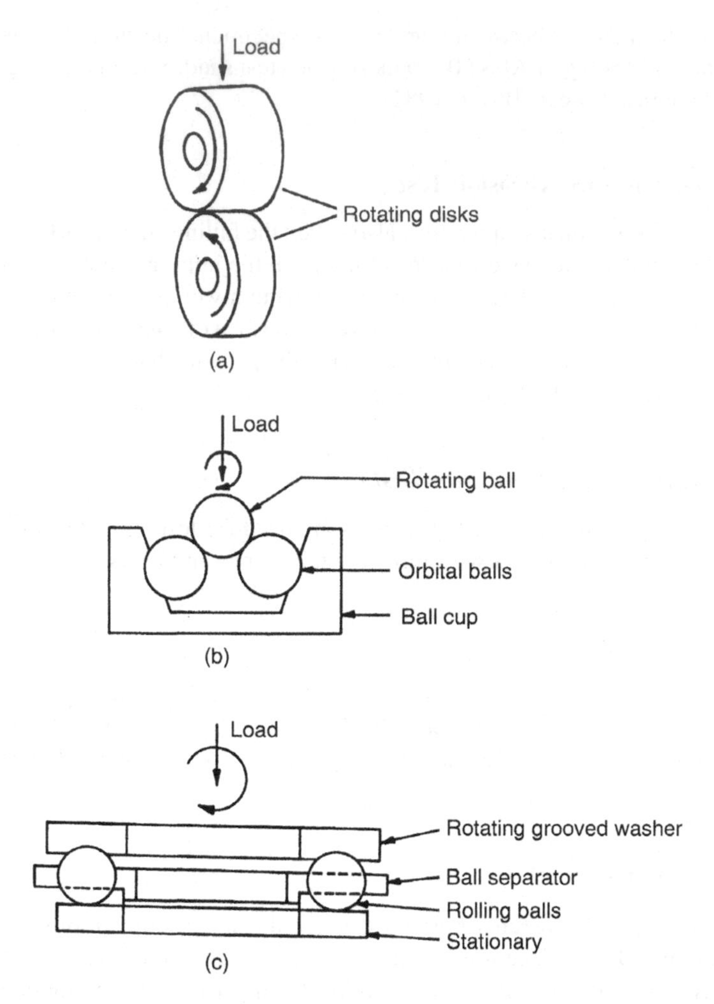

Figure 11.3.3 Schematic diagram of three types of rolling-contact fatigue test apparatus: (a) disk-on-disk, (b) rotating four ball, (c) balls-on-flat.

hopper into a stream of gas. A known amount of eroding particles is directed onto one or more test specimens. The weight loss of the test specimens is used as a measure of erosive wear (Bayer, 1976).

11.3.5 Corrosion Tests

Corrosion can occur because of electrochemical or chemical interactions with the environment. Electrochemical corrosion, also called electrolytic corrosion (EC), and accelerated business environment (ABE) tests are used to test specimens. The EC test, where the test time is a few minutes, is used to rank specimens that corrode by electrochemical means. These tests are useful during early development. In the ABE test, the test time can be from a fraction of a day

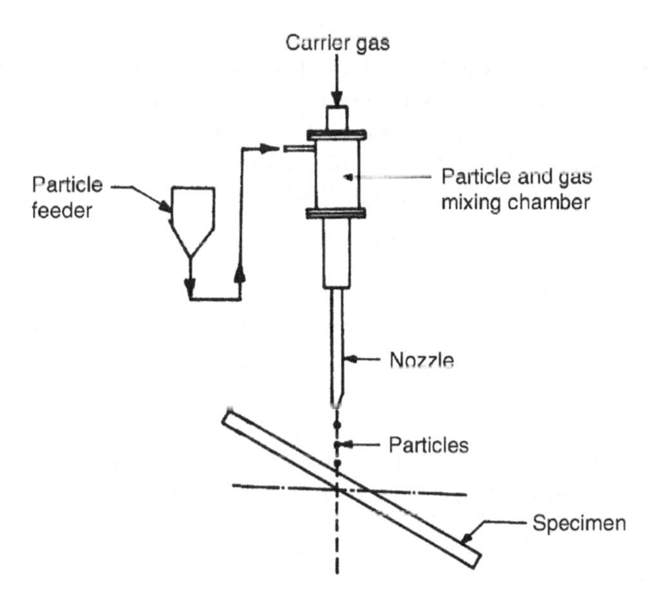

Figure 11.3.4 Schematic diagram of solid particle erosion test apparatus.

to several days depending on the environmental conditions. A properly simulated ABE test is generally used to predict the component life under actual conditions (Bhushan, 1996).

11.3.5.1 Electrochemical (EC) Test

In this test, two- or three-electrode cells are normally used to measure (1) the corrosion potential to determine the practical nobility of a material, and (2) the corrosion current density to determine the corrosion rate of two dissimilar materials to determine the corrosion rate of a material couple (Bhushan, 1996). A number of electrochemical instruments are commercially available for EC tests (Dean *et al.*, 1970).

11.3.5.2 Accelerated Business Environment (ABE) Test

The test samples are exposed to a controlled accelerated corrosive environment representative of the business environment. The degradation of the test specimen in the corrosive environmental test is measured by various methods, such as measuring weight loss, quantifying dimensional changes, noting changes in physical or chemical properties, measuring the size and number of defects on the surface using optical or scanning electron microscopes (visual methods), determining the total defect density by light-scattering techniques, determining the atomic concentration of substrate material on a coated surface (chemical analysis), and quantifying performance degradation (if measurable). Commonly used corrosion tests that approximate the corrosion produced in service are discussed here. Frequently, combinations of corrosive environments are used.

Salt Spray (Fog)

The neutral salt-spray (fog) test utilizes a box of suitable size, from about 2 m^3 to walk-in size, into which a 5% NaCl solution is aspirated with air. A common testing time is 72 hours, although exposure duration can vary considerably (ASTM B117-73). This test is commonly used for zinc coatings (Saur, 1975). For corrosion tests of gas-turbine components, salts such as Na_2SO_4 and NaCl are added in air (Nicoll, 1983).

Seawater

The test samples are partially or completely submerged in natural or synthetic (ASTM D1141-52) seawater for a fraction of a day to several months. This method is commonly used for marine applications (Bhushan and Dashnaw, 1981; Bhushan and Winn, 1981).

Corrosive Gases

In this test, the test specimen is exposed to an accelerated corrosive gas environment (with constituents representative of the business environment). The corrosive gases may consist of small fractions of Cl_2, NO_2, H_2S and SO_2 (such as air with 5 ppb Cl_2, 500 ppb NO_2, 35 ppb H_2S, and 275 ppb SO_2 at 70% relative humidity and 25°C). An exposure of a fraction of a day to few days is sufficient (Bhushan, 1996).

Temperature/Humidity (T/H)

In this test, test specimens are exposed to high temperature (T) and/or high humidity (H) for a fraction of a day to several days (Nicoll, 1983; Bhushan, 1996).

11.4 Closure

The screening tests include accelerated friction and wear tests and functional tests. Accelerated tests are required to reduce the number of options in the design of a machine component and/or materials (typically from one to three). Then functional tests are conducted on a smaller sample set. Accelerated tests reduce the testing time and cost. The accelerated tests serve a necessary function; however, these should be designed so that they properly simulate wear mechanisms and at the same time accelerate the wear process. If these tests are properly simulated, an acceleration factor between the simulated test and the functional test can be empirically determined to predict the component life based on the accelerated tests.

References

Bayer, R.G. (1976), *Selection and Use of Wear Tests for Metals*, STP-615, ASTM, Philadelphia, Pennsylvania.

Bayer, R.G. (1979), *Wear Tests for Plastics: Selection and Use*, STP-701, ASTM, Philadelphia, Pennsylvania.

Bayer, R.G. (1982), *Selection and Use of Wear Tests for Coatings*, STP-769, ASTM, Philadelphia, Pennsylvania.

Benzing, R.J., Goldblatt, I., Hopkins, V., Jamison, W., Mecklenburg, K., and Peterson, M.B. (1976), *Friction and Wear Devices*, Second edition, ASLE, Park Ridge, Illinois.

Bhushan, B. (1980), "Stick-Slip Induced Noise Generation in Water-Lubricated Compliant Rubber Bearings," *ASME J. Tribol.* **102**, 201–212.

Bhushan, B. (1987), "Overview of Coating Materials, Surface Treatments, and Screening Techniques for Tribological Applications Part 2: Screening Techniques," In *Testing of Metallic and Inorganic Coatings* (W.B. Harding and G.A. DiBari, eds), STP-947, pp. 310–319, ASTM, Philadelphia, Pennsylvania.

Bhushan, B. (1996), *Tribology and Mechanics of Magnetic Storage Devices*, Second edition, Springer-Verlag, New York.

Bhushan, B. (1999), *Handbook of Micro/Nanotribology*, Second edition, CRC Press, Boca Raton, Florida.

Bhushan, B. (2001), *Modern Tribology Handbook, Vol. 1: Principles of Tribology*, CRC Press, Boca Raton, Florida.

Bhushan, B. and Dashnaw, F. (1981), "Material Study for Advanced Stern-tube Bearings and Face Seals," *ASLE Trans.* **24**, 398–409.

Bhushan, B. and Gupta, B.K. (1997), *Handbook of Tribology: Materials, Coatings, and Surface Treatments*, McGraw-Hill, New York (1991); Reprint edition, Krieger, Malabar, Florida (1997).

Bhushan, B. and Lowry, J.A. (1995), "Friction and Wear Studies of Various Head Materials and Magnetic Tapes in a Linear Mode Accelerated Test Using a New Nano-Scratch Wear Measurement Technique," *Wear*. **190**, 1–15.

Bhushan, B. and Martin, R.J. (1988), "Accelerated Wear Test Using Magnetic-Particle Slurries," *Tribol. Trans.* **31**, 228–238.

Bhushan, B. and Sibley, L.B. (1982), "Silicon Nitride Rolling Bearings for Extreme Operating Conditions," *ASLE Trans.* **25**, 417–428.

Bhushan, B. and Winn, L.W. (1981), "Material Study for Advanced Stern-tube Lip Seals," *ASLE Trans.* **24**, 410–422.

Bhushan, B., Nelson, G.W., and Wacks, M.E. (1986), "Head-Wear Measurements by Autoradiography of Worn Magnetic Tapes," *ASME J. Tribol.* **108**, 241–255.

Clauss, F.J. (1972), *Solid Lubrication and Self-Lubricated Solids*, Academic, New York.

Dean, S.W., France, W.D., and Ketcham, S.J. (1970), "Electrochemical Methods of Testing," paper presented at the symposium on State of the Art in Corrosion Testing Methods, ASTM Annual Meeting, Toronto, Canada.

Doeblin, E.O. (1990), *Measurement Systems: Application and Design*, Fourth edition, McGraw-Hill, New York.

Dunkin, J.E. and Kim, D.E. (1996), "Measurement of Static Friction Coefficient Between Flat Surfaces," *Wear*. **193**, 186–192.

Nicoll, A.R. (1983), "A Survey of Methods Used for the Performance Evaluation of High Temperature Coatings," In *Coatings for High Temperature Applications* (E. Lang. ed), pp. 269–339, Applied Science Publishers, London.

Rabinowicz, E. and Tabor, D. (1951), "Metallic Transfer Between Sliding Metals: An Autoradiographic Study," *Proc. R. Soc. Lond. A* **208**, 455–475.

Saur, R.L. (1975), "Corrosion Testing: Protective and Decorative Coatings," In *Properties of Electrodeposits: Their Measurements and Significance* (R. Sard, H. Leidheiser, and F. Ogburn, eds), pp. 170–186, The American Electrochemical Society, Princeton, New Jersey.

Yust, C.S. and Bayer, R.G. (1988), *Selection and Use of Wear Tests for Ceramics*, STP-1010, ASTM, Philadelphia, Pennsylvania.

Further Reading

Bayer, R.G. (1976), *Selection and Use of Wear Tests for Metals*, STP-615, ASTM, Philadelphia, Pennsylvania.

Bayer, R.G. (1979), *Wear Tests for Plastics: Selection and Use*, STP-701, ASTM, Philadelphia, Pennsylvania.

Bayer, R.G. (1982), *Selection and Use of Wear Tests for Coatings*, STP-769, ASTM, Philadelphia, Pennsylvania.

Benzing, R.J., Goldblatt, I., Hopkins, V., Jamison, W., Mecklenburg, K., and Peterson, M.B. (1976), *Friction and Wear Devices*, Second edition, ASLE, Park Ridge, Illinois.

Bhushan, B. (1987), "Overview of Coating Materials, Surface Treatments, and Screening Techniques for Tribological Applications Part 2: Screening Techniques," In *Testing of Metallic and Inorganic Coatings* (W.B. Harding and G.A. DiBari, eds), STP-947, pp. 310–319, ASTM, Philadelphia, Pennsylvania.

Bhushan, B. (1996), *Tribology and Mechanics of Magnetic Storage Devices*, Second edition, Springer-Verlag, New York.

Bhushan, B. (2001), *Modern Tribology Handbook, Vol. 1: Principles of Tribology*, CRC Press, Boca Raton, Florida.

Bhushan, B. (2011), *Nanotribology and Nanomechanis II*, Third edition, Springer-Verlag, Heidelberg, Germany.

Bhushan, B. and Gupta, B.K. (1997), *Handbook of Tribology: Materials, Coatings, and Surface Treatments*, McGraw-Hill, New York (1991); Reprint edition, Krieger, Malabar, Florida (1997).

Clauss, F.J. (1972), *Solid Lubrication and Self-Lubricated Solids*, Academic, New York.

Nicoll, A.R. (1983), "A Survey of Methods Used for the Performance Evaluation of High Temperature Coatings," In *Coatings for High Temperature Applications* (E. Lang. ed), pp. 269–339, Applied Science Publishers, London.

Yust, C.S. and Bayer, R.G. (1988), *Selection and Use of Wear Tests for Ceramics*, STP-1010, ASTM, Philadelphia, Pennsylvania.

12

Tribological Components and Applications

12.1 Introduction

Common tribological components which are used in industrial applications include sliding-contact and rolling-contact bearings, seals, gears, cams and tappets, piston rings, electrical brushes, and cutting and forming tools. More recently, micro/nanoelectromechanical systems (MEMS/NEMS), also called micro/nanodevices or micro/nanocomponents are being produced using micro/nanofabrication techniques (Bhushan, 2010). Some of the common industrial applications include material processing, internal combustion engines for automotive applications, gas turbine engines for aerospace applications, railroads, and magnetic storage devices (Bhushan, 2001a, 2001b). For a component's desired performance and life, its friction and wear need to be minimized, or optimized, for a given application. The relevant friction and wear mechanisms are dependent upon the device and the operating conditions.

This chapter presents descriptions, relevant wear mechanisms, and typical materials for common tribological components, microcomponents, material processing and other industrial applications.

12.2 Common Tribological Components

Tribological components that operate at low to moderate contact stresses (on the order of 5 MPa) include sliding-contact bearings, seals, piston rings and electrical brushes (Bhushan, 2001a, 2001b). The components that operate at high Hertzian stresses (on the order of 500 MPa) include rolling-contact bearings, gears, cams, and tappets (Bhushan, 2001a).

12.2.1 Sliding-Contact Bearings

The machine elements that support a moving shaft against a stationary housing are called bearings. In general, we can classify bearings as either sliding-contact or rolling-contact bearings. In sliding-contact bearings, also known as sliding or plain bearings or bushings,

Introduction to Tribology, Second Edition. Bharat Bhushan.
© 2013 John Wiley & Sons, Ltd. Published 2013 by John Wiley & Sons, Ltd.

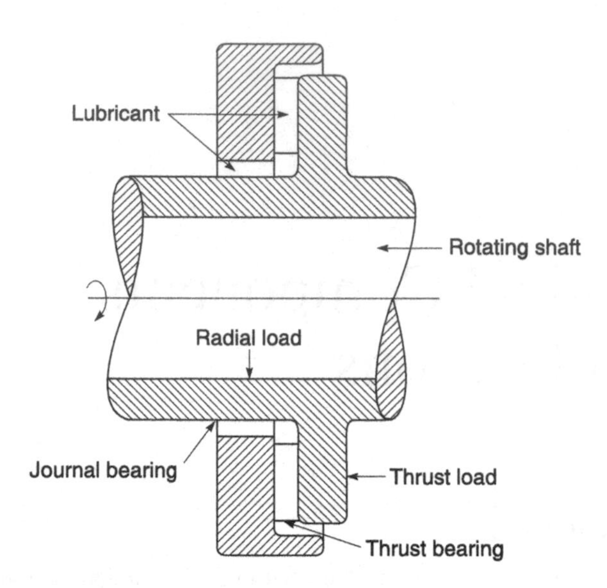

Figure 12.2.1 Schematic of a rotating shaft supported by a thrust and a journal bearing.

the load is transmitted between moving parts by sliding contact. The motion can be a planar motion (e.g., plane, curved, step, and composite sliders and pivoted-pad sliders) or a rotational motion (e.g., full and partial journal bearings, foil bearings, floating-ring bearings) (Bisson and Anderson, 1964; Neale, 1973; Booser, 1984; Fuller, 1984). (Also see Chapter 9 for examples.) Sliding bearings can be lubricated with a film of air, water, oil, grease, or the process fluid. Thrust and journal bearings are perhaps the most familiar and most widely used of all bearings types. Thrust bearings are used to support thrust (or axial) loads in a rotating machinery, Figure 12.2.1. These consist of multiple pads, either fixed or pivoted. Journal bearings are used to support radial (or normal) loads. These consist of a sleeve of bearing material wrapped partially or completely around a rotating shaft or journal, and are designed to support a radial load, Figure 12.2.1.

The bearings are generally lubricated with a liquid lubricant or grease. In self-acting bearings, lubrication is accomplished via hydrodynamic lubrication. Wear mechanisms are dependent upon the bearing materials and operating conditions. Sliding bearings most commonly fail by adhesive, abrasive, and/or chemical (or corrosive) wear mechanisms.

The selection of materials for sliding bearings is a multifunctional optimization problem. In general, the standard requirements are as follows: comformability, embeddability, compressive strength, fatigue strength, thermal conductivity, wear resistance, corrosive resistance, and cost. Bearing materials fall into two major categories: metals and nonmetals (Ku, 1970; Neale, 1973; Booser, 1984; Fuller, 1984; Glaeser, 1992; Bhushan and Gupta, 1997). Included in the metals are several types of soft metals – precious metals, tin- and lead-based alloys (babbitts), copper-based alloys (brasses and bronzes), aluminum-based alloys, cast iron, and porous metals. The nonmetals include wood, carbon-graphites, plastics, elastomers, ceramics, cermets, and several other proprietary materials. These materials can be used as a bulk material or as a lining on a bearing surface. In a lined bearing, the bearing material is bonded to a stronger backing material such as steel. The thickness of the liner material usually ranges from 0.25 mm to as high as 10 mm. Many soft metals, carbon-graphites, plastics, and elastomers are used as one of the slider materials against a sliding member such as stainless steel under unlubricated conditions. In many bearing applications, hard and soft materials are used to coat bearing substrates by various deposition techniques.

12.2.2 Rolling-Contact Bearings

Rolling-contact or rolling-element or antifriction bearings employ a number of balls or rollers between two surfaces known as inner and outer races or rings. The inner race is carried by the rotating shaft or journal and the outer race, mounted on the machine casing or bearing housing, is often stationary. The balls or rollers, also called rolling elements, are held in an angularly spaced relationship by a cage, also called a retainer or separator. The rolling elements accommodate relative motion between surfaces primarily by the action of rolling with a small slip (sliding) rather than pure sliding so that the frictional forces acting between the surfaces are primarily due to rolling resistance. Rolling bearings have much less friction than sliding bearings and, therefore, are also called antifriction bearings. The load capacity and stiffness of rolling bearings is much larger than that of sliding-contact bearings. Because of the use of balls or rollers, the actual area of contact is reduced to near zero, therefore, contact stresses are very high (Hertzian stresses), typically on the order of 500 MPa or more. There are many different kinds of bearings including radial ball bearings, angular-contact ball bearings, cylindrical roller bearings, and tapered roller bearings, Figure 12.2.2 (Bisson and Anderson, 1964; Neale, 1973; Booser, 1984; Harris, 1991, Zaretsky, 1992). (Also see Chapter 9). Angular-contact and tapered roller bearings are used to support radial and/or thrust loads. Roller bearings will carry a greater load than ball bearings of same size because of the greater contact area. However, they have the disadvantage of requiring almost perfect geometry of the raceways and rollers.

Rolling bearings are generally lubricated with a liquid lubricant or grease via elastohydrodynamic lubrication. Some bearings for operation in vacuum and high temperatures must be self-lubricated. The classic rolling-bearing failure mode is fatigue spalling, in which a sizable piece of the contact surface is dislodged during operation by fatigue cracking in the bearing metal under cyclic contact stressing (Tallian *et al.*, 1974; Zaretsky, 1992; Summers-Smith, 1994) (See also Chapter 7). When the surface motion at rolling-bearing contacts contains substantial sliding, the surface damage and wear can change abruptly from mild to severe adhesive wear, commonly called scuffing or smearing (Tallian, 1967; Scott, 1977).

The bearing industry has used SAE 52100 steel as a standard material since the 1920s. This is a high-carbon chromium steel that also contains small amounts of Mn, Si, Ni, Cu, and Mo. A minimum tolerable hardness for bearing components is about 58 HRC. SAE 52100 steels are generally used up to temperatures of about 200°C. Molybdenum air-hardening steels, known as high-speed steels (a class of tool steels), e.g., M-1, M-2, and M-50, are generally used up to temperatures of about 320°C. For a corrosive environment, AISI 440C stainless steel with a hardness of about 60 HRC is used for temperatures up to 250°C. Carburized or case-hardened steels such as AISI 4320 and AISI 4360, are also used for rolling bearings. The room temperature hardness of most carburized bearing steels is roughly 58–63 HRC, with a core hardness of 25–40 HRC (Neale, 1973; Bamburger, 1980; Zaretsky, 1992; Bhushan and Gupta, 1997). Some high-performance applications dictate the need for bearings that operate

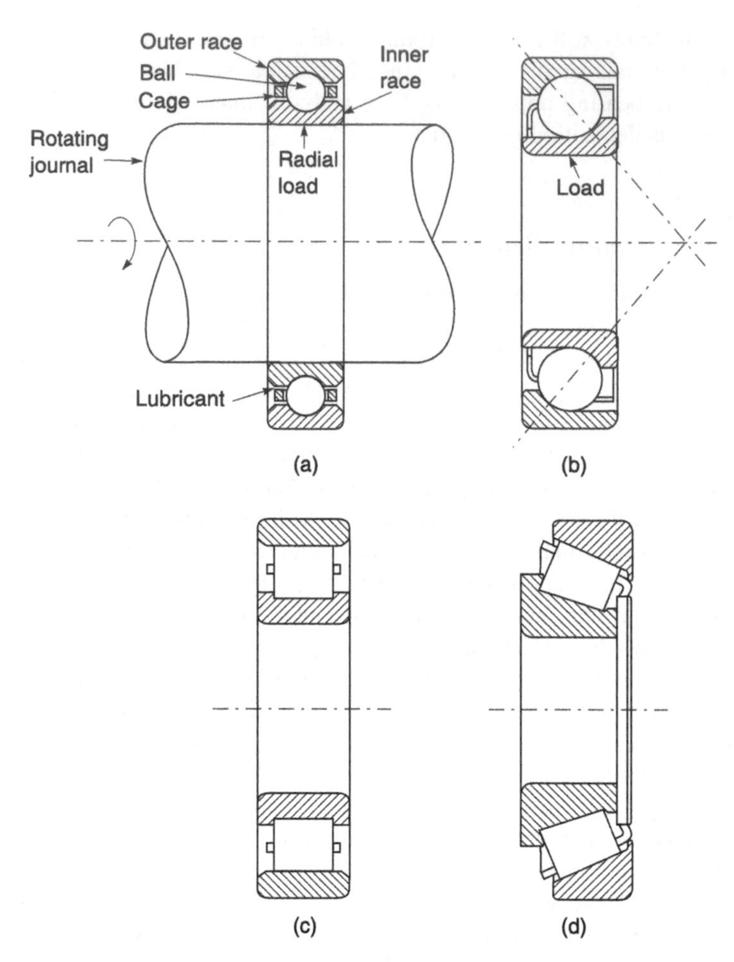

Figure 12.2.2 Schematics of (a) a radial ball bearing, (b) an angular-contact ball bearing, (c) a cylindrical roller bearing, and (d) a tapered roller bearing.

at high speeds and/or high temperatures (up to 1200°C) with high precision. Since high temperatures are beyond the range in which most ferrous materials are incapable of operating, more refractory materials and compounds are considered. Silicon nitride is the favorite ceramic material for high-performance applications. It has two to three times the hardness and one-third the dry friction coefficient of bearing steels, has good fracture toughness compared with other ceramic materials, and maintains its strength and oxidation resistance up to 1200°C, which makes it a promising high temperature rolling-bearing material (Bhushan and Sibley, 1982). Si_3N_4 has been used for bearing elements as well as for complete bearing configurations. Silicon nitride bearings are used in the aerospace/defense, tool spindle, chemical processing, nuclear, and automotive industries. Various hard and soft coatings and surface treatments have been developed for lightly loaded, long-life rolling-element bearings for applications in high vacuum and/or high temperatures, especially in applications with no external lubrication (Bhushan and Gupta, 1997).

Cages or retainers maintain the proper distance between rolling elements. In conventional rolling bearings, both metallic and nonmetallic retainers are used. Under normal temperatures, a large percentage of ball and roller bearings use stamped retainers of low-carbon steel or machined retainers of copper-based alloys such as iron-silicon bronze or a leaded brass. In applications where marginal lubrication exists during operation, silver-plated bronze or PTFE-based cage materials are used (Ku, 1970). For high-temperature and marginally lubricated or unlubricated conditions, such as in aerospace applications, potential self-lubricated cage materials are phenolic, polyimide, graphite and composites, Ga-In-WS_2 compact, Ta-Mo-MoS_2 compact, and metals containing MoS_2-graphite (Bhushan and Sibley, 1982; Gardos and McConnell, 1982). Carbon fiber-reinforced polyimide composite with a solid-lubricant additive has exhibited low friction and wear when tested against an Ni-based alloy (Rene 41) at a normal stress of about 175 MPa and a temperature of 315°C (Gardos and McConnell, 1982).

12.2.3 Seals

The primary function of seals, called fluid seals, is to limit loss of lubricant or process fluid (liquid or gas) from systems and to prevent contamination of systems by the operating environment. The seals are divided into two main classes: static and dynamic seals. Static seals are gaskets, O-ring joints, packed joints, and similar devices used to seal static connections or openings. A dynamic seal is used to restrict fluid flow through an aperture closed by relative moving surfaces. Dynamic seals include fixed clearance type (labyrinth seals, floating ring seals and ferrofluidic seals) and surface-guided type (mechanical face seals, lip seals and abradable seals) (Lebeck, 1991). The labyrinth seal shown in Figure 12.2.3a relies primarily on creating a high-loss leakage path to minimize leakage. In the ferrofluidic seal shown in Figure 12.2.3b, a magnetic fluid is held in place by magnets. A small pressure difference can be maintained by the fluid before it is pushed out of the gap. The magnetic fluid is a suspension of magnetic particles in a liquid. These seals have zero leakage. Figure 12.2.3c shows a schematic of a mechanical face seal. Generally there are two rings that mate at some annular surface. Usually one of the rings, called a mating ring, is rigidly mounted, and the other, called the primary ring, is flexibly mounted, so as to allow axial and angular freedom of the seal so as to self-align and be surface guided. Sliding of surfaces occurs in a direction normal to the leakage flow. Figure 12.2.3d shows a typical lip seal. It is made of a compliant material and contacts over a small axial length.

Lubrication of the sealing interface varies from hydrodynamic to no lubrication (e.g., in gas-path components such as a turbine or compressor blade tips). Adhesive wear is the dominant type of wear in well-designed seals. Other wear modes are abrasive wear, corrosive wear, fatigue wear, and blistering (Neale, 1973; Johnson and Schoenherr, 1980; Stair, 1984).

Which materials are used for seals depends on the seal design and the operating requirements. Mechanical face seals employed in oil-lubricated applications range from very hard combinations such as ceramics (e.g., alumina, tungsten carbide, silicon carbide, boron carbide) and cermets (e.g., cemented carbides) to babbitts, bronzes, carbon-graphites, thermoplastic resins, and elastomers. Carbon-graphites and polymers are also used in the presence of process fluids that are poor lubricants or under unlubricated conditions. A large number of mating materials have been used, such as Niresist iron, tool steel, hard-faced steel, nickel-copper-based

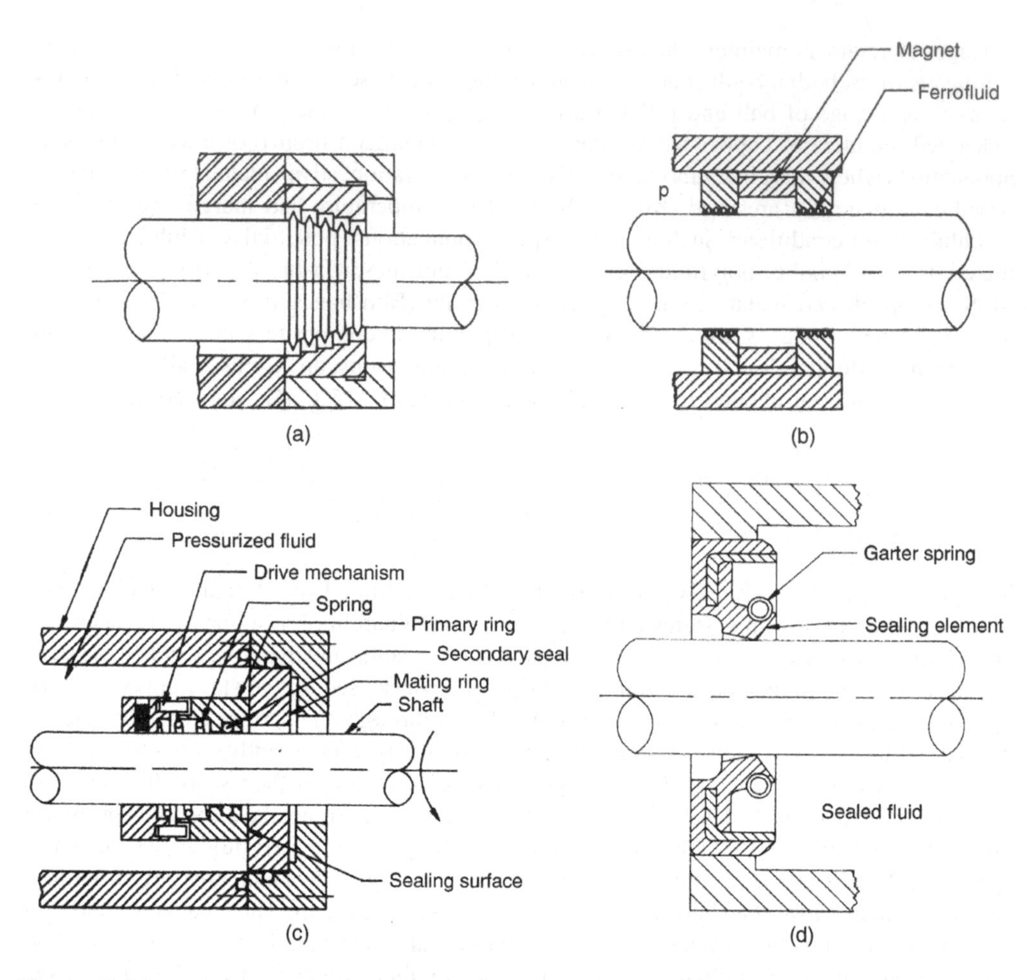

Figure 12.2.3 Schematics of (a) labyrinth seal, (b) ferrofluidic seal, (c) a mechanical face seal with outside pressurized, rotating primary ring and fixed mating ring, and (d) a lip seal on a cylindrical surface. Reproduced with permission from Lebeck, A.O. (1991), *Principles and Design of Mechanical Face Seals*, Wiley, New York. Copyright 1991. Wiley.

materials (e.g., Monel), nickel-molybdenum alloys (e.g., Hastelloy B or Hastelloy C), Cobalt-based alloys (e.g., Stellite), tungsten carbide, boron carbide, alumina, or plasma-sprayed coatings of various ceramics. For example, many merchant ships employ asbestos-filled phenolic in oil-lubricated face-seal configurations as the stationary-face insert and the rotating mating shaft is made of Niresist iron or chrome-nickel steel. Carbon-graphites and plastics such as PTFE, polyimide, poly(amide-imide), or phenolic composites are used in water-lubricated face seals (Paxton, 1979; Stair, 1984; Bhushan and Gupta, 1997).

Lip seals are made of compliant materials-elastomers. The elastomers most commonly used are butadiene-acrylonitrile (Buna N), polyacrylate, and vinylidene fluoride-hexafluoropropylene (Viton) at high temperatures (up to about 170°C). Other lip-seal materials that are used for specialized applications include silicone, fluorosilicone, and perfluoroelastomer (Kalrez). The main prerequisites for the mating surfaces of lip seals are hardness and high abrasion resistance. The liner materials include chrome-nickel steels, case-hardened cast iron, and plasma-sprayed coatings of various ceramics, such as Cr_2O_3, Al_2O_3, or mixtures of oxides with other materials to achieve specific characteristics (Bhushan and Gupta, 1997).

Abradable (rub-tolerant) seals are used in the compressor and turbine sections of aircraft gas turbine engines (Meetham, 1981). Abradable seals are used at the rotor–stator interface to maintain the close tolerance without catastrophic failure. One of the sliding members is supposed to abrade against another if there is any interference. Abradable coatings are applied to compressor and turbine castings in some engines. In some cases, wear-resistant coatings are applied to the tip regions of rotating blades to minimize wear. Many of the commonly used abradable materials are sintered ceramics and plasma-sprayed coatings of Ni–Cr bonded chrome carbide and tungsten carbide sliding against rub-tolerant materials such as felt metal and plasma-sprayed coatings of ceramics such as chrome oxide.

12.2.4 Gears

Gears are toothed wheels used for transmission of rotary motion from one shaft to another and a change in rotational speed (Dudley, 1964; Merritt, 1971; Shigley and Mischke, 1989). There are different types of gear including spur, helical, bevel and worm gears. Spur gears shown in Figure 12.2.4a are used to transmit rotary motion between parallel shafts and helical gears (Figure 12.2.4b) are used to transmit rotary motion between parallel and nonparallel shafts. The smaller of two mating gears is known as a pinion and the larger as a gear. To transmit motion at a constant angular-velocity ratio, an involute tooth profile is used. In the spur gears, the teeth are straight and parallel to the axis of rotation, whereas in helical gears, teeth are not parallel to the axis of rotation. The helix angle in helical gears is the same on each gear, but one gear must have a right hand helix and the other a left-hand helix. In spur gears, the line of contact is parallel to the axis of rotation; in helical gears the line is diagonal across the face of the tooth. The initial contact of spur-gear teeth is a line extending all the way across the face of the tooth, whereas the initial contact of helical-gear teeth is a point which changes into a line as the teeth come into more engagement. It is this gradual engagement of the teeth and the smooth transfer of load from one tooth to another which give helical gears the ability to transmit heavy loads at high speeds.

In the case of bevel gears, the rotational axes are not parallel to each other. Although bevel gears are usually made for a shaft angle of 90°, they can be produced for almost any angle. Figure 12.2.4c shows a straight bevel gear and a pinion. Worm gears are used to transmit motion between nonparallel, non-intersecting shafts. Figure 12.2.4d shows a worm and a worm gear used to transmit motion between nonparallel, non-intersecting shafts.

Contact occurs on lines or points, resulting in high Hertzian contact stresses, similar to that in rolling-contact bearings. The gear motion is associated primarily with rolling and some sliding motions. Gear teeth may operate under boundary, mixed, and fluid-film (elastohydrodynamic) lubrication regimes. Typical failure modes of gears are surface fatigue, scoring, pitting, scuffing (severe form of adhesive wear), abrasion, corrosive wear, and tooth breakage. The dominant failure mode for a well-lubricated gear pair is surface fatigue. Because of higher slip or sliding in gears as compared to rolling element bearings, the failure mode is generally surface fatigue

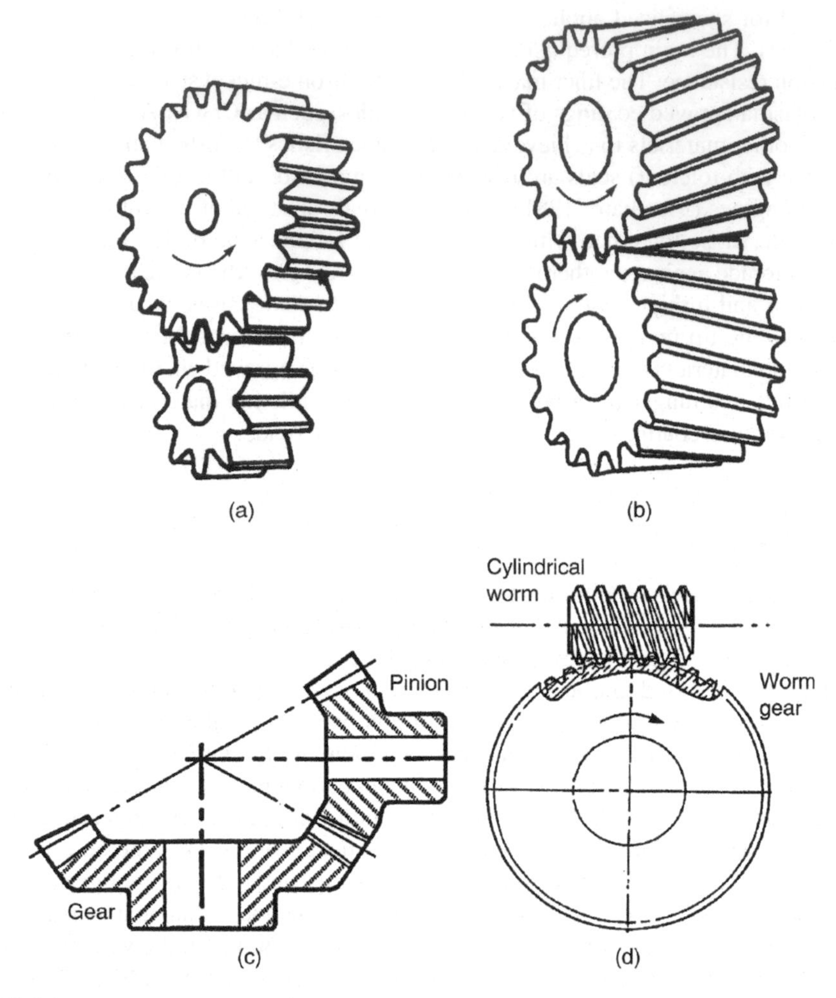

Figure 12.2.4 Schematics of (a) pair of spur gears used to transmit rotary motion between parallel shafts, (b) pair of helical gears used to transmit motion between parallel shafts, (c) pair of straight bevel gears used to transmit motion between intersecting shafts, and (d) worm and worm gears used to transmit motion between nonparallel, non-intersecting shafts.

rather than subsurface fatigue. When the elastohydrodynamic lubricant film is not sufficiently thick, metal-to-metal contact occurs, leading to other failure modes such as scuffing, a severe form of adhesive wear. The essential material requirements for gears are adequate bending fatigue strength and resistance to surface fatigue, adequate toughness to withstand the impact loads, adequate resistance to scuffing, and adequate resistance to abrasive wear (Dudley, 1964, 1980; Coleman, 1970; Neale, 1973; Anonymous, 1989a; Lee and Cheng, 1991).

For gears of useful load capacity, hard materials are generally required because high Hertzian stresses occur at contact spots (Dudley, 1964, 1980; Merritt, 1971; Neale, 1973; Anonymous, 1989b; Bhushan and Gupta, 1997). Gear wear is reduced by heat treatments or thermochemical treatments or by the application of coatings. The most commonly used gear material for power transmission is steel. Cast iron, bronze, and some nonmetallic materials are also used. In instrument gears, toys, and gadgets, such materials as aluminum, brass, zinc, and plastics (such as Nylon) are often used. Nonmetallic gears made of plastics are generally used against steel gears. The through-hardened steels for gears should have about 0.4% C. Their hardnesses range from 24 to 43 HRC. Cast-iron gears are generally not used above 250 Brinell (24 HRC). Cast-iron gears have unusually good wear resistance for their low hardness.

Case-hardening (carburizing) methods are used to put a thin, hard case on a medium-hard core of ferrous metals for gears. When the hardness is over about 400 Brinell (43 HRC), gear teeth become too brittle when through-hardened. The case-hardening treatments, when properly done, make a gear tooth that is tough and has good to excellent strength. Case-carburized gears at 55–63 HRC have the best capability when the surface layer of metal is around 0.8–0.9% C. Some improvements in wear resistance may occur if the outer layer is 1.0–1.1% C at the expense of a reduction in tooth-breaking strength. Other surface treatments that are used for steel gears include nitriding, sulfidizing, and phosphating to reduce friction and wear.

The performance limits of each gear of a pair can be rated in terms of maximum allowable transmitted power, which is proportional to allowable bending stress for the gear material for one-way bending. The allowable power rating also should be checked for each gear of a pair in terms of the risk of tooth pitting, which is denoted by allowable contact stress for the gear material. In other words, the performance limits of various gear materials can be compared in terms of allowable bending stress and allowable contact stress.

It is common practice, except when both members of a gear pair are surface-hardened, to have the two members of different strengths and usually of different alloys. This has been found to reduce the likelihood of scuffing. The smaller gear (pinion), which is the high-speed member and has the more arduous and frequent duty, is usually the harder member. Table 12.2.1 presents common material combinations suitable for the principal types of gears.

Table 12.2.1 Common material combinations suitable for the principal types of gears.

Gear duty	Material combination	
Motion only	Plastics, brass, mild steel, stainless steel in any combination	
Light power	Carbon steel	Brass Plastics Cast iron Steel
Worm drives	Alloy steel	Cast iron Phosphor bronze
High duty, industrial and marine	Alloy steel	Carbon steel Alloy steel
	Nitrided alloy steel	Nitrided alloy steel Alloy steel
Automotive	Carburized case-hardened alloy steel	Carburized case-hardened alloy steel
Aircraft and high duty	Carburized case-hardened and ground alloy steel	Carburized case-hardened and ground alloy steel

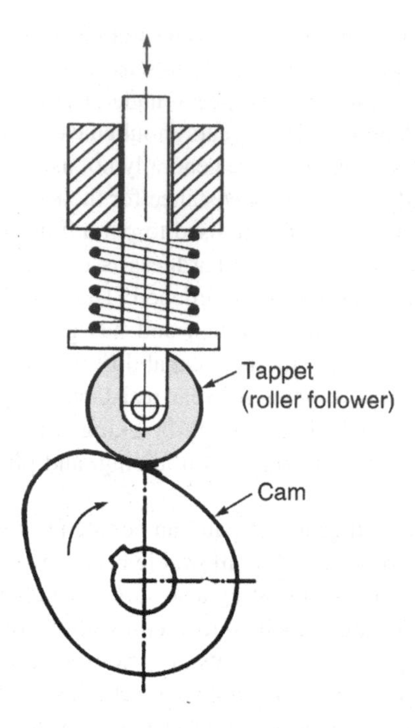

Figure 12.2.5 Schematic of a cam and a translating roller follower (tappet).

12.2.5 Cams and Tappets

Cams and tappets (or cam follower systems) are extensively employed in engineering machines to transform rotary motion to reciprocating sliding motion or vice versa, e.g., in automotive valve trains and textile machines. The cam follower can be a flat follower or a roller follower (Figure 12.2.5). The contact conditions are nominal points or line contacts which under load lead to elliptical and rectangular contact areas, respectively. There is always a rolling motion through the contact, accompanied by some sliding in the direction of rolling motion. The wear modes for cams and tappets are very similar to those for gears. Under heavy duty, cams and tappets suffer from burnishing (due to adhesive/abrasive wear processes), scuffing (due to severe adhesive wear processes), and pitting (due to fatigue wear processes) (Neale, 1973; Lee and Cheng, 1991).

The wear of cams and tappets can be reduced considerably by selecting hard material combinations or by hardening the cam material by heat treatments or thermochemical treatments or by applying coatings. Tappet materials are usually through-hardened high C, Cr, or Mo types of carburized low-alloy steels. The most common tappet material in automotive applications is gray hardenable cast iron containing Cr, Mo, and Ni or chilled cast iron.

Coatings and surface treatments are also used for cams and tappets. Running-in coatings include phosphate coatings, chemically produced oxide coatings on ferrous metals, and electrochemically deposited Sn and Al. Surfaces of cams can be hardened through diffusion treatment such as carburizing, nitriding, and Tufftriding. Several hard coatings such as TiN and TiC, applied by PVD and CVD, also can be applied on cams and tappets to achieve low coefficients of friction and low wear.

12.2.6 Piston Rings

Piston rings are mechanical sealing devices used for sealing pistons, piston plungers, reciprocating rods, etc., inside cylinders. In gasoline and diesel engines and lubricated reciprocating-type compressor pumps, the rings are generally split-type compression metal rings. When they are placed in the grooves of the piston and provided with a lubricant, a moving seal is formed between the piston and the cylinder bore. Piston rings are divided into two categories: compression rings and oil-control rings. Compression rings, generally two or more, are located near the top of the piston to block the downward flow of gases from the combustion chamber. Oil rings, generally one or more, are placed below the compression rings to prevent the passage of excessive lubricating oil into the combustion chamber yet provide adequate lubrication for the piston rings. In typical lubricated situations, the piston skirt is in direct contact with the cylinder and acts as a bearing member that supports its own weight and takes thrust loads. In unlubricated arrangements, it is necessary to keep these two surfaces separated because they are not frictionally compatible. This is usually accomplished with a rider ring that supports the piston, (Figure 12.2.6) (Neale, 1973).

An ideal piston-ring material must meet the following requirements: low friction and wear losses, superior scuffing resistance, tolerances for marginal lubrication and rapidly varying environments, good running-in wear behavior, long term reliability and consistency of performance, long maintenance-free life, and low production cost.

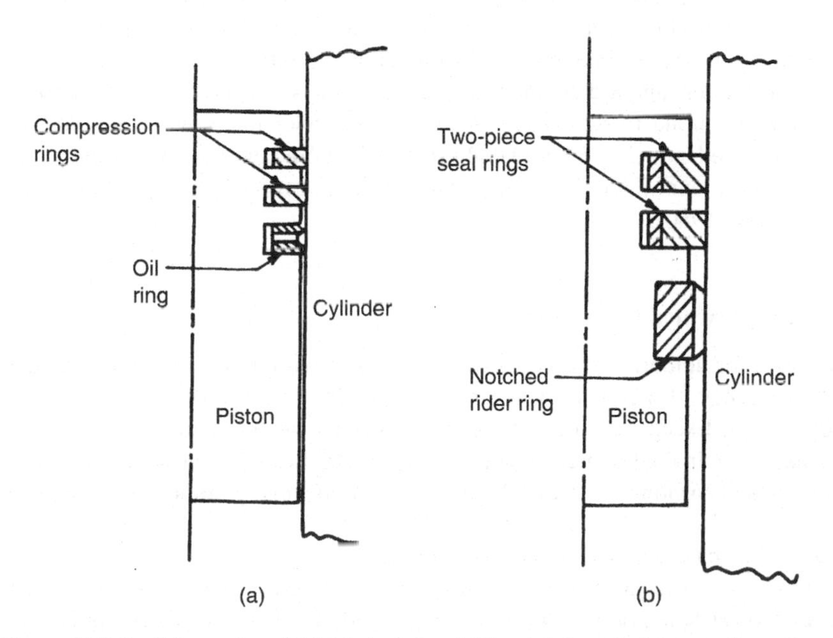

Figure 12.2.6 Schematics of (a) lubricated and (b) unlubricated piston configurations.

12.2.6.1 Lubricated Piston Rings

Gray cast iron with a hardness ranging from 200 to 400 Brinell is probably the most commonly used material for compression and oil rings. Pearlitic gray iron produced by either centrifugal or sand-cast methods is most widely used. In the heavier-duty engine applications, chromium-molybdenum alloy iron, spheroidal graphite iron, and carbidic malleable iron are used. Harder materials such as carbon steel or even En31 ball-bearing steel are also used.

Relatively thick coatings (up to 0.2 mm) of plated chromium on the ring periphery provide the best compromise between scuffing, wear, and corrosion resistance and low friction and resistance to oxidation at high temperatures; only one mating surface is coated. Generally, the use of chromium-plated top rings (with a hardness of 700–900 HV) run against cast-iron cylinder liners can reduce the ring and liner wear by a factor of 2 to 3. Rings coated with flame- or plasma-sprayed coatings of molybdenum (in thicknesses up to 0.25 mm) with a hardness of over 1000 HV are believed to have higher scuffing resistance than chromium-plated rings. The main limitation of molybdenum ring coatings is that they are subject to oxidation at 500°C, and at 730°C the oxide volatizes. Several plasma-sprayed coatings of composites, such as Mo-Cr-Ni alloy, and ceramics, such as chromium oxide, have been developed for achieving improved scuffing resistance under conditions of marginal lubrication. A wide variety of running-in coatings have been used on piston rings in order to reduce the scuffing resistance. For a review, see Neale (1973), Scott *et al.* (1975), Taylor and Eyre (1979), Ting (1980), and Bhushan and Gupta (1997).

Most cylinder liners are made of gray case iron. To increase the liner mechanical strengths, nickel, chromium, copper, molybdenum, titanium, and vanadium are added. Steel cylinder liners also have been used, and their advantage is that the walls can be made much thinner. They have to be hardened to at least 400 Brinell; however, for satisfactory resistance to wear and scuffing, they are hard-chromium plated. It should be noted that chromium-plated liners should not run against the chromium-plated piston rings, but rather against plain cast-iron rings or rings with other types of coatings. The sulfidizing and nitriding treatments also have been used in cylinder liners and are claimed to be comparable with chromium plating for scuffing resistance and as an aid in running-in. Aluminum cylinder liners with high silicon content have also been used, which are lightweight materials. Some engine test results show that the liner wear of aluminum liners is lower than that of conventional cast-iron liners (Taylor and Eyre, 1979; Ting, 1980).

12.2.6.2 Unlubricated Piston Rings

Materials used for unlubricated piston rings are almost without exception nonmetals. This is due to the tendency of metals to weld under dry sliding (Scott *et al.*, 1975; Fuchsluger and Vandusen, 1980). Exceptions are metals impregnated with lubricants or coated with wear-resistant materials. In the nonmetal category, plastics, carbons, and ceramics are the materials most widely used. Among plastics, filled PTFE is used most commonly. Another family of plastics consists of filled polyimides. Polyimides, while not as chemically resistant or as low in friction as PTFE, have a high temperature limit (315–370°C) and are more rigid. Some other rigid plastics that find use as piston rings are poly(amide-imides), polyphenylene sulfides, and aramids. For lower-temperature (below 150°C), lightly loaded, low-speed applications, rings of Nylon, acetal, ultrahigh-molecular-weight polyethylene, etc., have found use.

Carbon is another commonly used ring material. It is the material most commonly used in the temperature range from 370 to 540°C. Below 370°C, carbon is limited by its reaction to varying degrees of humidity and its fragility. Ceramics have low wear rates, but virtually all ceramics are expensive to fabricate and are subject to thermal shock. Consequently, the use of ceramics for piston ring applications has been limited to ceramic coatings for metal rings.

The most commonly used mating liner materials for nonmetallic piston rings are cast irons and hardened steels. When other materials, such as the 300 and 400 series stainless steels, must be used, a chrome-plated or nitrided wear surface frequently provides the compatibility needed. When neither course of action is open, a solid-lubricant coating is frequently used. Carbon-graphite is reported to have a lower wear rate in combination with Nickel-resist or nitrided steel than with chrome plate.

12.2.7 Electrical Brushes

Machines that utilize electrical brushes can be broadly classified into two groups. In the first group, the machines require a commutator. In these machines, the brushes must be capable of transferring the load current to the external circuit as well as assisting the commutation function. Within this class of machines are DC motors and generators. In the second class of machines, brushes are used only to transmit electric power from a stationary source to a moving component by means of a slip ring, Figure 12.2.7. Examples of slip-ring applications are AC generators, motors, and special applications. Brush wear is believed to be due to adhesion and particle transfer, while fatigue has been identified in some circumstances. A further wear mechanism that can occur is fracture caused by mechanical impact between the brush and the slip ring.

The need to transfer electric current efficiently across a sliding interface complicates the situation compared with normal wear. For example, it is not possible to consider reducing

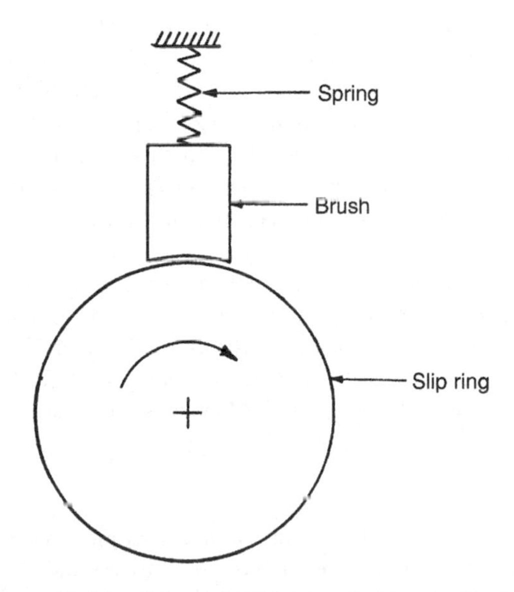

Figure 12.2.7 Schematic of an electrical brush-slip ring.

wear by using boundary lubrication or low-friction coatings (e.g., PTFE) on the surfaces, since this would cause unacceptably high electrical losses. For all brush applications, the brushes themselves are chosen as the sacrificial elements, so their wear rates exceed those of the machine slip ring or commutator by a factor of at least 10.

Most brushes are made of graphite-based materials such as electrographite, natural graphite, resin-bonded natural graphite, carbon and carbon-graphite, and metal-graphite (Shobert, 1965; Armington and Amey, 1979). Graphite is picked for its low friction and wear and high electrical conductivity. Electrographites are synthetic graphites characterized by a low coefficient of friction but a medium to high contact drop. Brushes of this material are among the most widely used because they exhibit good strength and controlled quality. Natural graphites have low density and low friction but exhibit a relatively high contact drop. They are particularly recommended for brushes operating on high-speed slip rings or commutators. Resin-bonded natural graphite is based on natural graphite that has been bonded with a phenolic or other resin. This produces high electrical resistivity and improved strength and gives good commutating ability for low operating current densities. Carbon-graphites, which are prepared by blending carbon and graphite and bonding them with a pitch or resin binder prior to baking, possess significant abrasive properties.

Metal-graphites are produced either by a powder technology or by infiltration of porous graphite with metals. By far the most common metal constituent is copper, although a good range of silver-graphite brushes is also manufactured. Small amounts of other metals (e.g., lead and tin) may also be added to provide improved bonding or reduced friction. Because of the metal constituent, these brushes have an appreciably lower electrical resistance than pure graphite or carbon-graphite grades. However, this is achieved at the expense of an increase in friction. Copper-graphite brushes are used where a high current-carrying capability is required.

Copper alloys and steel are most commonly used in slip-ring commutators or counter surfaces against which the brush is operated, although noble and rare metals are sometimes used for small-scale or special applications. Copper is chosen because of its good electrical and thermal properties. High-conductivity copper is most commonly used for commutators, although silver-bearing copper (<0.1% silver), chrome-copper (<1% chrome), and zirconium-copper (0.25% zirconium) also may be used, especially where higher strength at elevated temperatures is required. Common materials are bronze (copper-tin), phosphor bronze (copper-tin-phosphorus), gun metal (copper-tin-zinc), cupronickel (copper, 4% nickel), and Monel (nickel, 25–40% copper). Steel slip rings are used in applications demanding high sliding speeds, which make copper alloys unusable because of their lower mechanical strength and high wear.

Electroplating or other surface treatments can be used to provide an improved current transfer on a higher-strength substrate (e.g., copper on steel or high-strength aluminum) and will give a performance close to that observed for the bulk material of the surface coatings. Good commutating performance can be achieved by using a sintered-copper facing on steel commutator bars (Armington and Amey, 1979; McNab and Johnson, 1980).

12.3 MEMS/NEMS

Microelectromechanical systems (MEMS) refer to microscopic devices that have a characteristic length of less than 1 mm but more than 100 nm and combine electrical and mechanical components. Nanoelectromechanical systems (NEMS) refer to nanoscopic devices that have a characteristic length of less than 100 nm and combine electrical and mechanical components. In mesoscale devices, if the functional components are on micro- or nanoscale, they may be referred to as MEMS or NEMS, respectively (Bhushan, 2010). These are referred to as an intelligent miniaturized system comprising of sensing, processing, and/or actuating functions and combine electrical and mechanical components. The acronym MEMS originated in the USA. The term commonly used in Europe is microsystem technology (MST), and in Japan it is micromachines. Another term generally used is micro/nanodevices. MEMS/NEMS terms are also now used in a broad sense and include electrical, mechanical, fluidic, optical, and/or biological functions. MEMS/NEMS for optical applications are referred to as micro/nanooptoelectromechanical systems (MOEMS/NOEMS). MEMS/NEMS for electronic applications are referred to as radio-frequency-MEMS/NEMS or RF-MEMS/RF-NEMS. MEMS/NEMS for biological applications are referred to as BioMEMS/BioNEMS.

To put the characteristic dimensions of MEMS/NEMS and BioNEMS into perspective, see Figure 12.3.1. NEMS and BioNEMS shown in the figure range in size from 2 to 300 nm, and the size of MEMS is 12,000 nm. For comparison, individual atoms are typically a fraction of a nanometer in diameter, deoxyribonucleic acid (DNA) molecules are about 2.5 nm wide, biological cells are in the range of thousands of nm in diameter, and human hair is about 75 μm in diameter. NEMS can be built with weight as low as 10^{-20} N with cross sections of about 10 nm, and a micromachined silicon structure can have a weight as low as 1 nN. For comparison, the weight of a drop of water is about 10 μN and the weight of an eyelash is about 100 nN.

Micro/nanofabrication techniques include top-down methods, in which one builds down from the large to the small, and the bottom-up methods, in which one builds up from the small to the large (Bhushan, 2010). Top-down methods include micro/nanomachining methods and methods based on lithography as well as nonlithographic miniaturization for mostly MEMS and few NEMS fabrication. In the bottom-up methods, also referred to as nanochemistry, the devices and systems are assembled from their elemental constituents for NEMS fabrication, much like the way nature uses proteins and other macromolecules to construct complex biological systems. The bottom-up approach has the potential to go far beyond the limits of top-down technology by producing nanoscale features through synthesis and subsequent assembly. Furthermore, the bottom-up approach offers the potential to produce structures with enhanced and/or completely new functions. It allows a combination of materials with distinct chemical composition, structure, and morphology.

Tribological issues are important in MEMS/NEMS and BioMEMS/BioNEMS requiring intended and/or unintended relative motion (Bhushan, 1998, 2010, 2011). In these devices, various forces associated with the device scale down with the size. When the length of the machine decreases from 1 mm to 1 μm, the surface area decreases by a factor of a million, and the volume decreases by a factor of a billion. As a result, surface forces such as adhesion, friction, meniscus forces, and viscous forces that are proportional to surface area, become a thousand times larger than the forces proportional to the volume, such as inertial and electromagnetic forces. In addition to the consequence of a large surface-to-volume ratios, the small tolerances that these devices are designed for, make physical contacts more likely, thereby making them particularly vulnerable to adhesion between adjacent components. Slight particulate or chemical contamination present at the interface can be detrimental. Further, the small start-up forces and the torques available to overcome retarding forces are small, and the increase in resistive forces such as adhesion and friction become a serious tribological concern

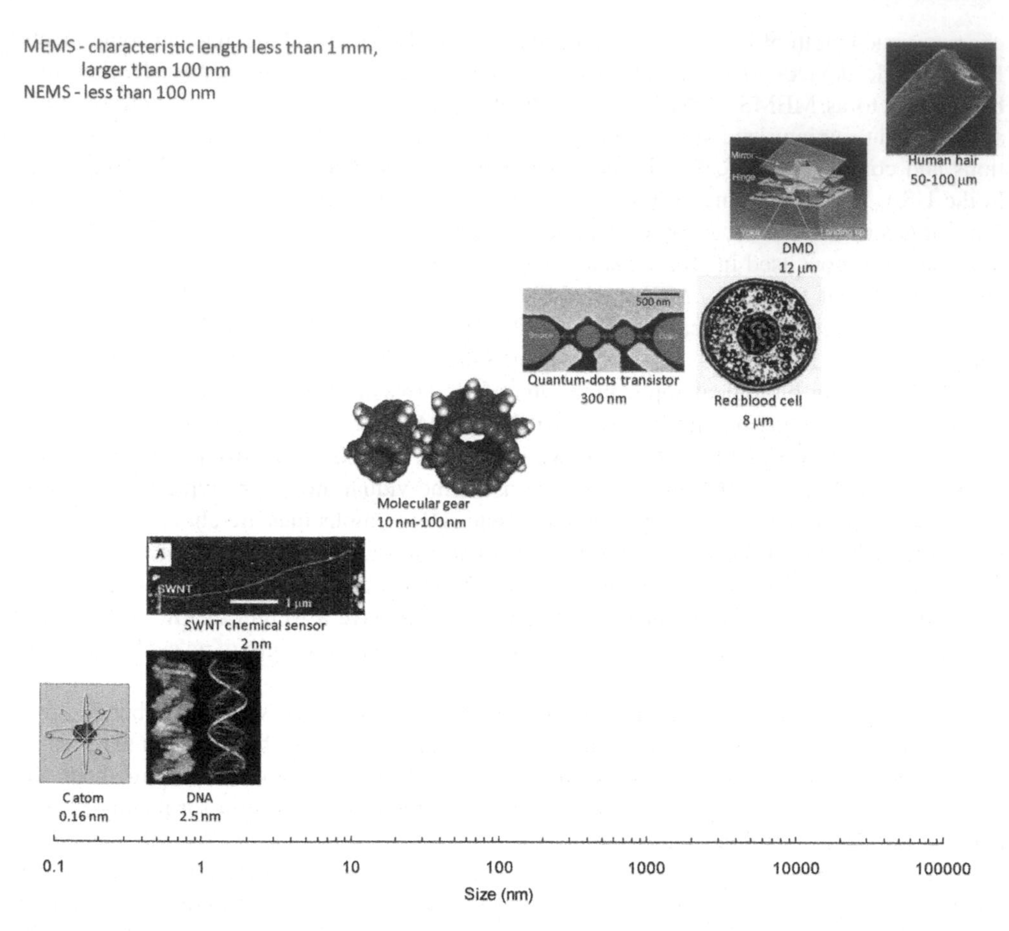

Figure 12.3.1 Characteristic dimensions of MEMS/NEMS and BioNEMS in perspective. Examples shown are of a single walled carbon nanotube (SWNT) chemical sensor (Chen *et al.*, 2004), molecular dynamic simulations of carbon-nanotube based gears (Srivastava, 2004), quantum-dot transistor obtained from van der Wiel *et al.* (2003), and DMD obtained from www.dlp.com. For comparison, dimensions and weights of various biological objects found in nature are also presented.

that limits the durability and reliability of MEMS/NEMS (Bhushan, 1998, 2010, 2011). A large lateral force required to initiate relative motion between two surfaces, large static friction, is referred to as "stiction," which has been studied extensively in the tribology of magnetic storage systems (Bhushan, 1996a, 1999, 2001a, 2003, 2011). The source of stiction is generally liquid mediated adhesion with the source of liquid being process fluid or capillary condensation of the water vapor from the environment. Adhesion, friction/stiction (static friction), wear, and surface contamination affect MEMS/NEMS and BioMEMS/BioNEMS performance and in some cases, can even prevent the devices from working.

Nanomechanical properties are scale dependent, therefore these should be measured at relevant scales.

Electrostatic micromotor
(Tai et al.,1989)

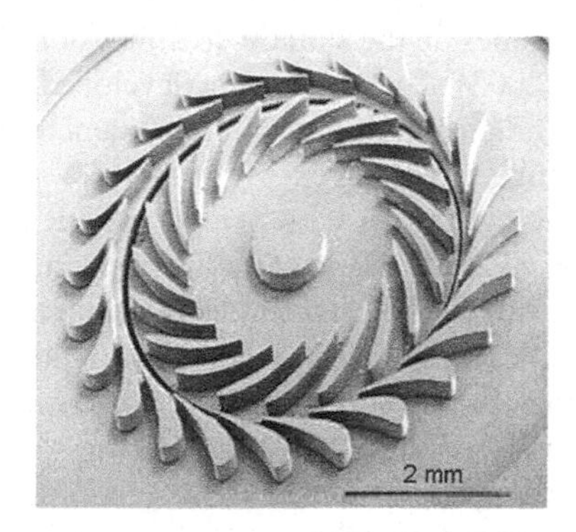

Microturbine bladed rotor and
nozzle guide vanes on the stator
(Spearing and Chen, 2001)

Six-gear chain
(www.sandia.gov)

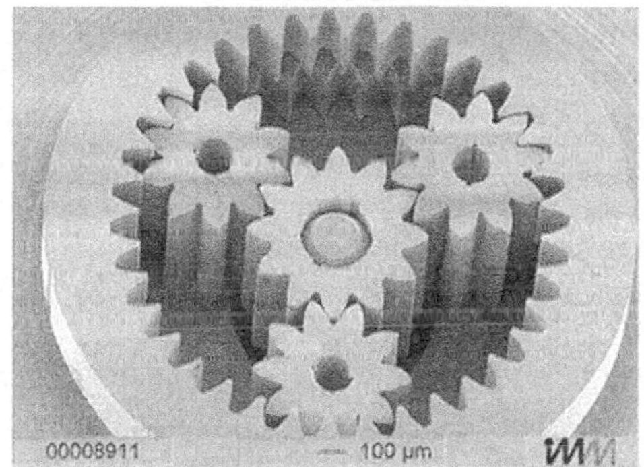

Ni-Fe wolfrom-type gear system
by LIGA (Lehr et al., 1996)

Figure 12.3.2 Examples of MEMS devices and components that experience tribological problems.

The following are some examples of MEMS/NEMS and BioMEMs and a microfabrication process that experience tribological issues.

12.3.1 MEMS

Figure 12.3.2 shows examples of several microcomponents that can encounter tribological problems. The polysilicon electrostatic micromotor has 12 stators and a 4-pole rotor and is produced by surface micromachining. The rotor diameter is 120 μm, and the air gap between the rotor and stator is 2 μm (Tai *et al.*, 1989). It is capable of continuous rotation up to speeds of 100,000 rpm. The intermittent contact at the rotor-stator interface and physical contact at the rotor-hub flange interface result in wear issues, and high stiction between the contacting surfaces limits the repeatability of operation or may even prevent the operation altogether. Next, a bulk micromachined silicon stator/rotor pair is shown with bladed rotor and nozzle guide vanes on the stator with dimensions less than a mm (Spearing and Chen, 2001; Frechette *et al.*, 2005). These are being developed for a high-temperature micro-gas turbine engine with rotor dimension of 4 to 6 mm in diameter and an operating speed of up to 1 million rpm (with a sliding velocity in excess of 500 m/s, comparable to velocities of large turbines operating at high velocities) to achieve high specific power, up to a total of about 10 W. Erosion of blades and vanes and design of the microbearings required to operate at extremely high speeds used in the turbines are some of the concerns. The ultra-short, high-speed micro hydrostatic gas journal bearings with a length to diameter ratio (L/D) of less than 0.1 are being developed for operation at surface speeds on the order of 500 m/s, which offer unique design challenges (Liu and Spakovszky, 2005). Microfabrica Inc. in the USA is developing microturbines with an outer diameter as low as 0.9 mm to be used as power sources for medical devices. They use precision ball bearings.

Next in Figure 12.3.2 is a scanning electron microscopy (SEM) micrograph of a surface micromachined polysilicon six-gear chain from Sandia National Lab. (For more examples of early version, see Mehregany *et al.*, 1988.) As an example of non-silicon components, a milligear system produced using the LIGA process for a DC brushless permanent magnet millimotor (diameter = 1.9 mm, length = 5.5 mm) with an integrated milligear box (Lehr *et al.*, 1996, 1997; Michel and Ehrfeld, 1998) is also shown. The gears are made of metal (electroplated Ni-Fe) but can also be made from injected polymer materials (e.g., Polyoxymethylene or POM) using the LIGA process. Even though the torque transmitted at the gear teeth is small, on the order of a fraction of nN m, because of the small dimensions of gear teeth, the bending stresses are large where the teeth mesh. Tooth breakage and wear at the contact of gear teeth is a concern.

Figure 12.3.3 shows an optical micrograph of a microengine driven by an electrostatically-activated comb drive connected to the output gear by linkages, for operation in kHz frequency range, which can be used as a general drive and power source to drive micromechanisms (Garcia and Sniegowski, 1995). Parts are fabricated from polysilicon. A microgear unit is used to convert reciprocating motion from a linear actuator into circular motion. Another drive linkage oriented at 90° to the original linkage, driving by another linear actuator, allows it to maintain continuous motion. The linkages are connected to the output gear through pin joints that allow relative motion.

One inset shows a polysilicon, multiple microgear speed reduction unit and its components after laboratory wear tests conducted for 600,000 cycles at 1.8% relative humidity (RH) (Tanner *et al.*, 2000). Wear of various components is clearly observed in the figure. Humidity was shown to be a strong factor in the wear of rubbing surfaces. In order to improve the wear characteristics of rubbing surfaces, vapor deposited self-assembled monolayers of fluorinated (dimethylamino) silane have been used (Hankins *et al.*, 2003). The second inset shows a comb drive with a deformed frame, which results in some fingers coming in contact. The contacting fingers can result in stiction.

Commercially available MEMS devices also exhibit tribological problems. Figure 12.3.4 shows an integrated capacitive-type silicon accelerometer fabricated using surface micromachining by Analog Devices, a couple of mm in dimension, which is used for the deployment of airbags in automobiles, and more recently for various other consumer electronics market (Core *et al.*, 1993; Sulouff, 1998). The central suspended beam mass (about 0.7 μg) is supported on the four corners by spring structures. The central beam has interdigitated cantilevered electrode fingers (about 125 μm long and 3 μm thick) on all four sides that alternate with those

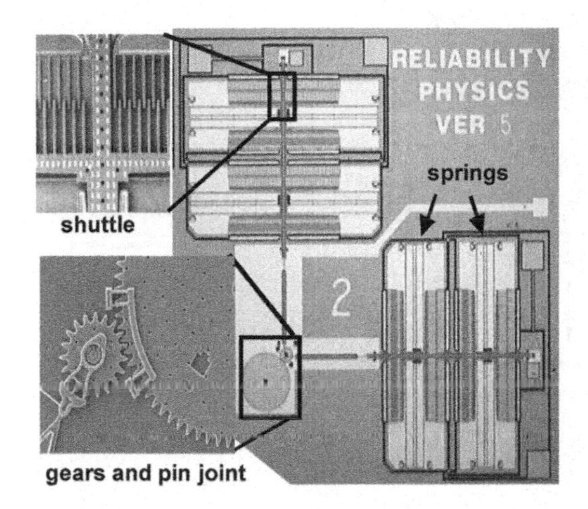

Microengine driven by electrostatically-actuated comb drive
Sandia Summit Technologies (www.mems.sandia.gov)

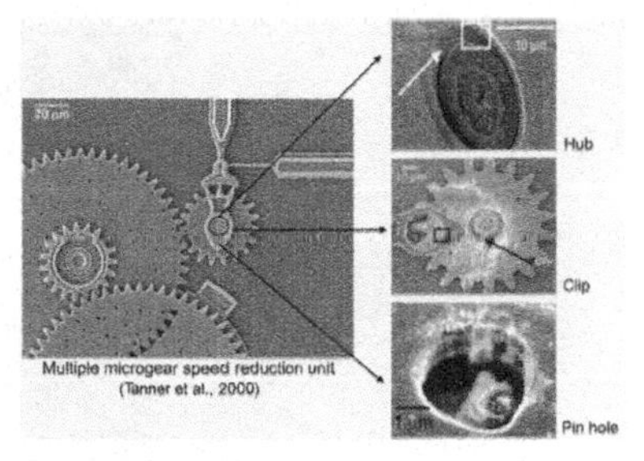

Microgear unit can be driven at speeds up to 250,000 RPM. Various sliding components are shown after wear test for 600k cycles at 1.8% RH (Tanner et al., 2000)

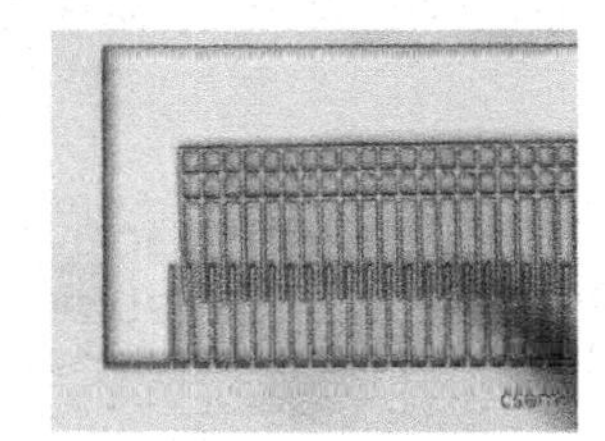

Stuck comb drive

Figure 12.3.3 Optical micrograph of a microengine driven by an electrostatically-actuated comb drive (microengine) fabricated by Sandia Summit Technologies. Reproduced with permission from Garcia, E.J. and Sniegowski, J.J. (1995), "Surface Micromachined Microengine," *Sensors and Actuators A* **48**, 203–214. Copyright 1995. Elsevier. One inset shows a polysilicon microgear speed reduction unit after laboratory wear test for 600,000 cycles at 1.8% relative humidity (Tanner *et al.*, 2000). The second inset shows a stuck comb drive (CSEM).

of the stationary electrode fingers as shown, with about a 1.3 μm gap. Lateral motion of the central beam causes a change in the capacitance between these electrodes, which is used to measure the acceleration. Stiction between the adjacent electrodes as well as stiction of the beam structure with the underlying substrate, under isolated conditions, is detrimental to the operation of the sensor (Core *et al.*, 1993; Sulouff, 1998). Wear during unintended contacts of these polysilicon fingers is also a problem. A molecularly thick diphenyl siloxane lubricant film, resistant to high temperatures and oxidation, is applied by a vapor deposition process on the electrodes to reduce stiction and wear (Martin and Zhao, 1997). As sensors are required to sense low g accelerations, they need to be more compliant and stiction becomes even a bigger concern.

Figure 12.3.4 also shows a cross-sectional view of a typical piezoresistive type pressure sensor, which is used for various applications including manifold absolute pressure (MAP) and tire pressure measurements in automotive applications, and disposable blood pressure measurements. The sensing material is a diaphragm formed on a silicon substrate, which bends with applied pressure (Smith, 1997; Parsons, 2001). The deformation causes a change in the band structure of the piezoresistors that are placed on the diaphragm, leading to a change in the resistivity of the material. The MAP sensors are subjected to drastic conditions – extreme

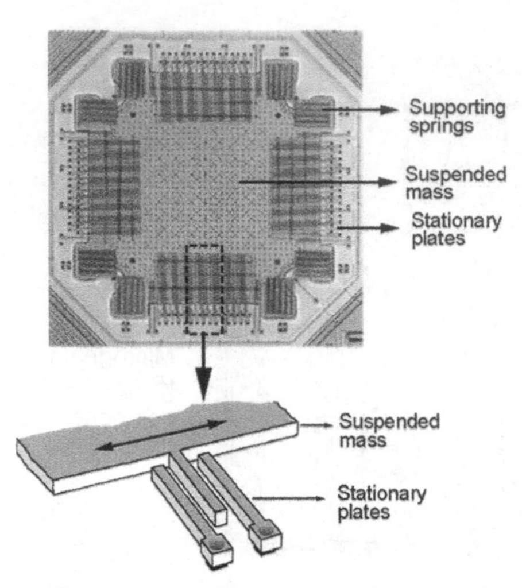

Capacitive type silicon accelerometer for automotive sensory applications (Sulouff, 1998)

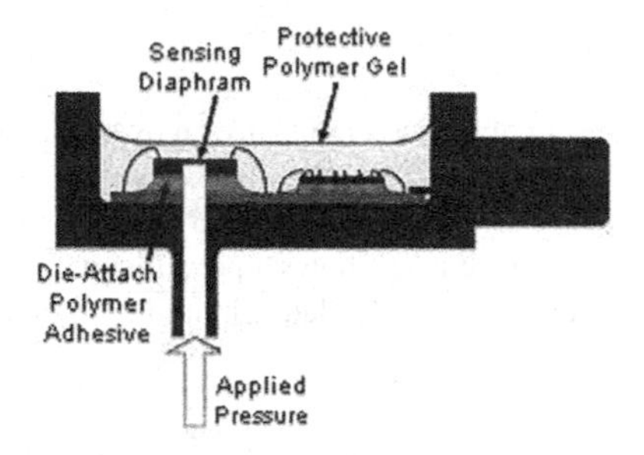

Piezoresistive type pressure sensor (Parsons, 2001)

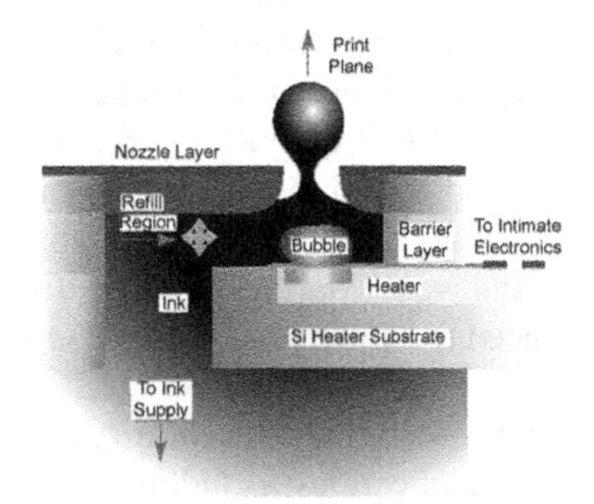

Thermal inkjet printhead (Baydo and Groscup, 2001)

Figure 12.3.4 Examples of commercial MEMS that experience tribological problems.

temperatures, vibrations, sensing fluid, and thermal shock. Fluid under extreme conditions could cause corrosive wear. Fluid cavitation could cause erosive wear. The protective gel encapsulent generally used can react with sensing fluid and result in swelling or dissolution of the gel. Silicon cannot deform plastically, therefore any pressure spikes leading to deformation past its elastic limit will result in fracture and crack propagation. Pressure spikes could also cause the diaphragm to delaminate from the support substrate. Finally, cyclic loading of the diaphragm during use can lead to fatigue and wear of the silicon diaphragm or its delamination.

The bottom schematic in Figure 12.3.4 shows a cross-sectional view of a thermal printhead chip (on the order of 10 to 50 cm^3 in volume) used in inkjet printers (Baydo and Groscup, 2001). They comprise of a supply of ink and an array of elements with microscopic heating resistors on a substrate mated to a matching array of ink-injection orifices or nozzles (about 70 μm in diameter) (Aden *et al.*, 1994; Le, 1998; Lee, 2003). In each element, a small chamber is heated by the resistor where a brief electrical impulse vaporizes part of the ink and creates a tiny bubble. The heaters operate at several kHz and are therefore capable of high-speed printing. As the bubble expands, some of the ink is pushed out of the nozzle onto the paper. When the bubble pops, a vacuum is created and this causes more ink from the cartridge to move into the printhead. Clogged ink ports are the major failure mode. There are various tribological concerns (Aden *et al.*, 1994). The surface of the printhead where the ink is shot out towards the paper can get scratched and damaged as a result of countless trips back and forth across the pages, which are somewhat rough. As a result of repeated heating and cooling, the heated resistors expand and contract. Over time, these elements will experience fatigue and may eventually fail. Bubble formation in the ink reservoir can lead to cavitation erosion of the chamber, which occurs when bubbles formed in the fluid become unstable and implode against the surface of the solid and impose impact energy on that surface. Fluid flow through nozzles may cause erosion and ink particles may also cause abrasive wear. Corrosion of the ink reservoir surfaces can also occur as a result of exposure of ink at high temperatures as well as due to ink pH. The substrate of the chip consists of silicon with a thermal barrier layer followed by a thin film of resistive material and then conducting material. The conductor and resister layers are generally protected by an overcoat layer of a plasma-enhanced chemical vapor deposition (PECVD) α-SiC:H layer, 200–500 nm thick (Chang *et al.*, 1991).

Figure 12.3.5 shows two digital micromirror devices (DMD) pixels used in digital light processing (DLP) technology for digital projection displays in computer projectors, high definition television (HDTV) sets, and movie projectors (Hornbeck and Nelson, 1988; Hornbeck, 1999, 2001). The entire array (chip set) consists of a large number of oscillating aluminum alloy micromirrors as digital light switches which are fabricated on top of a complementary metal-oxide-semiconductor (CMOS) static random access memory integrated circuit. The surface micromachined array consists of half a million to more than two million of these independently controlled reflective micromirrors, each about 12 μm square and with 13 μm pitch which flip backward and forward at a frequency of on the order of 5000 to 7000 times a second as a result of electrostatic attraction between the micromirror structure and the underlying electrodes. For binary operation, the micromirror/yoke structure mounted on torsional hinges is oscillated $\pm 10^\circ$ (with respect to the plane on the chip set), and is limited by a mechanical stop. Contact between cantilevered spring tips at the end of the yoke (four present on each yoke) with the underlying stationary landing sites is required for true digital (binary) operation. Stiction and wear during contact between aluminum alloy spring tips and landing

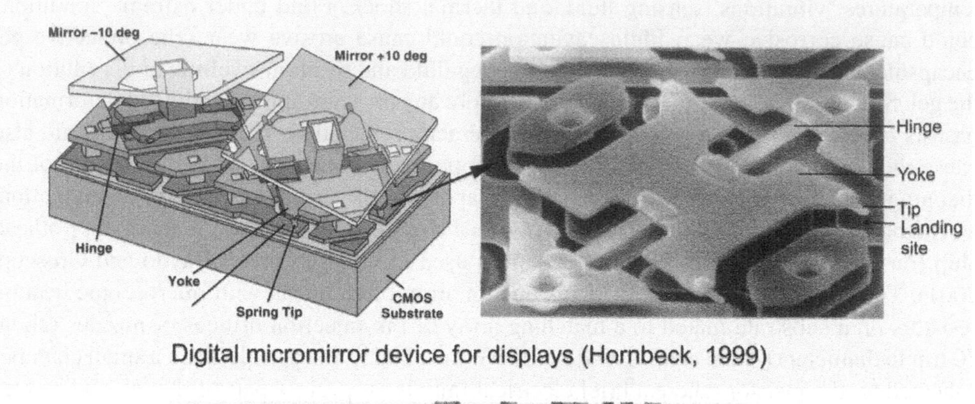

Digital micromirror device for displays (Hornbeck, 1999)

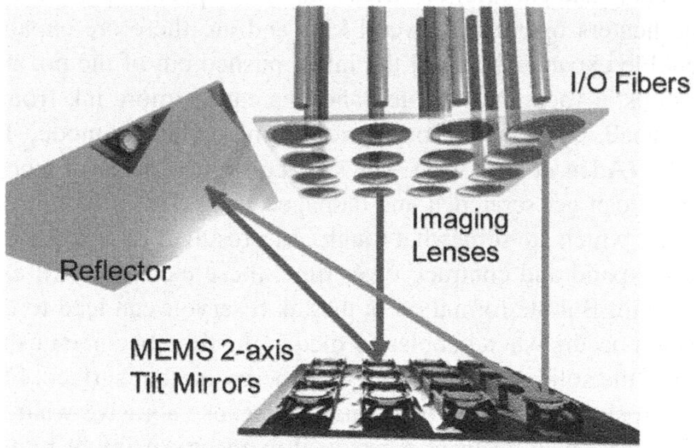

Tilt mirror arrays for switching optical signal in input and output fiber arrays in optical crossconnect for telecom (Aksyuk et al., 2003).

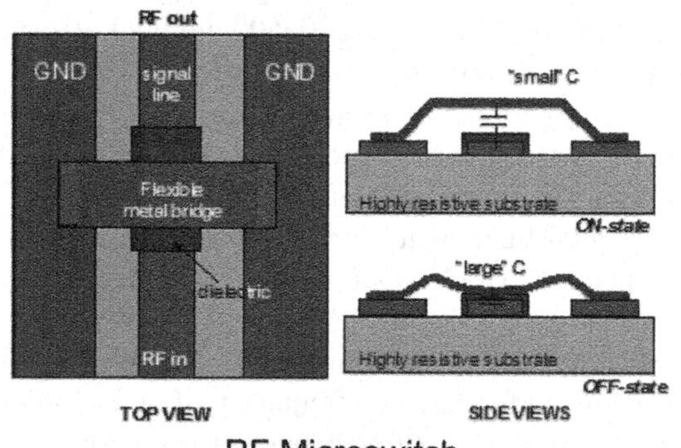

RF Microswitch
(Courtesy IMEC, Belgium)

Figure 12.3.5 Examples of two commercial MOEMS and one RF-MEMS device that experience tribological problems.

sites, hinge memory (metal creep at high operating temperatures), hinge fatigue, shock and vibration failure, and sensitivity to particles in the chip package and operating environment are some of the important issues affecting the reliable operation of a micromirror device (Henck, 1997; Douglass, 1998, 2003; Liu and Bhushan, 2004a, b). A vapor phase deposited self-assembled monolayer of the fatty acid perfluorodecanoic acid (PFDA) on surfaces of tip and landing sites is used to reduce stiction and wear (Hornbeck, 1997; Robbins and Jacobs, 2001). However, these films are susceptible to moisture, and to keep moisture out and create a background pressure of PFDA, a hermetic chip package is used. The spring tip is used in order to use the spring stored energy to pop up the tip during pull-off. A lifetime estimate of over one hundred thousand operating hours with no degradation in image quality is the norm. At a mirror modulation frequency of 7 kHz, each micromirror element needs to switch about 2.5 trillion cycles.

Figure 12.3.5 also shows a schematic of a 256 × 256-port large optical cross-connects, introduced in 2000 by Glimmerglass, Hayward, California, for optical telecommunication networks in order to be able to rapidly manipulate a larger number of optical signals (Aksyuk *et al.*, 2003). This optical microswitch uses 256 or more movable mirrors on a chip for switching a light beam from an input fiber to a few output fibers. The mirrors are made of gold-coated polysilicon and are about 500 μm in diameter. The reliability concerns are the same as those just described for DMDs. To minimize stiction, the chipset is hermetically sealed in dry nitrogen (90% N_2, 10% He).

Figure 12.3.5 also shows a schematic of an electrostatically-actuated capacitive-type RF microswitch for switching of RF signals at microwave and low frequencies (DeWolf and van Spengen, 2002). It is a membrane type and consists of a flexible metal (Al) bridge that spans the RF transmission line in the center of a coplanar waveguide. When the bridge is up, the capacitance between the bridge and RF transmission line is small, and the RF signal passes without much loss. When a DC voltage is applied between the RF transmission line and the bridge, the latter is pulled down until it touches a dielectric isolation layer. The large capacitance thus created shorts the RF signal to the ground. The failure modes include creep in the metal bridge, fatigue of the bridge, charging and degradation of the dielectric insulator, and stiction of the bridge to the insulator (DeWolf and van Spengen, 2002; Suzuki, 2002). The stiction occurs due to capillary condensation of water vapor from the environment, van der Waals forces, and/or charging effects. If the restoring force in the bridge of the switch is not large enough to pull the bridge up again after the actuation voltage has been removed, the device fails due to stiction. Humidity-induced stiction can be avoided by hermetically sealing the microswitch. Some roughness of the surfaces reduces the probability of stiction. Selected actuation waveforms can be used to minimize charging effects.

12.3.2 NEMS

Probe-based data recording technologies have been explored for ultra-high areal density recording where the probe tip (with a radius of about 5 nm) is expected to be scanned at velocities up to 100 mm/s. Major techniques include – thermomechanical (Vettiger *et al.*, 1999), phase change (Bhushan and Kwak, 2007), and ferroelectric recording (Bhushan and Kwak 2008; Kwak and Bhushan, 2008).

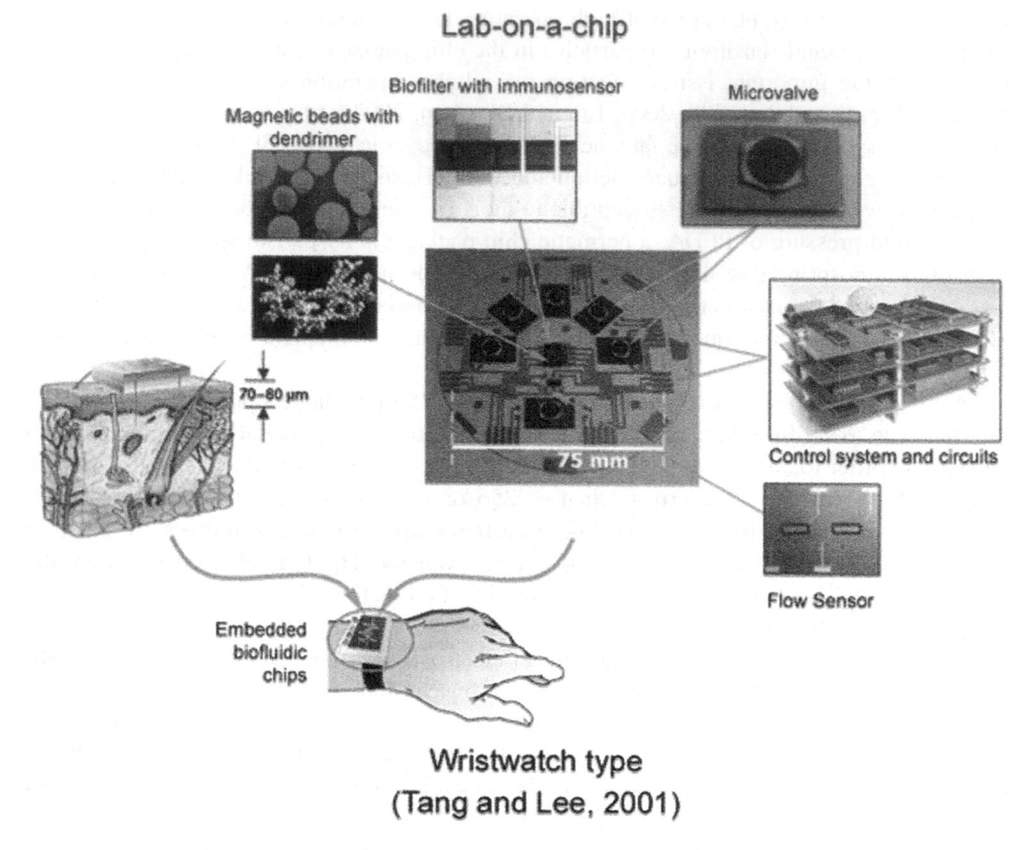

Figure 12.3.6 MEMS based biofluidic chip, commonly known as a lab-on-a-chip, that can be worn like a wristwatch.

12.3.3 BioMEMS

An example of a wristwatch type biosensor based on microfluidics referred to as a lab-on-a-chip system is shown in Figure 12.3.6 (Tang and Lee, 2001; van der Berg, 2003). These systems are designed to either detect a single or a class of (bio)chemical(s), or for system-level analytical capabilities for a broad range of (bio)chemical species known as a micro total analysis system (μTas), and have the advantage of incorporating sample handling, separation, detection, and data analysis onto one platform. The chip relies on microfluidics and involves the manipulation of tiny amounts of fluids in microchannels using microvalves. The test fluid is injected into the chip generally using an external pump or syringe for analysis. Some chips have been designed with an integrated electrostatically-actuated diaphragm type micropump. The sample, which can have volume measured in nanoliters, flows through microfluidic channels via an electric potential and capillary action using microvalves (having various designs including membrane type) for various analyses. The fluid is preprocessed and then analyzed using a biosensor.

If the adhesion between the microchannel surface and the biofluid is high, the biomolecules will stick to the microchannel surface and restrict flow. In order to facilitate flow, microchannel surfaces with low bioadhesion are required. Fluid flow in polymer channels can produce triboelectric surface potential, which may affect the flow.

12.3.4 Microfabrication Processes

In addition to in-use stiction, stiction issues are also present in some processes used for the fabrication of MEMS/NEMS (Bhushan, 2010). For example, the last step in surface micromachining involves the removal of sacrificial layer(s) called release since the microstructures are released from the surrounding sacrificial layer(s). The release is accomplished by an aqueous chemical etch, rinsing, and drying processes. Due to meniscus effects as a result of wet processes, the suspended structures can sometimes collapse and permanently adhere to the underlying substrate as shown in Figure 12.3.7 (Guckel and Burns, 1989). Adhesion is caused by water molecules adsorbed on the adhering surfaces and/or because of formation of adhesive bonds by silica residues that remain on the surfaces after the water has evaporated. This so-called release stiction is overcome by using dry release methods, such as CO_2 critical point drying or sublimation drying (Mulhern *et al.*, 1993). CO_2 at high pressure is in a supercritical state and becomes liquid. Liquid CO_2 is used to remove wet etchant, and then it is converted back to gas phase.

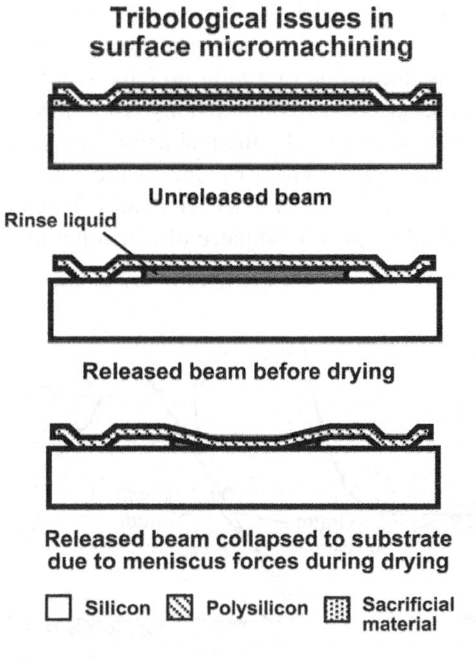

Figure 12.3.7 Microfabrication via surface micromachining.

12.4 Material Processing

The desired shape and accuracy of machine parts is obtained by the removal of material (material cutting) or by plastic deformation of material (metal forming) (Schey, 1977; Samuels, 1982; Booser, 1984; Anonymous, 1989c; Bhushan, 1996a, 2001a, Shaw, 1996, 1997). In material cutting, material is removed either by a cutting tool in the form of relatively large chips or by abrasives in the form of relatively small chips. The material-cutting processes that involve cutting tools include turning to produce cylindrical surfaces; milling to produce flat surfaces and surfaces of complex geometry; and drilling, boring, and reaming to produce round holes. The material-cutting processes that involve abrasives include grinding and free-abrasive and fixed-abrasive lapping or polishing. In all of these material-removal processes, the general removal mechanisms at the tip of the cutting edge are the same in all processes. The metal-forming processes include forging, rolling, drawing of wire, bar or tube, extrusion, and sheet-metal working. In most material processing, cutting fluids are used (Bastian, 1951; Braithwaite, 1967; Booser, 1984; Shaw, 1996, 1997). These cutting fluids also reduce friction at the cutting interface.

12.4.1 Cutting Tools

Cutting tools are used to cut, shape, and form bars, plates, sheets, castings, forgings, etc., to produce engineering components (Shaw, 1997). Figure 12.4.1 shows a schematic of a partially formed chip produced by moving a workpiece against a stationary tool. Tool wear usually takes place on the face or the flank of a cutting tool. Face or crater wear results from a chip moving across the face of the tool, whereas flank wear results from the rubbing action on the freshly formed surface of the job. The extent and location of crater wear are considerably affected by the formation of a built-up edge composed of highly strained and hardened fragments of material (Tipnis, 1980; Anonymous, 1989c; Shaw, 1997). Tool wear occurs by adhesive, abrasive, chemical (by thermal diffusion), and electrochemical wear. In general, the life of a cutting tool is judged by one of the following criteria: (1) complete failure of the tool; (2) cutting time for material removal to a predetermined crater depth or flank-wear land width; and (3) loss of workpiece dimensional tolerance and degradation of

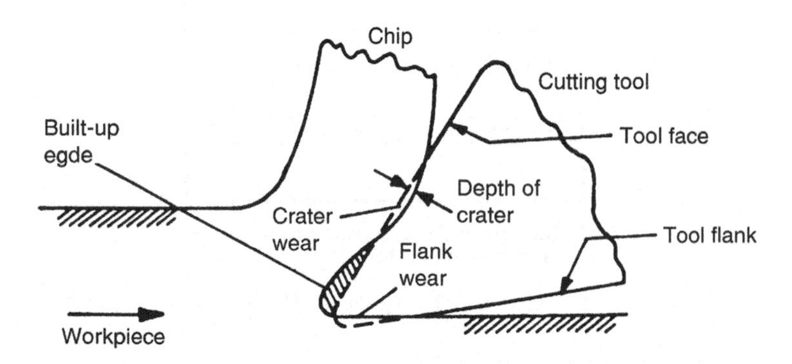

Figure 12.4.1 Schematic of a partially formed chip and crater (or face) and flank wear in cutting tools.

surface finish. The most important properties of a cutting tool material are its hot hardness (i.e., resistance to softening under temperatures generated at the cutting edge of the tool), toughness, and chemical stability and reactivity. Other relevant properties are elastic modulus, rupture strength, compressive strength, and coefficient of thermal expansion. Cutting-tool life is a most important factor in the economics of production.

Tools can be made from anything from an elastomer to a diamond, but a few tool materials dominate. Of the different cutting-tool materials in use today, about 40% are high-speed steels (HSS) (a class of tool steel), about 10% are cemented carbides (cermets), about 30% are carbon, alloy, and stainless steels, 5% are ceramics including diamonds and cubic boron nitride, and the rest are other materials (cast alloys, cast irons, nonferrous metals, and elastomers) (Budinski, 1980; Shaw, 1997). According to some estimates, 60% of all carbide tools are coated grades.

A comparison of hot hardness (hardness as a function of temperature) for commonly used tool materials is presented in Figure 12.4.2. Carbon steels are confined to hand tools, light-duty woodworking tools, and a few minor industrial applications on metals. HSS and cemented-carbide tools are primarily used for most metal cutting and processing of hard materials. Various ceramics are reserved for specialized applications. Coated tools exhibit considerable improvement in tool life and permit faster feed rates, which has led to their use becoming

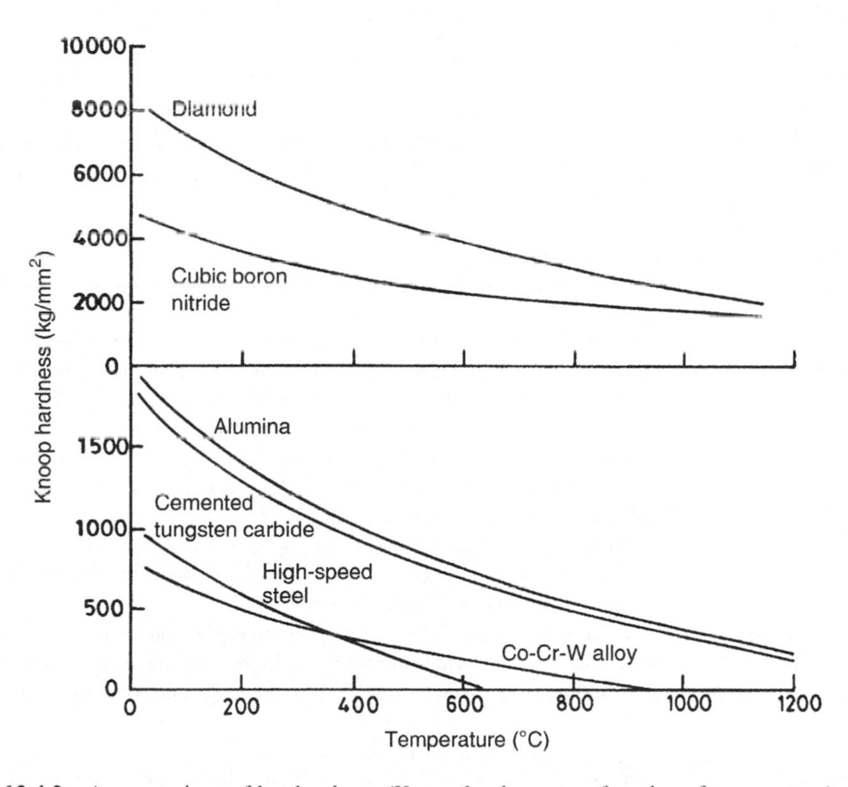

Figure 12.4.2 A comparison of hot hardness (Knoop hardness as a function of temperature) of high-speed steel, cemented tungsten carbide, alumina, cast Co-Cr-W alloy, cubic boron nitride, and diamond tool materials.

Table 12.4.1 Approximate grouping of tool materials for machining operations.

Tool class	Machining operation						
	Turning, facing, boring	Forming, grooving	Planing, shaping, breaching	Milling, hobbing	Drilling, reaming, tapping	Tapping	Sawing
High-speed steel (HSS)	√	+	+	+	+	+	+
Cast Co-Cr-W alloys	Δ	Δ	Δ	Δ	Δ	0	0
Cemented carbide	+	√	√	√	**X**	0	Δ
Alumina/composite	**X**	0	0	Δ	0	0	0
Cubic boron nitride/diamond	Δ	0	0	0	0	0	0

Key: most commonly used, +; frequently used, √; occasionally used, **X**; rarely used, Δ; and not used, 0.

more and more common. Each machining operation and type of work material places a specific demand on a cutting tool. An approximate grouping of bulk tool materials applicable to different machining operations is presented in Table 12.4.1, and general uses for various tool material classes are given in Table 12.4.2.

Carbon and alloy steels usually contain less than 5% total alloy content. This, combined with lower quality-control standards in their manufacture, makes them lower in cost than high-speed tool steels. However, these steels simply cannot equal the hardening and wear characteristics of tool steels. Low-carbon steels (such as AISI 1010 and AISI 1020) can be hardened by cold

Table 12.4.2 General uses of tool material classes.

Tool class	Use
High-speed steel (HSS)	Major tool material-relative toughness plus retention of hardness at relatively high temperatures allows high cutting speeds, fine cutting edges, and use in rough conditions; relatively low cost
Cast Co-Cr-W alloys	Retains hardness to higher temperatures than HSS; less tough; cut more difficult materials or at higher speed; higher cost than HSS
Cemented carbide	Major tool material-higher hardness plus retention of hardness at much higher temperatures allows much higher cutting speeds and lower wear rates; less tough but tough enough for rough usage if edge not too fine; higher cost and used as tool tip only
Alumina/composite	Retains hardness to higher temperature than cemented carbide and much lower wear rate which allows much higher cutting speeds; much less tough-use restricted to tools with strong edge and to good cutting conditions; main use turning and facing cast iron at 8 to 10 m/s; cost similar to cemented carbide
Cubic boron nitride/diamond	Much higher hardness and less tough; very low wear rates; very high cost; mainly high-speed finishing of nonferrous metals for good surface finish and very close dimensional tolerances

working to less than 200 Brinell. Medium-carbon grades (such as AISI 1040 and AISI 1060) can be hardened to 500–600 Brinell by direct-hardening or flame-hardening techniques. Alloy steels can be carburized (such as AISI 4320 and AISI 4620) or direct-hardened (such as AISI 4130 and AISI 4140) to a hardness no more than about 58 HRC.

Nearly 90% of the HSS tools are M series (with molybdenum as the major alloying element); the remaining are T series (with tungsten as the major alloying element). Although T-HSS are somewhat superior in wear resistance, they are more difficult to grind. Cobalt-containing M-HSS can be heat-treated to a higher hardness, 900–940 HV (67–70 HRC) versus 800–860 HV (64–66 HRC) for the non-cobalt-containing series. When tools are used in a corrosive environment, it is often necessary to use stainless steels, such as in the food industry and the chemical process industry. Of the various stainless steels, Cr-Ni steels with a martensitic structure, such as 440C, are the most useful as tool materials. Type 440C is capable of being quench-hardened to 58–60 HRC, and its wear characteristics are similar to those of air-hardening cold-working tool steel such as A2.

Cast Co-Cr-W alloys often contain some molybdenum and boron. In addition, vanadium, tantalum, and columbium as alloying elements and manganese and silicon as deoxidizers are generally present. The hardness of cast alloys ranges from 650 to 800 HV. The cast alloys can withstand higher temperatures than HSS and hence provide properties that are in between those of HSS and sintered carbides.

Sintered or cemented carbide tool materials are made of finely divided carbide particles of tungsten, titanium, tantalum, niobium, and other refractory metals bonded with cobalt, nickel, nichrome, molybdenum, or even steel alloys and are produced by powder metallurgy processes. The most commonly used cemented carbide is tungsten carbide with about 6 wt % cobalt binder. The most important property of cemented carbides is their hardness. The carbide particles that make up the major portion of these composites are harder than any metal. For example, the hardness of tungsten carbide, which is the most commonly used carbide, is about 2000 HV.

Ceramic tool materials primarily consist of polycrystalline alumina (Al_2O_3), cubic boron nitride (CBN), and diamonds. The ceramic composite tool materials typically are composites of Al_2O_3 and TiC or WC. Oxide ceramics and cermets are manufactured by either sintering or hot pressing. The principal elevated-temperature properties of the oxide ceramic tool materials are high hardness, chemical inertness, and wear resistance. However, these materials are relatively brittle (low transverse rupture strength) compared with HSS and cemented carbides. The hardness of CBN is two to three times that of sintered carbides. These materials can withstand very high temperatures without appreciable loss of hardness. This makes it possible to machine high-temperature Ni-based alloy work materials at tenfold the speeds normally employed with carbide tools. However, CBN is relatively brittle. Thin CBN wafers are compacted on carbide substrates to lend strength and to minimize cost. Natural and synthesized polycrystalline diamond compacts are used as tool materials. Diamond tools consist of 0.5 mm thick layers of sintered polycrystalline diamond bonded on a cemented tungsten carbide substrate or true diamond coating material. These tools are not suitable for machining ferrous alloys because at high temperatures the diamond tends to react with the carbon in steel and cast irons.

Coatings (typically 2–25 μm thick) of various ceramic materials, such as TiC, TiN, TiC_xN_y, TiO_xN_y, Al_2O_3, HfC, ZrC, TaC, HfN, and ZrN, on high-speed-steel and cemented-carbide substrates (usually WC-Co or WC-TiC-TaC-Co compacts), have been deposited successfully to increase tool life. The coatings have been deposited by various vapor-deposition techniques: activated reactive evaporation (ARE); ion plating; sputtering; and chemical vapor deposition (CVD). The use of CVD techniques for such coatings (e.g., TiC) on cemented tungsten carbide tools is well established. However, the CVD process is not suited to coat high-speed tools because of the high deposition temperature (1000–1200°C) of the process, which results in metallurgical changes and distortion of the tool. The low deposition temperatures of evaporation, ion plating, and sputtering are particularly attractive. Various surface treatments such as carburizing and nitriding of some steel tools and ion implantation are also found to improve tool life considerably.

12.4.2 Grinding and Lapping

Grinding is a versatile process that is used to manufacture parts that require a good surface finish (on the order of 1 μm peak to valley) and dimensional accuracy. Grinding is performed with small, extremely hard abrasive particles (grits) usually bonded together in the form of a wheel in the presence of a cutting fluid (Shaw, 1996, 1997; Bhushan, 1996a). The wheel can be either vitrified or resin or metal bonded. The most frequently used abrasives are MgO, SiO_2, Al_2O_3, SiC, cubic boron nitride (CBN), and diamond. Figure 12.4.3a shows a surface grinding operation and an undeformed shape of chip.

Lapping (finishing or polishing) is a fine finishing process and is usually the last stage in the finishing of a component (Samuels, 1982; Bhushan, 1996a). It is used to produce surfaces of

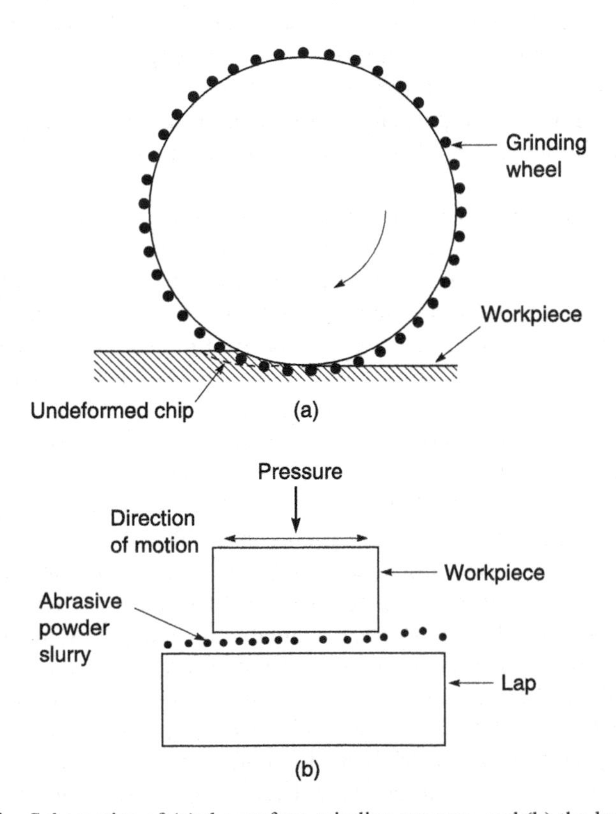

Figure 12.4.3 Schematics of (a) the surface grinding process, and (b) the lapping process.

extremely good finish (5–100 nm peak to valley) and flatness. In the free abrasive lapping, the workpiece is usually mounted on a steel puck with adhesive and moved across the face of the lap under normal ambient pressure in the presence of abrasive powder slurry, Figure 12.4.3b. The lap is usually made of soft materials such as bonze, tin, copper, or cast iron. The abrasive is suspended in a liquid carrier (e.g., ethylene glycol or a lubricant such as olive oil). Some of the abrasive gets embedded into the lap. Common abrasives used in lapping are Cr_2O_3, Al_2O_3, SiC, and diamond of various grit sizes ranging from about 0.05 μm to several μm.

In fixed-abrasive lapping, a lapping tape (abrasive impregnated tape) is rubbed against the workpiece. The lapping tapes are flexible and conform well to the workpiece. Lapping tapes normally use Cr_2O_3, Al_2O_3, and SiC of various grit sizes ranging from about 1.5 μm to 14 μm held in an organic binder. The tape substrate is typically acetate or polyester film of about 25 μm in thickness with a total thickness of about 40-45 μm (Bhushan, 1996a).

12.4.3 Forming Processes

Schematics of basic forms of wire, bar, and tube drawing operation, extrusion, and shearing (punching, blanking, or slitting) are shown in Figure 12.4.4 (Schey, 1977; Booser, 1984). The performance of forming punches and dies is usually dictated by the amount of wear, which is influenced by the kind and size of workpiece, the sharpness of the die radii, die construction and finish, lubrication, and hardness of the die material.

The conventional materials used for making dies and punches include alloy cast irons (typical carbon content 2.8–3.5% C), tool steels (typically D2, W1, O2, and A2 grades), and sintered carbides (typically WC-Co) (Neale, 1973). The application of hard coatings and surface treatments results in marked improvements in the useful life of forming tools. Examples include, a TiN coating applied by ion plating on punches, dies, and taps, and TiC, TiN, or TiC/TiN coatings by CVD to the chromium steel punches in punch and die assemblies. TiB_2 coating by CVD to A-6 and H-11 steel injection molding dies have been shown to improve their wear life (Bhushan and Gupta, 1997). Surface treatments such as carburizing, nitriding, boriding, and aluminizing of steels and ion implantation also have been used to improve tool life (Bhushan and Gupta, 1997). Surface treatment by boriding has been found to lead to significant improvements in wear life for deep drawing tools. Ion nitriding, boriding, and aluminizing treatments are helpful in extending the wear life of barrels and extrusion screws in polymer injection-molding machines. Dies of cemented carbide used for the drawing of copper or steel wire for metal-forming operations exhibit longer life as a result of ion implantation with carbon or nitrogen.

12.4.4 Cutting Fluids

A cutting tool generates high temperatures by the deformation of metal and by friction between the chip and the tool. The temperature rise typically ranges from 350 to 1000°C or even higher. The primary function of any cutting fluid is to dissipate the frictional heat away to keep the interface cool, especially in high-speed cutting operations, such as in turning (Bastian, 1951; Braithwaite, 1967; Booser, 1984; Shaw, 1996, 1997). The other function is to provide lubrication. For efficient cooling, the heat transfer properties of the cutting fluid should be good. Other considerations associated with the cutting fluid include lubricity, corrosion prevention, and health and safety hazards.

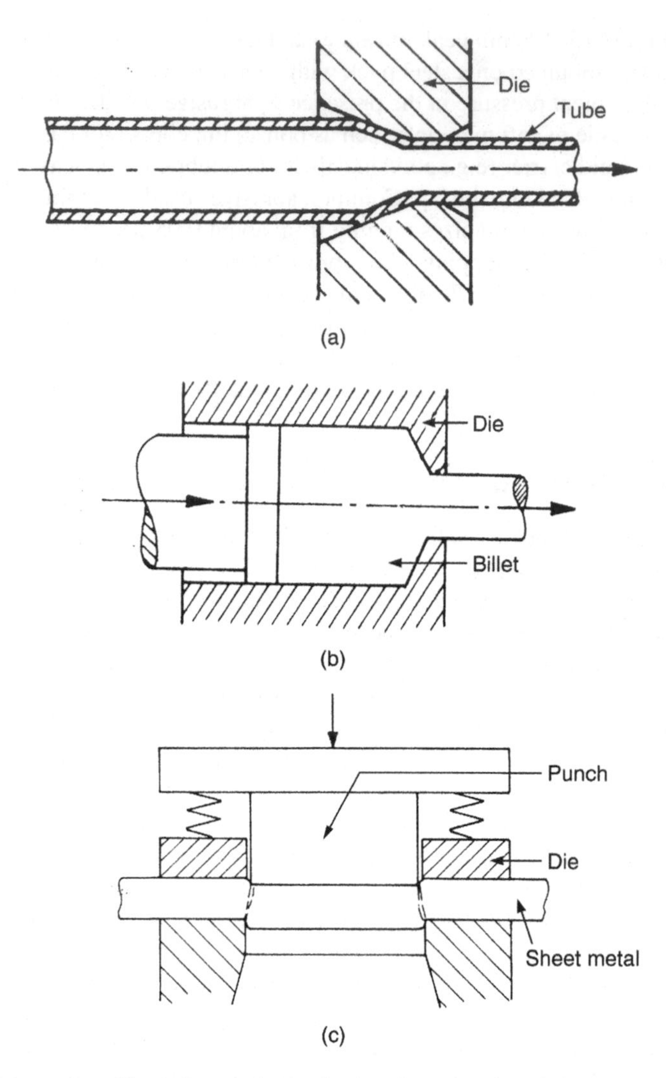

Figure 12.4.4 Schematics of basic forms of: (a) tube drawing operation; (b) extrusion; and (c) punching of a metal sheet of billet through the die operation.

Cutting fluids contain mineral oils, fatty oils, or a combination of these, mixtures of mineral oil and emulsifiers (soluble oils) added to water or synthetic fluids (organic and inorganic salts dissolved in water).

12.5 Industrial Applications

Tribology is extremely important in numerous industrial applications requiring relative motion, for example, automobiles, aircrafts, railroads and magnetic storage devices (O'Connor *et al.*, 1968; Neale, 1973; Peterson and Winer, 1980; Booser, 1983; Bhushan, 1996a, 2001a;

Bhushan and Gupta, 1997). Since the 1980s, microdevices involving mechanical components have been produced using microfabrication technologies. The above applications are briefly described next.

12.5.1 Automotive Engines

Internal combustion (IC) engines are the almost exclusive choice for use in automobiles (Rogowski, 1953; Taylor, 1966, 1968; Crouse, 1970; Judge, 1972; Heywood, 1988; Anonymous, 1997). In reciprocating IC engines, fuel is burned and its combustion power is converted from a linear reciprocation motion of the piston in its cylinder through a connecting rod to a rotating motion in the crankshaft. The crankshaft is connected to the drive shaft and a transmission which provides rotary motion at a desired speed to the rubber tires.

As just stated, in reciprocating IC engines, the piston moves back and forth in a cylinder and transmits power through a connecting rod and crank mechanism to the drive shaft. The cyclical piston motion produces a steady rotation in the crankshaft. The piston comes to rest at the top-center (TC) crank position and bottom-center (BC) crank position when the cylinder volume is a minimum or maximum, respectively. The majority of reciprocating engines operate on a four-stroke cycle, whereas some engines operate on a two-stroke cycle. In the four-stroke cycle, each cylinder requires four strokes of its piston, Figure 12.5.1, two revolutions of the crankshaft to complete the sequence of events which provide a power stroke (Rogowski, 1953). In the intake stroke, which starts with the piston at TC and ends with the piston at BC, fresh mixture is drawn into the cylinder by downward movement of the piston with the inlet

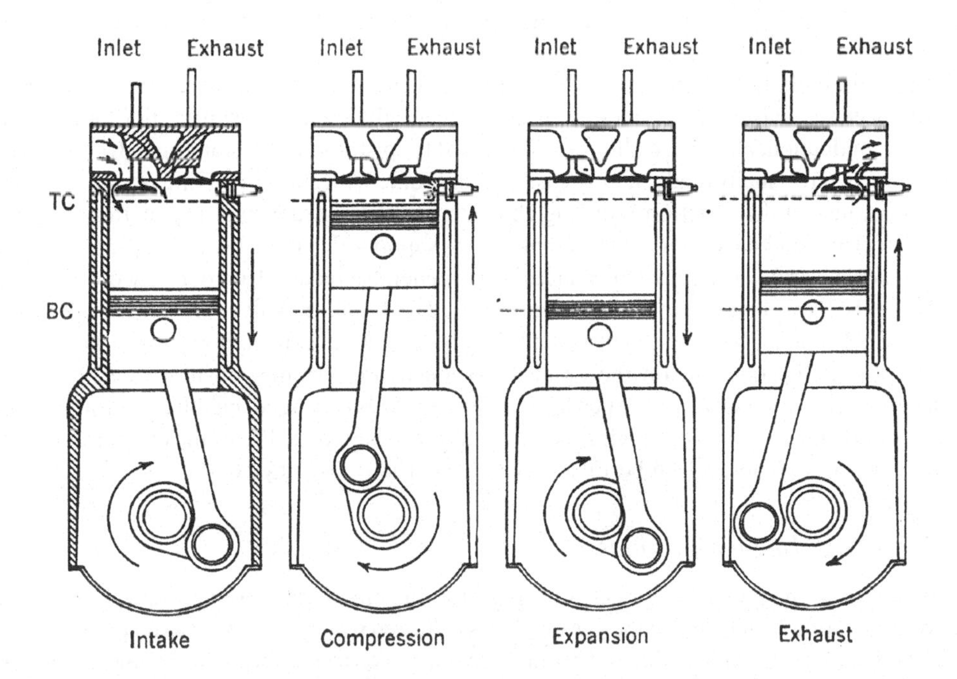

Figure 12.5.1 Schematic of the four-stroke operation cycle for an IC engine.

valve open and the exhaust valve closed. In the compression stroke, with all valves closed, the piston reverses direction from the intake stroke and compresses the trapped mixture. In the power stroke or expansion stroke, as the piston approaches the top of its compression stroke, the air-fuel mixture burns and develops the high pressure which powers the piston downward. Piston rings provide a seal against the cylinder wall to minimize gas leakage. In the exhaust stroke, as the piston approaches the end of its downward stroke at BC, the exhaust valve opens. The piston reverses direction and starts to move up towards TC and exhaust gases are expelled through the exhaust system. After this, the cycle starts again. Two types of IC engines commonly used in automobiles include spark ignition (SI) gasoline (or petrol) and compression ignition (CI) diesel engines. In the case of the SI gasoline engine, air-fuel mixture is drawn into the intake stroke and the air-fuel mixture is ignited by the spark plug in the power stroke. The compression ratio is about 8:1. In the case of the CI diesel engine, electrical ignition is not used and the intake system delivers only air. Fuel is injected directly into the combustion chamber. The compression ratio is very high, about 20:1. Air compression alone generates heat large enough to ignite the fuel on injection.

Figure 12.5.2 shows a schematic of a four-stroke and four-cylinder SI engine used in passenger cars (Weertman and Dean, 1981). With most water-cooled engines, the cylinders are cast into a single block containing cooling passages, lubricant passages, and supporting structure for the crankshaft, camshaft and other components. The cylinders are enclosed at the combustion end with a cylinder head containing the intake and exhaust valves as well as cooling passages. One end of the connecting rod is attached to the piston and the other end to the crankshaft which transforms the reciprocating motion of the piston to the rotating motion of the crankshaft. The mechanical components of a valve train which actuate the intake and exhaust valves include a cam, camshaft, spring-loaded valves, valve guides and oil seals. The cam actuates a hydraulic lifter or a manually adjusted tappet which actuates a push rod, rocker arm and finally a valve.

The engine cylinders are cast into a single block, generally from cast iron and passages for the cooling water are cast into the block. Pistons are generally made of cast iron. For light weight and heat conduction, aluminum alloy is widely used for pistons. The piston rings, which form a movable seal, are usually made of gray cast iron and some with chrome plating or molybdenum fillings because of their friction-reducing characteristics. The crankshaft is generally made of forged steel and nodular cast iron. The crankshaft is supported in main bearings. Cams and cam tappets, rocker arms, and camshaft materials are generally made of hardenable cast irons or forged steels. Phosphate coatings are commonly used for break-in. Valves are made from forged alloy steel. The main bearings, the camshaft bearings, and the connecting rod bearing (which support the connecting rod on the crankshaft) are hydrodynamic, oil-lubricated bearings, made of hard alloys such as copper-lead alloys, tin babbitts, lead babbitts and aluminum alloys. Lubrication of various moving components is the key to the life of the engine. For more details, see, for example, Chamberlin and Saunders (1983) and Heywood (1988).

12.5.2 Gas Turbine Engines

A schematic of the major components of a gas turbine engine is presented in Figure 12.5.3. Compression of the air entering the gas turbine is achieved by centrifugal or axial compressors. The combustion system heats the air from the compressor to the required turbine entry temperature. Hot gases from the combustion chamber are accelerated and directed by stationary

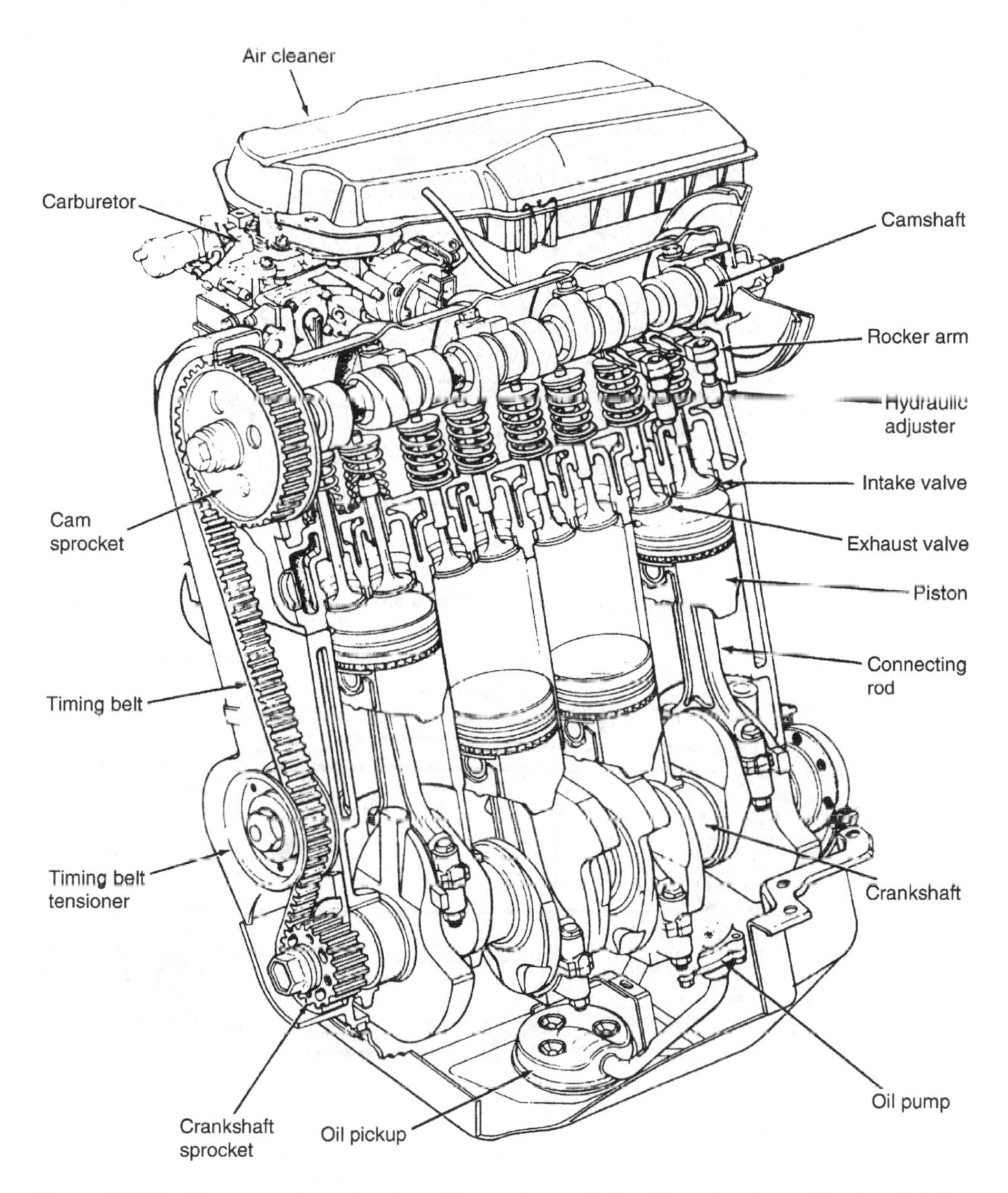

Figure 12.5.2 Cutaway schematic of an SI type four-cylinder IC engine from Chrysler (2.2 liter displacement, bore = 87.5 mm, stroke = 92 mm, compression ratio = 8.9, power = 65kW at 5000 RPM) (*Source*: Weertman and Dean, 1981).

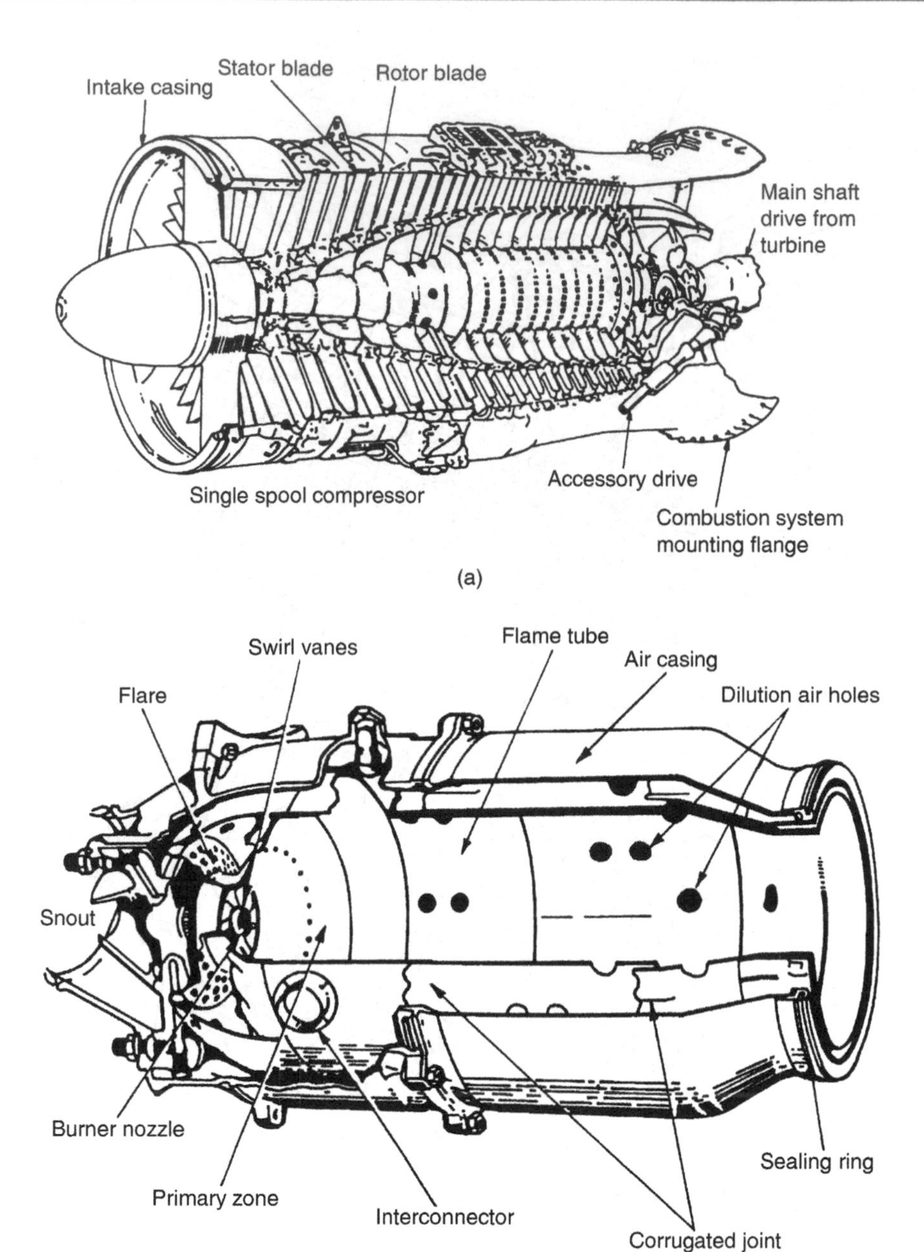

Figure 12.5.3 Schematic of gas turbine engine components: (a) a typical axial-flow compressor; (b) a typical combustion chamber; and (c) a typical turbine assembly (*Source*: Meetham, 1981). (*Continued*)

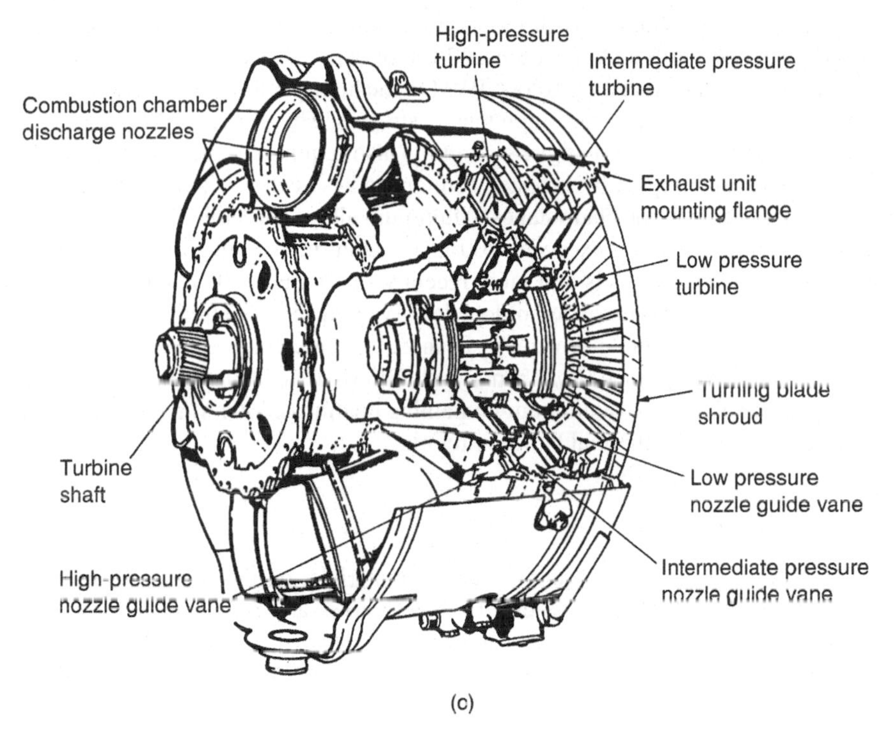

Figure 12.5.3 (*Continued*)

nozzle guide vanes into the turbine assembly, which provides the power to drive the compressor. For high-performance gas turbine engines, turbine entry temperatures should be as high as possible. Turbine temperatures can be as high as 1100°C for some aerospace applications (Meetham, 1981). Gas turbine engines are used in a wide variety of applications. The most demanding of these in terms of materials durability and reliability requirements under relatively severe conditions are aircraft propulsion, marine propulsion, and electric power generation.

The hot-section (turbine end) airfoils in such engines are required to retain mechanical and surface integrity for thousands of hours under conditions of very high stress at elevated temperatures. The hot gases are highly oxidizing and may contain contaminants such as chlorides and sulfates, which can lead to hot corrosion, and also can contain erosive media. Erosion may be caused by ingested sand. Temperature transients occurring during engine operation can cause thermal fatigue. Today, nickel-based superalloys are widely used because of their outstanding strength and oxidation resistance over the temperature range encountered. Nickel-based superalloys such as Nimonic 90, 105 and 115 and Udimet 500 and 700, containing Cr, Co, Ti, Al, and Mo, are commonly used for turbine blades, turbine and compressor casings, and combustion-chamber liners (Anonymous, 1967; Meetham, 1981). A casting process is used to produce turbine blades for economic design reasons. Sheet materials are used for turbine and compressor casings, combustion chambers, and various pipes. Various coatings are commonly used for protection. The mainline shafts are made of low-alloy steels or maraging steels. High-speed steel (18-4-1) and M-50 steels are the common materials in mainstream rolling-contact bearings. The cages are made of silver-plated En steel. Bearings are normally lubricated with low-viscosity synthetic oils. It should be noted that since the early 1970s, prototypes of stationary and rotating components of gas turbine engines for high-temperature (>1000°C) applications have been built making successful use of ceramic materials such as SiC.

With rotational speeds exceeding 10,000 RPM in the high-pressure compressors of large engines, compressor blades experience high tensile stresses. Therefore, specific tensile strength is a basic material requirement. Fatigue strength is also required to resist cyclic stresses. Resistance to erosion and impact by ingested foreign bodies such as sand, stones, and birds is important in early compressor-blade stages. Aluminum alloys were initially used for compressor blades. As exit temperatures exceeded 200°C, the aluminum alloys were superseded by the 12% chromium martensitic steels and more recently by titanium alloys (such as Ti-6Al-4V) because of their attractive combination of low density, high specific strength, good fatigue and creep resistance, and excellent corrosion resistance. Ceramic coatings are applied to provide corrosion and erosion resistance (Meetham, 1981; Bhushan and Gupta, 1997). Coatings generally used include aluminide (aluminum-containing coatings), MCrAlY (M = Fe, Co, Ni, or combinations thereof), and thermal barrier zirconia.

12.5.3 Railroads

Railroads are used for the transportation of bulk materials, on the order of 30 tons per car axle, at low energy costs. Diesel and electric locomotives are most widely used today. The tribological interface of most interest is the flanged steel wheel on steel rails with rotary motion. Figure 12.5.4 shows a schematic of a two-axle freight car truck having three main

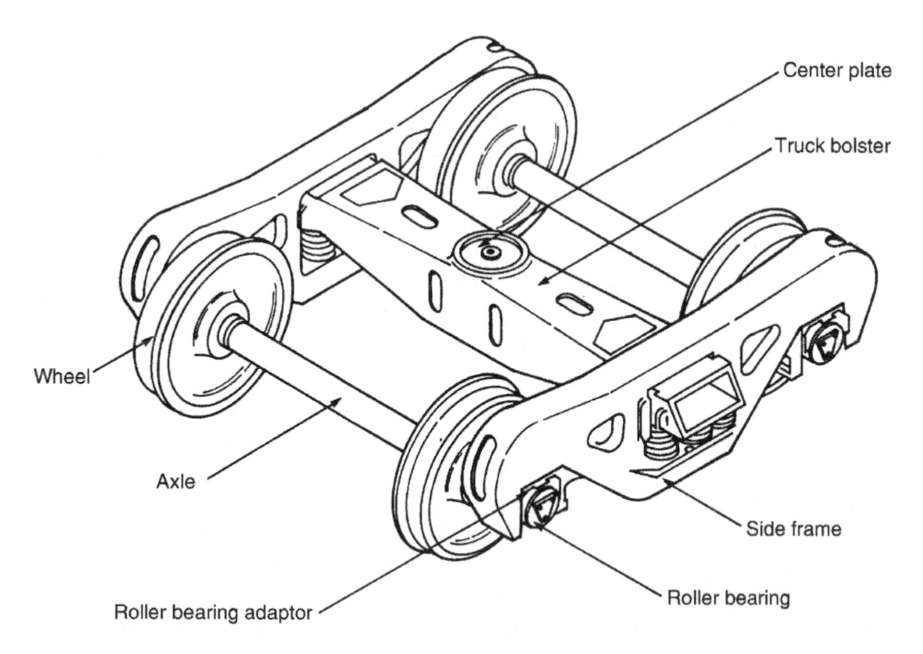

Figure 12.5.4 Schematic of a two-axle freight car truck. Reproduced with permission from Jackson, D.R. (1983), "Railroads" in *Handbook of Lubrication – Theory and Practice of Tribology Vol. I Application and Maintenance* (E.R. Booser, ed.), pp. 269–288, CRC Press, Boca Raton, Florida. Copyright 1983. Taylor and Francis.

components – a bolster and two side frames (Jackson, 1983). These assemblies mostly use grease-lubricated rolling bearings, with some using plain bearings. The wear of both the tread and the flange of the wheels and wear of the roller or plain bearings are some of the tribological issues.

12.5.4 Magnetic Storage Devices

Magnetic storage devices used for audio, video, and data-processing (computer) applications are tape and rigid disk drives. Magnetic recording and playback are accomplished by the relative motion between the magnetic medium (tape or disk) and a stationary or rotating read-write magnetic head (Mee and Daniel, 1996). Tape drives include drives with stationary head and linear tape motion (known as linear drives) and drives with rotary head and linear or helical tape motion (known as rotary drives). Figure 12.5.5 shows a schematic of a linear, data-processing IBM 3490 tape drive. After loading the cartridge into the drive, the tape leader is threaded into the drive to the take-up reel by a pentagon threading mechanism. A decoupler column placed near the cartridge entrance decouples any tape vibrations that may occur inside the cartridge. The tension is sensed and controlled by a tension transducer. Figure 12.5.6 shows a schematic of a rigid disk drive. A stack of disks is mounted on a sealed, grease-lubricated ball bearing or a hydrodynamic air bearing spindle. The disks are rotated by a DC motor at speeds ranging from few thousand to a maximum speed of about 10,000 RPM. A head slider

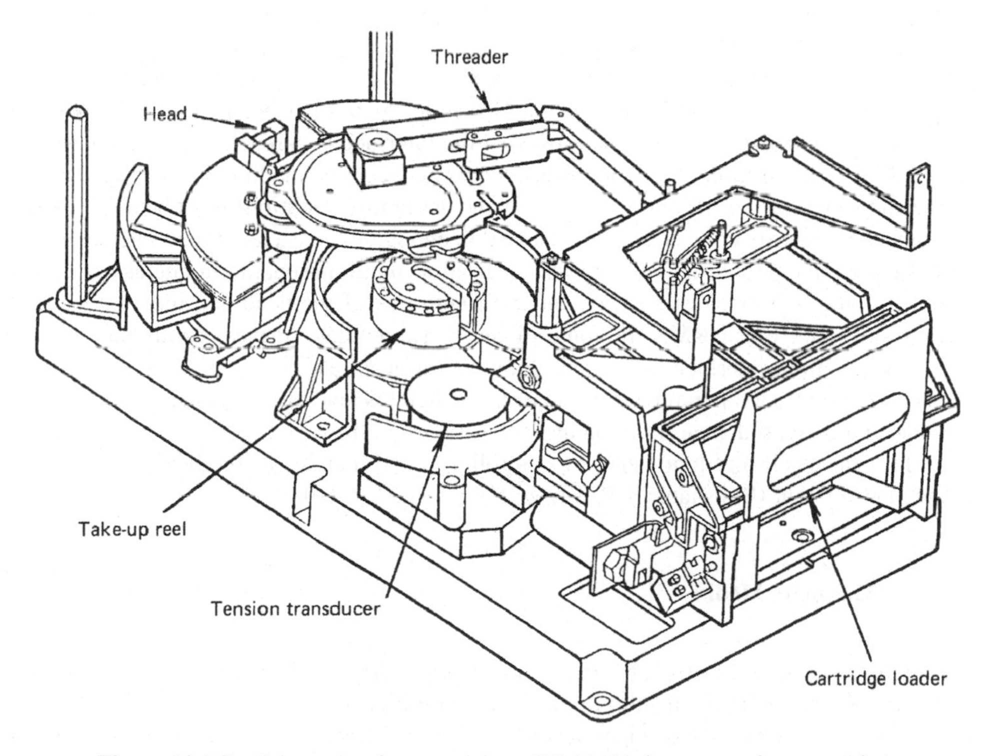

Figure 12.5.5 Schematic of tape path in an IBM 3490 data-processing tape drive.

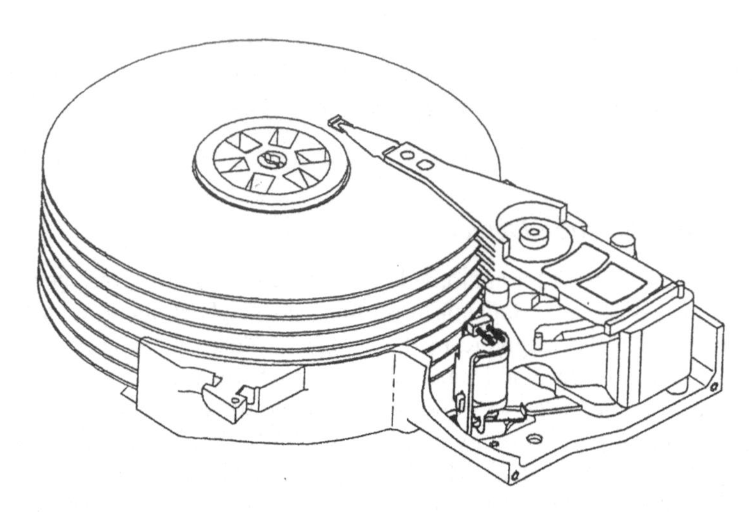

Figure 12.5.6 Schematic of a rigid disk drive.

is supplied for each disk surface. The slider-suspension assembly is actuated by a stepping motor or a voice-coil motor for read-write operation (Bhushan, 1992, 1996a).

For high areal recording density, the linear flux density (number of flux reversals per unit distance) and the track density (number of tracks per unit distance) should be as high as possible. The reproduced (read back) signal amplitude decreases with a decrease in the recording wavelength and/or track width. The signal loss occurs with an increase in head-to-medium spacing (clearance or flying height), requiring a short spacing. In order to minimize damage to the interface, the head-medium interface is designed such that under steady operating conditions, a load-carrying air film is formed at the interface as a result of hydrodynamic or elastohydrodynamic action. Physical contact occurs between the medium and the head during starting and stopping. The developed air film at the operating conditions must be thick enough to mitigate any asperity contacts, yet it must be thin enough to give a large read-back magnetic signal. In modern tape and rigid disk drives, the head-to-medium separation ranges from about 0.1–0.2 μm and 3–5 nm, respectively, and the roughness of the head and medium surfaces ranges from 1 to 10 nm RMS. In some of the consumer tape drives, continuous contact at the head-medium interface may occur. Smooth surfaces lead to an increase in adhesion, friction and interface temperatures, and closer flying heights lead to occasional rubbing of high asperities and increased wear. Friction and wear issues are resolved by appropriate selection of interface materials and lubricants, by controlling the dynamics of the head and medium, and the environment (Bhushan, 1992, 1996a, 1996b, 1999, 2001b).

12.5.4.1 Magnetic Media

Magnetic media fall into two categories: (a) particulate media, where magnetic particles are dispersed in a polymeric matrix and coated onto a polymeric substrate for tapes; and (b) thin-film media, where a continuous film of magnetic material is deposited by vacuum techniques onto a polymeric substrate for tapes or onto a rigid substrate such as aluminum, glass or glass ceramics for rigid-disks. Requirements of higher recording densities with low error rates have resulted in increased use of thin-film media which are smoother and with considerably thinner magnetic coating than the particulate media. Thin-film media are exclusively used for rigid disks and are used for high-density audio/video and data processing tapes along with particulate media.

Tapes

Cross sectional views of a particulate and a thin-film (evaporated) metal tape are shown in Figure 12.5.7. The base film for tapes is mostly polyethylene terephthalate (PET) film, followed

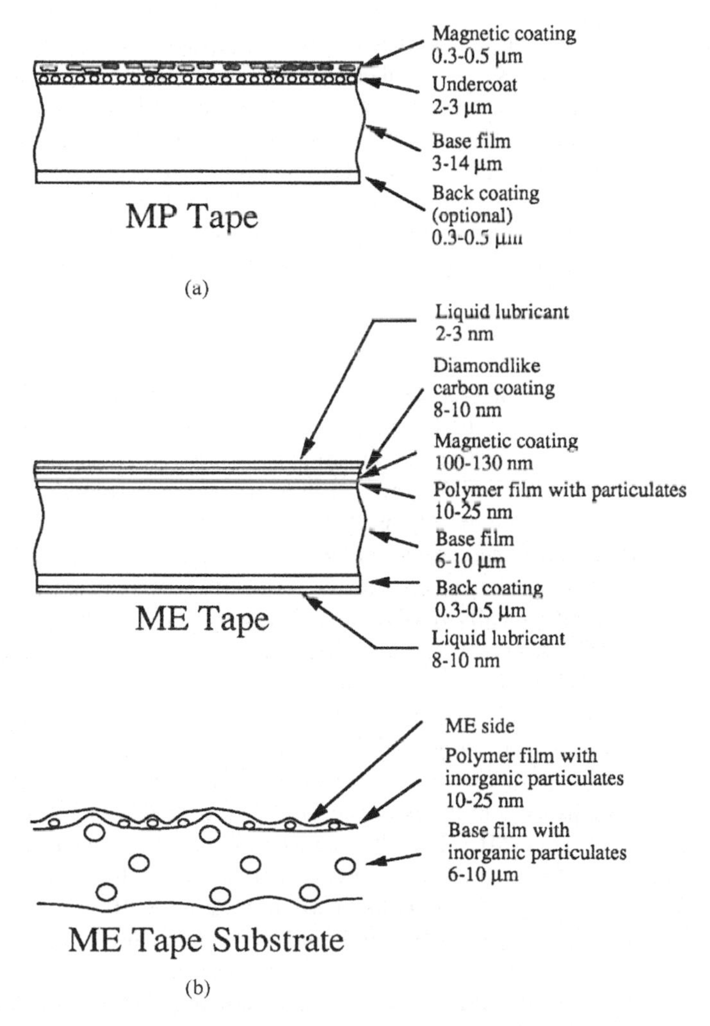

Figure 12.5.7 Sectional views of: (a) a particulate tape; and (b) a metal-evaporated (ME) magnetic tape and coated PET substrate for ME tapes.

by polyethylene naphthalate (PEN) polymers. Tapes use a 6.35–36.07 μm thick PET substrate with RMS roughness of about 2–5 nm and peak-to-valley (P-V) distance of 50–100 nm for particulate media and 1.5–2 nm RMS for the recording side of thin-film media. Particulates such as silica or titania with a bimodal distribution of sizes with mean diameters on the order of 0.5 μm and 2 μm, are added in the substrate as anti-slip agents (Bhushan, 1992).

The base film is coated on one side of a tape with a magnetic coating, typically 1–4 μm thick and containing 70–80% by weight (or 43–50% by volume) of submicron and acicular magnetic particles (such as γ-Fe_2O_3, Co-modified γ-Fe_2O_3, CrO_2 (only for tapes), and metal particles or hexagonal platelets of barium ferrite). These magnetic particles are held in polymeric binders such as polyester-polyurethane, polyether-polyurethane, nitrocellulose, poly(vinyl chloride), poly(vinyl alcohol-vinyl acetate), poly(vinylidene chloride), VAGH, phenoxy, and epoxy. To reduce friction, the coating consists of 1–7% by weight of lubricants (mostly fatty acid esters, e.g., tridecyl stearate, butyl stearate, butyl palmitate, butyl myristate, stearic acid, myrstic acid). Finally, the coating contains a cross linker or curing agent (such as lecithin) and solvents (such as tetrahydrofuran and methyl isobutyl ketone). In some media, carbon black is added for antistatic protection if the magnetic particles are highly insulating, and abrasive particles (such as Al_2O_3 and Cr_2O_3) are added as a head cleaning agent to improve wear resistance. The coating is calendered to a surface roughness of 5–15 nm RMS. For antistatic protection and for improved tracking, most magnetic tapes have a 1–3 μm thick backcoating of polyester-polyurethane binder containing a conductive carbon black and TiO_2, typically 10% and 50% by weight, respectively.

Thin-film (also called metal-film or ME) tapes consist of a polymer substrate (PET or polyimide) with an evaporated film of Co-Ni (with about 18% Ni) and experimental evaporated/sputtered Co-Cr (with about 17% Cr) (for perpendicular recording) which is typically 100–130 nm thick. Since the magnetic layer is very thin, the surface of the thin-film medium is greatly influenced by the surface of the substrate film. Therefore, an ultra-smooth PET substrate film (RMS roughness $\sim$ 1.5–2 nm) is used to obtain a smooth tape surface. A 10–25 nm thick precoat composed of polymer film with additives is generally applied to the recording side of the PET substrate to provide controlled topography, Figure 12.5.7b. This film generally contains inorganic particulates (typically SiO_2 with a particle size of 100–200 nm diameter and areal density of typically 10,000/mm^2). The polymer precoat is applied to reduce the roughness (mostly P-V distance) in a controlled manner from that of the PET surface, and to provide good adhesion with the ME films. The particles are added to the precoat to control the real area of contact and consequently the friction. A continuous magnetic coating is deposited on the polymer film. The polymer film is wrapped on a chill roll during deposition, which keeps the film at a temperature of 0 to –20°C. $Co_{80}Ni_{20}$ material is deposited on the film by a reactive evaporation process in the presence of oxygen; oxygen increases the hardness and corrosion resistance of the ME film. The deposited film, with a mean composition of $(Co_{80}Ni_{20})_{80}O_{20}$ consists of very small Co and Co-Ni crystallites which are primarily intermixed with oxides of Co and Ni. Diamondlike carbon in about 8–10 nm thickness is used to protect against corrosion and wear. A topical liquid lubricant (typically perfluoropolyether with reactive polar ends) is then applied to the magnetic and back coatings by rolling. The topical lubricant enhances the durability of the magnetic coating, and also inhibits the highly reactive metal coating from reacting with ambient air and water vapor. A backcoating is also applied to balance stresses in the tape, and for anti-static protection.

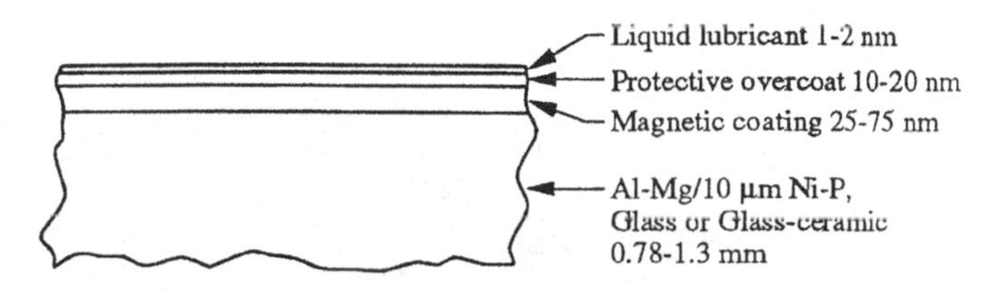

Figure 12.5.8 Sectional view of a thin-film magnetic rigid disk.

Rigid Disks

Figure 12.5.8 shows a sectional view of a thin-film rigid disk. The substrate for rigid disks is either non-heat-treatable aluminum-magnesium alloy AISI 5086 (95.4% Al, 4% Mg, 0.4% Mn, and 0.15% Cr) electroless plated with nickel-phosphorus (90–10 wt %) layer to improve its surface hardness to 600–800 kg/mm^2 (Knoop) and smoothness, chemically strengthened glass, or glass-ceramic. To minimize static friction at the head–disk interface, the start-stop zone or the entire disk surface of substrates is textured. Disks are textured either by mechanical texturing techniques using either free or fixed abrasives in the circumferential or random orientation to a typical RMS roughness of 4–8 nm, or a laser texturing technique to create bumps on the disk surface. The finished substrate is coated with a magnetic film 25–75 nm thick. Some metal films require a Cr undercoat (10–50 nm thick) as a nucleation layer to improve magnetic properties, such as coercivity. Magnetic films used are metal films of cobalt-based alloys (e.g., Co-Pt-Cr, Co-Pt-Ni). These metallic magnetic films have weak durability and poor corrosion resistance. Protective overcoats with a liquid lubricant overlay are generally used to provide low friction, low wear, and corrosion resistance. The protective coating is typically sputtered diamond-like carbon (DLC). In most cases, a thin layer of perfluoropolyether lubricant with reactive polar ends is used. The trend is to use partially bonded lubricant film consisting of an unbonded or lubricant film. The unbonded top layer would heal any worn areas on the disk surface where the lubricant may have been removed, and the bonded underlayer provides lubricant persistence. Furthermore, the bonded layer does not contribute to meniscus effects in stiction.

12.5.4.2 Magnetic Heads

Magnetic heads used to date consist either of conventional inductive or of thin-film inductive and magnetoresistive (MR) types. Film-head design capitalizes on semiconductor-like processing technology to reduce fabrication costs, and thin-film technology allows the production of high-track density heads with accurate track positioning control and high reading sensitivity. If an MR head design is used, it is only for read purposes. Inductive heads consist of a body forming the air bearing (referred to as the air bearing surface or ABS) and a magnetic ring core carrying the wound coil with a read-write gap. In film heads, the core and coils or MR stripes are deposited by thin-film technology. The body of a thin-film head is made of magnetic ferrites or nonmagnetic Al_2O_3-TiC and the head construction includes coatings of soft magnetic alloys, insulating oxide, and bonding adhesives.

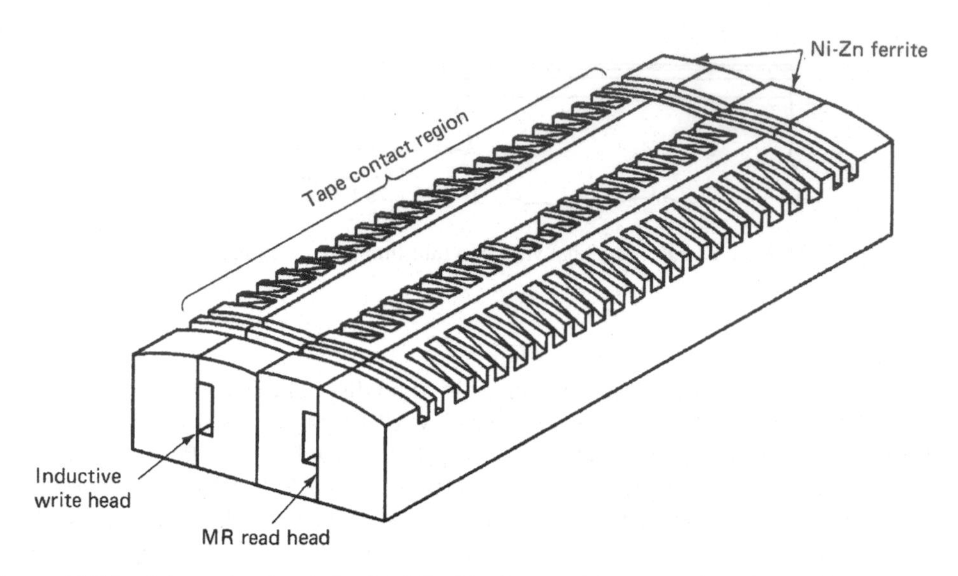

Figure 12.5.9 Schematic of an inductive/MR thin-film head (with a radius of cylindrical contour of 20 mm) in an IBM 3490 data processing tape drive.

Air-bearing surfaces of tape heads are cylindrical in shape. The tape is slightly underwrapped over the head surface to generate hydrodynamic lift during read-write operations. For inductive-coil tape heads, the core materials have been typically Permalloy and Sendust. However, since these alloys are good conductors, it is sometimes necessary to laminate the core structure to minimize losses due to eddy currents. The air-bearing surfaces of most inductive-coil type heads consist of plasma-sprayed coatings of hard materials such as Al_2O_3-TiO_2 and ZrO_2. MR read and inductive write heads in modern tape drives (such as IBM 3490) are miniaturized using thin-film technology, Figure 12.5.9. Film heads are generally deposited on Ni-Zn ferrite (11 wt % NiO, 22 wt % ZnO, 67 wt % Fe_2O_3) substrates.

The head sliders used in rigid disk drives are either a two- or three-rail, taper-flat design supported by a nonmagnetic 300 series steel leaf spring (flexure) suspension to allow motion along the vertical, pitch and roll axes (Figure 12.5.10a). The front taper pressurizes the air lubricant, while some air leaking over the side boundaries of the rail results in a pitch angle. The inductive type or inductive/MR type thin-film read-write elements used in high-end disk drives (e.g., IBM 3390) are integrated in the Al2O3-TiC (70–30 wt %) slider at the trailing edge of the center rail in the three-rail design where the lowest flying height occurs (Figure 12.5.10b). The suspension supplies a vertical load typically ranging from 30 mN (3 g) to 50 mN (5g) (apparent contact pressure ~35 kPa), dependent upon the size of the slider, which is balanced by the hydrodynamic load when the disk is spinning. As the drives get smaller and flying height decreases, the size and mass of the sliders and their gram load decreases. Lower mass results in higher air-bearing frequency, which reduces dynamic impact. Smaller size and required lower gram load results in an improvement in stiction and start-stop and flyability lifetimes. The surface roughness of the air-bearing rails is typically 1.5–2 nm RMS. The stiffness of the suspension (~25 mN/mm) is several orders of magnitude lower

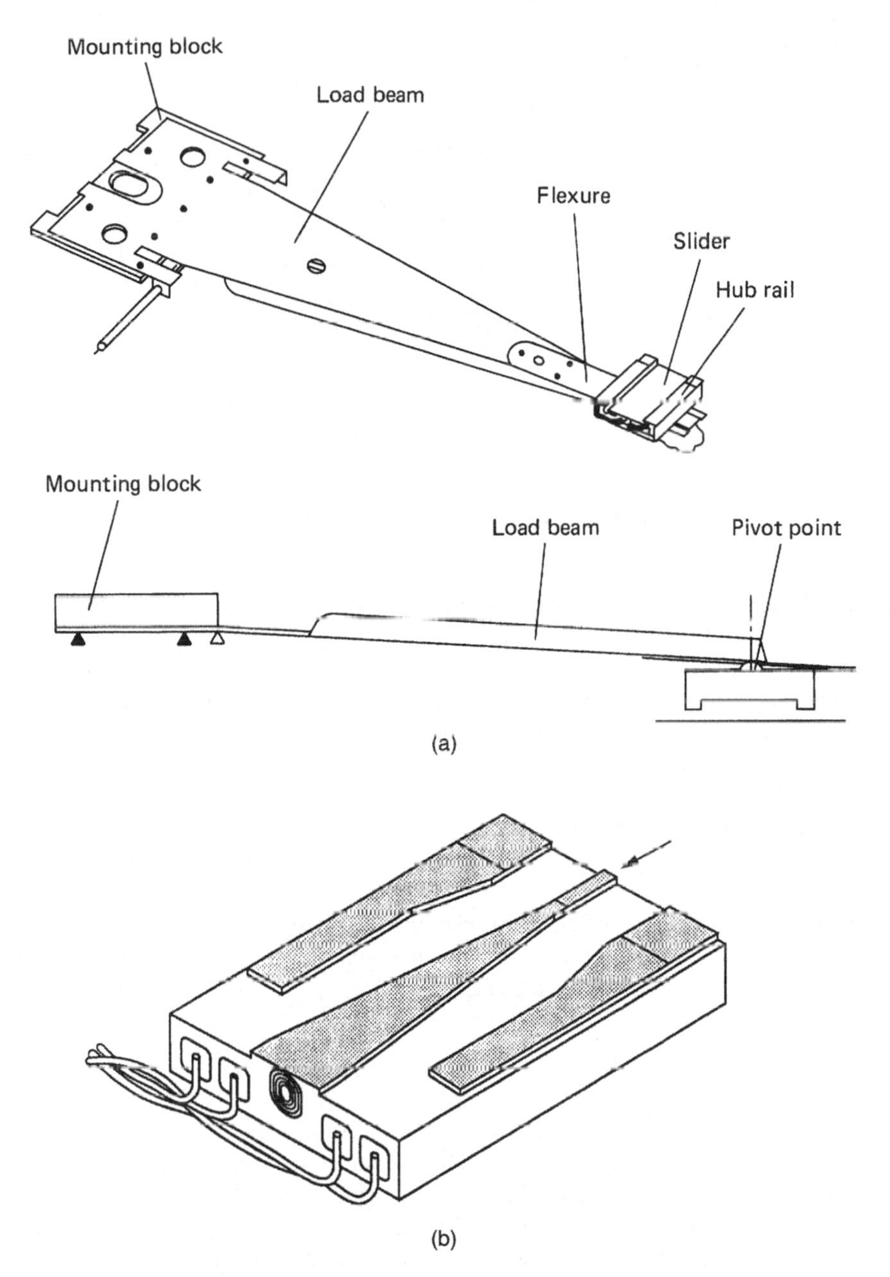

Figure 12.5.10 Schematics of: (a) an IBM 3370/3380/3390 type suspension-slider assembly; and (b) an IBM tri-rail thin-film nanoslider (direction of disk rotation is shown by an arrow).

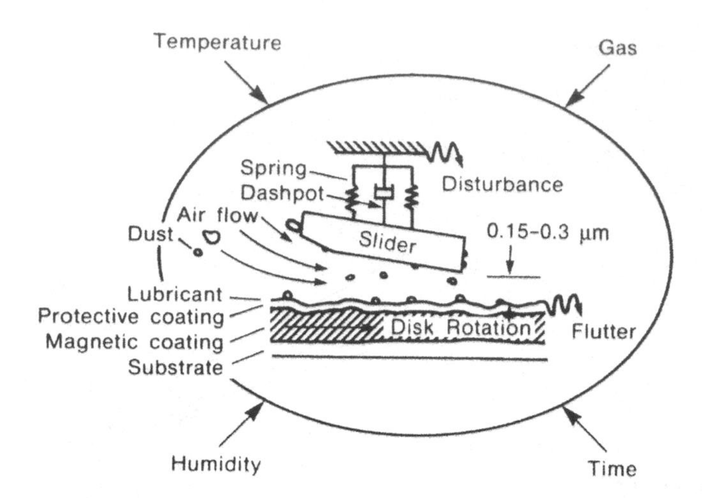

Figure 12.5.11 Schematic of the head-rigid disk interface.

than that of the bearing (~0.5 kN/mm), so that most dynamic variations are taken up by the suspension without degrading the air bearing.

Small disk drives also use inductive-coil type heads of one or two types: minimonolithic (mini-Winchester) and minicomposite. A minimonolithic head slider consists of a slider body and a core piece carrying the coil, both consisting of monolithic magnetic material (typically Mn-Zn ferrite). The taper-flat bearing area is provided by the outer rails of a tri-rail design. The center rail defines the width of the magnetic element in the trailing edge where a ferrite core is formed. A minimonolithic head slider consists of an Mn-Zn ferrite core and read-write gap, glass bonded into the air-bearing surface of a nonmagnetic, wear-resistant slider (typically calcium titanate).

A schematic representation of the head-disk interface is shown in Figure 12.5.11. Environment, usage time, and contamination (external and wear debris) play a significant role in the reliability and usable lifetime of the interface.

12.6 Closure

A variety of tribological components are used to accommodate relative motion. Common components include sliding-contact and rolling-contact bearings, seals, gears, cams and cam tappets, piston rings, electrical brushes, and cutting and forming tools. Tribological components also include various MEMS/NEMS devices. Some industrial applications include material processing, automotives, aerospace, railroads, and magnetic storage devices.

References

Aden, J.S., Bohorquez, J.H., Collins, D.M., Crook, M.D., Garcia, A., and Hess, U.E. (1994), "The Third-Generation HP Thermal Inkjet Printhead," *HP Journal* **45**(1), 41–45.

Aksyuk, V.A., Pardo, F., Carr, D., Greywall, D, Chan, H.B., Simon, M.E., Gasparyan, A., Shea, H., Lifton, V., Bolle, C., Arney, S., Frahm, R., Paczkowski, M., Haueis, M., Ryf, R., Neilson, D.T., Kim, J., Giles, C.R., and Bishop,

D. (2003), "Beam-Steering Micromirrors for Large Optical Cross-Connects," *J. Lightwave Technol.* **21**, 634–642.

Anonymous (1967), *Hot Corrosion Problems Associated with Gas Turbines*, STP-421, ASTM, Philadelphia, PA.

Anonymous (1989a), *Nomenclature of Gear Tooth Failure Modes*, ANSI/AGMA 110.04-1980, Amer. Gear Manuf. Assoc., 1901 North Fort Myer Drive, Arlington, VA 22209.

Anonymous (1989b), *Gear Material and Heat Treatment Manual*, ANSI/AGMA 2004-B89, Amer. Gear Manuf. Assoc., 1901 North Fort Myer Drive, Arlington, VA 22209.

Anonymous (1989c), *ASM Handbook-Ninth Edition, Vol. 16: Machining*, ASM International, Metals Park, OH.

Anonymous (1997), *SAE Handbook*, Vols. 1 to 3, SAE Inc., Warrandale, PA.

Armington, R.E. and Amey, D. (ed.) (1979), *Proc. Twenty-Fifth Holm Conf.*, Illinois Institute of Technology, Chicago, IL.

Bamburger, E.N. (1980), "Materials for Rolling Element Bearings," in *Bearing Design: Historical Aspects, Present Technology and Future Problems*, pp. 1–46, ASME, New York.

Bastian, E.L.H. (1951), *Mealworking Lubricants*, McGraw-Hill, New York.

Baydo, R. and Groscup, A. (2001), "Getting to the Heart of Ink Jet: Printheads," *Beyond Recharger*, May 10, 10–12. Also visit http://64.78.37.127.pdf/baydo.pdf.

Bhushan, B. (1992), *Mechanics and Reliability of Flexible Magnetic Media*, Springer-Verlag, New York.

Bhushan, B. (1996a), *Tribology and Mechanics of Magnetic Storage Devices*, Second edition, Springer-Verlag, New York.

Bhushan, B. (1996b), "Tribology of the Head-Medium Interface," in *Magnetic Recording Technology* (C.D. Mee and E. D. Daniel, eds.), Second edition, pp. 7.1–7.66, McGraw Hill, New York.

Bhushan, B. (1998), *Tribology Issues and Opportunities in MEMS*, Kluwer Academic, Dordrecht, Netherlands.

Bhushan, B. (1999), *Handbook of Micro/Nanotribology*, Second edition, CRC Press, Boca Raton, Florida.

Bhushan, B. (2001a), *Modern Tribology Handbook, Vol. 2: Materials, Coatings and Industrial Applications*, CRC Press, Boca Raton, FL.

Bhushan, B. (2001b), *Fundamentals of Tribology and Bridging the Gap Between the Macro- and Micro/Nanoscales*, NATO Science Series II. Mathematics, Physics and Chemistry Vol. 10, Kluwer Academic Publishers, Dordrecht, The Netherlands.

Bhushan, B. (2003), "Adhesion and Stiction: Mechanisms, Measurement Techniques, and Methods for Reduction," *J. Vac. Sci. Technol. B* **21**, 2262–2296.

Bhushan, B. (2010), *Springer Handbook of Nanotechnology*, Third edition, Springer-Verlag, Heidelberg, Germany.

Bhushan, B. (2011), *Nanotribology and Nanomechanics I – Measurement Techniques and Nanomechanics, II – Nanotribology, Biomimetics, and Industrial Applications*, Springer-Verlag, Heidelberg, Germany.

Bhushan, B. and Gupta, B.K. (1997), *Handbook of Tribology: Materials, Coatings and Surface Treatments*, McGraw-Hill, New York (1991), reprint ed., Krieger Publishing Co., Malabar, FL.

Bhushan, B. and Kwak, K.J. (2007), "Platinum-Coated Probes Sliding at up to 100 mm s^{-1} Against Coated Silicon Wafers for AFM Probe-Based Recording Technology," *Nanotechnology* **18**, Art. # 345504.

Bhushan, B. and Kwak, K.J. (2008), "Noble Metal-Coated Probes Sliding at up to 100 mm s^{-1} Against PZT Films for AFM Probe-Based Ferroelectric Recording Technology," (invited) *J. Phys.: Condens. Matter* **20**, Art. # 225013.

Bhushan, B. and Sibley, L.B. (1982), "Silicon Nitride Rolling Bearings for Extreme Operating Conditions," *ASLE Trans.* **25**, 417–428.

Bisson, E.E. and Anderson, W.J. (1964), *Advanced Bearing Technology*, Special Publication SP-38, NASA, Washington, DC.

Booser, E.R. (ed.) (1983), *CRC Handbook of Lubrication, Vol. I, Application and Maintenance*, CRC Press, Boca Raton, FL.

Booser, E.R. (ed.) (1984), *CRC Handbook of Lubrication, Vol. II, Theory and Design*, CRC Press, Boca Raton, FL.

Braithwaite, E.R. (1967), *Lubrication and Lubricants*, Elsevier, Amsterdam.

Budinski, K.G. (1980), "Tool Materials," in *Wear Control Handbook* (M.B. Peterson and W. O Winer, eds.), pp. 931–985, ASME, New York.

Chamberlin, W.B. and Saunders, J.D. (1983), "Automobile Engines," in *Handbook of Lubrication – Theory and Practice of Tribology Vol. I Application and Maintenance* (E.R. Booser, ed.), pp. 3–44, CRC Press, Boca Raton, FL.

Chang, L.S., Gendler, P.L., and Jou, J.H. (1991), "Thermal Mechanical and Chemical Effects in the Degradation of the Plasma-Deposited α-SC:H Passivation Layer in a Multlayer Thin-Film Device," *J. Mat. Sci* **26**, 1882–1290.

Chen, R.J., Choi, H.C., Bangsaruntip, S., Yenilmez, E., Tang, X., Wang, Q., Chang, Y.L., and Dai, H. (2004), "An Investigation of the Mechanisms of Electrode Sensing of Protein Adsorption on Carbon Nanotube Devices," *J. Am. Chem. Soc.* **126**, 1563–1568.

Coleman, W. (1970), "Gear Design Considerations," in *Interdisciplinary Approach to the Lubrication of Concentrated Contacts* (P. M. Ku, ed.), pp. 551–589, Special Publication SP-237, NASA, Washington, DC.

Core, T.A., Tsang, W.K., and Sherman, S.J. (1993), "Fabrication Technology for an Integrated Surface-Micromachined Sensor," *Solid State Technol.* **36** (Oct), 39–47.

Crouse, W.H. (1970), *Automotive Engine Design*, McGraw-Hill, New York.

DeWolf, I. and van Spengen, W.M. (2002), "Techniques to Study the Reliability of Metal RF MEMS Capacitive Switches," *Microelectronics Reliability* **42**, 1789–1794.

Douglass, M.R. (1998), "Lifetime Estimates and Unique Failure Mechanisms of the Digital Micromirror Devices (DMD)," *Proc. 36th Annual Inter. Reliability Phys. Symp.*, pp. 9–16, IEEE, New York.

Douglass, M.R. (2003), "DMD Reliability: A MEMS Success Story," in *Reliability, Testing, and Characterization of MEMS/MOEMS II*, Proc. of SPIE Vol. 4980, pp. 1–11, SPIE, Bellingham, WA.

Dudley, D.W. (1964), *Gear Handbook*, McGraw-Hill, New York.

Dudley, D.W. (1980), "Gear Wear," in *Wear Control Handbook* (M. B. Peterson and W.O. Winer, eds.), pp. 755–830, ASME, New York.

Frechette, L.G., Jacobson, S.A., Breuer, K.S., Ehrich, F.F., Ghodssi, R., Khanna, R., Wong, C.W., Zhang, X., Schmidt, M.A., and Epstein, A.H. (2005), "High-Speed Microfabricated Silicon Turbomachinery and Fluid Film Bearings," *J. MEMS* **14**, 141–152.

Fuchsluger, J.H. and Vandusen, V.L. (1980), "Unlubricated Piston Rings," in *Wear Control Handbook* (M.B. Peterson and W.O. Winer, eds.), pp. 667–698, ASME, New York.

Fuller, D.D. (1984), *Theory and Practice of Lubrication for Engineers*, Wiley, New York.

Garcia, E.J. and Sniegowski, J.J. (1995), "Surface Micromachined Microengine," *Sensors and Actuators A* **48**, 203–214.

Gardos, M.N. and McConnell, B.D. (1982), *Development of High Speed, High Temperature Self-Lubricating Composites*, Special Publication SP-9, STLE, Park Ridge, IL.

Glaeser, W.A. (1992), *Materials for Tribology*, Elsevier, Amsterdam, The Netherlands.

Guckel, H. and Burns, D.W. (1989), "Fabrication of Micromechanical Devices from Polysilicon Films with Smooth Surfaces," *Sensors and Actuators* **20**, 117–122.

Hankins, M.G., Resnick, P.J., Clews, P.J., Mayer, T.M., Wheeler, D.R., Tanner, D.M., and Plass, R.A. (2003), "Vapor Deposition of Amino-Functionalized Self-Assembled Monolayers on MEMS," *Proc. SPIE* **4980**, pp. 238–247, SPIE, Bellingham, WA.

Harris, T.A. (1991), *Rolling Bearing Analysis*, Third edition, Wiley, New York.

Henck, S.A. (1997), "Lubrication of Digital Micromirror Devices," *Tribol. Lett.* **3**, 239–247.

Heywood, J.B. (1988), *Internal Combustion Engine Fundamentals*, McGraw-Hill, New York.

Hornbeck, L.J. (1997), "Low Surface Energy Passivation Layer for Micromechanical Devices," U.S. Patent No. 5,602,671, Feb. 11.

Hornbeck, L.J. (1999), "A Digital Light Processing™ Update – Status and Future Applications," *Proc. Soc. Photo-Opt. Eng.* **3634**, *Projection Displays V*, 158–170.

Hornbeck, L.J. (2001), "The DMD™ Projection Display Chip: A MEMS-Based Technology," *MRS Bulletin*, **26**, 325–328.

Hornbeck, L.J. and Nelson, W.E. (1988), "Bistable Deformable Mirror Device," *OSA Technical Digest Series Vol. 8: Spatial Light Modulators and Applications*, 107–110.

Jackson, D.R. (1983), "Railroads," in *Handbook of Lubrication – Theory and Practice of Tribology Vol. I Application and Maintenance* (E.R. Booser, ed.), pp. 269–288, CRC Press, Boca Raton, FL.

Johnson, R.L. and Schoenherr, K. (1980), "Seal Wear," in *Wear Control Handbook* (M.B. Peterson and W.O. Winer, eds.), pp. 727–753, ASME, New York.

Judge, A.W. (1972), *Automotive Engines – In Theory, Design, Construction, Operation and Testing*, Robert Bentley, Cambridge, MA.

Ku, P.M. (ed.) (1970), *Interdisciplinary Approach to the Lubrication of Concentrated Contacts*, Special Publication SP-237, NASA, Washington, DC.

Kwak, K.J. and Bhushan, B. (2008), "Platinum-Coated Probes Sliding at up to 100 mm/s Against Lead Zirconate Titanate Films for Atomic Force Microscopy Probe-Based Ferroelectric Recording Technology," *J. Vac. Sci. Technol. A* **26**, 783–793.

Le, H. (1998), "Progress and Trends in Ink-jet Printing Technology," *J. Imaging Sci. Technol.* **42**, 49–62.
Lebeck, A.O. (1991), *Principles and Design of Mechanical Face Seals*, Wiley, New York.
Lee, E.R. (2003), *Microdrop Generation*, CRC Press, Boca Raton, FL.
Lee, S.C. and Cheng, H.S. (1991), "Scuffing Theory Modelling and Experimental Correlations," *ASME J. Trib.* **113**, 327–333.
Lehr, H., Abel, S., Doppler, J., Ehrfeld, W., Hagemann, B., Kamper, K.P., Michel, F., Schulz, Ch., and Thurigen, Ch. (1996), "Microactuators as Driving Units for Microrobotic Systems," *Proc. Microrobotics: Components and Applications* (A. Sulzmann, ed.), Vol. 2906, pp. 202–210, SPIE, Bellingham, WA.
Lehr, H., Ehrfeld, W., Hagemann, B., Kamper, K.P., Michel, F., Schulz, Ch., and Thurigen, Ch. (1997), "Development of Micro-Millimotors," *Min. Invas. Ther. Allied Technol.* **6**, 191–194.
Liu, H. and Bhushan, B. (2004a), "Nanotribological Characterization of Digital Micromirror Devices Using an Atomic Force Microscope," *Ultramicroscopy* **100**, 391–412.
Liu, H. and Bhushan, B. (2004b), "Investigation of Nanotribological and Nanomechanical Properties of the Digital Micromirror Device by Atomic Force Microscope," *J. Vac. Sci. Technol. A* **22**, 1388–1396.
Liu, L.X. and Spakovszky, Z.S. (2005), "Effect of Bearing Stiffness Anisotropy on Hydrostatic Micro Gas Journal Bearing Dynamic Behavior," *Proceedings of ASME Turbo Expo 2005*, Paper No. GT-2005-68199, ASME, New York.
Martin, J.R. and Zhao, Y. (1997), "Micromachined Device Packaged to Reduce Stiction," U.S. Patent No. 5,694,740, Dec. 9.
McNab, I.R. and Johnson, J. L. (1980), "Brush Wear," in *Wear Control Handbook* (M. B. Peterson and W.O. Winer, eds.), pp. 1053–1101, ASME, New York.
Mee, C.D. and Daniel, E.D. (1996), *Magnetic Recording Technology*, Second edition, McGraw-Hill, New York.
Meetham, G.W. (ed.) (1981), *The Development of Gas Turbine Materials*, Applied Science Publishers, London, UK.
Mehregany, M, Gabriel, K.J., and Trimmer, W.S.N. (1988), "Integrated Fabrication of Polysilicon Mechanisms," *IEEE Trans. Electronic Dev.* **35**, 719–723.
Merritt, H.E. (1971), *Gear Engineering*, Pitman, London, UK.
Michel, F. and Ehrfeld, W. (1998), "Microfabrication Technologies for High Performance Microactuators" in *Tribology Issues and Opportunities in MEMS* (B. Bhushan, ed.), pp. 53–72, Kluwer Academic, Dordrecht, Netherlands.
Mulhern, G.T., Soane, D.S., and Howe, R.T. (1993), "Supercritical Carbon Dioxide Drying of Microstructures," *Proc. Int. Conf. on Solid-State Sensors and Actuators*, pp. 296–299, IEEE, New York.
Neale, M.J. (ed.) (1973), *Tribology Handbook*, Newnes-Butterworth, UK.
O'Connor, J.J., Boyd, J., and Avallone, E.A. (1968), *Standard Handbook of Lubrication Engineers*, McGraw-Hill, New York.
Parsons, M. (2001), "Design and Manufacture of Automotive Pressure Sensors," *Sensors* **18**, 32–46.
Paxton, R.R. (1979), *Manufactured Carbon: A Self-Lubricating Material for Mechanical Devices*, CRC Press, Boca Raton, FL.
Peterson, M.B. and Winer, W.O. (eds.) (1980), *Wear Control Handbook*, ASME, New York.
Robbins, R.A. and Jacobs, S.J. (2001), "Lubricant Delivery for Micromechanical Devices," U.S. Patent No. 6,300,294 B1, Oct. 9.
Rogowski, A.R. (1953), *Elements of Internal Combustion Engines*, McGraw-Hill, New York.
Samuels, L.E. (1982), *Metallographic Polishing by Mechanical Methods*, Third edition, ASM International, Metals Park, OH.
Schey, J.A. (1977), *Introduction to Manufacturing Processes*, McGraw-Hill, New York.
Scott, D. (1977), "Lubricant Effects on Rolling Contact Fatigue – A Brief Review, Performance, Testing of Lubricants," *Proc. Symp on Rolling Contact Fatigue* (R. Tourret and E. P. Wright, eds.), pp. 3–15, 39–44, Heyden, London.
Scott, D., Smith, A.I., Tait, J., and Tremain, G.R. (1975), "Metals and Metallurgical Aspects of Piston Ring Scuffing – A Literature Survey," *Wear* **33**, 293–315.
Shaw, M.C. (1996), *Principles of Abrasive Processing*, Oxford University Press, Oxford, UK.
Shaw, M.C. (1997), *Metal Cutting Principles*, Second edition, Clarendon Press, Oxford, UK.
Shigley, J.E. and Mischke, C.R. (1989), *Mechanical Engineering Design*, Fifth edition, McGraw-Hill, New York.
Shobert, E.I. (1965), *Carbon Brushes*, Chemical Publishing Corp., New York.
Smith, G. (1997), "The Application of Microtechnology to Sensors for the Automotive Industry," *Microelectronics J.* **28**, 371–379.
Spearing, S.M. and Chen, K.S. (2001), "Micro-Gas Turbine Engine Materials and Structures," *Ceramic Eng. and Science Proc.* **18**, 11–18.

Srivastava, D. (2004), "Computational Nanotechnology of Carbon Nanotubes," in *Carbon Nanotubes: Science and Applications* (M. Meyyappan, ed.), pp. 25–63, CRC Press, Boca Raton, FL.
Stair, W.K. (1984), "Dynamic Seals," in *Handbook of Lubrication: Theory and Practice of Tribology, Vol. 2: Theory and Design* (E. R. Booser, ed.), pp. 581–622, CRC Press, Boca Raton, FL.
Sulouff, R.E. (1998), "MEMS Opportunities in Accelerometers and Gyros and the Microtribology Problems Limiting Commercialization," in *Tribology Issues and Opportunities in MEMS* (B. Bhushan, ed.), pp. 109–120, Kluwer Academic, Dordrecht, The Netherlands.
Summers-Smith, J.D. (1994), *An Introductory Guide to Industrial Tribology*, Mech. Eng. Publications Ltd., London, UK.
Suzuki, K. (2002), "Micro Electro Mechanical Systems (MEMS) Micro-Switches for Use in DC, RF, and Optical Applications," *Jpn. J. Appl. Phys.* **41**, 4335–4339.
Tai, Y.C., Fan, L.S., and Muller, R.S. (1989), "IC-processed Micro-Motors: Design, Technology and Testing," *Proc. IEEE Micro Electro Mechanical Systems*, 1–6.
Tallian, T.E. (1967), "On Competing Failure Modes in Rolling Contact," *ASLE Trans.* **10**, 418–439.
Tallian, T.E., Baile, G.H., Dalal, H., and Gustafsson, O.G. (1974), *Rolling Bearing Damage*, SKF Industries Inc., King of Prussia, PA.
Tang, W.C. and Lee, A.P. (2001), "Defense Applications of MEMS," *MRS Bulletin* **26**, 318-319. Also see www.darpa.mil/mto/mems.
Tanner, D.M., Smith, N.F., Irwin, L.W., Eaton, W.P., Helgesen, K.S., Clement, J.J., Miller, W.M., Walraven, J.A., Peterson, K.A., Tangyunyong, P., Dugger, M.T., and Miller, S.L. (2000), *MEMS Reliability: Infrastructure, Test Structures, Experiments, and Failure Modes*, SAND2000-0091, Sandia National Laboratories, Albuquerque, New Mexico. Download from http://www.prod.sandia.gov/cgi-bin/techlib/access-control.pl/2000/000091.pdf.
Taylor, B.J. and Eyre, T.S. (1979), "A Review of Piston Rings and Cylinder Liner Materials," *Tribol. Int.* **12**, 79–89.
Taylor, C.F. (1966 and 1968), *The Internal Combustion Engine in Theory and Practice*, Vol. 1 (1966), Vol. 2 (1968), MIT Press, Cambridge, MA.
Ting, L. L. (1980), "Lubricated Piston Rings and Cylinder Bore Wear," in *Wear Control Handbook* (M.B. Peterson and W.O. Winer, eds.), pp. 609–665, ASME, New York.
Tipnis, V.A. (1980), "Cutting Tool Wear," in *Wear Control Handbook* (M.B. Peterson and W.O. Winer, eds.), pp. 891–930, ASME, New York.
van der Berg, A. (ed.) (2003), *Lab-on-a-Chip: Chemistry in Miniaturized Synthesis and Analysis Systems*, Elsevier, Amsterdam.
van der Wiel, W.G., De Franceschi, S., Elzerman, J.M., Fujisawa, T., Tarucha, S., and Kouwenhoven, L.P. (2003), "Electron Transport Through Double Quantum Dots," *Rev. Modern Phys.* **75**, 1–22.
Vettiger, P., Brugger, J., Despont, M., Drechsler, U., Duerig, U., Haeberle, W., Lutwyche, M., Rothuizen, H., Stutz, R., Widmer, R., and Binnig, G. (1999), "Ultrahigh Density, High Data-Rate NEMS Based AFM Data Storage System," *Microelec. Eng.* **46**, 11–17.
Weertman, W.L. and Dean, S.W. (1981), "Chrysler Corporation's New 2.2 Liter 4 Cylinder Engine," SAE paper 810007.
Zaretsky, E.V. (1992), *Life Factors for Rolling Bearings*, STLE, Park Ridge, IL.

Further Reading

Bhushan, B. (2001a), *Modern Tribology Handbook, Vol. 2: Materials, Coatings and Industrial Applications*, CRC Press, Boca Raton, FL.
Bhushan, B. (2001b), *Fundamentals of Tribology and Bridging the Gap Between the Macro- and Micro/Nanoscales*, NATO Science Series II. Mathematics, Physics and Chemistry – Vol. 10, Kluwer Academic Publishers, Dordrecht, The Netherlands.
Bhushan, B. (2010), *Springer Handbook of Nanotechnology*, Third edition, Springer-Verlag, Heidelberg, Germany.
Bhushan, B. and Gupta, B.K. (1997), *Handbook of Tribology: Materials, Coatings and Surface Treatments*, McGraw-Hill, New York (1991), reprint ed., Krieger Publishing Co., Malabar, FL.
Booser, E.R. (ed.) (1983), *CRC Handbook of Lubrication, Vol. I Application and Maintenance*, CRC Press, Boca Raton, FL.
Booser, E.R. (ed.) (1984), *CRC Handbook of Lubrication, Vol. II, Theory and Design*, CRC Press, Boca Raton, FL.

Neale, M.J. (ed.) (1973), *Tribology Handbook*, Newnes-Butterworth, UK.

O'Connor, J.J., Boyd, J., and Avallone, E.A. (1968), *Standard Handbook of Lubrication Engineers*, McGraw-Hill, New York.

Peterson, M.B. and Winer, W.O. (eds.) (1980), *Wear Control Handbook*, ASME, New York.

Totten, G.E. (2006), *Handbook of Lubrication and Tribology: Vol. 1 – Applications and Maintenance*, Second edition, CRC Press, Boca Raton, Florida.

13

Green Tribology and Biomimetics

13.1 Introduction

The ecological or green tribology is a relatively new field. Green tribology is defined as the science and technology of the tribological aspects which provide ecological balance and minimize environmental and biological impacts (Bartz, 2006; Nosonovsky and Bhushan, 2010, 2012). Reduction in consumption of energy resources is also an important aspect of green tribology. Energy or environmental sustainability and whatever has an impact upon today's environment should be emphasized. Green tribology requires the use of environmentally friendly materials, lubricants, and processes. The first scientific volume on green tribology was published in 2010 in *Philosophical Transaction of the Royal Society A* (Nosonovsky and Bhushan, 2010) and later a book came out in 2012 (Nosonovsky and Bhushan, 2012).

Tribological aspects are important in various applications. Since the early 2000s, there has been significant interest in renewable energy production, such as wind turbines, tidal turbines, or solar panels. Many of these energy production technologies present their unique tribological challenges. Environmentally friendly tribological components, materials and surfaces can be fabricated by mimicking nature, a field referred to as biomimetics. In this chapter, we introduce green tribology and biomimetics and its applications in tribology.

13.2 Green Tribology

Green tribology can be viewed in the broader context of two "green" areas: green engineering and green chemistry (Nosonovsky and Bhushan, 2010). The US Environmental Protection Agency (EPA) defined green engineering as "the design, commercialization and use of processes and products that are technically and economically feasible while minimizing (i) generation of pollution at the source (ii) risk to human health and the environment" (Anonymous, 2010).

Green chemistry, also known as sustainable chemistry, is defined as "the design of chemical products and processes that reduce or eliminate the use or generation of hazardous substances" (Anonymous, 2010). Based on Nosonovsky and Bhushan (2010), the focus of green chemistry is on minimizing the hazards and maximizing the efficiency of any chemical choice. It is distinct

Introduction to Tribology, Second Edition. Bharat Bhushan.
© 2013 John Wiley & Sons, Ltd. Published 2013 by John Wiley & Sons, Ltd.

from environmental chemistry which focuses on chemical phenomena in the environment. While environmental chemistry studies the natural environment as well as pollutant chemicals in nature, green chemistry seeks to reduce and prevent pollution at its source. Green chemistry technologies provide a number of benefits, including reduced waste, eliminating costly end-of-the-pipe treatments, safer products, reduced use of energy and resources, and improved competitiveness of chemical manufacturers and their customers. Green chemistry consists of chemicals and chemical processes designed to reduce or eliminate negative environmental impacts. The use and production of these chemicals may involve reduced waste products, non-toxic components, and improved efficiency.

The principles of green chemistry are applicable to green tribology as well. However, since tribology involves not only the chemistry of surfaces but also other aspects related to the mechanics and physics of surfaces, there is a need to modify these principles.

13.2.1 Twelve Principles of Green Tribology

Twelve principles of green tribology have been proposed by Nosonovsky and Bhushan (2010). Some principles are related to the design and manufacturing of tribological applications (iii–x), while others belong to their operation (i–ii and xi–xii).

(i) *Minimization of heat and energy dissipation.* Friction is the primary source of energy dissipation. According to some estimates, about one-third of the energy consumption in the United States is spent overcoming friction. Most energy dissipated by friction is converted into heat and leads to heat pollution of the atmosphere and the environment. The control of friction and friction minimization, which leads to both energy conservation and the prevention of damage to the environment due to the heat pollution, are a primary task of tribology. It is recognized that for certain tribological applications (e.g., car brakes and clutches) high friction is required; however, ways of effective use of energy for these applications should be sought as well.

(ii) *Minimization of wear* is the second most important task of tribology which has relevance to green tribology. In most industrial applications wear is undesirable. It limits the lifetime of components and therefore creates the problem of their recycling. Wear can also lead to catastrophic failure. In addition, wear creates debris and particles which contaminate the environment and can be hazardous for humans in certain situations. For example, wear debris generated after human joint replacement surgery is the primary source of long-term complications in patients.

(iii) *Reduction or complete elimination of lubrication and self-lubrication.* Lubrication is a focus of tribology since it leads to the reduction of friction and wear. However, lubrication can also lead to environmental hazards. It is desirable to reduce lubrication or achieve the self-lubricating regime, when no external supply of lubrication is required. Tribological systems in living nature often operate in the self-lubricating regime. For example, joints form essentially a closed self-sustainable system.

(iv) *Natural lubrication* (e.g., vegetable oil-based) should be used in cases when possible, since it is usually environmentally friendly.

(v) *Biodegradable lubrication* should also be used when possible to avoid environmental contamination. In particular, water lubrication is an area which has attracted the attention of researchers in recent years. Natural oil (such as canola) lubrication is another option, especially discussed in the developing countries.

(vi) *Sustainable chemistry and green engineering principles* should be used in the manufacturing of new components for tribological applications, coatings, and lubricants.

(vii) *Biomimetic approach* should be used whenever possible. This includes biomimetic surfaces, materials, and other biomimetic and bio-inspired approaches, since they tend to be more ecologically friendly.

(viii) *Surface texturing* should be applied to control surface properties. Conventional engineered surfaces have random roughness, and the randomness is the factor which makes it extremely difficult to overcome friction and wear. On the other hand, many biological functional surfaces have complex structures with hierarchical roughness, which defines their properties. Surface texturing provides a way to control many surface properties relevant to making tribo-systems more ecologically friendly.

(ix) *Environmental implications of coatings* and other methods of surface modification (texturing, depositions, etc.) should be investigated and taken into consideration.

(x) *Design for degradation* of surfaces, coatings, and tribological components. Similar to green chemistry applications, the ultimate degradation/utilization should be taken into consideration during design.

(xi) *Real-time monitoring, analysis, and control* of tribological systems during their operation should be implemented to prevent the formation of hazardous substances.

(xii) *Sustainable energy applications* should become the priority of tribological design as well as engineering design in general.

13.2.2 Areas of Green Tribology

Important areas for green tribology include biodegradable and environmentally-friendly lubricants and materials and tribology of renewable and/or sustainable sources of energy (Nosonovsky and Bhushan, 2010, 2012). Bio-inspired materials and surfaces can developed for green tribology applications and will be discussed in a following section.

13.2.2.1 Biodegradable Lubricants and Materials

Natural (e.g., vegetable-oil based or animal-fat based) biodegradable lubricants, are the oils that can be used for engines, hydraulic applications, and metal cutting applications (Nosonovsky and Bhushan, 2010, 2012). In particular, corn, soybean, coconut oils have been used so far (the latter is of particular interest in tropical countries such as India). These lubricants are potentially biodegradable, although in some cases chemical modification or additives for best performance are required. Vegetable oils can have good lubricity, comparable to that of mineral oil. In addition, they have a very high viscosity index and high flash/fire points. However, natural oils often lack sufficient oxidative stability, which means that the oil will oxidize rather quickly during use, becoming thick, and will polymerize to a plastic-like consistency. Chemical modification of vegetable oils and/or the use of antioxidants can address this problem (Mannekote and Kailas, 2009).

Ionic liquids (ILs) have been explored as lubricants for various device applications due to their good electrical conductivity and thermal conductivity, where the latter allows frictional heating dissipation (Palacio and Bhushan, 2010). Since they do not emit volatile organic compounds, they are regarded as "green" lubricants. It has been shown that some ILs can match or even exceed the tribological behavior of high performance lubricants.

Powder lubricants and, in particular, boric acid lubricants tend to be more ecologically friendly than the traditional liquid lubricants. Boric acid and MoS_2 powder can also be used as an additive to the natural oil. Friction and wear experiments show that the nanoparticles of boric acid additive exhibited superior friction and wear performance with respect to various lubricants (Lovell *et al.*, 2010).

It has been suggested that environmental aspects should become an integral part of brake design (Yun *et al.*, 2010). Preliminary data obtained with animal experiments revealed that inhaled metallic particles remain deposited in the lungs of rats six months after exposure. The presence of inhaled particles had a negative impact on health and led to emphysema (destroyed alveoli), inflammatory response, and morphological changes of the lung tissue.

13.2.2.2 Renewable Energy

The tribology of renewable sources of energy is a relatively new field of tribology (Nosonovsky and Bhushan, 2010, 2012). Based on the US President's proposed clean energy standard, about 80% of electricity will come from clean energy sources by 2035. There are a number of renewable energy production sources whose usage continues to grow. Important renewable energy systems include conversion of wind or tidal streams to rotational motion for generating electricity and the use of solar panels to harness solar energy. Wind and tidal turbines include bearings and gears with unique challenges because of the high loads and their size and the need for field service.

Figure 13.2.1 shows a photograph of wind turbine blades. Wind turbines consist of a rotor with wing-shaped blades that are attached to a hub. The hub is attached to the nacelle which

Figure 13.2.1 Photograph of a wind turbine.

Table 13.2.1 Typical specifications of a commercial three-bladed, upwind, horizontal-axis wind turbine for power generation of 1.5 MW (GE Energy 2.5 MW Series TC3).

Parameter	Machine data
Number of Blades	3
Rotor diameter	103 m (swept area = 8328 m^2)
Tower	Height: 85–100 m, weight: ~60,000 kg
Rotational speed	25–60 RPM
Components	**Machine data**
Nacelle (which houses gear box)	Size of a school bus and weight ~ 50,000 kg
Gear box	Three-stage planetary/helical gears with a gear ratio of 1:78. Weight: ~ 20,000 kg
Blade-pitch bearings	Dual, four-point ball bearings
Main shaft bearings	Double-row spherical roller bearings
Lubrication system	Forced lubrication

houses the gear box, the drive train, the support bearings, and the generator. Table 13.2.1 shows the typical specifications of a commercial 2.5 MW wind turbine. Figure 13.2.2 shows a turbine Nacelle layout. Figure 13.2.3 shows a configuration of a three-stage gear box for a 2–3 MW wind turbine. Since the turbine power is proportional to the area swept by the blades (the square of the rotor diameter), the size of the rotor blades has increased dramatically

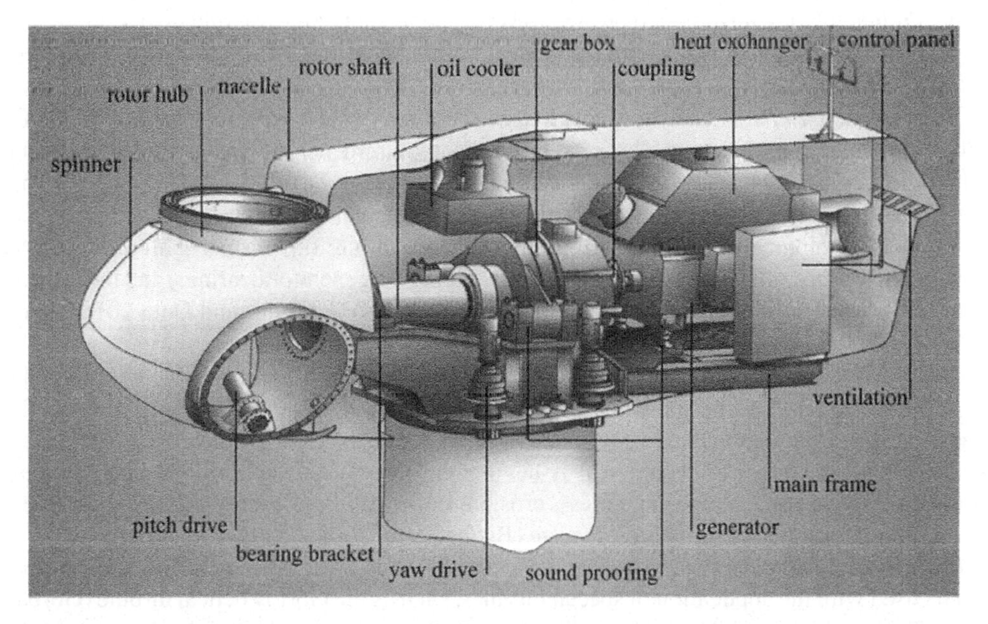

Figure 13.2.2 Wind turbine Nacelle layout which houses gear box, drive train, support bearings and electric generator (GE Energy 2.5 MW).

Figure 13.2.3 Configuration of three-stage gear box for 2-3 MW wind turbine. Reproduced with permission from Hau, E. (2006), *Wind Turbines*, Springer-Verlag, Berlin, Germany. Copyright 2006. Springer.

(Hau, 2006). Furthermore, to take advantage of less turbulent but faster wind, up to 100 m above ground, blades are mounted on high towers. The weight of the rotating blades dominates over inertial loads which puts enormous demand on the bearings. To minimize the weight of the rotor blades, they are made of fiber composites. The gear box is designed to function as a speed increaser and transmit power from the 25–60 RPM turbine rotor to the 1000–1800 RPM electric generator. The gear box ratio requirements are rather large. Since the tribological components need to be serviced on-site (on-shore or off-shore) with components located at high elevations, reliability of rather large components becomes a major tribological challenge. Tribological issues in wind turbines include failure of the mainshaft and gearbox bearings and gears, water contamination, electric arcing on generator bearings, and the erosion of blades (due to solid particles, cavitation, rain, hail stones) (Kotzalas and Doll, 2010; Wood *et al.*, 2010; Terrell *et al.*, 2012).

Tidal power turbines are another important method of producing renewable energy. Tidal power turbines are especially popular in Europe (particularly, in the UK), which remains the leader in this area, although several potential sites in North America have been suggested. There are several specific tribological issues related to tidal power turbines, such as their lubrication (seawater, oils, and greases), erosion, corrosion, and biofouling, as well as the interaction between these modes of damage (Batten *et al.*, 2008; Wood *et al.*, 2010).

Besides tidal, the ocean water flow and wave energy and river flow energy (without dams) can be used with the application of special turbines, such as the Gorlov helical turbine (Gorban *et al.*, 2001), which provides the same direction of rotation independent of the direction of the current flow. These applications also involve specific tribological issues.

Geothermal energy plants are used in the United States (in particular, on the Pacific coast and Alaska); however, their use is limited to the geographical areas at the edges of tectonic plates (Rybach, 2007). In 2007, they produced 2.7 GW of energy in the US, with the Philippines (2.0 GW) and Indonesia (1.0 GW) in second and third place (Bertani, 2007). There are many issues related to the tribology of geothermal energy sources.

13.3 Biomimetics

Biomimetics means mimicking biology or living nature. Biomimetics allows biologically inspired design, adaptation, or derivation from nature. The word biomimetics was coined by the polymath Otto Schmitt in 1957, who, in his doctoral research, developed a physical device that mimicked the electrical action of a nerve. Biomimetics is derived from the Greek word biomimesis. Other words used include bionics (coined in 1960 by Jack Steele of Wright-Patterson Air Force Base in Dayton, OH), biomimicry, and biognosis. The word "biomimetics" first appeared in *Webster's Dictionary* in 1974 and is defined as: "the study of the formation, structure or function of biologically produced substances and materials (as enzymes or silk) and biological mechanisms and processes (as protein synthesis or photosynthesis) especially for the purpose of synthesizing similar products by artificial mechanisms which mimic natural ones." The field of biomimetics is highly interdisciplinary. It involves the understanding of biological functions, structures, and principles of various objects found in living nature by biologists, physicists, chemists, and material scientists, and the biologically-inspired design and fabrication of various materials and devices of commercial interest by engineers, material scientists, chemists, biologists, and others (Bhushan, 2009, 2012).

Nature has evolved over the 3.8 billion years since life is estimated to have appeared on the Earth (Gordon, 1976). Biological materials are highly organized from the molecular to the nano-, micro-, and macroscales, often in a hierarchical manner with an intricate nanoarchitecture that ultimately makes up a myriad of different functional elements (Alberts *et al.*, 2008). Nature uses commonly found materials. Properties of materials and surfaces result from a complex interplay between surface structure and morphology and physical and chemical properties. Many materials, surfaces, and objects in general provide multi-functionality.

Bio-inspired materials and surfaces are eco-friendly or green which have generated significant interest and are helping to shape green science and technology. Many of the biological objects exhibit controlled adhesion, low friction, wear resistance, good lubrication and high mechanical properties.

The objective of biomimetics research is to develop biologically-inspired materials and surfaces of commercial interest (Bhushan, 2012). The approach is threefold:

1. Objects are selected from living nature that provide functionality of commercial interest.
2. The objects are characterized to understand how a natural object provides functionality. Then it is modeled and structures are generally fabricated in the lab using nature's route to verify one's understanding. Modeling is used to develop optimum structures.
3. Nature has a limited toolbox and uses rather basic materials and routine fabrication methods; it capitalizes on hierarchical structures. Once one understands how nature does it, one can then fabricate optimum structures using smart materials and fabrication techniques to provide functionality of interest.

13.3.1 Lessons from Nature

The understanding of the functions provided by objects and processes found in nature can guide us to design and produce nanomaterials, nanodevices, and processes (Bhushan, 2009, 2012). There are a large number of objects, including bacteria, plants, land and aquatic animals, and seashells, with properties of commercial interest. Figure 13.3.1 provides an overview of various objects from nature and their selected functions (Bhushan, 2009, 2012). These include bacteria (Jones and Aizawa, 1991), plants (Koch *et al.*, 2008, 2009), insects/spiders/lizards/frogs (Autumn *et al.*, 2000; Gorb, 2001; Bhushan, 2007, 2010), aquatic animals (Bechert *et al.*, 1997, 2000; Dean and Bhushan, 2010), birds (Jakab, 1990; Bechert *et al.*, 2000), seashells/bones/teeth (Lowenstam and Weiner, 1989; Sarikaya and Aksay, 1995; Mann, 2001; Alexander and Diskin, 2004; Meyers *et al.*, 2008), spiders' web (Jin and Kaplan, 2003; Bar-Cohen, 2011), moth-eye effect (Genzer and Efimenko, 2006; Mueller, 2008) and structure coloration (Parker, 2009), the

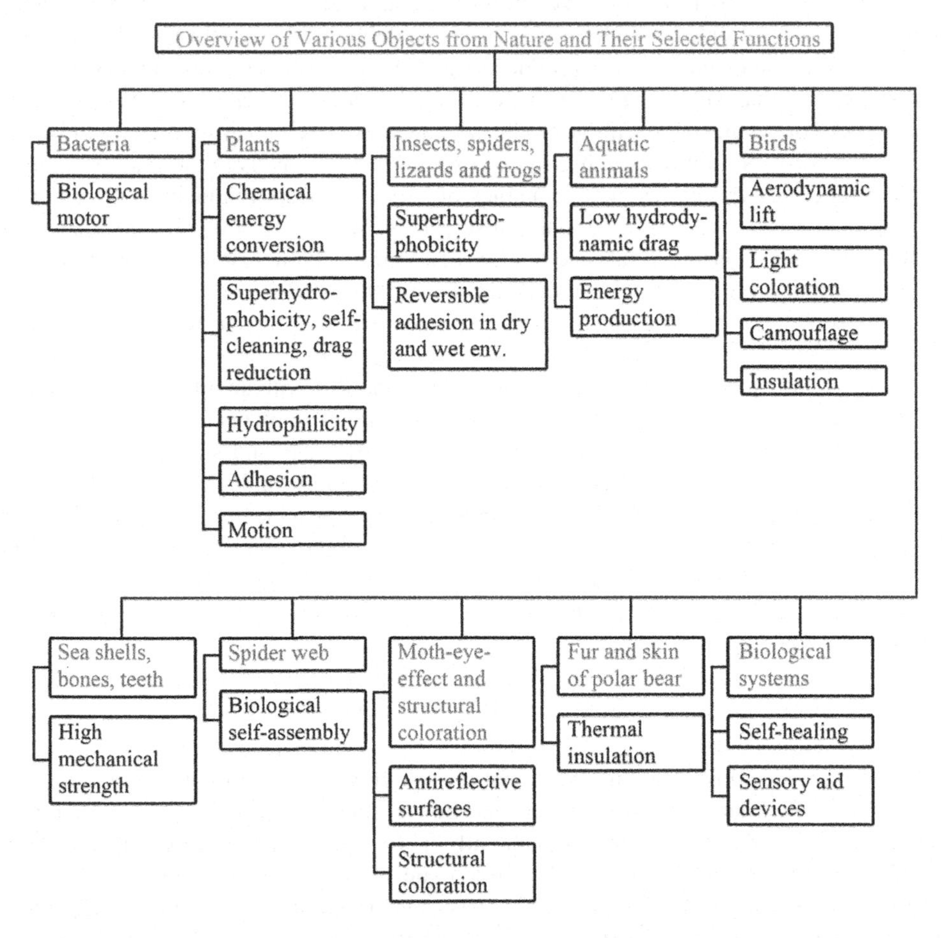

Figure 13.3.1 An overview of various objects from nature and their selected function. Reproduced with permission from Bhushan, B. (2009), "Biomimetics: Lessons from Nature – An Overview," *Phil. Trans. R. Soc. A* **367**, 1445–1486, by permission of the Royal Society.

fur and skin of polar bears (Stegmaier *et al.*, 2009), and biological systems with self-healing capacity (Fratzl and Weinkamer, 2007; Nosonovsky and Bhushan, 2009), and sensory-aid devices (Barth *et al.*, 2003; Bar-Cohen, 2011).

Figure 13.3.2 shows a montage of some examples from nature (Bhushan, 2009, 2012). Some leaves of water-repellent plants, such as *Nelumbo nucifera* (Lotus), are known to be superhydrophobic, self-cleaning, and antifouling, due to their hierarchical roughness (microbumps superimposed with a nanostructure) and the presence of a hydrophobic wax coating (Neinhuis and Barthlott, 1997; Barthlott and Neinhuis, 1997; Wagner *et al.*, 2003; Burton and Bhushan, 2006; Bhushan and Jung, 2006, 2011; Bhushan, 2009, 2011; Koch *et al.*, 2008, 2009). Water droplets on these surfaces readily sit on the apex of nanostructures because air bubbles fill in the valleys of the structure under the droplet. Therefore, these leaves exhibit considerable superhydrophobicity, Figure 13.3.2(a). Two strategies used for catching insects by plants for digestion are having sticky surfaces or sliding structures. As an example, for catching insects using sticky surfaces, the glands of the carnivorous plants of the genus *Pinguicula* (butterworts) and *Drosera* (sundew), shown in Figure 13.3.2(b), secrete adhesives and enzymes to trap and digest small insects, such as mosquitoes and fruit flies (Koch *et al.*, 2009). Water striders (*Gerris remigis*) have the ability to stand and walk upon a water surface without getting wet, Figure 13.3.2(c). Even the impact of rain droplets with a size greater than the water strider's size does not immerse it in the water. Gao and Jiang (2004) showed that the special hierarchical structure of the water strider's legs, which are covered by large numbers of oriented tiny hairs (microsetae) with fine nanogrooves and covered with cuticle wax, makes the leg surfaces superhydrophobic, is responsible for the water resistance, and enables them to stand and walk quickly on the water surface.

A gecko is the largest animal that can produce high (dry) adhesion to support its weight with a high factor of safety. Gecko skin is comprised of a complex hierarchical structure of lamellae, setae, branches, and spatula (Autumn *et al.*, 2000; Gao *et al.*, 2005; Bhushan, 2007). The attachment pads on two feet of the Tokay gecko have an area of approximately 220 mm^2, Figure 13.3.2(d). Approximately 3×10^6 setae on their toes that branch off into about three billion spatula on two feet can produce the clinging ability of approximately 20 N (the vertical force required to pull a lizard down a nearly vertical (85°) surface) and allow them to climb vertical surfaces at speeds of over 1 m/s, with the capability to attach or detach their toes in milliseconds (Bhushan, 2007).

Shark skin, which is a model from nature for a low drag surface, is covered by very small individual tooth-like scales called dermal denticles (little skin teeth), ribbed with longitudinal grooves (aligned parallel to the local flow direction of the water). These grooved scales lift vortices to the tips of the scales, resulting in water moving efficiently over their surface (Bechert *et al.*, 2000; Dean and Bhushan, 2010). The spacing between these dermal denticles is such that microscopic aquatic organisms have difficulty adhering to the surface, making the skin surface antifouling (Carman *et al.*, 2006; Genzer and Efimenko, 2006; Kesel and Liedert, 2007; Ralston and Swain, 2009; Bixler and Bhushan, 2012). An example of scale structure on the right front of a Galapagos shark (*Carcharhinus galapagensis*) is shown in Figure 13.3.2(e) (Jung and Bhushan, 2010).

Birds consist of several consecutive rows of covering feathers on their wings, which are flexible, Figure 13.3.2(f). These movable flaps develop the lift. When a bird lands, a few feathers are deployed in front of the leading edges of the wings, which help to reduce the drag on the wings.

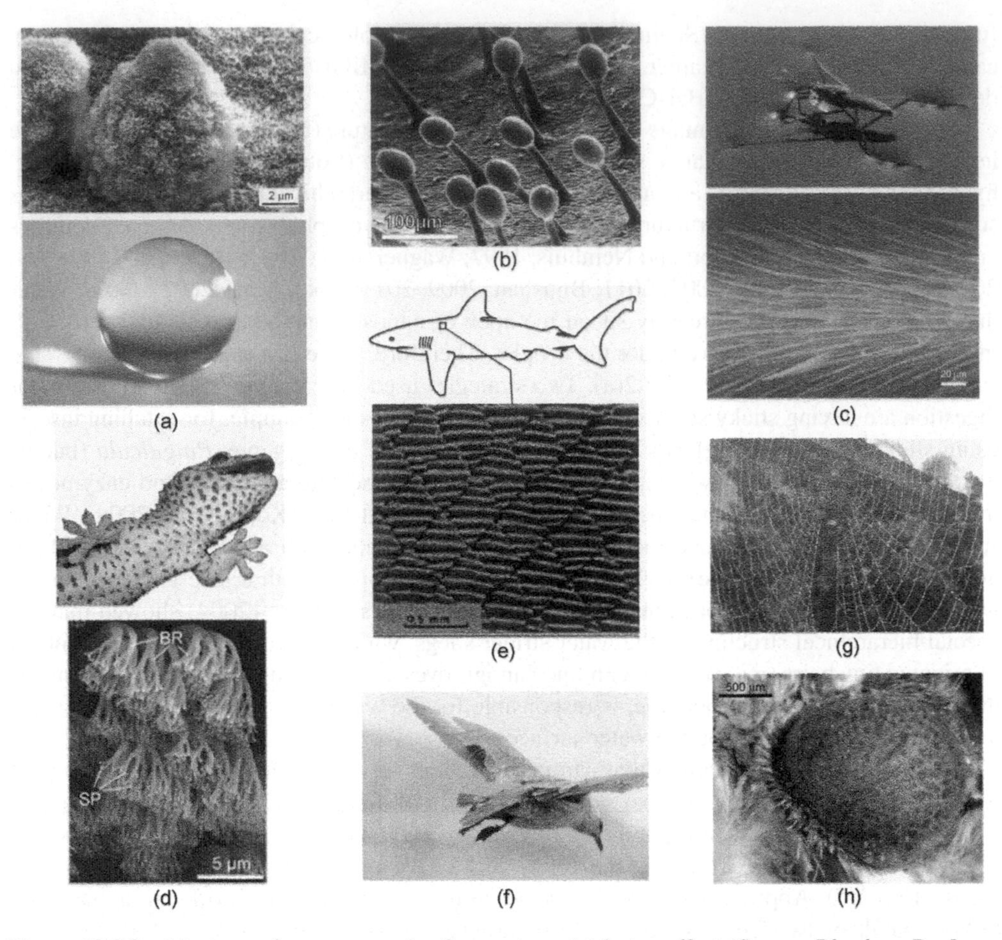

Figure 13.3.2 Montage of some examples from nature: (a) Lotus effect (*Source*: Bhushan, B., Jung, Y.C., and Koch, K. (2009), "Micro-, Nano- and Hierarchical Structures for Superhydrophobicity, Self-Cleaning and Low Adhesion," *Phil. Trans. R. Soc. A* **367**, 1631–1672, by permission of the Royal Society), (b) glands of carnivorous plant secrete adhesive to trap insects (Reproduced with permission from Koch, K., Bhushan, B., and Barthlott, W. (2009), "Multifunctional Surface Structures of Plants: An Inspiration for Biomimetics," *Prog. Mater. Sci.* **54**, 137–178. Copyright 2009. Elsevier), (c) water strider walking on water (Reproduced with permission from Gao, X. F. and Jiang, L. (2004), "Biophysics: Water-repellent Legs of Water Striders," *Nature* **432**, 36. Copyright 2004. Nature Publishing Group), (d) gecko foot exhibiting reversible adhesion (Reproduced with permission from Gao, H., Wang, X., Yao, H., Gorb, S., and Arzt, E. (2005), "Mechanics of Hierarchical Adhesion Structures of Geckos," *Mech. Mater.* **37**, 275–285. Copyright 2005. Elsevier), (e) scale structure of shark reducing drag (Reproduced with permission from Jung, Y. C. and Bhushan, B. (2010), "Biomimetic Structures for Fluid Drag Reduction in Laminar and Turbulent Flows," J. *Phys.: Condens. Matter* **22**, 035104. Copyright 2010. IOP Science), (f) wings of a bird in landing approach, (g) spiderweb made of silk material (Reproduced with permission from Bar-Cohen, Y. (2011), *Biomimetics: Nature-Based Innovation*, CRC Press, Boca Raton, FL. Copyright 2011. Taylor and Francis), and (h) antireflective moth's eye (Reproduced with permission from Genzer, J. and Efimenko, K. (2006), "Recent Developments in Superhydrophobic Surfaces and Their Relevance to Marine Fouling: A Review," *Biofouling* **22**, 339–360. Copyright 2006. Taylor and Francis).

The spider generates silk fiber and has a sufficient supply of raw material for its silk to span great distances (Jin and Kaplan, 2003; Bar-Cohen, 2011). Spiderweb is a structure built of a one-dimensional fiber, Figure 13.3.2(g). The fiber is very strong and continuous and is insoluble in water. The web can hold a significant amount of water droplets, and it is resistant to rain, wind, and sunlight (Sarikaya and Aksay, 1995; Bar-Cohen, 2011).

The eyes of moths are antireflective to visible light and consist of hundreds of hexagonally organized nanoscopic pillars, each approximately 200 nm in diameter and height, which result in a very low reflectance for visible light, Figure 13.3.2(h) (Genzer and Efimenko, 2006; Mueller, 2008). These nanostructures' optical surfaces make the eye surface nearly antireflective in any direction.

13.3.2 Industrial Significance

The word biomimetics is relatively new; however, our ancestors looked to nature for inspiration and development of various materials and devices many centuries ago (Ball, 2002; Bar-Cohen, 2011; Vincent *et al.*, 2006; Anonymous, 2007; Meyers *et al.*, 2008; Bhushan, 2012). For example, the Chinese tried to make artificial silk some 3000 years ago. Leonardo da Vinci, a genius of his time, studied how birds fly and proposed designs of flying machines. In the twentieth century, various products, including the design of aircraft, have been inspired by nature. Since the 1980s, the artificial intelligence and neural networks in information technology have been inspired by the desire to mimic the human brain. The existence of biocells and deoxyribonucleic acid (DNA) serves as a source of inspiration for nanotechnologists who hope one day to build self-assembled molecular-scale devices. In molecular biomimetics, proteins are being utilized in controlling materials formation in practical engineering towards self-assembled, hybrid, functional materials structure (Grunwald *et al.*, 2009; Tamerler and Sarikaya, 2009). Since the mid-1990s, the so-called Lotus effect has been used to develop a variety of surfaces for superhydrophobicity, self-cleaning, low adhesion, and drag reduction in fluid flow, as well as antifouling (Bhushan *et al.*, 2009; Bhushan, 2011; Bhushan and Jung, 2011). Replication of the dynamic climbing and peeling ability of geckos has been carried out to develop treads of wall-climbing robots (Cutkosky and Kim, 2009). Replication of shark skin has been used to develop moving objects with low drag, for example, wholebody swimsuits (Dean and Bhushan, 2010). Nanoscale architecture used in nature for optical reflection and anti-reflection has been used to develop reflecting and anti-reflecting surfaces. In the field of biomimetic materials, there is an area of bio-inspired ceramics based on seashells and other biomimetic materials. Inspired by the fur of the polar bear, artificial furs and textiles have been developed. Self-healing of biological systems found in nature is of interest for self-repair. Biomimetics is also guiding in the development of sensory-aid devices.

Various features found in nature objects are on the nanoscale. The major emphasis on nanoscience and nanotechnology since early 1990s has provided a significant impetus in mimicking nature using nanofabrication techniques for commercial applications (Bhushan, 2010). Biomimetics has spurred interest across many disciplines.

13.4 Closure

Green tribology is a novel area of science and technology. It is related to other areas of tribology as well as other "green" disciplines, namely, green engineering and green chemistry.

The twelve principles of green tribology are formulated. The field of biomimetics offers many examples in nature of materials and surfaces which can be exploited in green tribology.

References

Alberts, B., Johnson, A., Lewis, J., Raff, M. Roberts, K., and Walter, P. (eds.) (2008), *Molecular Biology of the Cell*, Garland Science, New York.

Alexander, R. M. and Diskin, A. (2004), *Human Bones: A Scientific and Pictorial Investigation*, Pi Press, New York.

Anonymous (2007), *Biomimetics: Strategies for Product Design Inspired by Nature*, Dept. of Trade and Industry, London, UK.

Anonymous (2010), *Green Engineering*, http://www.epa.gov/oppt/greenengineering/.

Autumn, K., Liang, Y.A., Hsieh, S.T., Zesch, W., Chan, W.P., Kenny, T.W., Fearing, R., and Full, R.J. (2000), "Adhesive Force of a Single Gecko Foot-Hair," *Nature* **405**, 681–685.

Ball, P. (2002), "Natural Strategies for the Molecular Engineer," *Nanotechnology*. **13**, R15–R28.

Bar-Cohen, Y. (2011), *Biomimetics: Nature-Based Innovation*, CRC Press, Boca Raton, FL.

Barth, F.G., Humphrey, J.A.C., and Secomb, T.W. (2003), *Sensors and Sensing in Biology and Engineering*, Springer-Verlag, New York.

Barthlott, W. and Neinhuis, C. (1997), "Purity of the Sacred Lotus, or Escape from Contamination in Biological Surfaces," *Planta* **202**, 1–8.

Bartz, W.J. (2006), "Ecotribology: Environmentally acceptable tribological practices," *Tribol. Int.* **39**, 728–733.

Batten, W.M.J., Bahaj, A.S., Molland, A.F., and Chaplin, J.R. (2008), "The Prediction of the Hydrodynamic Performance of Marine Current Turbines," *Renew. Energ.* **33**, 1085–1096.

Bechert, D.W., Bruse, M., Hage, W., Van Der Hoeven, J.G.T., and Hoppe, G. (1997), "Experiments on Drag-Reducing Surfaces and Their Optimization with an Adjustable Geometry," *J. Fluid Mech.* **338**, 59–87.

Bechert, D.W., Bruse, M., and Hage, W. (2000), "Experiments with Three-Dimensional Riblets as an Idealized Model of Shark Skin," *Exp. Fluids* **28**, 403–412.

Bertani, E.R. (2007), "World Geothermal Generation in 2007," *Geo-Heat Center Bulletin* **28**, 8–19.

Bhushan, B. (2007), "Adhesion of Multi-Level Hierarchical Attachment Systems in Gecko Feet," *J. Adhes. Sci. Technol.* **21**, 1213–1258.

Bhushan, B. (2009), "Biomimetics: Lessons from Nature – An Overview," *Phil. Trans. R. Soc. A* **367**, 1445–1486.

Bhushan, B. (2010), *Springer Handbook of Nanotechnology*, Third edition, Springer, Heidelberg, Germany.

Bhushan, B. (2011), "Biomimetics Inspired Surfaces for Drag Reduction and Oleophobicity/philicity," *Beilstein J. Nanotechnol.* **2**, 66–84.

Bhushan, B. (2012), *Biomimetics: Bioinspired Hierarchical-Structured Surfaces for Green Science and Technology*, Springer-Verlag, Heidelberg, Germany.

Bhushan, B. and Jung, Y.C. (2006), "Micro and Nanoscale Characterization of Hydrophobic and Hydrophilic Leaf Surface," *Nanotechnology* **17**, 2758–2772.

Bhushan, B. and Jung, Y.C. (2011), "Natural and Biomimetic Artificial Surfaces for Superhydrophobicity, Self-Cleaning, Low Adhesion, and Drag Reduction," *Prog. Mater. Sci.* **56**, 1–108.

Bhushan, B., Jung, Y.C., and Koch, K. (2009), "Micro-, Nano- and Hierarchical Structures for Superhydrophobicity, Self-Cleaning and Low Adhesion," *Phil. Trans. R. Soc. A* **367**, 1631–1672

Bixler, G.D. and Bhushan, B. (2012), "Biofouling: Lessons from Nature," *Phil. Trans. R. Soc. A* **370**, 2381–2471.

Burton, Z. and Bhushan, B. (2006), "Surface Characterization and Adhesion and Friction Properties of Hydrophobic Leaf Surfaces," *Ultramicroscopy* **106**, 709–719.

Carman, M.L., Estes, T.G., Feinburg, A.W., Schumacher, J.F., Wilkerson, W., Wilson, L.H., Callow, M.E., Callow, J.A., and Brennan, A.B. (2006), "Engineered Antifouling Microtopographies—Correlating Wettability with Cell Attachment," *Biofouling* **22**, 11–21.

Cutkosky, M. R. and Kim, S. (2009), "Design and Fabrication of Multi-Materials Structures for Bio-Inspired Robots," *Phil. Trans. R. Soc. A* **367**, 1799–1813.

Dean, B. and Bhushan, B. (2010), "Shark-Skin Surfaces for Fluid-Drag Reduction in Turbulent Flow: A Review," *Phil. Trans. R. Soc. A* **368**, 4775–4806; 368, 5737.

Fratzl, P. and Weinkamer, R. (2007), "Nature's Hierarchical Materials," *Prog. Mat. Sci.* **52**, 1263–1334.

Gao, X. F. and Jiang, L. (2004), "Biophysics: Water-repellent Legs of Water Striders," *Nature* **432**, 36.

Gao, H., Wang, X., Yao, H., Gorb, S., and Arzt, E. (2005), "Mechanics of Hierarchical Adhesion Structures of Geckos," *Mech. Mater.* **37**, 275–285.

Genzer, J. and Efimenko, K. (2006), "Recent Developments in Superhydrophobic Surfaces and Their Relevance to Marine Fouling: A Review," *Biofouling* **22**, 339–360.

Gorb, S. (2001), *Attachment Devices of Insect Cuticle*, Kluwer Academic, Dordrecht, Netherlands.

Gorban, A.N., Gorlov, A.M., and Silantyev, V.M. (2001), "Limits of the Turbine Efficiency for Free Fluid Flow," *ASME J. Energ. Resour.* **123**, 311–317.

Gordon, J.E. (1976), *The New Science of Strong Materials, or Why You Don't Fall Through the Floor*, Second edition, Penguin, London, UK.

Grunwald, I., Rischka, K., Kast, S.M., Scheibel, T., and Bargel, H. (2009), "Mimicking Biopolymers on a Molecular Scale: Nano(bio)technology Based on Engineering Proteins," *Phil. Trans. R. Soc. A* **367**, 1727–1747.

Hau, E. (2006), *Wind Turbines*, Springer-Verlag, Berlin, Germany.

Jakab, P.L. (1990), *Vision of a Flying Machine*, Smithsonian Institution Press, Washington D.C.

Jin, H.-J. and Kaplan, D.L. (2003), "Mechanism of Silk Processing in Insects and Spiders," *Nature* **424**, 1057–1061.

Jones, C.J. and Aizawa, S. (1991), "The Bacterial Flagellum and Flagellar Motor: Structure, Assembly, and Functions," *Adv. Microb. Physiol.* **32**, 109–172.

Jung, Y.C. and Bhushan, B. (2010), "Biomimetic Structures for Fluid Drag Reduction in Laminar and Turbulent Flows," *J. Phys.: Condens. Matter* **22**, 035104.

Kesel, A. and Liedert, R. (2007), "Learning from Nature: Non-Toxic Biofouling Control by Shark Skin Effect," *Comp. Biochem. Physiol. A* **146**, S130.

Koch, K., Bhushan, B., and Barthlott, W. (2008), "Diversity of Structure, Morphology, and Wetting of Plant Surfaces," *Soft Matter* **4**, 1943–1963.

Koch, K., Bhushan, B., and Barthlott, W. (2009), "Multifunctional Surface Structures of Plants: An Inspiration for Biomimetics," *Prog. Mater. Sci.* **54**, 137–178.

Kotzalas, M. and Doll, G.L. (2010), "Tribological Advancements for Reliable Wind Turbine Performace," *Phil. Trans. R. Soc. A* **368**, 4829–4850.

Lovell, M.R., Kabir, M.A., Menzes, P.L., and Higgs III, C.F. (2010), "Influence of Boric Acid Additive Size on Green Lubricant Performance," *Phil. Trans R. Soc. A* **368**, 4851–4868.

Lowenstam, H.A. and Weiner, S. (1989), *On Biomineralization*, Oxford University Press, Oxford, UK.

Mann, S. (2001), *Biomineralization*, Oxford University Press, Oxford, UK.

Mannekote, J.K. and Kailas, S.V. (2009), "Performance Evaluation of Vegetable Oils as Lubricant in a Four Stroke Engine," *Proc. Fourth World Tribology Congress*, p. 331, Kyoto, Japan, September 6–11.

Meyers, M.A., Chen, P., Lin, A.Y.M., and Seki, Y. (2008), "Biological Materials: Structure and Mechanical Properties," *Prog. Mater. Sci.* **53**, 1–206.

Mueller, T. (2008), "Biomimetics Design by Natures," *National Geographic* April 2008, 68–90.

Neinhuis, C. and Barthlott, W. (1997), "Characterization and Distribution of Water-Repellent, Self-Cleaning Plant Surfaces," *Annals of Botany* **79**, 667–677.

Nosonovsky, M. and Bhushan, B. (2009), "Thermodynamics of Surface Degradation, Self-Organization, and Self-Healing for Biomimetic Surfaces," *Phil. Trans. R. Soc. A* **367**, 1607–1627.

Nosonovsky, M. and Bhushan, B. (2010), "Green Tribology: Principles, Research Areas and Challenges," *Phil. Trans R. Soc. A* **368**, 4677–4694.

Nosonovsky, M. and Bhushan, B. (2012), *Green Tribology: Biomimetics, Energy Conservation and Sustainability*, Springer-Verlag, Heidelberg, Germany.

Palacio, P. and Bhushan, B. (2010), "A Review of Ionic Liquids for Green Molecular Lubrication in Nanotechnology," *Tribol. Lett.* **40**, 247–268.

Parker, A.R. (2009), "Natural Photonics for Industrial Applications," *Phil. Trans. R. Soc. A* **367**, 1759–1782.

Ralston, E. and Swain, G. (2009), "Bioinspiration – the Solution for Biofouling Control?" *Bioinsp. Biomim.* **4**, 1–9.

Rybach, L. (2007), "Geothermal Sustainability," *Geo-Heat Centre Quarterly Bulletin* **28**, 2–7.

Sarikaya, M. and Aksay, I.A. (1995), *Biomimetic Design and Processing of Materials*, American Institute of Physics, Woodbury, New York.

Stegmaier, T., Linke, M., and Planck, H. (2009), "Bionics in Textiles: Flexible and Translucent Thermal Insulations for Solar Thermal Applications," *Phil. Trans. R. Soc. A* **367**, 1749–1758.

Tamerler, C. and Sarikaya, M. (2009), "Molecular Biomimetics: Nanotechnology and Molecular Medicine Utilizing Genetically Engineered Peptides," *Phil. Trans. R. Soc A* **367**, 1705–1726.

Terrell, E.J., Needleman, W.M., and Kyle, J.P. (2012), "Wind Tribology," in *Green Tribology: Biomimetices, Energy Conservation and Sustainability* (eds. M. Nosonovsky and B. Bhushan), pp. 483–530, Springer-Verlag, Heidelberg, Germany.

Vincent, J.F.V., Bogatyreva, O.A., Bogatyrev, N.R., Bowyer, A., and Pahl, A.K. (2006), "Biomimetics: Its Practice and Theory," *J. Royal Soc. Interf.* **3**, 471–482.

Wagner, P., Furstner, R., Barthlott, W., and Neinhuis, C. (2003), "Quantitative Assessment to the Structural Basis of Water Repellency in Natural and Technical Surfaces," *J. Exper. Botany* **54**, 1295–1303.

Wood, R.J.K., Bahaj, A.S., Turnock, S.R., Wang, L., and Evans, M. (2010), "Tribological Design Constraints of Marine Renewable Energy Systems," *Phil. Trans. R. Soc. A* **368**, 4807–4827.

Yun, R., Lu, Y., and Filip, P. (2010), "Application of Extension Evaluation Method in Development of Novel Eco-friendly Brake Materials," *SAE Int. J. Mater. Manuf.* **2**, 1–7.

Further Reading

Bhushan, B. (2009), "Biomimetics: Lessons from Nature – An Overview," *Phil. Trans. R. Soc. A* **367**, 1445–1486.

Bhushan, B. (ed.) (2009), Special Journal Issue on Biomimetics I: Functional Biosurfaces and II: Fabrication and Applications., *Phil. Trans. R. Soc. A* **367**, No. 1893 and 1894.

Bhushan, B. (2012), *Biomimetics: Bioinspired Hierarchical-Structured Surfaces for Green Science and Technology*, Springer-Verlag, Heidelberg, Germany.

Nosonovsky, M. and Bhushan, B. (2010), Special Journal Issue on Green Tribology, *Phil. Trans. R. Soc A* **368**, No. 1929.

Nosonovsky, M. and Bhushan, B. (2012), *Green Tribology: Biomimetics, Energy Conservation and Sustainability*, Springer-Verlag, Heidelberg, Germany.

Units, Conversions, and Useful Relations

A.1 Fundamental Constants

Constant	Symbol	SI units
Avogadro's constant	AN	6.022×10^{23} mol^{-1}
Boltzmann's constant	k	1.381×10^{-23} J/ K
Molar gas constant	R = AN k	8.315 J/(K mol) (or Pa m^3/(K mol))
Electronic charge	$-e$	1.602×10^{-19} C (A s)
Permittivity of free space	..	8.854×10^{-12} C^2/(J m)
Mass of $\frac{1}{12}$ of ^{12}C atom (atomic mass unit)	*amu*	1.661×10^{-27} kg
Mass of electron	m_e	9.109×10^{-31} kg
Gravitational constant	G	6.670×10^{-11} N m^2/kg^2
Gravitational acceleration (New York)	g	9.807 m/s^2
Speed of light in vacuum	c	2.998×10^8 m/s

Introduction to Tribology, Second Edition. Bharat Bhushan.
© 2013 John Wiley & Sons, Ltd. Published 2013 by John Wiley & Sons, Ltd.

A.2 Conversion of Units

1 nm = 10 A (angstrom)
1 liter (ℓ) = 10^{-3} m^3 = 1000 cm^3 (cc)
1 gallon (US) = (1/7.4805) ft^3 = 3.78 ℓ
1 N = 10^5 dyne = (1/9.807) kgf = (1/4.448) 1b
1 m N/m = 1 dyne/cm = 1 erg/cm^2 = 1 mJ/m^2 (unit of surface tension)
1 Pa (N/m^2) = 10 dyne/cm^2 = 10^{-5} bar = (1/6894) psi
1 atm = 760 mm Hg = 1.013×10^5 Pa (N/m^2) = 1.013×10^6 dyne/cm^2 = 1.013 bar
1 torr = 1 mm Hg = 1.316×10^{-3} atm = 133.3 Pa (N/m^2)
1 J = 1 N m = 1 W s = 10^7 dyne cm = 10^7 ergs
1 cal = 4.187 J (Mechanical equivalent of heat)
1 BTU = 778.2 ft 1b = 252 cal
1 W (N m/s) = 1 J/s = 10^7 dyne cm/s
1 hp = 550 ft 1b/s = 0.746 kW
1 kT = 4.114×10^{-14} erg = 4.114×10^{-21} J at 298 K (~25°C)
1 kT per molecule = 0.592 kcal/mol = 2.478 kJ/mol at 298 K
1 eV = 1.602×10^{-12} erg = 1.602×10^{-19} J
1 eV per molecule = 23.06 kcal/mol = 96.48 kJ/mol
1 poise (P) = 1 dynes s/cm^2 = 10^{-1} kg/(m s) = 10^{-1} Pa s (unit of absolute or dynamic viscosity)
1 Reyn = 1 1b s/in^2 = 68,750 Poise
1 Stoke (St) = 10^2 mm^2/s (unit of kinematic viscosity = absolute viscosity/density)
1 kg/m^3 = 62.43×10^{-3} 1b/ft^3 (density)
°C = (°F – 32)/1.8
0°C = 273.15 K (triple point of water)
1 W/(m K) = (1/4.187 × 10^2) cal/(s cm °C) = 57.79 BTU/(h ft °F) (thermal conductivity)

A.3 Useful Relations

The energy equivalent, mc^2, of one atomic mass unit = 1.492×10^{-10} J

The mass of an atom or molecule = (molecular weight) (1.661×10^{-27} kg)

The kgf (in metric units) is the force required to support a standard kilogram (kg) body against gravity in a vacuum; or the force applied to give a body the standard acceleration. The word kilogram is used for the unit of mass,

$$1 \text{ kgf} = (1 \text{ kg})(9.807 \text{ m/s}^2)$$

The Newton is the force (in SI units) which, if applied to a standard kilogram body, would give that body an acceleration of 1 m/s^2,

$$1 \text{ N} = 1 \text{ kg m/s}^2 = 1/9.807 \text{ kgf}$$

The dyne is the force which if applied to a standard gram body, would give that body an acceleration of 1 cm/s^2, i.e.,

$$1 \text{ dyne} = 1 \text{ g cm/s}^2 = 10^{-5} \text{ N} = (1/980.7) \text{ gf}$$

The specific gravity of a solid or liquid is the ratio of the mass of the body to the mass of an equal volume of water at some standard temperature (typically 4°C in physics and 15.6°C in engineering); density of water at 4°C = 10^3 kg/m^3. The specific gravity of gases is usually expressed in terms of that of hydrogen or air.

Index

Introduction to Tribology, Second Edition. Bharat Bhushan.

© 2013 John Wiley & Sons, Ltd. Published 2013 by John Wiley & Sons, Ltd.

Lightning Source UK Ltd.
Milton Keynes UK
UKOW07n1931200917
309560UK00004B/5/P

9 781119 944539